LIST OF THE ELEMENTS WITH THEIR SYMBOLS AND ATOMIC WEIGHTS

Element	Symbol	Atomic number	Atomic weight	Element	Symbol	Atomic number	Atomic weight	Element	Symbol	Atomic number	Atomic weight
Actinium	Ac	89	227.0278	Hahnium b	Ha	105	(262)	Protactinium	Pa	91	231.0359
Aluminum	Al	13	26.98154	Helium	He	2	4.00260	Radium	Ra	88	226.0254
Americium	Am	95	(243)a	Holmium	Ho	67	164.9304	Radon	Rn	86	(222)
Antimony	Sb	51	121.75	Hydrogen	H	1	1.0079	Rhenium	Re	75	186.207
Argon	Ar	18	39.948	Indium	In	49	114.82	Rhodium	Rh	45	102.9055
Arsenic	As	33	74.9216	Iodine	I	53	126.9045	Rubidium	Rb	37	85.4678
Astatine	At	85	(210)	Iridium	Ir	77	192.22	Ruthenium	Ru	44	101.07
Barium	Ba	56	137.33	Iron	Fe	26	55.847	Rutherfordium b	Rf	104	(261)
Berkelium	Bk	97	(247)	Krypton	Kr	36	83.80	Samarium	Sm	62	150.36
Beryllium	Be	4	9.01218	Lanthanum	La	57	138.9055	Scandium	Sc	21	44.9554
Bismuth	Bi	83	208.9804	Lawrencium	Lr	103	(260)	Selenium	Se	34	78.96
Boron	B	5	10.81	Lead	Pb	82	207.2	Silicon	Si	14	28.0855
Bromine	Br	35	79.904	Lithium	Li	3	6.941	Silver	Ag	47	107.8682
Cadmium	Cd	48	112.41	Lutetium	Lu	71	174.967	Sodium	Na	11	22.98977
Calcium	Ca	20	40.08	Magnesium	Mg	12	24.305	Strontium	Sr	38	87.62
Californium	Cf	98	(251)	Manganese	Mn	25	54.9380	Sulfur	S	16	32.06
Carbon	C	6	12.011	Mendelevium	Md	101	(258)	Tantalum	Ta	73	180.9479
Cerium	Ce	58	140.12	Mercury	Hg	80	200.59	Technetium	Tc	43	(98)
Cesium	Cs	55	132.9054	Molybdenum	Mo	42	95.94	Tellurium	Te	52	127.60
Chlorine	Cl	17	35.453	Neodymium	Nd	60	144.24	Terbium	Tb	65	158.9254
Chromium	Cr	24	51.996	Neon	Ne	10	20.179	Thallium	Tl	81	204.383
Cobalt	Co	27	58.9332	Neptunium	Np	93	237.0482	Thorium	Th	90	232.0381
Copper	Cu	29	63.546	Nickel	Ni	28	58.69	Thulium	Tm	69	168.9342
Curium	Cm	96	(247)	Niobium	Nb	41	92.9064	Tin	Sn	50	118.69
Dysprosium	Dy	66	162.50	Nitrogen	N	7	14.0067	Titanium	Ti	22	47.88
Einsteinium	Es	99	(254)	Nobelium	No	102	(259)	Tungsten	W	74	183.85
Erbium	Er	68	167.26	Osmium	Os	76	190.2	Unnilennium b	Une	109	(266)
Europium	Eu	63	151.96	Oxygen	O	8	15.9994	Unnilhexium b	Unh	106	(263)
Fermium	Fm	100	(257)	Palladium	Pd	46	106.4	Unnilseptium b	Uns	107	(262)
Fluorine	F	9	18.998403	Phosphorus	P	15	30.97376	Uranium	U	92	238.0289
Francium	Fr	87	(223)	Platinum	Pt	78	195.08	Vanadium	V	23	50.9415
Gadolinium	Gd	64	157.25	Plutonium	Pu	94	(244)	Xenon	Xe	54	131.29
Gallium	Ga	31	69.72	Polonium	Po	84	(209)	Ytterbium	Yb	70	173.04
Germanium	Ge	32	72.59	Potassium	K	19	39.0983	Yttrium	Y	39	88.9059
Gold	Au	79	196.9665	Praseodymium	Pr	59	140.9077	Zinc	Zn	30	65.38
Hafnium	Hf	72	178.49	Promethium	Pm	61	(145)	Zirconium	Zr	40	91.22

a Approximate values for radioactive elements are listed in parentheses.

b The official name and symbol have not been agreed to. The names for elements 106, 107, and 109 represent their atomic numbers, as in un (1) nil (0) hex (6) = unnilhexium (Unh) for element 106. Element 109 was reported in 1982; 108 is not known.

Essentials
of General,
Organic,
and Biological
Chemistry

Essentials of General, Organic, and Biological Chemistry

John McMurry
Cornell University

PRENTICE HALL, Englewood Cliffs, New Jersey 07632

Library of Congress Cataloging-in-Publication Data

McMurry, John.
    Essentials of general, organic, and biological chemistry/John McMurry.
        p.   cm.
    Includes index.
    ISBN 0-13-286261-1
    1, Chemistry. I. Title.
QD33.M138 1989
540—dc 19   88-26577
                                                             CIP

Development editors: Elizabeth Fletcher Foy and Dan Schiller
Production editor: Eleanor Henshaw Hiatt
Interior design: Judith A. Matz-Coniglio
Manufacturing buyer: Paula Massenaro
Page layout: Martin J. Behan
Cover design: Judith A. Matz-Coniglio
Cover photograph: A ribbon representation of the protein insulin (see Chapter
    15). The image was generated on the Evans & Sutherland PS 390 computer
    graphics system using Tripos® SYBYL computational chemistry software.

Photo credits and acknowledgments appear on pages 487–488, which constitute
    a continuation of the copyright page.

© 1989 by Prentice-Hall, Inc.
A Division of Simon & Schuster
Englewood Cliffs, New Jersey 07632

Printed in the United States of America

10 9 8 7 6 5 4 3 2 1

ISBN   0-13-286261-1

Prentice-Hall International (UK) Limited, *London*
Prentice-Hall of Australia Pty. Limited, *Sydney*
Prentice-Hall Canada Inc., *Toronto*
Prentice-Hall Hispanoamericana, S.A., *Mexico*
Prentice-Hall of India Private Limited, *New Delhi*
Prentice-Hall of Japan, Inc., *Tokyo*
Simon & Schuster Asia Pte. Ltd., *Singapore*
Editora Prentice-Hall do Brasil, Ltda., *Rio de Janeiro*

Brief Contents

v

Contents

All about atoms

3 The Structure of Matter: Chemical Bonds 46

4 Chemical Reactions 69

5 Solids, Liquids, and Gases 92

6 Solutions 112

7 Acids, Bases, and Salts 135

8 Introduction to Organic Chemistry: Alkanes 160

15   The Molecules of Life: Amino Acids and Proteins   317

16   The Molecules of Life: Enzymes, Vitamins, and Hormones   339

17 The Molecules of Life: Nucleic Acids 364

18 Metabolism I: The Generation of Biochemical Energy 389

19 Metabolism II: Catabolic and Anabolic Pathways 407

Preface

In writing this book, my goal was to create the finest available introductory text on general, organic, and biological chemistry. The book is designed for use in a one-term college course for students with no prior preparation in chemistry. It is also well suited for a two-quarter or two-semester course.

In addition to the careful pedagogy that should be expected of any introductory text, this book includes many special features that set it apart.

CONTENT AND EXPOSITION

Writing   Great care has gone into the writing of this book, and every attempt has been made to make it as lucid and readable as possible. Paragraphs begin with summary sentences, transitions between topics are smooth, and new concepts are introduced only as needed. Explanations are concise and to the point; the focus is on the essentials.

Applications   Each chapter includes a brief discussion of some special topic that shows how the material presented in the chapter is relevant to a specific biological or medical application.

Chapter 1   Measuring Percent Body Fat
Chapter 2   Chernobyl and Cesium
Chapter 3   Biologically Important Ions
Chapter 4   Mercury—Reactivity and Toxicity
Chapter 5   Inhaled Anesthetics
Chapter 6   Gout and Kidney Stones—Problems in Solubility
Chapter 7   Ulcers and Antacids
Chapter 8   Displaying Molecular Shapes
Chapter 9   The Chemistry of Vision
Chapter 10   Morphine Alkaloids
Chapter 11   Chemical Warfare among the Insects
Chapter 12   Thiol Esters—Biological Carboxylic Acid Derivatives
Chapter 13   Cell-Surface Carbohydrates
Chapter 14   Cholesterol and Heart Disease
Chapter 15   Protein and Nutrition
Chapter 16   Medical Use of Enzymes—Isoenzymes
Chapter 17   Viruses
Chapter 18   Barbiturates
Chapter 19   Diabetes, A Metabolic Disorder
Chapter 20   Medical Uses of Radioactivity

Interludes   Each chapter ends with a brief Interlude extending the material of the chapter into a broader context: ecological, clinical, social, or technological.

Chapter 1   Powers of Ten
Chapter 2   Are Atoms Real?
Chapter 3   Diamond and Graphite
Chapter 4   Regulation of Body Temperature
Chapter 5   Blood Pressure
Chapter 6   Dialysis
Chapter 7   Acid Rain
Chapter 8   Petroleum
Chapter 9   Alkene Polymers
Chapter 10   Chlorofluorocarbons and the Ozone Layer
Chapter 11   A Biological Aldol Reaction
Chapter 12   Nylons and Polyamides
Chapter 13   Sweetness
Chapter 14   Chemical Communication
Chapter 15   Determining Protein Structure
Chapter 16   Penicillin, the First Antibiotic
Chapter 17   Recombinant DNA
Chapter 18   Diets, Babies, and Hibernating Bears
Chapter 19   Exercise and Weight
Chapter 20   Archaeological Radiocarbon Dating

Currency   Coverage is up to date throughout the book—particularly in the biological chapters, where new breakthroughs seem to occur almost daily. DNA sequencing (Section 17.10), recombinant DNA technology (Interlude, Chapter 17), prostaglandins (Section 14.10), cell-surface carbohydrates (Application, Chapter 13), chlorofluorocarbons and the ozone layer (Interlude, Chapter 10), and diagnostic imaging techniques (Interlude, Chapter 20) are among the currently exciting topics treated in this book.

GRAPHICS

Color   Full color has been used throughout this book, not only to make it attractive but to serve important pedagogical purposes. Many topics, especially in organic and biological chemistry, become much clearer when the reacting portions of complex molecules are color coded.

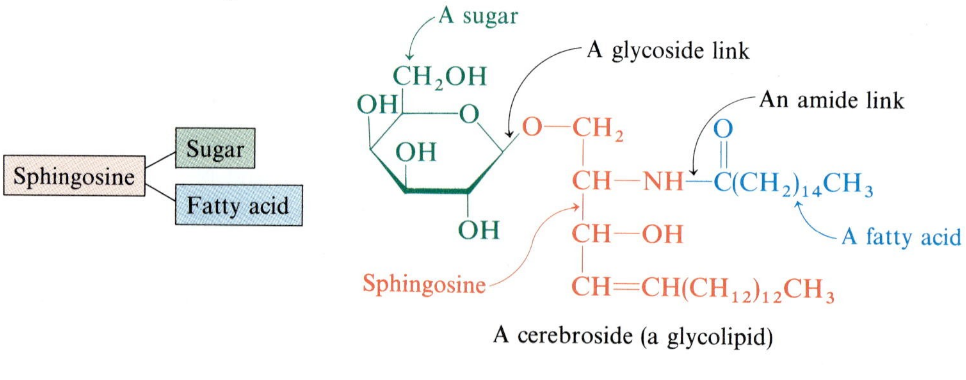

A cerebroside (a glycolipid)

Important groups, such as phosphate, are colored consistently throughout the text so that students can recognize them more quickly and relate material presented in different chapters. Moreover, in the chapters on bioenergetics (Chapters 18 and 19), energy rich forms of compounds, such as ATP and reduced coenzymes, are depicted in red, whereas their lower energy counterparts, such as ADP and oxidized coenzymes, are represented in blue.

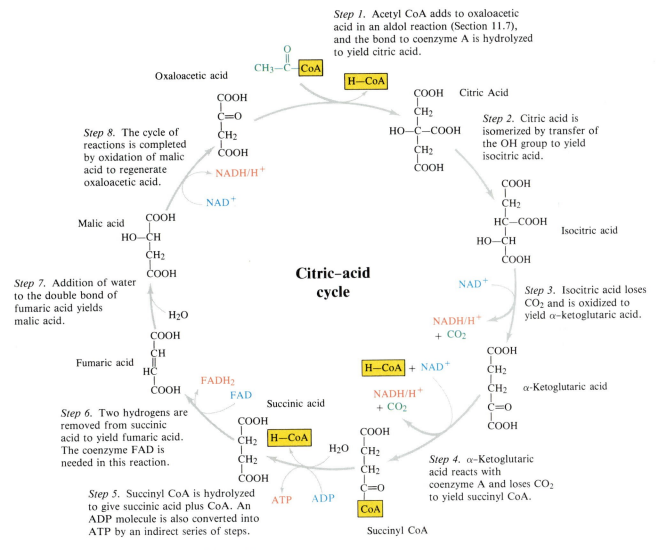

Figure 18.6
The citric-acid cycle, an eight-step series of reactions whose net effect is the metabolic breakdown of acetyl groups (from acetyl CoA) into two molecules of carbon dioxide plus energy.

Computer-Generated Structures   Computer-generated molecular models are used extensively in the organic and biological chapters, both for their accuracy in portraying the three-dimensional structures of molecules and for their visual appeal.

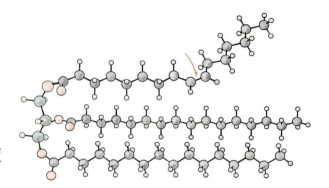

Figure 14.3
Molecular model of an unsaturated triglyceride. The double bond (arrow) prevents the molecule from adopting a regular shape and crystallizing easily.

PROBLEM SOLVING

Solved Problems   Eighty-two fully worked-out sample problems are used in the text to demonstrate methods of solving commonly encountered types of problems.

Practice Problems   Solved problems are always followed by related practice problems for students to solve. Most text sections conclude with additional practice problems to provide an immediate test of understanding. The book includes a total of 245 practice problems, answers to all of which are provided in a section at the end of the text.

Solved Problem 9.4   What product would you expect from the following reaction?

$$\overset{\displaystyle CH_3}{\underset{\displaystyle |}{CH_3CH_2C}}{=}CHCH_3 + HCl \longrightarrow \ ?$$

Solution   We know that reaction of an alkene with HCl leads to formation of an alkyl chloride addition product. The question here is which of two possible products will form. To make a prediction, look at the starting alkene, and count the number of hydrogens already attached to each double-bond carbon. Then write the product by attaching H to the carbon that already has more hydrogens and attaching Cl to the carbon that has fewer hydrogens.

$$\overset{\displaystyle CH_3}{\underset{\displaystyle |}{CH_3CH_2C}}{=}CHCH_3 + HCl \longrightarrow \overset{\displaystyle CH_3}{\underset{\displaystyle \underset{\displaystyle Cl}{|}}{\underset{|}{CH_3CH_2C}}}{-}CH_2CH_3$$

No hydrogens on this carbon, so —Cl attaches here.   One hydrogen already on this carbon, so —H attaches here.   3-Chloro-3-methylpentane

Practice Problems   **9.7**   What products would you expect from these following reactions?

(a) cyclohexene + HBr $\longrightarrow$?   (b) 1-hexene + HCl $\longrightarrow$?

(c) $(CH_3)_2CHCH{=}CH_2$ + HI $\longrightarrow$?   (d) 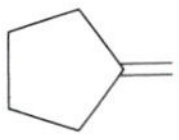 $\bigcirc{=}CH_2$ + HCl $\longrightarrow$?

9.8   What alkenes are the following alkyl halides likely to have been made from? (Be careful, there may be more than one answer.)
(a) 3-chloro-3-ethylpentane   (b) $(CH_3)_2CHCBr(CH_3)_2$

Ballpark Answers to Problems   In addition to the standard factor-label method used throughout for solving numerical calculations, the idea of first getting an approximate or "ballpark" answer is taught. This method serves not only as a test of understanding specific concepts but also as a useful device for developing general intellectual skills.

Solved Problem 7.6   When a 5.00 mL sample of household vinegar (dilute aqueous acetic acid) was titrated, 44.5 mL of 0.100 M NaOH solution were required to reach the end point. What is the acid concentration of vinegar?

Ballpark Solution   Since the volume of base required to neutralize the acid sample is about nine times the volume of the sample (44.5 mL versus 5 mL), the concentration of the acid is about nine times that of the base, or 0.9 M.

Solution   First, write the balanced equation for the neutralization to find the number of moles of base required to neutralize each mole of acid. In this example, the mole ratio of base to acid is 1:1.

$$CH_3COOH + NaOH \longrightarrow CH_3COO^-Na^+ + H_2O$$

Next, calculate how many moles of NaOH are consumed in the titration:

$$44.5 \, \text{mL} \times \frac{0.100 \text{ moles NaOH}}{\text{L}} \times \frac{1 \, \text{L}}{1000 \, \text{mL}} = 0.00445 \text{ moles NaOH}$$

Since the balanced equation says that each mole of NaOH reacts with one mole of acetic acid, the 5.00 mL sample of vinegar must contain 0.00445 moles of acetic acid:

$$\text{Acetic acid concentration} = \frac{0.00445 \text{ moles } CH_3COOH}{5.00 \, \text{mL}} \times \frac{1000 \, \text{mL}}{1 \text{ L}}$$

$$= 0.890 \text{ M}$$

Thus, the acetic acid concentration in vinegar is 0.890 M, very close to the ballpark answer.

Practice Problems   **7.9**   In order to determine the concentration of an old bottle of aqueous HCl whose label had become unreadable, a titration was carried out. What is the HCl concentration if 58.4 mL of 0.250 M NaOH was required to titrate a 20.0 mL sample of the acid?

7.10   How many mL of 0.150 M NaOH are required to neutralize 50.0 mL of 0.200 M H_2SO_4?

Review Problems   Nearly 1000 review problems are provided at the ends of chapters. These are classified by subject; in addition, there is a section of uncategorized problems for each chapter. Answers to selected review problems are given at the end of the book.

PEDAGOGY

Goals   Each chapter begins with a brief introductory overview, followed by a list of goals for the student to keep in mind while studying.

Marginal Definitions   Key terms are boldfaced in the text, and the definition of each term is repeated in the margin for easy review. In addition, a complete glossary of terms is provided at the end of the book.

Summaries   All chapters end with clear, concise summaries that recapitulate the key points in the text, with important terms in boldface.

Appendices   A review of exponential notation, a table of commonly used conversion factors, and a discussion of the concepts of accuracy and precision in measurement are provided in appendices.

SUPPLEMENTS

Study Guide and Solutions Manual   A carefully prepared *Study Guide and Solutions Manual* accompanies this text. Written by Susan McMurry, this companion volume answers all in-text and end-of-chapter problems and explains in detail how the answers are obtained. In addition, study hints and self-test materials for each chapter are included.

Instructor's Manual   The *Instructor's Manual*, written by the author, includes lecture suggestions and teaching tips, suggested readings, and approximately 500 test questions.

Laboratory Manual   The *Laboratory Manual*, developed by Scott Mohr and his associates at Boston University, has been class tested for 12 years.

Instructor's Manual to the Laboratory Manual   This supplement provides information on reagents and equipment needed for the laboratory experiments, along with answers to any questions in the *Laboratory Manual*.

Transparency Set   A set of 60 transparencies of the most important figures and tables in the text is available free with adoptions of 100 or more copies.

For more information on supplements, contact your local Prentice Hall sales representative.

ACKNOWLEDGMENTS

It's a pleasure to thank the many people whose help and suggestions were so valuable in preparing this book. First is my wife, Susan, herself a chemist, who not only read and improved the entire manuscript but also prepared the accompanying *Study Guide and Solutions Manual*. Next is the first-rate staff at Prentice Hall whose care and criticisms made this book so much better than it would otherwise have been—Dan Joraanstad, Elizabeth Foy, Eleanor Hiatt, Judy Matz-Coniglio, and especially Dan Schiller, for effort far beyond the call of duty. Finally, I thank the reviewers:

Dennis Berzansky
Westmoreland County Community College

Jennie Caruthers
University of Colorado, Boulder

Mary E. Castellion
Norwalk, Connecticut

Ronald E. DiStefano
Northampton Community College

Paula Getzin
Kean College

C. S. Jones
Southwest Missouri State University

Irving Lillien
Miami-Dade Community College

Howard Ono
California State University, Fresno

John Paparelli
San Antonio College

J. Graham Rankin
Shell Development Co.

Jack W. Timberlake
University of New Orleans

Rod S. Tracey
College of the Desert

Mary S. Vennos
Essex Community College

Robert W. Wallace
Bentley College

Anonymous to me at the time they read the manuscript, the persons listed above provided many excellent suggestions that I gladly incorporated in the final product. Needless to say, they cannot be blamed for any errors that remain. I would be deeply grateful for any suggestions and corrections from users of this book.

John McMurry

A Note to the Student

We have the same goals. Yours is to learn chemistry, and mine is to do everything possible to help you learn. It's going to take some work on your part, but the following suggestions should prove helpful.

Don't read the text immediately. As you begin each new chapter, look it over first. Read the introductory paragraphs and familiarize yourself with the chapter goals. Find out what topics will be covered, and then turn to the end of the chapter and read the summary. You'll be in a much better position to learn new material if you first have a general idea of where you're going.

Work the problems. There are no shortcuts here; working problems is the only way to learn chemistry. The sample problems show you how to approach the material, the in-chapter practice problems provide immediate practice, and the end-of-chapter problems provide additional drill. Answers to the in-chapter practice problems and selected review problems are given at the end of this book; full answers and explanations for all problems are given in the accompanying *Study Guide and Solutions Manual.*

Use the study guide. The *Study Guide and Solutions Manual* that accompanies this text gives complete solutions to all problems. It also provides chapter outlines, additional study hints, and self-test materials. This material can be extremely useful when you're working problems and when you're studying for an exam. Find out what's there now so you'll know where to find it when you need help.

Ask questions. Faculty members and teaching assistants are there to help you. Most of them will turn out to be genuinely nice people with a sincere interest in helping you learn.

Good luck. I sincerely hope you enjoy learning about chemistry and come to see the beauty and logic of its structure.

John McMurry

CHAPTER 1

Chemistry: Matter and Measurement

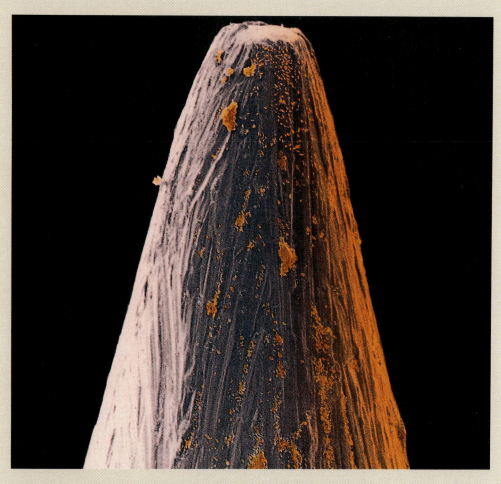

How many bacteria fit on the tip of a pin? We'll see in this chapter how objects are measured.

Look around you. Everything you see, touch, taste, and smell is made of chemicals. Many of these chemicals—those that make up rocks, trees, and your own body—occur naturally. Many others, however, are synthetic: The plastics, the fibers, and many of the medicines that are so important a part of modern life do not occur in nature but have been created in the chemical laboratory.

Just as everything you see is made of chemicals, many of the natural changes you see taking place around you are the result of chemical reactions: the change of one chemical into another. The flowering of plants in the spring, the color change of a leaf in the fall, and the growth and aging of a human body—all are the result of chemical reactions. To understand these or any other processes in life, you must have a basic understanding of chemistry.

As you might expect, the chemistry of many life processes is complex, and it's not possible to jump into their study without the right background. Thus, our general plan is to increase gradually in complexity, beginning in the first seven chapters with a firm grounding in the scientific fundamentals that govern all of chemistry, moving in the next five chapters to a look at organic molecules, and then coming in the next seven chapters to biological chemistry. We'll begin in this chapter by looking at these topics:

1. **What is matter, and what is chemistry?** The goal: you should learn the meaning of some important chemical terms.

2. **How is matter described?** The goal: you should become familiar with metric units of measure for describing mass, length, volume, and temperature.

3. **How good are the measurements we make?** The goal: you should learn the meaning of significant figures, should become familiar with using scientific notation, and should learn how to round off numbers.

4. **How can a quantity be converted from one unit of measure to another?** The goal: you should learn the techniques for solving numerical problems involving the conversion of quantities from one unit to another.

5. **What are heat, density, and specific gravity?** The goal: you should learn what these quantities are, why they are important, and how they can be used.

1.1 MATTER

■ **Chemistry** The study of the nature, properties, and transformations of matter.

■ **Matter** The physical material that makes up the universe; anything that has mass and occupies space.

■ **Property** A characteristic or trait useful for identifying a substance or object.

■ **Physical property** A property that does not involve a chemical change in a substance or object.

■ **Chemical property** A property that involves a chemical reaction of a substance.

Chemistry is the science of matter. **Matter**, in turn, is a catch-all word used to describe anything physically real—anything you can see, touch, taste, or smell. In more scientific terms, matter is anything that has mass (weight) and volume (occupies space). Thus, a light bulb contains matter, but the light coming from the bulb contains no matter.

How might we describe different kinds of matter more specifically? Any characteristic that can be used to describe or identify something is called a **property**. Size, color, and temperature are all well-known properties of matter. Still other properties include such characteristics as solubility and melting point. Thus, table salt (sodium chloride) dissolves in water, but a metal coin doesn't; an ice cube always melts at exactly 32°F, but a crystal of sodium chloride doesn't melt until it's heated to 1474°F.

Properties of matter can be divided into two types: physical properties and chemical properties. **Physical properties** are those characteristics that can be determined without altering the chemical make-up of the sample. For example, when you step on a scale, you can determine your weight. When you step off the scale, however, you're unchanged. Weight, therefore, is a physical property. In the same way, color, temperature, melting point, and many other characteristics are all physical properties.

In contrast to physical properties, **chemical properties** are those characteristics that *do* involve a change in the chemical make-up of the sample. For example, if you never wax your car and often leave it out in the rain, the metal parts are likely to rust. The iron of the car changes chemically, combining with oxygen from the air to give the new substance called iron oxide, or rust. Rusting, the chemical combination of iron with oxygen, is therefore a chemical property of iron (Figure 1.1).

Figure 1.1
The difference between physical and chemical properties.

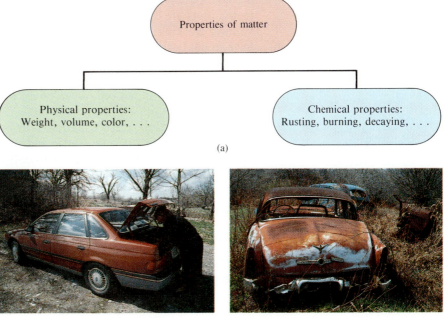

(a)

(b) The car weighs 1200 kg.   (c) The car rusts in the rain.

1.2 PHYSICAL QUANTITIES

■ **Physical quantity** A physical property that can be measured.

■ **Unit** A specific quantity, used for measurement.

■ **SI Unit** An internationally agreed-on unit of measure derived from the metric system.

■ **Kilogram (kg)** The SI unit of mass; equal to 2.205 pounds.

■ **Meter (m)** The SI unit of length; equal to 3.280 feet.

■ **Cubic meter (m³)** The SI unit of volume; equal to 264.2 gallons.

■ **Kelvin (K)** The SI unit of temperature; equal to 1.8 Fahrenheit degrees.

■ **Metric units** Units of a common system of measure used throughout the world.

■ **Gram (g)** The metric unit of mass; equal to 1/1000 kilogram.

■ **Liter (L)** The metric unit of volume; equal to 1.057 quarts.

■ **Celsius degree (C°)** The metric unit of temperature; equal to 1.8 Fahrenheit degrees.

Any physical property of matter that can be measured is called a **physical quantity**. Mass, volume, length, and temperature are common examples. *Mass* is a measure of the amount of matter in an object. (Mass and weight are related but aren't identical; we'll see the difference between the two in the next section.) *Volume* is a measure of the amount of space occupied by an object. *Length* is a measure of the distance an object extends in a particular direction. *Temperature* is a measure of how much heat energy an object contains. Other physical properties such as odor and color are not physical quantities, however, because they can't be measured.

Notice that all physical quantities are described by both a number and a label, or **unit**. A number alone isn't much good without a unit to define it. If you asked how much blood an accident victim had lost, the answer "3" wouldn't tell you much. 3 drops? 3 milliliters? 3 pints? 3 liters? (Let's hope it's not 3 liters; an adult human has only 5 to 6 liters of blood.)

By agreement among scientists in different countries, the standard units used to define physical quantities are those called **SI units** after the French name, *Système International d'Unités* (International System of Units). Thus, mass is measured in **kilograms** (abbreviated **kg**), length is measured in **meters** (**m**), volume is measured in **cubic meters** (**m³**), and temperature is measured in **kelvins** (**K**). These basic SI units and their common equivalents are shown in Table 1.1.

SI units are closely related to the more familiar **metric units** that are used in all industrialized nations of the world except the United States. If you compare the SI and metric units shown in Table 1.2, you'll find that the metric unit of mass is the **gram** (abbreviated **g**; 1 g = 1/1000 kg) rather than the kilogram. Similarly, the metric unit of volume is the **liter** (**L**) rather than the cubic meter (1 L = 1/1000 m³), and the metric unit of temperature is the **Celsius degree** (**C°**) rather than the kelvin (1 K = 1 C°). The meter (m) is the unit of length in both SI and metric systems. Although SI units are now preferred in chemical research, metric units are commonly used in biology and medicine. You'll probably find yourself working primarily with the metric system, but you may occasionally have to deal with SI units.

One problem with any system of measure is that the sizes of the units often turn out to be inconveniently large or small. A biologist describing the diameter of a red blood cell (0.000006 m) would find the meter to be an inconveniently large unit, but an astronomer measuring the average distance from the earth to the sun (150,000,000,000 m) would find the meter to be inconveniently small. For this reason, metric and SI fundamental units can be modified through the use of prefixes to refer to either smaller or larger quantities. For example, the prefix *milli-* means one thousandth. Thus, 1 millimeter (1 mm) is

Table 1.1 Some SI Units and Their Common Equivalents

Quantity	SI Unit	Symbol	Common Equivalent
Mass	Kilogram	kg	1 kilogram = 2.205 pounds
Length	Meter	m	1 meter = 3.280 feet
Volume	Cubic meter	m³	1 cubic meter = 264.2 gallons
Temperature	Kelvin	K	1 kelvin = 1.8 Fahrenheit degree

Table 1.2   Some Metric Units and Their Equivalents

Quantity	Metric Unit	Symbol	Common and SI Equivalent
Mass	Gram	g	1 gram = 0.002205 pound = 0.001 kilogram
Length	Meter	m	1 meter = 3.280 feet
Volume	Liter	L	1 liter = 1.057 quarts = 0.001 cubic meter
Temperature	Celsius degree	C°	1 Celsius degree = 1.8 Fahrenheit degree = 1 kelvin

1/1000 meter, and 1 milligram (1 mg) is equal to 1/1000 gram. Similarly, *kilo-* means one thousand, and a kilogram (1 kg) is therefore equal to 1000 grams. (Note that the SI unit for mass already contains the *kilo-* prefix.)

A list of prefixes is shown in Table 1.3, with the most commonly used ones displayed in bold type. *Mega-* (10^6), *kilo-* (10^3), *milli-* (10^{-3}), and *micro-* (10^{-6}) have exponents that are multiples of 3, whereas *deci-* (10^{-1}) and *centi-* (10^{-2}) do not. *Centi-* is used primarily for length measurements, where 1 centimeter (1 cm) equals 1/100 meter. *Deci-* is used primarily in clinical chemistry, where concentrations of components in blood are referred to in milligrams per deciliter (1 dL = 1/10 liter).

Table 1.3   Some Prefixes for Multiples of Metric and SI Units

Prefix	Symbol	Meaning[a]	Example
mega	**M**	$1,000,000 = 10^6$	1 megameter (Mm) = 10^6 m
kilo	**k**	$1,000 = 10^3$	1 kilogram (kg) = 10^3 g
hecto	h	$100 = 10^2$	1 hectogram (hg) = 100 g
deka	da	$10 = 10^1$	1 dekaliter (daL) = 10 L
deci	**d**	$0.1 = 10^{-1}$	1 deciliter (dL) = 0.1 L
centi	**c**	$0.01 = 10^{-2}$	1 centimeter (cm) = 0.01 m
milli	**m**	$0.001 = 10^{-3}$	1 milligram (mg) = 0.001 g
micro	**μ**	$0.000001 = 10^{-6}$	1 micrometer (μm) = 10^{-6} m
nano	n	$0.000000001 = 10^{-9}$	1 nanogram (ng) = 10^{-9} g
pico	p	$0.000000000001 = 10^{-12}$	1 picogram (pg) = 10^{-12} g

[a] The exponential method of writing large and small numbers (for example, 10^6 for 1,000,000) is explained in Section 1.7.

Practice Problems

1.1   Bottles of wine sometimes carry the notation "Volume − 75 cL." What does the unit cL stand for?

1.2   Identify the full names of these units:
(a) mL   (b) kg   (c) cm   (d) km   (e) μg

1.3   Write the abbreviation of each of the following units:
(a) liter   (b) microliter   (c) nanometer   (d) megameter

1.3   MASS AND ITS MEASUREMENT

■ **Mass**   The amount of matter in an object.

■ **Weight**   The measure of the gravitational force exerted on an object by the earth, moon, or other massive body.

The terms "mass" and "weight," although often used interchangeably, are really quite different. **Mass** is a measure of the amount of matter in an object, whereas **weight** is a measure of the gravitational force that the earth, moon, or other heavenly body exerts on the object. Clearly, the amount of matter (mass) in an object does not depend on its location. Whether you're standing on earth, riding in a space shuttle, or standing on the moon, the amount of matter (mass) in your body is the same. Just as clearly, however, the weight of an object *does* depend on its location. Your weight might be 140 lb on earth, 0 lb in a space shuttle, and 23 lb on the moon, since the pull of gravity is quite different in these locations.

At the same location, two objects of identical mass will have identical weights; that is, gravity will pull equally on them. Thus, the *mass* of an object can be determined on a chemical balance by comparing the *weight* of the object to the weight of a known reference standard. Much of the confusion between mass and weight is simply due to a language problem: We speak of "weighing" when we really mean that we're determining *mass* by *comparing* two weights. Figure 1.2 shows some of the instruments normally used for determining mass in the laboratory.

We saw in Section 1.2 that mass is measured by the kilogram, which is equal to 1000 grams. The standard kilogram, against which all other masses are compared, is defined as the mass of a cylindrical bar of platinum-iridium alloy stored in a vault just outside Paris, France. Forty copies of this bar are distributed throughout the world, with two (Numbers 4 and 20) now stored at the U.S. National Bureau of Standards near Washington, D.C.

■ **Milligram (mg)**   A unit of mass equal to 1/1000 gram.

■ **Microgram (μg)**   A unit of mass equal to 1/1000 milligram.

One kilogram is just over 2 lb, which is far too large for many purposes in chemistry and medicine. Thus, smaller units of mass such as the gram (g; 1/1000 kg), the **milligram** (**mg**; 1/1000 g), and the **microgram** (**μg**; 1/1000 mg; sometimes called a gamma, γ) are more commonly used. Table 1.4 shows the relationships for converting between metric and common measures.

Figure 1.2
Two common laboratory balances. (Left) The single-pan balance has a sliding weight that is adjusted until the weight of the object in the pan is just balanced. (Right) A modern electronic balance.

Table 1.4   Conversions Among Common Measures of Mass

Unit	Equivalent	Unit	Equivalent
1 kilogram (kg)	= 1000 gram = 2.205 pound	1 ton	= 2000 pound = 907.03 kilogram
1 gram (g)	= 0.001 kilogram = 1000 milligram = 0.03527 ounce	1 pound	= 16 ounce = 0.454 kilogram = 454 gram
1 milligram (mg)	= 0.001 gram = 1000 microgram	1 ounce	= 0.02835 kilogram = 28.35 gram = 28,350 milligram
1 microgram (μg)	= 0.000001 gram = 0.001 milligram		

1.4   MEASURING LENGTH

The meter is the standard measure of length in the metric system. Although originally defined in 1790 as one ten-millionth the distance from the equator to the North Pole, the meter was redefined in 1889 as the distance between two thin lines on a bar of platinum-iridium alloy stored near Paris, France. To accommodate an increasing need for precision, however, the meter was redefined once again in 1986 as equal to the distance traveled by light in a vacuum in 1/299,792,458 second. Although this definition isn't as easy to understand as the distance between two scratches on a bar, it has the great advantage that it can't be lost, damaged, or destroyed.

One meter is 39.37 inches or about 10 percent longer than an English yard, a length that is much too large for most measurements in chemistry and medicine. Other, more commonly used measures of length are the **centimeter** (**cm**; 1/100 meter) and the **millimeter** (**mm**; 1/1000 meter). One centimeter is a bit less than one-half inch, 0.3937 in. to be exact. Put another way, there are 2.54 centimeters in one inch. A millimeter, in turn, is 0.03937 in., or about the thickness of a dime. Table 1.5 lists the relationships of these units, and Figure 1.3 (page 8) shows a comparison between inches and centimeters.

■ **Centimeter (cm)**   A unit of length equal to 1/1000 meter, or about 0.3937 inch.

■ **Millimeter (mm)**   A unit of length equal to 1/1000 meter, or about the thickness of a dime.

Table 1.5   Conversions Among Different Measures of Length

Unit	Equivalent	Unit	Equivalent
1 kilometer (km)	= 1000 meter = 0.6214 mile	1 mile	= 1.609 kilometer = 1609 meter
1 meter (m)	= 100 centimeter = 1000 millimeter = 1.0936 yard = 39.37 inch	1 yard	= 0.9144 meter = 91.44 centimeter
		1 foot	= 0.3048 meter = 30.48 centimeter
1 centimeter (cm)	= 0.01 meter = 10 millimeter = 0.3937 inch	1 inch	= 2.54 centimeter = 25.4 millimeter
1 millimeter (mm)	= 0.001 meter = 0.1 centimeter		

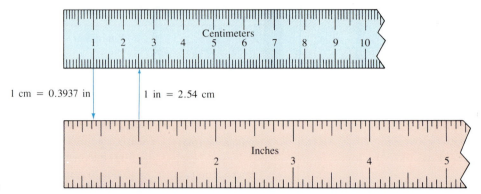

Figure 1.3
A comparison of a metric centimeter with an English inch.

1 cm = 0.3937 in

1 in = 2.54 cm

1.5 MEASURING VOLUME

■ **Milliliter (mL)** A unit of volume equal to 1/1000 liter.

■ **Cubic centimeter (cc or cm³)** An alternative name for a milliliter.

The metric liter (L; 0.001 m³) is the most commonly used measure of volume in chemistry and medicine. One liter has the volume of a cube 10 centimeters on each edge (Figure 1.4) and is just a bit larger than one U.S. quart. Each liter is further divided into 1000 **milliliters (mL)**, where one milliliter is the size of a cube one centimeter on each edge. In fact, the milliliter is often called a **cubic centimeter**, **cm³** or **cc**, in medical work. Table 1.6 shows the relationships among measures of volume.

Table 1.6 Conversions Among Different Measures of Volume

Unit	Equivalent	Unit	Equivalent
1 cubic meter (m³)	= 1000 liter	1 gallon	= 3.7856 liter
	= 264.2 gallons	1 quart	= 0.9464 liter
1 liter (L)	= 0.001 m³		= 946.4 milliliter
	= 1000 milliliter	1 ounce	= 29.57 milliliter
	= 1.057 quart		
1 milliliter (mL)	= 0.001 liter		
	= 1000 microliter		
1 microliter (μL)	= 0.001 milliliter		

Figure 1.4
(a) A large cube 10 cm on a side has a volume of 1000 cm³, or 1 liter. A smaller cube 1.0 cm on a side has a volume of 1 cm³, or 1 mL. Liquid volumes are often measured by using a graduated cylinder (b), a pipette (c), or a syringe (d).

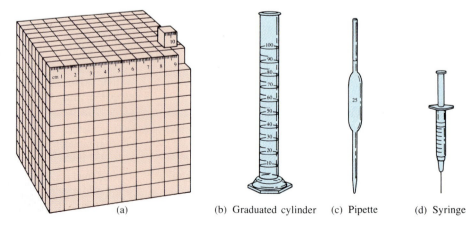

(a) (b) Graduated cylinder (c) Pipette (d) Syringe

1.6   MEASUREMENT AND SIGNIFICANT FIGURES

How much does a nickel weigh? If you put a nickel on an ordinary bathroom scale, the scale would probably register 0 lbs (or 0 kg if you've gone metric). If you placed the same nickel on a common laboratory scale, however, you might get a reading of 4.95 g. Trying again by placing the nickel on an expensive analytical balance like those found in clinical and research laboratories, you might find a weight of 4.94438 g. Clearly, the exactness of your answer depends on the equipment used for the measurement.

Every experimental measurement, no matter how exact, has a degree of uncertainty to it because there's always a limit to the number of digits that can be determined. Should we call the mass of the nickel measured on the laboratory scale 4.94 g, 4.95 g, or 4.96 g? Should the mass measured on the analytical balance be 4.94437 g, 4.94438 g, or 4.94439 g? Different people making the same measurement might come up with slightly different answers because the measuring equipment is being stretched to its limits.

In order to indicate the exactness of a measurement, the value recorded should use all the digits known with certainty, plus one additional estimated digit that is usually considered uncertain by plus or minus 1. The total number of digits used to express a value is called the number of **significant figures**. Thus, the quantity 4.95 g has three significant figures (4, 9, and 5), and the quantity 4.94438 g has six significant figures. Remember: All but one of the significant figures are known with certainty; the final significant figure is only an estimate accurate to plus or minus 1 (written as ± 1).

Finding the number of significant figures in a measurement is usually simple but can be troublesome when zeroes are involved. Depending on the circumstance, zeroes might be significant, or they might simply be space fillers to locate the decimal point. For example, how many significant figures does each of the following quantities have?

■ **Significant figures**   In describing a quantity, the total number of digits whose values are known with certainty, plus one estimated digit.

- 94.072 g   [five significant figures (9, 4, 0, 7, 2)]
- 0.0834 g   [three significant figures (8, 3, 4)]
- 23000 g   [*anywhere* from two (2, 3) to five (2, 3, 0, 0, 0) significant figures]
- 138.200 g   [six significant figures (1, 3, 8, 2, 0, 0)]

The following rules are helpful in cases where zeroes are present:

1. *Zeroes in the middle of a number are always significant.* Thus, 94.072 has five significant figures.
2. *Zeroes at the beginning of a number are not significant*; they serve only to locate the decimal point. Thus, 0.0834 has three significant figures.
3. *Zeroes at the end of a number but* before *the decimal point may or may not be significant.* We can't tell whether they are part of the measurement or whether they serve only to locate the decimal point. Thus, 23000 may have two, three, four, or five significant figures. (In the next section, we'll see a way to indicate how many figures are significant.)
4. *Zeroes at the end of a number but* after *the decimal point are significant.* (We assume that these zeroes wouldn't be indicated unless they were significant.) Thus, 138.200 has six significant figures.

One final point about determining significant figures: Some numbers are *exact*, and so the idea of significant figures is meaningless. Such exact numbers often occur in counting objects and in converting quantities from one unit to another. Thus, a class might have *exactly* 32 students (not 31.9, 32.0, or 32.1); similarly, there are *exactly* 100 cents to the dollar.

Solved Problem 1.1   How many significant figures does each of the following quantities have?
(a) 2730.78 m   (b) 0.0076 mL   (c) 3400 kg   (d) 3400.0 m^2

Solution   (a) six (rule 1)   (b) two (rule 2)
(c) two, three, or four (rule 3)   (d) five (rule 4)

Practice Problem   **1.4**   How many significant figures does each of the following quantities have? Explain your answers.
(a) 3.45 meters   (b) 0.1400 kilograms   (c) 10.003 liters   (d) 35 cents

---

1.7   SCIENTIFIC NOTATION

We often find ourselves having to work either with very large or with very small numbers. For example, the diameter of the red blood cell shown in Figure 1.5 is very small—approximately 0.000006 m (6 μm). On the other hand, the number of red blood cells in a normal adult is very large—approximately 200,000,000,000,000. Rather than write out such numbers in their entirety, it's much more convenient to express them using **scientific notation**. A number is expressed in scientific notation as the product of a number between 1 and 10 times the number 10 raised to a power. Thus, the number 215 can be written in scientific notation as 2.15×10^2:

■ **Scientific notation**   A way of representing a large or small number as a power of 10.

$$215 = 2.15 \times 100 = 2.15 \times (10 \times 10) = 2.15 \times 10^2$$

Notice that in this case, where the number to be expressed in scientific notation is *larger* than 1, the decimal point has been moved to the left until it's *next to* the first digit. The exponent on the 10 tells how many places we had to move the decimal point to get it next to the first digit:

$$2\ 1\ 5. = 2.15 \times 10^2$$

decimal point moved 2 places left, so exponent is 2

In order to express a number *smaller* than 1 in scientific notation, we have to move the decimal point to the right until it's just *past* the first digit. The number of places moved is the negative exponent of 10. For example, the number 0.00215 can be rewritten as 2.15×10^{-3}:

$$0.00215 = 2.15 \times \frac{1}{1000} = 2.15 \times \frac{1}{10 \times 10 \times 10} = 2.15 \times \frac{1}{10^3} = 2.15 \times 10^{-3}$$

$$0.0\ 0\ 2\ 1\ 5 = 2.15 \times 10^{-3}$$

decimal point moved 3 places right, so exponent is -3

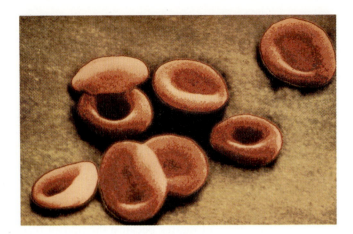

Figure 1.5
The red blood cells shown here have diameters of 6 μm, or 6 $\times$ 10^{-6} m.

The use of scientific notation is particularly helpful for indicating how many significant figures are present in a large number. We saw in the previous section that zeroes at the end of a number but to the left of a decimal point may or may not be significant. Thus, if we read that the distance from the earth to the sun is 93,000,000 miles, we don't really know how many significant figures are indicated by this number. Some of the zeroes might be significant, or they might merely act to locate the decimal point. Using scientific notation, however, we can indicate how many of the zeroes are significant. Rewriting 93,000,000 as 9.3 $\times$ 10^7 indicates two significant figures, whereas writing it as 9.300 $\times$ 10^7 indicates four significant figures. (Appendix A explains how to multiply and divide numbers expressed in scientific notation.)

Solved Problem 1.2   There are 1,760,000,000,000,000,000,000 molecules in 1 gram of sucrose (ordinary sugar). Use scientific notation to express this number to three significant figures.

Solution   Since we have to move the decimal point 21 places to the left to position it next to the first digit, the answer is 1.76 $\times$ 10^{21}.

Solved Problem 1.3   Use scientific notation to indicate the diameter of a sodium atom (0.000000000388 m).

Solution   Since we have to move the decimal point 10 places to the right to move it past the first significant figure, the answer is 3.88 $\times$ 10^{-10} m.

Practice Problems

1.5   Ordinary table salt, or sodium chloride, is made up of small particles called *ions*, which we discuss in Chapter 3. If the distance between a sodium ion and a chloride ion is 0.000000000278 m, what is this distance in scientific notation? How many pm (picometers) is this?

1.6   Convert these values to scientific notation:
(a) 58 g   (b) 46,792 m   (c) 0.0006720 cm   (d) 345.3 kg

1.7   Convert these values from scientific notation back into normal notation:
(a) 4.885 $\times$ 10^4 mg   (b) 8.3 $\times$ 10^{-6} m   (c) 4.00 $\times$ 10^{-2} mL

1.8   ROUNDING OFF NUMBERS

It often happens that a measurement appears to have more significant figures than are really justified by the precision of the work. You might read in a World Almanac, for example, that the population of the United States in 1980 was 226,549,448. Because of the errors inherent in census taking, it's very unlikely that there were *exactly* this number of people. More likely, the number you read is accurate only to two or three significant figures and should be **rounded off** to 227,000,000 people.

■ **Rounding off**   The procedure used to make sure that values are expressed with the correct number of significant figures.

There are only two rules for rounding off numbers:

1. *If the first digit you decide to remove is 4 or less, simply drop it and all following digits.* Thus, 2.4271 becomes 2.4 when rounded off to two significant figures because the first of the dropped digits (a 2) is 4 or less.

2. *If the first digit you decide to remove is 5 or greater, round it upward by adding a 1 to the digit to its left.* Thus, 4.5832 becomes 4.6 when rounded off to two significant figures because the first of the dropped digits (an 8) is 5 or greater.

The need to round off numbers is common when doing arithmetic calculations on physical quantities, where the results often contain more digits than are significant. (This is particularly true when using a hand-held calculator that shows eight-digit answers; just because a calculator gives you eight digits doesn't mean that they're all useful.) There are two rules for dealing with these situations:

1. *In carrying out a multiplication or division of two numbers, the result can't have more significant figures than either of the two original numbers.* Suppose you wanted to calculate the mileage your car was getting. If you drove 187.3 miles and used 5.2 gallons of gasoline, you could find your mileage by dividing the number of miles driven by the amount of gasoline used:

$$\text{Mileage} = \frac{\text{number of miles driven}}{\text{number of gallons used}}$$

four significant figures

$$= \frac{187.3 \text{ miles}}{5.2 \text{ gallons}}$$

two significant figures

$$= 36.019231 \text{ miles per gallon}$$

$$= 36 \text{ miles per gallon (rounded off)}$$

Since the number of gallons is known to only two significant figures, the calculated mileage must also be rounded off to two significant figures, giving a value of 36 miles per gallon.

2. *If two numbers are added or subtracted, the result can't have more digits to the right of the decimal point than either of the two original numbers.* If you

added 4.672 mL of a sterile saline solution to a container already holding 34.8 mL of solution, the total volume would have to be rounded off:

Volume of solution added =   4.672 mL   (three digits after decimal point

Volume already present   = 34.8   mL   (one digit after decimal point)

Total volume   = 39.472 mL   (unrounded)
                          = 39.5   mL   (rounded off to one digit after decimal point)

Rounding off the total volume is necessary because of the difference in precision of the two measurements. Since we don't know the volume of solution in the container past the first digit after the decimal point (it could be any value between 34.7 mL and 34.9 mL), we can't know the total of the combined volumes past the same digit.

Solved Problem 1.4   Suppose that you weighed 124 lb before dinner. How much would you weigh after dinner if you ate a total of 1.884 lb of food?

Solution   Your after-dinner weight is found by adding your original weight to the weight of the food consumed:

$$124 \text{ lb} + 1.884 \text{ lb} = 125.884 \text{ lb} \quad \text{(unrounded)}$$

Since the value of your original weight has no significant figures after the decimal point, your after-dinner weight also must have no significant figures after the decimal point. Thus, 125.884 lb must be rounded off to 126 lb.

Practice Problems   **1.8**   Round off each of the following quantities to the indicated number of significant figures:
(a) 2.304 g (3 significant figures)   (b) 188.3784 mL (5 significant figures)
(c) 0.00887 L (1 significant figure)   (d) 1.00039 kg (4 significant figures)

1.9   Carry out the following calculations, rounding each result to the correct number of significant figures:
(a) 4.87 mL + 46.0 mL   (b) 3.4 × 0.023 g
(c) 55 mg − 4.671 mg + 0.894 mg

1.9   CALCULATIONS: CONVERTING A QUANTITY FROM ONE UNIT TO ANOTHER

Many activities in the laboratory—measuring, weighing, preparing and diluting solutions, and so on—involve numerical calculations. It's often necessary to convert a quantity from one unit to another: "If there are 10 ounces of sodium bicarbonate in this box, how much is that in grams?"

Converting between units isn't mysterious; we all do it every day. If you run 9 laps around a quarter-mile track, you have to convert between the distance-unit "lap" and the distance-unit "mile" to find that you have run 2.25 miles (9 laps divided by 4 laps-per-mile). Converting from one scientific unit to another is just as easy as converting from laps into miles. It just seems harder because the units are less familiar.

The simplest way to carry out calculations involving different units is to use the **factor-label method**. In this method, a physical quantity described in one unit is converted into a quantity with different units by using a **conversion factor**:

■ **Factor-label method** A method of problem solving in which equations are set up so that unwanted units cancel and only the correct units remain.

■ **Conversion factor** A fraction that states the relationship between different units of measure.

$$\begin{pmatrix}\text{Quantity}\\ \text{in old units}\end{pmatrix} \times (\text{conversion factor}) = \begin{pmatrix}\text{Quantity}\\ \text{in new units}\end{pmatrix}$$

A conversion factor is simply any specific relationship between units, such as those given in Tables 1.4 through 1.6. For example, Table 1.5 shows that 1 km = 0.6214 mile. Writing this relationship as a fraction restates it in the form of a conversion factor, miles-per-kilometer or kilometers-per-mile:

$$\frac{0.6214 \text{ mile}}{1 \text{ km}} \quad \text{or} \quad \frac{1 \text{ km}}{0.6214 \text{ mile}}$$

These two conversion factors express the same relationship

The key to the factor-label method of problem solving is that units (labels) are treated in the same way as numbers: Units can be multiplied or divided, just as numbers can. In solving a problem, the idea is to set up an equation so that the unwanted units cancel, leaving only the desired units. Thus, if you wanted to find out how many kilometers there are in a marathon (26.22 miles), you could simply multiply the marathon distance in miles by the conversion factor in kilometers-per-mile:

$$\underset{\text{Starting quantity}}{26.22 \cancel{\text{ mile}}} \times \underset{\text{Conversion factor}}{\frac{1 \text{ km}}{0.6214 \cancel{\text{ mile}}}} = \underset{\text{New quantity}}{42.20 \text{ km}}$$

The unit "mile" cancels from the left side of the equation since it appears both above and below the division line. With "mile" now gone, "km" is the only remaining unit.

The factor-label method *always* gives the right answer if the equation is set up so that unwanted units cancel. If the equation is set up in any other way, the units will not cancel properly, and you won't get the right answer. Thus, if you were to multiply the marathon distance in miles by the incorrect conversion factor, miles-per-kilometer, you would end up with an incorrect answer expressed in meaningless units:

Incorrect $$26.22 \text{ mile} \times \frac{0.6214 \text{ mile}}{1 \text{ km}} = \frac{16.29 \text{ (miles)}^2}{\text{km}}$$ **Incorrect**

The main drawback to using the factor-label method is that it's easy to get the "right" answer without understanding what you're doing. It's therefore best when first approaching a problem to think through a rough ballpark answer in your head before doing the exact calculation. If your rough guess isn't close to the exact answer, then there's a misunderstanding somewhere. Let's try a few examples.

Solved Problem 1.5   A child is 21.5 inches long at birth. How long is this in centimeters?

Ballpark Solution   According to Table 1.5, 1 in. equals 2.54 cm. In other words, a centimeter is smaller than an inch, and it takes about two-and-one-half times more centimeters than inches to measure the same length. Thus, 21.5 in. ought to be about 2.5 times 20, or approximately 50 cm.

Exact Solution   Using the factor-label method, we can set up the following equation so that inch units cancel:

$$21.5 \text{ in.} \times \frac{2.54 \text{ cm}}{1 \text{ in.}} = 54.6 \text{ cm} \quad \text{(rounded off from 54.61)}$$

The ballpark solution and the exact solution agree closely.

Solved Problem 1.6   A standard aluminum soda can contains 12.0 ounces. How many mL is this?

Ballpark Solution   According to Table 1.6, 1 ounce equals 29.57 mL, or about 30 mL. Thus 12 ounces equals about 30 × 12, or 360 mL.

Exact Solution   Set up an equation so that "ounce" units cancel:

$$12.0 \text{ oz} \times \frac{29.57 \text{ mL}}{1 \text{ oz}} = 355 \text{ mL}$$

The ballpark answer and the exact answer agree.

Solved Problem 1.7   Administration of the digitalis preparations used to control atrial fibrillation in heart patients must be carefully controlled, because even a modest overdose can be fatal. To take differences between patients into account, dosages are usually prescribed in terms of μg per kg body weight. Thus, a child and an adult may differ greatly in weight, but both will receive the same dosage per kilogram body weight. At a dosage of 20 μg per kg body weight, how many mg of digitalis should a 160-lb patient receive?

Ballpark Solution   This is a complex problem requiring several conversions. We have to convert lb into kg, carry out a calculation, and then convert μg (micrograms) into mg. According to Table 1.4, a kilogram is a bit more than twice as large as a pound (1 kg = 2.205 lb). It therefore takes about one-half as many kg as lb to measure the same mass, and a 160-lb patient will weigh about 80 kg. At a dosage of 20 μg for each kg body weight, an 80-kg patient should receive 80 × 20 μg or about 1600 μg digitalis. A milligram is 1000 times larger than a microgram, so it will take only one thousandth as many mg as μg to equal the same dose. Our ballpark answer is about 1.6 mg digitalis.

Exact Solution   We have to set up an equation so that all unwanted units cancel, and only the unit "mg" remains:

$$160 \;\cancel{lb} \times \frac{1 \;\cancel{kg}}{2.205 \;\cancel{lb}} \times \frac{20 \;\cancel{\mu g} \text{ digitalis}}{1 \;\cancel{kg}} \times \frac{1 \text{ mg}}{1000 \;\cancel{\mu g}} = 1.5 \text{ mg digitalis}$$

(rounded off)

The ballpark solution and the exact solution agree.

Practice Problems   **1.10**   Carry out the necessary conversions to answer the following questions.
(a) How many kg does a 7.5-lb infant weigh?
(b) How many mL are in a 4.0-oz bottle of cough medicine?

1.11   Calculate the dosage in mg per kg body weight for a 135-lb adult who takes two aspirin tablets weighing 0.324 g each. Calculate the dosage for a 40-lb child who also takes two aspirin tablets.

1.10   MEASURING TEMPERATURE

Just when you were getting used to hearing daily temperatures given on the weather report in both Fahrenheit (°F) and Celsius (°C) units, along comes yet a third system. The SI unit for temperature degrees is the kelvin (K). (Note that we say only kelvin, not degrees kelvin.)

Both the kelvin and the Celsius degree are defined as one-hundredth of the interval between the freezing point of water (0°C) and the boiling point of water (100°C). Although the size of a kelvin and a Celsius degree are identical, the numbers assigned to various points on the scales differ. Whereas the Celsius scale defines the freezing point of water as 0°C, the kelvin scale defines the coldest possible temperature, $-273.15°C$ (sometimes called "absolute zero"), as 0 K. Thus, 0 K = $-273.15°C$, and 273.15 K = 0°C as the following equations indicate (for most purposes, rounding off to 273 is sufficient):

$$\text{Temperature in K} = \text{temperature in °C} + 273$$

$$\text{Temperature in °C} = \text{temperature in K} - 273$$

For practical applications in medicine and clinical chemistry, both the Fahrenheit scale and the Celsius scale are still used almost exclusively. The Fahrenheit scale defines the freezing point of water as 32°F and the boiling point of water as 212°F. Thus, it takes 180 Fahrenheit degrees to cover the same range encompassed by only 100 Celsius degrees, and a Celsius degree is therefore $180/100 = 9/5 = 1.8$ times as large as a Fahrenheit degree.

Converting between Fahrenheit and Celsius scales is similar to converting between different units of length or volume but is a bit more complex because *two* corrections need to be made—one to adjust for the difference in degree size, and one to adjust for the different zero points. The size correction is done by remembering that a Celsius degree is 9/5 the size of a Fahrenheit degree (or, conversely, that a Fahrenheit degree is 5/9 the size of a Celsius degree). The

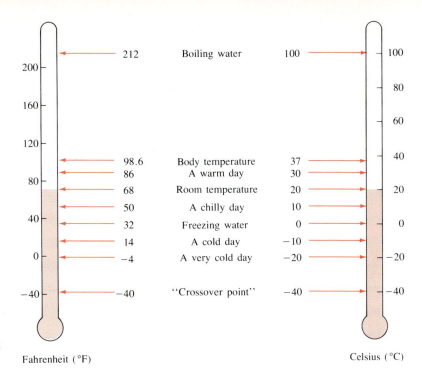

Figure 1.6
A comparison of Fahrenheit and Celsius scales.

zero point correction is done by remembering that the freezing point is higher by 32 on the Fahrenheit scale that it is on the Celsius scale. The following formulas show the conversion methods:

Celsius to Fahrenheit: $°F = (1.8 × °C) + 32°$

Fahrenheit to Celsius: $°C = \dfrac{(°F − 32°)}{1.8}$

The quickest way to become familiar with converting between Fahrenheit and Celsius scales is to remember a few convenient points and make rough estimates for values intermediate between these remembered points. Figure 1.6 lists some common temperatures in both Celsius and Fahrenheit.

Solved Problem 1.8   The highest land temperature ever recorded was 136°F in Al Aziziyah, Libya, on September 13, 1922. What is this temperature in °C?

Ballpark Solution   Celsius degrees are about twice as large as Fahrenheit degrees (1 C° = 1.8 F°), and it therefore takes about one-half as many of them to span the same range. A temperature of 136°F is about 100 Fahrenheit degrees above freezing (32°F), which would correspond to about 50 Celsius degrees above freezing. Since freezing on the Celsius scale is 0°C, our rough answer is that 136°F equals about 50°C.

Exact Solution   According to the formula given in Section 1.10, we can set up the following equation:

$$(136 − 32)\ °\!\!\!\!\diagup\!\!F \times \frac{1°C}{1.8\ °\!\!\!\!\diagup\!\!F} = 58°C \quad \text{(rounded off from 57.77778°C)}$$

The ballpark solution and the exact solution agree.

Practice Problems   *1.12*   What is the body temperature in °C of someone with a fever of 103°F?

1.13   The use of mercury thermometers is limited by the fact that mercury freezes at −38.9°C. What does this correspond to in °F?

1.11   HEAT AND ENERGY

■ **Heat**   A form of energy transferred from a hotter object to a colder one.

■ **Joule (J)**   The SI unit of energy; equal to 4.184 calories.

■ **Calorie (cal)**   A common unit of energy, defined as the amount of heat necessary to raise the temperature of one gram of water by one Celsius degree.

■ **Kilocalorie (Kcal)**   A unit of energy equal to 1000 calories.

■ **Specific heat**   The amount of heat that will raise the temperature of one gram of a substance by one Celsius degree.

What is heat? Everyone knows what it feels like, but few people really know what it is. In fact, **heat** is a form of energy that is transferred from a hotter object to a colder object when the two come in contact. Energy is measured in SI units by the **joule** (**J**; pronounced jool), but the metric **calorie** (**cal**) is still much more widely used. One calorie is the amount of heat necessary to raise the temperature of one gram of water by one Celsius degree (specifically, from 14.5°C to 15.5°C). A **kilocalorie** (**kcal**; pronounced **kay**-cal), often called a *large calorie* by nutritionists, is equal to 1000 cal

$$1 \text{ kcal} = 1000 \text{ cal}$$

$$1 \text{ cal} = 1/1000 \text{ kcal} = 4.184 \text{ J}$$

Not all substances have their temperatures raised to the same extent when equal amounts of heat are added. One calorie will raise the temperature of one gram of water by one Celsius degree, but the same amount of heat will raise the temperature of one gram of iron by 10 Celsius degrees. The amount of heat that will raise the temperature of one gram of a substance by one Celsius degree is called the substance's **specific heat**. Specific heats vary greatly from one substance to another, as shown in Table 1.7.

$$\text{Specific heat} = \frac{\text{calories}}{\text{gram °C}}$$

As indicated by Table 1.7, the specific heat of water is higher than that of most other substances. Thus, a large transfer of heat is required either to cool (remove heat from) or to warm (add heat to) a given amount of water. As a result, the human body, which is about 60 percent water, is able to maintain a steady internal temperature, even under changing outside conditions.

Table 1.7   Specific Heats of Some Common Substances

Substance	Specific Heat (cal/g °C)	Substance	Specific Heat (cal/g °C)
Water	1.00	Gold	0.031
Alcohol	0.59	Iron	0.106
Mercury	0.033	Sodium	0.293

Solved Problem 1.9	Taking a bath might use about 95 kg water. How much energy (in calories) is needed to heat the water from a cold 15°C to a warm 40°C?
Ballpark Solution	The water is being heated 25°C (from 15°C to 40°C), and it therefore takes 25 cal to heat each gram. Since 95 kg of water is 95,000 g, or about 100,000 g, it takes about 25 × 100,000 cal, or 2,500,000 calories to heat the entire tubfull of water.
Exact Solution	Set up an equation so that unwanted units cancel:

$$25°C \times \frac{1.0 \text{ cal}}{g \cdot °C} \times \frac{1000 \text{ g}}{kg} \times 95 \text{ kg} = 2,400,000 \text{ cal}$$

Practice Problems

1.14 Assuming that Coca Cola has the same specific heat as water, how much energy in calories is removed when 350 g of Coke (about the contents of one 12-oz can) is cooled from 25°C to 3°C?

1.15 If it takes 161 cal to raise the temperature of a 75-g bar of aluminum by 10°C, what is the specific heat of aluminum?

1.12 DENSITY

We've all heard the question: "Which weighs more, a pound of feathers or a pound of lead?" The answer, of course, is that they both weigh the same. A pound of feathers has far more *volume* than a pound of lead, but a pound is still a pound.

■ **Density** The mass of an object per unit of volume.

The physical property that relates the mass of an object to its volume is called **density**. Density, which is simply the mass of an object divided by its volume, is usually expressed in units of grams-per-milliliter (g/mL) for liquids and grams-per-cubic-centimeter (g/cm³) for solids. Thus, if we know the density of a substance, we know how much a given volume will weigh. The densities of some common materials are listed in Table 1.8.

$$\text{Density} = \frac{\text{mass (gm)}}{\text{volume (mL or cm}^3\text{)}}$$

Gases, as well as solids and liquids, can have their densities measured. As Table 1.8 shows, helium is much less dense than air, and helium-filled balloons therefore rise in the atmosphere.

Table 1.8 Densities of Some Common Materials

Substance	Density (g/mL)	Substance	Density (g/mL)
Ice (0°C)	0.917	Human fat	0.94
Water (4.0°C)	1.0000	Cork	0.22–0.26
Gold	19.3	Table sugar	1.59
Helium (25°C)	0.000194	Balsa wood	0.12
Air (25°C)	0.001185	Earth	5.54
Urine (25°C)	1.003–1.030	Blood plasma (25°C)	1.027

AN APPLICATION: MEASURING PERCENT BODY FAT

Much as we might complain about it, none of us would survive long without the layer of adipose tissue (body fat) lying just under the skin. Not only does this fat layer act as a shock absorber and as a thermal insulator to maintain body temperature, it also serves as a long-term energy storehouse. A normal adult body contains about 50 percent cellular protoplasm, 24 percent blood and other extracellular fluids, 7 percent bone, and 19 percent body fat. Overweight sedentary individuals have a higher fat percentage, while some world-class female gymnasts are reported to have as little as 3 percent body fat.

Percent body fat is measured by the underwater immersion method shown in the figure below. Suppose that a person weighs 60.0 kg on land. The individual then climbs into a pool of water, exhales air from the lungs to sink below the surface, and is found to weigh 2.9 kg while totally immersed. The underwater body weight is less than the land weight because water helps support the body by giving it buoyancy. Since fat is less dense than other body components, it makes a person's body particularly buoyant. The higher the percentage body fat, the more buoyant the person and the greater the difference between land weight and underwater body weight.

With body weight known, body volume is determined next. Physics tells us that the buoyancy of a submerged object (that is, the land weight minus the submerged weight) is equal to the weight of the water displaced by the object. Thus, our submerged person with a buoyancy of 57.1 kg (60.0 kg − 2.9 kg) must be displacing 57.1 kg water. Since we know that water has a density of 1.0 kg/L, we can find the volume of the submerged body:

Body volume

$$= \text{weight of water displaced} \times \text{density of water}$$

$$= (\text{land weight} - \text{underwater weight})\,\cancel{kg} \times \frac{1\ L}{\cancel{kg}}$$

$$= (60.0 - 2.9)\,\cancel{kg} \times \frac{1\ L}{\cancel{kg}} = 57.1\ L$$

With both body weight and body density now known, overall body density is calculated:

$$\text{Body density} = \frac{\text{body weight}}{\text{body volume}} = \frac{60.0\ \text{kg}}{57.1\ \text{L}}$$

$$= 1.05\ \text{kg/L} = 1.05\ \text{g/mL}$$

A chart is then used to correlate body density with percent body fat. Our individual with a density of 1.05 g/mL has about 20 percent body fat—not bad, but some more exercise might be a good idea.

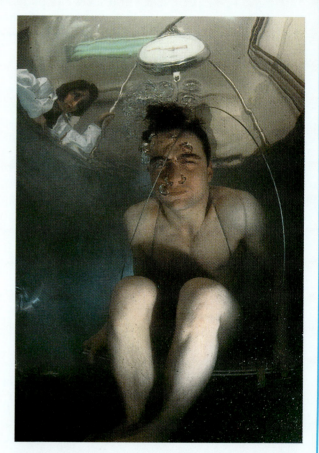

Measurement of percent body fat by the underwater immersion method.

Most substances change in volume when heated or cooled, and densities are therefore temperature dependent. For example, at 3.98°C, a 1-mL container will hold exactly 1.0000 g of water (density = 1.0000 g/mL). As the temperature is raised, however, the volume occupied by the water expands so that only 0.9584 g fits in the 1-mL container at 100°C (density = 0.9584 g/mL). When reporting a density, the temperature must also be specified.

Although most substances expand indefinitely when heated and contract indefinitely when cooled, water behaves differently. Water contracts normally when cooled from 100°C to 3.98°C, but below this temperature it begins to *expand* again. Thus, the density of liquid water is at its maximum of 1.0000 g/mL at 3.98°C, but decreases to 0.99987 g/mL at 0°C. When freezing occurs, the density drops still further to a value of 0.917 g/cm^3 for ice at 0°C. Ice and any other substance with a density less than that of water will float, but any substance with a density greater than that of water will sink.

Knowing the density of a substance can be very useful, since it's often easier to measure the volume of a liquid rather than its weight. Suppose, for example, that you needed 1.5 g of ethyl alcohol. Rather than use a dropper to weigh out exactly the right amount, it would be much easier to look up the density of ethyl alcohol (0.7893 g/mL at 20°C) and measure out the correct volume with a syringe or graduated cylinder. As the following calculation shows, density is really just a conversion factor for changing between mass (grams) and volume (milliliters):

$$1.5 \cancel{g} \text{ ethyl alcohol} \times \frac{1 \text{ mL}}{0.7893 \cancel{g}} = 1.9 \text{ mL ethyl alcohol}$$

(Note that calling the density 1 mL/0.7893 g expresses the same relationship as calling it 0.7893 g/mL.)

Solved Problem 1.10   A glass stopper weighing 16.8 g has a volume of 7.6 cm^3. What is the density of glass?

Solution   Density is equal to the mass of an object divided by its volume:

$$\text{Density} = \frac{\text{Mass}}{\text{Volume}} = \frac{16.8 \text{ g}}{7.6 \text{ cm}^3} = 2.2 \text{ g/cm}^3 \quad \begin{array}{l} \text{(rounded to 2} \\ \text{significant figures)} \end{array}$$

Solved Problem 1.11   What volume of isopropyl alcohol (rubbing alcohol) would you use if you needed 25.0 g? (The density of isopropyl alcohol is 0.7855 g/mL at 20°C.)

Ballpark Solution   Since 1 mL of isopropyl alcohol weighs a bit less than 1 g, it will take a bit more than 25 mL to get 25 g—perhaps about 30 mL.

Exact Solution   Use density as a conversion factor, and set up an equation so that the unwanted units cancel:

$$25.0 \cancel{g} \text{ isopropyl alcohol} \times \frac{1 \text{ mL}}{0.7855 \cancel{g}} = 31.8 \text{ mL isopropyl alcohol}$$

The ballpark solution and the exact solution agree.

Practice Problems

1.16 Which of the solids whose densities are given in Table 1.8 will float on water, and which will sink?

1.17 Chloroform, which was once used as an anesthetic agent, has a density of 1.474 g/mL. What volume would you use if you needed 12.37 g?

1.13 SPECIFIC GRAVITY

■ **Specific gravity** The density of a substance divided by the density of water at the same temperature.

For many purposes ranging from winemaking to medicine, it's more useful to know the **specific gravity** of a substance than to know its density. The specific gravity of any material is simply the density of the material divided by the density of water at the same temperature. Since all units cancel out, specific gravity has no units:

$$\text{Specific gravity} = \frac{\text{density of substance (g/mL)}}{\text{density of water (1 g/mL)}}$$

At normal temperatures, the density of water is very close to 1 g/mL. Thus, the specific gravity of a substance is numerically equal to its density.

■ **Hydrometer** A weighted bulb used to measure specific gravity of a liquid.

Specific gravities are usually measured only for liquids, where an instrument called a **hydrometer** is used. A hydrometer has a weighted bulb on the end of a calibrated glass tube, as shown in Figure 1.7. The depth to which the hydrometer sinks when placed in the fluid to be measured indicates its specific gravity.

Figure 1.7
(a) A hydrometer has a weighted bulb at the end of a calibrated glass tube. (b) The depth to which the hydrometer sinks in a fluid indicates the specific gravity of the fluid.

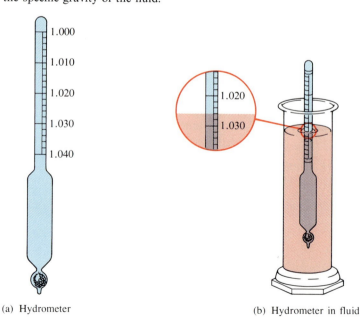

(a) Hydrometer

(b) Hydrometer in fluid

As a general rule, water that contains dissolved particles has a specific gravity higher than 1.00, because the dissolved material is usually denser than water. In winemaking, for example, the amount of fermentation taking place is gauged by observing the change in specific gravity on going from grape juice, which contains 20 percent dissolved sugar (sp. gr. = 1.082), to dry wine, which contains 12 percent alcohol (sp. gr. = 0.984). In medicine, a hydrometer called a urinometer is used to indicate the amount of solids dissolved in urine. Certain diseases, such as diabetes mellitus, are characterized by an abnormal specific gravity of the urine, giving an indication of the amount of wastes passing from the kidneys.

INTERLUDE: POWERS OF 10

It's not easy to grasp the enormous differences in size represented by numbers given as powers of 10 (exponential notation), but the following pictures might help. Imagine that you're looking at an ordinary household pin. Since the pin is only about 3 cm long, you can see little detail. Imagine, though, that the pin is magnified in size by about 10^2. That small change in the exponent on going from 1 to 10^2 means that the pin is now 100 times larger and would look something like that in part (a) of the figure below.

Now imagine that the pin is magnified by about 10^3 as in part (b) of the figure below. You see not only the bluntness of the tip (it's not as sharp as you thought) but also that the pin is covered by tiny bacteria about 5×10^{-5} cm in length. Increasing the magnification still further to about 10^5 gives you a picture of the internal details of a bacterium about to undergo cell division [part (c) of the figure below] and shows a cell wall that is only *10 nm* thick. The magnification has changed only from 1 to 10^5, but that small change in exponent has made an extraordinary difference in your view. Powers of 10 are powerful indeed.

Bacteria on a household pin at different magnifications. The individual pictures are explained in the text.

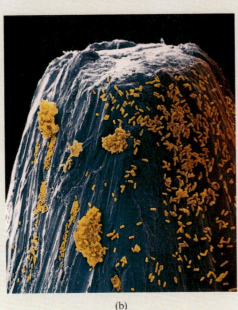
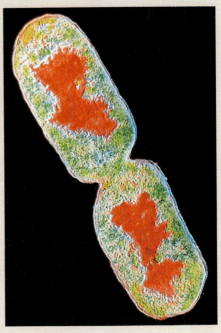

(a)   (b)   (c)

SUMMARY

Matter is anything that has mass and occupies space. Any characteristic that can be used to identify matter is called a **property**. **Physical properties**, such as color, mass, or length, can be determined without changing the chemical make-up of the sample. **Chemical properties**, such as rusting or burning, involve a change in the chemical make-up of the sample.

Physical quantities are properties that can be measured; they are described by both a number and unit. Among the most important physical quantities are mass, length, volume, and temperature. The units used to describe these properties are either those of the International System of Units (**SI units**) or of the **metric** system. **Mass** is the amount of matter an object contains; it is measured in **kilograms (kg)** or **grams (g)**. Length is measured in **meters (m)**. Volume is measured in cubic meters (m^3) in the SI system, and by **liters (L)** or milliliters (mL) in the metric system. Temperature is measured in **kelvins (K)** in the SI system and by **Celsius degrees (C°)** in the metric system.

When measuring physical quantities, it's important to indicate the exactness of the measurement by using the correct number of **significant figures**. All but one of the significant figures in a number should be known with certainty; the final digit is estimated to ± 1. Measurements of small and large quantities are often expressed in **scientific notation** as the product of a number between 1 and 10 times some power of 10. For example: $3562 = 3.562 \times 10^3$.

Many applications in science and medicine involve numerical calculations that require the manipulation and conversion of different units of measurement. The best way to solve such problems is to use the **factor-label method**, in which an equation is set up using **conversion factors** so that unwanted units cancel out and only the desired unit(s) remains.

Density is the physical property that relates mass to volume; it is usually expressed in units of g/mL (read as grams-per-milliliter) or g/cm^3 (grams-per-cubic-centimeter). If the same amount of mass is present in two samples, the sample of the denser substance, such as lead or mercury, will appear "heavier" because its mass is concentrated in a smaller volume. The **specific gravity** of an object—usually a liquid—is the object's density divided by the density of water at the same temperature. Since the density of water is approximately 1 g/mL, specific gravity and density have the same numerical value.

Heat is a form of energy that is transferred from hotter to cooler objects. The **specific heat** of a substance is the amount of heat necessary to raise the temperature of 1 gram of substance by 1°C. Water has an unusually high specific heat, which helps our bodies to maintain an even temperature.

REVIEW PROBLEMS

Definitions and Units

1.18   What is the difference between a physical property and a chemical property?

1.19   What is the difference between a physical quantity and a number?

1.20   Give four examples of physical properties.

1.21   What is the difference between mass and weight?

1.22   What are the basic units used in the metric system to measure mass, volume, and length?

1.23   What are the basic units used in the SI system to measure mass, volume, and length?

1.24 Which is larger, a Fahrenheit degree or a Celsius degree? By how much?

1.25 What is the difference between a kelvin and a Celsius degree?

1.26 What is the difference between density and specific gravity?

1.27 Is specific heat a chemical property or a physical property?

1.28 What do these abbreviations stand for? (a) cL (b) dm (c) μm (d) nL

1.29 How many picograms are in 1 mg? in 35 ng?

1.30 How many microliters are in 1 L? in 20 mL?

1.31 Which is smaller, a liter or a quart? By about how much?

Unit Conversions

1.32 The speed limit in Canada is 100 km/hour. How many miles/hour is this?

1.33 The speed limit in the United States is 55 miles/hour. How many km/hour is this?

1.34 Batrachotoxin, the active component of South American arrow poison obtained from the Golden frog (*Phyllobates terribilis*), is so toxic that a single frog contains enough poison (1100 μg) to kill 2200 people. How many micrograms would it take to kill one person? How many ounces?

1.35 Carry out the following conversions: (a) 550 mL expressed in L; in quarts (b) 3340 m expressed in cm; in km; in miles

1.36 What is the height in meters of a person 5′4″ tall?

1.37 How many liters are in a 10-mL syringe?

1.38 What is the temperature of the normal human body (98.6°F) in °C? in K?

1.39 If you were feeling sick and a thermometer measured your body temperature at 39.3°C, what would your temperature be in °F?

1.40 The spiny anteater of New Guinea has a normal body temperature of 22.2°C. What is the anteater's temperature in °F? in K?

1.41 The lowest land temperature ever recorded was −128.6°F at Vostok Station, Antarctica. What is this in °C? in K?

1.42 How many grams of meat are in a quarter-pound hamburger? How many kg?

1.43 The World Trade Center in New York has an approximate floor area of 406,000 m². How many square feet of floor does the building have?

1.44 If each story of the World Trade Center (Problem 1.43) is 8.0 ft high, what is the total volume of the building in cubic feet? in m³?

1.45 The longest recorded jump by a flea is 13.0 in. How far is this in mm?

1.46 A normal value for blood cholesterol is 200 mg/dL. If a normal adult has a total blood volume of 5 L, how much total cholesterol is present?

1.47 The white blood cell concentration (w.b.c.) in normal blood is approximately 5000 cells/mm³. How many white blood cells does a normal adult (Problem 1.46) have? Express the answer in scientific notation.

Scientific Notation and Significant Figures

1.48 Express the answers to Problems 1.43 and 1.44 in scientific notation assuming three significant figures.

1.49 The diameter of the earth at the equator is 7,926.381 miles. (a) Round off the earth's diameter to four significant figures; to two significant figures; to six significant figures. (b) Express the earth's diameter in scientific notation.

1.50 What is the earth's diameter (Problem 1.49) in km? Express the result in scientific notation to five significant figures.

1.51 Express the following numbers in scientific notation, assuming all figures are significant: (a) 2586 (b) 4957500 (c) 0.003870

1.52 Convert the following numbers from scientific notation into normal notation: (a) 4.87×10^3 (b) 5.501×10^6 (c) 2.540×10^{-3}

1.53 How many significant figures does each of the following numbers have? (a) 237401 (b) 0.300 (c) 3.01 (d) 244.4 (e) 50000

Heat, Density, and Specific Gravity

1.54 Calculate the specific heat of copper if it takes 23 cal to heat a 5.0-g sample from 25°C to 75°C.

1.55 Assume that 50 cal of heat is applied to a 15-g sample of sulfur at 20°C. What is the final temperature of the sample if the specific heat of sulfur is 0.175 cal/g °C?

1.56 Aspirin has a density of 1.40 g/cm³. What is the volume in cm³ of a tablet weighing 250 mg?

1.57 What is the density of isopropyl alcohol ("rubbing alcohol") if a 5.000-mL sample weighs 3.928 g at room temperature? What is the specific gravity of isopropyl alcohol?

1.58   Ethylene glycol (automobile antifreeze) has a specific gravity of 1.1088 at room temperature. What is the volume of 1 kg of ethylene glycol? What is the volume of 2 lb of ethylene glycol?

1.59   What is the mass of 500 mL of ethylene glycol (Problem 1.58)? What is the weight of 1 pint of ethylene glycol?

1.60   Which weighs more, a kilogram of feathers or a pound of lead?

1.61   In which liquid will a urinometer float higher, ethanol (sp. gr. = 0.7893) or chloroform (sp. gr. = 1.4832)?

Additional Problems

1.62   Gemstones are weighed in *carats*, where one carat = 200 mg. What is the mass in grams of the Hope Diamond, the world's largest blue diamond at 44.4 carats?

1.63   What is the weight in ounces of the Hope diamond (Problem 1.62)? How many significant figures should your answer have?

1.64   If you were cooking in an oven calibrated in Celsius degrees, what temperature would you use if the recipe called for 350°F?

1.65   Macau, on the southern coast of China, is reported to be the most densely populated territory in the world, with 293,000 people living in an area of 6.2 square miles. What is the population density of Macau in people/miles^2? Express your answer in scientific notation with the proper number of significant figures.

1.66   What is the population density of Macau (Problem 1.65) in people/km^2 expressed in scientific notation with the proper number of significant figures?

1.67   Bozo Miller is generally acknowledged as the world's greatest trencherman, having eaten 27 two-pound chickens in a single sitting. What was the total mass of the meal in grams, expressed in scientific notation with the proper number of significant figures?

1.68   What dosage in grams per kilogram of body weight does a 130-lb woman receive if she takes two 250-mg tablets of penicillin? How many 125-mg tablets would a 40-lb child take to receive the same dosage?

1.69   The density of air at room temperature is 1.3 g/L at 0°C. What is the mass of the air in a room that is 4.0-m long, 3.0-m wide, and 2.5-m high?

1.70   Approximately 75 mL of blood are pumped by the normal human heart at each beat. Assuming an average pulse of 72 beats per minute, how much blood is pumped in a day?

CHAPTER 2

Matter and Molecules

A crystal of pure silver grown in the laboratory.

Matter—what is it? We answered this question in the last chapter by saying that matter is anything that has mass and volume. In this chapter, we'll examine matter more closely to see what it's made of and how it can be described. We'll look at these topics:

1. **What are elements?** The goal: you should learn what an element is and what the names and symbols of the first 20 elements are.
2. **What are atoms, and what are they made of?** The goal: you should learn the three elementary particles that make up atoms and how electrons are distributed in shells around atoms.
3. **What is the periodic table?** The goal: you should learn how elements are arranged in the periodic table and how the position of an element in the periodic table relates to its electronic structure.
4. **What are molecules, compounds, and mixtures?** The goal: you should learn the classifications of matter and how to distinguish among them.

2.1 ELEMENTS

It was once thought that all matter was composed of four elemental substances: earth, air, fire, and water. Although the specifics were wrong, this early notion of elemental substances was correct in principle. Matter is in fact formed from a combination of one or more of 107 presently known elements. An **element** is a fundamental substance that cannot be chemically changed or broken down into anything simpler. Iron, gold, aluminum, and mercury are common examples.

Elements are numbered from 1 through 107 in order of increasing size and complexity. Only 89 of the 107 known elements occur naturally; the remaining 18 have been made artificially using high-energy particle accelerators. As indicated in Figure 2.1, which shows approximate elemental compositions of the earth's crust and the human body, the 89 naturally occurring elements are not equally abundant. Oxygen and silicon together account for 75 percent of the mass in the earth's crust. Oxygen, carbon, and hydrogen account for nearly all the mass in a human body.

■ **Element** One of the 107 presently known fundamental substances that cannot be broken down chemically into any simpler substance.

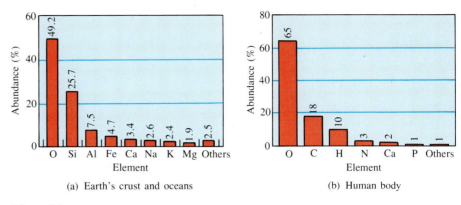

Figure 2.1
Elemental compositions of (a) the earth's crust, and (b) the human body.

2.2 NAMES AND SYMBOLS FOR ELEMENTS

Rather than write out the full names of elements, chemists have devised a timesaving shorthand in which elements are referred to by one- or two-letter symbols (Table 2.1). Most of the symbols are simply the first one or two letters of the element name, such as H (hydrogen) and Al (aluminum), but others, such as Na (natrium = sodium), are derived from Latin words. Note that all two-letter symbols have only their first letter capitalized; the second letter is always lowercase. A list of the most common elements is given in Table 2.1, and a complete alphabetical list of all 107 elements is given on the inside back cover.

Table 2.1   Names and Symbols for Some Common Elements

Al	Aluminum	Cl	Chlorine	Mn	Manganese	Cu	Copper (Cuprum)
Ar	Argon	F	Fluorine	N	Nitrogen	Fe	Iron (Ferrum)
Ba	Barium	He	Helium	O	Oxygen	Pb	Lead (Plumbum)
B	Boron	H	Hydrogen	P	Phosphorus	Hg	Mercury (Hydrargyrum)
Br	Bromine	I	Iodine	Si	Silicon	K	Potassium (Kalium)
Ca	Calcium	Li	Lithium	S	Sulfur	Ag	Silver (Argentum)
C	Carbon	Mg	Magnesium	Zn	Zinc	Na	Sodium (Natrium)

Although many of the elements may be unfamiliar to you and their names appear exotic, at least 20 are essential for human life. In addition to such well-known elements as carbon, hydrogen, oxygen, and nitrogen, less familiar elements such as molybdenum and selenium are also important (Table 2.2, page 30).

Practice Problems   **2.1**   Look at the alphabetical list on the inside back cover, and find the symbols for these elements:
(a) uranium (the fuel in nuclear reactors)
(b) titanium (the skin of jet fighters)
(c) tungsten (the filament in light bulbs)

2.2   What elements do these symbols represent?
(a) Na   (b) Ca   (c) Pd   (d) K   (e) Sr   (f) Sn

Table 2.2 Elements Essential for Human Life

Element	Symbol	Function
Carbon	C	These four elements are present throughout all living organisms.
Hydrogen	H	
Oxygen	O	
Nitrogen	N	
Calcium	Ca	Necessary for growth of teeth and bones
Chlorine	Cl	Necessary for maintaining salt balance in body fluids
Chromium	Cr	Aids in carbohydrate metabolism
Cobalt	Co	Component of vitamin B-12
Copper	Cu	Necessary to maintain blood chemistry
Fluorine	F	Aids in development of teeth and bones
Iodine	I	Necessary for thyroid function
Iron	Fe	Necessary for oxygen-carrying ability of blood
Magnesium	Mg	Necessary for bones, teeth, and muscle and nerve action
Manganese	Mn	Necessary for carbohydrate metabolism and bone formation
Molybdenum	Mo	Component of enzymes necessary for metabolism
Phosphorus	P	Necessary for growth of bones and teeth; present in DNA/RNA
Potassium	K	Component of body fluids; necessary for nerve action
Selenium	Se	Aids vitamin E action and fat metabolism
Sodium	Na	Component of body fluids; necessary for nerve and muscle action
Sulfur	S	Component of proteins; necessary for blood clotting
Zinc	Zn	Necessary for growth, healing, and overall health

2.3 ATOMIC THEORY

■ **Atom** The smallest and simplest piece that an element can be broken into while still maintaining the chemical properties of the element.

■ **Subatomic particle** An elementary particle from which atoms are made.

■ **Proton** A positively charged subatomic particle found in the nucleus of atoms.

■ **Neutron** An electrically neutral subatomic particle found in the nucleus of atoms.

■ **Electron** A negatively charged subatomic particle that orbits around the nucleus at a distance.

Take a piece of aluminum foil and cut it in two. Then take one of the pieces, cut *it* in two, and so on. Assuming that you had infinitely small scissors and infinite dexterity, how long could you keep dividing the foil? Is there a limit, or is matter infinitely divisible into ever smaller and smaller pieces? In fact, there is a limit. The smallest and simplest piece that an element can be broken into while still maintaining the element's properties is called an **atom** (derived from the Greek, *atomos*, meaning indivisible).

Atoms are extremely small, ranging from about 7.4×10^{-11} m in diameter for hydrogen to 5.24×10^{-10} m for cesium;* even the smallest speck of dust visible to the naked eye contains about 10^{16} atoms. It's difficult to appreciate just how small atoms really are, although it might help if you realize that a fine pencil line is about 3 *million* atoms across.

Although chemically indivisible, atoms are nevertheless made up of three small **subatomic particles**, called **protons**, **neutrons**, and **electrons**. Protons have a mass of 1.673×10^{-24} g, neutrons have a mass of 1.675×10^{-24} g, and electrons have a mass of 9.109×10^{-28} g. Since these masses are much too small

* Atomic diameters are often expressed in angstrom units (Å), where 1 Å $= 10^{-10}$ m. Thus, the diameter of a hydrogen atom is 0.74 Å, and the diameter of a cesium atom is 5.24 Å.

Table 2.3 A Comparison of Subatomic Particles

| | Mass | | Charge |
	In Grams	In amu	In Charge Units
Electron	9.109×10^{-28}	5.486×10^{-4}	-1
Proton	1.673×10^{-24}	1.007	$+1$
Neutron	1.675×10^{-24}	1.009	0

■ **Atomic mass unit (amu)** A convenient unit of mass on the atomic scale; approximately equal to the mass of a proton or neutron.

■ **Dalton** An alternate name for atomic mass unit.

for convenience, we describe them by a new unit called the **atomic mass unit** (**amu**), also called a **dalton**. One atomic mass unit is 1.661×10^{-24} g, or almost exactly the mass of a proton or a neutron. Thus, for all practical purposes, the masses of both protons and neutrons are 1 amu. An electron is so much lighter than either a proton or a neutron, however, that its mass is usually ignored (mass of electron = 0 amu).

The most important characteristic of protons and electrons is that they are electrically charged: protons have a positive charge ($+$), and electrons have a negative charge ($-$). Neutrons, however, have no electrical charge. Table 2.3 summarizes the data for all three subatomic particles.

Because they are charged, protons and electrons exert forces on each other. Just as the north pole of one magnet pushes away the north pole of another magnet but attracts an opposite (south) pole, particles with similar charges repel each other, and particles with opposite charges attract. Thus, one proton repels another, and one electron repels another, but a proton and an electron attract each other. It is this interplay of repulsive and attractive forces that explains much of chemistry.

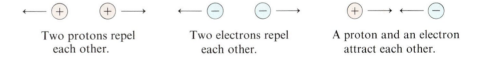

Two protons repel each other.   Two electrons repel each other.   A proton and an electron attract each other.

Atoms of different elements differ from each other according to how many protons they contain, a value called the *atomic number* of the element. Each atom contains an identical number of protons and electrons so that, overall, atoms are neutral; they have no net charge. Thus, the atomic number also equals the number of electrons an atom has. Hydrogen, atomic number 1, has only one proton (and one electron); helium, atomic number 2, has two protons (and two electrons); oxygen, atomic number 8, has eight protons (and eight electrons); and so on up to element number 107.

■ **Atomic number** The primary characteristic that distinguishes atoms of different elements; equal to the number of protons in an atom's nucleus.

Atomic number of an element = Number of protons in atom

■ **Mass number** The sum of an atom's protons and neutrons.

The sum of an atom's protons and neutrons is called its *mass number*. Most hydrogen atoms have one proton and no neutrons and therefore have mass number = 1; helium atoms, with two protons and two neutrons, have mass number = 4; oxygen atoms, with eight protons and eight neutrons have mass number = 16; and so on. In general, atoms contain at least as many neutrons as protons, but there is no simple way to predict how many neutrons a given atom will have.

Mass number of an atom = Number of protons + neutrons in atom

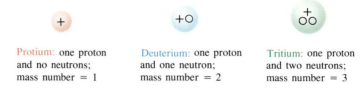

Protium: one proton and no neutrons; mass number = 1

Deuterium: one proton and one neutron; mass number = 2

Tritium: one proton and two neutrons; mass number = 3

Figure 2.2
The three isotopes of hydrogen. All have atomic number = 1.

■ **Isotopes** Atoms of the same element that have different numbers of neutrons in their nuclei. Isotopes have the same atomic number but different mass numbers.

Although all atoms of a given element have the same number of protons (the atomic number characteristic of that element) different atoms of an element can have different mass numbers, depending on how many neutrons they have. Such atoms with *identical atomic numbers* but *different mass numbers* are called **isotopes**. Hydrogen, for example, has three isotopes (Figure 2.2). All hydrogen atoms have one proton, but most have no neutrons. These hydrogens, called *protium*, have mass number = 1. A small percentage of hydrogen atoms have one neutron. This isotope, mass number = 2, is called *deuterium*. A third isotope, called *tritium*, has two neutrons and mass number = 3. Although tritium does not occur naturally, it can be made in nuclear reactors. We'll see in Chapter 20 how isotopes can be used in biology and medicine.

Different isotopes are represented by a format in which the mass number is given as a superscript and the atomic number is given as a subscript. Thus, protium is $^1_1$H, deuterium is $^2_1$H, and tritium is $^3_1$H.

Mass number (sum of protons and neutrons) → 3

$^3_1$H ← Symbol of element

Atomic number (number of protons) → 1

■ **Atomic weight** The average mass (in amu) of a large sample of an element's atoms.

If we were to take a large number of hydrogen atoms, we would find that 99.985 percent of them had mass number 1 (protium) and 0.015 percent had mass number 2. The *average* mass of a large number of an element's atoms is called the element's **atomic weight**. Thus, hydrogen has atomic weight 1.008, which is slightly higher than 1. Similarly, chlorine has two principal isotopes with mass numbers 35 (75.53 percent) and 37 (24.47 percent), and its atomic weight is 35.5. This value is calculated in the following way:

Contribution from ^{35}Cl	75.53% of 35	= 26.44
Contribution from ^{37}Cl	24.47% of 37	= 9.05
	Atomic weight	= 35.5

Solved Problem 2.1

Most phosphorus atoms, atomic number 15, have mass number 31. How many protons and how many neutrons are in phosphorus?

Solution

The atomic number of an atom tells how many protons the atom contains, and the mass number tells the total number of protons and neutrons. Thus, phosphorus has 15 protons and $31 - 15 = 16$ neutrons.

Practice Problems

2.3 The uranium used in nuclear reactors has atomic number 92 and mass number 235. How many protons and neutrons does this uranium have?

2.4 Chlorine, one of the atoms found in common table salt (sodium chloride), has two common isotopes with mass numbers 35 and 37. If chlorine has atomic number 17, how many protons and neutrons does each isotope have?

2.5 Show how both chlorine isotopes (Problem 2.4) can be symbolized using subscripts for atomic number and superscripts for mass number.

2.4 THE STRUCTURE OF ATOMS

■ **Nucleus** The dense, positively charged mass at the center of an atom where protons and neutrons are located.

Subatomic particles are not distributed at random throughout an atom. Rather, protons and neutrons are pressed together into a dense, positively charged mass called the **nucleus**, which is located at the center of the atom. Most of an atom is just empty space, with all of the mass concentrated in the center and with the electrons situated a great distance away from the nucleus. For comparison, if an atom had the diameter of a parking lot and the nucleus were a small car parked in the center, the edge of the atom would be more than 50 *miles* away (Figure 2.3).

2.5 ENERGY LEVELS OF ELECTRONS

How are the electrons distributed in an atom? Since they are in constant motion around the nucleus, it's impossible to define precisely the positions of specific electrons. It turns out, however, that electrons are not perfectly free to move about; they are confined to specific regions within the atom according to the amount of energy they have. Different electrons have different amounts of energy and thus occupy different regions within the atom.

Figure 2.3
The relative size of the nucleus within an atom. If the nucleus were the size of a small car, the atom would extend for about 50 miles in all directions.

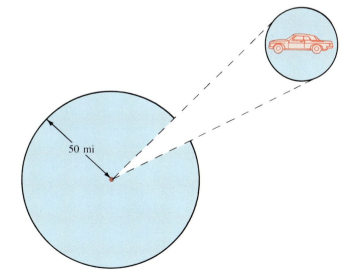

50 mi

Table 2.4   Distribution of Electrons into Shells

Number of Shell	Electron Capacity of Shell
First	2
Second	8
Third	18
Fourth	32
Fifth	50

■ **Shell**   An imaginary layer surrounding an atom's nucleus where electrons are located.

The electrons in an atom are grouped by energy into **shells**, which can be thought of as successive layers of electrons at increasing distances from the nucleus. The farther a shell is from the nucleus, the more electrons it can hold and the greater the energies of those electrons. For example, an atom's lowest-energy electrons occupy the first shell, which is nearest the nucleus and has a capacity of only two electrons. The second shell is farther from the nucleus and can hold eight electrons; the third shell is still farther from the nucleus and can hold eighteen electrons; and so on as indicated in Table 2.4.

■ **Subshell**   A subregion of a shell where electrons of the same energy level are located.

Within each shell, electrons are further grouped into various energy **subshells**, and within each subshell, electrons are grouped by pairs into orbitals. There are four different kinds of subshells, denoted *s*, *p*, *d*, and *f*. Of the four, we will be concerned only with *s* and *p* subshells, since most of the atoms found in living organisms use only these.

■ **Orbital**   A specifically shaped region of space around an atom, denoted *s*, *p*, *d*, or *f*, where electrons of a specific energy level are found.

Orbitals are different-shaped regions of space within the atom where electrons of a specific energy level are found. For example, orbitals in an *s* subshell are spherical with the atomic nucleus in the center, and orbitals in a *p* subshell are dumbbell-shaped, as shown in Figure 2.4. Note that we aren't saying anything about the *path* an electron follows as it moves about the nucleus; we are only defining a general region of space in which the electron moves about. It's often helpful to think of an orbital in terms of a time-lapse photograph of an electron's movement around the nucleus. In such a photograph, the orbital would appear as a kind of blurry cloud showing where the electron has been.

Different shells have different numbers and kinds of orbitals. The two electrons of the first shell occupy a single *s* orbital, designated 1*s*. (The "1" refers to the first shell, and the "*s*" refers to the spherical shape of the orbital.) The eight electrons of the second shell occupy one *s* orbital (designated 2*s*) and three different *p* orbitals (each designated 2*p*). The eighteen electrons of the third shell occupy an *s* orbital (3*s*), three *p* orbitals (3*p*), and five *d* orbitals (3*d*). These electron distributions are indicated in Figure 2.5.

Figure 2.4
The shapes of *s* and *p* orbitals: (a) *s* orbitals are spherical, and (b) *p* orbitals are dumbbell-shaped. Each orbital can hold only two electrons.

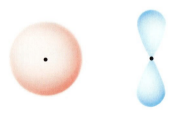

(a) An *s* orbital   (b) A *p* orbital

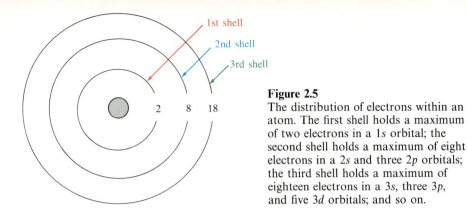

1st shell
2nd shell
3rd shell

2   8   18

Figure 2.5
The distribution of electrons within an atom. The first shell holds a maximum of two electrons in a $1s$ orbital; the second shell holds a maximum of eight electrons in a $2s$ and three $2p$ orbitals; the third shell holds a maximum of eighteen electrons in a $3s$, three $3p$, and five $3d$ orbitals; and so on.

2.6   ELECTRON CONFIGURATION OF ATOMS

■ **Electron configuration** The specific way that an atom's electrons are distributed into shells and subshells.

The specific way an atom's electrons are distributed throughout its shells and orbitals is called the atom's **electron configuration**. We can determine the electron configuration of any element by following a series of three rules:

1. Find the atomic number of the element to see how many electrons it has.
2. Begin assigning electrons to orbitals according to the order shown in Figure 2.6. Fill the lowest-energy orbitals first before moving on to higher levels. (Note that there is some overlap of energies between shells, with the $4s$ orbital lower in energy than the $3d$ orbitals.) Remember that each orbital can hold only two electrons.
3. If two or more orbitals have the same energy (for example, the three p orbitals or the five d orbitals), fill all orbitals halfway before completely filling any one of them.

Complete electron configurations for the first 20 elements are shown in Table 2.5. Since this list includes almost all of the important elements found in living organisms, we won't be concerned with heavier atoms. Notice that the number of electrons in each subshell is indicated in Table 2.5 by using superscripts. For example, the notation $1s^2$ for helium means that the $1s$ orbital in a helium atom has two electrons. The total number of electrons in each shell is found by adding the number of electrons in the subshells within that shell. Thus, the notation $1s^2\ 2s^2\ 2p^6\ 3s^2$ for magnesium means that magnesium has

Figure 2.6
Relative energy levels of orbitals and their capacities. In determining the electron configuration of an atom, the lowest-energy orbitals are filled first.

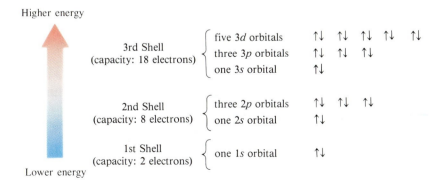

Higher energy

Lower energy

3rd Shell (capacity: 18 electrons)
- five $3d$ orbitals
- three $3p$ orbitals
- one $3s$ orbital

2nd Shell (capacity: 8 electrons)
- three $2p$ orbitals
- one $2s$ orbital

1st Shell (capacity: 2 electrons)
- one $1s$ orbital

Table 2.5   Electron Configurations of the First 20 Elements

	Element	Atomic Number	Electron Configuration		Element	Atomic Number	Electron Configuration
H	Hydrogen	1	$1s^1$	Na	Sodium	11	$1s^2\ 2s^2\ 2p^6\ 3s^1$
He	Helium	2	$1s^2$	Mg	Magnesium	12	$1s^2\ 2s^2\ 2p^6\ 3s^2$
Li	Lithium	3	$1s^2\ 2s^1$	Al	Aluminum	13	$1s^2\ 2s^2\ 2p^6\ 3s^2\ 3p^1$
Be	Beryllium	4	$1s^2\ 2s^2$	Si	Silicon	14	$1s^2\ 2s^2\ 2p^6\ 3s^2\ 3p^2$
B	Boron	5	$1s^2\ 2s^2\ 2p^1$	P	Phosphorus	15	$1s^2\ 2s^2\ 2p^6\ 3s^2\ 3p^3$
C	Carbon	6	$1s^2\ 2s^2\ 2p^2$	S	Sulfur	16	$1s^2\ 2s^2\ 2p^6\ 3s^2\ 3p^4$
N	Nitrogen	7	$1s^2\ 2s^2\ 2p^3$	Cl	Chlorine	17	$1s^2\ 2s^2\ 2p^6\ 3s^2\ 3p^5$
O	Oxygen	8	$1s^2\ 2s^2\ 2p^4$	Ar	Argon	18	$1s^2\ 2s^2\ 2p^6\ 3s^2\ 3p^6$
F	Fluorine	9	$1s^2\ 2s^2\ 2p^5$	K	Potassium	19	$1s^2\ 2s^2\ 2p^6\ 3s^2\ 3p^6\ 4s^1$
Ne	Neon	10	$1s^2\ 2s^2\ 2p^6$	Ca	Calcium	20	$1s^2\ 2s^2\ 2p^6\ 3s^2\ 3p^6\ 4s^2$

two electrons in its first shell, eight electrons in its second shell, and two electrons in its third shell.

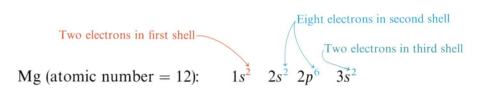

Two electrons in first shell — Eight electrons in second shell — Two electrons in third shell

$$\text{Mg (atomic number} = 12): \qquad 1s^2 \quad 2s^2\ 2p^6 \quad 3s^2$$

Solved Problem 2.2   Show how the electron configuration of oxygen can be assigned.

Solution   Oxygen, atomic number 8, has eight electrons to be placed in specific orbitals. Assignments are made by putting two electrons in each orbital, according to the order shown in Figure 2.6. When all electrons are assigned, the configuration is complete. For oxygen, the first two electrons are placed in the $1s$ orbital ($1s^2$), the next two electrons are placed in the $2s$ orbital ($2s^2$), and the next four electrons are spread over the three available $2p$ orbitals ($2p^4$). Thus, oxygen has the configuration $1s^2\ 2s^2\ 2p^4$

Practice Problem   2.6   Show how electron configurations can be assigned for these elements (check your answers in Table 2.5):
(a) C   (b) Na   (c) Cl   (d) Ca.

2.7   THE PERIODIC TABLE

By the early 1860s, 62 elements were known, and chemists had begun to look for ways of organizing their accumulated knowledge. Soon, certain similarities among groups of elements were noticed. For example, lithium, sodium, and potassium were all known to react violently with water; chlorine, bromine, and iodine were all known to have similar chemical behavior, as were calcium, strontium, and barium. Furthermore, graphs of the effect of atomic number on various physical properties tended to show a periodic pattern of rising and

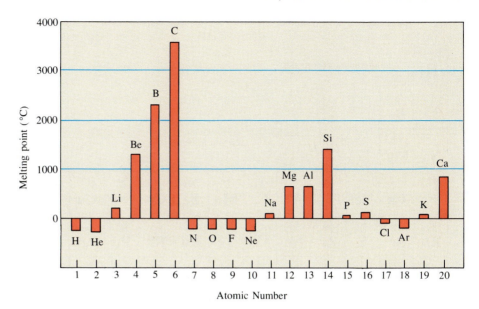

Figure 2.7
A graph of melting point versus atomic number for the first 20 elements shows a clear rise-and-fall pattern (periodicity).

■ **Periodic table** The standard chart displaying the elements in order of increasing atomic number so that elements with similar properties fall into groups.

■ **Period** A horizontal row of elements in the periodic table. Elements are listed in order of increasing atomic number.

■ **Group** A vertical column of elements in the periodic table. Elements in a group have similar chemical properties.

■ **Main group** An element group on the far right (Groups 3A–8A) or far left (Groups 1A–2A) of the periodic table.

■ **Transition metal group** An element group in the middle (1B–10B) of the periodic table.

■ **Alkali metal** An element in Group 1A of the periodic table (Li, Na, K, Rb, Cs, Fr).

falling. For example, the graph of atomic number versus melting point in Figure 2.7 shows a clear rise-and-fall periodicity.

In 1869, the Russian chemist Dmitri Mendeleev suggested a tabular way of classifying the elements according to the similarities of their properties. In modified form, Mendeleev's table of the elements has come down to us today as the **periodic table**, shown in Figure 2.8, page 38.

Beginning at the upper left corner of the table, elements are arranged in horizontal rows, or **periods**, in order of increasing atomic number. The first row contains only two elements, hydrogen and helium; the second and third rows each contain eight elements; the fourth and fifth rows each contain 18; the sixth row contains 32; and the seventh row contains 21. When this arrangement is used, the elements fall into 18 vertical columns, or **groups**, which have remarkably similar chemical properties.

Notice that not all groups (columns) have the same number of elements. The larger groups on the left and right of the table are usually called the **main groups**, while those in the middle are called the **transition metal groups**. Notice also that the two series of 14 elements following lanthanum and actinium are placed outside the table and are not given group numbers. The main reason for this arrangement is simply so that the table will fit conveniently on a page. The arrangement also makes chemical sense, however, because the 14 elements following lanthanum (called the *lanthanides*) are all quite similar, as are the 14 elements following actinium (called the *actinides*).

2.8 GROUP CHARACTERISTICS OF THE ELEMENTS

Groups of elements in the periodic table often show remarkable similarities in their chemical properties. Look at the following four groups:

● *Group 1A*—**Alkali metals:** Lithium, sodium, potassium, rubidium, and cesium are shiny, soft, low-melting metals. All react violently with water

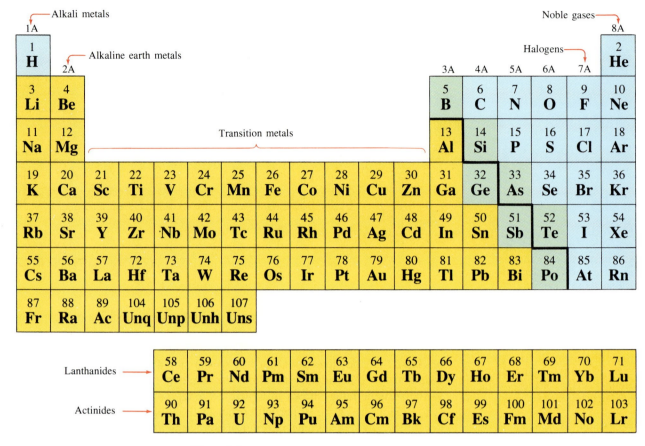

Figure 2.8
The periodic table of the elements. The elements cerium (Ce) through lutetium (Lu) follow lanthanum (La) in the table, and the elements thorium (Th) through lawrencium (Lr) follow actinium (Ac). Metallic elements are shown in yellow, metal-like elements (metalloids) are in green, and nonmetallic elements are in blue.

to form products that are extremely alkaline—hence the name *alkali metals*. Because of their high reactivity, the alkali metals are never found in nature in the pure state but only in chemical combination with other elements.

■ **Alkaline earth metal** An element in Group 2A of the periodic table (Be, Mg, Ca, Sr, Ba, Ra).

- *Group 2A*—**Alkaline earth metals:** Beryllium, magnesium, calcium, strontium, barium, and radium are also lustrous, silvery metals, but all are less reactive than their neighbors in Group 1.

■ **Halogen** An element in Group 7A of the periodic table (F, Cl, Br, I).

- *Group 7A*—**Halogens:** Fluorine, chlorine, bromine, and iodine are reactive, corrosive nonmetals. The halogens are found in nature only in combination with other elements, such as sodium in table salt (sodium chloride). In fact, their group name, halogen (**hal**-o-jen), derives from the Greek word *hals*, meaning salt.

■ **Noble gas** An element in Group 8A of the periodic table (He, Ne, Ar, Kr, Xe, Rn).

- *Group 8A*—**Noble gases:** Helium, neon, argon, krypton, xenon, and radon are gases of extremely low reactivity. Helium, neon, and argon don't react with any other element; krypton and xenon react with very few.

■ **Metal** A malleable element with a lustrous appearance that is a good conductor of heat and electricity.

■ **Nonmetal** An element at the right side of the periodic table that is a poor conductor of heat and electricity.

In addition to the group similarities mentioned, we can roughly divide the elements of the periodic table into metals and nonmetals. **Metals** usually have a lustrous appearance, are good conductors of heat and electricity, and are malleable rather than brittle. **Nonmetals,** by contrast, might be either gaseous, liquid, or solid, and are poor conductors of heat and electricity. The dividing line between the two groups follows a zigzag path angling from boron at the top middle to the lower right corner. Elements to the left of the line have metallic properties, while elements to the right are nonmetals.

Practice Problem **2.7** Which of the following elements would you expect to be metals?
(a) scandium (Sc)   (b) technetium (Tc)   (c) selenium (Se)

AN APPLICATION: CHERNOBYL AND CESIUM

Few people had heard of the element cesium until April 26, 1986, when an explosion destroyed the unit 4 reactor at the Chernobyl nuclear power plant in Russia. As a consequence of that accident, however, cesium came to be a household word throughout much of Europe.

Discovered in 1860 in mineral waters from the small town of Dürkheim, Germany, and named for the Latin *caesius*, meaning sky blue, naturally occurring cesium has only one isotope, $^{133}_{55}$Cs. Although this naturally occurring isotope is stable and harmless, two unstable isotopes, $^{134}_{55}$Cs and $^{137}_{55}$Cs, are formed as by-products in nuclear reactors and were released into the atmosphere at Chernobyl. Cesium-134 and cesium-137 are radioactive, undergoing decay slowly and producing harmful radiation. (We'll see exactly what this means in Chapter 20.) As dust from the accident slowly settled to the earth, radioactive cesium became widely dispersed on agricultural land throughout northern Europe and the European part of the Soviet Union.

None of this would matter quite so much except for cesium's position in the periodic table. As a Group-1A element with atomic number 55, cesium is just below potassium in the periodic table. It is chemically similar to potassium and can be mistaken for that element by living organisms. Potassium, an essential element in the human diet, is necessary for nerve action and is a constituent of all body tissues and fluids.

When crops were grown and animals were grazed on contaminated land, radioactive cesium moved into the human food chain, where its presence was detected at low levels. Careful monitoring by public health officials minimized the potential hazard, although it is estimated that an additional 28,000 cancers may occur worldwide in the next 50 years as a result of the accident. Over the same 50-year period, however, there will be an estimated 600 *million* spontaneous cancers.

The Chernobyl nuclear power plant after the devastating fire and explosion of April 26, 1986, that destroyed the unit 4 reactor.

2.9   ELECTRON CONFIGURATIONS AND THE PERIODIC TABLE

What makes it possible to classify elements in the periodic table? Why do many groups of elements have similar properties? The answers appear by looking at Table 2.6, which shows the electron configurations of elements within several groups. *Elements within each group (column) of the periodic table have similar electron configurations in their outermost electron shells.* For example, all the alkali metals have a single electron in their outer shell. Similarly, all the alkaline earth elements have two electrons in their outer shell; all the halogens have seven electrons in their outer shell; and all the noble gases have eight electrons in their outer shell except for helium, which has two (Figure 2.9).

We'll soon see that the chemistry these elements undergo is due largely to the number of electrons in their outer shells. Elements that have eight outer-shell electrons (electron *octets*) are unusually stable and unreactive, whereas elements with other numbers of electrons are less stable and more reactive.

Table 2.6   Electron Configurations for Four Groups of Elements

Group	Element	Atomic Number	Configuration by Shell 1 2 3 4 5 6	Group	Element	Atomic Number	Configuration by Shell 1 2 3 4 5
	Li (Lithium)	3	2 1		F (Fluorine)	9	2 7
Group 1A	Na (Sodium)	11	2 8 1	*Group 7A*	Cl (Chlorine)	17	2 8 7
Alkali	K (Potassium)	19	2 8 8 1	**Halogens**	Br (Bromine)	35	2 8 8 7
Metals	Rb (Rubidium)	37	2 8 18 8 1		I (Iodine)	53	2 8 18 8 7
	Cs (Cesium)	55	2 8 18 18 8 1				
	Be (Beryllium)	4	2 2		He (Helium)	2	2
Group 2A	Mg (Magnesium)	12	2 8 2	*Group 8A*	Ne (Neon)	10	2 8
Alkaline	Ca (Calcium)	20	2 8 8 2	**Noble**	Ar (Argon)	18	2 8 8
Earths	Sr (Strontium)	38	2 8 18 8 2	**Gases**	Kr (Krypton)	36	2 8 18 8
	Ba (Barium)	56	2 8 18 18 8 2		Xe (Xenon)	54	2 8 18 18 8

Practice Problem   **2.8**   Which group of elements in the periodic table do you suppose has six outer-shell electrons? Explain your answer.

Figure 2.9
Elements in Group 8A, such as neon, are gases. When excited by an electric current they give off light of characteristic colors.

2.10   ATOMS AND MOLECULES; COMPOUNDS AND MIXTURES

■ **Molecule**   A group of atoms bonded together in a discrete unit.

■ **Chemical reaction**   A chemical change brought about by making and breaking of bonds between atoms.

■ **Chemical compound**   A chemical substance formed by joining together atoms of different elements.

■ **Reactant**   A starting substance that undergoes change in a chemical reaction.

■ **Product**   A substance formed as the result of a chemical reaction.

■ **Chemical equation**   The written expression that describes a chemical reaction.

Atoms, as we've seen, are the smallest individual units of matter. **Molecules** are larger units formed by the chemical combining, or bonding together, of two or more atoms. For example, when one oxygen atom combines with two hydrogen atoms, the three bond tightly together to form one molecule of water, symbolized as H_2O. A chemical change that involves the making or breaking of bonds between atoms is called a **chemical reaction**, and the new substance formed (water) is a **chemical compound**. Chemical compounds are written by listing the element symbols of the various atoms and indicating the number of atoms of each element by a subscript.

Chemical reactions are always written by showing the starting **reactants** on the left, the **products** on the right, and an arrow between them to indicate a chemical transformation. Although this manner of writing reactions is sometimes called a **chemical equation**, the word "equation" is not really correct. The number and kinds of atoms must be the same on both sides of the reaction arrow, but the chemical substances themselves are not the same since a reaction is occurring.

Notice how chemical compounds differ from mixtures: *Mixtures* result simply from blending two or more substances together in any proportion without changing the individual substances in the mixture. Thus, the elements hydrogen and oxygen can be mixed together in any proportion without changing them, just as a spoonful of sugar and a spoonful of salt can be mixed. *Compounds*, however, result from the joining together of different elements to yield new molecules. Once a chemical reaction takes place, individual hydrogens and oxygens are no longer present. In their place are new molecules of water. All of the new water molecules are identically composed of two hydrogen atoms and one oxygen atom.

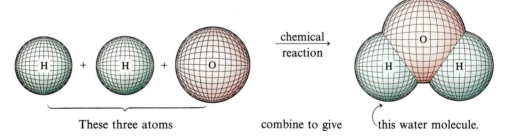

These three atoms   combine to give   this water molecule.

■ **Law of definite proportions**   Every chemical compound is formed by a combination of elements in a defined proportion.

The idea that atoms join together in definite ratios to make molecules is one of the cornerstones of chemistry. Called the *law of definite proportions*, it is applicable to all chemical compounds:

Law of Definite Proportions: Every chemical compound is formed by a combination of elements in a defined proportion.

Elements never combine in random proportions to make compounds. Thus *all* water molecules consist of two hydrogen atoms and one oxygen atom (H_2O); *all* ammonia molecules consist of one nitrogen atom and three hydrogen atoms (NH_3); *all* sucrose molecules (table sugar) consist of 12 carbon atoms, 22 hydrogen atoms, and 11 oxygen atoms ($C_{12}H_{22}O_{11}$); and so on.

Practice Problem   **2.9**   Glucose, also called dextrose, is a chemical compound made of six carbon atoms, six oxygen atoms, and twelve hydrogen atoms. What is the shorthand formula of glucose?

INTERLUDE: ARE ATOMS REAL?

All of chemistry rests on the belief that matter is composed of the tiny particles we call atoms. Every chemical reaction and every physical law that governs the behavior of matter is explained by chemists in terms of atomic theory.

But how do we know that atoms are real, rather than imaginary? How do we know that our explanations have a factual basis and are not just fanciful theories? The best answer is that we can now actually *see* individual atoms through the use of an extraordinary device called a *scanning tunneling microscope*. Invented in 1981 by a research team at the IBM Corporation, this instrument has achieved magnifications of up to ten million (!) allowing chemists for the first time to look directly at atoms themselves. The figure below for example, shows a computer-enhanced representation of individual atoms lined up on the face of a crystal of

silicon. There is little doubt that atoms are indeed real.

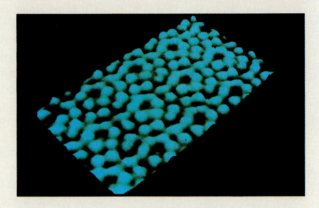

Individual atoms on the face of a crystal of silicon are seen at a magnification of ten million in this picture taken by a scanning tunneling microscope.

SUMMARY

All matter is formed from a combination of one or more of the 107 presently known **elements**. These elements, fundamental substances that cannot be chemically changed into anything simpler, are usually referred to by one- or two-letter shorthand symbols. For example, H is hydrogen, Ca is calcium, and Na is sodium.

An **atom** is the smallest and simplest unit into which matter can be chemically broken down. Atoms are made up of three small subatomic particles called **protons**, **neutrons**, and **electrons**. Protons have a positive electrical charge; electrons have a negative electrical charge; and neutrons are electrically neutral. As a result, protons and electrons attract each other. The protons and neutrons in an atom are pressed together into a dense, positively charged mass called the **nucleus**. Electrons are situated a great distance away from the nucleus, leaving most of the atom as empty space.

Elements differ according to the number of protons their atoms contain. All atoms of a given element have the same number of protons and an equal number of electrons—the **atomic number** characteristic of that element. Different atoms of the same elements may have different numbers of neutrons, however, and thus have different **mass numbers** (the total number of protons and neutrons). Atoms with identical numbers of protons and electrons but different numbers of neutrons are called **isotopes**. The **atomic weight** of an element is simply the weighted average mass of an element's naturally occurring isotopes.

The electrons surrounding an atom are grouped into layers, or **shells**, according to the amount of energy they have. Within each shell, electrons are grouped into **subshells**, and within each subshell into **orbitals**, regions of space to which they are confined. The s orbitals are spherical-shaped regions, and the p orbitals are dumbbell-shaped. Each orbital and each shell can hold only a specific number of electrons. The innermost shell can hold two electrons in an s orbital ($1s^2$); the second shell can hold eight electrons in one s and three p orbitals ($2s^2\,2p^6$); the third shell can hold 18 electrons in one s, three p, and five d orbitals ($3s^2\,3p^6\,3d^{10}$); and so on. **Electron configurations** of specific elements are determined by placing the elements' electrons into orbitals, beginning with the lowest-energy orbital.

The 107 elements are organized by atomic number into the **periodic table**. Elements within each of the 18 columns, or **groups**, in the table have the same number of electrons in their outermost shell and therefore have similar chemical properties.

Molecules are units of matter formed by the chemical combination of two or more atoms. Whenever a chemical reaction occurs, atoms combine in a specific proportion to yield products of defined composition—a fundamental principle of chemistry called the **law of definite proportions**.

REVIEW PROBLEMS

Elements and Their Symbols

2.10 What symbols identify these elements? (a) zinc (b) mercury (c) barium (d) gold (e) silicon (f) carbon (g) selenium

2.11 What elements are specified by these symbols? (a) N (b) O (c) K (d) Cl (e) Ca (f) P (g) Mg (h) Mn

2.12 The symbol CO stands for carbon monoxide, a chemical compound, but the symbol Co stands for cobalt, an element. Explain how the two can be told apart.

2.13 What is wrong with these statements? Correct them. (a) The symbol for bromine is BR. (b) The symbol for manganese is Mg. (c) The symbol for carbon is Ca. (d) The symbol for potassium is po.

2.14 What is the most abundant element in the earth's crust? In the human body?

2.15 Small amounts of the following elements in our diets are essential for good health. What is the chemical symbol for each? (a) iron (b) copper (c) cobalt (d) molybdenum (e) chromium (f) fluorine (g) sulfur

2.16 What is wrong with these statements? Correct them. (a) Water has the formula H2O. (b) Carbon has atomic number 12.

Atomic Structure

2.17 What are the names of the three subatomic particles? What are their masses in amu, and what electrical charge does each of the three have?

2.18 Give the names and symbols for elements that meet these descriptions. (a) an element that has atomic number 6 (b) an element that is likely to have an isotope with mass number 9 (c) an element that has 16 protons

2.19 There are two naturally occurring isotopes of carbon. How many protons does each have? How many electrons?

2.20 The two naturally occurring isotopes of carbon (Problem 2.19) have mass numbers of 12 and 13. How many neutrons does each have? Write the symbol for each isotope, indicating both atomic number and mass number.

2.21 The isotope of iodine with mass number 131 is often used in medicine as a radioactive tracer. Write the symbol for this isotope, indicating both mass number and atomic number.

2.22 What is the maximum number of electrons that can go into an orbital?

2.23 What is the maximum number of electrons that can go into the first shell? The second shell? The third shell?

2.24 Without looking at Table 2.5, write the electron configuration for magnesium, atomic number 12.

2.25 Without looking at Table 2.5, write the electron configuration for phosphorus, atomic number 15.

2.26 Write the electron configuration of the element that has atomic number 19.

2.27 How many electrons does the element with atomic number 12 have in its outer shell?

The Periodic Table

2.28 Thallium, a highly toxic element, has atomic number 81. Locate thallium in the periodic table, give its symbol, and predict whether you would expect thallium to be a metal or a nonmetal.

2.29 Americium, atomic number 95, is used in household smoke detectors. What is the symbol for americium? Is americium a metal or a nonmetal?

2.30 One isotope of americium (Problem 2.29) has mass number 241. How many protons and how many neutrons are in a nucleus of americium-241?

2.31 How many electrons would you expect to be in the outer shell of the group-4A elements? Explain your reasoning.

2.32 What other two elements in the periodic table would you expect to be most similar to sulfur?

2.33 What other element in the periodic table would you expect to be most similar to boron?

2.34 How many electrons would you expect fran-

cium, atomic number 87, to have in its outer shell? To what other elements is francium similar?

2.35 What elements in addition to lithium make up the alkali-metal family?

2.36 What elements in addition to fluorine make up the halogen family?

2.37 What elements in addition to helium make up the noble-gas family?

2.38 Cesium, atomic number 55, has only one more electron than xenon, atomic number 54, yet its chemical behavior is completely different. Explain.

Molecules and Compounds

2.39 What is the difference between an element and a compound? Between a compound and a mixture?

2.40 Classify each of the following as an element, a compound, or a mixture: (a) aluminum foil (b) table salt (c) water (d) air (e) a banana (f) diamond

2.41 The amino acid glycine has the shorthand formula $C_2H_5NO_2$. What elements are present in glycine? Of how many atoms does glycine consist?

2.42 Ribose, an essential part of RNA (ribonucleic acid), has the shorthand formula $C_5H_{10}O_5$. Of how many atoms does ribose consist?

2.43 What is the shorthand formula for penicillin V, whose molecules contain 16 carbons, 18 hydrogens, 2 nitrogens, 5 oxygens, and 1 sulfur?

2.44 State and explain the law of definite proportions.

Additional Problems

2.45 Define these terms: (a) isotope (b) atomic weight (c) electron configuration (d) orbital (e) mass number

2.46 What is the shape of an s orbital? A p orbital?

2.47 If you had one atom of hydrogen and one atom of carbon, which of the two would weigh more? Explain your answer.

2.48 If you had a pile of 10^{23} hydrogen atoms and another pile of 10^{23} carbon atoms, which of the two piles would weigh more? (See Problem 2.47.)

2.49 If your pile of hydrogen atoms in Problem 2.48 weighed about one gram, how much would you expect your pile of carbon atoms to weigh?

2.50 Based on your answer to Problem 2.49, how

much would you expect a pile of 10^{23} sodium atoms to weigh?

2.51 One of the many hazards of fallout from nuclear explosions is the production of radioactive strontium, which can be absorbed by the body and deposited in growing bones in place of calcium. Why do you suppose the body mistakes strontium for calcium?

2.52 An unidentified element is found to have an electron configuration by shell of 2 8 18 8 2. To what family (period) does this element belong? Is the element a metal or a nonmetal? How many protons does an atom of the element have? What is the name of the element?

2.53 An unidentified element is found to have an electron configuration by shell of 2 8 18 7. To what family (group) does this element belong? Is the element a metal or a nonmetal? How many protons does an atom of the element have? What is the name of the element?

2.54 Titanium, atomic number 22, is used in building jet fighter planes because of its combination of high strength and light weight. If titanium has an electron configuration by shell of 2 8 10 2, in what orbital are the outer-shell electrons?

2.55 Zirconium, atomic number 40, is directly beneath titanium (Problem 2.54) in the periodic table. What electron configuration by shell would you expect zirconium to have? Is zirconium a metal or a nonmetal?

CHAPTER 3

The Structure of Matter: Chemical Bonds

A computer-generated model of a molecule of cholesterol—the fatty substance that plays a role in heart disease (see Chapter 14). Carbon atoms are shown in green and hydrogens in blue. The pink sphere is an oxygen atom.

We ended the last chapter with a brief look at molecules—units of matter made up of two or more atoms linked together in a definite proportion. For example, every molecule of water is made of two hydrogen atoms and one oxygen atom, and every molecule of methane (natural gas) is made of one carbon atom and four hydrogen atoms:

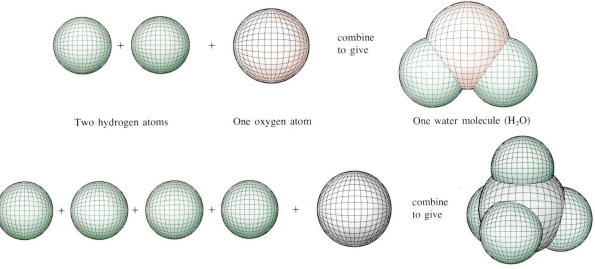

Two hydrogen atoms   One oxygen atom   combine to give   One water molecule (H_2O)

Four hydrogen atoms   One carbon atom   combine to give   One methane molecule (CH_4)

Clearly, there must be something that holds atoms together in molecules. Otherwise, molecules would simply fly apart into atoms, and no chemical compounds could exist. We call the links that join atoms together *chemical bonds*. In this chapter, we'll examine the following questions about bonds:

1. **How are bonds described electronically?** The goal: you should learn why bonds form and how an atom's electrons are involved in bonding.

2. **What kinds of chemical bonds are there?** The goal: you should learn how to distinguish between the two fundamental types of chemical bonds.

3. **What happens to bonds in a chemical reaction?** The goal: you should learn how the large class of reactions called *oxidation/reduction* reactions occur.

4. **What shapes do molecules have?** The goal: you should learn how to describe the three-dimensional shapes of some simple molecules.

3.1   THE OCTET RULE

Look again at the electron configurations shown in Table 2.6 for several families of elements in the periodic table. Elements in Group 1A (the alkali-metal family) have a single electron in their outermost shell; elements in Group 7A (the halogen family) have seven electrons in their outermost shell; and elements in Group 8A (the noble-gas family) have eight electrons in their outermost shell. Of the three groups, both the alkali metals and the halogens are extremely reactive, while the noble gases are inert and unreactive. Evidently there is something special about having eight outer-shell electrons (filled s and p subshells) that leads to unusual stability and lack of chemical reactivity.

■ **Octet rule**   Atoms undergo reactions in order to attain a noble-gas electronic configuration with eight outer-shell electrons.

According to the **octet rule**, atoms undergo reactions in order to obtain eight outer-shell electrons. In other words, atoms seek to attain a highly stable noble-gas electron configuration with filled s and p subshells. But how can this happen? How can an alkali metal, which has one outer-shell electron, attain a noble-gas configuration? Similarly, how can a halogen, which has seven outer-shell electrons, attain a noble-gas configuration?

One way that atoms can attain a noble-gas configuration is by either gaining or losing an appropriate number of electrons. For example, look back at the electron-configurations given for sodium and chlorine in Tables 2.5 and 2.6. Sodium has a total of two first-shell electrons, eight second-shell electrons, and one third-shell electron ($1s^2\ 2s^2\ 2p^6\ 3s^1$), and chlorine has a total of two first-shell electrons, eight second-shell electrons, and seven third-shell electrons ($1s^2\ 2s^2\ 2p^6\ 3s^2\ 3p^5$). The simplest way that sodium can attain an outer-shell octet is by *losing* the single electron in its $3s$ orbital, and the simplest way that chlorine can attain an outer-shell octet is by *gaining* an additional electron in its $3p$ orbital. In so doing, the sodium atom acquires a positive charge since it now has one less electron than it has protons, and the chlorine atom gains a negative charge since it now has one more electron than it has protons:

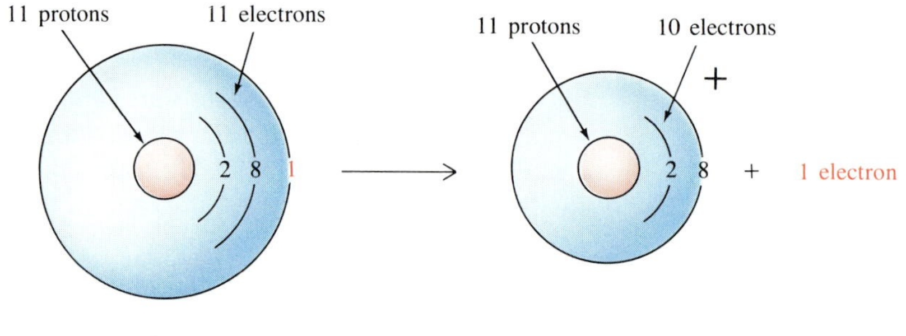

| A sodium atom | A positively charged sodium ion |

| A chlorine atom | A negatively charged chloride ion |

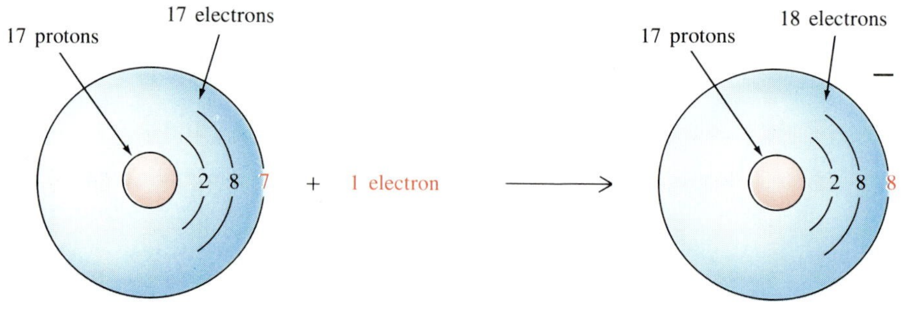

Practice Problems

3.1 Write the electron configuration of potassium, atomic number 19, and show how potassium can attain a noble-gas configuration.

3.2 How many electrons do you think magnesium, atomic number 12, would have to lose to attain a noble-gas configuration?

3.3 Oxygen, atomic number 8, has six outer-shell electrons. How do you think oxygen can most easily attain a noble-gas config·ation?

3.2 IONS AND IONIC BONDING

When a neutral chlorine atom gains a negatively charged electron to reach a noble-gas configuration, it must become negatively charged (Cl^-). Similarly, when a neutral sodium atom *loses* a negatively charged electron to reach a noble-gas configuration, it must become positively charged (Na^+). Atoms that are electrically charged because they contain unequal numbers of protons and electrons are called **ions**. Positively charged ions are called **cations** (**cat**-ions), and negatively charged ions are called **anions** (**an**-ions).

If sodium comes in contact with chlorine, an instantaneous reaction occurs to yield sodium chloride, a compound quite unlike either of the elements from which it's formed. Sodium is a soft, silvery metal that reacts violently with water, and chlorine is a corrosive and poisonous green gas, but sodium chloride is a stable, white, crystalline substance necessary for life. Sodium chloride is formed when a sodium atom transfers an electron to a chlorine atom to form a sodium cation (Na^+) and a chloride anion (Cl^-). Since opposite electrical charges strongly attract each other, we say that the positive sodium ion and negative chloride ion are held together by an **ionic bond**.

■ **Ion** An electrically charged atom or group of atoms.

■ **Cation** A positively charged ion.

■ **Anion** A negatively charged ion.

■ **Ionic bond** The electrical attraction between an anion and a cation

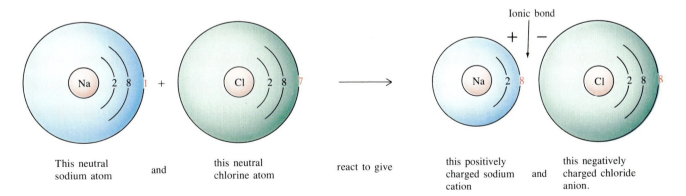

| This neutral sodium atom | and | this neutral chlorine atom | react to give | this positively charged sodium cation | and | this negatively charged chloride anion. |

When a billion billion sodium atoms transfer electrons to a billion billion chlorine atoms, a visible crystal of sodium chloride results. In this crystal, equal numbers of sodium cations and chloride anions are packed together in a regular arrangement. Each positively charged sodium ion is surrounded by six negatively charged chloride ions, and each chloride ion is surrounded by six sodium ions (Figure 3.1). This packing arrangement allows each ion to be stabilized by the

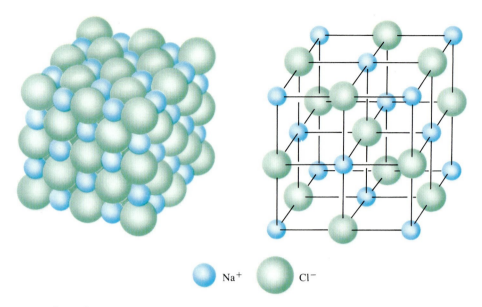

Na⁺   Cl⁻

Figure 3.1
The packing arrangement of ions in a sodium chloride crystal. Each positively charged sodium ion is surrounded by six negatively charged chloride ions and vice versa. The crystal is held together by ionic bonds.

attraction of unlike charges on its six nearest-neighbor ions while being as far as possible from like-charged ions.

Because of the three-dimensional packing arrangement of ions in a sodium chloride crystal, we can't speak of specific ionic bonds between specific pairs of ions. That is, there's really no such thing as an individual NaCl "molecule." Rather, there are many ionic bonds between an ion and its nearest neighbors, and we can speak only of the whole crystal as being an **ionic solid**.

■ **Ionic solid** A chemical compound held together by ionic bonds between anions and cations.

3.3 IONIC COMPOUNDS

What is true for sodium and chlorine is also true for many other elements. In general, metallic elements on the far left side of the periodic table are able to give up electrons and become positively charged cations, while nonmetallic elements on the far right side of the periodic table are able to accept electrons and become negatively charged anions. Thus, a large number of ionic compounds are possible.

Group-1A elements: All the alkali metals—lithium, sodium, potassium, rubidium, and cesium—have a single electron in their outermost shell that they can give up to attain noble-gas configurations. Thus, the positively charged ions Li⁺, Na⁺, K⁺, Rb⁺, and Cs⁺ are all well-known. (The lithium ion, of course, does not have an outer-shell octet. Lithium metal, with the electron configuration $1s^2\,2s^1$, loses its 2s electron to attain the helium noble-gas configuration $1s^2$ in the lithium ion.)

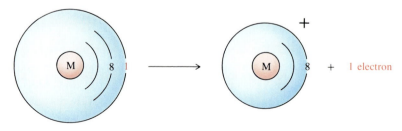

where M = an alkali metal (Na, K, Rb, or Cs)

- **Group-2A elements:** All the alkaline-earth metals—beryllium, magnesium, calcium, strontium, barium, and radium—have *two* electrons in their outermost shell. If both are given up, the alkali metals can attain noble-gas configurations. Thus, the *doubly* positive ions Be^{2+}, Mg^{2+}, Ca^{2+}, Sr^{2+}, Ba^{2+}, and Ra^{2+} are all well-known. (Beryllium, $1s^2\,2s^2$, achieves the helium configuration $1s^2$ in the Be^{2+} ion.)

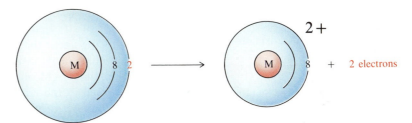

where M = an alkaline earth metal (Mg, Ca, Sr, Ba, or Ra)

- **Group-7A elements:** All the halogens—fluorine, chlorine, bromine, and iodine—have seven electrons in their outermost shell. By gaining one more electron, all can attain noble-gas configurations. Thus, the negatively charged ions F^-, Cl^-, Br^-, and I^- are all well-known.

where X = a halogen (F, Cl, Br, I)

Table 3.1 shows the number of charges on some of the most commonly encountered anions and metal cations.

Table 3.1   Some Common Cations and Anions

Group	Element	Ion	Group	Element	Ion
1A	Hydrogen	H^+	6A	Oxygen	O^{2-}
1A	Lithium	Li^+	6A	Sulfur	S^{2-}
1A	Sodium	Na^+	7A	Fluorine	F^-
1A	Potassium	K^+	7A	Chlorine	Cl^-
2A	Magnesium	Mg^{2+}	7A	Bromine	Br^-
2A	Calcium	Ca^{2+}	7A	Iodine	I^-
2A	Barium	Ba^{2+}			
3A	Aluminum	Al^{3+}			

3.4 OXIDATION/REDUCTION REACTIONS

■ **Oxidation** The loss of electrons by a reactant in a chemical reaction.

■ **Reduction** The gain of electrons by a reactant in a chemical reaction.

■ **Oxidizing agent** The reactant that causes an oxidation by taking electrons.

■ **Reducing agent** The reactant that causes a reduction by giving electrons.

In describing the formation of ions by the gain or loss of electrons, we often use the terms *oxidation* and *reduction*. An **oxidation** is a loss of electrons and a **reduction** is a gain of electrons. A substance that causes an oxidation by taking electrons is called an **oxidizing agent**, and a substance that causes a reduction by giving electrons is called a **reducing agent**.

Clearly, oxidations and reductions are linked. Whenever one substance is oxidized (has electrons removed from it), another substance must be reduced (have electrons added to it). In the reaction of sodium with chlorine, for example, sodium is the reducing agent that is oxidized by chlorine; that is, sodium gives up an electron to chlorine. Conversely, chlorine is the oxidizing agent that is reduced by sodium; that is, chlorine accepts an electron from sodium. Of course, the compound formed in the reaction, Na^+Cl^-, is electrically neutral since it contains an equal number of positive and negative ions.

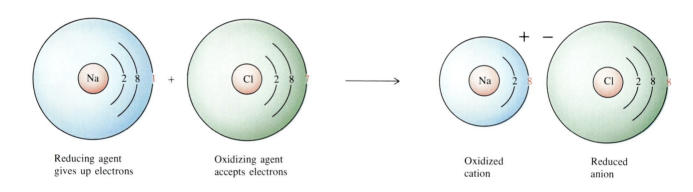

Reducing agent gives up electrons

Oxidizing agent accepts electrons

Oxidized cation

Reduced anion

■ **Oxidation state** The charge on an ion.

■ **Redox reaction** A general term for a reaction in which oxidations and reductions occur.

The charge on an ion is referred to as the ion's **oxidation state**. Thus, the sodium cation Na^+ has an oxidation state of $+1$, the magnesium cation Mg^{2+} has an oxidation state of $+2$, and the chloride anion Cl^- has an oxidation state of -1.

Oxidation/reduction reactions, often termed **redox reactions**, are not limited to just the alkali and alkaline-earth metals. Many of the transition metal elements in the middle of the periodic table (Figure 2.8) can also form cations by giving up electrons. For example, aluminum, a Group-3A element with three outer-shell electrons, gives up all three electrons in reacting with chlorine to form aluminum chloride, $AlCl_3$. The resultant aluminum ion, Al^{+3}, has an oxidation state of $+3$, and each of the three chloride ions has an oxidation state of -1 (see page 53).

Some metals in the middle of the periodic table can form more than one kind of ion. Thus, iron can react with chlorine to give either $FeCl_2$ or $FeCl_3$. Since a chloride ion always has a -1 oxidation state, the iron atom in $FeCl_2$ must have a $+2$ oxidation state (Fe^{2+}), whereas that in $FeCl_3$ must have a $+3$ oxidation state (Fe^{3+}).

In general, it's possible to tell the oxidation state of a metal *cation* by looking at the number and oxidation state of the *anions* in the compound. The following solved problems give examples of how this can be done.

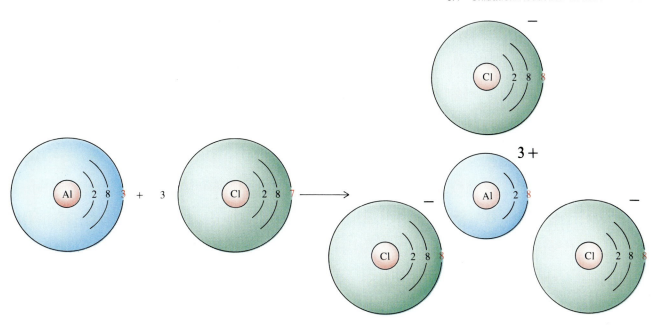

Solved Problem 3.1   What is the oxidation state of the titanium atom in $TiCl_4$?

Solution   First look at the anion (chloride), and then work backward. Since there are four chlorides in $TiCl_4$, each of which must have an oxidation state of -1, the oxidation state of titanium must be Ti^{4+}.

Solved Problem 3.2   What is the oxidation state of iron in iron oxide (rust, Fe_2O_3)?

Solution: First look at the anion. Since oxygen, a Group-6A element, must gain two electrons to achieve a noble-gas electron configuration, an oxygen anion must have a -2 oxidation state, O^{2-}. Furthermore, there are three oxygens in Fe_2O_3, for a total negative charge of $2- \times 3 = 6-$. Since Fe_2O_3 is neutral overall, however, the two iron atoms must have a combined *positive* charge of $6+$, or $3+$ on each. Thus, the iron atoms in iron oxide have $+3$ oxidation states.

$$\text{iron oxide} = Fe_2O_3 = 2\,Fe^{3+} \text{ and } 3\,O^{2-}$$

Practice Problems

3.4   Potassium, a silvery metal, reacts with bromine, a corrosive reddish liquid, to yield potassium bromide, a white solid. Identify the oxidizing and reducing agents, and tell the oxidation state of the product ions.

3.5   Identify the oxidizing and reducing agents in these reactions, and tell the oxidation state of each product ion.
(a) calcium + chlorine $\longrightarrow CaCl_2$ (b) barium + oxygen $\longrightarrow BaO$
(c) magnesium + iodine $\longrightarrow MgI_2$

3.6   What is the oxidation state of the metal atom in these compounds?
(a) VCl_3 (b) MgO (c) $SnCl_4$ (d) CrO_3

AN APPLICATION: BIOLOGICALLY IMPORTANT IONS

The human body requires many different ions for proper functioning. Some of these ions, such as Ca^{2+}, are used as structural material in bones and teeth. Others, such as Fe^{2+} found in blood hemoglobin, are required for specific chemical reactions in the body. And still others, such as K^+ and Na^+, are generally utilized throughout the body for many functions. Some of the most important ions and their functions are shown in the following table.

Some Biologically Important Ions	Ion	Location	Function	Dietary Source
	Ca^{2+}	Outside cell; 99% of Ca^{2+} is in bones and teeth as $Ca_3(PO_4)_2$ and $CaCO_3$	Component of bone and teeth; necessary for blood clotting and muscle contraction	Milk, whole grains, leafy vegetables
	Fe^{2+}	Blood hemoglobin	Transports oxygen from lungs to cells	Liver, red meat, vegetables
	K^+	Inside cells	Maintains osmotic pressure in cells; regulates insulin release	Milk, oranges, bananas, meat
	Na^+	Extracellular fluids	Protects against fluid loss; involved in muscle contraction	Table salt, seafood
	Mg^{2+}	Outside cells; bone	Present in many enzymes; muscle contraction	Leafy green plants
	Cl^-	Extracellular fluid; gastric juice	Maintains fluid balance in cells; helps transfer CO_2 from blood to lungs	Table salt, seafood
	HCO_3^-	Extracellular fluid	Controls acid/base balance in blood	From CO_2 in breathing
	HPO_4^{2-}	Inside cells; bones and teeth	Controls acid/base balance in cells; bone and teeth structure	Fish, poultry, milk

3.5 COVALENT BONDS

Although ionic bonding accounts for the formation of such compounds as sodium chloride, it doesn't account for the formation of many other compounds. For example, the elements hydrogen, oxygen, nitrogen, fluorine, chlorine, bromine, and iodine all exist, *not* as individual atoms, but as **diatomic molecules** in which two atoms are bonded together—H_2, O_2, N_2, F_2, Cl_2, Br_2, and I_2.

A moment's thought indicates that the bonds in these molecules can't be ionic. Take the chlorine molecule, Cl_2, for example. If one chlorine atom with seven outer-shell electrons gave one electron to another chlorine atom, Cl^+ and

■ **Diatomic molecule** A molecule that consists of two atoms bonded together.

Cl⁻ ions would result. The Cl⁻ ion would have a stable electron octet, but the Cl⁺ ion would have only *six* outer-shell electrons, making it worse off (less stable) than when it was neutral.

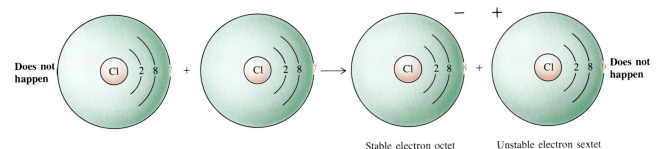

Does not happen + → + **Does not happen**

Stable electron octet   Unstable electron sextet

If the Cl—Cl bond in the Cl_2 molecule is not ionic, what is it? How can a bond form so that *both* chlorine atoms can achieve stable outer-shell electron octets? The answer is that the two chlorine atoms must *share* electrons. As shown earlier in Table 2.5, each individual chlorine atom has the outer-shell electron configuration $3s^2 3p^5$. Two of the three $3p$ orbitals are filled by two electrons each (four total), while the third $3p$ orbital holds only one electron. In schematic form, we can represent the situation by showing the chlorine atom surrounded by seven electrons (dots). Six of the dots come in three pairs, representing the filled $3s$ and the two filled $3p$ orbitals, while the seventh dot is unpaired (the half-filled $3p$ orbital):

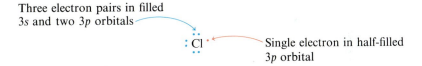

Three electron pairs in filled $3s$ and two $3p$ orbitals

Single electron in half-filled $3p$ orbital

When two chlorine atoms approach each other, the unpaired $3p$ electrons from each are *shared* by both atoms. Thus, each chlorine "owns" six outer-shell electrons and "shares" two more, for the necessary total of eight (Figure 3.2).

How does electron sharing between atoms lead to bonding? The best analogy is to think of two people who both want the same object. If both people

Figure 3.2
The formation of a Cl_2 molecule from two Cl atoms by sharing electrons. Both atoms in the resultant Cl_2 molecule have outer-shell electron octets (represented by dots). Inner-shell electrons are not shown.

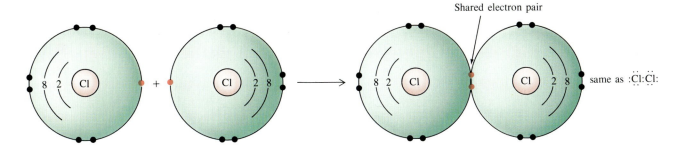

Shared electron pair

same as

■ **Covalent bond** A bond that results when two atoms share one or more pairs of electrons.

hold on to the object tightly, they are bound together. Neither person can walk away from the other as long as they both hold on. Similarly, when both chlorine atoms hold on to the shared electrons, the atoms are bonded together.

We call the shared-electron bond a **covalent** bond (co-**vale**-ent). Of the two sorts of bonding, ionic and covalent, covalent bonding is by far the more common. Almost all molecules found in living organisms are held together by covalent bonds.

Practice Problem   *3.7*   The hydrogen molecule, H_2, is held together by a covalent bond. Show how the H—H bond might arise. What noble-gas electron configuration do the two hydrogen atoms achieve?

3.6 COVALENT BONDS BETWEEN UNLIKE ATOMS

Covalent bonds can form between unlike atoms as well as between like atoms. Water consists of two hydrogen atoms joined by covalent bonds to an oxygen atom—H_2O; ammonia consists of three hydrogen atoms covalently bonded to a nitrogen atom—NH_3; and methane consists of four hydrogen atoms covalently bonded to a carbon atom—CH_4. Notice that the different atoms in these examples—hydrogen, oxygen, nitrogen, and carbon—all form different numbers of covalent bonds when they join to other atoms in neutral covalent molecules. Hydrogen forms one bond, oxygen forms two bonds, nitrogen forms three bonds, and carbon forms four bonds.

■ **Covalence** The number of bonds formed by an atom in a molecule.

The total number of bonds that an element forms is called its **covalence** and is equal to the number of electrons that the element needs to gain a noble-gas configuration. For example, hydrogen, with one outer-shell electron, needs one more electron to achieve the helium configuration ($1s^2$). Hydrogen can gain the needed electron by bonding to one other atom, and the covalence of hydrogen is therefore one. Similarly, oxygen, with six outer-shell electrons, needs two more electrons to reach a stable octet. Oxygen can gain the two needed electrons by getting one from each of two other atoms and thus has a covalence of two.

In the same way, nitrogen has five outer-shell electrons and a covalence of three, and carbon has four outer-shell electrons and a covalence of four. The covalences of these and other common elements are shown in Table 3.2. Note

Table 3.2 Covalences of Some Common Elements

Element	Number of Outer-Shell Electrons	Covalence	Element	Number of Outer-Shell Electrons	Covalence
H	1	1	S	6	2
F	7	1	N	5	3
Cl	7	1	P	5	3
Br	7	1	C	4	4
I	7	1	Si	4	4
O	6	2			

that the number of outer-shell electrons and the covalence of an atom always total eight, except for hydrogen where the total is 2.

Water, ammonia, and methane can be represented as shown in Figure 3.3. Also shown in Figure 3.3 are the standard ways used to depict covalent bonds. The representations of molecules that use dots to indicate outer-shell electrons are called **Lewis structures** (after G. N. Lewis of the University of California). The advantage of using Lewis structures is that they make clear how many outer-shell electrons surround each atom. Because it's much too time-consuming to show electron dots every time a covalent bond is drawn, however, covalent bonds are usually indicated simply by drawing a line between atoms (**line-bond structures**). *Whenever two atoms are shown connected by a line, a covalent bond is indicated.* Thus, these line-bond structures are a convenient way of showing the atom-to-atom connections in a molecule.

■ **Lewis structure**   A way of representing molecules by using dots to represent outer-shell electrons.

■ **Line-bond structure**   A way of representing molecules using lines between atoms to represent covalent bonds.

Solved Problem 3.3   Hydrogen sulfide, the "rotten-egg smell" of decaying matter, has the formula H_2S. Draw H_2S in both Lewis and line-bond structures.

Solution   Sulfur, a Group-6A element directly beneath oxygen in the periodic table, has six outer-shell electrons and a covalency of two. Four of the six electrons are in two pairs, and two of the six are unpaired:

$$:\dot{S}\cdot \; + \; H\cdot \; + \; H\cdot \; \longrightarrow \; :\ddot{S}:H \qquad :\ddot{S}\text{—}H \qquad \text{Hydrogen sulfide}$$

Practice Problems

3.8   What are the covalences of each atom in these molecules?
(a) PH_3 (b) H_2Se

3.9   Draw Lewis and line-bond structures for each of the molecules listed in Problem 3.8.

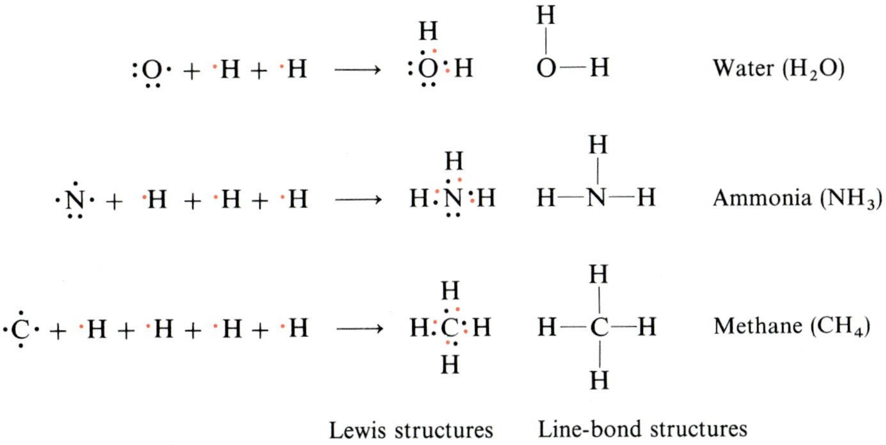

Figure 3.3
Covalent bonds in water, ammonia, and methane. The outer-shell electrons of hydrogen are indicated by colored dots; those of O, N, and C by black dots. Lewis structures are shown in the middle, and line-bond structures are shown on the right side of the figure.

$$:\dot{O}\cdot \; + \; \cdot H \; + \; H \; \longrightarrow \; :\ddot{O}:H \qquad O\text{—}H \qquad \text{Water } (H_2O)$$

$$\cdot\dot{N}\cdot \; + \; \cdot H \; + \; H \; + \; \cdot H \; \longrightarrow \; H:\ddot{N}:H \qquad H\text{—}N\text{—}H \qquad \text{Ammonia } (NH_3)$$

$$\cdot\dot{C}\cdot \; + \; \cdot H \; + \; \cdot H \; + \; \cdot H \; + \; \cdot H \; \longrightarrow \; H:\ddot{C}:H \qquad H\text{—}C\text{—}H \qquad \text{Methane } (CH_4)$$

Lewis structures   Line-bond structures

3.7 COVALENT BONDS IN ORGANIC MOLECULES

We'll see when we begin a study of organic chemistry in Chapter 8 that most molecules found in living organisms contain carbon. Carbon, with a covalence of four, has the ability to covalently bond to other carbon atoms, forming a fantastically diverse array of molecules. In fact, more than nine *million* different carbon-containing compounds are known (and several more were probably made in the time it's taken you to read this paragraph).

Covalent bonds form between carbon atoms in the same way they form between other atoms—by the sharing of electrons. For example, in ethane, C_2H_6, three of the four covalencies of each carbon atom are used in bonds to hydrogen, and the fourth covalency is used in a carbon-carbon bond. There is no other arrangement in which the covalencies of all eight atoms can be satisfied.

$$2 \cdot \overset{\cdot}{\underset{\cdot}{C}} \cdot + 6\,H\cdot \longrightarrow \quad \underset{\underset{H}{|}}{\overset{\overset{H}{|}}{H{:}\overset{..}{C}{:}\overset{..}{C}{:}H}} \quad = \quad \underset{\underset{H}{|}\;\underset{H}{|}}{\overset{\overset{H}{|}\;\overset{H}{|}}{H{-}C{-}C{-}H}} \quad = \quad CH_3CH_3 \qquad \text{Ethane}$$

Lewis structure   Line-bond structure   Condensed structure

■ **Condensed structure** A shorthand way of representing molecules without showing individual bonds.

Even line-bond structures become awkward when discussing large molecules, and **condensed structures** are often used. In its condensed form, ethane is written as CH_3CH_3, meaning that each carbon atom has three hydrogen atoms bonded to it (CH_3) and the two CH_3 units are bonded to each other. You'll get a lot more practice with condensed structures in Chapter 8.

Solved Problem 3.4   Methyl alcohol, or "wood alcohol," has the formula CH_4O. Show Lewis, line-bond, and condensed structures of methyl alcohol.

Solution   First write out the atoms with their outer-shell electrons indicated, and then piece the atoms together so that all covalencies are satisfied. There is only one way this can be done:

$$\cdot \overset{\cdot}{\underset{\cdot}{C}} \cdot + \cdot \overset{..}{\underset{..}{O}} \cdot + 4\,H\cdot \longrightarrow \quad \underset{\underset{H}{|}}{\overset{\overset{H}{|}}{H{:}\overset{..}{C}{:}\overset{..}{O}{:}H}} \quad = \quad \underset{\underset{H}{|}}{\overset{\overset{H}{|}}{H{-}C{-}O{-}H}} \quad = \quad CH_3OH \qquad \text{Methyl alcohol}$$

Lewis structure   Line-bond structure   Condensed structure

Practice Problems   **3.10**   Methylamine, CH_5N, is responsible for the characteristic odor of fish. Show Lewis, line-bond, and condensed structures of methylamine.

3.11   Propane (LP gas), used to heat homes in rural areas, has the formula C_3H_8. Show Lewis, line-bond, and condensed structures of propane.

3.8 MULTIPLE COVALENT BONDS

■ Single bond A covalent bond that results from sharing two electrons between atoms.

The sharing of two electrons between atoms results in what is called a **single bond**. Thus, the Cl—Cl bond in chlorine, the O—H bonds in water, the N—H bonds in ammonia, and the C—H and C—C bonds in ethane, are all single covalent bonds. Many molecules cannot be explained so simply, however. For example, the oxygen atoms in the O_2 molecule and the nitrogen atoms in the N_2 molecule could not have electron octets if single bonds were present:

$$2\,:\!\overset{\displaystyle .}{\underset{\displaystyle .}{O}}\!\cdot \;\longrightarrow\; :\!\overset{\displaystyle .}{O}\!:\!\overset{\displaystyle .}{O}\!: \qquad \text{Each oxygen has only } \textit{seven} \text{ electrons.} \quad \textbf{Unstable}$$

$$2\,:\!\overset{\displaystyle .}{N}\!\cdot \;\longrightarrow\; :\!\overset{\displaystyle .}{N}\!:\!\overset{\displaystyle .}{N}\!: \qquad \text{Each nitrogen has only } \textit{six} \text{ electrons.} \quad \textbf{Unstable}$$

■ Double bond A covalent bond that results from sharing four electrons between atoms.

■ Triple bond A covalent bond that results from sharing six electrons between atoms.

The only way that the individual O and N atoms in the O_2 and N_2 molecules can have outer-shell electron octets is if they share more than two electrons, resulting in the formation of *multiple* covalent bonds. If the two oxygen atoms share *four* electrons, each oxygen then has a stable outer-shell octet of electrons, and a **double bond** results. Similarly, if two nitrogen atoms share *six* electrons, each nitrogen then has an outer-shell electron octet, and a **triple bond** results.

$$2\,:\!\overset{\displaystyle .}{\underset{\displaystyle ..}{O}}\!\cdot \;\longrightarrow\; :\!\overset{}{\underset{\displaystyle ..}{O}}\!:\!:\!\overset{}{\underset{\displaystyle ..}{O}}\!: \;=\; O{=}O$$

Each oxygen now has eight electrons, a double bond.

$$2\,:\!\overset{\displaystyle .}{\underset{\displaystyle .}{N}}\!\cdot \;\longrightarrow\; :\!\overset{}{\underset{\displaystyle ..}{N}}\!:\!:\!:\!\overset{}{\underset{\displaystyle ..}{N}}\!: \;=\; N{\equiv}N$$

Each nitrogen now has eight electrons, a triple bond.

Multiple covalent bonding is also common in organic molecules. For example, ethylene, a simple molecule used commercially as a plant hormone to induce ripening in fruit, has the formula C_2H_4. The only way that all six atoms can have their necessary covalencies is for the two carbon atoms to share four electrons in a carbon-carbon double bond:

$$2\cdot\!\overset{\displaystyle .}{C}\!\cdot + 4\,H\cdot \;\longrightarrow\; \begin{matrix} H & & H \\ :\!C & :\!: & C\!: \\ H & & H \end{matrix} \;=\; \begin{matrix} H \\ \diagdown \\ H \diagup \end{matrix}C{=}C\begin{matrix} H \\ \diagup \\ \diagdown H \end{matrix} \;=\; H_2C{=}CH_2 \;\; \text{or} \;\; CH_2{=}CH_2$$

Ethylene—the carbons share four electrons in a double bond.

As another example, acetylene, the gas used in welding, has the formula C_2H_2. Thus, the two acetylene carbons must share six electrons in a carbon-carbon triple bond:

$$2\cdot\!\overset{\displaystyle .}{C}\!\cdot + 2\,H\cdot \;\longrightarrow\; H\!:\!C\!:\!:\!:\!C\!:\!H \;=\; H{-}C{\equiv}C{-}H \;=\; HC{\equiv}CH$$

Acetylene—the carbons share six electrons in a triple bond.

Notice that multiple bonds are indicated as double or triple lines in condensed formulas. Ethylene is $H_2C{=}CH_2$ and acetylene is $HC{\equiv}CH$.

Practice Problems

3.12 Formaldehyde, the substance used as a preservative for biological specimens, has the formula CH_2O. What kind of bond exists between carbon and oxygen? Draw formaldehyde in Lewis and line-bond structures.

3.13 Hydrogen cyanide, a deadly poisonous gas, has the formula HCN. What kind of bond exists between carbon and nitrogen? Show hydrogen cyanide in Lewis and line-bond structures.

3.9 THE SHAPES OF COVALENT MOLECULES

What shapes do covalent molecules have? Are the two O—H bonds in water (H—O—H) randomly oriented, or is there a specific relationship between them? In fact, all covalent molecules have an exact three-dimensional shape. Thus, the water molecule is bent, with a **bond angle** of 104.5° between O—H bonds. Similarly, the ammonia molecule (NH_3) is pyramid-shaped, with an angle of 107° between any two N—H bonds, and the methane molecule (CH_4) is tetrahedral, with an angle of 109.5° between any two C—H bonds (Figure 3.4).

The simplest way to explain the shapes of these molecules is to keep in mind the general rule that electrons always repel each other and seek to be as

■ **Bond angle** The angle formed between any two adjacent covalent bonds.

Figure 3.4
The shapes of water, ammonia and methane molecules. All have bond angles of approximately 109°. In these three-dimensional drawings, normal lines are in the plane of the paper, dashed lines recede into the paper, and heavy wedged lines come out of the paper toward the viewer.

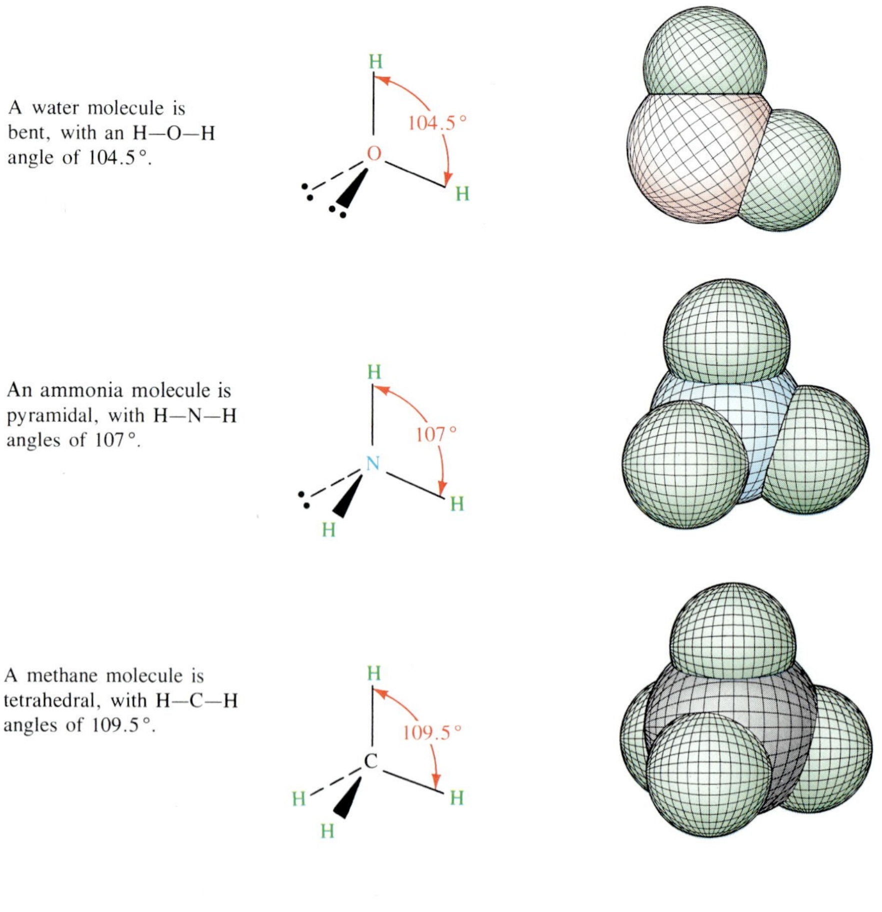

A water molecule is bent, with an H—O—H angle of 104.5°.

A methane molecule is tetrahedral, with H—C—H angles of 109.5°.

An ammonia molecule is pyramidal, with H—N—H angles of 107°.

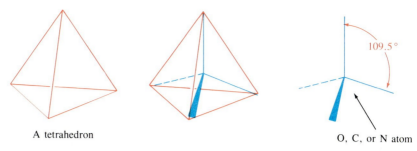

A tetrahedron   O, C, or N atom

Figure 3.5
The tetrahedral geometry of water, ammonia, and methane. The O, N, or C atom is located at the center of the tetrahedron, while the four outer-shell electron pairs are oriented toward the four corners. The angle between any two pairs is 109.5°. In water, two of the pairs are used for covalent bonding to hydrogens, and two are not; in ammonia, three of the pairs are used for covalent bonding to hydrogens, and one is not; and in methane, all four pairs are used for covalent bonding to hydrogens.

■ **Lone pair** A pair of outer-shell electrons not used by an atom for forming bonds.

■ **Tetrahedron** A geometrical figure with four identical triangular faces.

far away from each other as possible. In water, for example, there are eight outer-shell electrons on oxygen. These eight electrons are grouped into four pairs, two of which are used in the two H—O covalent bonds and two of which (called **lone pairs**) are not used for bonding. These four electron pairs can be as far away from each other as possible if they move toward the four corners of an imaginary tetrahedron. (A **tetrahedron** is a regular geometric figure whose four identical faces are equilateral triangles.) The angle between two lines drawn from the center to two corners of the tetrahedron is 109.5° (Figure 3.5).

The situations in ammonia and methane are similar to that in water. In all cases, an atom sits in the center of an imaginary tetrahedron with its four electron pairs oriented toward the four corners. The pairs may either be used for forming covalent bonds to other atoms, or may be nonbonding lone pairs. In all cases, the angle formed by two hydrogens and the central atom is approximately 109.5° as indicated in Figure 3.4.

Practice Problem   **3.14**   Phosphine, PH_3, has a shape similar to that of ammonia. Sketch the molecule in three dimensions, and predict the H—P—H bond angle.

3.10   COVALENT BONDS IN POLYATOMIC IONS

Imagine a neutral water molecule, H—O—H. What would happen if you were to remove one hydrogen but leave behind the electron pair of the former O—H bond? The hydrogen you removed would bear a positive charge, since it consists of only a proton (the H nucleus) without any electrons. In other words, it would be an H^+ ion. Conversely, the O—H fragment would have to bear a negative charge in order to maintain overall electrical neutrality, and would be an OH^- ion.

$$H\!:\!\overset{..}{\underset{..}{O}}\!:\!H \longrightarrow H^+ + {}^-\!:\!\overset{..}{\underset{..}{O}}\!:\!H$$

The oxygen atom has an outer-shell octet of electrons and carries a negative charge.

The hydrogen atom has *no* electrons and carries a positive charge.

Table 3.3   Some Common Polyatomic Ions

Name	Formula	Name	Formula
Ammonium ion	NH_4^+	Phosphate ion	PO_4^{3-}
Hydronium ion	H_3O^+	Nitrate ion	NO_3^-
Hydroxide ion	OH^-	Acetate ion	$CH_3CO_2^-$
Sulfate ion	SO_4^{2-}	Cyanide ion	CN^-
Bisulfate ion	HSO_4^-	Permanganate ion	MnO_4^-
Carbonate ion	CO_3^{2-}	Dichromate ion	$Cr_2O_7^{2-}$
Bicarbonate ion	HCO_3^-		

■ **Hydroxide ion**   The HO^- ion formed by loss of H^+ from water.

To convince yourself that the OH^- ion (called a **hydroxide ion**) must have a negative charge, you can simply count the total number of protons and electrons present and then subtract. The OH^- ion has nine protons (eight in the oxygen nucleus and one in the hydrogen nucleus) and ten electrons (two in the first shell of oxygen, six unshared electrons in the outer shell of oxygen, and two shared between oxygen and hydrogen in a covalent bond). Ten electrons and nine protons result in an overall charge of -1.

■ **Polyatomic ion**   An ion that contains two or more atoms linked by covalent bonds.

The hydroxide ion is an example of a **polyatomic ion**, a charged particle made up of two or more atoms linked by covalent bonds. Table 3.3 lists other common polyatomic ions, both positive and negative. In all cases, the charge on the ion can be calculated by carrying out a "bookkeeping" count of the numbers of protons and electrons present.

Polyatomic ions form ionic compounds in the same way that simple monoatomic ions do (Section 3.3). For example, household lye is sodium hydroxide, NaOH, an ionic compound of positive sodium ions and negative hydroxide ions; baking soda is sodium bicarbonate, $NaHCO_3$, an ionic compound of positive sodium ions and negative bicarbonate ions; and most common lawn fertilizer is ammonium nitrate, NH_4NO_3, an ionic compound of positive ammonium ions and negative nitrate ions. Some other important compounds formed from polyatomic ions are shown in Table 3.4. (See Figure 3.6.)

Table 3.4   Some Important Compounds Formed from Polyatomic Ions

Formula	Name	Use
$AgNO_3$	Silver nitrate	Antiseptic, germicide, astringent
$CaSO_4$	Calcium sulfate	Plaster casts
Li_2CO_3	Lithium carbonate	Treatment of depression
$(NH_4)_2CO_3$	Ammonium carbonate	"Smelling salts"
$KMnO_4$	Potassium permanganate	Antiseptic, disinfectant

Figure 3.6
Crystals of the mineral barite consist largely of the ionic compound barium sulfate, $BaSO_4$.

Solved Problem 3.5 The nitrate ion has the formula NO_3^- and the Lewis structure shown. Tell how many protons and how many electrons are present in the ion, and then explain why it has a negative charge.

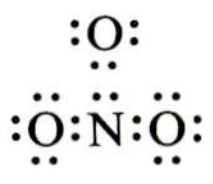

$$:\overset{\cdot\cdot}{\underset{}{O}}:$$
$$:\overset{\cdot\cdot}{\underset{\cdot\cdot}{O}}:\overset{\cdot\cdot}{N}:\overset{\cdot\cdot}{\underset{\cdot\cdot}{O}}: \qquad \text{Nitrate ion}$$

Solution Nitrogen (atomic number 7) has seven protons, and each of the three oxygens (atomic number 8) has eight protons, for a total of $[7 + (3 \times 8)] = 31$ protons. Each of the four atoms has two inner-shell electrons ($1s^2$) for a total of eight; there are 16 electrons in lone pairs; and there are eight electrons in bonds, for a total of $8 + 16 + 8 = 32$ electrons. Thus, a nitrate ion has one more electron than proton and carries a negative charge.

Practice Problems **3.15** Carbonate ion has the following Lewis structure. Count the number of protons and electrons in carbonate ion, and tell why it has two negative charges.

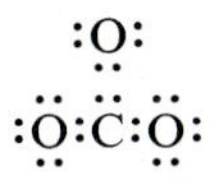

$$:\overset{\cdot\cdot}{\underset{\cdot\cdot}{O}}:$$
$$:\overset{\cdot\cdot}{\underset{\cdot\cdot}{O}}:\overset{\cdot\cdot}{C}:\overset{\cdot\cdot}{\underset{\cdot\cdot}{O}}: \qquad \text{Carbonate ion}$$

3.16 The ammonium ion has the formula NH_4^+. How many protons and electrons are present in the ion? What shape would you predict for the ammonium ion (see Section 3.9).

3.11 NAMING CHEMICAL COMPOUNDS

Different kinds of compounds are named by different rules. Ordinary table salt is named *sodium chloride* because of its formula, NaCl, but the food preservative BHT that you see listed on most food labels is properly named *2,6-di-*tert-*butyl-4-methylphenol* (now you see why it has a nickname) because of special rules for organic compounds. We'll see in this section how ionic compounds are named and then introduce additional rules in later chapters as the need arises.

As the compounds listed in Table 3.4 indicate, ionic compounds are named by first identifying the cation and then the anion. A positively charged metal cation simply takes the name of the metal itself, while a negatively charged anion is given an -*ide* ending. Sodium chlor*ide* (NaCl), potassium brom*ide* (KBr), and aluminum chlor*ide* ($AlCl_3$) are examples.

It can be a little confusing to use the same name for both a metal and its ion, and you sometimes have to stop and think about what's being referred to. For example, a medical technician might talk about sodium or potassium in the bloodstream. Since both sodium and potassium *metal* react violently with water, however, they are unlikely to be present in blood. Thus, the technician must be referring to the inert and harmless sodium and potassium *ions*.

In naming ionic compounds of metals that can have more than one oxidation state, the metal's oxidation state is noted in parentheses. Thus, $FeCl_2$

is called iron(II) chloride and $FeCl_3$ is called iron(III) chloride. This situation occurs primarily in the transition metals—those in the middle part of the periodic table—but if you're in doubt about whether to include the oxidation state in a name, it's safer to put it in than leave it out.

Solved Problem 3.6   Magnesium carbonate is used as an ingredient in Bufferin tablets. Write its formula.

Solution   Look at the cation and the anion parts of the compound separately: Magnesium, a Group-2A element, forms the doubly positive Mg^{2+} cation; carbonate ion is doubly negative, $CO_3{}^{2-}$. Since magnesium carbonate must be neutral overall, its formula is $MgCO_3$.

Solved Problem 3.7   Give systematic names for these compounds:
(a) KF   (b) $MgCl_2$ (c) $TiCl_3$ (d) Fe_2O_3

Solution
(a) Potassium fluoride: No Roman numeral is necessary.
(b) Magnesium chloride: No Roman numeral is necessary since magnesium (left side of periodic table) forms only Mg^{2+}.
(c) Titanium(III) chloride: The Roman numeral III is necessary to specify the $+3$ charge on titanium (middle of periodic table).
(d) Iron(III) oxide: Since the three oxide anions (O^{2-}) have a total negative charge of -6, the two iron cations must total $+6$. Thus, each is Fe(III).

Practice Problems

3.17   Barium sulfate is an ionic compound swallowed by patients before X-raying the gastrointestinal tract. Write its formula. How many barium ions are present?

3.18   The compound Ag_2S is partly responsible for the tarnish found on silverware. Name Ag_2S, and indicate the oxidation state of the silver.

3.19   Name each of these compounds:
(a) CuO   (b) BF_3 (c) $NaNO_3$ (d) Cu_2SO_4 (e) Li_3PO_4

3.20   Write formulas for these compounds:
(a) barium hydroxide   (b) copper(II) carbonate
(c) magnesium bicarbonate   (d) chromium(VI) oxide

SUMMARY

Chemical bonds are the links that hold atoms together in molecules. There are two kinds of chemical bonds: **ionic bonds**, and **covalent bonds**. Ionic bonds result from the electrical attraction of oppositely charged particles, called **ions**. Ions are formed by the transfer of electrons between atoms as each atom seeks to achieve a stable, noble-gas electron configuration by obtaining an outer-shell

INTERLUDE: DIAMOND AND GRAPHITE

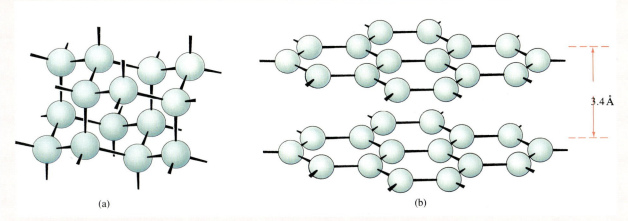

(a)   (b)

The structures of (a) diamond and (b) graphite.

Diamond and graphite are both made of pure carbon, but what a difference there is between them. Diamond, a rare colorless gemstone, is one of the hardest known materials. Graphite, a flaky black substance, is so soft that it's used as a lubricant and as the "lead" in lead pencils.

The differences between diamond and graphite are due to their different structures (see drawing). A crystal of diamond is essentially a single enormous molecule. Each carbon atom in the crystal is covalently bonded to four neighboring carbons with exactly the same tetrahedral geometry as that in methane (Section 3.9). The result is an immense three-dimensional array of interlocked atoms that gives diamond its great strength and hardness. Graphite, however, consists of two-dimensional sheets of carbon atoms stacked on top of each other in layers. Each carbon atom is joined to only three other carbons and has a double bond like that in ethylene (Section 3.8). Because the sheets are not bonded together, they can slip and slide over one another, making graphite a good lubricant.

Scientists have known for nearly 40 years that diamonds can be made synthetically by subjecting graphite to very high pressures at temperatures around 1800°C (see photograph). More recently, though, methods have been discovered for depositing an extremely thin film of diamond onto a variety of other materials to give these materials a nearly indestructible protective coat. Coated knives and scalpels should remain forever sharp, coated glass lenses should remain forever scratchproof, and coated bearings should never wear out. These and other applications may soon become reality.

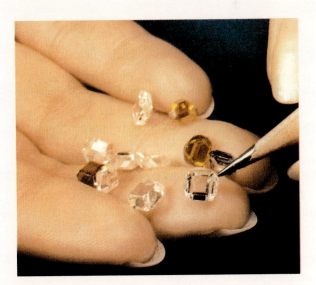

Synthetic diamonds.

octet of electrons. When sodium metal reacts with chlorine gas, sodium acts as a **reducing agent** by donating one of its outer-shell electrons to chlorine. At the same time, chlorine acts as an **oxidizing agent** by accepting an extra electron into its outer shell. The result is the formation of the **ionic solid**, $Na^+ Cl^-$.

In general, metallic elements on the far left side of the periodic table are able to give up electrons and become positively charged ions (**cations**). Non-metallic elements on the far right side of the periodic table are able to accept electrons and become negatively charged ions (**anions**).

Covalent bonds result from a sharing of electron pairs between atoms. By sharing electrons, the bonded atoms can achieve stable noble-gas electron configurations. For example, the chlorine molecule, Cl_2, results when two Cl atoms share two electrons.

The number of covalent bonds formed by an atom, called its **covalence**, is equal to the number of electrons the atom needs to share in order to achieve an outer-shell electron octet. Hydrogen has a covalence of one, oxygen has a covalence of two, nitrogen has a covalence of three, and carbon has a covalence of four. **Multiple covalent bonds** form when an atom shares more than two electrons. The ethylene molecule, $H_2C=CH_2$, has a carbon-carbon **double bond** formed by sharing four electrons between carbons. The acetylene molecule, $HC\equiv CH$, has a carbon-carbon **triple bond** formed by sharing six electrons.

Covalent bonds in molecules have a specific, rather than random, orientation. In methane, for example, the four covalent carbon-hydrogen bonds are oriented toward the four corners of an imaginary tetrahedron, with the carbon atom at the center.

REVIEW PROBLEMS

Ions and Ionic Compounds

3.21 Define these terms: (a) octet rule (b) ion (c) anion (d) cation (e) ionic bond (f) noble-gas electron configuration

3.22 What common feature do the alkali metals have that allows them to form positively charged ions?

3.23 Strontium is a Group-2 element with atomic number 38. How many electrons must strontium lose to achieve a noble-gas configuration?

3.24 How many electrons and how many protons does a strontium atom (Problem 3.23) have? How many electrons and protons does the strontium ion have?

3.25 Write out the electron configuration of calcium (atomic number 20), and show how calcium can achieve a noble-gas configuration.

3.26 Would you expect elements in Group 4 to gain electrons or lose electrons when they form ions?

3.27 Count the number of protons and electrons in the sulfate ion, and tell why it has the indicated charge.

Sulfate ion, SO_4^{2-}

$$\begin{array}{c} \ddot{\text{O}} \\ \ddot{\text{O}} : \ddot{\text{S}} : \ddot{\text{O}} \\ \ddot{\text{O}} : \end{array}$$

Oxidation/Reduction Reactions

3.28 Stannous (tin) fluoride, the toothpaste additive used to prevent cavities, has the formula SnF_2. What is the oxidation state of the stannous ion?

3.29 In the reaction of tin with fluorine to make stannous fluoride (Problem 3.28), what is the oxidizing agent, and what is the reducing agent? How many electrons does each of the three atoms involved in the reaction gain or lose?

$$Sn + F_2 \longrightarrow SnF_2$$

3.30 Titanium dioxide, TiO_2, is used as the pigment in white paint. What is the oxidation state of the titanium atom in titanium dioxide?

3.31 What is the oxidation state of the metal atom in these compounds? (a) CsF (b) $FeBr_3$ (c) HgO (d) Na_2S

3.32 Name the compounds listed in Problem 3.31.

3.33 Some ionic compounds consist of more than two kinds of ions. What is the oxidation state of the metal in vanadium oxychloride, $VOCl_3$?

3.34 Which of the following formulas is most likely to be correct for arsenic chloride: AsCl? As_2Cl_3? $AsCl_3$? $AsCl_4$? Explain.

Covalent Bonds

3.35 What are the covalences of these atoms? (a) carbon (b) oxygen (c) nitrogen (d) iodine (e) sulfur

3.36 Explain why the sum of an atom's covalence and the number of its outer-shell electrons usually totals eight.

3.37 Look up tellurium (atomic number 52) in the periodic table, and predict its probable covalence. Explain your reasoning.

3.38 Germanium (atomic number 32) is an element used in the manufacture of transistors. Judging from its position in the periodic table, what is the probable covalence of germanium?

Structures of Covalent Molecules

3.39 Define the following terms: (a) covalence (b) covalent bond (c) Lewis structure (d) line-bond structure

3.40 Chloroform, $CHCl_3$, was at one time used as an anesthetic agent. Draw a Lewis electron-dot structure and a line-bond structure for chloroform.

3.41 Sketch a probable three-dimensional shape for chloroform (Problem 3.40).

3.42 Which of the following two structures shown for formaldehyde, CH_2O, is more likely to be correct? Explain.

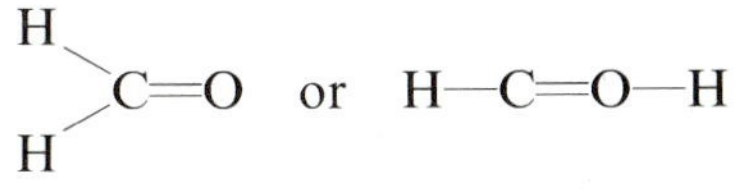

$$\underset{H}{\overset{H}{\diagdown}}C{=}O \quad \text{or} \quad H{-}C{=}O{-}H$$

3.43 Tetrachloroethylene, C_2Cl_4, is used commer-

cially as a dry-cleaning solvent. Propose a line-bond structure for tetrachloroethylene. What kind of carbon-carbon bond is present in the molecule?

3.44 Draw the line-bond formula of carbon dioxide, CO_2. What kinds of carbon-oxygen bonds are present?

3.45 Ethanol, or "grain alcohol," has the formula C_2H_5OH. Propose a structure for ethanol that is consistent with the rules of covalence.

3.46 Isopropyl alcohol, or "rubbing alcohol," has the formula C_3H_7OH. Propose a structure for isopropyl alcohol.

3.47 If a research paper appeared reporting the structure of a new molecule with formula C_2H_8, most chemists would be highly skeptical. Why?

3.48 Acetic acid (vinegar) has the formula $C_2H_4O_2$. Propose a suitable line-bond structure for acetic acid. What kind of carbon-oxygen bond(s) are present?

3.49 Hydrazine, a substance used to make rocket fuel, has the formula N_2H_4. Propose a line-bond structure for hydrazine.

Naming Chemical Compounds

3.50 Maalox™, an over-the-counter antacid, is a mixture of magnesium hydroxide and aluminum hydroxide. Write the formulas of each.

3.51 Write formulas for the following substances: (a) sodium bicarbonate (baking soda) (b) potassium nitrate (a backache remedy) (c) calcium carbonate (an antacid)

3.52 Write the names of these substances: (a) K_2CO_3 (b) $MgCO_3$ (c) $Ca(CH_3CO_2)_2$ (d) AgCN (e) $Na_2Cr_2O_7$ (f) $AlPO_4$

3.53 Indicate which of the following structures is most likely to be correct for calcium phosphate: (a) Ca_2PO_4 (b) $CaPO_4$ (c) $Ca_2(PO_4)_3$ (d) $Ca_3(PO_4)_2$

Additional Problems

3.54 Which of the following formulas is most likely to be correct for acetone, a common solvent: C_3H_6O or C_2H_7O? Explain.

3.55 If acetone (Problem 3.54) contains a carbon-oxygen double bond, propose two possible line-bond structures.

3.56 Which of the following formulas is most likely to be correct for dimethylamine, a chemical used in making detergents: C_2H_6N or C_2H_7N?

3.57 Propose a line-bond structure for dimethylamine (Problem 3.56).

3.58 What is the oxidation state of manganese in the permanganate ion, MnO_4^-?

3.59 The following formulas are all unlikely to be correct. What is wrong with each? (a) $NaBr_2$ (b) CCl_3 (c) K_2O_3 (d) N_2H_5

3.60 What is a likely formula for potassium permanganate, a purple substance used as a disinfectant? (See Problem 3.58.)

3.61 How many sodium ions are likely to be present in sodium phosphate?

3.62 Write the formulas of these substances: (a) molybdenum(V) chloride (b) cobalt(III) bromide (c) scandium(III) oxide

3.63 The cyanide ion, CN^-, has the following structure. Count the number of protons and electrons present, and show why the cyanide ion must have a negative charge.

$$^-:C:::N:\qquad \text{Cyanide ion}$$

3.64 Count the number of protons and electrons present in the acetate ion, and show why it bears a negative charge.

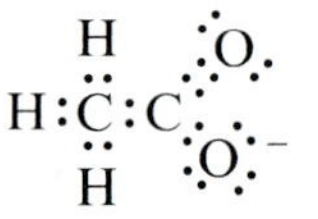

$$\begin{array}{cc} H & :\overset{..}{O}. \\ H:\overset{..}{\underset{..}{C}}:\overset{..}{C}.^{} & \\ H & :\overset{..}{\underset{..}{O}}:^- \end{array}\qquad \text{Acetate ion}$$

3.65 Common household bleach contains sodium hypochlorite, NaClO. What is the charge on the hypochlorite ion? Propose a structure for this ion.

3.66 The active ingredient in Rolaids™ has the formula $Al(OH)_2NaCO_3$. Identify the anions and cations present in this substance, and explain why the molecule is neutral overall.

CHAPTER 4

Chemical Reactions

This rock is in an unstable state, but it won't fall unless it gets an energetic push. We'll see in this chapter that the same is true for most chemical reactions.

We've seen numerous examples in past chapters of chemical reactions—the changes that occur when two or more substances react together to form a new compound. In this chapter, we'll look at reactions in more detail to learn how they are described, what their requirements are, and what chemists can do to modify them. In particular, we'll examine the following points:

1. **What is a chemical equation?** The goal: you should learn how to write and interpret simple chemical equations, and how to balance them.
2. **What are the quantitative relationships in chemical reactions?** The goal: you should learn how to derive the proper weight relationships among substances in a reaction, should understand the concept of a mole, and should learn how to carry out mole-to-gram and gram-to-mole conversions.
3. **Why and how do chemical reactions occur?** The goal: you should learn how energy changes take place in chemical reactions and how to interpret these changes using reaction energy diagrams.
4. **What factors affect reactions?** The goal: you should learn how changes in temperature, concentration, and catalyst can affect reactions.
5. **What is a chemical equilibrium?** The goal: you should learn what it means when a reaction reaches equilibrium, and what a reversible reaction is.

4.1 CHEMICAL EQUATIONS

We saw in Chapter 2 that *chemical equations* are shorthand statements used to represent chemical reactions. Starting reagents (called *reactants*) are shown on the left and the new substances formed (called *products*) on the right, with an arrow between them to indicate the direction of the chemical transformation. For example, hydrogen reacts with oxygen to form water, H_2O:

$$2 H_2 + O_2 \longrightarrow 2 H_2O \quad \text{(water)}$$

Look carefully at the way this equation is written. Since we know from Section 3.5 that hydrogen and oxygen exist as diatomic H_2 and O_2 *molecules*, rather than as individual H and O atoms, we must write them as such in the equation.

Now count the numbers of atoms on each side of the reaction arrow. On the starting-material side are four hydrogen atoms (two atoms in each of two H_2 molecules) and two oxygen atoms (two atoms in one O_2 molecule); on the product side are also four hydrogen atoms (two atoms in each of two H_2O molecules) and two oxygen atoms (one in each of two H_2O molecules). The number of atoms of each kind is the same on both sides of the equation.

■ **Balanced** Describing an equation in which the numbers and kinds of atoms on both sides of the reaction arrow are the same.

When the numbers and kinds of *atoms* on both sides of an equation are the same, we say that the equation is **balanced**. The numbers placed before each substance—the "2" in front of H_2, for example—are called **coefficients**; they tell how many molecules of that substance are necessary to balance the equation. *All chemical equations must be balanced, because atoms are neither created nor destroyed in reactions.*

■ **Coefficients** The numbers placed before each substance in a chemical equation to tell how many units of that substance are required to balance the equation.

Practice Problem   **4.1**   Count the numbers and kinds of atoms on both sides of the reaction arrows, and tell which of these equations are balanced:

(a) $HCl + KOH \longrightarrow H_2O + KCl$

(b) $CH_4 + Cl_2 \longrightarrow CH_2Cl_2 + 2\,HCl$

(c) $H_2O + MgO \longrightarrow Mg(OH)_2$

(d) $Na + H_2O \longrightarrow NaOH + H_2$

(e) $Al(OH)_3 + H_3PO_4 \longrightarrow AlPO_4 + 2H_2O$

4.2 BALANCING CHEMICAL EQUATIONS

Balancing chemical equations involves four steps:

1. Write an unbalanced equation using the correct formulas for all substances involved. For example, hydrogen and oxygen must be written as H_2 and O_2, rather than as H and O, since we know that hydrogen and oxygen exist as diatomic molecules. *These subscripts must never be changed in balancing an equation, since doing so would mean a change in the reaction itself.*

2. Find the proper coefficients that tell how many of each molecule are required to balance the equation. This step usually takes a mix of common sense along with trial and error.

3. When the equation is balanced, make sure the coefficients are reduced to their lowest whole-number values.

4. Check your answer to make sure the numbers and kinds of atoms on both sides of the equation are the same.

Let's work through some solved problems to see how a chemical equation is balanced.

Solved Problem 4.1   Natural gas (methane, CH_4) burns in oxygen to yield water and carbon dioxide (CO_2). Write a balanced equation for the reaction.

Solution   Step 1. *Write the unbalanced equation using correct formulas for all substances:*

$$CH_4 + O_2 \longrightarrow CO_2 + H_2O \qquad \text{(Unbalanced)}$$

Step 2. *Find coefficients to balance the equation:* Look first at the unbalanced equation and note that there is one carbon on each side of the arrow. Thus, the equation is already balanced with respect to carbon. Next, look at the hydrogens and note that there are four hydrogens on the left (in CH_4) and only two on the right (in H_2O). If we place a 2 before the H_2O, the number of hydrogens on both sides is the same:

$$CH_4 + O_2 \longrightarrow CO_2 + 2H_2O \qquad \text{(balanced for C and H)}$$

Now look at the number of oxygens. There are two oxygens on the left (in O_2) but four on the right (two in CO_2 and one in each of two H_2O's). If we place a 2 before the O_2, the number of oxygens on both sides is the same:

$$CH_4 + 2O_2 \longrightarrow CO_2 + 2H_2O \qquad \text{(completely balanced for C, H, and O)}$$

Step 3. *Make sure the coefficients are reduced to their lowest whole-number values:* In fact, our answer is already correct, but we might, through trial and error, have arrived at a different answer for the balanced equation:

$$2CH_4 + 4O_2 \longrightarrow 2CO_2 + 4H_2O$$

Although the previous equation is balanced, the coefficients are not yet in their lowest whole numbers. Dividing all the coefficients by 2 is necessary to reach the final equation.

Solved Problem 4.2   Potassium chlorate, $KClO_3$, decomposes when heated to yield potassium chloride and oxygen, a reaction used to provide oxygen for the emergency breathing masks in airliners. Balance the equation.

Solution   First, write the unbalanced equation:

$$KClO_3 \longrightarrow KCl + O_2$$

Next, find the proper coefficients. We might start by noticing that the equation is already balanced for K and Cl but is unbalanced for O. In order to balance for O, it's necessary to put a coefficient of 2 before $KClO_3$ and a coefficient of 3 before O_2, giving 6 O's on both sides of the equation:

$$2KClO_3 \longrightarrow KCl + 3O_2 \qquad \text{(balanced for O)}$$

Unfortunately, when we balanced the equation for O, we *unbalanced* it for K and Cl. Thus, we have to rebalance for K and Cl by placing a coefficient of 2 before KCl. The final equation is therefore:

$$2KClO_3 \longrightarrow 2KCl + 3O_2 \qquad \text{(balanced for K, Cl, and O)}$$

One further point about balancing equations: When any of the substances involved in the reaction contain polyatomic covalent ions of the sort described in Section 3.10, the ions are treated as whole units. Thus, reaction of sulfuric acid (H_2SO_4) with sodium hydroxide (NaOH) yields sodium sulfate and water:

$$H_2SO_4 + 2\,NaOH \longrightarrow Na_2SO_4 + H_2O$$

The sulfate ion, $SO_4{}^{2-}$, is treated as a whole unit for purposes of balancing the equation, since it appears unchanged on both sides of the equation. On the left there is one $SO_4{}^{2-}$ unit (in H_2SO_4), and on the right there is one $SO_4{}^{2-}$ unit (in Na_2SO_4).

Practice Problems

4.2 Write a balanced equation for the reaction of metallic sodium with chlorine to yield sodium chloride.

4.3 Ozone (O_3) is formed in the earth's upper atmosphere by action of solar radiation on oxygen molecules. Write a balanced equation for the formation of ozone from oxygen.

4.4 Balance these equations:
(a) $Ca(OH)_2 + HCl \longrightarrow CaCl_2 + H_2O$
(b) $Al + O_2 \longrightarrow Al_2O_3$
(c) $Ag_2O + HCl \longrightarrow AgCl + H_2O$
(d) $CH_3CH_3 + O_2 \longrightarrow CO_2 + H_2O$

4.3 AVOGADRO'S NUMBER AND THE MOLE

Imagine that you wanted to do a laboratory experiment combining hydrogen and oxygen to form water. How much hydrogen and how much oxygen should you use? According to the balanced equation describing the reaction, it takes two molecules of hydrogen and one molecule of oxygen to make two molecules of water.

$$2\,H_2 + O_2 \longrightarrow 2\,H_2O$$

In reality, of course, individual molecules are so tiny that we can't see them even with the most powerful microscopes. Thus, a *visible* chemical reaction must involve reacting many *quadrillions* of hydrogen molecules in a 2:1 ratio with many quadrillions of oxygen molecules to yield many quadrillions of water molecules.

2-to-1 ratio of H_2 to O_2

$2\,H_2$ molecules $+\ 1\,O_2$ molecule $\longrightarrow 2\,H_2O$ molecules

$1000\,H_2$ molecules $+\ 500\,O_2$ molecules $\longrightarrow 1000\,H_2O$ molecules

$2 \times 10^{20}\,H_2$ molecules $+\ 1 \times 10^{20}\,O_2$ molecules $\longrightarrow 2 \times 10^{20}\,H_2O$ molecules

How, though, can you be sure that you have the correct ratio of hydrogen and oxygen molecules in your reaction vessel? Clearly, it's impossible to count out the right number of molecules, and you might therefore try to weigh the proper amounts of both substances on a chemical balance.

But the weighing approach leads to yet another problem. How many molecules are there in a gram of a substance? The answer depends on what the substance is, because different atoms and different molecules have different masses. Just as the *atomic* weight of an element is the average mass of individual atoms of the element (Section 2.3), the **formula weight** of a compound is the sum of the individual atomic weights of all atoms in the formula of the compound. For example, the *atomic weight* of a hydrogen *atom* is 1.0 amu, but the *formula weight* of an H_2 *molecule* is 2×1.0 amu, or 2.0 amu. Similarly, the formula weight of an O_2 molecule is 2×16.0 amu, or 32.0 amu. (The term *molecular weight*, abbreviated mol wt, is often used in place of formula weight when referring to covalent molecules rather than ionic compounds.)

The sum of individual atomic weights for all atoms in a molecule or ion.

$$= \text{sum of atomic weights of atoms in compound}$$

Since the weight ratio of a *single* H_2 molecule to a *single* O_2 molecule is 2:32, it follows that the weight ratio of *any* given number of H_2 molecules to the same number of O_2 molecules will always be 2:32. In other words, *whenever* the weight ratio of an H_2 sample to an O_2 sample is 2:32, the samples will contain the same number of molecules (Figure 4.1).

One particularly easy way to use the correlation between formula weights and numbers of molecules is simply to weigh out amounts in grams that are numerically equal to the formula weights of each sample. For example, the number of hydrogen molecules in 2.0 grams of H_2 is the same as the number of oxygen molecules in 32.0 grams of O_2. In fact, *there will always be the same number of molecules in any sample whose weight in grams is numerically equal to the formula weight of the substance in the sample.* This amount of any substance is called a **mole**, abbreviated **mol** and derived from the latin word *moles*, meaning "pile." Thus, a mole of atoms is a very big pile of them.

An amount of a substance in grams equal to the formula weight of the substance.

$$= \text{formula weight of substance in grams}$$

The exact number of molecules in a mole of substance can be found by a brief calculation. To take H_2 as an example, we know both the number of grams of H_2 per mole (2.02 g/mol) and the average mass per H_2 molecule (2.02 amu). Since we also know from Section 2.3 that 1 amu $= 1.66 \times 10^{-24}$ gram, we can

Figure 4.1
Equal numbers of H_2 and O_2 molecules always have a weight ratio equal to the ratio of their formula weights (2:32).

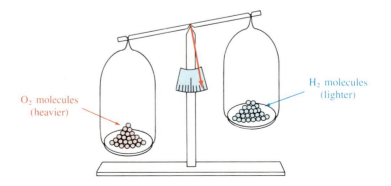

O_2 molecules (heavier)

H_2 molecules (lighter)

■ **Avogadro's number**   The number of particles (atoms, ions, or molecules) in one mole (6.02×10^{23}).

calculate a value of 6.02×10^{23} molecules per mole. This value is called **Avogadro's number**, after the Italian scientist who first conceived the idea. *One mole of any substance always contains Avogadro's number of molecules:*

$$\frac{2.02 \ \text{g H}_2}{1 \ \text{mol H}_2} \times \frac{1 \ \text{H}_2 \ \text{molecule}}{2.02 \ \text{amu}} \times \frac{1 \ \text{amu}}{1.66 \times 10^{-24} \ \text{g H}_2} = \frac{6.02 \times 10^{23} \ \text{H}_2 \ \text{molecules}}{\text{mol H}_2}$$

It helps to understand the idea of a mole if you think about it as simply a very large number of something. One mole of ping pong balls is 6.02×10^{23} ping pong balls, one mole of bowling balls is 6.02×10^{23} bowling balls, and one mole of molecules is 6.02×10^{23} molecules. A mole of bowling balls is much *heavier* than a mole of ping pong balls, but the *number* of balls in one mole of each is the same. Similarly, one mole of oxygen molecules is heavier than one mole of hydrogen molecules (32.0 g versus 2.0 g), but the number of molecules is 6.02×10^{23} in each.

$$\text{One mol H}_2 \ = \ 2.0 \ \text{g H}_2 \ = 6.02 \times 10^{23} \ \text{H}_2 \ \text{molecules}$$
$$\text{One mol O}_2 \ = 32.0 \ \text{g O}_2 \ = 6.02 \times 10^{23} \ \text{O}_2 \ \text{molecules}$$
$$\text{One mol Na} \ = 23.0 \ \text{g Na} \ = 6.02 \times 10^{23} \ \text{Na atoms}$$
$$\text{One mol CH}_4 = 16.0 \ \text{g CH}_4 = 6.02 \times 10^{23} \ \text{CH}_4 \ \text{molecules}$$

How big is Avogadro's number? Our minds can't really conceive of the magnitude of a number like 6.02×10^{23}, but one useful analogy is that if you were to measure the total amount of water in all the world's oceans, the answer would be approximately Avogadro's number of milliliters.

Solved Problem 4.3   What is the formula weight of ethyl alcohol, C_2H_6O?

Solution   First list the elements present in the molecule, and then look up the atomic weight of each:

$$\text{C (12.0 amu)} \qquad \text{H (1.0 amu)} \qquad \text{O (16.0 amu)}$$

Next, multiply the atomic weight of each element by the number of times the atom is present, and total the result:

$$2 \, \text{C} = 2 \times 12.0 \ \text{amu} = 24.0 \ \text{amu}$$
$$6 \, \text{H} = 6 \times 1.0 \ \text{amu} \ = \ \ 6.0 \ \text{amu}$$
$$\text{O} = 16.0 \ \text{amu}$$
$$\text{Formula weight} = 46.0 \ \text{amu}$$

Solved Problem 4.4   How many molecules are there in 3.0 mol of ethyl alcohol?

Solution   Since one mol of *any* substance (ethyl alcohol included) always contains 6.02×10^{23} molecules, 3.0 mol of any substance contains $3.0 \times (6.02 \times 10^{23})$ molecules.

$$3.0 \ \text{mol ethyl alcohol} \times \frac{6.02 \times 10^{23} \ \text{molecules}}{1 \ \text{mol ethyl alcohol}} = 1.8 \times 10^{24} \ \text{molecules}$$

Practice Problems   **4.5**   Calculate the formula weights of these substances:
(a) Ammonia, NH_3 (b) Water, H_2O (c) Rust, Fe_2O_3

4.6   Sucrose, or common table sugar, has the formula $C_{12}H_{22}O_{11}$. What is the formula weight of sucrose? How many molecules are there in 1.7 mol of sucrose?

4.7   How many molecules are in each of these samples?
(a) 2.0 mol of glucose   (b) 0.25 mol of penicillin   (c) 18.0 g of water

4.4   GRAM/MOLE CONVERSIONS

We saw in the previous section that the coefficients in a balanced equation describe how many moles of each substance are needed for a reaction. In order to do actual laboratory work, however, it's necessary to deal in grams rather than in moles, since it's relatively easy to weigh out specific amounts of reactants. This means, however, that there has to be a convenient way of converting between grams and moles to be sure that correct ratios of reactants are used. Thus, if you needed to know how many moles of water were in a 9.0-gram sample, you would first have to calculate the formula weight of H_2O to find how many grams are in 1 mol. Using the formula weight to find how many grams are in one mole, you could then find how many moles are in 9.0 grams. The following solved problem shows how this is done.

Solved Problem 4.5   How many moles of water are in 9.0 grams?

Solution   First calculate the formula weight of water:

$$\text{formula weight of } H_2O = (2 \times 1.0 \text{ amu}) + 16.0 \text{ amu} = 18.0 \text{ amu}$$

Since the formula weight tells how many grams of water are in 1 mol, it can be used as a conversion factor:

$$\text{Number of moles water} = 9.0 \text{ grams water} \times \left\{ \frac{1.0 \text{ mol water}}{18 \text{ grams water}} \right. = 0.50 \text{ moles water}$$

formula weight
used as conversion factor

As another example of how a gram/mole conversion might be used, the anesthetic agent ethyl chloride (C_2H_5Cl) is prepared by reaction of ethylene (C_2H_4) with hydrogen chloride. How many moles of HCl is needed to convert a given amount of ethylene into ethyl chloride? To find out, you first have to write a balanced equation and then convert grams into moles to see how many moles of ethylene reactant you have. The following solved problem shows how this can be done.

Solved Problem 4.6   How many moles of HCl are needed to convert 5.0 g of ethylene, C_2H_4, into ethyl chloride, C_2H_5Cl?

Solution   First, write the balanced equation:

$$C_2H_4 + HCl \longrightarrow C_2H_5Cl$$

According to the coefficients in this equation, one mole of HCl is required for each mole of ethylene. To find out how much HCl is required for 5.0 g of ethylene, we must find how many moles of ethylene are in 5.0 g. We do this by calculating the formula weight of ethylene:

Formula weight of C_2H_4 = $(2 \times 12.0 \text{ amu}) + (4 \times 1.0 \text{ amu}) = 28.0 \text{ amu}$

We then use the formula weight as the conversion factor to change grams into moles (remember to use the right number of significant figures!):

$$\text{Number of moles of } C_2H_4 = 5.0 \text{ g } C_2H_4 \times \frac{1 \text{ mole } C_2H_4}{28.0 \text{ g } C_2H_4} = 0.18 \text{ mol } C_2H_4$$

Since we have 0.18 mol of ethylene, we also need 0.18 mol of HCl to make ethyl chloride.

We have just used a gram-to-mole calculation to conclude that 5.0 g of ethylene (0.18 mol) will react with 0.18 mol HCl. To find out how many *grams* of HCl we need for the reaction, however, we have to carry out one more calculation—a mole-to-gram conversion.

Solved Problem 4.7   How many grams of HCl are in 0.18 mol?

Solution   First, calculate the formula weight of HCl:

Formula weight of HCl = 1.0 amu + 35.5 amu = 36.5 amu

Next, use the formula weight of HCl as a conversion factor to find out how many grams of HCl are in 0.18 mol:

$$\text{Number of grams of HCl} = 0.18 \text{ mol HCl} \times \frac{36.5 \text{ g HCl}}{1 \text{ mol HCl}} = 6.6 \text{ g HCl}$$

Look carefully at the sequence of steps used in the solved problems. The important point is that the coefficients of a balanced equation tell the relative numbers of *moles* needed for a reaction but that *grams* are used to weigh actual amounts of reactants in the laboratory. Moles tell *how many molecules* of each reactant we have to use, but grams tell exactly *how much mass* of each reactant we need.

Moles: numbers of molecules

Grams: mass of reactants

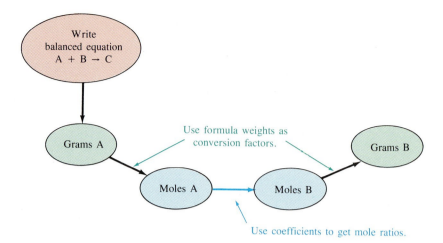

Figure 4.2
A flow diagram summarizing conversions between moles and grams for chemical reactions.

The diagram in Figure 4.2 illustrates the necessary conversions. Note that you can't go directly from the number of grams of one reactant to the number of grams of another reactant. You *must* first convert to moles.

Practice Problems

4.8 How many moles are in a 10.0-g sample of ethyl alcohol, C_2H_6O? How many grams are in a 0.10 mol sample of ethyl alcohol?

4.9 How many moles are in 25.0 g of each of these substances?
(a) NaCl   (b) $NaHCO_3$ (sodium bicarbonate)
(c) NaOCl (household bleach)   (d) $C_6H_{12}O_6$ (glucose)

4.10 How many grams are in these samples?
(a) 1.33 mol of iron   (b) 0.025 mol of aspirin, $C_9H_8O_4$

4.11 How many grams of ethylene, C_2H_4, and how many grams of HCl would it take to prepare 10.0 g of ethyl chloride, C_2H_5Cl?

4.5 REACTIONS WITH LIMITING AMOUNTS OF REACTANTS

What would happen if five people showed up to play tennis but only one court was available? Since only two people at a time can play on a court (at least in singles), only two people get to play. The other three sit and watch, because the number of players is limited by the number of available courts.

$$5 \text{ people} + 1 \text{ tennis court} \longrightarrow 2 \text{ players} + 3 \text{ spectators}$$

Now imagine what would happen if five H_2 molecules came in contact with one O_2 molecule. Since only two H_2 molecules react with each O_2 molecule, only two H_2 molecules undergo a chemical reaction. The other three sit and watch, because the number of H_2 molecules that react is limited by the number of available O_2 molecules.

$$5 H_2 + 1 O_2 \longrightarrow 2 H_2O + 3 H_2 \quad \text{(unreacted)}$$

What is true for five molecules of H_2 and one molecule of O_2 is also true for five *moles* of H_2 and one *mole* of O_2: There will be 2 mole of H_2O formed and 3 mole of H_2 left over. *Whenever the numbers of reactant molecules are different from those required for a balanced reaction, some excess reactant will be left over.* The extent to which a chemical reaction takes place always depends on the reactant present in limiting amount—the **limiting reactant**. The following solved problem shows how to tell if there is a limiting reactant and how to calculate the amount of product formed in such a case.

■ **Limiting reactant**   The reactant present in limiting amount that restricts the extent to which a reaction can occur.

Solved Problem 4.8   How much sodium chloride can be formed from 10.0 g of sodium and 10.0 g of chlorine? Which reactant is limiting? How much of the other reactant will be left over?

Solution   First, write the balanced equation:

$$2\,Na + Cl_2 \longrightarrow 2\,NaCl$$

Second, calculate the formula weights of the reactants and products:

$$Na = 23.0 \text{ amu} \qquad Cl_2 = 2 \times 35.5 \text{ amu} = 71.0 \text{ amu}$$

$$NaCl = 23.0 \text{ amu} + 35.5 \text{ amu} = 58.5 \text{ amu}$$

Third, determine how many moles of each reactant are present:

$$\text{Moles Na} = 10.0 \text{ g Na} \times \frac{1 \text{ mol Na}}{23.0 \text{ g Na}} = 0.435 \text{ mol Na}$$

$$\text{Moles } Cl_2 = 10.0 \text{ g } Cl_2 \times \frac{1 \text{ mol } Cl_2}{71.0 \text{ g } Cl_2} = 0.141 \text{ mol } Cl_2$$

According to the balanced equation, Na and Cl_2 react in a 2:1 ratio. Therefore 0.435 mol of Na requires $0.435/2 = 0.218$ mol of Cl_2 for a complete reaction. Since only 0.141 mol of Cl_2 is available, however, this is the limiting reactant. Thus, 0.141 mol of Cl_2 can react with only $2 \times 0.141 = 0.282$ mol of Na, yielding only 0.282 mol of NaCl product. Converting these mole amounts into grams, we get:

$$\text{grams } Cl_2 \text{ reacted} = 0.141 \text{ mol } Cl_2 \times \frac{71.0 \text{ g } Cl_2}{1 \text{ mol } Cl_2} = 10.0 \text{ g } Cl_2 \text{ reacted}$$

$$\text{grams Na reacted} = 0.282 \text{ mol Na} \times \frac{23.0 \text{ g Na}}{1 \text{ mol Na}} = 6.48 \text{ g Na reacted}$$

$$\text{grams Na remaining} = 10.0 \text{ g Na} - 6.48 \text{ g Na} = 3.5 \text{ g Na remaining}$$

$$\text{grams NaCl formed} = 0.282 \text{ mol NaCl} \times \frac{58.5 \text{ g NaCl}}{1 \text{ mol NaCl}}$$

$$= 16.5 \text{ g NaCl formed}$$

The answer can be checked by observing that the masses of Na and Cl_2 reacted total the calculated mass of NaCl formed $(10.0 \text{ g} + 6.48 \text{ g} = 16.5 \text{ g})$.

Practice Problems

4.12 Methane (CH_4) reacts with oxygen to yield CO_2 and water. If 2 mol of oxygen and 5 mol of methane are present, which reactant is limiting? How many moles of excess reactant remain after reaction? How many moles of each product are formed?

4.13 If 20.0 g of methane is allowed to react with 50.0 g of oxygen, which reactant is limiting? How many grams of excess reactant remain after reaction? How many grams of each product are formed?

4.6 WHY DO CHEMICAL REACTIONS OCCUR?

If you think about it for a minute, you'll conclude that just because you can write a balanced chemical equation doesn't mean the reaction will take place. For example, it's easy to write a balanced chemical equation for the spontaneous decomposition of rust (Fe_2O_3) to give iron and oxygen, yet common sense tells you that rusted metal doesn't do this. Similarly, it's easy to write a balanced equation for the conversion of table salt into sodium and chlorine, but no one has ever observed toxic green chlorine gas emanating from a salt shaker.

$$2\,Fe_2O_3 \longrightarrow 4\,Fe + 3\,O_2$$
$$2\,NaCl \longrightarrow 2\,Na + Cl_2$$

These reactions don't take place even though the equations are balanced.

Clearly, just being able to describe a reaction by a balanced equation isn't enough. There must be some other requirement that has to be met before a reaction can actually take place. This additional requirement involves *stability*. In order for a chemical reaction to be favored, the products of the reaction must be more stable than the starting reactants. But what is "stability"? What does it mean to say that one chemical is more stable than another?

In general, the stability of a substance refers to the amount of potential energy the substance contains. **Potential energy** is stored energy—the capacity to give off energy under the proper conditions. Substances with a large amount of potential energy are less stable and more reactive, whereas substances with a small amount of potential energy are more stable and less reactive.

A good analogy for the relationship between potential energy and chemical stability is that of a rock poised near the top of a hill. The rock at the top of the hill has a large amount of potential energy stored in it because of its unstable position. When it rolls downhill, however, it releases its stored energy until it reaches a low-energy, stable position at the bottom. In the same way, the energy level in a chemical reaction goes downhill as the potential energy stored in the chemical bonds of an unstable, high-energy reactant is released and a low-energy, stable product is formed (Figure 4.3).

Whenever a high-energy reactant is converted into a low-energy product during a chemical reaction, energy is released to the surroundings, and the reaction is said to be **exothermic**. (Conversely, if a reaction absorbs energy from the surroundings, it is said to be **endothermic**.) This energy release usually appears

■ **Potential energy** Energy that is stored because of the position or composition of an object.

■ **Exothermic reaction** A reaction that gives off heat to the surroundings.

■ **Endothermic reaction** A reaction that absorbs heat from the surroundings.

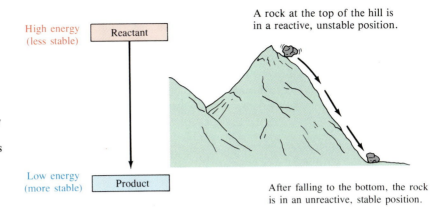

Figure 4.3
The relationship between energy and stability. Just like a rock near the top of a hill, substances that are high in energy are unstable; they seek to drop "downhill" in energy to reach a low-energy, stable product.

as heat, as when methane (natural gas) reacts with oxygen during combustion in a home furnace. Occasionally, however, light (as in fireflies) and sound (as in dynamite explosions) are also produced by chemical reactions.

The exact amount of heat released during a chemical reaction is called the **heat of reaction** and is a quantity that can be measured. During the combustion of methane, for example, 213 kilocalories of heat are released for every mole of methane that reacts. Since one calorie is defined (Section 1.11) as the amount of energy necessary to heat one gram of water one Celsius degree, 213 kcal is enough energy to heat 213 *kilo*grams (470 lbs!) of water one Celsius degree. All of that energy is packed into just one mole (16 grams) of methane and two moles (64 grams) of oxygen, and is released when these two compounds react.

■ **Heat of reaction** The exact amount of heat released or absorbed during a chemical reaction.

$$1 \text{ mol } CH_4 + 2 \text{ mol } O_2 \longrightarrow CO_2 + 2 H_2O + 213 \text{ kcal heat}$$

Practice Problems

4.14 If 213 kcal heat is released during the combustion of 1 mol methane, how much heat is released during the combustion of 0.20 mol of methane? During the combustion of 8.0 g of methane?

4.15 How much energy is required to heat a cup of coffee, assuming that the cup holds 200 g of water and that the water is heated 80°C (from 20°C to 100°C)? How much methane must be burned to obtain this much energy?

4.7 HOW DO CHEMICAL REACTIONS OCCUR?

Although you might think of molecules as being rigid and static, they're actually in constant motion at normal temperatures. If the molecule is in a crystal, its overall position is fixed, but individual bonds within the molecule stretch, contract, bend, and rotate. If, however, the molecule is in a liquid or a gas, then the entire molecule moves about more or less at random, often colliding with its neighbors. At low temperatures, molecular motions are relatively sluggish; at higher temperatures, motions become more vigorous, and collisions become more frequent. In fact, that's really what "temperature" is—a measure of the amount of molecular motion in a substance.

For a chemical reaction to occur, reactant atoms, ions, or molecules have to collide. Some chemical bonds then have to break, while new ones form. One of the simplest ways to think about what happens during a collision is to return to the analogy of a rock near the top of a hill. Although the rock would be much more stable at the bottom, let's assume that it's trapped behind a barrier in a small depression and isn't able to fall spontaneously. In order to fall, it has to be lifted over the barrier and given a shove. In other words, before the rock can *release* its potential energy in a fall, we first have to "collide" with it, putting energy *into* the rock to start it falling (Figure 4.4).

A similar situation occurs in chemical reactions. Just because a chemical reaction yields a stable product doesn't mean the reaction will occur all by itself; it may need a "shove" to get it started. We've all seen the flame on a gas stove

AN APPLICATION: MERCURY—REACTIVITY AND TOXICITY

As the only metallic element that is liquid at room temperature (mp = $-38.9°C$), mercury has fascinated people since ancient times. Containers of mercury have been found buried with kings in Egyptian pyramids, and even its symbol, *Hg*, from the Latin *hydrargyrum*, meaning "liquid silver," hints at mercury's uniqueness.

Much of the recent interest in mercury has concerned its toxicity, but here too there are surprises. For example, mercury(I) chloride (better known as *calomel*) is nontoxic and has a long history of medical use as a bowel purgative, yet mercury(II) chloride is used as a fungicide and rat poison. Similarly, an alloy of 50 percent elemental mercury with 50 percent silver-tin has been used safely by dentists for many years to fill tooth cavities, yet exposure to elemental mercury vapors for prolonged periods leads to headaches, tremors, and loss of hair and teeth.

Why is mercury toxic in some forms but not in others? It turns out that the toxicity of mercury and its derivatives is related to reactivity; only soluble Hg(II) compounds are toxic. The toxicity arises because Hg(II) compounds are transported through the bloodstream to all parts of the body, where they react with different enzymes and interfere with various biological processes. Elemental mercury and Hg(I) compounds become toxic only when converted into Hg(II) compounds, a reaction that is extremely slow in the body. Calomel, for example, passes through the body long before it is converted into any Hg(II) compounds. The dental use of mercury alloys is safe, because mercury does not evaporate from the alloys and neither reacts with nor dissolves in saliva; thus it does not enter the body.

Of particular current concern with regard to mercury toxicity is the environmental danger posed by pollution from both natural and industrial sources. Microorganisms present in lakes and streams are able to convert many mercury-containing wastes into a soluble and highly toxic Hg(II) compound called methylmercury. Methylmercury is concentrated to high levels in fish, particularly shark and swordfish, which are then poisonous when eaten. Although commercial fishing is now monitored carefully, more than fifty deaths from eating contaminated fish were recorded in Minimata, Japan, during the 1950s before the cause of the problem was realized.

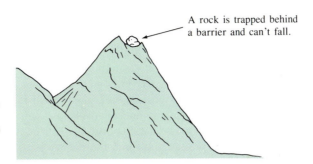

A rock is trapped behind
a barrier and can't fall.

Figure 4.4
A rock behind a barrier near the top of a hill could release its energy by falling to a more stable position at the bottom. Before it can do so, however, energy has to be put *into* it to raise it over the barrier.

■ **Reaction energy diagram** A pictorial way of representing the energy changes that occur during a chemical reaction.

as methane releases energy in its reaction with oxygen, but we know that the reaction doesn't begin all by itself. Methane and oxygen don't react until we give them a shove by holding up a lighted match. The heat of the match provides energy to start the reaction. Once started, the reaction sustains itself by giving off enough energy to shove other molecules over the barrier.

A helpful way to visualize the energy situation in a chemical reaction is to use a **reaction energy diagram** like that shown in Figure 4.5. On this diagram, the potential energy levels of molecules are shown at all stages of the reaction from beginning (on the left) to end (on the right). At the beginning, reactant molecules have a certain amount of energy as indicated on the left side of the diagram. Like the rock near the top of the hill, though, the molecules are unable to react; they simply don't have enough energy to surmount the barrier that keeps them from going downhill toward stable products.

As reactants approach each other and begin to collide, their total potential energy *rises* because the negatively charged electron clouds surrounding the two molecules bump together and repel each other. If the collision is a soft one, the molecules simply spring apart and the energy level drops back to what it was at the beginning. If the collision is a particularly violent one, however, the energy level of the molecules rises until it is sufficient to get over the barrier. The reactants momentarily join together, a reaction occurs as some bonds break and others form, and the energy level drops as heat is released when the stable product forms.

Figure 4.5
A reaction energy diagram, used to describe what happens during a chemical reaction. Reaction begins on the left and proceeds to the right. Since energy is released in exothermic reactions, the product energy level is lower than that of reactants. The height of the barrier between reactant energy level and the peak is the activation energy, E_{act}; the difference between reactant and product energy levels is the heat of reaction.

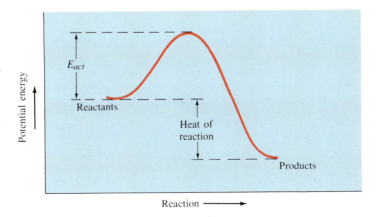

■ **Activation energy (E_{act})** The amount of energy that reactants must have in order to surmount the energy barrier to reaction.

The amount of energy necessary for reactants to surmount the energy barrier is called the reaction's **activation energy**, E_{act}, as indicated on the diagram. Reactions that have a high activation energy occur very slowly since few collisions take place with enough energy to raise reactants over the high barrier. Reactions with a low activation energy, however, take place rapidly since almost all collisions are vigorous enough to push reactants over the barrier.

The difference in energy levels between reactants and products is the heat of reaction, discussed in Section 4.6. For exothermic reactions, in which heat is released to the surrounding, the energy level of the products is lower than the energy level of the reactants. Notice that the size of the activation energy and the size of the heat of reaction are unrelated. A reaction might take place very slowly (have a large E_{act}) even though it goes far downhill once it does react (has a large heat of reaction).

Solved Problem 4.9   Draw a reaction energy diagram for a reaction that is very fast but releases only a small amount of heat.

Solution   A very fast reaction is one that has only a small E_{act}, and a reaction that releases only a small amount of heat is one that has a small heat of reaction. Thus, the diagram must show a small energy barrier and a small energy difference between starting materials and products (Figure 4.6).

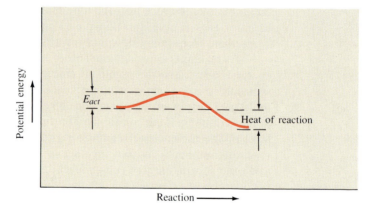

Figure 4.6
A reaction energy diagram for Solved Problem 4.9. Both E_{act} and the heat of reaction are small.

Practice Problem   **4.16**   Draw a reaction energy diagram for a reaction that is very slow and releases a large amount of heat.

4.8 EFFECT OF TEMPERATURE, CONCENTRATION, AND CATALYSTS ON REACTIONS

■ **Temperature** A measure of the amount of random motion of particles in a substance.

There are several things we can do to help reactants over an activation barrier and thereby speed up a reaction. One possibility is to add energy to the reactants by raising the **temperature**. With more energy in the system, collisions between reactant molecules are more violent and more likely to lead to a successful reaction. Thus, a bicycle rusts faster on a warm day in summer than on a cold day in winter. In general, a 10°C rise in temperature leads to a doubling or tripling of the reaction's speed.

Increase in temperature ⟹ Increase in reaction rate

■ **Concentration**   A measure of the amount of dissolved substance per unit volume of solvent.

Another possibility for speeding up a reaction is to increase the **concentration** of the reactants. With molecules more crowded together, collisions become more frequent and reactions more likely. Although different reactions respond differently to concentration changes, we often find that doubling or tripling the concentration also doubles or triples the reaction rate. Thus, flammable materials burn much more rapidly in pure oxygen than in air because the concentration of O_2 molecules is higher (air is approximately 20% oxygen and 80% nitrogen). For this reason, hospitals must take extraordinary precautions to be sure that no flames are used near patients receiving oxygen.

Increase in concentration ⟹ Increase in reaction rate

■ **Catalyst**   A substance that speeds up a chemical reaction without itself undergoing change.

A third possibility for speeding up a reaction is to add a catalyst. **Catalysts** are substances that accelerate chemical reactions without themselves undergoing change. For example, ammonia is produced for use as an agricultural fertilizer by reaction of hydrogen with nitrogen:

$$3\,H_2 + N_2 \xrightarrow[\text{catalyst}]{\text{Fe}} 3\,NH_3$$

Although a mixture of hydrogen and nitrogen is completely inert in the absence of a catalyst, reaction occurs when a small amount of iron (Fe) catalyst is added.

Catalysts increase the rate of a reaction by lowering the height of the activation-energy barrier. To return once more to the analogy of a rock behind a barrier at the top of a hill, a catalyst functions by finding some alternative *low* point in the barrier for the rock to be lifted over, rather than by lifting it directly over the *top* of the barrier. On a reaction energy diagram, the catalyzed reaction has a lower E_{act} as in Figure 4.7.

Figure 4.7
A reaction energy diagram for a reaction in the presence (green line) and absence (red line) of a catalyst. The catalyzed reaction has a lower E_{act} because it uses an alternate pathway with a lower energy barrier.

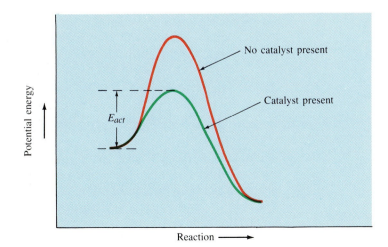

■ **Enzyme** A large protein molecule that acts as a catalyst in biological reactions.

The chemical reactions in all living organisms are catalyzed by large molecules called **enzymes**. The human body alone is estimated to have some 50,000 different enzymes, each of which catalyzes a specific chemical reaction necessary for the myriad functions of life. When even one of these enzymes malfunctions, drastic consequences to health can result.

Although we'll study more about enzymes in Chapter 16, a brief example now is worth mentioning. The two major sources of carbohydrates in nature—*starch* and *cellulose*—both consist of a large number (up to 2000) of glucose units bonded together. (The exact manner of bonding is slightly different in the two cases.) Those carbohydrates that we eat—the ones in grains and cereals—are composed of *starch*, whereas those we can't eat—the ones in grass and leaves—are composed of cellulose. In order to be useful as a human food source, a reaction with water is necessary to break the bonds between units and to release individual glucose molecules. It turns out, however, that the uncatalyzed reaction of water with either starch or cellulose has a high E_{act} and can't take place under normal conditions.

Fortunately, human saliva and gastric secretions contain enzymes that are able to catalyze the reaction of starch with water, lowering the E_{act} for the process so that glucose molecules are released in the mouth and stomach for further digestion. We don't, however, have any enzymes for catalyzing the reaction of cellulose with water, and we therefore can't digest grasses and fibers.

$$\text{Starch} + \text{H}_2\text{O} \quad \underset{\substack{\text{enzymes in}\\ \text{saliva}}}{\overset{\text{no catalyst}}{\diagdown\diagup}} \quad \begin{matrix} \text{No reaction} \\ \\ \text{Many glucose molecules} \end{matrix}$$

$$\text{Cellulose} + \text{H}_2\text{O} \longrightarrow \text{No reaction}$$

4.9 REVERSIBLE REACTIONS AND CHEMICAL EQUILIBRIUM

Many chemical reactions seem to go in only one direction because the products are so much more stable than the reactants. Thus, sodium and chlorine react to give sodium chloride, but the reverse reaction of sodium chloride decomposing to give sodium and chlorine does not occur:

$$\underset{\text{(less stable)}}{2\,\text{Na} + \text{Cl}_2} \quad \underset{\text{does not occur}}{\overset{\text{takes place readily}}{\rightleftharpoons}} \quad \underset{\text{(more stable)}}{2\,\text{NaCl}}$$

What happens, though, when the reactants and products are of approximately *equal* stability? This is the case in the reaction of acetic acid (vinegar) with ethyl alcohol to yield ethyl acetate (a solvent used in nail-polish remover). The products (ethyl acetate and water) and the reactants (acetic acid and ethyl alcohol) are of similar stability:

$$CH_3\overset{\displaystyle O}{\overset{\displaystyle \|}{C}}OH + HOCH_2CH_3 \underset{\text{or this direction?}}{\overset{\text{this direction?}}{\rightleftharpoons}} CH_3\overset{\displaystyle O}{\overset{\displaystyle \|}{C}}OCH_2CH_3 + H_2O$$

Acetic acid　　　Ethyl acetate　　　　　　　Ethyl alcohol　　　Water

■ **Reversible reaction** A reaction that can proceed in either the forward or the reverse direction because the reactants and products are of similar stability.

Imagine the situation if you took a mixture of acetic acid and ethyl alcohol. The two would begin to react to form ethyl acetate and water. But as soon as ethyl acetate and water formed, *they* would begin to react to go back to acetic acid and ethyl alcohol. When the products and reactants are of similar stability, the reaction can go in either direction and is thus said to be a **reversible reaction**. Ultimately, a **chemical equilibrium** is established. Both products and reactants are present at equilibrium (although not necessarily in equal amounts) and the forward and reverse reactions both occur at the same rate, so that the product and reactant concentrations remain constant.

■ **Chemical equilibrium** The point in a reversible reaction at which forward and reverse reactions take place at the same rate, so that the concentrations of products and reactants no longer change.

4.10 CHEMICAL REACTIONS IN THE BODY

All the characteristics discussed in this chapter apply just as well to reactions taking place in a living organism as to those taking place in a test tube. All reactions in living organisms must be balanced; all have a certain heat of reaction; and all have a certain activation energy that must be surmounted.

Chemical reactions in the human body take place at 37°C (98.6°F), the normal body temperature of a healthy person. This temperature is maintained by the heat released when food is "burned" in the body to yield water, carbon dioxide, and waste products. In fact, the "caloric value" of a given food is really just a heat of reaction telling how much energy is released when the food is burned in oxygen in the laboratory. Just as one gram of methane releases 13.3 kcal when burned (Section 4.6), one gram of table sugar releases 4.1 kcal, one gram of fat releases 9.4 kcal, and so on (Table 4.1).

$$\text{Food} + \text{oxygen} \longrightarrow CO_2 + H_2O + \text{heat}$$

If the body's thermostat is unable to maintain a temperature of 37°C, the rates of the many thousands of chemical reactions that take place constantly in the body will change accordingly. If, for example, a hiker in the mountains were to be trapped overnight by a sudden storm and was unable to keep warm, **hypothermia** could result. Hypothermia is a dangerous state that occurs when the body is unable to generate enough heat to maintain normal temperature.

■ **Hypothermia** The medical condition that results from uncontrolled loss of body temperature.

Table 4.1　Caloric Values of Some Foods

Substance (1 g Sample)	Caloric Value (kcal)
Protein	5.6
Carbohydrate	4.1
Fat	9.4
Alcohol	7.1

All chemical reactions in the body slow down because of the lower temperature, metabolism drops, and death can result when the body is no longer able to fuel itself.

Conversely, a marathon runner on a hot, humid day might become overheated, and **hyperthermia** could result. Hyperthermia, also called *heat stroke*, is an uncontrolled rise in temperature as the result of the body's inability to lose sufficient heat. All chemical reactions in the body are accelerated at higher temperatures, metabolism speeds up, and brain damage can result.

■ **Hyperthermia**   The medical condition that results from an uncontrolled rise in body temperature.

INTERLUDE: REGULATION OF BODY TEMPERATURE

Maintenance of normal body temperature is crucial if the thousands of different chemical reactions taking place in the body are to occur at the right rate. Body temperature is maintained both by the thyroid gland and by the hypothalamus, which act together to regulate metabolic rate see the figure below. When the body's environment changes, temperature receptors in the skin, spinal cord, and abdomen send signals to the hypothalamus gland in the brain. The hypothalamus contains both heat-sensitive neurons, which detect an increase in temperature, and cold-sensitive neurons, which detect a decrease in temperature.

Stimulation of the heat-sensitive neurons on a hot day causes a variety of effects: Impulses are sent to stimulate the sweat glands, dilate the blood vessels of the skin, decrease muscular activity, and reduce metabolic rate. Sweating cools the body through evaporation; approximately 540 calories of heat are removed by evaporation of 1 gram of sweat. Dilated blood vessels cool the body by allowing more blood to flow close to the surface of the skin, where heat is removed by contact with air. Decreased muscular activity and a reduced metabolic rate cool the body by lowering internal heat production.

Stimulation of the cold-sensitive neurons on a cold day also causes a variety of effects: The hormone epinephrine is released to stimulate metabolic rate; peripheral blood vessels contract to decrease blood flow to the skin and prevent heat loss; and muscular contractions increase to produce more heat, resulting in shivering and "goosebumps."

One further comment: Drinking alcohol to warm up on a cold day actually has exactly the opposite of the intended effect. Alcohol causes blood vessels to dilate, resulting in a warm feeling as blood flow to the skin increases. Although the warmth feels good temporarily, body temperature ultimately drops as heat is lost through the skin at an increased rate.

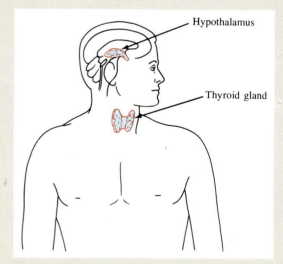

Location of the hypothalamus and thyroid, the body's heat-regulation centers.

SUMMARY

All chemical equations must be **balanced**. That is, the numbers and kinds of atoms must be the same on both sides of the equation. Many quadrillions of molecules are involved when carrying out a chemical reaction on a visible scale, but the *ratio* of reactant molecules is always that described by the balanced equation.

The **formula weight** of a substance is the total of the atomic weights of all atoms in the substance. The amount of any chemical compound whose weight in grams is equal to its formula weight is called a **mole**. One mole of any substance always contains **Avogadro's number** of molecules, 6.02×10^{23}. Conversions between moles and grams are necessary to calculate the amounts of starting materials needed to carry out a reaction.

In order for a reaction to be favorable, the products must be more stable than the starting reactants. In chemical terms, "stability" refers to the amount of **potential energy** a molecule contains. High-energy molecules are unstable and reactive, whereas low-energy molecules are stable and unreactive. Thus, chemical reactions take place when a high-energy reactant yields a low-energy product and gives off heat in the process. The exact amount of heat given off is called the **heat of reaction**.

Chemical reactions occur when molecules collide with sufficient energy for them to fuse together momentarily and go on to yield products. The exact amount of collision energy necessary is called the **activation energy**, E_{act}. A high activation energy results in a slow reaction because few collisions occur with sufficient vigor, whereas a low activation energy results in a fast reaction. Reaction rates can be increased either by raising the temperature, by raising the concentrations of the reactants, or by adding a **catalyst**.

If reactants and products are of similar stability, the reaction is **reversible** and can take place in either direction. Reversible reactions ultimately reach a **chemical equilibrium** where both forward and reverse reactions take place at the same rate and where both reactants and products are present.

REVIEW PROBLEMS

Balancing Chemical Equations

4.17 What is meant by the term "balanced equation"?

4.18 Large amounts of ammonia, NH_3, are prepared industrially for use as a fertilizer by reaction of N_2 with H_2. Write a balanced equation for the reaction.

4.19 According to the balanced equation you wrote in Problem 4.18, how many moles of hydrogen are needed to react with one mole of nitrogen? How many moles of H_2 react with 3 moles of N_2? How many moles of N_2 react with 3 moles of H_2?

4.20 How many moles of ammonia would you obtain (Problem 4.18) if you allowed 5 moles of H_2 to react with 1 mol of N_2? What is the limiting reactant? How much of the nonlimiting reactant would remain after reaction?

4.21 Write a balanced equation for the reaction of sodium bicarbonate ($NaHCO_3$) with sulfuric acid (H_2SO_4) to yield CO_2, Na_2SO_4, and H_2O.

4.22 Spoilage of wine into vinegar is caused by reaction of ethyl alcohol (C_2H_6O) with oxygen to yield acetic acid ($C_2H_4O_2$) and water. Write a balanced equation for the reaction.

4.23 Photosynthesis in green leaves converts carbon dioxide and water into glucose and oxygen. Write a balanced equation for the process:

$$CO_2 + H_2O \longrightarrow C_6H_{12}O_6 + O_2$$

4.24 Balance the following equations:
(a) $Mg + HCl \longrightarrow MgCl_2 + H_2$
(b) $AgNO_3 + CaCl_2 \longrightarrow AgCl + Ca(NO_3)_2$
(c) $CH_4 + Cl_2 \longrightarrow CHCl_3$ (chloroform) $+ HCl$

Formula Weights and Moles

4.25 What is a mole of a substance, and how many molecules does it have?

4.26   Vitamin A has the formula $C_{20}H_{30}O$. What is its formula weight?

4.27   What is the formula weight of diazepam (trade name, Valium), $C_{16}H_{13}ClN_2O$?

4.28   Calculate the formula weights of these substances:
(a) Benzene, C_6H_6
(b) Sodium bicarbonate, $NaHCO_3$
(c) Chloroform, $CHCl_3$
(d) Penicillin V, $C_{16}H_{18}N_2O_5S$

4.29   How many moles are present in 5.00-g samples of each of the molecules listed in Problem 4.28?

4.30   How many grams are present in 0.050-mol samples of each of the molecules listed in Problem 4.28?

4.31   Iron(II) sulfate, $FeSO_4$, is used in the clinical treatment of iron-deficiency anemia. What is the formula weight of $FeSO_4$? How many moles of $FeSO_4$ are in a standard 300-mg tablet?

4.32   Ethyl acetate, $C_4H_8O_2$, is often used as nail-polish remover. Calculate the formula weight of ethyl acetate.

4.33   How many molecules are present in 0.30 mol of ethyl acetate (Problem 4.32)? How many molecules are present in 10.0 g of ethyl acetate?

4.34   Ethyl acetate (Problem 4.32) will undergo a reaction with H_2 in the presence of a catalyst to yield ethyl alcohol: $C_4H_8O_2 + H_2 \rightarrow C_2H_6O$. Write a balanced equation for the reaction. How many moles of ethyl alcohol are produced by reaction of 1.5 mol of ethyl acetate?

4.35   How many grams of ethyl alcohol are produced by reaction of 1.5 mol of ethyl acetate with H_2 (Problem 4.34)?

4.36   How many grams of ethyl alcohol are produced by reaction of 12.0 g of ethyl acetate with H_2 (Problem 4.34)?

4.37   Calculate the formula weight of caffeine, $C_8H_{10}N_4O_2$.

4.38   An average cup of coffee contains approximately 125 mg of caffeine (Problem 4.37). How many moles of caffeine are in one cup? How many molecules?

4.39   How many moles of aspirin, $C_9H_8O_4$, are in a 500-mg tablet?

4.40   How many molecules of aspirin are in a 500 mg tablet (Problem 4.39)?

Chemical Reactions

4.41   What is meant by the "stability" of a molecule?

4.42   In a favorable reaction, which are more stable: products or reactants?

4.43   What is a heat of reaction?

4.44   What is the activation energy of a reaction?

4.45   List three ways in which you might be able to increase the rate of a chemical reaction.

4.46   What is a catalyst, and what effect does it have on the activation energy of a reaction?

4.47   Which reaction is faster: one with $E_{act} = 10$ kcal/mol or one with $E_{act} = 5$ kcal/mol? Explain your answer.

4.48   Draw reaction energy diagrams for reactions that meet these descriptions:
(a) a very slow reaction that has a small heat of reaction
(b) a very fast reaction with a high heat of reaction

4.49   Draw a reaction energy diagram for a reaction whose products are exactly as stable as its reactants. What is the heat of reaction in this case?

4.50   What does it mean in terms of chemical reactions when the caloric value of a food is given?

4.51   Combustion of a 1.0-g sample of glucose, $C_6H_{12}O_6$ gives off 4.1 kcal of heat $C_6H_{12}O_6 + O_2 \rightarrow CO_2 + H_2O + $ heat.   (a) Write a balanced equation for the reaction.   (b) How much heat is given off during combustion of 10.0 g of glucose?   (c) What is the formula weight of glucose?   (d) What is the heat of reaction for combustion of 1.0 mol of glucose.

4.52   Write a balanced chemical equation for the combustion (chemical reaction with oxygen) of ethyl alcohol, C_2H_6O.

4.53   What is the heat of reaction produced by burning 1.0 mol of ethyl alcohol (Problem 4.52) if burning 5.0 g gives off 35.5 kcal of heat?

4.54   How many moles of ethyl alcohol must be burned (Problem 4.53) to obtain 75 kcal of heat?

4.55   How many grams of ethyl alcohol must be burned (Problem 4.53) to obtain 400 kcal of heat?

4.56   How many grams of ethyl alcohol must be burned (Problem 4.53) to heat 500 mL of water from room temperature (20°C) to 100°C?

Additional Problems

4.57   Dichloromethane, CH_2Cl_2, the solvent used to decaffeinate coffee beans, is prepared by reaction of CH_4 with Cl_2. Write the balanced reaction. (HCl is also formed.)

4.58 How many grams of Cl_2 are needed to react with 50.0 g of CH_4 in the preparation of dichloromethane (Problem 4.57)?

4.59 How many grams of dichloromethane could be prepared by reaction of 50 g of CH_4 and 50 g of Cl_2 (Problem 4.57)? What is the limiting reagent?

4.60 How many grams of which reagent would be left over after carrying out the reaction in Problem 4.59?

4.61 Zinc metal reacts with hydrochloric acid (HCl) according to the equation: $Zn + 2\,HCl \rightarrow ZnCl_2 + H_2$. How many grams of hydrogen would be produced if 150 grams of zinc reacted?

4.62 Lithium oxide is used aboard the space shuttle to remove water from the atmosphere according to the equation: $Li_2O + H_2O \rightarrow 2\,LiOH$. How many grams of Li_2O must be carried on board to remove 80 kg of water?

4.63 Batrachotoxin, $C_{31}H_{42}N_2O_6$, the active component of South American arrow poison, is so toxic that 0.5 μg can kill a person. How many molecules is this?

CHAPTER **5**

Solids, Liquids, and Gases

It will take a long time before this iceberg can absorb enough heat from the ocean to melt. We'll see why in this chapter.

One of the ways that scientists use to simplify a subject is to group similar things together. Biologists, for example, have learned to make sense of many thousands of different life forms by using a careful system of classification whereby similar organisms are grouped together. By studying the general properties that make members of a group similar, we can learn something about *all* members of the group without having to study each one individually.

In this chapter, we'll see how matter can be classified and what the characteristics of the different classes are. We'll answer these questions:

1. **What are the three states of matter?** The goal: you should learn the general characteristics of solids, liquids, and gases.

2. **How can the behavior of gases be explained?** The goal: you should learn what the kinetic theory of gases is and how it accounts for the physical properties of ordinary gases.

3. **What is gas pressure?** The goal: you should learn what gas pressure is and how to calculate the effect of temperature and volume changes on gas pressure.

4. **What laws govern the behavior of gases?** The goal: you should learn what the four main gas laws are and how they can be used to predict the behavior of gases.

5. **What happens when a liquid boils?** The goal: you should learn what takes place when a liquid changes to a gas and what a heat of vaporization is.

5.1 STATES OF MATTER

■ **State of matter** The physical state of a substance as a solid, a liquid, or a gas.

■ **Solid** A substance that has a definite shape and volume.

■ **Liquid** A substance that has a definite volume but that changes shape to fill the container it's placed in.

■ **Gas** A substance that has neither a definite volume nor a definite shape.

Matter is usually classified into three groups familiar to everyone: solids, liquids, and gases. These three groups, called **states of matter**, are classified according to their physical similarities. **Solids** have a definite volume and a definite shape; neither the volume nor the shape of a solid changes when it's placed in a different container. **Liquids** have a definite volume but not a definite shape; the volume of liquid doesn't change when it's poured into a different container, but its shape changes to conform to the shape of the new container. **Gases** have neither a definite volume nor a definite shape; a gas will expand to fill the volume and to take the shape of any container it's placed in (Figure 5.1).

Why and how does one state of matter differ from another? Although we've been concerned primarily with isolated individual molecules up to this point, molecules aren't isolated in reality. Visible amounts of substances contain

(a) Ice: A solid has a definite volume and a definite shape independent of its container.

(b) Water: A liquid has a definite volume but a variable shape that depends on its container.

(c) Steam: A gas has both variable volume and shape that depend on its container.

Figure 5.1
Characteristics of solids, liquids, and gases.

vast numbers of molecules that can exert attractive forces on each other, thereby affecting the properties of the sample as a whole.

The forces attracting molecules together are weak electrical interactions that vary in strength depending on molecular structure. If the attractive forces are negligible, molecules are free to move about independently of one another, and the substance is a gas. If the attractive forces are stronger, molecules draw more closely together but are still free to slip and slide over one another, and the resulting substance is a liquid. If, however, the attractive forces are sufficiently strong, the molecules are rigidly locked together in an ordered way, and the substance is a solid (Figure 5.2, page 95).

5.2   GASES AND THE KINETIC THEORY

■ **Kinetic theory of gases**   A set of four assumptions for explaining the general behavior of gases.

Molecules in the gas state are not attracted to their neighbors and are thus able to move about freely, filling any container they're placed in. According to the **kinetic** (kih-**net**-ic) **theory of gases**, the behavior of gases can be explained by several assumptions:

1. A gas consists of a great many tiny particles moving about at random with no attractive forces between molecules.

2. Gas particles move in straight lines with an energy proportional to the temperature. Thus, gas molecules move faster as the temperature increases. (In fact, gas molecules move much faster than you might suspect. The average speed of a helium atom at room temperature and atmospheric pressure is close to that of a high-speed rifle bullet.)

3. The amount of space actually occupied by gas molecules is much smaller than the amount of space between molecules. Thus, most of the volume taken up by a gas is just empty space.

4. When molecules collide, they spring apart elastically without reacting, and their total energy is conserved.

Let's look at some of the properties of gases to see how the kinetic theory explains their behavior.

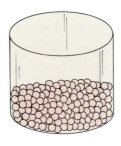

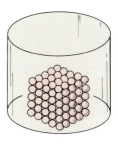

Figure 5.2
Gases, liquids, and solids differ in the amount of freedom that molecules have to move around as a result of the different strengths of attraction between molecules.

(a) A gas: The individual molecules feel little attraction for one another and are free to move about.

(b) A liquid: The individual molecules are attracted to one another but can slide over each other.

(c) A solid: The individual molecules are strongly attracted to one another and cannot move around.

5.3 PRESSURE

We're all familiar with the effects of air pressure. When you climb a mountain or fly in an airplane, the change in air pressure against your eardrum as you climb or descend can cause a painful "popping." When you pump up a bicycle tire, you increase the pressure of air against the inside walls to keep the tire firm.

In scientific terms, **pressure** is defined as a force per unit area pushing against a surface. In the bicycle tire, for example, you measure the force of air molecules pressing against the inside walls of the tire. The units you probably use for tire pressure are pounds-per-square-inch (psi), where 1 psi is equal to the force exerted by a 1-pound object sitting on a 1-square-inch surface. We on earth are under pressure from the atmosphere, the blanket of air pressing down on us, as shown in Figure 5.3. Atmospheric pressure is not constant, however; it varies slightly from day to day depending on the weather and depending on altitude. Air pressure is about 14.7 psi at sea level but only about 4.9 psi on the summit of Mt. Everest.

■ **Pressure** The force per unit area exerted on a surface.

Figure 5.3
A column of air weighing 14.7 lb presses down on each square inch of the earth's surface at sea level, resulting in what we call atmospheric pressure.

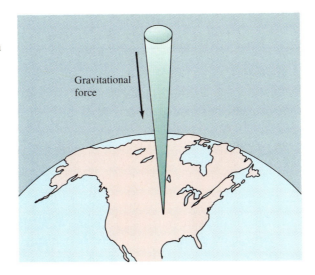

Gravitational force

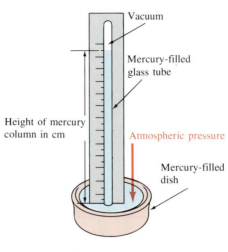

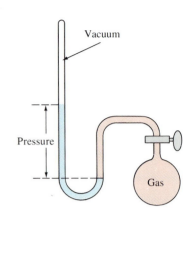

(a) Mercury barometer   (b) Manometer

Figure 5.4
(a) A mercury barometer, used to read atmospheric pressure, and (b) a mercury-filled manometer, used to read the pressure of a gas sample. Both devices work by measuring the height of a column of mercury forced by gas pressure into a sealed tube.

■ **Millimeter of mercury (mm Hg)**   A common unit of pressure equal to the force exerted by a 1-mm column of mercury. Standard atmospheric pressure is 760 mm Hg.

■ **Torr**   Alternate name for mm Hg.

■ **Mercury barometer**   A mercury-filled glass tube used to measure atmospheric pressure.

■ **Pascal (Pa)**   The SI unit of pressure; equal to 0.007500 mm Hg.

The most commonly used unit of pressure is the **millimeter of mercury**, abbreviated **mm Hg** and often called **torr** after the Italian physicist Evangelista Torricelli. This rather unusual unit, which dates back to the early 1600s, is measured by a device called a **mercury barometer**, illustrated in Figure 5.4. A thin tube, sealed at one end, is filled with mercury and then inverted into a dish of mercury. As some mercury runs from the tube into the dish, a vacuum is created at the sealed end of the tube. Since the pressure in the vacuum is zero but the pressure forcing mercury from the dish up into the tube is that of the atmosphere, a column of mercury is held in the tube. At sea level, the pressure of the atmosphere is sufficient to hold up a column of mercury 760 mm high—hence the use of mm Hg as a unit of pressure.

Pressure is measured in nonmedical laboratories by an SI unit called the **pascal (Pa)**, where 1 Pa = .007500 mm Hg (or 1 mm Hg = 133.3224 Pa). Measurements in pascals are becoming more common, and many clinical laboratories may make the switchover in the next few years.

One atmosphere (1 atm) = 760 mm Hg = 14.7 psi = 101,325 Pa

1 mm Hg = 1 torr = 133.3221 Pa

■ **Manometer**   A mercury-filled U-tube used to measure pressure.

Gas pressures in a container are usually measured by a device called a **manometer** (ma-**nah**-me-ter), also shown in Figure 5.4. A manometer consists simply of a U-tube filled with mercury. One end of the tube is sealed with a vacuum, and one end is open to the gas-filled container. The pressure of the gas forces mercury into the sealed end, and the amount of pressure can be read by measuring the difference between mercury levels at the two ends.

Practice Problem 5.1   The air pressure outside a jet plane flying at 35,000 feet is about 220 mm Hg. How many atmospheres is this? How many psi? How many pascals?

5.4 PARTIAL PRESSURE; DALTON'S LAW

The kinetic theory of gases says that each molecule acts independently of all others because they are so far apart. As far as any one molecule is concerned, the chemical identity of other molecules is irrelevant. Thus, *mixtures* of different gases act just the same as pure substances and obey the same laws.

■ **Partial pressure** The contribution to total gas pressure caused by each individual component of a mixture of gases.

Let's take a sample of dry air as an example. Dry air is a mixture of about 21 percent oxygen by volume and 79 percent nitrogen; in other words, 21 of every 100 molecules in air are oxygen, and 79 of every 100 are nitrogen. Thus, 21 percent of atmospheric air pressure is caused by the oxygen molecules, and 79 percent is caused by the nitrogen molecules. We therefore say that, at a *total* air pressure of 760 mm Hg, the **partial pressure** caused by the contribution of oxygen is (0.21×760) mm Hg, or 160 mm Hg. Similarly, the partial pressure of nitrogen in air is (0.79×760) mm Hg, or 600 mm Hg.

■ **Dalton's law of partial pressure** The total pressure exerted by a mixture of gases is equal to the sum of the partial pressures exerted by each individual gas.

According to *Dalton's law of partial pressure*, the total pressure exerted by *any* gas mixture is the sum of the individual pressures of the components in the mixture:

Dalton's law of partial pressure: $P_{total} = P_{gas\ 1} + P_{gas\ 2} + \cdots$

Solved Problem 5.1 Humid air on a warm summer day is approximately 20 percent oxygen, 75 percent nitrogen, and 5 percent water vapor. What is the partial pressure of each component if atmospheric pressure that day is 750 mm Hg?

Solution The partial pressure of any gas is obtained by multiplying the concentration of the gas by the total gas pressure in the sample.

Oxygen partial pressure: 0.20×750 mm Hg = 150 mm Hg
Nitrogen partial pressure: 0.75×750 mm Hg = 560 mm Hg
Water vapor partial pressure: 0.05×750 mm Hg = 40 mm Hg

Practice Problems

5.2 Assuming a total pressure of 9.5 atm, what is the partial pressure of each component in a mixture of 98 percent helium and 2.0 percent oxygen breathed by deep-sea divers?

5.3 How does the partial pressure of oxygen in diving gas (Practice Problem 5.2) compare with its partial pressure in normal air?

5.5 GAS LAWS

Gases, unlike liquids and solids, have remarkably similar behavior regardless of their molecular structures. Helium and chlorine, for example, are vastly different in their *chemical* properties, but they are extremely similar in much of their physical behavior. Observations of scientists in the 1700s led to the formulation of what are now called the **gas laws**. The four gas laws allow us to predict the influence of pressure, volume, and temperature on any gas or mixture of gases.

■ **Gas laws** A series of laws that describe the behavior of gases under conditions of differing pressure, volume, and temperature.

Boyle's Law: The Relation Between Pressure and Volume   Imagine that you have a sample of gas inside a cylinder with a plunger at one end (Figure 5.5). What would happen if you were to halve the volume of the gas by pushing the plunger halfway down? Since the gas molecules have only half as much room to move around in, the number of collisions with the walls of the cylinder increases. Thus, the pressure of the gas in the cylinder increases. According to *Boyle's law, the pressure of a gas at constant temperature is inversely proportional to its volume.* As volume goes up or down, pressure goes down or up, a relationship that can be expressed in the following way:

■ **Boyle's law**   The pressure of a gas at constant temperature is inversely proportional to its volume.

$$\text{Boyle's law:} \qquad \text{Pressure} \propto \frac{1}{\text{Volume}} \quad \text{(the sign } \propto \text{ means proportional to)}$$

or: $\qquad$ Pressure × volume = PV = a constant value when temperature (T) and number of moles (n) are fixed

Since $P \times V$ is always a constant value for a given amount of gas at constant temperature, we can restate Boyle's law by saying that the starting pressure (P_1) times the starting volume (V_1) is equal to the final pressure (P_2) times the final volume (V_2):

$$\text{Boyle's law restated:} \qquad P_1 \times V_1 = P_2 \times V_2$$

As an example of Boyle's-law behavior, think about what happens when you squeeze a balloon too hard. By squeezing, you decrease the volume but increase the pressure inside the balloon. If you squeeze too hard, the balloon bursts.

Solved Problem 5.2   In a typical automobile engine, the gas mixture in a cylinder is compressed from 1 atm to 9.5 atm. If the uncompressed volume of the cylinder is 750 mL, what is the volume when fully compressed?

Ballpark Solution   Since the pressure increases by about ten times (from 1 atm to 9.5 atm), the volume must decrease by about ten times from 750 mL to 75 mL.

Exact Solution   According to Boyle's law, the pressure of a sample times its volume is constant:

$$P_1 \times V_1 = P_2 \times V_2$$

So: $\qquad$ (1 atm) × (750 mL) = (9.5 atm) × V_2

Solving for V_2, the final volume, gives:

$$V_2 = \frac{1 \text{ atm} \times 750 \text{ mL}}{9.5 \text{ atm}} = 79 \text{ mL}$$

The final calculated volume of 79 mL agrees closely with the ballpark answer.

Practice Problems

5.4   A cylinder of compressed oxygen used to aid breathing has a volume of 5.0 L at 90 atm pressure. What volume would the same amount of oxygen have if the pressure were 1.0 atm, assuming constant temperature?

5.5   A sample of hydrogen gas at 273 K has a volume of 2.5 L at 5.0 atm pressure. What is its pressure if its volume is changed to 10.0 L? to 0.20 L?

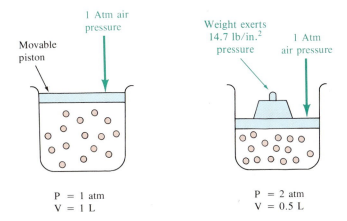

1 Atm air pressure

Movable piston

P = 1 atm
V = 1 L

Weight exerts 14.7 lb/in.² pressure

1 Atm air pressure

P = 2 atm
V = 0.5 L

Figure 5.5
Decreasing the volume of a gas sample increases crowding of the molecules and thereby increases the pressure (Boyle's law).

Charles' Law: The Relation Between Volume and Temperature   Imagine that you again have a sample of gas inside a cylinder with a plunger at one end. What would happen if you were to double the sample's temperature (in kelvins) while letting the plunger move freely to keep the pressure constant? The gas molecules would move with twice as much energy and need twice as much room to move around in. Thus, the volume of the gas in the cylinder would double. According to *Charles' law, the volume of a gas at constant pressure is directly proportional to its kelvin temperature.* As temperature goes up or down, volume also goes up or down:

■ **Charles' law**   The volume of a gas at constant pressure is directly proportional to its kelvin temperature.

Charles' law: Volume ∝ Temperature

or: $\dfrac{\text{Volume}}{\text{Temperature}} = V/T =$ a constant value when pressure (P) and number of moles (n) are fixed

or: $\dfrac{V_1}{T_1} = \dfrac{V_2}{T_2}$

As an example of Charles'-law behavior, think of what happens when a hot-air balloon is filled. Heating causes the air inside to expand and fill the bag (Figure 5.6).

Figure 5.6
Hot air balloons illustrate Charles' law: the volume of a gas increases as its temperature rises.

Solved Problem 5.3   An average adult inhales a volume of 0.50 L air into the lungs with each breath. If the air is warmed from room temperature (20°C = 293 K) to body temperature (37°C = 310 K) while in the lungs, what is the volume of the air when exhaled?

Ballpark Solution   The temperature of the air rises about 5 percent (from 293 K to 310 K) in the lungs so the volume of the air should also rise about 5 percent from 0.50 L to 0.53 L.

Exact Solution   According to Charles' law, the volume of the gas divided by its temperature is constant:

$$\frac{V_1}{T_1} = \frac{V_2}{T_2} \quad \text{so} \quad \frac{0.50 \text{ L}}{293 \text{ K}} = \frac{V_2}{310 \text{ K}}$$

Solving for V_2: $V_2 = \dfrac{V_1 \times T_2}{T_1} = \dfrac{(0.50 \text{ L}) \times (310 \text{ K})}{293 \text{ K}} = 0.53 \text{ L}$

The final calculated volume of 0.53 L agrees with the ballpark answer.

Practice Problem   **5.6**   A sample of chlorine gas has a volume of 0.30 L at 273 K and 1 atm pressure. What is its volume at 350 K and 1 atm pressure? at 500°C?

Gay-Lussac's Law: The Relation Between Pressure and Temperature   Imagine next that you have a fixed volume of gas in a sealed container. What would happen if you were to double its temperature (in kelvins)? The gas molecules move about with twice as much energy and collide with the walls of the container twice as hard. Thus, the pressure in the container doubles. According to *Gay-Lussac's law, the pressure of a gas at a constant volume is directly proportional to its kelvin temperature.* As temperature goes up or down, pressure also goes up or down:

■ **Gay-Lussac's law**   The pressure of a gas at a constant volume is directly proportional to its kelvin temperature.

Gay-Lussac's law: Pressure ∝ Temperature

or: $\dfrac{\text{Pressure}}{\text{Temperature}} = P/T = $ a constant value when volume (V) and number of moles (n) are fixed

or: $\dfrac{P_1}{T_1} = \dfrac{P_2}{T_2}$

As an example of Gay-Lussac's-law behavior, think of what happens when an aerosol can is thrown into an incinerator. Pressure builds up, and the can explodes (hence the warning statement on aerosol cans).

Solved Problem 5.4 What would the inside pressure become if an aerosol can with an initial pressure of 4.5 atm were heated in a fire from room temperature (20°C) to 600°C?

Ballpark Solution Since 20°C is about 300 K and 600°C is about 900 K, the temperature in the can triples when it is placed in the fire. According to Gay-Lussac's law, a tripling in temperature leads to a tripling in pressure, giving a value of about 3×4.5 atm $= 14$ atm in the can.

Exact Solution Gay-Lussac's law says that pressure divided by temperature is a constant:

$$\frac{P_1}{T_1} = \frac{P_2}{T_2} \quad \text{or} \quad \frac{4.5 \text{ atm}}{293 \text{ K}} = \frac{P_2}{873 \text{ K}}$$

Solving for P_2: $P_2 = \frac{P_1 \times T_2}{T_1} = \frac{(4.5 \text{ atm}) \times (873 \text{ K})}{293 \text{ K}} = 13 \text{ atm}$

Practice Problem **5.7** What final temperature is required for the pressure inside an automobile tire to increase from 30 psi at 0°C to 45 psi assuming the volume remains constant?

Avogadro's Law: The Relation Between Volume and Moles of Sample

Imagine finally that you have two equal volumes of different gases in sealed containers at the same temperature and pressure. How many molecules does each sample contain? Since molecules in a gas are so tiny compared to the empty space surrounding them, the chemical identity of the molecules doesn't matter. According to *Avogadro's law, equal volumes of gases at the same temperature and pressure contain equal numbers of molecules.* One volume of oxygen gas contains as many molecules as one volume of chlorine gas, which contains as many molecules as one volume of carbon dioxide gas, and so on.

■ **Avogadro's law** Equal volumes of gases at the same temperature and pressure contain equal numbers of molecules.

Avogadro's law: Volume $\propto n$ (number of moles)

or: $\dfrac{\text{Volume}}{n} = \dfrac{V}{n} =$ a constant value at fixed pressure (P) and temperature (T)

or: $\dfrac{V_1}{n_1} = \dfrac{V_2}{n_2}$

Notice that the *values* of temperature and pressure for the gas samples don't matter according to Avogadro's law. It's only necessary that the temperatures and pressure of the two samples be the same. For comparison purposes, however, it's convenient to define 0°C (273 K) and one atmosphere pressure (760 mm Hg) as **standard temperature and pressure (STP)** of the gas samples.

■ **Standard temperature and pressure (STP)** Standard conditions for a gas, defined as 0°C and 1 atm pressure.

Because equal volumes of gases at the same temperature and pressure contain equal numbers of molecules, they also contain equal numbers of moles. In other words, one mole (6.02×10^{23} molecules) of any gas has the same volume as one mole of any other gas at the same temperature and pressure. Called the **standard molar volume**, one mole of any gas has a volume of 22.4 liters at STP.

A summary of all four gas laws is given in Table 5.1 (page 102).

■ **Standard molar volume** The volume of one mole of a gas at standard temperature and pressure (22.4 L).

Table 5.1 A Summary of the Four Gas Laws

Gas Law		Variable Quantities	Constant Quantities
Boyle's law:	$P_1 \times V_1 = P_2 \times V_2$	Pressure, volume	Temperature, number of moles
Charles's law:	$V_1/T_1 = V_2/T_2$	Volume, temperature	Pressure, number of moles
Gay-Lussac's law:	$P_1/T_1 = P_2/T_2$	Pressure, temperature	Volume, number of moles
Avogadro's law:	$V_1/n_1 = V_2/n_2$	Volume, number of moles	Temperature, pressure

---

Solved Problem 5.5 How many molecules do you inhale in an average breath, assuming a volume of 0.50 L at STP?

Ballpark Solution One mole of particles in a gas (6.02×10^{23} molecules) has a volume of 22.4 L at STP. Since 0.50 L is about 2% of 22.4 L (0.5/22.4), each breath contains about 2% of Avogadro's number of molecules, or 10^{22}.

Exact Solution Use Avogadro's law as a conversion factor to change volume into numbers of molecules:

$$\text{Number of molecules} = 0.50\,\cancel{L} \times \frac{6.02 \times 10^{23} \text{ molecules}}{22.4\,\cancel{L}}$$

$$= 1.3 \times 10^{22} \text{ molecules}$$

The calculated answer of 1.3×10^{22} molecules agrees with the ballpark answer.

Practice Problem 5.8 How many moles of methane gas, CH_4, are in a 100,000-L storage tank at STP? How many grams of methane is this? How many grams of carbon dioxide gas, CO_2, could the same tank hold?

5.6 THE UNIVERSAL GAS LAW

■ **Universal gas law** A law that relates the effects on a gas sample of temperature, pressure, volume, and molar amount ($PV = nRT$).

All four of the gas laws just discussed can be combined into a single *universal gas law* that relates the effects of temperature, pressure, volume, and molar amount. Any time we know the value of three of the four quantities, we can calculate the value of the fourth.

$$\text{Universal gas law:} \qquad \frac{PV}{nT} = R \quad \text{(a constant)}$$

where: P = pressure of gas in mm Hg

V = volume of gas in liters (L)

n = number of moles of gas

R = 62.3 mm Hg L/mol K (a constant value, valid for all gases)

T = temperature of gas in kelvins (K)

The universal gas law can be rewritten in different ways depending on which of the four quantities we wish to calculate. For example, it's often written as $PV = nRT$.

Different formulations of the universal gas law:

$$P = nRT/V \qquad n = PV/RT$$
$$T = PV/nR \qquad V = nRT/P$$
$$R = PV/nT \qquad PV = nRT$$

The following solved problem shows how to use the universal gas law.

Solved Problem 5.6 How many moles of air are there in the lungs of an average person with a total lung capacity of 3.8 L? Assume that the person is at sea level and has a normal body temperature of 37°C.

Solution The problem asks for a value of n when P, V, and T are known. We first carry out whatever conversions are necessary to put the quantities in their correct units (V in liters; P in mm Hg; and T in kelvins) and then set up the universal gas law to solve for n:

$$n = PV/RT$$

where: $P = 760$ mm Hg (1 atm)
$V = 3.8$ L
$T = 310$ K (37°C)

$$n = \frac{(760 \text{ mm Hg}) \times (3.8 \text{ L})}{62.3 \dfrac{\text{mm Hg L}}{\text{mol K}} \times (310 \text{ K})} = 0.15 \text{ mol air}$$

There are 0.15 moles of air in the lungs of an average person.

Practice Problems

5.9 An aerosol spray-deodorant can with a volume of 350 mL contains 3.2 g of propane gas (C_3H_8) as propellant. What is the pressure of gas in the can at 20°C?

5.10 A helium gas cylinder of the sort used to fill balloons has a volume of 180 L and a pressure of 2200 psi (150 atm) at 25°C. How many moles of helium are in the tank? How many grams?

AN APPLICATION: INHALED ANESTHETICS

William Morton's demonstration in 1846 of ether-induced anesthesia during surgery must surely rank as one of the most important medical break-throughs of all time. Before that date, all surgery had been carried out with the patient fully conscious. Many other inhaled anesthetic agents have now been introduced, including nitrous oxide, halothane, and enflurane to name a few.

$$
\begin{array}{c}
\quad\ \text{H}\ \ \text{H}\ \ \ \ \ \ \ \ \text{H}\ \ \text{H}\\
\quad\ \ |\ \ \ |\ \ \ \ \ \ \ \ \ |\ \ \ |\\
\text{H}-\text{C}-\text{C}-\text{O}-\text{C}-\text{C}-\text{H}\\
\quad\ \ |\ \ \ |\ \ \ \ \ \ \ \ \ |\ \ \ |\\
\quad\ \text{H}\ \ \text{H}\ \ \ \ \ \ \ \ \text{H}\ \ \text{H}
\end{array}
\qquad \text{N}_2\text{O}
$$

Ether	Nitrous oxide

$$
\begin{array}{cc}
\ \ \text{F}\ \ \ \text{Br} & \ \ \ \text{F}\ \ \ \ \text{F}\ \ \ \ \ \ \ \text{F}\\
\ \ |\ \ \ \ |\ & \ \ \ \ |\ \ \ \ \ |\ \ \ \ \ \ \ \ |\\
\text{F}-\text{C}-\text{C}-\text{H} & \text{H}-\text{C}-\text{C}-\text{O}-\text{C}-\text{H}\\
\ \ |\ \ \ \ |\ & \ \ \ \ |\ \ \ \ \ |\ \ \ \ \ \ \ \ |\\
\ \ \text{F}\ \ \ \text{Cl} & \ \ \ \text{Cl}\ \ \ \text{F}\ \ \ \ \ \ \text{F}
\end{array}
$$

Halothane	Enflurane

Despite their great importance and use, surprisingly little is known about how inhaled anesthetics work in the body. The potency of different anesthetics correlates well with their solubility in vegetable oils, leading many scientists to believe that they act by dissolving in the fatty membranes surrounding nerve cells. The resultant change in the nerve-cell membranes apparently decreases the ability of sodium ions to pass into the cells, thereby blocking the firing of nerve impulses.

The depth of anesthesia is determined by the concentration of anesthetic agent that reaches the brain. Concentration in the brain, in turn, depends on the solubility and transport of the anesthetic agent in blood and on its partial pressure in inhaled air. Nitrous oxide must be administered at a partial pressure of up to 500 mm Hg to induce loss of consciousness, but the more soluble halothane can be administered at a partial pressure of 20 mm Hg.

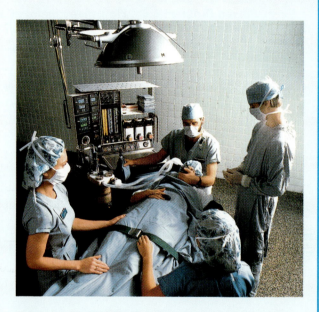

Doctors use sophisticated equipment to monitor the vital signs of a patient undergoing surgery and administer the precise amount of anesthetic that is safe and effective.

5.7   LIQUIDS

As in gases, molecules in the liquid state are in constant motion. They are free to slide about and move over one another, allowing the liquid to take the shape of whatever container it's in. Unlike the situation in gases, however, molecules in liquids are always in contact with their neighbors. Thus, there is little empty space between molecules, and liquids are far denser and less compressible than gases.

Let's take a sample of liquid water and think about what happens to an individual molecule as it moves randomly about. If the molecule happens to be

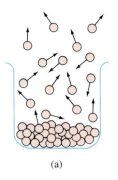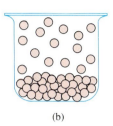

(a)   (b)

Figure 5.7
The transfer of molecules between liquid and gas states.
(a) Molecules that escape in an open container drift away
until the liquid has entirely evaporated. (b) Molecules in a
closed container reach an equilibrium; that is, the number
of molecules going from the liquid to the gas is the same
as the number of molecules going from the gas back to
the liquid.

■ **Evaporation** The
spontaneous conversion of a
liquid to a gas.

near the surface, it might occasionally move upward with enough energy to
break free of the liquid and escape into the gas state. In an open container, the
now-gaseous water molecule will wander away from the liquid, and the process
will continue until all water molecules have escaped (Figure 5.7a). This, of course,
is exactly what happens during **evaporation**. We're all familiar with seeing a
puddle of water evaporate after a rainstorm.

If the water sample is in a closed container, however, the situation is
different. In a closed container, the gaseous water molecules can never escape
very far from the liquid. Thus, there is a good chance that the random motion
of a gaseous water molecule will occasionally cause it to reenter the liquid. After
the concentration of water molecules in the gas state has built sufficiently high,
the number of molecules reentering the liquid becomes equal to the number of
molecules escaping from the liquid. At this point, an equilibrium exists (Figure
5.7b).

■ **Vapor** The gaseous state of
a substance that is normally a
liquid.

■ **Vapor pressure** The
pressure of a vapor at
equilibrium with its liquid.

Once water molecules have escaped from the liquid to the gas state, they
are subject to all of the gas laws previously discussed. In a closed vessel at
equilibrium, for example, the gaseous water molecules (called **vapor**) make their
own contribution to the total pressure of the gas above the liquid according to
Dalton's law of partial pressure. We call this contribution the **vapor pressure** of
the liquid.

■ **Boiling point (bp)** The
temperature at which the vapor
pressure of a liquid is equal to
atmospheric pressure.

The vapor pressure of a liquid depends on both the temperature and the
identity of the liquid. As the temperature rises, molecules become more energetic
and more likely to escape into the gas state. Thus, vapor pressure rises with
increasing temperature, as shown in Figure 5.8 (page 106). Ultimately, the **boiling
point (bp)** of the liquid is reached, and the entire liquid is converted into gas. At
the boiling point, the vapor pressure of the liquid is equal to the pressure of
the atmosphere. Bubbles of vapor therefore form under the surface of the liquid
and force their way to the top, giving rise to the violent action seen during a
vigorous boil.

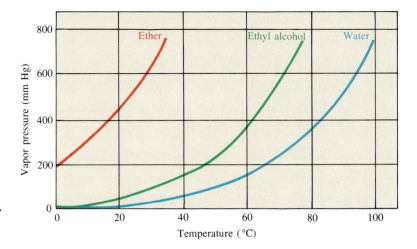

Figure 5.8
A plot showing the change of vapor pressure with temperature for ether, ethyl alcohol, and water. At a liquid's boiling point, its vapor pressure is equal to atmospheric pressure.

If atmospheric pressure is higher or lower than normal, the boiling point of a liquid changes accordingly. At high altitudes, for example, atmospheric pressure is lower than at sea level, and boiling points are also lower. On top of Mt. Everest (29,028 ft), the boiling temperature of water is approximately 71°C. In a pressure cooker, however, pressure is higher and boiling points are also higher (Figure 5.9).

A liquid doesn't become a gas instantaneously on reaching its boiling point. Rather, a liquid first reaches its boiling point, and then even *more* heat is required to convert it fully into a gas *without further raising its temperature*. The amount of heat required to vaporize one gram of a liquid after reaching its boiling point is called the liquid's **heat of vaporization**. As indicated in Table 5.2, which gives the heats of vaporization of several common liquids, water has a particularly high value. Thus, water evaporates more slowly than many other liquids and takes a long time to boil away.

■ **Heat of vaporization** The amount of heat necessary to convert one gram of liquid into a gas when the liquid is at its boiling point.

Figure 5.9
Water in hot subterranean springs is often under such high pressure that it does not boil even when its temperature exceeds 100°C. At the surface, subjected only to atmospheric pressure, it quickly vaporizes. Then, as the vapor cools, it condenses again into tiny droplets, forming steam. (Yellowstone National Park.)

Table 5.2 Heats of Vaporization of Some Common Substances

Liquid	Boiling Point (°C)	Heat of Vaporization (cal/g)
Ammonia	−33.4	327
Butane	−0.5	91.5
Ether	34.6	84.0
Ethyl alcohol	78.3	204
Water	100.0	540

5.8 SOLIDS

Molecules in the solid state are rigidly locked into ordered positions. Although individual bonds within a molecule can stretch and bend, the molecule as a whole cannot move past its neighbors. But what happens if we raise the temperature of the solid? As more and more energy is put in, the molecule begins to stretch, bend, and vibrate more and more vigorously. Finally, if the temperature is raised high enough and molecular motions become vigorous enough, the molecules break free from one another, and the substance becomes liquid.

■ **Melting point (mp)** The temperature at which a solid turns into a liquid.

The temperature at which a solid turns into a liquid is called the **melting point (mp)** of the solid. Ice, for example, has a melting point of 0°C. Different substances have different melting points depending on their structures and on the strength of the attractive forces between molecules or ions. As you might expect, ionic substances like sodium chloride tend to have much higher melting points than covalent substances like glucose since the electrical attractions between oppositely charged ions are very strong. Melting points of some common substances are given in Table 5.3.

Notice in Table 5.3 that even oxygen, a gas at room temperature, has a melting point. Every substance becomes a solid if cooled to a low enough temperature to decrease molecular motions, and therefore every substance has some point at which its solid form will melt and become liquid (assuming that no heat-induced decomposition occurs first).

■ **Heat of fusion** The amount of heat necessary to convert one gram of solid into a liquid when the solid is at its melting point.

Solids don't become liquid instantaneously on reaching their melting points. Rather, a substance first reaches its melting point while still solid, and additional heat then begins the melting process *without further raising the temperature of the sample*. Thus, both solid and liquid forms coexist in the sample while melting is occurring. For example, solid ice has a temperature of 0°C when it just begins to melt, and liquid water is also at 0°C after the ice has just finished melting. The amount of heat necessary to melt one gram of any solid after reaching the solid's melting point is called its **heat of fusion**. For ice, the heat of fusion is 80 cal/g.

Table 5.3 Melting Points of Some Common Substances

Substance	Melting Point (°C)	Substance	Melting Point (°C)
Sodium metal	97.8	Glucose	146
Sodium chloride	801	Morphine	254
Sodium hydroxide	318	Cholesterol	148.5
Oxygen	−218	Benzene	5.5

INTERLUDE: BLOOD PRESSURE

Having your blood pressure measured is a quick and easy way to get an indication of the state of your circulatory system. Although blood pressure varies with age, a normal adult male has a reading near 120/80, and a normal adult female has a reading near 110/70. Abnormally high values signal an increased risk of heart attack and stroke.

Pressure varies greatly in different types of blood vessels. Usually, though, arterial pressure in the upper arm is measured as the heart goes through a full cardiac cycle. **Systolic pressure** is the maximum pressure developed in the artery just after contraction as the heart forces the maximum amount of blood into the artery. **Diastolic pressure** is the minimum pressure that occurs at the end of the heart cycle.

Blood pressure is measured by a *sphygmomanometer*, a device consisting of a squeeze bulb, a flexible cuff, and a mercury manometer. The cuff is placed around the upper arm over the brachial artery and inflated by the bulb to about 200 mm Hg pressure, an amount great enough to squeeze the artery shut and prevent blood flow. Air is then slowly released from the cuff, and pressure drops. As cuff pressure reaches the systolic pressure, blood spurts through the artery, creating a turbulent tapping sound that can be heard through a stethoscope. The pressure registered on the manometer at the moment the first sounds are heard is the systolic blood pressure. Sounds continue until the pressure in the cuff becomes low enough to allow diastolic blood flow. At this point, blood flow becomes smooth, no sounds are heard, and a diastolic blood pressure reading is recorded on the manometer. Readings are usually recorded as systolic/diastolic—for example, 120/80. The figure below shows the sequence of events during measurement.

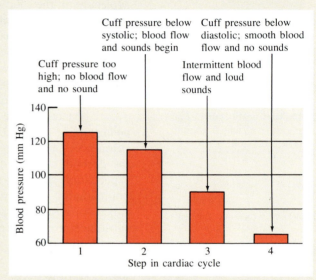

The sequence of events during blood pressure measurement, including the sounds heard.

SUMMARY

There are three states of matter: solids, liquids, and gases. According to the **kinetic theory of gases**, the physical behavior of gases can be explained by assuming that they consist of tiny particles moving at random and separated from other molecules by great distances. Four laws have been formulated to explain gas behavior:

1. **Boyle's law** states that the pressure of a gas at constant temperature is inversely proportional to its volume: $P \propto 1/V$.
2. **Charles' law** states that the volume of a gas at constant pressure is directly proportional to its temperature in kelvins: $V \propto T$.
3. **Gay-Lussac's law** states that the pressure of a gas at constant volume is directly proportional to its temperature in kelvins: $P \propto T$.
4. **Avogadro's law** states that equal volumes of gases at the same temperature

contain the same number of molecules. At 0°C and 1 atmosphere pressure, called **standard temperature and pressure (STP)**, one mole of any gas (6.02×10^{23} molecules) occupies a volume of 22.4 liters.

The four gas laws can be combined into a single **universal gas law**, $PV = nRT$, that relates the effects of temperature, pressure, volume, and molar amount. Any time the values of three of the quantities are known, the fourth can be calculated.

Molecules are in much closer contact in liquids than in gases. Occasionally, however, molecules escape from the surface of the liquid into the gas state. As the temperature of a liquid rises, molecules become more energetic and more likely to escape. At the liquid's **boiling point**, the entire liquid is converted into gas. The amount of heat necessary to vaporize one gram of liquid after reaching its boiling point is called its **heat of vaporization**.

Molecules in solids are locked into fixed and ordered positions. As the temperature is raised, molecular motions increase in vigor until, at the solid's **melting point**, molecules break free from one another, and the substance becomes liquid. The amount of heat necessary to melt one gram of a substance after reaching its melting point is called its **heat of fusion**.

REVIEW PROBLEMS

Gases and Pressure

5.11 How is one atmosphere of pressure defined?

5.12 What four variables are used to describe the physical state of a gas?

5.13 What is meant by partial pressure?

5.14 What is Dalton's law of partial pressure?

5.15 What are the four assumptions made by the kinetic theory of gases?

5.16 How does the kinetic theory of gases explain the phenomenon of gas pressure?

5.17 Convert these values into mm Hg:
(a) standard pressure (b) 0.25 atm (c) 7.5 atm
(d) 28.0 inches Hg

5.18 Atmospheric pressure at the top of Mt. Whitney, the highest point in the lower 48 states, is 440 mm Hg. How many atmospheres is this?

5.19 If the partial pressure of oxygen in air at 1.0 atm is 160 mm Hg, what is the partial pressure on the summit of Mt. Whitney (Problem 5.18)?

Gas Laws

5.20 What conditions are defined as standard temperature and pressure (STP)?

5.21 What is Boyle's law, and what variable must be kept constant in order for the law to hold?

5.22 What is Charles' law, and what variable must be kept constant in order for the law to hold?

5.23 What is Gay-Lussac's law, and what variable must be kept constant in order for the law to hold?

5.24 What is Avogadro's law?

5.25 How many liters does a mole of gas occupy at STP?

5.26 How does the kinetic theory of gases explain Charles' law?

5.27 What is the universal gas law?

5.28 Oxygen gas is commonly sold in 50-L steel containers at a pressure of 150 atm. What volume would the gas occupy at a pressure of 1 atm if its temperature remained unchanged?

5.29 What would be the pressure in the oxygen cylinder of Problem 5.28 if its temperature were raised from 20°C to 75°C?

5.30 A compressed-air tank carried by scuba divers has a volume of 8 L and a pressure of 140 atm at 20°C. What is the volume of air in the tank at STP?

5.31 What is the partial pressure of oxygen in the scuba tank (Problem 5.30) assuming that air is 21 percent oxygen?

5.32 Which sample contains more molecules: 1.0 L of O_2 at STP, or 1.0 L of H_2 at STP?

5.33 Which of the two samples in Problem 5.32 weighs more?

5.34 Which sample contains more molecules? 2.0 L of Cl_2 at STP, or 3.0 L of CH_4 at 300 K and 1.5 atm? Which sample weighs more?

5.35 Which sample contains more molecules: 2.0 L of CO_2 at 300 K and 500 mm Hg, or 1.5 L of N_2 at 57°C and 760 mm Hg? Which sample weighs more?

5.36 What's the effect on the pressure of a gas if you simultaneously:
(a) halve its volume and double its kelvin temperature?
(b) double its volume and halve its kelvin temperature?

5.37 How many mL of Cl_2 gas would you measure out to obtain 0.20 g at STP?

5.38 If 15.0 g of CO_2 gas has a volume of 0.30 L at 300 K, what is its pressure in mm Hg?

5.39 If 20.0 g of N_2 gas has a volume of 0.40 L and a pressure of 6.0 atm, what is its temperature?

5.40 If 18.0 g of O_2 gas has a temperature of 350 K and a pressure of 550 mm Hg, what is its volume?

5.41 How many moles of a gas will it take to occupy a volume of 0.55 L at a temperature of 347 K and a pressure of 2.5 atm?

5.42 What is the total mass of the oxygen in a typical room measuring 4.0 meters long, 5.0 meters wide, and 2.5 meters high? Assume that the gas in the room is at STP and that air contains 21 percent oxygen and 79 percent nitrogen.

5.43 What is the total mass of nitrogen in the room described in Problem 5.42? What is the total mass of air in the room?

Liquids

5.44 What is the vapor pressure of a liquid?

5.45 How does the arrangement of molecules in a liquid differ from that in a gas?

5.46 What is the value of a liquid's vapor pressure at the liquid's boiling point?

5.47 What is the effect of pressure on a liquid's boiling point?

5.48 What is a liquid's heat of vaporization?

5.49 What does it mean when we say that a liquid and its vapor are in equilibrium at a given temperature?

Solids

5.50 Why do ionic compounds often have higher melting points than covalent compounds?

5.51 How does the arrangement of molecules in a solid differ from that in a liquid?

5.52 What is a solid's heat of fusion?

Additional Problems

5.53 Hydrogen and oxygen react according to the equation $2 H_2 + O_2 \rightarrow 2 H_2O$. How many moles of hydrogen are required to react with each mole of oxygen? According to Avogadro's law, how many liters of hydrogen are required to react with 2.5 L of oxygen at STP?

5.54 If 3.0 L of hydrogen and 1.5 L of oxygen at STP reacted to yield water (Problem 5.53), how many moles of water would be formed? What gas volume would the water have at a temperature of 100°C and 1 atm pressure?

5.55 Approximately 240 mL-per-minute of carbon dioxide (CO_2) gas is exhaled by an average adult at rest. Assuming a temperature of 37°C, how many moles of CO_2 is this?

5.56 How many grams of CO_2 are exhaled by an average resting adult in 24 hours (Problem 5.55)?

5.57 The average oxygen content of arterial blood is approximately 200 mL per liter of blood. Assuming a body temperature of 37°C, how many moles of oxygen are transported by each liter of arterial blood? How many grams?

5.58 Imagine that you have a balloon full of air connected to the open end of a mercury manometer. Describe the effect on the mercury column in the manometer of: (a) squeezing the balloon (b) warming up the balloon (c) cooling the balloon

5.59 Imagine that you have two identical vessels, one containing hydrogen at STP and the other containing oxygen at STP. How could you tell which was which without opening them?

5.60 According to Avogadro's law, one mole of any gas has a volume of 22.4 L at STP. What is the formula weight of each of the following gases, and what are the densities in grams-per-liter at STP?
(a) CH_4 (b) CO_2 (c) O_2 (d) UF_6

5.61 Gas pressure outside the space shuttle is approximately 1×10^{-14} mm Hg at a temperature of approximately 1 K. If the gas is almost entirely hydrogen atoms (H, not H_2), what volume of space is occupied by one mole of atoms? What is the density of H gas in atoms-per-liter?

5.62 Ammonia is prepared commercially by reacting hydrogen with nitrogen in the presence of a catalyst: $H_2 + N_2 \rightarrow NH_3$.
(a) Write a balanced equation for the reaction.
(b) If one mole of nitrogen and three moles of hydrogen are mixed at a total pressure of 100 atm and allowed to react, what will the final pressure be when reaction is complete? Assume that temperature and volume are constant throughout.

5.63 If two moles of nitrogen and two moles of hydrogen at a total pressure of 80 atm are allowed to react to make ammonia (Problem 5.62), what will the final pressure be when reaction is complete?

5.64 What is the molecular weight of a gas if a 1.50 g sample has a volume of 1.12 L at STP?

CHAPTER 6

Solutions

Everyone knows that the ocean is salty. But seawater is a complex solution—it contains dissolved salts other than sodium chloride, and atmospheric gases as well.

Up to this point, we've been concerned primarily with pure substances, both elements and compounds. In day-to-day life, however, most of the materials we come in contact with are *mixtures*. Air, for example, is a gaseous mixture of oxygen and nitrogen; blood is a liquid mixture of many different components; and rocks are solid mixtures of different minerals. We'll look closely in this chapter at the characteristics and properties of solutions and other mixtures.

1. **What are mixtures, and what are solutions?** The goal: you should learn to distinguish between mixtures and solutions.

2. **How is the concentration of a solution expressed?** The goal: you should learn the most common ways of expressing concentrations of substances dissolved in solution, and you should learn how to convert between different units.

3. **How are dilutions carried out?** The goal: you should learn how to make up solutions of different concentration.

4. **Why is water a solvent?** The goal: you should learn how water and other solvents are able to dissolve substances.

5. **What is an electrolyte?** The goal: you should learn what an electrolyte is and how to express the concentration of an electrolyte solution.

6. **What is osmosis?** The goal: you should learn the principle of osmosis and its application to dialysis.

6.1 MIXTURES

We saw in Section 2.10 that a mixture is any combination of two or more substances that retain their individual chemical identities. Mixtures are classified according to their visual appearance as either heterogeneous or homogeneous. **Heterogeneous mixtures** are those in which the mixing of components is obviously nonuniform. Most rocks, for example, show a grainy character due to the heterogeneous mixing of different mineral particles. **Homogeneous mixtures** are those in which the mixing *is* uniform, at least to the naked eye. Seawater, a homogeneous mixture of NaCl in water, is an obvious example.

There are three kinds of homogeneous mixtures—solutions, colloids, and suspensions—which differ according to the size of their particles. The divisions among the three groups aren't always clear, but the concepts are useful none-

■ **Heterogeneous mixture**
A mixture that is visually nonuniform.

■ **Homogeneous mixture**
A mixture that is uniform to the naked eye.

■ **Solution** A homogeneous mixture containing molecule-sized particles uniformly dispersed in another material.

■ **Colloid** A homogeneous mixture that contains particles larger than a typical molecule, uniformly dispersed in another material.

■ **Suspension** A homogeneous mixture containing particles just large enough to be visible to the naked eye.

theless. **Solutions**, the most important class of homogeneous mixtures, contain particles the size of a typical covalent molecule (0.2–15 nm diameter). **Colloids** are also homogeneous in appearance but contain larger particles than solutions (15–1000 nm diameter). These particles might be either very large individual molecules, such as proteins, or aggregations of smaller molecules. **Suspensions** contain still larger particles that have diameters greater than 1000 nm—the size of a speck just visible to the naked eye.

Solutions, colloids, and suspensions can be differentiated in several ways. For example, solutions are transparent, although they may be colored. Colloids and suspensions, however, generally have a murky or opaque appearance because their larger particles block out light. In addition, solutions and colloids don't separate on standing, but suspensions usually settle into layers over time. Table 6.1 gives some examples of these three types of homogeneous mixtures and lists some of their characteristics.

Table 6.1   Three Kinds of Homogeneous Mixtures: Solutions, Colloids, and Suspensions

Kind of Mixture	Particle Diameter	Examples	Characteristics
Solution	0.2–2.0 nm	Sea water, air, vinegar	Transparent to light, nonfilterable, does not separate on standing
Colloid	2.0–1000 nm	Butter, milk, fog, pearls	Murky or opaque to light, nonfilterable, does not separate on standing
Suspension	> 1000 nm	Blood, paint, aerosol sprays	Murky or opaque to light, can be separated by filtration or on standing

6.2   SOLUTIONS

Although we usually think only of a solid dissolved in a liquid when talking about solutions, there are many other kinds of solutions, as indicated in Table 6.2. Even solutions of one solid with another are common, as in metal alloys such as 14-karat gold (58 percent gold and 42 percent silver). These solid/solid solutions can't be made by simply grinding two solids together, however, because

Table 6.2   Some Different Kinds of Solutions

Solution Type	Example
Gas in gas	Air (oxygen and nitrogen)
Gas in liquid	Carbonated water (CO_2 in water)
Liquid in liquid	Vinegar (acetic acid in water)
Solid in liquid	Saline solution (NaCl in water)
Solid in solid	Gold alloy

grinding can never be done finely enough to mix particles at the molecular level. They must be made by first melting the two solids, then mixing them to form a liquid solution, and then allowing the liquid mixture to cool and solidify.

■ **Solute** The minor substance in a homogeneous mixture.

■ **Solvent** The major substance in a homogeneous mixture.

For solutions in which a gas or solid is dissolved in a liquid, the dissolved substance is called the **solute** (**soll**-yute), and the liquid is called the **solvent**. When one liquid is dissolved in another, however, the distinction between solute and solvent is less clear. Usually, though, the minor liquid is considered the solute, and the major liquid is the solvent. Thus, ethyl alcohol is the solute and water the solvent in a mixture that is 10 percent ethyl alcohol and 90 percent water, but water is the solute and ethyl alcohol the solvent in a mixture that is 90 percent ethyl alcohol and 10 percent water.

As we saw in Section 6.1, liquid solutions are easily distinguished from other kinds of mixtures by the following properties:

1. Solutions are transparent. Although they may be colored, you can see through solutions because they don't have large particles that block out light.
2. Solutions don't separate into their components on standing. A saline solution, for example, never separates into salt and water; it always remains a transparent solution.
3. Solutions can't be separated by filtration. Both solvent and solute molecules pass through filter paper.

Practice Problem **6.1** Which of the following liquid mixtures are true solutions?
(a) milk (b) seawater (c) blood (d) gasoline

6.3 SOLUBILITY

Imagine that you were asked to prepare 100 mL of a saline solution. You might measure out the water, add solid sodium chloride, and stir until it dissolves. But how much NaCl should you add? Can you dissolve any amount you choose in water, or is there a limit? In fact, there is a limit. At 20°C, a maximum of 35.8 grams of NaCl will dissolve in 100 mL of water. Any amount added above this limit will simply sink to the bottom of the container and sit there.

■ **Solubility** The amount of a substance that can be dissolved in a given volume of solvent.

■ **Saturated solution** A solution in which the solute has reached its solubility limit.

The maximum amount of a substance that will dissolve in a given amount of a liquid, usually expressed in g/100 mL, is called its **solubility**. When a solution has reached its solubility limit, we say that it is **saturated**. Like melting point and boiling point, solubility is a physical property characteristic of a specific substance. Different substances have greatly differing solubilities, as shown in Table 6.3 (page 116). Also indicated in Table 6.3 is the fact that the solubility of most substances increases as the temperature is raised. The effect of temperature on solubility is different for every substance and can be either large (as with ammonium nitrate) or small (as with sodium chloride).

Table 6.3 Solubilities of Some Common Substances in Water

Substance	Solubility in Water	(g/100 mL)
	at 20°C	at 100°C
Ammonium nitrate (NH_4NO_3)	178	871
Sodium chloride (NaCl)	35.8	39.1
Sodium bicarbonate ($NaHCO_3$)	9.6	23.6
Sodium hydroxide (NaOH)	109	347
Glucose	92.3	—
Sucrose	204	487

Table 6.4 Solubilities of Some Gases in Water

Gas	Solubility in Water (g/100 mL)
	at 20°C
Ammonia (NH_3)	51.8
Carbon dioxide (CO_2)	0.169
Hydrogen chloride (HCl)	77.1
Oxygen (O_2)	0.0043

Gases, as well as solids, can be soluble in liquids. For example, both oxygen and carbon dioxide gas dissolve to a slight extent in water (Table 6.4). According to *Henry's law*, the solubility of a gas in a liquid varies with its pressure if temperature is held constant. If the pressure of the gas doubles, solubility doubles; if the gas pressure is halved, solubility is halved.

■ **Henry's law** The solubility of a gas in a liquid varies with its pressure if temperature is held constant.

Henry's law: Gas solubility $\propto$ Gas pressure

$$\frac{\text{Solubility}}{\text{Pressure}} = \text{A constant value at fixed temperature}$$

Henry's law is what accounts for the fizzing in a suddenly opened bottle of soft drink or champagne. The bottle is sealed under more than one atmosphere of carbon dioxide pressure, causing some of the CO_2 to dissolve. When the bottle is opened, however, CO_2 pressure drops, and dissolved gas comes bubbling out of solution.

Table 6.5 Solubilities of Some Liquids in Water

Liquid	Solubility in Water (g/100 mL)
	at 20°C
Acetic acid ($C_2H_4O_2$)	Miscible
Chloroform ($CHCl_3$)	0.71
Diethyl ether (ether, $C_4H_{10}O$)	6.9
Ethyl alcohol (C_2H_6O)	Miscible
Phenol (carbolic acid, C_6H_6O)	9.1

AN APPLICATION: GOUT AND KIDNEY STONES: PROBLEMS IN SOLUBILITY

One of the major pathways for the breakdown of nucleic acids in the body is by their conversion to uric acid. About 0.5 g per day of uric acid is excreted in the urine of normal people. Unfortunately, the solubility of uric acid in water is fairly low—only about 7 mg/mL at 37°C. When too much uric acid is produced by the body, its concentration in blood and urine rises, and excess uric acid sometimes comes out of solution to be deposited in the joints and kidneys.

Gout, a disorder of nucleic-acid metabolism that primarily affects middle-aged men, is characterized by excessive uric-acid production leading to the deposit of uric acid crystals in soft tissue around the joints. The big toe and the hands are often affected. Deposition of the crystals causes an acute and painful inflammation that can lead ultimately to arthritis and to bone destruction.

Similarly with kidney stones. Excess uric acid in the urine can lead to the formation and deposit of crystals in the kidney. Although often quite small, kidney stones cause excruciating pain when passed through the ureter. In some cases, complete blockage of the ureter can occur.

Treatment of excessive uric acid production involves both dietary modification and drug therapy. Foods rich in nucleoproteins, such as liver, sardines, and asparagus, must be avoided, and drugs such as allopurinol are taken to lower production of uric acid.

Liquids, too, can be soluble in other liquids (Table 6.5). Some liquids, particularly organic compounds such as chloroform ($CHCl_3$) and diethyl ether ($C_4H_{10}O$), are only slightly soluble in water. When one of these liquids is mixed with water, a small amount dissolves but the remainder stays undissolved and forms a separate liquid layer apart from the water. (We're all familiar with seeing oil "float" on water.) Other liquids such as ethyl alcohol are completely water soluble in all proportions. The word **miscible** (**miss**-uh-bul) is used to describe such behavior.

■ **Miscible**   Soluble in all proportions without limit.

6.4   UNITS FOR EXPRESSING CONCENTRATION

Although we might speak in casual conversation of a solution as being either "dilute" or "concentrated," laboratory work usually requires an exact knowledge of a solution's **concentration**—the amount of solute dissolved in a given amount of solvent. As indicated in Table 6.6, there are several common methods for expressing concentration. The units differ, but all of the methods describe how much solute is present per amount of solvent.

■ **Concentration**   The amount of solute per unit volume of solvent.

Table 6.6   Some Methods for Expressing Concentration

Solute Measure	Solvent Measure	Concentration Measure
Weight	Volume	Weight/volume percent (w/v)%
Weight	Weight	Weight/weight percent (w/w)%
Volume	Volume	Volume/volume percent (v/v)%
Number of moles	Volume	Moles/liter, **molarity (M)**

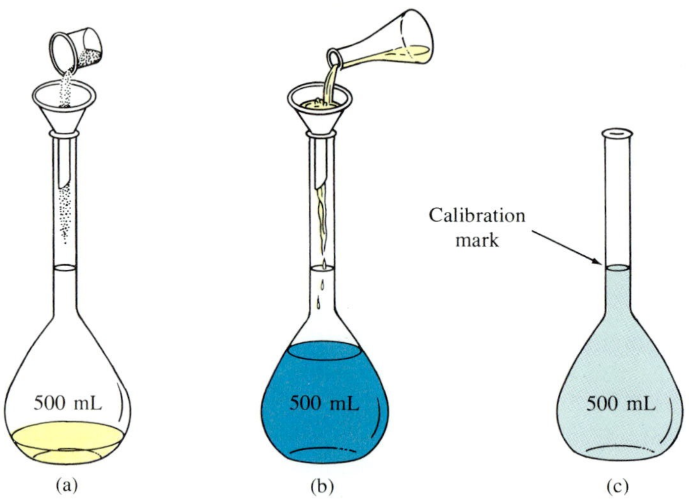

Figure 6.1
Preparation of a weight/volume percent solution. (a) A weighed amount of solute is placed in a volumetric flask containing a small amount of solvent. (b) More solvent is added to dissolve the solute. (c) Still more solvent is added to fill the flask to the known calibration mark in the neck.

Calibration mark

500 mL   500 mL   500 mL

(a)   (b)   (c)

■ **Weight/volume percent concentration (w/v)%**
Concentration expressed as the number of grams of solute dissolved in 100 mL of solution.

Weight/Volume Percent Concentration   One of the most common methods for expressing concentration is to give the number of grams (weight) of the solute per 100 mL (volume) of final solution: *weight/volume percent concentration* (w/v)%. (In clinical laboratories, g/dL is often used rather than g/100 mL, since 1 dL = 100 mL.) For example, if 15 g of glucose is dissolved in enough water to give 100 mL of solution, the glucose concentration is 15 g/100 mL, or 15% (w/v). (Remember that *percent* means number of parts per hundred. Thus, 15% means 15 parts out of 100, or 15/100.)

$$\textbf{(w/v)\% concentration} = \frac{\text{grams of solute}}{100 \text{ mL of solution}}$$

Notice that a weight/volume solution is made by dissolving the weighed solute in just enough solvent to give a *final* volume of 100 mL, not by dissolving it in a *starting* volume of 100 mL solvent. After all, if the solute were dissolved in 100 mL of solvent, the final volume of the solution would be a bit larger than 100 mL, since you would have added the volume of the solute. In practice, a weight/volume solution is made by weighing out the appropriate amount of solute and placing it with a small amount of solvent in a **volumetric flask** of the sort shown in Figure 6.1. Enough solvent is then added to dissolve the solute, and further solvent is added until a precisely calibrated final volume is reached.

■ **Volumetric flask**   A flask whose volume has been precisely calibrated.

Solved Problem 6.1   How much NaCl is needed to prepare 250 mL of a 1.5% (w/v) solution?

Ballpark Solution   A 1.5% (w/v) solution contains 1.5 g of solute per 100 mL. Thus, 250 mL of the solution contains 2.5 times 1.5 g, or about 3.5 g NaCl.

Exact Solution   Use the concentration of the solution as a conversion factor:

$$1.5\% \text{ (w/v) means:} \quad \frac{1.5 \text{ g NaCl}}{100 \text{ mL solution}} \quad \text{or} \quad \frac{100 \text{ mL solution}}{1.5 \text{ g NaCl}}$$

Then set up an equation to convert the amount of solution to the amount of NaCl needed:

$$\text{grams NaCl} = 250 \text{ mL solution} \times \frac{1.5 \text{ g NaCl}}{100 \text{ mL solution}} = 3.8 \text{ g NaCl}$$

Thus, 3.8 g of NaCl is needed to prepare 250 mL of a 1.5% (w/v) solution.

6.2   How many grams of solute are needed to prepare these solutions?
(a) 100 mL of 12% (w/v) glucose ($C_6H_{12}O_6$)   (b) 75 mL of 2.0% (w/v) KCl

6.3   How many mL of 5.0% (w/v) NaOH solution can be made from 8.0 g NaOH?

■ **Weight/weight percent concentration, (w/w)%**
Concentration expressed as the number of grams of solute per 100 grams of solution.

Weight/Weight Percent Concentration   The *weight/weight percent concentration*, (w/w)%, of a solution gives the number of grams of solute per number of grams of solution. For example, if 2.5 g of NaCl is dissolved in enough water to give a final solution weighing 100 g, the NaCl concentration is 2.5 g/100 g = 2.5% (w/w). Another way of looking at this method is to say that 2.5 percent of the weight of the solution (2.5 grams of every 100 grams of solution) is NaCl, and 97.5 percent of the weight of the solution is water.

$$(w/w)\% \text{ concentration} = \frac{\text{grams of solute}}{100 \text{ grams of solution}}$$

There is no particular advantage to expressing concentrations as weight/weight percents, and the method is not often used.

■ **Volume/volume percent concentration, (v/v)%**
Concentration expressed as the number of mL solute per 100 mL solution.

Volume/Volume Percent Concentration   Solutions of one liquid in another are often expressed as *volume/volume percent concentrations*, (v/v)%, that give volume of liquid solute per 100 mL of solution. For example, if 10.0 mL of ethyl alcohol is dissolved in enough water to give 100 mL of solution, the ethyl alcohol concentration is 10 mL/100 mL = 10% (v/v). A volumetric flask is used to prepare volume/volume solutions, just as for weight/volume solutions.

$$(v/v)\% \text{ concentration} = \frac{\text{mL of solute}}{100 \text{ mL of solution}}$$

Solved Problem 6.2   How much methyl alcohol is needed to prepare 75 mL of a 5.0% (v/v) solution?

Ballpark Solution   A 5.0% (v/v) solution contains 5 mL of solute per 100 mL. Thus, 75 mL of the solution contains 75/100 times 5 mL, or about 4 mL.

Exact Solution   First use the concentration of the solution as a conversion factor:

5.0% (v/v) means: $\dfrac{5.0 \text{ mL methyl alcohol}}{100 \text{ mL solution}}$ or $\dfrac{100 \text{ mL solution}}{5.0 \text{ mL methyl alcohol}}$

Then set up an equation to convert the amount of solution to the amount of methyl alcohol needed:

$$\text{mL methyl alcohol} = 75 \text{ mL solution} \times \frac{5.0 \text{ mL methyl alcohol}}{100 \text{ mL solution}}$$

$$= 3.8 \text{ mL methyl alcohol}$$

Thus, 3.8 mL of methyl alcohol is needed.

6.4   Describe how you would use a 500-mL volumetric flask to prepare a 7.5% (v/v) solution of acetic acid in water.

6.5   How many mL of solute are needed to prepare these solutions?
(a) 100 mL of 22% (v/v) ethyl alcohol   (b) 150 mL of 12% (v/v) acetic acid

Mole/Volume Concentration: Molarity   Although useful for many purposes, the three methods just discussed for expressing concentration are awkward when it comes to using them in chemical calculations. The problem is that the three percentage methods give only *weights* or *volumes* of solute, whereas we need to know *numbers of molecules* (really, numbers of moles) to do chemical calculations (Sections 4.3–4.5).

Problems involving mole calculations are most easily handled when solute concentrations are expressed as **molarity** (**M**), the number of moles of solute per liter of solution. For example, a solution made by dissolving 1.0 mol (36.5 g) of HCl in a final volume of 1.0 L of water has a concentration of 1.0 mol/L, or 1.0 M.

■ **Molarity (M)**   Concentration expressed as the number of moles of solute per liter of solution.

$$\text{Molarity (M)} = \frac{\text{moles of solute}}{1 \text{ L of solution}}$$

Solutions of a given molarity are prepared in volumetric flasks by the same method used to prepare weight/volume solutions (Figure 6.1). The only difference is that the amount of solute used is expressed in *moles* instead of grams. The following practice problems show how calculations involving solutions of known molarity are carried out.

Solved Problem 6.3   How would you prepare a 250 mL solution of 0.50 M NaCl?

Solution   First determine how many moles of NaCl are needed by using the given molarity as a conversion factor:

$$0.50 \text{ M NaCl means:} \quad \frac{0.50 \text{ mole NaCl}}{1.0 \text{ L solution}} \quad \text{or} \quad \frac{1.0 \text{ L solution}}{0.50 \text{ mole NaCl}}$$

$$\text{moles NaCl} = 250 \text{ mL solution} \times \frac{0.50 \text{ mole NaCl}}{1000 \text{ mL solution}} = 0.125 \text{ mole NaCl}$$

Next, convert moles of NaCl into grams by using the formula weight of NaCl as a conversion factor:

formula weight of NaCl = 23.0 amu (Na) + 35.5 amu (Cl) = 58.5 amu

Thus, there are 58.5 grams NaCl per mole.

$$\text{grams NaCl} = 0.125 \text{ mole NaCl} \times \frac{58.5 \text{ g NaCl}}{1 \text{ mole NaCl}} = 7.31 \text{ g NaCl}$$

To prepare the desired solution, we would place 7.31 g of NaCl in a 250-mL volumetric flask, add enough water to dissolve it, and then add a further amount of water until the total solution volume is 250 mL.

Practice Problems

6.6   How many moles of solute are present in these solutions?
(a) 125 mL of 0.20 M $NaHCO_3$ (b) 650 mL of 2.5 M H_2SO_4?

6.7   How many grams of solute would you use to prepare these solutions?
(a) 500 mL of 1.25 M NaOH   (b) 1.5 L of 0.25 M glucose ($C_6H_{12}O_6$)

Solved Problem 6.4 Stomach acid, a dilute solution of HCl in water, can be neutralized by reaction with sodium bicarbonate, $NaHCO_3$, according to the equation:

$$HCl + NaHCO_3 \longrightarrow NaCl + H_2O + CO_2$$

Assuming that you had 1.80 g of pure HCl, how many mL of a 0.125 M $NaHCO_3$ solution would be needed to neutralize it?

Solution Since chemical calculations must be done in moles rather than in grams, we first have to find how many moles of HCl are in 1.80 g. We do this by using the formula weight of HCl as a conversion factor:

HCl formula weight = 1.0 amu (H) + 35.5 amu (Cl) = 36.5 amu

Thus, there are 36.5 grams HCl per mole.

$$\text{Moles HCl} = 1.80 \text{ g HCl} \times \frac{1 \text{ mole HCl}}{36.5 \text{ g HCl}} = 0.0493 \text{ mole HCl}$$

Next, we check the coefficients of the balanced equation to find that each mole of HCl reacts with one mole of $NaHCO_3$, and we find how many mL of 0.125 M $NaHCO_3$ solution contains 0.0493 moles.

$$\text{mL solution} = 0.0493 \text{ mole HCl} \times \frac{1000 \text{ mL solution}}{0.125 \text{ mole HCl}} = 395 \text{ mL solution}$$

Thus, 395 mL of the 0.125 M $NaHCO_3$ solution is needed to neutralize 1.80 g of HCl.

Practice Problems

6.8 How many mL of a 0.200 M glucose ($C_6H_{12}O_6$) solution are needed to provide a total of 25.0 g glucose?

6.9 The concentration of cholesterol ($C_{27}H_{46}O$) in normal blood is approximately 0.005 M. How many grams of cholesterol are in 750 mL of blood?

6.5 DILUTION

■ **Dilution** A decrease in the concentration of a solution caused by addition of solvent.

Many solutions, from orange juice to chemical reagents, are prepared and stored in high concentrations and are then **diluted** before use. For example, you might make up one-half gallon of orange juice by adding water to a canned concentrate.

Orange juice concentrate + water $\longrightarrow$ dilute orange juice

In the same way, you might buy a medicine or chemical reagent in concentrated solution and then dilute it before use.

Concentrated chemical + solvent $\longrightarrow$ dilute solution

The key fact to remember when diluting a concentrated solution is that the amount of *solute* remains constant; only the *volume* is changed by adding more solvent. In other words:

$$\text{Moles of solute} = \mathbf{M_{orig}} \times V_{orig} = \mathbf{M_{final}} \times V_{final}$$

| | [original concentration] | [original volume] | [final concentration] | [final volume] |

which we can rewrite as:

$$\mathbf{M}_{final} = \mathbf{M}_{orig} \times \frac{V_{orig}}{V_{final}}$$

where: $\dfrac{V_{orig}}{V_{final}}$ is the *dilution factor*

■ **Dilution factor**   The ratio of original-to-final volumes of a solution being diluted.

This equation says that the final concentration after dilution ($\mathbf{M}_{final}$) can be found by multiplying the original concentration ($\mathbf{M}_{orig}$) by a dilution factor. The **dilution factor** is simply the ratio of the original and final volumes (V_{orig}/V_{final}). If the volume *increases* by a factor of five from 10 mL to 50 mL, then the concentration must *decrease* by a factor of five because the dilution factor is 10 mL/50 mL, or 1:5. Solved Problem 6.5 shows how to use this relationship for calculating the effect of dilution.

Solved Problem 6.5   What is the final concentration if 75 mL of a 3.5 M glucose solution is diluted to a volume of 400 mL?

Ballpark Solution   Since the volume of the solution *increases* about five times (from 75 mL to 400 mL), the concentration must *decrease* about five times from 3.5 M to approximately 0.7 M.

Exact Solution   Use the ratio of original-to-final volume as a dilution factor to calculate the final concentration:

$$\mathbf{M}_{final} = \mathbf{M}_{orig} \times \frac{V_{orig}}{V_{final}} = 3.5 \text{ M} \times \frac{75 \text{ mL}}{400 \text{ mL}} = 0.66 \text{ M}$$

The calculation yields a final concentration of 0.66 M for the glucose solution, which is close to the ballpark answer.

Practice Problem   **6.10**   Hydrochloric acid is normally purchased at a concentration of 12.0 M. What is the final concentration if 100 mL of 12.0 M HCl is diluted to 500 mL?

The relationship between concentration and volume can also be used to determine how much of a dilution to make in order to prepare a solution of known concentration. We know that:

$$M_{orig} \times V_{orig} = M_{final} \times V_{final}$$

Therefore:
$$V_{orig} = V_{final} \times \frac{M_{final}}{M_{orig}}$$

This equation says that we can determine what original volume (V_{orig}) to dilute in order to reach a desired final concentration by multiplying the final volume (V_{final}) by the ratio of the final and original concentrations (M_{final}/M_{orig}). For example, if we want to decrease the original concentration by a factor of five from 5.0 M to 1.0 M, then the volume we start with must be five times less than the volume we want to end up with (5.0 M/1.0 M). Solved Problem 6.6 shows how to use this relationship for calculating the effect of dilution.

Solved Problem 6.6   Sodium hydroxide (NaOH) solution can be purchased at a concentration of 1.0 M. How would you prepare 1.0 L of 0.25 M NaOH?

Ballpark Solution   Since the concentration of the NaOH has decreased by a factor of four in the dilution (from 1.0 M to 0.25 M), the volume must have increased by a factor of four. Thus, if we end up with 1.0 L, we must have started with one-fourth of 1 L, or 250 mL.

Exact Solution   Using the foregoing equation, we can solve to find the original volume of concentrated NaOH:

$$V_{orig} = V_{final} \times \frac{M_{final}}{M_{orig}} = 1.0 \text{ L} \times \frac{0.25 \text{ M}}{1.0 \text{ M}} = 0.25 \text{ L} = 250 \text{ mL}$$

Practice Problem   **6.11**   Concentrated ammonia solution, NH_3 in H_2O, is commercially available at a concentration of 16.0 M. How much would you use to prepare 500 mL of a 1.25 M solution?

6.6   WATER AND ITS STRUCTURE: POLAR COVALENT BONDS

Water is the most abundant and most carefully studied chemical compound on earth. More than 75 percent of the earth's surface is covered by water, totaling an estimated 1.4×10^9 km³ (1.4 billion cubic kilometers!). Even the human body is about 50–60 percent water.

We saw in Section 3.7 that a water molecule consists of an oxygen atom covalently bound to two hydrogen atoms. We also saw in Chapter 3 that there are two fundamental kinds of bonds between atoms: ionic bonds and covalent bonds. An ionic bond such as that in NaCl results from the *donation* of an electron from one atom to another to yield charged ions, Na$^+$ and Cl$^-$, whereas

■ **Electronegativity** The ability of an atom to attract electrons.

■ **Polar covalent bond** A covalent bond in which one atom attracts bonding electrons more strongly than the other atom.

■ **Polarized** Having a partial positive or negative charge as the result of being in a polar covalent bond.

■ **Hydrogen bond** A weak attraction between a hydrogen and a nearby oxygen, nitrogen, or fluorine atom.

a covalent bond such as that in water results from the *sharing* of two electrons between atoms.

Although not pointed out previously, the electrons in a covalent bond are not necessarily shared *equally* between atoms. Elements at the upper right of the periodic table—oxygen, nitrogen, fluorine, and chlorine—are said to be **electronegative** because they have a much greater ability to attract electrons than do most other elements. We've already seen that these electronegative elements are the ones most likely to form ionic bonds with Group 1A and Group 2A elements from the left side of the periodic table (Section 3.3).

When oxygen bonds to hydrogen to form the covalent O—H bond of water, the electronegative oxygen atom attracts the electron pair more strongly than hydrogen does. The result is a bond *intermediate in type between perfectly ionic and perfectly covalent*. The oxygen atom gains a *partial* (but not a full) negative charge, while the hydrogen atom is left with a partial (but not a full) positive charge. The partial positive charge is indicated δ^+, and the partial negative charge δ^-, where δ is the lowercase Greek letter, delta. We call such a bond a **polar covalent bond** (Figure 6.2), and we say that the individual oxygen and hydrogen atoms are **polarized**.

One consequence of having polar covalent bonds is that water molecules are attracted to each other by weak electrical forces. In the liquid state, water molecules orient so that a positive hydrogen of one molecule points toward the negative oxygen of another molecule, forming what is called a **hydrogen bond**. Each water molecule can be joined by hydrogen bonds to four of its neighbors, as shown in Figure 6.3.

Hydrogen bonding is not limited to water. It can occur whenever an O—H bond or N—H bond is present in a molecule, as in ethyl alcohol, CH_3CH_2O—H, or ammonia, H_2N—H. Although only about 5 percent as strong as normal covalent bonds, hydrogen bonds are nevertheless responsible for controlling much of the behavior of proteins and nucleic acids, as we'll see in later chapters.

Figure 6.2
The continuity in bonding type from (a) perfectly ionic, where one atom owns both bonding electrons, resulting in the formation of positive and negative ions; to (b) a polar covalent bond, where the two bonding electrons are shared unequally, resulting in the development of partial positive (δ^+) and partial negative (δ^-) charges on the atoms; to (c) a perfectly covalent bond, where the two bonding electrons are shared equally. The O—H bonds in water are polar-covalent, with oxygen having a partial negative charge (δ^-) and hydrogen a partial positive charge (δ^+).

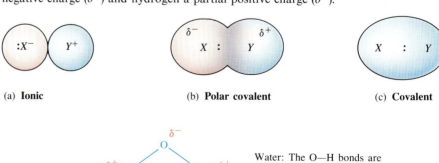

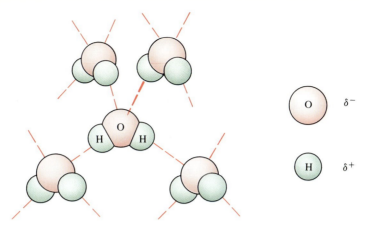

Figure 6.3
Hydrogen bonds in water, formed as the result of electrical attractions between positively polarized hydrogen atoms (δ^+) and negatively polarized oxygen atoms (δ^-). Each water molecule can bond to four neighboring molecules.

6.7 WATER AS A SOLVENT

Why does water dissolve salt but not oil, while ether dissolves oil but not salt? The rule of thumb about solubility is that *like dissolves like*. A strongly polar molecule like water dissolves the ionic compound NaCl, whereas a nonpolar organic molecule like ether dissolves the nonpolar organic molecules in oil.

When NaCl crystals are put into water, ions at the crystal surface come into contact with polar water molecules. Positively charged sodium ions are attracted to the negatively polarized oxygen end of water, while negatively charged chloride ions are attracted to the positively polarized hydrogen end. The combined forces of attraction between an ion and several water molecules are enough to pull the ion away from the crystal, exposing a fresh surface, until ultimately the crystal dissolves (Figure 6.4).

Once in solution, Na^+ and Cl^- ions are completely surrounded by solvent molecules, a phenomenon called **solvation**. The water molecules form a loose shell around the ions, stabilizing them by electrical attraction, as shown in Figure 6.4.

■ **Solvation** The surrounding of a solute ion or molecule by solvent molecules.

Figure 6.4
A salt crystal dissolves when polar water molecules surround the individual Na^+ and Cl^- ions, pulling them from the crystal surface into solution.

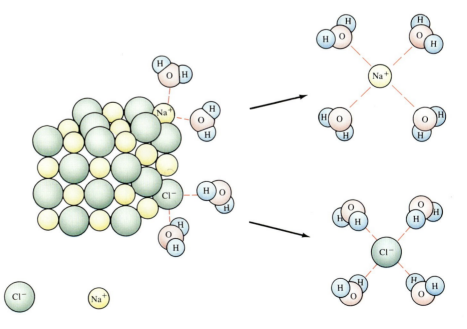

Solubility is an unpredictable and complex matter that is not well understood by scientists, even today. In spite of what you might expect, not all ionic compounds dissolve in water: Silver chloride (AgCl), calcium carbonate ($CaCO_3$, limestone), barium sulfate ($BaSO_4$), and many others are practically insoluble. Many polar covalent molecules, however, including sugars, amino acids, and even some proteins, *do* dissolve in water. Water solvates these polar covalent molecules in the same way it solvates ions—by surrounding the solute molecules and orienting to maximize electrical attractions.

6.8 HYDRATION OF SOLIDS

■ **Hydrate** A solid substance that has water molecules included in its crystals.

Some ionic compounds attract water strongly enough to hold onto water molecules even when crystalline. Such compounds are called solid **hydrates**. For example, the plaster of Paris used to make casts for broken limbs is calcium sulfate monohydrate, $CaSO_4 \cdot H_2O$. The raised dot between $CaSO_4$ and H_2O in the formula is used to indicate that one molecule of water is present for each $CaSO_4$ but that the two are merely attracted together, not chemically bonded.

$$CaSO_4 \cdot H_2O \qquad \text{A hydrate}$$

■ **Hygroscopic** Having the ability to attract water vapor from the air.

Still other ionic compounds attract water so strongly that they literally pull water vapor from the surrounding atmosphere to become hydrated. Compounds that show this behavior, such as $CuSO_4$ and $MgSO_4$, are called **hygroscopic**; they are often used as drying agents. You might have noticed a small bag of a hygroscopic compound (probably silica, SiO_2) included in the packing material of a new stereo or VCR to keep humidity low during shipping.

6.9 IONS IN SOLUTION: ELECTROLYTES

Try a simple experiment. Take a flashlight battery, a bulb, and three short pieces of wire, and tape them together as shown in Figure 6.5 so that the wires are in contact with the battery ends and with the bulb. Next take a glass of water, dissolve as much table salt in it as you can, and immerse the ends of the wires

Figure 6.5
A simple experiment to show that electricity can flow through a solution of ions.

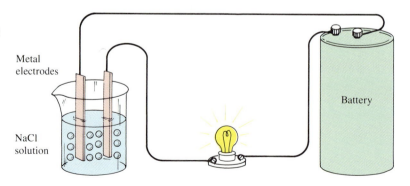

Metal electrodes

NaCl solution

Battery

in the salt water. The bulb lights! Evidently, electricity flows from the battery through the salt water to complete the circuit.

Substances such as NaCl that conduct electricity when dissolved in water are called **electrolytes**. Conduction occurs because negatively charged Cl^- anions migrate through the solution toward the wire connected to the positive end of the battery (the **anode**), while at the same time positively charged Na^+ cations migrate toward the wire connected to the negative end of the battery (the **cathode**).

As you might expect, the more ions there are in solution, the better the solution is able to conduct electricity. Thus, distilled water contains no ions of any sort, is a **nonelectrolyte**, and is nonconducting; ordinary tap water contains low concentrations of dissolved ions (mostly Na^+, K^+, Mg^{2+}, and Cl^-), is a weak electrolyte, and is weakly conducting; and 1 M NaCl has a high concentration of ions, is a strong electrolyte, and is highly conducting.

■ **Electrolyte** A substance that conducts electricity when dissolved in water.

■ **Anode** The positive electrode of a battery.

■ **Cathode** The negative electrode of a battery.

■ **Nonelectrolyte** A substance that does not conduct electricity when dissolved in water.

Practice Problems

6.12 How can you account for the fact that 1 M HCl is a strong electrolyte? What ions do you think are present in solution?

6.13 Unlike HCl (Problem 6.12), 1 M HF is a weak electrolyte. Which do you think forms ions in solution more easily, HCl or HF?

6.10 EQUIVALENTS, MILLIEQUIVALENTS, AND BODY ELECTROLYTES

Imagine dissolving both NaCl and KBr in the same solution. Since the cations (K^+ and Na^+) and anions (Cl^- and Br^-) are all mixed together, an identical solution could just as well be made from KCl + NaBr. Thus, we can no longer speak of having an NaCl + KBr solution; we can only speak of having a solution with four different ions in it.

A similar situation exists for blood and other body fluids, which contain many different anions and cations, often in small concentration. Since they're all mixed together, we can't "assign" specific cations to specific anions, and we can't talk about specific ionic compounds. To discuss such mixtures, we need a new term, called *equivalents* of ions.

One **equivalent** (**Eq**) of an ion is the amount in grams that contains Avogadro's number of charges (or one mole of charges). If the ion has a charge of $+1$ or -1, one equivalent is simply equal to the formula weight of the ion in grams. Thus, one equivalent of Na^+ is 23 grams, and one equivalent of Cl^- is 35.5 grams. If the ion has a charge of $+2$ or -2, however, one equivalent is equal to the ion's formula weight in grams divided by two. Thus, one equivalent of Mg^{2+} is $24.3/2 = 12.15$ grams, and one equivalent of CO_3^{2-} is $(12.0 + 16.0 + 16.0 + 16.0)/2 = 30.0$ grams.

■ **Equivalent (Eq)** The amount of an ion in grams that contains Avogadro's number of charges.

$$\text{One equivalent of ion} = \frac{\text{formula weight of ion in grams}}{\text{number of charges on ion}}$$

Table 6.7 Concentrations of Common Ions in Blood

Ion	Concentration (mEq/L)
Ca^{2+}	4.5–6.0
Cl^-	98–106
$HCO_3{}^-$	25–29
Mg^{2+}	3
K^+	3.5–5.0
Na^+	136–145

■ **Milliequivalent (mEq)** The amount of an ion equal to one-thousandth of an equivalent.

Because ion concentrations in body fluids are often low, clinical chemists find it more convenient to talk about *milliequivalents* of ions. One **milliequivalent (mEq)** of an ion is one thousandth of an equivalent. For example, the normal concentration of Na^+ in blood is 0.14 Eq/L, or 140 mEq/L.

$$1 \text{ mEq} = 0.001 \text{ Eq} \qquad 1 \text{ Eq} = 1000 \text{ mEq}$$

Normal blood levels of other common ions are shown in Table 6.7.

Solved Problem 6.7   The normal concentration of Ca^{2+} in blood is 5.0 mEq/L. How many milligrams of Ca^{2+} are in 1.00 L of blood?

Solution   First calculate how many grams are in 1 Eq of Ca^{2+}. The formula weight of Ca^{2+} is 40.1 amu, but because Ca^{2+} has two charges, we must divide the formula weight by 2. Therefore,

$$1 \text{ Eq Ca}^{2+} = \frac{40.1 \text{ g Ca}^{2+}}{2} = 20.0 \text{ g Ca}^{2+}$$

Next, calculate how many grams are in 5.0 mEq/L Ca^{2+}:

$$\text{g/L Ca}^{2+} = \frac{20.0 \text{ g Ca}^{2+}}{1 \text{ Eq Ca}^{2+}} \times \frac{1 \text{ Eq Ca}^{2+}}{1000 \text{ mEq}} \times \frac{5.0 \text{ mEq}}{1.00 \text{ L}} = 0.10 \text{ g/L Ca}^{2+}$$

Finally, convert grams into milligrams:

$$\text{milligrams Ca}^{2+} = \frac{0.10 \text{ g Ca}^{2+}}{1 \text{ L}} \times \frac{1000 \text{ mg}}{1 \text{ g}} = 100 \text{ mg/L Ca}^{2+}$$

Thus, there are 100 mg of Ca^{2+} in 1 liter of blood.

Practice Problems

6.14   How many grams are in one equivalent of each of the following?
(a) K^+ (b) Br^- (c) Mg^{2+} (d) $SO_4{}^{2-}$

6.15   How many grams are in 1 mEq of each ion in Problem 6.14?

6.16   Look at the data in Table 6.7, and calculate how many milligrams of Mg^{2+} are in 250 mL of blood.

6.11 OSMOSIS AND OSMOTIC PRESSURE

■ **Semipermeable membrane**
A thin membrane that allows water or other small solvent molecules to pass through but that blocks the flow of larger solute molecules or ions.

Certain materials, including those that make up the membranes around living cells, are semipermeable. A **semipermeable membrane** is one that allows water or other small solvent molecules to pass through but blocks the passage of larger solute molecules or ions. It's as if the membrane has tiny holes in it through which only small molecules can pass. Larger molecules and bulky solvated ions evidently can't fit through the holes.

When two solutions of different concentration are separated by a semipermeable membrane, water passes from the more dilute side (the one with less solute but more solvent) to the more concentrated side (the one with more solute but less solvent). Called **osmosis** (ozz-**mo**-sis), this passage of solvent has the effect of bringing the concentrations on the two sides closer to equality. As a result of osmosis, the water level rises on the side with more solute and drops on the side with less solute (Figure 6.6). The taller column of water pushes down with more weight than the shorter column, and pressure therefore develops on the membrane. Eventually, this pressure, called **osmotic pressure**, becomes great enough to halt the flow of more solvent across the membrane.

■ **Osmosis** The passage of solvent molecules across a semipermeable membrane from a more dilute solution to a more concentrated solution.

■ **Osmotic pressure** The amount of external pressure that must be applied to halt the passage of solvent molecules across a semipermeable membrane.

In the special case where pure water is present on one side of the semipermeable membrane and a solution on the other, the concentrations on the two sides can never become equal no matter how much water passes from one side to the other, because all of the solute remains on the one side. Osmotic pressure can be extremely high in such situations, even for rather dilute solutions. For example, the osmotic pressure of a 0.15 M NaCl solution is 7.4 atm, a value high enough to support a difference in water level of approximately 230 ft!

Osmotic pressure is proportional to the difference in molar concentrations of solute particles on the two sides of the osmotic membrane. For compounds that yield ions, the concentration of *particles* (ions) is often a factor of two or more greater than the molar concentration of the compound. Thus, a 1 M solution of glucose has one mole of particles per liter of solution, but a 1 M solution of NaCl has *two* moles of particles per liter—one mole of Na^+ ions and one mole of Cl^- ions.

■ **Osmolarity (Osmol)** The number of moles of dissolved solute particles (ions or molecules) per liter of solution.

We use the term **osmolarity (Osmol)** to describe the number of particles in solution. The osmolarity of a solution is equal to the molarity times the number of particles produced by each solute molecule. Thus, a 0.2 M glucose solution has an osmolarity of 0.2 Osmol, but a 0.2 M solution of NaCl has an osmolarity of 0.4 Osmol.

Osmolarity = molarity × number of particles per solute unit

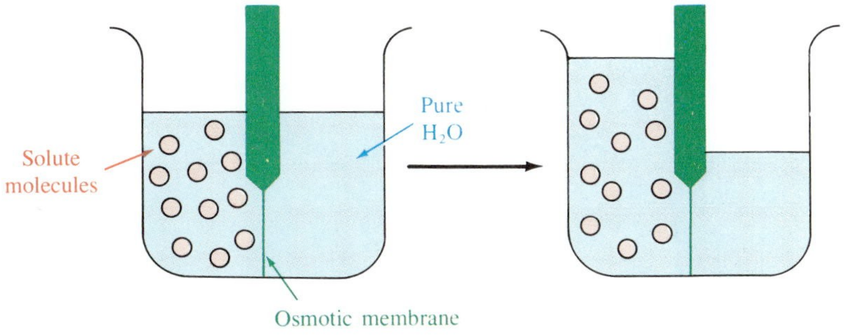

Figure 6.6
The phenomenon of osmosis. Water molecules pass through the semipermeable membrane from the side of the less concentrated solution to the side of the more concentrated one. As a result of this passage, the water level rises on the more concentrated side, and pressure develops.

■ **Plasma**   The fluid surrounding blood cells.

■ **Isotonic**   Having the same osmolarity as another solution, usually blood (0.30 Osmol).

■ **Hypotonic**   Having a lower osmolarity than another solution, usually blood.

■ **Hemolysis**   Bursting of red blood cells due to a buildup of pressure in the cell.

■ **Hypertonic**   Having a higher osmolarity than another solution, usually blood.

■ **Crenation**   The shriveling of red blood cells due to a loss of water.

Osmosis is particularly important in living organisms, because the membranes around cells are semipermeable. The fluids both in and out of cells must therefore have the same osmolarity to prevent buildup of osmotic pressure and rupture of the cell membrane.

In blood, the fluid (**plasma**) surrounding red blood cells has an osmolarity of 0.30 Osmol and is **isotonic** with (that is, has the same osmolarity as) the cell contents. If the cells are removed from plasma and placed in an isotonic saline solution (0.15 M NaCl = 0.30 Osmol), they are unharmed. If, however, red blood cells are placed in pure water or in any solution with an osmolarity *lower* than the cell contents (a **hypotonic** solution), water will pass through the membrane into the cell, causing the cell to swell up and burst, a process called **hemolysis**.

Finally, if red blood cells are placed in a solution having an osmolarity *greater* than the cell contents (a **hypertonic** solution), water will pass out of the cell into the surrounding solution, causing the cell to shrivel, a process called **crenation**. Figure 6.7 shows photographs of human red blood cells under all three conditions—isotonic, hypotonic, and hypertonic. Clearly, it's critical that solutions used for intravenous feeding be isotonic to prevent red blood cells from being destroyed.

Practice Problems

6.17   What is the osmolarity of these solutions?
(a) 0.35 M KBr   (b) 0.15 M glucose

6.18   Human blood has an osmolarity of 0.30 Osmol. What concentration of salt (saline) solution in percent (*w/v*) has the same osmolarity?

Figure 6.7
Red blood cells in (a) an isotonic solution (normal in appearance), (b) a hypotonic solution (swollen), and (c) a hypertonic solution (shriveled). Notice that the intact cell in (b) has lost its normal concave shape. The other cells have already burst open from the pressure of the water they absorbed, leaving behind their empty cell membranes.

(a)

(b)

(c)

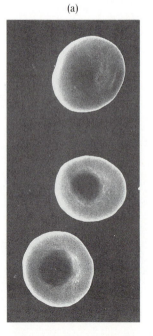
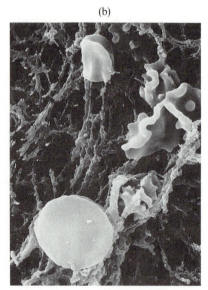

INTERLUDE: DIALYSIS

Osmotic membranes of the sort discussed in Section 6.11 have pores through which only small solvent molecules can pass. Other kinds of semipermeable membranes, however, have larger pores that allow the passage of ions and small covalent molecules while preventing passage of large molecules such as proteins and DNA. (The exact dividing line between a "small" molecule and a "large" one is, of course, imprecise.)

The passage of ions and small molecules through a membrane, a process called *dialysis* (di-**al**-uh-sis), is exactly analogous to osmosis. Solvent and small solute molecules pass through the membrane toward the side where their respective concentrations are lower, but large colloidal particles cannot pass.

Perhaps the most important medical use of dialysis is in artificial kidney machines, where *hemodialysis* is used to cleanse the blood of patients whose kidneys malfunction. Blood is taken from the body and pumped through a long cellophane dialysis tube suspended in an isotonic solution containing glucose, NaCl, $NaHCO_3$ and other essential materials. Small waste materials pass from the blood through the dialysis membrane where they are washed away, but cells, proteins, and other important blood components are prevented by their size from passing through the membrane. Purified blood is then returned to the body (see figure).

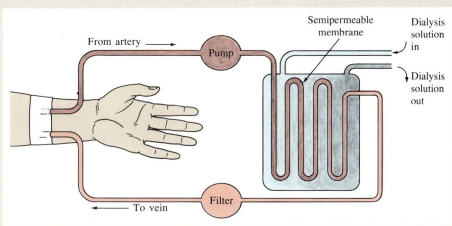

Schematic operation of a hemodialysis unit used for purifying blood. Blood is pumped from an artery through a coiled semipermeable membrane of cellophane. Small waste products pass through the membrane and are washed away by an isotonic dialysis solution.

SUMMARY

Mixtures are classified according to their visual appearance as either **heterogeneous**, where the mixing is nonuniform, or **homogeneous**, where the mixing is uniform. **Solutions**, the most important class of homogeneous mixtures, contain particles the size of a typical covalent molecule (0.2–0.15 nm diameter), whereas **colloids** and **suspensions** contain larger particles.

The maximum amount of one substance (the **solute**) that can be dissolved in another (the **solvent**) is called the substance's **solubility**. The **concentration** of a solution can be expressed in several ways. In the **weight/volume percent** ($w/v\%$) method, concentration is expressed as the number of grams of solute per

100 mL of solution. In the **molarity** method, concentration is expressed as the number of moles of solute per liter of solution. Solubilities of solids generally increase with temperature.

Water, the most common solvent, contains **polar covalent bonds** because of the greater **electronegativity** of oxygen over hydrogen. One consequence of their polar structure is that water molecules orient in the liquid state so that a positively polarized (δ^+) hydrogen of one molecule points toward the negatively polarized (δ^-) oxygen of another molecule, forming a **hydrogen bond**. Another consequence of their polar structure is that water molecules can dissolve ionic compounds by surrounding and stabilizing the dissolved ions in solution.

Substances that form ions when dissolved in water and whose water solutions therefore conduct electricity are called **electrolytes**. Body fluids contain small concentrations of many different electrolytes.

Certain materials, including those that make up the membranes around living cells, are **semipermeable**. These membranes contain pores that allow water or other small solvent molecules to pass but block the passage of solute ions and molecules, a phenomenon called **osmosis**. When the number of dissolved particles on the two sides of the membrane is different, water passes from the more dilute side to the more concentrated side, resulting in the buildup of **osmotic pressure**. A similar effect is noted when membranes of larger pore-size are used. In **dialysis**, the membrane allows the passage of solvent and small dissolved molecules but prevents passage of colloidal and larger particles.

REVIEW PROBLEMS

Solutions and Solubility

6.19   What is the difference between a homogeneous mixture and a heterogeneous one?

6.20   How can you tell a solution from a colloid? A colloid from a suspension?

6.21   How can you explain the observation that opening a warm can of soda produces much more foaming than opening a cold can?

6.22   What characteristic of water allows it to dissolve ionic solids?

6.23   If a single 5-g block of salt is placed in water, it dissolves slowly, but if 5 g of powdered salt is placed in water, it dissolves rapidly. How can you explain this difference?

6.24   Suppose you had a mixture of sand and sugar. How could you use solubility to separate the mixture into its two components?

6.25   Suppose you had a mixture of salt (soluble in water) and aspirin (soluble in chloroform). How could you use the differences in solubility behavior to separate the mixture into its two components?

Concentrations of Solutions

6.26   How is weight/volume percent concentration defined?

6.27   How is molarity defined as a means of expressing concentration?

6.28   How is volume/volume percent concentration defined?

6.29   How much water would you add to 100 mL of orange juice concentrate if you wanted the final juice to be 20 percent the strength of the original?

6.30   Describe how you would prepare 500 mL of a 5.0% (v/v) ethyl alcohol solution.

6.31   A dilute solution of boric acid, $B(OH)_3$, is often used as an eyewash. How would you prepare 500 mL of a 0.50% (w/v) boric acid solution?

6.32   What is the molarity of the boric acid solution in Problem 6.31?

6.33   Describe how you would prepare 250 mL of a 0.10 M NaCl solution.

6.34   The sterile saline solution used to rinse contact lenses has a concentration of 0.40% (w/v). How would you prepare 50 mL of solution if you wanted to save money by not buying the rinse from the drugstore?

6.35 Describe how you would prepare 1.0 L of a 7.5% (w/v) KBr solution.

6.36 Which of the following solutions is more concentrated?
(a) 0.5 M KCl or 5% (w/v) KCl
(b) 2.5% (w/v) $NaHSO_4$ or 0.025 M $NaHSO_4$

6.37 What is the weight/volume percent concentration of the following solutions?
(a) 5.0 g KCl in 75 mL water
(b) 15 g sucrose in 350 mL water
(c) 4.5 g acetic acid in 35 mL water
(d) 0.75 g cholesterol in 150 mL chloroform

6.38 How many grams of each substance are needed to prepare these solutions?
(a) 50 mL of 8% (w/v) KCl
(b) 200 mL of 7.5% (w/v) acetic acid

6.39 If you had only 23 g KOH remaining in a bottle, how many mL of 10% (w/v) solution could you prepare? How many mL of 0.25 M solution?

6.40 The concentration of glucose in blood is approximately 90 mg/100 mL. What is the weight/volume percent concentration of glucose?

6.41 Assuming a glucose concentration of 90 mg/100 mL and a blood volume of 5 L, how many grams of glucose are present in the blood of an average adult?

6.42 What is the molarity of these solutions?
(a) 12.5 g $NaHCO_3$ in 350 mL water
(b) 45.0 g H_2SO_4 in 300 mL water

6.43 How many moles of solute are in these solutions?
(a) 200 mL of 0.30 M acetic acid, CH_3COOH
(b) 1.50 L of 0.25 M NaOH
(c) 750 mL of 2.5 M nitric acid, HNO_3

6.44 How many grams of solute are in each of the solutions in Problem 6.43?

6.45 How many mL of a 0.75 M HCl solution would you need to obtain 0.0040 mol?

6.46 A flask containing 450 mL of 0.50 M H_2SO_4 was accidentally knocked to the floor. How many grams of $NaHCO_3$ would you need to put on the spill to neutralize the acid according to the following equation?

$$H_2SO_4 + 2\,NaHCO_3 \longrightarrow$$
$$Na_2SO_4 + 2\,H_2O + 2\,CO_2$$

6.47 Nalorphine, a relative of morphine, is used to combat withdrawal symptoms in narcotics victims. How many mL of a 0.40% (w/v) solution of nalorphine must be injected to obtain a dose of 1.5 mg?

Electrolytes

6.48 What is an electrolyte? Give an example.

6.49 What does it mean when we say that the concentration of Ca^{2+} in blood is 3.0 mEq/L?

6.50 Which of the following solutions is the strongest electrolyte, and which is the weakest?
(a) 1.0 M NaCl   (b) 1.0 M Na_2SO_4 (c) 1.0 M NH_3

6.51 Kaochlor, a 10% (w/v) KCl solution, is an oral electrolyte supplement administered for potassium deficiency. How many mEq of K^+ are in a 30 mL dose?

6.52 Look up the concentration of Cl^- ion in blood (Table 6.7), and calculate how many grams of Cl^- are in 100 mL blood.

Osmosis

6.53 Explain why a red blood cell will swell up and burst when placed in pure water.

6.54 What does it mean when we say that a 0.15 M NaCl solution is isotonic with blood, whereas distilled water is hypotonic?

6.55 Describe briefly what would happen if a 1.0 M NaCl solution were separated from distilled water by an osmotic membrane.

6.56 Why does a 0.10 M KBr solution have an osmolarity of 0.20 Osmol, but a 0.10 M glucose solution has an osmolarity of 0.10 Osmol?

6.57 Which of the following solutions has the higher osmolarity?
(a) 0.25 M KBr or 0.40 M Na_2SO_4
(b) 0.30 M NaOH or 3% (w/v) NaOH

6.58 What is the difference between an osmotic membrane and a dialysis membrane?

6.59 What would happen if a 0.40 M glucose solution were separated from a 0.40 M NaCl solution by an osmotic membrane?

Additional Problems

6.60 Emergency treatment of cardiac arrest victims sometimes involves injection of calcium chloride solutions directly into the heart muscle. How many grams of $CaCl_2$ are administered in an injection of

5 mL of a 5% solution? How many mEq Ca^{2+} does this correspond to?

6.61 Nitric acid, HNO_3, is available commercially at a concentration of 16 M. How much would you use to prepare 750 mL of a 0.20 M solution?

6.62 How much 16 M nitric acid (Problem 6.61) would react with 5.50 g of KOH according to the following equation?

$$HNO_3 + KOH \longrightarrow H_2O + KNO_3$$

6.63 Hydrochloric acid, HCl, is purchased commercially at a concentration of 12.0 M. How would you prepare the following solutions?
(a) 250 mL of 0.40 M solution
(b) 1.60 L of 0.050 M solution

6.64 An old bottle of 12.0 M hydrochloric acid (Problem 6.63) has only 25 mL left in it. What would the HCl concentration be if 500 mL water were added?

6.65 *Ringer's solution*, used in the treatment of burns and wounds, is prepared by dissolving 8.6 g NaCl, 0.30 g KCl, and 0.33 g $CaCl_2$ in water and diluting to a volume of 1.00 L. What is the molarity of each of the three components?

6.66 What is the osmolarity of Ringer's solution (Problem 6.65)? Is it hypotonic, isotonic, or hypertonic with blood plasma (0.30 Osmol)?

6.67 In most states, a person with a blood alcohol concentration of 0.10% (*v/v*) is considered legally drunk. How much total alcohol does this concentration represent, assuming a blood volume of 5.0 L?

6.68 What does it mean when we say that sodium sulfate is hygroscopic?

6.69 When colorless copper(II) sulfate, $CuSO_4$, is dissolved in water, a blue solution of copper sulfate pentahydrate forms. What is the formula of this pentahydrate?

6.70 The carbon-chlorine bond in chloromethane, CH_3Cl, is polar because a chlorine atom attracts electrons more strongly than a carbon atom. Draw chloromethane using the standard δ^+/δ^- notation to indicate bond polarity.

6.71 Draw the structure of an ammonia molecule, NH_3, and indicate the polarization of the N—H bonds by using the δ^+/δ^- notation.

6.72 In light of your answer to Problem 6.71, show how a hydrogen bond can form between two ammonia molecules.

CHAPTER 7

Acids, Bases, and Salts

This cross section of a fir tree from the Black Forest in Germany shows clear evidence of the effects of acid rain. Each ring represents a year's growth; the tiny outermost rings (just under the thick bark layer) indicate that for the last 20 years growth was severely stunted. You'll learn more about acid rain and the other properties of acids—beneficial and harmful—in this chapter.

Acids! The word calls up images of dangerous, corrosive liquids that eat away everything they touch. Although a few well-known substances such as sulfuric acid, H_2SO_4, do indeed fit this description, most acids are relatively harmless. In fact, many acids, such as ascorbic acid (vitamin C), are necessary for life.

As we'll see in this chapter, the reaction of an acid with a base to yield a product called a *salt* is one of the most common and fundamental reactions in all of chemistry. We'll cover these topics:

1. **What are acids, bases, and salts?** The goal: you should learn how to recognize acids and bases, and how to write equations for their reactions to yield salts.

2. **What is the pH scale for measuring acidity?** The goal: you should learn what the pH scale means and how it is used for measuring the strengths of acid solutions.

3. **How is acidity determined?** The goal: you should learn methods for determining acid strength in the laboratory.

4. **How are acid concentrations expressed?** The goal: you should learn the common methods used to express acid concentrations in solution.

5. **What is a buffer?** The goal: you should learn the concept of a buffer and how to explain a buffer's effect on acid strength.

7.1   ACIDS

■ **Acid** A substance that is able to donate a hydrogen ion, H^+.

What exactly is an acid? According to a definition suggested in 1923 by the Danish chemist, Johannes Brønsted, an **acid** is a substance that is able to give away a hydrogen ion, H^+, to another molecule or ion. Since a hydrogen *atom* consists of a proton and an electron (Section 2.3), a hydrogen *ion*, H^+, is simply a proton. For example, H_2SO_4 (sulfuric acid), **HCl** (hydrochloric acid), HNO_3 (nitric acid), CH_3COOH (acetic acid), and many others are all capable of giving up a proton, H^+, as shown in Table 7.1. Of course, when a neutral acid gives up H^+, an anion is also formed.

Two points about Table 7.1 bear mentioning. The first is that water is included in Table 7.1 as an example of an acid, even though you might not think of it that way. When water gives up H^+, the hydroxide ion, OH^-, results. A second point about Table 7.1 is that different acids can have different numbers of hydrogen atoms available for giving away. Thus, HCl and HNO_3 are both

Table 7.1 Some Common Acids and Their Anions

Acid Name	Formulas			Anion Name
Acetic acid	CH_3COOH	$\rightleftharpoons$	$H^+ + CH_3COO^-$	Acetate ion
Hydrochloric acid	HCl	$\rightleftharpoons$	$H^+ + Cl^-$	Chloride ion
Nitric acid	HNO_3	$\rightleftharpoons$	$H^+ + NO_3^-$	Nitrate ion
Phosphoric acid	H_3PO_4	$\rightleftharpoons$	$H^+ + H_2PO_4^-$	Dihydrogen phosphate ion
Sulfuric acid	H_2SO_4	$\rightleftharpoons$	$H^+ + HSO_4^-$	Bisulfate ion
Water	H_2O	$\rightleftharpoons$	$H^+ + OH^-$	Hydroxide ion

■ **Monoprotic acid**   A substance that has one acidic hydrogen atom.

■ **Diprotic acid**   A substance that has two acidic hydrogen atoms.

■ **Triprotic acid**   A substance that has three acidic hydrogen atoms.

monoprotic acids since each has only one proton to give up; H_2SO_4 is a **diprotic acid** since it has two protons; and H_3PO_4 is a **triprotic acid** since it has three protons. Acetic acid (CH_3COOH), an example of an organic acid, actually has a total of four protons, but only the one attached to oxygen is acidic. The three protons bonded to carbon are not given up easily.

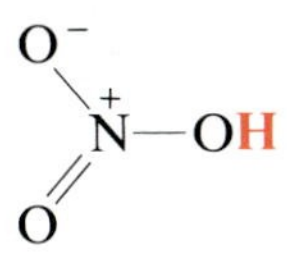

Nitric acid 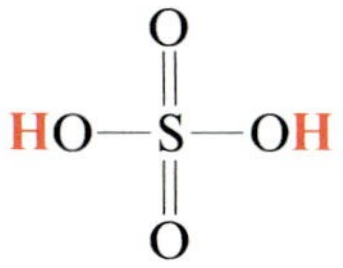   Sulfuric acid 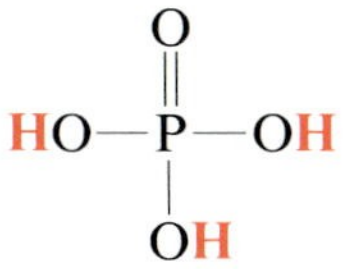   Phosphoric acid

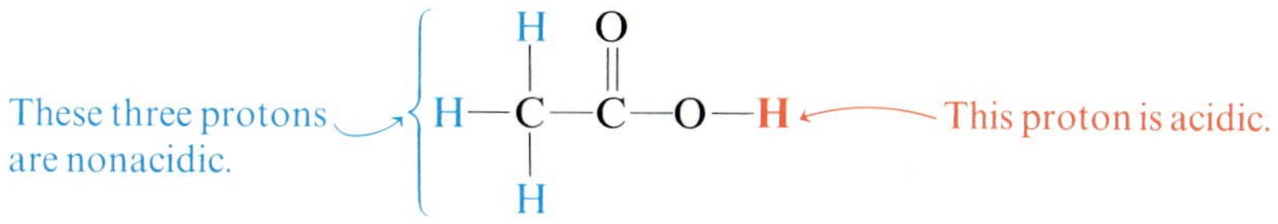

Acetic acid

7.2 BASES

■ **Base**   A substance that can accept a hydrogen ion (H^+) from an acid.

If an acid is any substance that donates a proton, what is a base? According to the Brønsted definition, a **base** is any substance that accepts a proton from an acid. Thus, the ideas of acids and bases are intertwined: An acid reacts with a base, and a base reacts with an acid. During the reaction, a proton is transferred. Acid/base reactions are always written as equilibrium processes using a forward-and-backward double arrow (Section 4.9). The position of the equilibrium in a specific reaction can lie either on the right or on the left depending on the exact nature of the reactants and on the experimental conditions. We can formulate the general reaction as follows:

These electrons from the base are
used to make new B—H bond.

$$B: + H{-}A \rightleftharpoons B{-}H + A^-$$

$$B:^- + H{-}A \rightleftharpoons B{-}H + A^-$$

where: B: or B:$^-$ = a base

H—A = an acid

Table 7.2   Some Common Bases

Base Name	Formulas			Product Name
Acetate ion	$CH_3COO^- + H^+$	$\rightleftharpoons$	CH_3COOH	Acetic acid
Ammonia	$NH_3 + H^+$	$\rightleftharpoons$	$^+NH_4$	Ammonium ion
Bicarbonate ion	$HCO_3^- + H^+$	$\rightleftharpoons$	H_2CO_3	Carbonic acid
Carbonate ion	$CO_3^{2-} + H^+$	$\rightleftharpoons$	HCO_3^-	Bicarbonate ion
Hydroxide ion	$OH^- + H^+$	$\rightleftharpoons$	H_2O	Water
Water	$H_2O + H^+$	$\rightleftharpoons$	H_3O^+	Hydronium ion

Notice in these acid/base reactions that both electrons in the resulting B—H bond must come from the base when a bond forms to H^+. In fact, it's the availability of two unshared electrons for bonding that allows a substance to act as a base. Called a *lone pair of electrons* (Section 3.7), a base must have these two unshared electrons in order to accept H^+ from an acid.

A base can be either neutral (B:) or negatively charged (B:$^-$). If the base is neutral, then the protonated product has a positive charge ($^+$B—H); if the base is negatively charged to begin with, then the protonated product is neutral (B—H). Table 7.2 gives some examples of common bases.

Notice that some of the same substances appear in both Table 7.1 (acids) and Table 7.2 (bases), but on different sides of the reaction arrow. For example, when acetic acid acts as an *acid* by donating a proton, it yields the *base* product, acetate ion. Looked at from the other side, the *base* acetate ion can accept a proton to yield the *acid* product, acetic acid. The two reactions are, of course, just the forward and reverse directions of the same acid/base equilibrium.

Acetic acid (acid)   Acetate ion (base)

Notice also that water can act *either* as an acid or as a base. When water acts as an acid, it donates a proton and becomes OH^-; when it acts as a base, it accepts a proton and becomes H_3O^+. How water reacts in any specific instance depends on the experimental conditions and on the nature of other reagents present.

Water as an acid
(H$^+$ donor): $H-O-H \rightleftharpoons H^+ + OH^-$

Water as a base
(H$^+$ acceptor): $H-\overset{..}{\underset{..}{O}}-H + H^+ \rightleftharpoons H-\overset{..}{O}^+-H = H_3O^+$
                                                                                      |
                                                                                      H

7.3   THE NATURE OF ACIDS AND BASES IN AQUEOUS SOLUTION

■ **Dissociation**   The splitting apart of a substance to yield two or more ions.

Although it's sometimes convenient to write a chemical equation that portrays the splitting apart (**dissociation**) of an acid into a proton and an anion, this dissociation doesn't occur spontaneously in practice. A bare proton is too unstable to exist by itself, and acids therefore only give up a proton when there is a base present to accept it.

$$H—A \quad \xrightarrow{\;\;\;\not\rightarrow\;\;\;} \quad H^+ + A^-$$

The dissociation of a pure acid doesn't occur spontaneously.

$$H—A + Base \quad \longrightarrow \quad H—Base^+ + A^-$$

Instead, a base is necessary to accept the proton from the acid.

■ **Hydronium ion** H_3O^+, the species formed when an acid is dissolved in water.

When an acid is dissolved in water, water acts as a base to accept a proton from the acid and generate the **hydronium ion** (H_3O^+). If HCl gas is dissolved in water, for example, a solution of H_3O^+ and Cl^- ions results. (Historically, in fact, the first way of defining an acid—the **Arrhenius acid definition**—was as any substance that dissolves in water to yield H_3O^+ ions.) In writing such equations, we indicate the state of the reactants and products by using a label in parenthesis—g for gaseous, l for liquid, s for solid, and aq for **aqueous** (water solution). Often, H_3O^+ and H^+ (aq) are used interchangeably, since both refer to the same species.

■ **Arrhenius acid**   A substance that increases the concentration of H_3O^+ when dissolved in water.

■ **Aqueous**   Refers to a solution with water as the solvent.

Hydrochloric acid

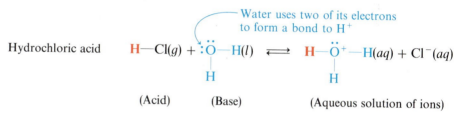

(Acid)   (Base)   (Aqueous solution of ions)

Just as the Arrhenius definition of an acid is a substance that yields H_3O^+ when dissolved in water, the Arrhenius definition of a base is a substance that yields the hydroxide ion, OH^-, when dissolved in water. Solid NaOH, for example, dissociates to yield Na^+ and OH^- ions.

$$NaOH(s) \quad \xrightarrow[H_2O]{\text{dissolve in}} \quad Na^+(aq) + OH^-(aq)$$

Aqueous solutions of all acids show similar chemical properties because all contain the hydronium ion, H_3O^+; only the anion differs from one acid solution to another. Thus, all aqueous acid solutions turn litmus paper from blue to red, and all are able to react with bases. Similarly, aqueous solutions of all bases show similar chemical properties because all contain the hydroxide ion, HO^-. All aqueous base solutions turn litmus paper from red to blue, and all are able to react with acids.

7.4   REACTIONS OF ACID WITH BASES: NEUTRALIZATION

■ **Neutralization reaction**   The reaction of an acid with a base to yield water and a salt.

Since acids are proton donors and bases are proton acceptors, it's not surprising that they react with each other. For example, the acid HCl reacts with the base NaOH to yield NaCl and water. The overall process is called a **neutralization reaction** because both acid and base are used up and the products that result are neutral (that is, are neither acidic nor basic).

A neutralization reaction:

$$HCl(aq) + NaOH(aq) \quad \longrightarrow \quad NaCl(aq) + H_2O(l)$$

An acid   A base   Neutral products

Acid/base neutralizations are among the most important reactions in all life processes and, indeed, in all of chemistry. Let's look at several kinds of these reactions.

Reactions of Acids with Metal Hydroxides   As we've seen, metal hydroxides such as KOH and NaOH are ionic solids that dissociate in water to give solutions containing the hydroxide ion, OH^-. Acids react with these metal-hydroxide bases to yield water and a salt. For example, HCl reacts with KOH to give water and potassium chloride. The H^+ from HCl and the OH^- from KOH join to yield neutral H_2O, and the K^+ and Cl^- are left to yield KCl:

$$HCl(aq) + KOH(aq) \longrightarrow H_2O(l) + KCl(aq)$$

   An acid   A base   Water   A salt

■ **Salt** An ionic substance composed of a positive ion other than H^+ and a negative ion other than OH^-.

Although we usually think specifically of sodium chloride, table salt, when the word salt is mentioned, a **salt** is more generally defined as any ionic substance composed of a positive ion (cation) other than H^+ and a negative ion (anion) other than OH^-. Thus, KCl, NaBr, Na_2SO_4, $LiNO_3$, and many other ionic compounds are all called salts. When a salt is formed in an acid/base reaction, the cation comes from the base and the anion comes from the acid. Some further examples of salt formed by neutralization reactions between acids and bases are shown:

$$HNO_3(aq) + KOH(aq) \longrightarrow H_2O(l) + KNO_3(aq)$$
$$H_2SO_4(aq) + 2\,LiOH(aq) \longrightarrow 2\,H_2O(l) + Li_2SO_4(aq)$$
$$2\,HCl(aq) + Mg(OH)_2(aq) \longrightarrow 2\,H_2O(l) + MgCl_2(aq)$$

Some salts

Solved Problem 7.1   Show the products from the neutralization reaction of HBr with $Ca(OH)_2$.

Solution   All neutralizations involve reaction of an acid with a base to yield water and a salt. The identity of the salt can be found by taking the cation from the base [Ca^{2+} from $Ca(OH)_2$] and the anion from the acid (Br^- from HBr). As the formulas indicate, however, HBr provides only one H^+ (aq) when dissolved in water, but $Ca(OH)_2$ provides *two* OH^- ions when similarly dissolved.

$$HBr(g) \rightleftharpoons H^+(aq) + Br^-(aq)$$
$$Ca(OH)_2(s) \rightleftharpoons Ca^{2+}(aq) + 2\,OH^-(aq)$$

Thus, the balanced equation requires two HBr molecules to react with $Ca(OH)_2$, yielding $CaBr_2$ and two water molecules:

$$2\,HBr(aq) + Ca(OH)_2(aq) \longrightarrow 2\,H_2O(l) + CaBr_2(aq)$$

Practice Problems

7.1   Show the products you would obtain from these acid/base reactions:
(a) $HNO_3 + Mg(OH)_2 \rightarrow$?   (b) $H_2SO_4 + Ba(OH)_2 \rightarrow$?

7.2   Maalox, an over-the-counter antacid, contains aluminum hydroxide, $Al(OH)_3$. Write a balanced equation for the reaction of $Al(OH)_3$ with stomach acid (HCl).

Reactions of Acids with Metal Bicarbonates and Metal Carbonates   What products would you expect from reaction of an acid such as HBr with a metal bicarbonate such as $NaHCO_3$? Since the bicarbonate anion, HCO_3^-, is a base (Table 7.2), we might expect it to react with H^+ to yield H_2CO_3. This is exactly what happens. It turns out, however, that H_2CO_3 (carbonic acid) is not a very stable molecule; much of it spontaneously decomposes to yield water and carbon dioxide gas, which bubbles out of the reaction mixture:

$$H^+(aq) + HCO_3^-(aq) \rightleftharpoons [H_2CO_3(aq)] \rightleftharpoons H_2O(l) + CO_2(g)$$

An acid   Bicarbonate   Carbonic acid   Water   Carbon
             ion   (unstable)   dioxide

Thus, the reaction of an acid with a metal bicarbonate yields water, a salt, and carbon dioxide gas. The net effect of the reaction is to neutralize the acid, just as reaction with a metal hydroxide neutralizes it. For example, the reaction of HBr with $NaHCO_3$ in aqueous solution proceeds in this way:

$$HBr(aq) + NaHCO_3(aq) \longrightarrow H_2O(l) + CO_2(g) + NaBr(aq)$$

Acids react with metal carbonates such as Na_2CO_3 in exactly the same way they react with metal bicarbonates. Water, a salt, and carbon dioxide gas are the products. For example:

$$2\,HCl(aq) + Na_2CO_3(aq) \longrightarrow H_2O(l) + CO_2(g) + 2\,NaCl(aq)$$

The only difference in the reactions of carbonate and bicarbonate is that a carbonate ion (CO_3^{2-}) can neutralize twice as much acid as a bicarbonate ion (HCO_3^-) because it reacts with two hydrogen ions rather than one.

Although most metal carbonates are insoluble in water (marble is almost pure calcium carbonate, $CaCO_3$), they nevertheless react easily with aqueous acid. In fact, geologists often test for carbonate-bearing rocks simply by putting a few drops of aqueous HCl on the rock and watching to see if bubbles of CO_2 form.

$$CaCO_3(s) + 2\,HCl(aq) \longrightarrow H_2O(l) + CO_2(g) + CaCl_2(aq)$$

Solved Problem 7.2   Write the products from reaction of HI with $LiHCO_3$.

Solution   Acids react with metal bicarbonate to yield water, a salt, and carbon dioxide. The cation of the salt comes from the metal bicarbonate, and the anion of the salt comes from the acid. Thus the balanced reaction is:

$$HI(aq) + LiHCO_3(aq) \longrightarrow H_2O(l) + LiI(aq) + CO_2(g)$$

Practice Problem   **7.3**   Write the products of these reactions:
(a) $KHCO_3 + HNO_3 \rightarrow$?   (b) $MgCO_3 + H_2SO_4 \rightarrow$?

Reactions of Acids with Ammonia   Aqueous solutions of ammonia are usually labeled "ammonium hydroxide," but this is really a misnomer. Only about one percent of the ammonia molecules in aqueous solution are protonated by water to yield NH_4^+ and OH^- ions; the remaining 99 percent are unprotonated. In writing the equation, we use forward and backward arrows of *unequal* length to indicate the position of the equilibrium.

$$NH_3 + H-O-H \rightleftharpoons NH_4^+(aq) + HO^-(aq)$$

99% of ammonia molecules are unprotonated.

1% of ammonia molecules are protonated by water

■ **Ammonium salt** An ionic substance formed by reaction of ammonia with an acid.

Acids react with ammonia to yield **ammonium salts** such as ammonium chloride, NH_4Cl. Like metal salts, ammonium salts are soluble in water and dissociate into NH_4^+ and Cl^- ions.

$$HCl(aq) + :NH_3(aq) \longrightarrow NH_4Cl(aq)$$

Ammonium chloride

Look at the way the reaction of ammonia with HCl has been written, with the lone pair of electrons on nitrogen specifically indicated. Even though ammonia is not negatively charged like hydroxide ion, the nitrogen atom can still use its lone electron pair to form a covalent bond to the hydrogen ion. In so doing, nitrogen effectively gives an electron to the H^+ and thus gains the positive charge for itself. We can indicate the direction of this electron donation by showing a curved arrow moving from the nitrogen to the hydrogen ion.

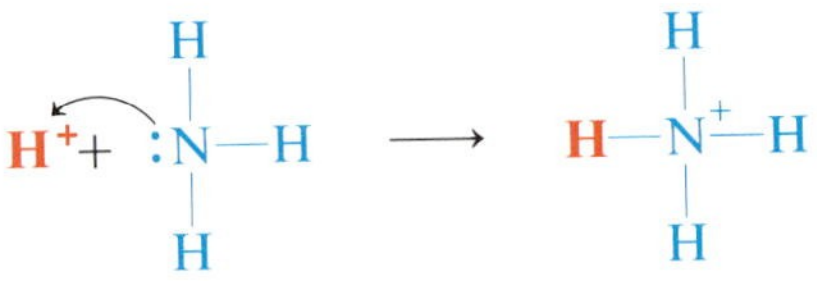

Ammonia   Ammonium ion

Living organisms contain a group of compounds called *amines*, which contain ammonia-like nitrogen atoms and which can undergo neutralization reactions with acids just as ammonia can. Methylamine, for example, is the compound primarily responsible for the distinctive odor of fish.

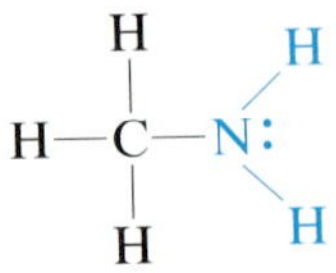

Methylamine, a naturally occurring base

> **Solved Problem 7.3**   Show the products from reaction of ammonia with HNO_3.
>
> **Solution**   Ammonia reacts with acids to yield ammonium salts containing the ammonium cation (NH_4^+) and the anion from the acid. Thus, the balanced equation is: $NH_3(aq) + HNO_3(aq) \rightarrow NH_4NO_3(aq)$

Practice Problems

7.4   What products would you expect from reaction of ammonia with sulfuric acid? $H_2SO_4 + NH_3 \rightarrow ?$

7.5   Show how methylamine, the compound responsible for the odor of fish, can react with HCl to give a methylammonium salt.

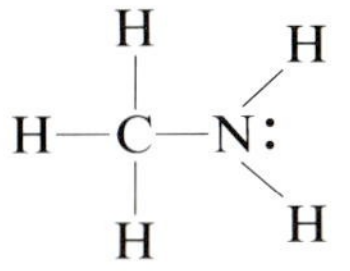

Methylamine

7.5   ACID AND BASE STRENGTH

Different acids differ in their ability to give up a proton. Compounds such as HCl and H_2SO_4, which give up a proton easily, are called *strong acids*. Those compounds such as CH_3COOH that don't give up a proton easily are called *weak acids*. Some of the more common acids are listed in Table 7.3 in order of strength. The five acids shown at the top of the table are very strong; phosphoric acid is intermediate in strength; and those shown at the bottom of the table are weak.

Strong acids are almost entirely dissociated in water solution to yield H_3O^+ and an anion. In other words, the position of the dissociation equilibrium for a strong acid lies far toward the right side of the equation. Weak acids, however, dissociate only to a slight extent, and their dissociation equilibria therefore lie toward the left side of the equation. For example, only about one

Table 7.3   Relative Strengths of Some Common Acids and Bases

	Acid			Base	
Strong acid	Perchloric acid	$HClO_4$	ClO_4^-	Perchlorate ion	Weak base
	Sulfuric acid	H_2SO_4	HSO_4^-	Bisulfate ion	
	Hydrobromic acid	HBr	Br^-	Bromide ion	
	Hydrochloric acid	HCl	Cl^-	Chloride ion	
	Nitric acid	HNO_3	NO_3^-	Nitrate ion	
	Phosphoric acid	H_3PO_4	$H_2PO_4^-$	Dihydrogen phosphate ion	
	Acetic acid	CH_3COOH	CH_3COO^-	Acetate ion	
	Carbonic acid	H_2CO_3	HCO_3^-	Bicarbonate ion	
	Water	H_2O	OH^-	Hydroxide ion	
Weak acid	Methyl alcohol	CH_3OH	CH_3O^-	Methoxide ion	Strong base
	Ammonia	NH_3	NH_2^-	Amide ion	

percent of the CH_3COOH molecules in a 0.10 M aqueous solution of acetic acid are dissociated into CH_3COO^- and H_3O^+; the remaining 99 percent are neutral and undissociated.

Hydrochloric acid (strong):

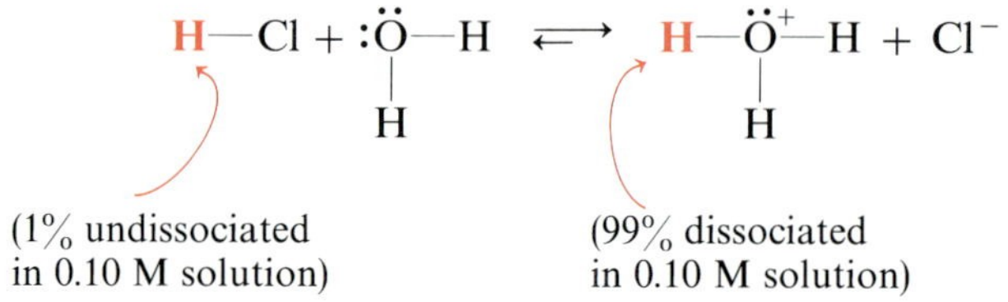

$$H-Cl + :\overset{..}{O}-H \rightleftharpoons H-\overset{..}{\overset{+}{O}}-H + Cl^-$$

(1% undissociated in 0.10 M solution)   (99% dissociated in 0.10 M solution)

Acetic acid (weak):

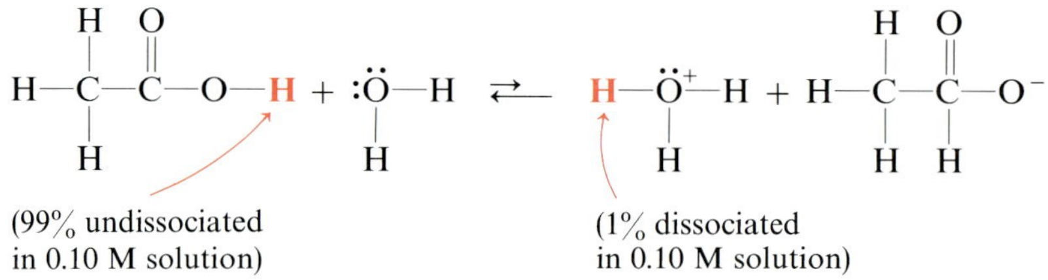

(99% undissociated in 0.10 M solution)   (1% dissociated in 0.10 M solution)

Diprotic acids such as sulfuric acid can undergo two stepwise dissociations in water. The first dissociation of H_2SO_4 to yield HSO_4^- (bisulfate ion) occurs to the extent of nearly 100 percent, but the second dissociation of HSO_4^- to yield SO_4^{2-} (sulfate ion) is much more difficult and takes place to a much lesser extent because it involves further loss of a positively charged H^+ from an ion that already carries one negative charge.

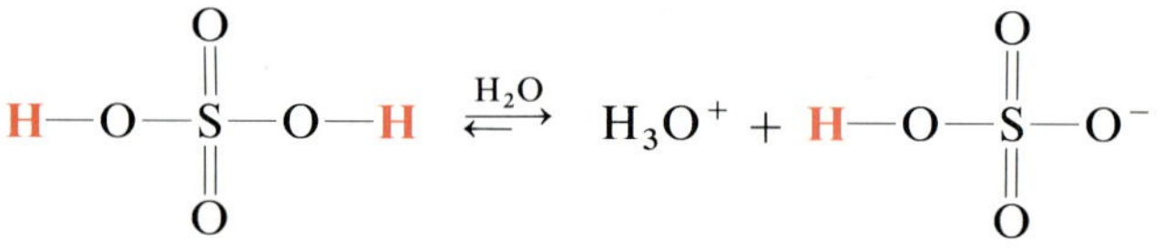

Sulfuric acid (a diprotic acid)   Bisulfate ion

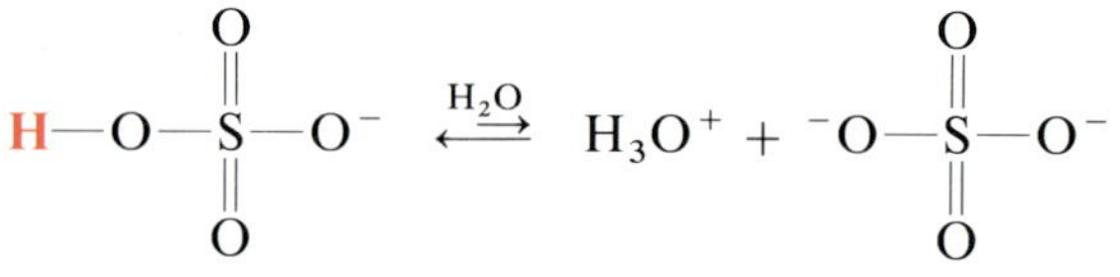

Bisulfate ion   Sulfate ion

Just as acids differ in their ability to *donate* a proton, bases differ in their ability to *accept* a proton. Strong bases such as hydroxide ion accept and hold a proton tightly, but weak bases such as chloride ion have little affinity for a proton. Table 7.3 also lists some of the more common bases in order of strength.

The most striking feature of Table 7.3 is the relationship between acid and base strength. The anion from dissociation of a strong acid is a weak base, and the anion from dissociation of a weak acid is a strong base. Conversely, when a strong base accepts a proton, the product is a weak acid; when a weak base accepts a proton, the product is a strong acid.

Why is there a relationship between acid and base strength? The easiest way to answer this question is to think about what it means for an acid or base to be strong or weak. By definition, a strong acid is one that gives up a proton readily. In other words, the anion formed by dissociation of a strong acid has little affinity for the proton; otherwise it wouldn't have given up the proton in the first place. But this is exactly the definition of a weak base—a substance that has little affinity for a proton.

$$H-A + H_2O \rightleftharpoons H_3O^+ + A^-$$

If this is a strong acid because it gives up the proton readily . . .

. . . then this anion is a weak base because it has little affinity for the proton

For example:

$$HCl + H_2O \rightleftharpoons H_3O^+ + Cl^-$$

(HCl—a strong acid)

(Cl$^-$—a weak base)

Similarly, a weak acid is one that holds a proton tightly and gives it up with difficulty. In other words, the anion formed by dissociation of a weak acid is not stable but would instead prefer to bond to the proton. This is exactly the definition of a strong base—a substance that has a high affinity for a proton.

$$A^- + H_3O^+ \rightleftharpoons H-A + H_2O$$

If this anion is a strong base because it has a high affinity for a proton . . .

. . . then this is a weak acid because it will not readily give up the proton.

For example

$$H-\underset{\underset{H}{|}}{\overset{\overset{H}{|}}{C}}-O^- + H_3O^+ \rightleftharpoons H-\underset{\underset{H}{|}}{\overset{\overset{H}{|}}{C}}-O-H + H_2O$$

(Methoxide ion—a strong base)

(Methyl alcohol—a weak acid)

7.6 THE ACIDITY OF WATER

We saw in Section 7.1 that even pure water can dissociate to a slight extent into ions. Since each dissociation yields one H_3O^+ ion and one OH^- ion, the concentrations of the two ions are identical; at 25°C, the concentration of each is 1.00×10^{-7} M. (In the following expressions of concentration, brackets [ ] indicate that the concentration of the enclosed species is given in moles per liter. Thus, we don't need to use the molarity unit M.)

$$2\,H_2O \rightleftarrows H_3O^+ + OH^-$$

At 25°C:

$$[H_3O^+] = [OH^-] = 1.00 \times 10^{-7}$$

Ion product constant (K_w) The product of H_3O^+ and OH^- molar concentrations in water ($K_w = 10^{-14}$).

It turns out to be very useful to define a quantity called the **ion product constant (K_w)**, which is obtained by multiplying the molar concentrations of hydronium and hydroxide ions. Since both $[H_3O^+]$ and $[OH^-]$ equal 1.00×10^{-7}, K_w is equal to 1.00×10^{-14}.

Ion product constant (K_w):
$$K_w = [H_3O^+] \times [OH^-]$$
$$= (1.00 \times 10^{-7}) \times (1.00 \times 10^{-7})$$
$$= 1.00 \times 10^{-14}$$

K_w *always has the same value for any aqueous solution,* regardless of whether or not an acid, base, or salt is present. If an acid is present, making the hydronium-ion concentration high, then the hydroxide-ion concentration of the solution is low. By using K_w, an unknown H_3O^+ concentration can be found from a known OH^- concentration, and vice versa. For example, in 0.10 M HCl solution, where $[H_3O^+] = 0.10$, $[OH^-]$ must be 10^{-13}. Since $K_w = [H_3O^+] \times [OH^-]$, then

$$[OH^-] = \frac{K_w}{[H_3O^+]}$$

For example, if $[H_3O^+] = 0.10$, then

$$[OH^-] = \frac{K_w}{[H_3O^+]} = \frac{10^{-14}}{0.10} = 10^{-13}$$

Similarly, if a base is present so that the hydroxide-ion concentration is high, then the hydronium-ion concentration of the solution is low. For example, in 0.10 M NaOH solution, where $[OH^-] = 0.10$, $[H_3O^+]$ must be 10^{-13}. Since $K_w = [H_3O^+] \times [OH^-]$, then

$$[H_3O^+] = \frac{K_w}{[OH^-]}$$

For example, if $[OH^-] = 0.10$, then

$$[H_3O^+] = \frac{K_w}{[OH^-]} = \frac{10^{-14}}{0.10} = 10^{-13}$$

In general, any solution with a hydronium-ion concentration greater than 10^{-7} M is considered acidic; any solution with a hydronium-ion concentration less than 10^{-7} M is considered basic; and any solution with a hydronium-ion concentration near that of pure water (10^{-7} M) is considered neutral.

Acidic solution: $[H_3O^+] > 10^{-7}$

Neutral solution: $[H_3O^+] \approx 10^{-7}$

Basic solution: $[H_3O^+] < 10^{-7}$

Solved Problem 7.4   What is $[OH^-]$ in milk if $[H_3O^+] = 4.5 \times 10^{-7}$? Is milk acidic or basic?

Solution   Since the product of $[H_3O^+]$ and $[OH^-]$ for any solution is 10^{-14}, we can set up the following equation:

$$[H_3O^+][OH^-] = (4.5 \times 10^{-7}) \times [OH^-] = 10^{-14}$$

Solving this equation for $[OH^-]$, we get:

$$[OH^-] = \frac{10^{-14}}{4.5 \times 10^{-7}} = 2.2 \times 10^{-8}$$

Thus, the OH^- concentration of milk is 2.2×10^{-8} M. Since $[H_3O^+]$ is higher than $[OH^-]$, milk is slightly acidic.

Practice Problem   **7.6**   What is $[OH^-]$ in the following solutions? Identify each as either acidic or basic.
(a) beer; $[H_3O^+] = 3.2 \times 10^{-5}$ (b) household ammonia; $[H_3O^+] = 3.1 \times 10^{-12}$

7.7 MEASURING ACIDITY IN AQUEOUS SOLUTIONS: pH

It's often necessary to know the exact concentration of H_3O^+ or OH^- ions in solution. In medicine, for example, it's crucial that the hydronium-ion concentration of blood be very close to 4.0×10^{-8} M. If the acidity of blood changes only slightly from this value, death can result.

Although perfectly correct, it's nevertheless awkward to refer to low concentrations of H_3O^+ by giving their molarity. For instance, if you were asked which concentration is higher, 9.0×10^{-8} M or 3.5×10^{-7} M, you'd probably have to stop and think for a minute before answering. Fortunately, there's an easier way to express and compare $[H_3O^+]$—the pH method.

■ **pH**   A number that describes the acidity of an aqueous solution. Mathematically, pH is equal to the negative logarithm of a solution's H_3O^+ concentration.

The **pH** of any aqueous solution is simply a number, usually between 0 and 14, that tells the solution's acid strength. A pH below 7 corresponds to an acid solution, with acid strength increasing at lower pH; a pH above 7 corresponds to a base solution, with base strength increasing at higher pH; and a pH of exactly 7 corresponds to a neutral solution. The pH scale and pH values of some common substances are shown in Figure 7.1.

Figure 7.1
The pH scale and the pHs of some common substances. A low pH value corresponds to a strongly acidic solution; a high pH value corresponds to a strongly basic solution; and a pH of 7 corresponds to a neutral solution.

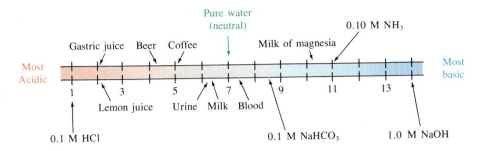

Since pH and $[H_3O^+]$ both tell the concentration of hydronium ions in a solution, they must be related. Mathematically, the relationship can be expressed in the following way:

$$[H_3O^+] = 10^{-pH}$$

This equation says that the hydronium concentration of a solution, $[H_3O^+]$, is equal to the number 10 raised to the negative power of the solution's pH (hence the derivation of the term pH: small "p" for power and large "H" for hydrogen ion). For example, in neutral water, where $[H_3O^+] = 10^{-7}$, the pH is **7**; in a strong acid solution, where $[H_3O^+] = 10^{-1}$, the pH is **1**; and in a strong base solution, where $[H_3O^+] = 10^{-14}$, the pH is **14**. Figure 7.2 summarizes the relationships.

It's important to realize that the pH scale covers an enormous range of acidities. Because the expression for pH involves powers of ten, a change of only one pH unit means a *tenfold* change in $[H_3O^+]$; a change of two pH units means a *hundredfold* change in $[H_3O^+]$; and a change of 12 pH units, from a strong acid solution with pH 1 to a strong base solution with pH 13, means a change of 10^{12} (a million million) in $[H_3O^+]$.

It might help give you feeling for the size of quantities involved by thinking of a backyard swimming pool, which contains about 100,000 L water. You would have to add only 0.1 mole of HCl (3.7 g) to lower the pH of the pool from 7.0 (neutral) to 6.0, but you would have to add 10,000 moles of HCl (370 kg!) to lower the pH of the pool to 1.0.

Figure 7.2
The relationship between pH, $[H_3O^+]$, and $[OH^-]$. A high pH corresponds to a low $[H_3O^+]$ and a high $[OH^-]$, while a low pH corresponds to a high $[H_3O^+]$ and a low $[OH^-]$.

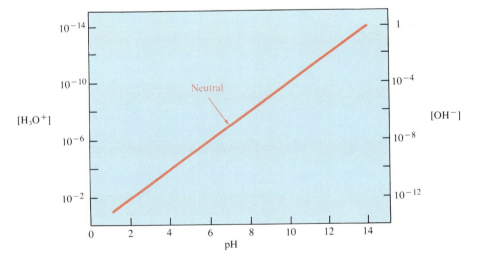

The conversion between pH and $[H_3O^+]$ is simple when the pH is a whole number, but what does it mean when we say that the pH of blood is 7.40? One answer is that if pH = 7.40, then $[H_3O^+] = 10^{-7.40}$. Although correct, it's nevertheless awkward to use numbers like $10^{-7.40}$, which can be better written as 4.0×10^{-8}. But how can we convert $10^{-7.40}$ into 4.0×10^{-8}? If pH = 7.40, then

$$[H_3O^+] = 10^{-7.40} = 4.0 \times 10^{-8}$$

Converting between fractional powers of ten like $10^{-7.40}$ and their equivalent values like 4.0×10^{-8} requires a knowledge of how to work with logarithms.* For most purposes, though, it's sufficient just to realize that a pH value intermediate between two whole numbers must correspond to a hydronium-ion concentration intermediate between the $[H_3O^+]$ values for those same two numbers. For example, if the pH of blood is between 7 and 8, then the hydronium-ion concentration of blood must be between 10^{-7} M and 10^{-8} M.

If pH = 7.40, then $[H_3O^+] = 10^{-7.40} = 4.0 \times 10^{-8}$

A value between
7 and 8

A value between
10^{-7} and 10^{-8}

Solved Problem 7.5   The pH of Coca Cola is approximately 3.3. Tell whether Coke is acidic or basic, and estimate its hydronium-ion concentration.

Ballpark Solution   A pH value lower than 7 corresponds to an acid solution. Thus, Coca Cola is acidic. Since a pH of 3.3 is intermediate between 3 and 4, $[H_3O^+]$ must be intermediate between 10^{-3} and 10^{-4}—say about 4×10^{-4}. (The exact value is 5.0×10^{-4}.)

Practice Problems   **7.7**   Identify the following solutions as acidic or basic, and rank them in order of increasing acidity:
(a) saliva, pH = 6.5   (b) pancreatic juice, pH = 7.9   (c) orange juice, pH = 3.7   (d) wine, pH = 3.5

7.8   Estimate $[H_3O^+]$ values for each of the solutions in Problem 7.7.

7.8 LABORATORY DETERMINATION OF ACIDITY

We've talked at length about what it means for a solution to be acidic or basic, but we've not yet touched on the most important practical matter: How do you determine the pH of a solution?

---

* If you're familiar with using logarithms, you can use the equation pH = $-\log[H_3O^+]$ to convert between pH and $[H_3O^+]$. Even if you're not familiar with logarithms, you can try a simple test if your calculator has a "Log" key: Enter the number 4.0×10^{-8} (the value of $[H_3O^+]$ for blood), and then press the Log key. The answer -7.40 appears. Changing the sign then gives 7.40, the pH of blood.

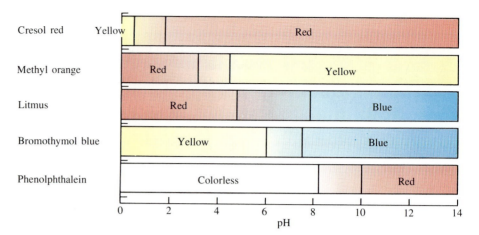

Figure 7.3
Some acid/base indicators and their color changes. The area between indicated colors is the range in which transition occurs.

■ **Indicator** A dye that changes color to indicate the pH of a solution.

There are two common ways to measure the pH of a solution. The simpler but less accurate method is to use an **indicator**, a dye that changes color depending on pH. For example, the well-known dye *litmus* is red below pH 4.8 but blue above pH 7.8; phenolphthalein (fee-nol-**thay**-lean) is colorless below pH 8.2 but red above pH 10; and so on as shown in Figure 7.3. Paper strips ("pH paper") impregnated with a combination of different indicators are available, which allow one to determine pH simply by putting a drop of solution on the paper and comparing the color that appears to the color on a calibration chart. The pH value obtained with these strips is accurate to within several tenths of a pH unit.

A second and more accurate method of determining pH is to use an electronic pH meter, like the one shown in Figure 7.4. Electrodes are dipped into the solution, and the meter provides a pH reading accurate to the second decimal point.

AN APPLICATION: ULCERS AND ANTACIDS

Although their causes are unknown, peptic ulcers are one of the hazards of fast-paced modern life—nature's way of telling us to slow down. Ulcers occur either in the stomach itself or in the upper part of the small intestine (the *duodenum*) when the protective mucosal lining is penetrated and gastric juices begin to dissolve the stomach or intestinal wall. The resultant lesion causes considerable pain and can lead to eventual perforation of the wall.

Much of the tissue damage done by ulcers is due to the fact that gastric secretions are strongly acidic, reaching a maximum acid strength approximately 60 minutes after eating. Thus, one of the simplest methods for controlling ulcer damage is to lower stomach acidity through the use of antacids. As you might expect, all the common over-the-counter antacid remedies are bases, either hydroxides, carbonates, or bicarbonates. Since aluminum

and calcium salts tend to cause constipation, most formulations also contain magnesium salts, which have a counteracting laxative effect. Sodium bicarbonate ($NaHCO_3$, baking soda) is less suitable than other bases because it tends to enter the bloodstream too rapidly.

Ingredients of Some Antacid Preparations

Trade Name	Active Ingredients
Alka-Seltzer	$NaHCO_3 + KHCO_3$
Bisodol	$CaCO_3 + Mg(OH)_2$
DiGel	$Al(OH)_3 + MgCO_3$
Gelusil	$Al(OH)_3 + Mg(OH)_2$
Maalox	$Al(OH)_3 + Mg(OH)_2$
Rolaids	$NaAl(OH)_2CO_3$
Tums	$CaCO_3$

Figure 7.4
Using a pH meter to obtain an accurate reading of pH.

7.9 TITRATION

Determining the pH of a solution with a meter or with indicator paper tells the solution's hydronium-ion concentration but doesn't necessarily tell its total acid concentration (neutralizing capacity). For example, in a 0.10 M solution of acetic acid, the total acid concentration is 0.10 M, yet the hydronium-ion concentration is only 0.0013 M (pH = 2.9) because acetic acid is only about one percent dissociated (Section 7.5).

In order to measure the total acid concentration of a solution, and thereby find out the neutralizing capacity of the solution, you have to carry out a **titration** (tie-**tray**-shun). Titrating an acid solution to find its concentration consists simply of adding a known amount of base until the acid is exactly neutralized. Similarly, titrating a basic solution consists of adding a known amount of acid until the solution is neutralized.

■ **Titration** An experimental method for determining acid (or base) concentration by neutralizing a sample with a base (or acid) of known concentration.

Figure 7.5
Carrying out a titration. (a) A known volume of the acid (or base) solution to be analyzed is placed in the flask along with an indicator. (b) Base (or acid) of known concentration is then added drop by drop until (c) the color change of the indicator shows that neutralization is complete (the end point).

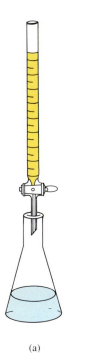
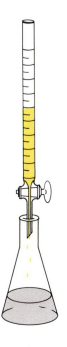

(a)   (b)   (c)

The experimental procedure for carrying out a titration is shown in Figure 7.5. A measured volume of solution of unknown acidity or basicity is placed in a flask, and an indicator is added. If the unknown solution is an acid, then a solution of known base concentration is slowly added until the indicator changes color to signal neutralization. If the unknown solution is a base, then a solution of known acid concentration is slowly added until neutralization is complete (the **end point**). By determining how much acid (or base) is needed to neutralize the unknown base (or acid) solution, the concentration of the unknown can be calculated. The following solved problem shows how the calculation can be done. (It might help you to review Sections 6.4 and 6.5.)

■ **End point**   The point at which a titration is complete.

Solved Problem 7.6   When a 5.00 mL sample of household vinegar (dilute aqueous acetic acid) was titrated, 44.5 mL of 0.100 M NaOH solution were required to reach the end point. What is the acid concentration of vinegar?

Ballpark Solution   Since the volume of base required to neutralize the acid sample is about nine times the volume of the sample (44.5 mL versus 5 mL), the concentration of the acid is about nine times that of the base, or 0.9 M.

Solution   First, write the balanced equation for the neutralization to find the number of moles of base required to neutralize each mole of acid. In this example, the mole ratio of base to acid is 1:1.

$$CH_3COOH + NaOH \longrightarrow CH_3COO^-Na^+ + H_2O$$

Next, calculate how many moles of NaOH are consumed in the titration:

$$44.5\ \cancel{mL} \times \frac{0.100\ \text{moles NaOH}}{\cancel{L}} \times \frac{1\ \cancel{L}}{1000\ \cancel{mL}} + 0.00445\ \text{moles NaOH}$$

Since the balanced equation says that each mole of NaOH reacts with one mole of acetic acid, the 5.00 mL sample of vinegar must contain 0.00445 moles of acetic acid:

$$\text{Acetic acid concentration} = \frac{0.00445\ \text{moles CH}_3\text{COOH}}{5.00\ \cancel{mL}} \times \frac{1000\ \cancel{mL}}{1\ \text{L}}$$

$$= 0.890\ \text{M}$$

Thus, the acetic acid concentration in vinegar is 0.890 M, very close to the ballpark answer.

Practice Problems

7.9   In order to determine the concentration of an old bottle of aqueous HCl whose label had become unreadable, a titration was carried out. What is the HCl concentration if 58.4 mL of 0.250 M NaOH was required to titrate a 20.0 mL sample of the acid?

7.10   How many mL of 0.150 M NaOH are required to neutralize 50.0 mL of 0.200 M H_2SO_4?

7.10 EQUIVALENTS OF ACIDS AND BASES: NORMALITY

We saw in the last chapter (Section 6.10) that it's sometimes useful to think in terms of ion equivalents, where one equivalent of an ion is defined as the formula weight of the ion in grams divided by the number of charges on the ion.

$$\text{One equivalent of ion} = \frac{\text{formula weight of ion in grams}}{\text{number of charges on ion}}$$

■ **Acid (base) equivalent** The amount in grams of an acid (or base) that can donate one mole of H^+ (or OH^-) ions.

Since some acids have more than one H^+ per molecule and some bases have more than one OH^- per molecule, it's often helpful to think of **acid** or **base equivalents**. One equivalent of an acid is equal to the formula weight of the acid in grams, divided by the number of H^+ ions produced. Thus, one equivalent of an acid is the weight in grams that can donate one mole of H^+ ions. Similarly, one equivalent of a base is the weight in grams that can produce one mole of OH^- ions:

$$\text{One equivalent of acid} = \frac{\text{formula weight of acid in grams}}{\text{number of } H^+ \text{ ions produced}}$$

$$\text{One equivalent of base} = \frac{\text{formula weight of base in grams}}{\text{number of } OH^- \text{ ions produced}}$$

One equivalent of the monoprotic acid HCl is 36.5 g, the formula weight of the acid in grams, but one equivalent of the diprotic acid H_2SO_4 is 49 g, the formula weight of the acid in grams (98 g) divided by 2.

$$\text{One equiv. HCl} = \frac{\text{form. wt. HCl}}{1} = \frac{(1.0 + 35.5)\text{ g}}{1} = 36.5\text{ g}$$

Since HCl is monoprotic . . . , we divide by 1.

$$\text{One equiv. } H_2SO_4 = \frac{\text{form. wt. } H_2SO_4}{2} = \frac{[(2 \times 1.0) + 32 + (4 \times 16)]\text{ g}}{2} = 49\text{ g}$$

Since H_2SO_4 is diprotic . . . , we divide by 2.

The advantage to using acid and base equivalents is that we often care only about the total acidity or basicity of a solution rather than about the exact structures of the acids or bases present. If we need one mole of H^+ ions for a titration, it doesn't matter whether we use HCl, HNO_3, or H_2SO_4. All that matters is that we use one equivalent of H^+. *One equivalent of any acid neutralizes one equivalent of any base.*

Because the idea of acid/base equivalents is so useful, we sometimes give acid and base concentrations in normality (N) rather than in molarity. The **normality** of an acid or base solution is defined as the number of equivalents of acid or base per liter of solution. For example, a solution made by dissolving 1.0 equivalent (36.6 g) of HCl in 1.0 L of water has a concentration of 1.0 equiv/L, or 1.0 N.

■ **Normality (N)** A measure of acid (or base) concentration expressed as the number of acid (or base) equivalents per liter of solution.

$$\text{Normality (N)} = \frac{\text{equivalents of acid or base}}{1\text{ L solution}}$$

Note that the values of molarity (M) and normality (N) are the same for monoprotic acids such as HCl but are not the same for diprotic or triprotic acids. Thus, a solution made by dissolving 1.0 equivalent (49 g; 0.50 mole) of the diprotic acid H_2SO_4 in 1.0 L of water has a *normality* of 1.0 equiv/L, or 1.0 N, but a *molarity* of 0.50 mol/L, or 0.50 M. For any acid (or base), normality is always equal to molarity times the number of H^+ (or OH^-) ions produced.

$$\text{Normality of acid} = (\text{molarity of acid}) \times (\text{number of } H^+ \text{ ions produced})$$

$$\text{Normality of base} = (\text{molarity of base}) \times (\text{number of } OH^- \text{ ions produced})$$

Solved Problem 7.7   What is the normality of the solution made by diluting 6.5 g of H_2SO_4 to a volume of 200 mL?

Solution   First calculate how many equivalents of H_2SO_4 are in 6.5 g. The formula weight of H_2SO_4 is 98, and the acid is diprotic. Thus, one equivalent of H_2SO_4 is 49 g. This value can be used as a conversion factor to find how many equivalents are in 6.5 g.

$$6.5 \text{ g } H_2SO_4 \times \frac{1 \text{ eq } H_2SO_4}{49 \text{ g } H_2SO_4} = 0.13 \text{ eq } H_2SO_4$$

Next, determine the concentration of the acid:

$$\frac{0.13 \text{ eq } H_2SO_4}{200 \text{ mL}} \times \frac{1000 \text{ mL}}{1 \text{ L}} = 0.65 \text{ eq } H_2SO_4/L = 0.65 \text{ N}$$

Thus, the concentration of the sulfuric acid solution is 0.65 N

Practice Problems

7.11   How many equivalents are in the following samples?
(a) 5.0 g HNO_3 (b) 12.5 g $Ca(OH)_2$ (c) 4.5 g H_3PO_4

7.12   What would the normalities of the solutions be if each of the samples in Problem 7.11 were dissolved in water and diluted to a volume of 300 mL?

7.11   BUFFER SOLUTIONS

Much of the body's chemistry is dependent on pH. If the pH inside cells changes only slightly from normal values, body chemistry is altered and death can result. How, though, does the body maintain the pH of various fluids with such precise control?

The pH of any fluid or solution can be kept relatively constant through the use of buffers. A **buffer** is a combination of substances that act together to prevent a drastic change in the pH of a solution. If acid is added to a buffer solution, the added hydronium ions are neutralized; if base is added to a buffer, the added hydroxide ions are neutralized.

How does a buffer work? Most common buffers are mixtures of a weak acid and the anion of that acid. For example, the *carbonate buffer* system in

■ **Buffer**   A combination of substances, usually a weak acid and its anion, that act together to prevent a large change in the pH of a solution.

blood consists of a combination of carbonic acid, H_2CO_3, and the bicarbonate ion, HCO_3^-. If too much base enters the bloodstream, it's removed by reaction with H_2CO_3 to yield more HCO_3^-. The only change in blood chemistry is a slight increase in HCO_3^- concentration, which has little effect on pH.

$$H_2CO_3(aq) + OH^-(aq) \rightleftharpoons H_2O(l) + HCO_3^-(aq)$$

A strong base . . . is converted into . . . a weak base.

If too much acid enters the bloodstream, it's removed by reaction with HCO_3^- to yield more H_2CO_3. The only change in blood chemistry is a slight increase in carbonic-acid concentration.

$$HCO_3^-(aq) + H_3O^+(aq) \rightleftharpoons H_2CO_3(aq) + H_2O(l)$$

A strong acid . . . is converted into . . . a weak acid.

As we saw in Section 7.4, carbonic acid is unstable. Thus, an additional step in the carbonate buffer system is to cleanse the blood of excess carbonic acid by chemical breakdown into carbon dioxide and water. The CO_2 is then removed from the body in breathing.

$$H_2CO_3(aq) \rightleftharpoons H_2O(l) + CO_2(g)$$

■ **Acidosis** The medical condition that results when blood pH drops below 7.35.

The regulation of blood pH by the carbonate buffer is particularly important in preventing the medical conditions called *acidosis* and *alkalosis*. **Acidosis**, the condition that results when blood pH drops below 7.35, leads in mild cases to fainting and depression of the nervous system, and in more severe cases to coma. It can be brought on either by the difficulty in breathing that accompanies asthma and emphysema or by metabolic irregularities due to fasting. Breathing difficulties lower blood pH by drastically raising blood CO_2 concentration, while fasting lowers blood pH by causing the overproduction of blood-soluble acidic substances.

■ **Alkalosis** The medical condition that results when blood pH rises above 7.45.

Alkalosis, the condition that results when blood pH rises above 7.45, leads to stimulation of the nervous system and to subsequent convulsions. Alkalosis arises either from heavy breathing (**hyperventilation**) that removes large amounts of CO_2 from the blood or from metabolic abnormalities.

■ **Hyperventilation** Heavy breathing that depletes carbon dioxide from the lungs and blood.

Another important buffer in the body is the *phosphate buffer* system present inside cells. Based on the weak acid, dihydrogen phosphate ion ($H_2PO_4^-$), and its anion, HPO_4^{2-}, the phosphate buffer system works on the same principle as the carbonate buffer. An increase in hydroxide-ion concentration is neutralized by formation of more HPO_4^{2-}, and an increase in hydronium-ion concentration is neutralized by formation of more $H_2PO_4^-$.

If base is added: $H_2PO_4^-(aq) + OH^-(aq) \rightleftharpoons HPO_4^{2-}(aq) + H_2O(l)$

If acid is added: $HPO_4^{2-}(aq) + H_3O^+(aq) \rightleftharpoons H_2PO_4^-(aq) + H_2O(l)$

Practice Problem

7.13 A mixture of HCl and NaCl is a very poor buffer system. Propose an explanation, taking into account the base strength of the Cl^- ion.

INTERLUDE: ACID RAIN

The problem of acid rain has emerged as one of the most important environmental issues of recent times. Both the causes and the effects of acid rain are well understood. The problem is what to do about it.

As the water that evaporates from oceans and lakes condenses into raindrops, it dissolves small quantities of gases from the atmosphere. Under normal conditions, rain is slightly acidic (pH 5.6) because of dissolved CO_2. In recent decades, however, the acidity of rainwater in many industrialized areas of the world has increased by a factor of over 100 to a pH of 3 to 3.5.

The culprit behind acid rain is industrial and automotive pollution. Each year in the United States and Canada, large coal-burning power plants and smelters pour millions of tons of sulfur dioxide (SO_2) gas into the atmosphere, where some is oxidized by air to produce sulfur trioxide (SO_3). Sulfur oxides then dissolve in rain to form sulfurous acid and sulfuric acid:

$$SO_2 + H_2O \longrightarrow H_2SO_3 \quad \text{Sulfurous acid}$$
$$SO_3 + H_2O \longrightarrow H_2SO_4 \quad \text{Sulfurous acid}$$

Nitrogen oxides produced both by coal-burning plants and by automobiles make a further contribution to the problem. Nitrogen dioxide (NO_2) dissolves in water to form nitric (HNO_3) and nitrous (HNO_2) acids:

$$2\,NO_2 + H_2O \longrightarrow HNO_3 + HNO_2$$

Although a pH of 3.5 doesn't sound particularly acidic, many processes in nature require such a fine balance that they are dramatically upset by the shift that has occurred in the pH of rain. Many thousands of lakes in the Adirondack region of upper New York state and in southeastern Canada have becomes so acidic that all fish life has disappeared; massive tree die-offs have occurred throughout Central Europe; and countless statues throughout Europe are being slowly dissolved away. What should be done?

This statue in a Kentucky cemetery is being rapidly eaten away by acidic rainfall.

The combustion of coal produces oxides of sulfur and nitrogen that dissolve in rainwater, giving rise to acid rain.

SUMMARY

Acids are substances that donate hydrogen ions (a proton; H^+); **bases** are substances that accept hydrogen ions. Thus, the generalized reaction of an acid with a base involves the reversible transfer of a proton:

$$A—H + :B \;\rightleftharpoons\; A^- + H—B^+$$

where: $A—H$ = a generalized acid

$:B$ = a generalized base

The reaction of an acid with a base is called a **neutralization reaction**. Reaction of an acid with a metal-hydroxide base such as KOH yields water and a **salt**; reaction with bicarbonate ion ($HCO_3{}^-$) or carbonate ion ($CO_3{}^{2-}$) yields water, a salt, and carbon dioxide gas; and reaction with ammonia yields an ammonium salt:

$$HCl(aq) + KOH(aq) \longrightarrow H_2O(l) + KCl(l) \quad \text{(a salt)}$$

$$HCl(aq) + NaHCO_3(aq) \longrightarrow H_2O(l) + NaCl(aq) + CO_2(g)$$

$$HCl(aq) + NH_3(aq) \longrightarrow NH_4Cl(aq) \quad \text{(an ammonium salt)}$$

Different acids and bases differ in their ability to give up or accept a proton. **Strong acids** give up a proton easily and are 100 percent dissociated in aqueous solution; **weak acids** give up a proton reluctantly and are only slightly **dissociated** in water. Similarly, **strong bases** accept and hold a proton readily, whereas **weak bases** have little affinity for a proton.

Acids do not spontaneously dissociate into H^+ and an anion, because a bare hydrogen ion is too unstable. In aqueous solution, however, water can accept a proton from an acid to generate a **hydronium ion, H_3O^+**. The strength of an aqueous acid solution is measured by its **pH**, which is related to its hydronium-ion concentration, $[H_3O^+]$. A pH below 7 means an acidic solution; a pH near 7 means a neutral solution; and a pH above 7 means a basic solution.

Acid and base concentrations are often expressed in units of **normality** (N) rather than molarity. The normality of a solution tells how many **equivalents** of H^+ or OH^- ions are present per liter of solution. One equivalent of an acid (or base) is the formula weight of the acid (or base) in grams, divided by the number of H^+ (or OH^-) ions that the acid (or base) produces. Acid (or base) concentrations are determined in the laboratory by **titrating** a solution of unknown concentration with a base (or acid) solution of known strength until an indicator signals that neutralization is complete.

The pH of a solution can be regulated through the use of a **buffer** that acts to remove either added H^+ ions or added OH^- ions. The **carbonate buffer** present in blood and the **phosphate buffer** present in cells are particularly important examples.

REVIEW PROBLEMS

Acids, Bases, and Salts

7.14   What are the molecular formulas of the following substances?
(a) sulfuric acid   (b) nitric acid

(c) magnesium hydroxide

7.15   What happens when a strong acid such as HBr is dissolved in water?

7.16   What happens when a weak acid such as CH_3COOH is dissolved in water?

7.17   What happens when a strong base such as KOH is dissolved in water?

7.18   What is the difference between a monoprotic acid and a diprotic acid? Give an example of each.

7.19   List four strong acids and two weak acids.

7.20   List two strong bases and two weak bases.

7.21   What do all aqueous acid solutions have in common?

7.22   Write the equilibrium expressions for the three successive dissociations of phosphoric acid, H_3PO_4, in water.

7.23   Tums, a drugstore remedy for acid indigestion, contains $CaCO_3$. Write an equation for the reaction of Tums with gastric juice (HCl).

7.24   Formulate the reaction of HNO_3 with Na_2CO_3.

7.25   Alka-Seltzer, a drugstore antacid, contains a mixture of $NaHCO_3$, aspirin, and citric acid, $C_6H_5O_7H_3$. Why do you suppose Alka-Seltzer foams and bubbles when dissolved in water?

Acid and Base Strength: pH

7.26   What does pH measure?

7.27   How is pH defined?

7.28   What is an indicator, and why is it used?

7.29   How does indicator paper help you to distinguish between acids and bases?

7.30   The electrode of a pH meter was placed in a sample of urine, and a reading of 7.9 was obtained. Is urine acidic, basic, or neutral?

7.31   Why isn't water acidic even though it dissociates to give H_3O^+?

7.32   Normal gastric juice has a pH of about 2. Assuming that gastric juice is primarily aqueous HCl, what is the HCl concentration?

7.33   How is K_w defined? What is its numerical value?

7.34   What is the approximate pH of a 0.10 M solution of a strong acid?

7.35   A 0.10 N solution of the deadly poison hydrogen cyanide, HCN, has a pH of 5.2. Is HCN acidic or basic? Is it strong or weak?

7.36   What are the H_3O^+ concentrations of solutions with the following pHs?
(a) pH $= 4$ (b) pH $= 11$ (c) pH $= 0$

7.37   What is the OH^- concentration of each solution in Problem 7.36?

7.38   Human spinal fluid has a pH of 7.4. Approximately what is the H_3O^+ concentration of spinal fluid?

7.39   Approximately what pHs do these H_3O^+ concentrations correspond to?
(a) fresh egg white: $[H_3O^+] = 2.5 \times 10^{-8}$
(b) apple cider: $[H_3O^+] = 5.0 \times 10^{-4}$

7.40   The pH of a 0.10 N ascorbic acid (vitamin C) solution is 2.6. Is ascorbic acid a weak or a strong acid? Explain.

7.41   What color would phenolphthalein indicator be at these pHs? (Figure 7.3)
(a) pH $= 4.5$ (b) pH $= 11.2$ (c) pH $= 7.1$

7.42   What color would bromothymol blue indicator be at each of the pHs given in Problem 7.41?

Buffers

7.43   What are the two components of a buffer system? Tell in your own words how a buffer works to hold pH nearly constant.

7.44   Which system would you expect to be a better buffer: $HNO_3 + NaNO_3$, or $CH_3COOH + CH_3COO^-Na^+$? Explain.

7.45   The pH of a buffer solution containing 0.1 M acetic acid and 0.1 M sodium acetate is 5.60. Write the equation for reaction of this buffer with a small amount of HNO_3.

Concentrations of Acid and Base Solutions

7.46   What does it mean when we talk about acid and base equivalents?

7.47   How is normality defined as a means of expressing acid or base concentration?

7.48   How many equivalents are in 500 mL of 0.50 M HNO_3? of 0.50 M H_3PO_4?

7.49   How many equivalents of NaOH are required to react with 0.035 equivalents of the triprotic acid, H_3PO_4?

7.50   How many mL of 0.0050 N KOH are required to neutralize 25 mL of 0.0050 N H_2SO_4? to neutralize 25 mL of 0.0050 N HCl?

7.51   Explain how you would prepare 250 mL of a 0.10 N HCl solution starting from 12.0 N HCl.

7.52   A 0.10 N solution of trisodium phosphate,

Na_3PO_4, used as a cleaning agent has a pH of 12.0. Is this solution acidic or basic?

7.53 What are the H_3O^+ and OH^- concentrations of the solution in Problem 7.52?

7.54 What is the molarity of a 0.10 N H_3PO_4 solution?

7.55 How many grams of HCN are required to make 250 mL of 0.10 N solution?

7.56 Since hydrogen cyanide is a gas, it's easier to measure by volume than by weight (Section 5.6). How many liters of HCN at STP are required to make 250 mL of 0.10 N solution?

7.57 How many equivalents are there in each of the following?
(a) 0.25 mol $Mg(OH)_2$ (b) 2.5 g $Mg(OH)_2$
(c) 15 g CH_3COOH

7.58 What is the normality of these solutions?
(a) 0.75 M H_2SO_4 (b) 0.13 M $Ba(OH)_2$
(c) 1.4 M HF

7.59 What is the normality of a solution made by dissolving 5.0 g $Ca(OH)_2$ in enough water to make 400 mL of solution?

7.60 What is the normality of a solution made by dissolving 25 g of citric acid ($C_6H_5O_7H_3$; a triprotic acid) in 750 mL water?

Additional Problems

7.61 How many mL of 0.50 N NaOH solution is required to titrate 40 mL of a 0.10 N H_2SO_4 solution to an end point?

7.62 How many equivalents do each of these following samples contain? How many milliequivalents?
(a) 20 mL of 0.015 N HNO_3
(b) 170 mL of 0.025 N H_2SO_4

7.63 How many mL of 0.12 N HCl is required to titrate 25 mL of a 0.20 N ammonia solution to an end point?

7.64 A 0.20 N solution of $NaHCO_3$ is used to titrate 50 mL of an acetic acid solution of unknown concentration. If 37.2 mL of $NaHCO_3$ is required to reach the end point, what is the concentration of the acetic acid solution?

7.65 A 0.15 N solution of HCl is used to titrate 30 mL of a $Ca(OH)_2$ solution of unknown concentration. If 140 mL of HCl is required, what is the concentration of the $Ca(OH)_2$ solution?

7.66 Why doesn't pure water conduct electricity, even though it dissociates into H_3O^+ and OH^- ions?

7.67 Which solution contains more acid: 50 mL of a 0.20 N HCl solution, or 50 mL of a 0.20 N acetic acid solution? Which has a higher hydronium-ion concentration? Which has the lower pH?

7.68 A 0.010 N solution of aspirin has pH = 3.3. Is aspirin a strong or a weak acid?

7.69 Oxalic acid, **HOOCCOOH**, is a toxic diprotic acid found in spinach leaves. What is the normality of a solution made by dissolving 12.0 g of oxalic acid in 400 mL of water?

7.70 How much 0.10 N KOH would be required to titrate 25 mL of the oxalic acid solution made in Problem 7.69 to an end point?

CHAPTER 8

Introduction to Organic Chemistry: Alkanes

Drilling for offshore oil in the ocean. Crude oil is a mixture of many different hydrocarbons—mostly alkanes, the compounds you'll be reading about in this chapter. From this mixture a great variety of petroleum products can be refined.

As knowledge of chemistry slowly evolved in the 1700s, mysterious differences were noted between compounds obtained from animals and those from minerals. Chemicals from animal sources were often more difficult to isolate, to purify, and to work with than those from mineral sources. To express this difference, the term *organic chemistry* was introduced to mean the study of compounds from living organisms, while *inorganic chemistry* was used to refer to the study of compounds from minerals.

Today we know that there aren't any fundamental differences between organic and inorganic compounds; the same scientific principles are applicable to both. The only common characteristic of compounds from living sources is that all contain the element carbon. Thus, *organic chemistry* is now defined as the study of carbon compounds.

Why is carbon special? The answer to this question involves the unique ability of carbon atoms to bond together, forming long chains and rings. Of all the elements, only carbon is able to form an immense array of compounds, from methane with one carbon atom, to deoxyribonucleic acid (DNA) with tens of billions of carbon atoms. In this and the next four chapters, we'll look at the chemistry of organic compounds, beginning with an exploration of these topics:

1. **What are the structures of organic molecules?** The goal: you should learn the importance of chemical structure in organic chemistry and should become familiar with the idea of isomers.

2. **How are organic molecules classified?** The goal: you should learn how organic molecules are classified into functional-group families.

3. **What shapes do organic molecules have?** The goal: you should learn the importance of three-dimensionality to organic molecules.

4. **How are organic molecules depicted?** The goal: you should become familiar with the shorthand conventions for portraying organic molecules.

5. **How are organic molecules named?** The goal: you should become familiar with the rules for naming simple organic compounds.

6. **What is the chemistry of alkanes?** The goal: you should become familiar with the chemical behavior of alkanes.

8.1 THE NATURE OF ORGANIC MOLECULES

Let's review what we've seen in earlier chapters about the structures of organic molecules:

1. Carbon is always *tetravalent*; it always forms four bonds (Section 3.7). In methane, carbon is connected to four hydrogen atoms:

$$
\text{Methane, CH}_4 \qquad
\begin{array}{c}
\quad\;\; \text{H} \\
\quad\;\; | \\
\text{H} - \text{C} - \text{H} \\
\quad\;\; | \\
\quad\;\; \text{H}
\end{array}
$$

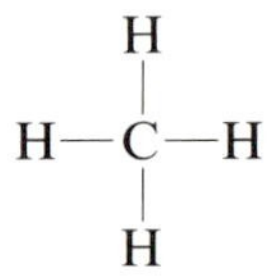

2. Organic molecules contain *covalent bonds* (Section 3.8). In ethane, the bonds result from the sharing of two electrons, either between C and C or between C and H:

$$
\text{Ethane, C}_2\text{H}_6 \qquad
\begin{array}{c}
\text{H}\;\,\text{H} \\
\text{H}\!:\!\ddot{\text{C}}\!:\!\ddot{\text{C}}\;\text{H} \\
\text{H}\;\,\text{H}
\end{array}
\;=\;
\begin{array}{c}
\text{H}\;\;\;\text{H} \\
|\quad\; | \\
\text{H}-\text{C}-\text{C}-\text{H} \\
|\quad\; | \\
\text{H}\;\;\;\text{H}
\end{array}
$$

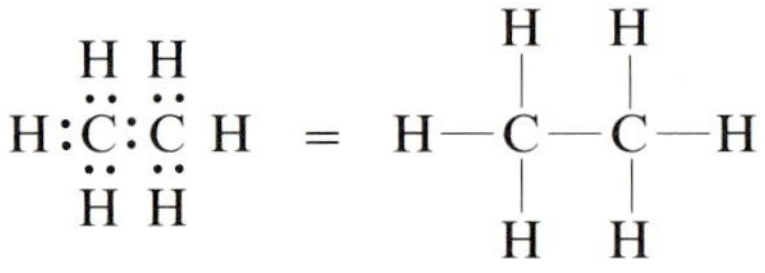

3. Organic molecules contain *polar covalent bonds* when carbon bonds to an element on the far right or far left of the periodic table (Section 6.6). In chloromethane, the electronegative chlorine atom attracts electrons more strongly than carbon, resulting in polarization of the carbon-chlorine bond so that carbon has a partial positive charge, δ^+:

$$
\text{Chloromethane, CH}_3\text{Cl} \qquad
\begin{array}{c}
\text{H} \\
\text{H}\!:\!\ddot{\text{C}}\!:\!\text{Cl} \\
\text{H}
\end{array}
\;=\;
\begin{array}{c}
\quad\;\; \text{H} \\
\quad\;\; | \\
\text{H}-\text{C}^{\delta^+}\!-\text{Cl}^{\delta^-} \\
\quad\;\; | \\
\quad\;\; \text{H}
\end{array}
$$

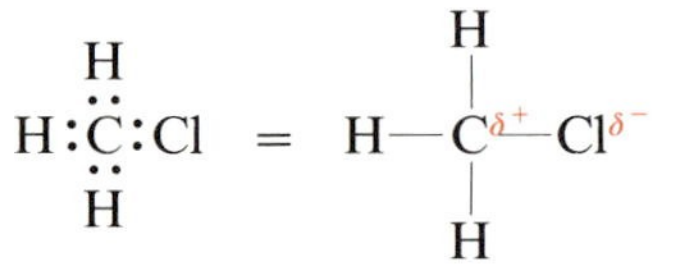

In methyllithium, the lithium attracts electrons less strongly than carbon, resulting in polarization of the carbon-lithium bond so that carbon has a partial negative charge, δ^-:

$$
\text{Methyllithium, CH}_3\text{Li} \qquad
\begin{array}{c}
\text{H} \\
\text{H}\!:\!\ddot{\text{C}}\!:\!\text{Li} \\
\text{H}
\end{array}
\;=\;
\begin{array}{c}
\quad\;\; \text{H} \\
\quad\;\; | \\
\text{H}-\text{C}^{\delta^-}\!-\text{Li}^{\delta^+} \\
\quad\;\; | \\
\quad\;\; \text{H}
\end{array}
$$

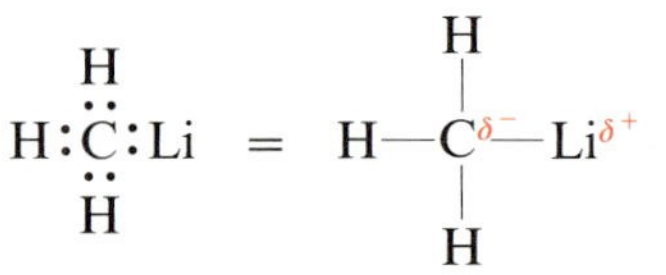

4. Carbon can form *multiple covalent bonds* by sharing more than two electrons with a neighboring atom (Section 3.8). In ethylene, the two carbon

atoms share four electrons in a double bond; in acetylene, the two carbons share six electrons in a triple bond:

Ethylene, C_2H_4

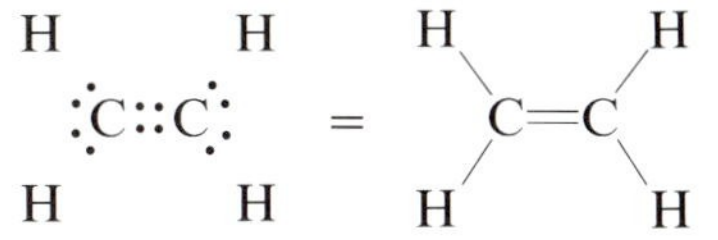

Acetylene, C_2H_2 H:C:::C H = H—C≡C—H

5. Covalently bonded molecules have specific three-dimensional *shapes* (Section 3.9). When carbon is bonded to four atoms as in methane, CH_4, the bonds are oriented toward the four corners of an imaginary tetrahedron with carbon in the center:

Methane (tetrahedral geometry)

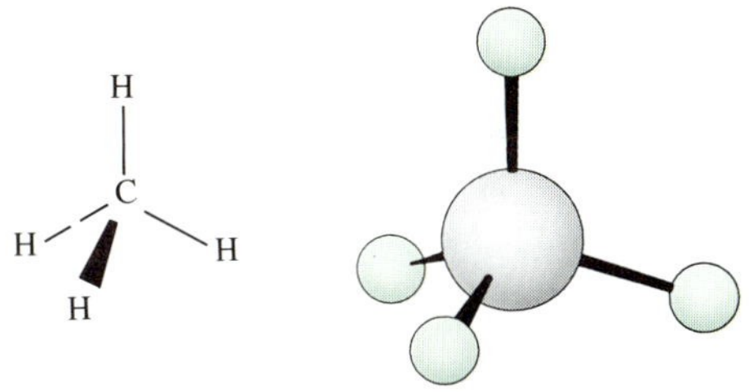

As a result of their covalent bonding, organic compounds have properties quite different from those of inorganic salts. For example, many inorganic compounds have high melting points and boiling points because they consist of large collections of oppositely charged ions held together by strong electrical attractions (Section 3.2). Organic compounds, by contrast, consist of small, individual molecules held together by covalent bonds. The forces between molecules are fairly weak, and organic compounds therefore have lower melting and boiling points. In fact, many simple organic compounds are liquid at room temperature.

Solubility and electrical conductivity are other important differences between organic and inorganic compounds. Whereas many inorganic salts dissolve in water to yield ionic solutions capable of conducting electricity (Section 6.9), most organic compounds are insoluble in water and do not conduct electricity. Water dissolves ionic compounds by solvation of ions but cannot dissolve the great majority of nonpolar organic compounds. Only a few small polar organic molecules such as glucose and ethyl alcohol dissolve in water. This lack of water solubility has important practical consequences in digestion where fats and other nonpolar organic molecules consumed in food must be chemically broken down and made soluble in water before they can be used by the body. In diseases such as sprue, fats are not digested properly and therefore can't be absorbed by the body.

8.2 FAMILIES OF ORGANIC MOLECULES: FUNCTIONAL GROUPS

At last count, there were more than nine million organic compounds described in the scientific literature. Each has unique physical properties such as melting point and boiling point, and each has unique chemical properties. The situation isn't as hopeless as it sounds, however, because chemists have learned through experience that organic compounds can be classified into families according to their structural features, and that the chemical behavior of the members of a family is often predictable. Instead of nine million compounds with random chemical reactivity, there are just a few general families of organic compounds whose chemistry is predictable.

The structural features that allow us to class compounds together are called functional groups. A **functional group** is a part of a larger molecule and is composed of an atom or group of atoms that has characteristic chemical behavior. A given functional group undergoes the same reactions in every molecule it's a part of. For example, the carbon-carbon double bond is one of the simplest functional groups. Thus, ethylene (C_2H_4), the simplest compound with a double bond, undergoes many chemical reactions similar to those of cholesterol ($C_{27}H_{46}O$), a far larger and more complex compound. Both, for example, react with hydrogen in the same manner (Figure 8.1).

The example shown in Figure 8.1 is typical; the chemistry of an organic molecule, regardless of size and complexity, is determined by the functional

■ **Functional group** A part of a larger molecule composed of an atom or group of atoms that has characteristic chemical behavior.

Figure 8.1
The reactions of (a) ethylene and (b) cholesterol with hydrogen. The carbon-carbon double bond functional group adds two hydrogen atoms in both cases, regardless of the complexity of the rest of the molecule.

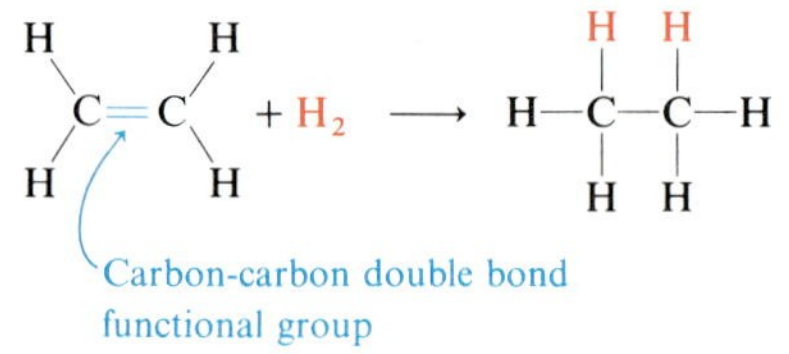

(a) Ethylene

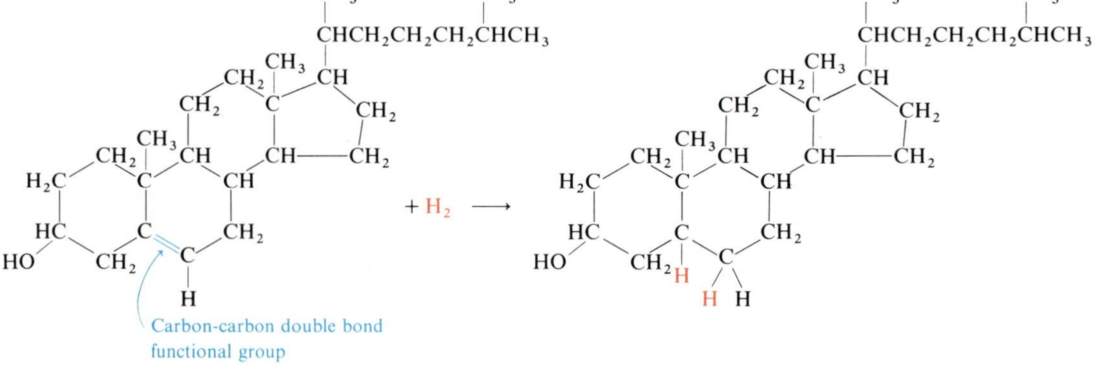

(b) Cholesterol

groups it contains. Table 8.1 lists some of the most common functional groups found in organic molecules and gives examples of their occurrence. Much of the chemistry we'll study in the next four chapters is the chemistry of these functional groups, so it's best to memorize their names and structures now.

Table 8.1   Some Important Families of Organic Molecules

Family Name	Functional Group Structure[a]	Simple Example	Name Ending
Alkane	(contains only C—H and C—C single bonds)	CH_3CH_3 ethane	-ane
Alkene	C=C	$H_2C{=}CH_2$ ethylene	-ene
Alkyne	—C≡C—	H—C≡C—H acetylene (ethyne)	-yne
Arene	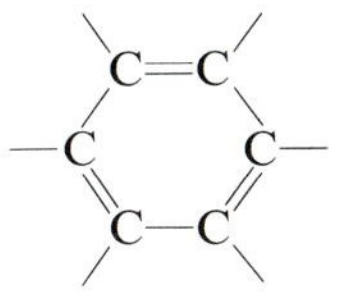 C=C / —C C— / C—C (benzene ring)	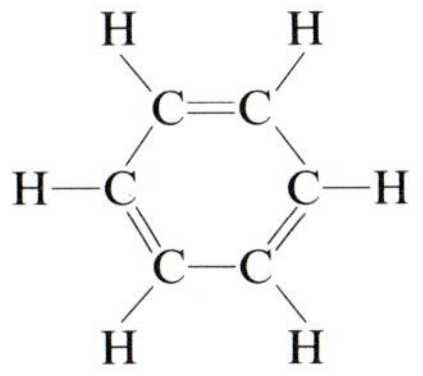 benzene	none
Alcohol	—C—O—H	CH_3—OH methyl alcohol (methanol)	-ol
Ether	—C—O—C—	CH_3—O—CH_3 dimethyl ether	none
Amine	—N—H, —N—H, —N— with H	CH_3—NH_2 methylamine	-amine
Aldehyde	—C(=O)—H	CH_3—C(=O)—H acetaldehyde (ethanal)	-al
Ketone	C—C(=O)—C	CH_3—C(=O)—CH_3 acetone	-one
Carboxylic acid	—C(=O)—OH	CH_3—C(=O)—OH acetic acid	-ic acid
Ester	—C(=O)—O—	CH_3—C(=O)—O—CH_3 methyl acetate	-ate
Amide	—C(=O)—NH_2, —C(=O)—N—H, —C(=O)—N—	CH_3—C(=O)—NH_2 acetamide	-amide

[a] The bonds whose connections aren't specified are assumed to be attached to carbon or hydrogen atoms in the rest of the molecule.

Practice Problems   **8.1**   Locate and identify the functional groups in these molecules:
(a) lactic acid, from sour milk   (b) styrene, used to make polystyrene

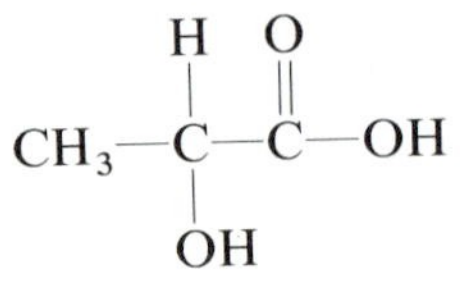

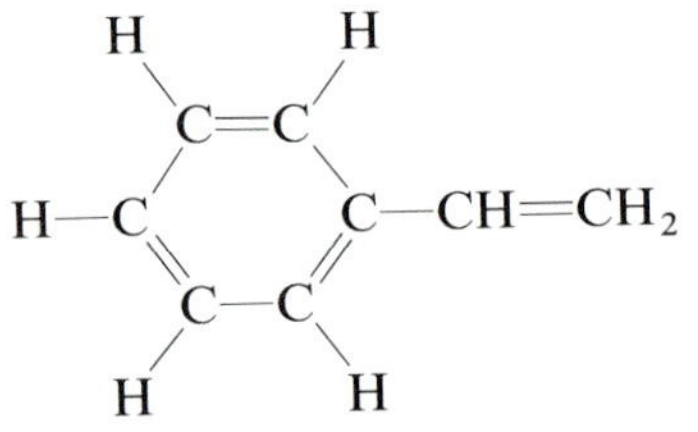

$$CH_3-\underset{\underset{\displaystyle OH}{|}}{\overset{\overset{\displaystyle H}{|}}{C}}-\overset{\overset{\displaystyle O}{\|}}{C}-OH$$

$$H-C\overset{\overset{\displaystyle H}{|}}{\underset{}{=}}C-CH=CH_2$$

8.2   Propose structures for molecules that fit these descriptions:
(a) C_2H_4O containing an aldehyde functional group
(b) $C_3H_6O_2$ containing a carboxylic acid functional group

8.3   THE STRUCTURE OF ORGANIC MOLECULES: ALKANES AND THEIR ISOMERS

How can there be so many different organic compounds? The answer is simply that a relatively small number of atoms can bond together in a great many different ways to form a great many different compounds. Take molecules that contain only carbon and hydrogen and that have only single bonds, for example. Such compounds belong to the family of organic molecules called **alkanes**.

If we imagine ways that one carbon and four hydrogens can combine, there is only one possibility: methane, CH_4. If we imagine ways that two carbons and six hydrogens can combine, only ethane, CH_3CH_3, is possible; and if we imagine the combination of three carbons with eight hydrogens, only propane, $CH_3CH_2CH_3$, is possible. (Remember from Section 3.7 that elements form a specific number of bonds in covalent molecules. Carbon forms four bonds, and hydrogen forms one.)

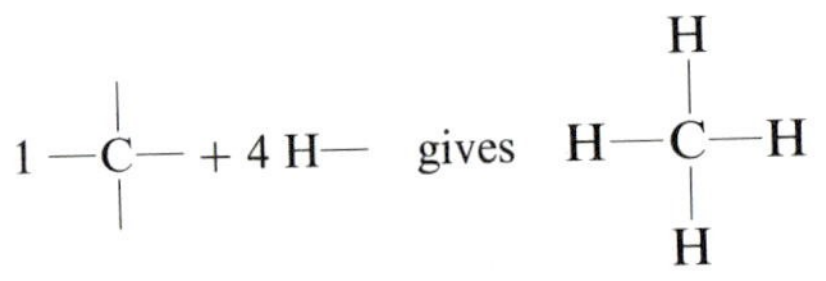

■ **Alkane**   A molecule that contains only carbon and hydrogen and that has only single bonds.

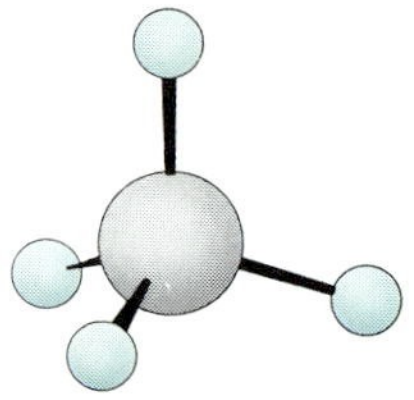

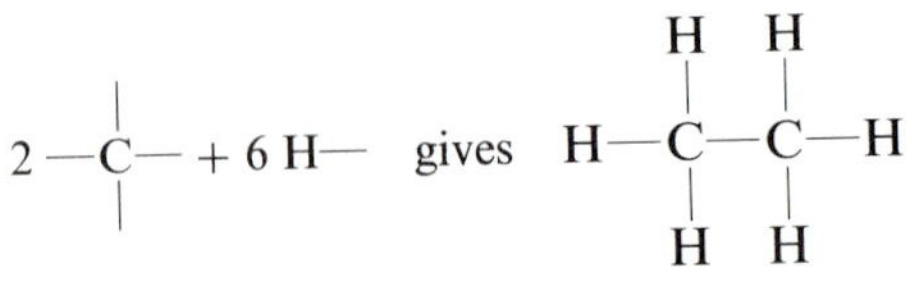

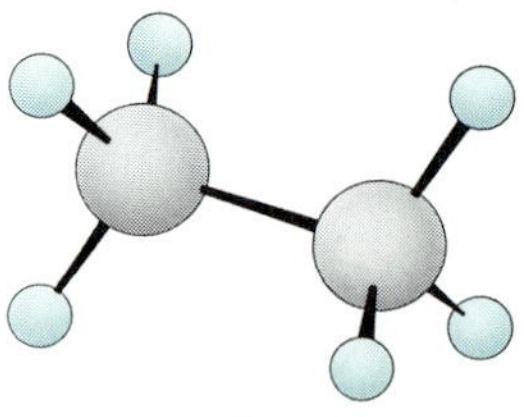

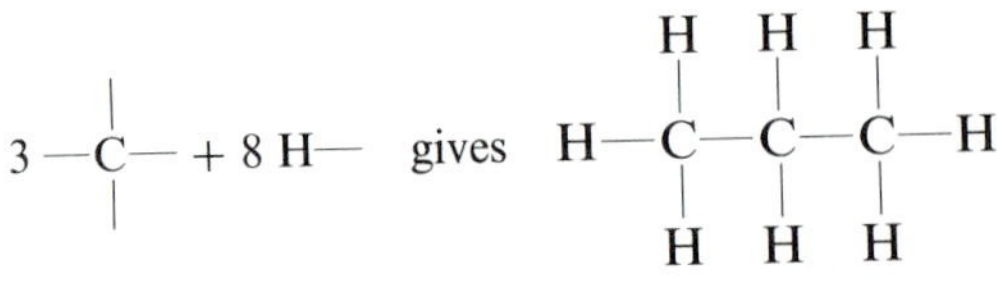

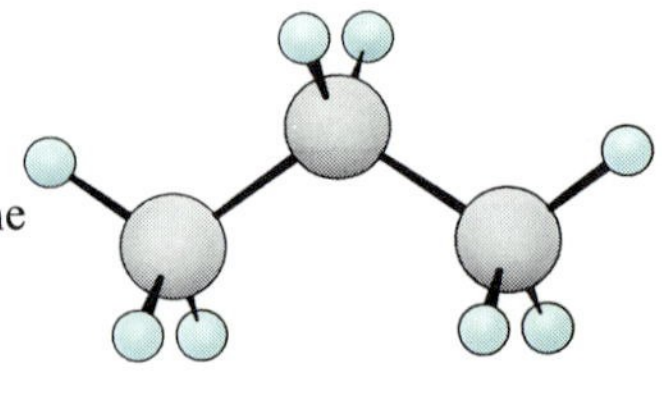

If larger numbers of carbons and hydrogens combine, *more than one kind of molecule can be formed*. There are two ways in which molecules with the formula C_4H_{10} can be formed: The four carbons can either be in a row, or they can have a branched arrangement. Similarly, there are *three* ways in which molecules with the formula C_5H_{12} can be formed; and so on for larger alkanes. Compounds with all their carbons connected in a row are called **straight-chain alkanes**, whereas those with a branching connection of carbons are called **branched-chain alkanes**. Note that in a straight-chain alkane, you can draw a line through all the carbon atoms without lifting your pencil from the paper. In a branched-chain alkane, however, you must either lift your pencil from the paper or retrace your steps in order to draw a line through all carbons.

■ **Straight-chain alkane** An alkane that has all its carbon atoms connected in a row

■ **Branched-chain alkane** An alkane that has a branching connection of carbon atoms along its chain.

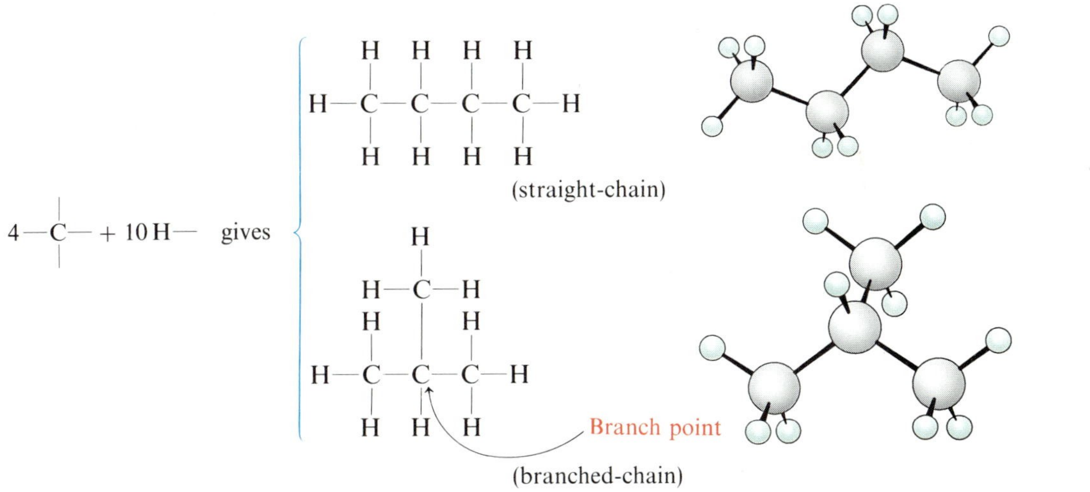

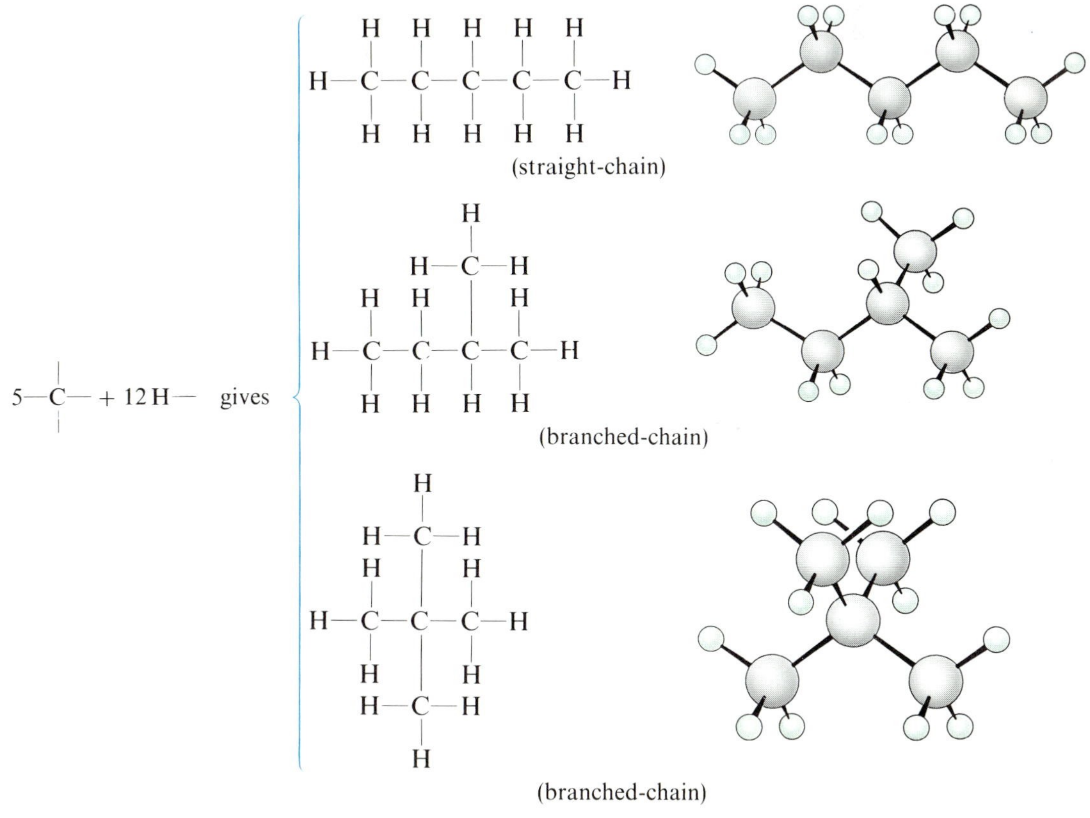

(straight-chain)

(branched-chain)

Branch point

$4 -\!\!C\!\!- + 10\,H\!\!-$ gives

(straight-chain)

(branched-chain)

(branched-chain)

$5 -\!\!C\!\!- + 12\,H\!\!-$ gives

Table 8.2   Numbers of Possible Alkane Isomers

Formula	Number of Isomers	Formula	Number of Isomers
C_6H_{14}	5	$C_{10}H_{22}$	75
C_7H_{16}	9	$C_{20}H_{42}$	366,319
C_8H_{18}	18	$C_{30}H_{62}$	4,111,846,763
C_9H_{20}	35	$C_{40}H_{82}$	62,491,178,805,831

■ **Isomers** Compounds with the same molecular formula but with different connections between atoms.

Compounds like the two different molecules with formula C_3H_8 and the three different molecules with formula C_4H_{10} are called *isomers* (**eye**-so-mers). **Isomers** are compounds that have the same molecular formula but have different connections between their atoms. As Table 8.2 shows, the number of possible alkane isomers grows rapidly as the number of carbon atoms increases.

It's important to realize that different isomers are completely different chemical compounds. They have different structures, different physical properties such as melting point and boiling point, and potentially different physiological properties. For example, ethyl alcohol and dimethyl ether both have the formula

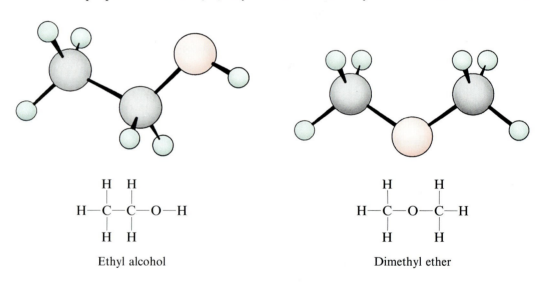

Ethyl alcohol   Dimethyl ether

Table 8.3   Some Properties of Ethyl Alcohol and Dimethyl Ether

Name	Formula	Structure	Boiling Point	Melting Point	Physiological Activity
Ethyl alcohol	C_2H_6O	H—C—C—O—H	78.5°C	−117.3°C	Central-nervous-system depressant
Dimethyl ether	C_2H_6O	H—C—O—C—H	−23°C	−138.5°C	General anesthetic

C_2H_6O. Yet ethyl alcohol is a liquid central-nervous-system depressant with a boiling point of 78.5°C, whereas dimethyl ether is a gaseous general anesthetic with a boiling point of −23°C (Table 8.3).

Practice Problem   **8.3**   Draw the straight-chain isomer with the formula C_7H_{16}.

8.4 WRITING ORGANIC STRUCTURES

■ **Condensed structures** A shorthand way of drawing an organic structure in which carbon-carbon and carbon-hydrogen bonds are "understood" rather than explicitly shown.

As you might imagine, it's both time-consuming and awkward to draw all the bonds and all the atoms in organic compounds, even for relatively small molecules like C_4H_{10}. Thus, a shorthand way of drawing condensed structures is often used. In **condensed structures**, carbon-hydrogen and carbon-carbon single bonds aren't shown; rather, they're "understood." If a carbon atom has three hydrogens bonded to it, we write CH_3; if the carbon has two hydrogens bonded to it, we write CH_2; and so on. For example, the four-carbon straight-chain compound (called *butane*) and its branched-chain isomer (called *2-methylpropane*) can be written as the following condensed structures:

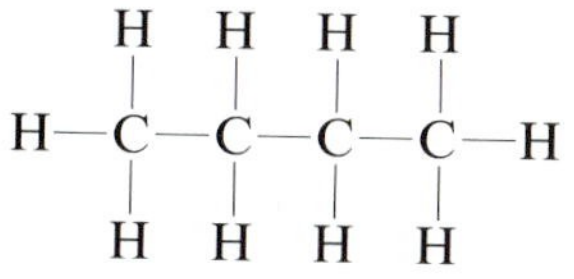

$$\text{Butane} \qquad CH_3CH_2CH_2CH_3$$

$$\text{2-Methylpropane} \qquad CH_3CHCH_3$$

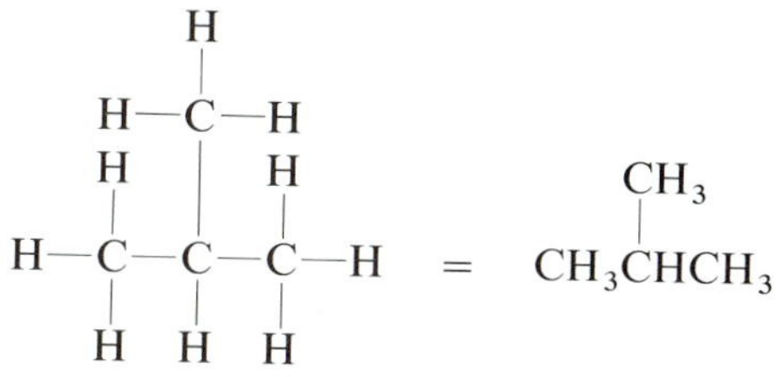

Note that the horizontal bonds between carbons aren't shown—the CH_3 and CH_2 units are simply placed next to each other—but that the vertical bond in 2-methylpropane is shown for clarity.

Practice Problem   **8.4**   Draw the three isomers of C_5H_{12} as condensed structures.

(a) 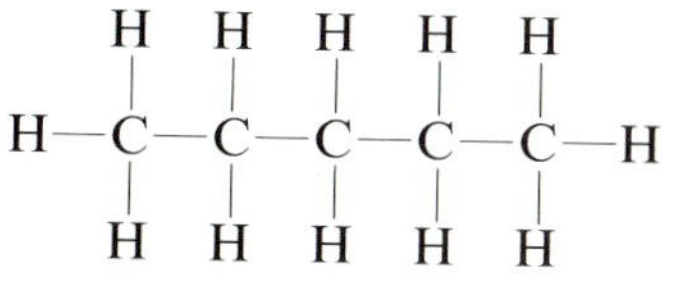 Pentane

(b) 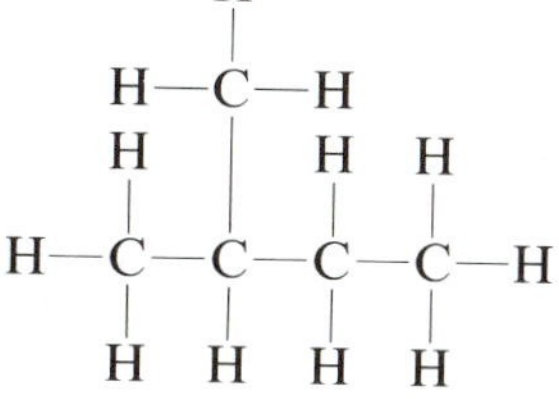 2-Methylbutane

(c) 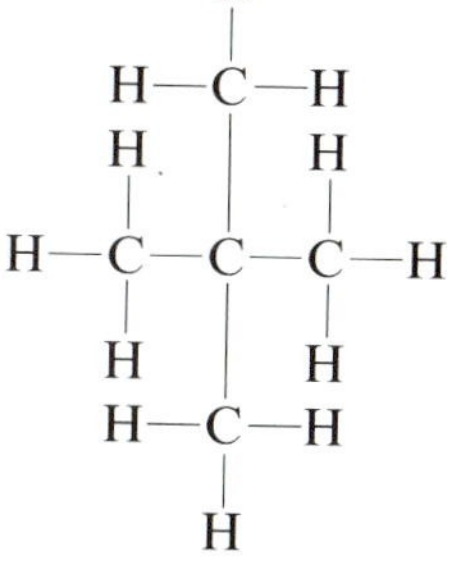 2,2-Dimethylpropane

8.5 THE SHAPES OF ORGANIC MOLECULES

Although every carbon atom in an alkane has its four bonds pointing toward the four corners of a tetrahedron, chemists don't usually worry about exact three-dimensional shapes when writing the condensed structure of an organic molecule. Thus, a molecule can be arbitrarily shown in a great many ways when its structure is written. For example, the straight-chain, four-carbon alkane butane might be represented by any of the structures shown in Figure 8.2. These structures don't imply any particular three-dimensional shape for butane; they only indicate the *connections* between atoms without specifying geometry.

In fact, butane has no one single shape because *rotation* is possible around carbon-carbon single bonds. The two parts of a molecule joined by a carbon-carbon single bond are free to spin around the bond, giving rise to an infinite number of possible three-dimensional structures, or **conformations**. A given butane molecule might be fully extended at one instant [Figure 8.3 (a)] but be

■ **Conformation** The exact three-dimensional structure of a molecule.

AN APPLICATION: DISPLAYING MOLECULAR SHAPES

Molecular shape is critical to the proper functioning of all biological molecules. The tiniest difference in shape between two compounds can cause them to behave differently or to have different physiological effects in the body. It's therefore critical that chemists have techniques available both for determining molecular shape with great precision and for visualizing those shapes in useful and manageable ways.

Three-dimensional shapes of molecules are determined by *X-ray crystallography*, a technique that allows us to "see" molecules in a crystal using

X-ray waves rather than light waves. The molecular "picture" obtained by X-ray crystallography looks at first like a series of regularly spaced dark spots on a photographic film. After computerized manipulation of the data, however, recognizable molecules like those in the following figure emerge. Relatively small molecules like morphine are usually displayed on paper, but enormous biological molecules like enzymes are best displayed on computer terminals, where their structures can be enlarged, rotated, and otherwise manipulated for the best view.

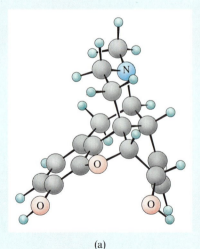

(a)

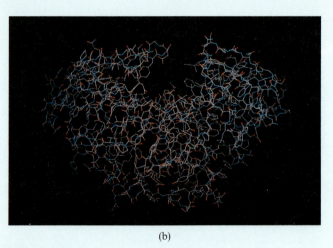

(b)

Computer-generated structures of (a) morphine and (b) penicillopepsin enzyme taken from X-ray crystallographic data.

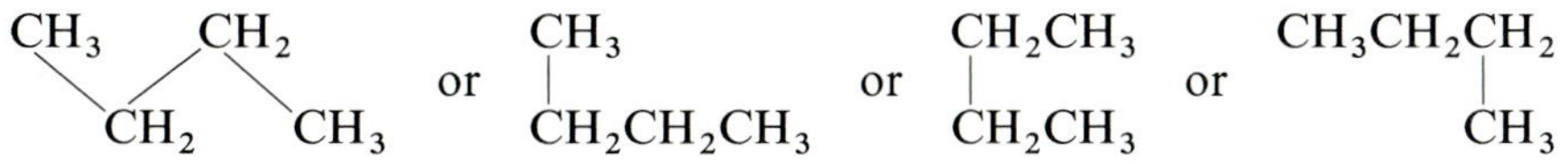

$$CH_3 \quad CH_2 \qquad CH_3 \qquad\qquad CH_2CH_3 \qquad CH_3CH_2CH_2$$
$$\diagdown \;\; \diagup \qquad \text{or} \qquad | \qquad \text{or} \qquad | \qquad \text{or} \qquad |$$
$$CH_2 \quad CH_3 \qquad CH_2CH_2CH_3 \qquad CH_2CH_3 \qquad\qquad CH_3$$

Figure 8.2
Some representations of butane, C_4H_{10}. The molecule is the same regardless of how it's drawn. These structures imply only that butane has a continuous chain of four carbon atoms.

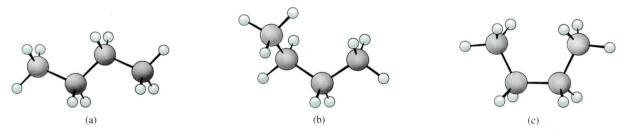

(a)   (b)   (c)

Figure 8.3
Some possible conformations of butane. There are many other conformations as well.

twisted an instant later [Figure 8.3 (b) or 8.3 (c)]. An actual sample of butane contains a great many molecules that are constantly changing their shape. At any given instant, however, most of the molecules have the less crowded extended conformation shown in Figure 8.3 (a). The same is true for all other alkanes.

Solved Problem 8.1   The following molecules have the same formula, C_7H_{16}. Which of the structures represent the same molecule?

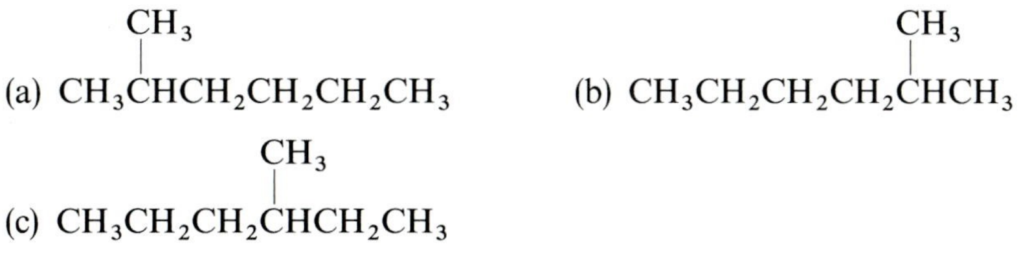

$$\qquad\qquad CH_3 \qquad\qquad\qquad\qquad\qquad\qquad\qquad\qquad CH_3$$
$$\qquad\qquad | \qquad\qquad\qquad\qquad\qquad\qquad\qquad\qquad\qquad |$$
(a) $CH_3CHCH_2CH_2CH_2CH_3$ (b) $CH_3CH_2CH_2CH_2CHCH_3$

$$\qquad\qquad\qquad\qquad CH_3$$
$$\qquad\qquad\qquad\qquad |$$
(c) $CH_3CH_2CH_2CHCH_2CH_3$

Solution   The important point in determining whether two structures are identical is to pay attention to the order of connection between atoms. Don't get confused by the apparent differences caused by writing a structure right-to-left versus left-to-right. In this example, molecule (a) has a straight chain of six carbons with a $-CH_3$ branch on the second carbon from the end. Molecule (b) also has a straight chain of six carbons with a $-CH_3$ branch on the second carbon from the end and is therefore identical to (a). The only difference between (a) and (b) is that one is written "forward" and one is written "backward." Molecule (c), by contrast, has a straight chain of six carbons with a $-CH_3$ branch on the *third* carbon from the end and is therefore an isomer of (a) and (b).

Practice Problems

8.5 Which of the following three structures are identical?

(a)
$$CH_3 \quad CH_3$$
$$CH_2CH_2CHCH_2CH_3$$

(b)
$$CH_3$$
$$CH_3CH_2CH_2CCH_3$$
$$CH_3$$

(c)
$$CH_3$$
$$CH_3CH_2CHCH_2CH_2CH_3$$

8.6 According to Table 8.2, there are five isomers with the formula C_6H_{14}. Draw as many as you can.

8.7 Using toothpicks for bonds and two marshmallows to represent the carbon atoms, build a model of an ethane molecule, CH_3CH_3. (Hydrogen atoms can be "understood" to be at the ends of the toothpicks coming out of the marshmallows.) Rotate the model around a C—C bond, and note the changing relationships between hydrogens on neighboring carbons.

8.6 NAMING ALKANES

In earlier times when relatively few pure organic chemicals were known, new compounds were named at the whim of their discoverer. Thus, urea is a crystalline substance first isolated from urine, and the barbiturates are a group of tranquilizing agents named by their discoverer in honor of his friend, Barbara. As more and more compounds became known, however, the need for **nomenclature**—a systematic method of naming organic compounds—became apparent.

■ **Nomenclature** The system for naming molecules.

The system of nomenclature now used is that devised by the International Union of Pure and Applied Chemistry—the IUPAC system (pronounced as **eye**-you-pac). In the IUPAC system, a chemical name has three parts: parent, suffix, and prefix. The parent name specifies the overall size of the molecule by telling how many carbon atoms are present in the longest continuous chain; the suffix identifies what family the molecule belongs to; and the prefix specifies the location of the functional groups and other substituents on the chain:

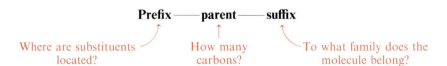

Prefix —— **parent** —— **suffix**

Where are substituents located?   How many carbons?   To what family does the molecule belong?

Straight-chain alkanes are named simply by counting the number of carbon atoms in the chain and adding the family suffix *-ane*. With the exception of the first four compounds—methane, ethane, propane, and butane—whose names have historical origins, the alkanes are named from Greek numbers according to the number of carbons present. Thus, *pent*ane is the five-carbon alkane, *hex*ane is the six-carbon alkane, and so on as shown in Table 8.4.

Table 8.4   Names of Straight-Chain Alkanes

No. of Carbons	Structure	Name	No. of carbons	Structure	Name
1	CH_4	*Methane*	6	$CH_3CH_2CH_2CH_2CH_2CH_3$	*Hexane*
2	CH_3CH_3	*Ethane*	7	$CH_3CH_2CH_2CH_2CH_2CH_2CH_3$	*Heptane*
3	$CH_3CH_2CH_3$	*Propane*	8	$CH_3CH_2CH_2CH_2CH_2CH_2CH_2CH_3$	*Octane*
4	$CH_3CH_2CH_2CH_3$	*Butane*	9	$CH_3CH_2CH_2CH_2CH_2CH_2CH_2CH_2CH_3$	*Nonane*
5	$CH_3CH_2CH_2CH_2CH_3$	*Pentane*	10	$CH_3CH_2CH_2CH_2CH_2CH_2CH_2CH_2CH_2CH_3$	*Decane*

Branched chain alkanes are named by the following four steps.

Step 1. Name the main chain.   Find the *longest continuous chain of carbons* present in the molecule, and use the name of that chain as the parent name. The longest chain may not always be obvious from the manner of writing; you may have to "turn corners" to find it.

$$CH_3-CH_2$$
$$CH_3-CH-CH_2-CH_3$$

Named as a substituted pentane, not as a substituted butane, because the *longest* chain has five carbons.

Step 2. Number the carbon atoms in the main chain.   Beginning at the end nearer the first branch point, number each carbon atom in the parent chain:

$$CH_3$$
$$CH_3-CH-CH_2-CH_2-CH_3$$
$$1 \quad 2 \quad 3 \quad 4 \quad 5$$

The first (and only) branch occurs at C2 if we start numbering from the left, but would occur at C4 if we started from the right by mistake.

Step 3. Identify and number the branching substituent.   Assign a number to each branching substituent on the parent chain according to its point of attachment.

$$CH_3$$
$$CH_3-CH_2-CH_2-CH-CH_3$$
$$5 \quad 4 \quad 3 \quad 2 \quad 1$$

The main chain is a pentane. There is a —CH_3 substituent group connected to C2 of the chain.

If there are two substituents on the same carbon, assign the same number to both of them. There must always be as many numbers in the name as there are substituents.

$$CH_2-CH_3$$
$$CH_3-CH_2-C-CH_2-CH_2-CH_3$$
$$1 \quad 2 \quad 3 \quad 4 \quad 5 \quad 6$$
$$CH_3$$

The main chain is a hexane. There are two substituents, a—CH_3 and a—CH_2CH_3, both connected to C3 of the chain.

■ **Alkyl group**   The part of an alkane that remains when one hydrogen atom is removed.

The —CH_3 and —CH_2CH_3 substituents that branch off the main chain in the previous two compounds are called **alkyl groups**. An alkyl group can be thought of as the part of an alkane that remains when one hydrogen is removed.

■ **Methyl group** —CH_3, the alkyl group derived from methane.

■ **Ethyl group** —CH_2CH_3, the alkyl group derived from ethane.

For example, removal of one hydrogen from methane gives the **methyl group**, —CH_3, and removal of one hydrogen from ethane gives the **ethyl group**, —CH_2CH_3. Notice that these alkyl groups are named simply by replacing the -*ane* ending of the parent alkane with a -*yl* ending.

Methane

$$H-\underset{\underset{H}{|}}{\overset{\overset{H}{|}}{C}}-H \xrightarrow{\text{remove one H}} -\underset{\underset{H}{|}}{\overset{\overset{H}{|}}{C}}-H = -CH_3 \quad \text{(a methyl group)}$$

Ethane

$$H-\underset{\underset{H}{|}}{\overset{\overset{H}{|}}{C}}-\underset{\underset{H}{|}}{\overset{\overset{H}{|}}{C}}-H \xrightarrow{\text{remove one H}} -\underset{\underset{H}{|}}{\overset{\overset{H}{|}}{C}}-\underset{\underset{H}{|}}{\overset{\overset{H}{|}}{C}}-H = -CH_2CH_2 \quad \text{(an ethyl group)}$$

Methane and ethane are special because each has only one kind of hydrogen. It doesn't matter which of the four methane hydrogens is removed because all four are equivalent. Thus, there is only one possible kind of methyl group. Similarly, it doesn't matter which of the six ethane hydrogens is removed because all six are equivalent and only one kind of ethyl group is possible.

The situation is more complex for larger alkanes that contain more than one kind of hydrogen. For example, propane has two different kinds of hydrogens, whose removal can give two different kinds of propyl groups. Removal of any one of the six hydrogens attached to an end carbon yields a straight-chain propyl group called ***n*-propyl**, whereas removal of one of the two hydrogens attached to the central carbon yields a branched-chain propyl group called **isopropyl**. (The "*n*" prefix in *n*-propyl stands for *normal*, meaning straight-chain.)

■ ***n*-Propyl group** —$CH_2CH_2CH_3$, the alkyl group derived by removing a hydrogen atom from an end carbon of propane.

■ **Isopropyl group** —$CH(CH_3)_2$, the alkyl group derived by removing a hydrogen atom from the central carbon of propane.

$$-\underset{\underset{H}{|}}{\overset{\overset{H}{|}}{C}}-\underset{\underset{H}{|}}{\overset{\overset{H}{|}}{C}}-\underset{\underset{H}{|}}{\overset{\overset{H}{|}}{C}}-H = -CH_2CH_2CH_3$$

n-Propyl group (straight-chain)

Propane

$$H-\underset{\underset{H}{|}}{\overset{\overset{H}{|}}{C}}-\underset{\underset{H}{|}}{\overset{\overset{H}{|}}{C}}-\underset{\underset{H}{|}}{\overset{\overset{H}{|}}{C}}-H$$

remove H from outside carbon

remove H from inside carbon

$$H-\underset{\underset{H}{|}}{\overset{\overset{H}{|}}{C}}-\underset{\underset{H}{|}}{\overset{}{C}}-\underset{\underset{H}{|}}{\overset{\overset{H}{|}}{C}}-H = CH_3CHCH_3 \text{ or } (CH_3)_2CH-$$

Isopropyl group (branched-chain)

It's important to realize that alkyl groups themselves are not compounds and that the "removal" of a hydrogen from an alkane is just a way of looking at things, not a chemical reaction. Alkyl groups are simply parts of molecules that help us to name compounds.

Step 4. Write the name as a single word. Use hyphens to separate the different prefixes, and use commas to separate numbers if necessary. If two or more different substituent groups are present, cite them in alphabetical order. If two or more identical substituents are present, use one of the prefixes *di-*, *tri-*, *tetra-*, and so forth, but don't use these prefixes for alphabetizing purposes.

Let's look at some examples:

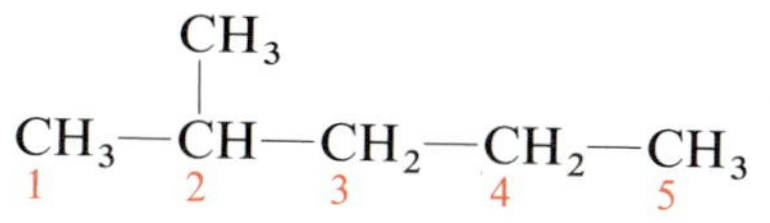

$$CH_3-\underset{2}{\underset{|}{\overset{CH_3}{C}H}}-\underset{3}{CH_2}-\underset{4}{CH_2}-\underset{5}{CH_3}$$
(1)

2-Methylpentane (a five-carbon main chain with a 2-methyl substituent)

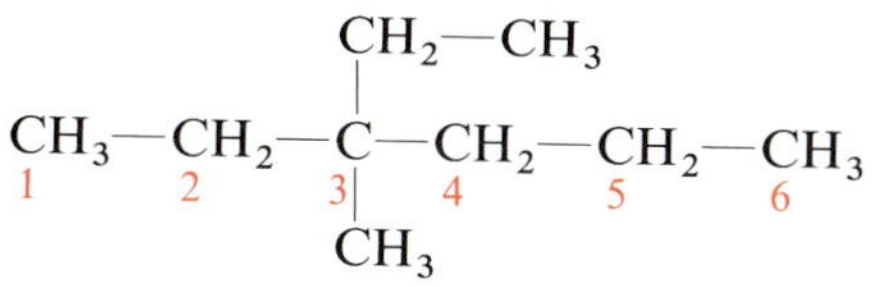

$$\underset{1}{CH_3}-\underset{2}{CH_2}-\underset{3}{\overset{CH_2-CH_3}{\underset{|}{C}}}-\underset{4}{CH_2}-\underset{5}{CH_2}-\underset{6}{CH_3}$$
$$CH_3$$

3-*Ethyl-3-methylhexane* (a six-carbon main chain with 3-ethyl and 3-methyl substituents cited alphabetically)

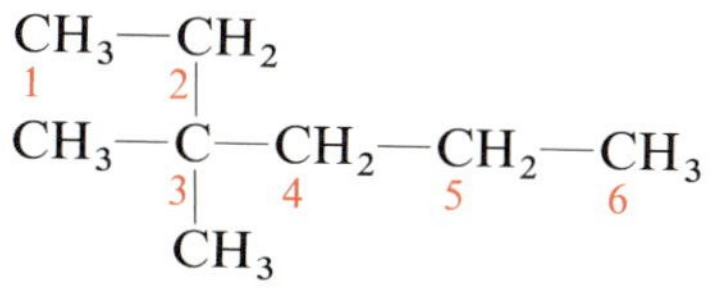

$$\underset{1}{CH_3}-\underset{2}{\overset{CH_2}{\underset{|}{C}}}$$
$$\underset{3}{CH_3}-\underset{}{\overset{}{\underset{|}{C}}}-\underset{4}{CH_2}-\underset{5}{CH_2}-\underset{6}{CH_3}$$
$$CH_3$$

3,3-*Dimethylhexane* (a six-carbon main chain with two 3-methyl substituents)

■ **Primary (1°) carbon** A carbon atom that is bonded to one other carbon.

■ **Secondary (2°) carbon** A carbon atom that is bonded to two other carbons.

■ **Tertiary (3°) carbon** A carbon atom that is bonded to three other carbons.

■ **Quaternary (4°) carbon** A carbon atom that is bonded to four other carbons.

One further word of explanation about alkanes and alkyl groups: It's sometimes useful to think about the *number of other carbon atoms attached* to a given carbon atom. There are four possible substitution patterns for carbon, called *primary, secondary, tertiary,* and *quaternary.* As indicated by the following structures, a **primary (1°) carbon atom** has one other carbon attached to it, a **secondary (2°) carbon atom** has two other carbons attached to it, a **tertiary (3°) carbon atom** has three other carbons attached to it, and a **quaternary (4°) carbon atom** has four other carbons attached to it.

$$\underset{\begin{array}{c}H\end{array}}{\overset{H}{R-\underset{|}{\overset{|}{C}}-H}}$$

$$\underset{\begin{array}{c}H\end{array}}{\overset{R}{R-\underset{|}{\overset{|}{C}}-H}}$$

$$\underset{\begin{array}{c}R\end{array}}{\overset{R}{R-\underset{|}{\overset{|}{C}}-H}}$$

$$\underset{\begin{array}{c}R\end{array}}{\overset{R}{R-\underset{|}{\overset{|}{C}}-R}}$$

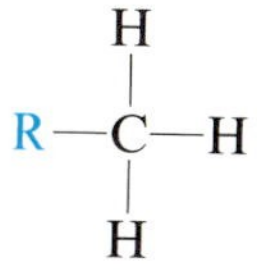

Primary carbon (1°) has one other carbon attached.

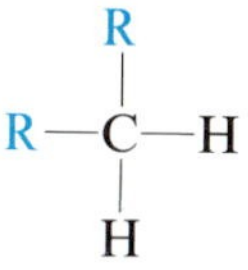

Secondary carbon (2°) has two other carbons attached.

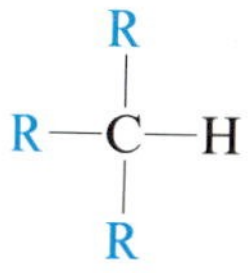

Tertiary carbon (3°) has three other carbons attached.

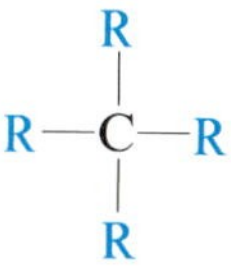

Quaternary carbon (4°) has four other carbons attached.

■ **R** The general symbol for an alkyl group.

The symbol **R** is used here and in later chapters to represent a *generalized* alkyl group. The group R may stand for methyl, ethyl, propyl, or any other alkyl group. For example, the generalized formula R—OH for an alcohol might refer to CH_3OH, CH_3CH_2OH, or to any of a great many other possibilities.

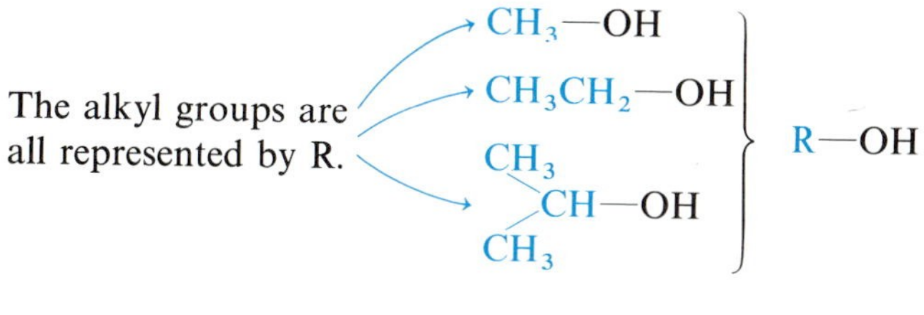

The alkyl groups are all represented by R.

$$CH_3-OH$$
$$CH_3CH_2-OH$$
$$\underset{CH_3}{\overset{CH_3}{\underset{|}{C}H-OH}}$$

$$\Bigg\}\ R-OH$$

Solved Problem 8.2   What is the IUPAC name of this alkane?

$$CH_3 \qquad\qquad CH_3$$
$$CH_3-CH-CH_2-CH_2-CH-CH_2-CH_3$$

Solution   First, find and name the longest continuous chain of carbon atoms (in this case seven carbons, or heptane). Second, number the main chain beginning at the end nearer the first branch (on the left in this case).

$$CH_3 \qquad\qquad CH_3$$
$$CH_3-CH-CH_2-CH_2-CH-CH_2-CH_3$$
$$1\quad\;\;2\quad\;\;3\quad\;\;4\quad\;\;5\quad\;\;6\quad\;\;7$$

Third, identify and number the substituents (a 2-methyl and a 5-methyl in this case). Fourth, write the name as one word using the prefix *di-* since there are two methyl groups. Separate the two numbers by a comma, and use a hyphen between the numbers and the word: 2,5-dimethylheptane.

Solved Problem 8.3   Identify the carbon atoms in this molecule as primary, secondary, tertiary, or quaternary.

$$CH_3 \qquad\quad CH_3$$
$$CH_3CHCH_2CH_2CCH_3$$
$$CH_3$$

Solution   Look at each carbon atom in the molecule, count the number of others carbon atoms attached, and make the assignment.

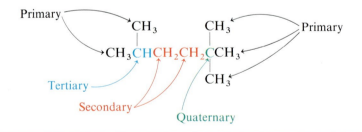

Practice Problems   **8.8**   What are the IUPAC names of these alkanes?

$$CH_2-CH_3 \qquad\qquad CH_3$$
(a) $CH_3-CH-CH_2-CH_2-CH_2-CH-CH_3$

$$CH_2-CH_3$$
(b) $CH_3-CH_2-CH_2-CH_2-C-CH_2-CH_3$
$$CH_2-CH_3$$

8.9 Draw structures corresponding to these IUPAC names:
(a) 3-methylhexane (b) 3,4-dimethyloctane (c) 2,2,4-trimethylpentane

8.10 Identify the carbon atoms in the molecules shown in Problem 8.8 as primary, secondary, tertiary, or quaternary.

8.11 Draw and name alkanes that meet these descriptions:
(a) an alkane with a tertiary carbon atom
(b) an alkane that has both a tertiary and a quaternary carbon atom

8.7 CYCLIC ORGANIC MOLECULES

■ **Acyclic alkane** An alkane that contains no rings.

■ **Cycloalkane** An alkane that contains a ring of carbon atoms.

The compounds we've been dealing with thus far have all been *open-chain* or **acyclic alkanes**—compounds that do not contain rings. **Cycloalkanes**, which contain rings of carbon atoms, are also well known and widespread throughout nature. Compounds of all ring sizes from three through thirty carbons and beyond have been prepared. The four simplest cycloalkanes having three carbons (cyclopropane), four carbons (cyclobutane), five carbons (cyclopentane), and six carbons (cyclohexane) are shown.

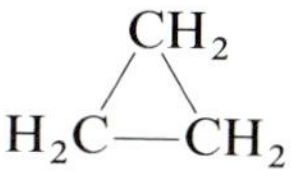

Cyclopropane

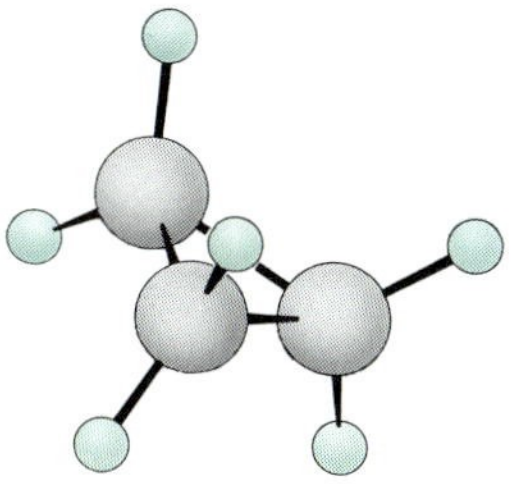

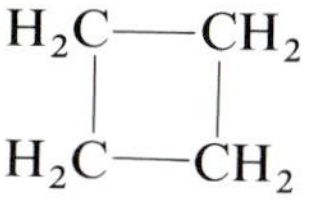

Cyclobutane

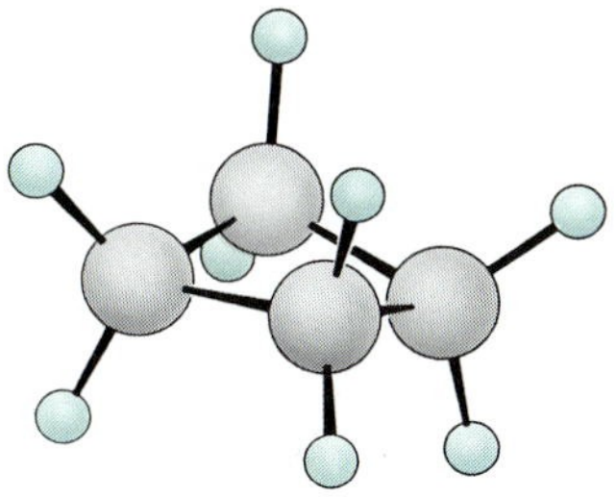

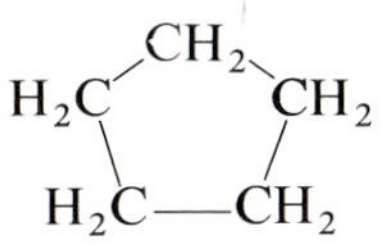

Cyclopentane

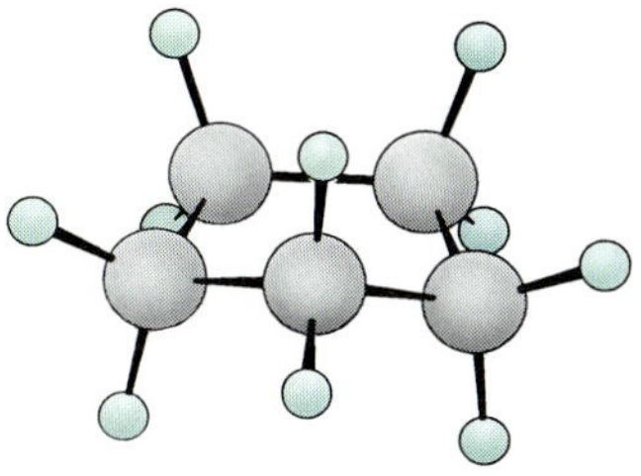

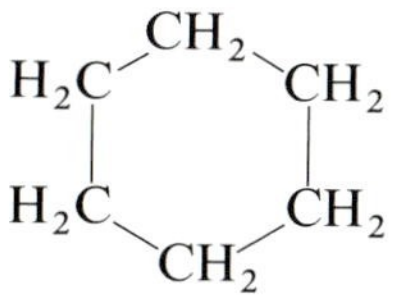

Cyclohexane

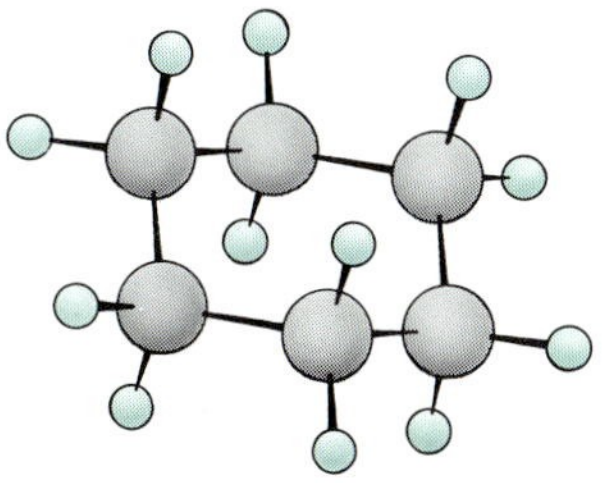

Cyclic and acyclic alkanes are similar in many ways but differ in one important respect. Because of their cyclic structures, cycloalkanes are more rigid and less flexible than their open-chain counterparts. Rotation is not possible around the carbon-carbon bonds in cycloalkanes without breaking open the ring.

8.8 DRAWING AND NAMING CYCLOALKANES

Even condensed structures become cluttered and awkward when working with large molecules (see the cholesterol structure in Figure 8.1, for example). Thus, an even more streamlined way of drawing structures is often used in which cycloalkanes are represented simply by polygons. A triangle represents cyclopropane, a square represents cyclobutane, a pentagon represents cyclopentane, and so on.

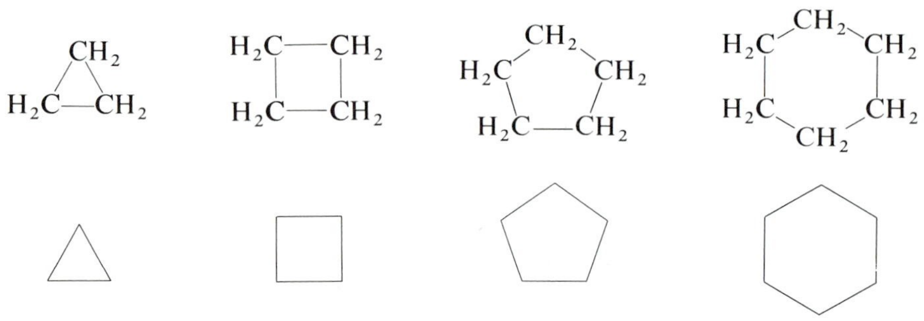

$$CH_2$$
$$H_2C—CH_2$$

$$H_2C—CH_2$$
$$H_2C—CH_2$$

$$CH_2$$
$$H_2C \qquad CH_2$$
$$H_2C—CH_2$$

$$CH_2$$
$$H_2C \qquad CH_2$$
$$H_2C \qquad CH_2$$
$$CH_2$$

Cyclopropane   Cyclobutane   Cyclopentane   Cyclohexane

■ **Line structure** A shorthand way of representing cycloalkane structures as polygons without showing individual carbon atoms.

Notice that carbon and hydrogen atoms aren't even shown in these **line structures**. A carbon atom is simply "understood" to be at every intersection of two lines, and the proper number of hydrogen atoms necessary to fill out carbon's valency is supplied mentally. Methylcyclohexane looks like this in a line structure:

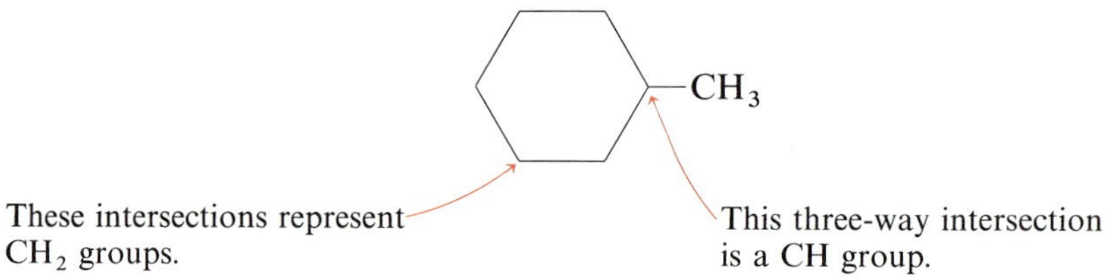

$$—CH_3$$

These intersections represent CH_2 groups.

This three-way intersection is a CH group.

Cycloalkanes are named by a straightforward extension of the rules for open-chain alkanes. In most cases, only two rules are needed:

1. **Use the cycloalkane name as the parent.** That is, compounds should be named as alkyl-substituted cycloalkanes rather than as cycloalkyl-substituted alkanes. If there is only one substituent on the ring, we don't even need to number it since all ring positions are equivalent.

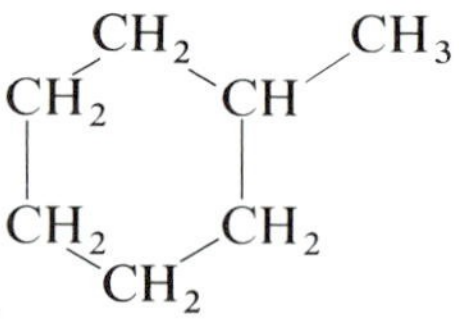

$$CH_2 \qquad CH_3$$
$$CH_2 \qquad CH$$
$$CH_2 \qquad CH_2$$
$$CH_2$$

is the same as

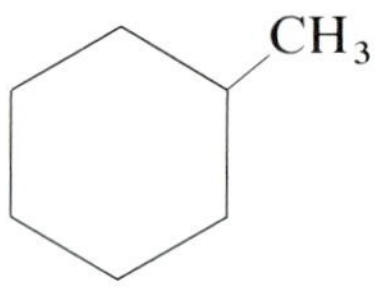

$$CH_3$$

Methylcyclohexane (not cyclohexylmethane)

2. **Number the substituents.** Start numbering at the group that has alphabetical priority, and proceed around the ring in the direction that gives the second substituent the lowest possible number.

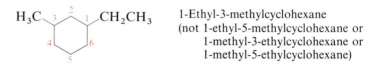

$$H_3C \qquad CH_2CH_3$$

1-Ethyl-3-methylcyclohexane (not 1-ethyl-5-methylcyclohexane or 1-methyl-3-ethylcyclohexane or 1-methyl-5-ethylcyclohexane)

Solved Problem 8.4   What is the IUPAC name of this cycloalkane?

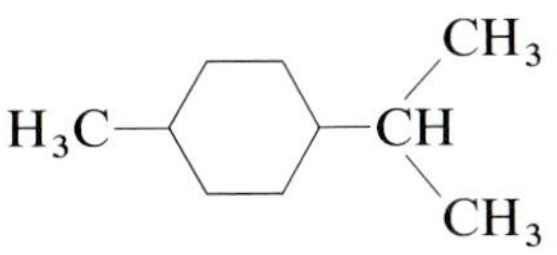

$$H_3C \!-\!\!\bigcirc\!\!-\! \underset{\underset{CH_3}{\big|}}{\overset{\overset{CH_3}{\big|}}{CH}}$$

Solution   First, identify the parent cycloalkane (cyclohexane in this case). Second, identify the two substituents (a methyl group and an isopropyl group). Third, number the compound beginning at the group having alphabetical priority (isopropyl rather than methyl), and proceed around the ring in a direction that gives the second group the lowest possible number.

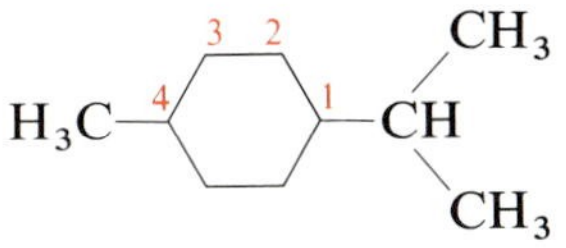

$$H_3C \!-\!\!\overset{4}{\underset{3 \quad 2}{\bigcirc}}\!\!\!\overset{1}{-}\! \underset{\underset{CH_3}{\big|}}{\overset{\overset{CH_3}{\big|}}{CH}} \qquad \text{1-Isopropyl-4-methylcyclohexane}$$

Solved Problem 8.5   Draw a line structure for 1,4-dimethylcyclohexane.

Solution   First draw a hexagon to represent a cyclohexane ring, and then attach a —CH_3 (methyl) group at an arbitrary position that becomes C1. Then count around the ring to C4, and attach another —CH_3 group (written here as H_3C— since it is attached on the left side of the ring).

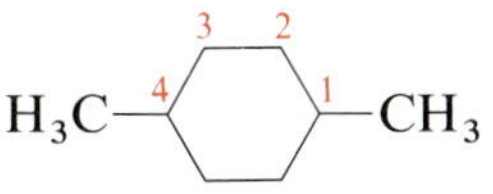

$$H_3C \!-\!\!\overset{4}{\underset{3 \quad 2}{\bigcirc}}\!\!\!\overset{1}{-}\! CH_3 \qquad \text{1,4-Dimethylcyclohexane}$$

Practice Problems   **8.12**   What are the IUPAC names of these cycloalkanes?

(a) 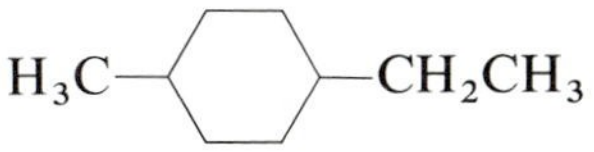 $H_3C \!-\!\!\bigcirc\!\!-\! CH_2CH_3$

(b) 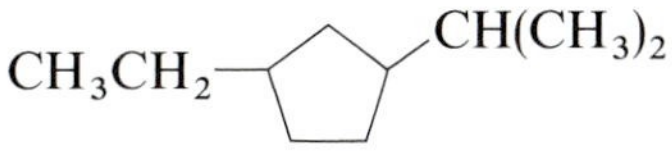 $CH_3CH_2 \!-\!\!\bigpentagon\!\!-\! CH(CH_3)_2$

8.13   Draw structures representing these IUPAC names. Use simplified line structures rather than condensed structures.
(a) 1,1-diethylcyclohexane   (b) 1,3,5-trimethylcycloheptane

8.9 CHEMISTRY OF ALKANES

■ **Paraffin** A mixture of waxy alkanes having 20 to 36 carbon atoms.

If you've ever done any home canning, you might have used paraffin wax to seal the jar. Chemically, what you've used is a mixture of straight-chain alkanes in the C_{20} to C_{36} range. The word **paraffin**, derived from the Latin *parum affinis* meaning slight affinity, is often applied to alkanes because of their lack of chemical reactivity. Alkanes show little chemical affinity for other molecules and are inert to acids, bases, and most other common laboratory reagents. Their only reaction of real importance is their combination with oxygen.

■ **Combustion** The burning of an organic substance with oxygen.

The chemical reaction of alkanes with oxygen occurs during **combustion** in an engine or furnace when the alkane is burned as fuel. Carbon dioxide and water are formed as products, and a large amount of heat is released. For example, methane, the principal component of natural gas, reacts with oxygen to release 213 kcal of heat for each mole of methane that burns:

$$CH_4 + 2\,O_2 \longrightarrow CO_2 + 2\,H_2O + 213 \text{ kcal/mol}$$

Propane (the LP gas used in campers and rural homes), gasoline (a mixture of C_5-to-C_{11} alkanes), kerosene (a mixture of C_{11}-to-C_{14} alkanes), and other alkanes burn similarly.

INTERLUDE: OCCURRENCE AND USES OF ALKANES: PETROLEUM

Although many alkanes occur naturally throughout the plant and animal world, natural gas and petroleum deposits provide the most abundant supply. Laid down eons ago, these deposits are largely derived from the decomposition of marine organic matter.

Natural gas consists chiefly of methane, with smaller amounts of ethane, propane, and butane also present. *Petroleum* is a complex mixture of hydrocarbons that must be separated, or *refined*, into different fractions before it can be used (see figure below). Petroleum refining begins with distillation to separate the crude oil into three main fractions according to boiling points: straight-run gasoline (bp 30–200°C), kerosene (bp 175–300°C), and gas oil (bp 275–400°C). The residue is then further distilled under reduced pressure to recover lubricating oils, waxes, and asphalt, as shown in the table.

Distillation of petroleum is just the beginning of the process for making automobile fuel. It's long been known that straight-chain alkanes burn far less smoothly than branched-chain compounds, a quality measured by determining a compound's *oc-*

The Principal Products of Petroleum Refining

Product	Boiling-point Range (°C)	Alkane Size
Asphalt, tar	Nonvolatile	
Lubricating oil, wax	Nonvolatile	C_{20}–C_{35}
Gas oil, diesel fuel	275–400	C_{14}–C_{25}
Kerosene	175–300	C_{11}–C_{14}
Straight-run gasoline	20–200	C_5–C_{11}
Natural gas	Below 20	C_1–C_4

tane number. Heptane, a particularly bad fuel, is assigned an octane rating of 0, and 2,2,4-trimethylpentane (known as isooctane) is given a rating of 100. Straight-run gasoline, with its high percentage of unbranched alkanes, is thus a poor fuel. Petroleum chemists, however, have devised sophisticated methods to remedy the problem. One of these methods, *catalytic cracking*, involves taking the kerosene cut (C_{11}-C_{14}) and "cracking" it into smaller C_3-C_5 molecules at high temperature. These small hydrocarbons are then catalytically recombined to yield C_7-C_{10} branched-chain molecules that are perfectly suited for use as automobile fuels.

SUMMARY

Organic chemistry is the study of carbon compounds. Carbon has the unique ability to form strong bonds to other carbon atoms, resulting in the formation of rings or long chains, and giving rise to many millions of possible structures. Organic compounds can be classified into various families according to the functional groups they contain. A **functional group** is a part of a larger molecule and is composed of an atom or group of atoms that has characteristic chemical reactivity. A given functional group undergoes the same chemical reactions in every molecule where it occurs.

Compounds of carbon and hydrogen that contain only single bonds are called **alkanes**. **Straight-chain alkanes** have all their carbons connected in a row; **branched-chain alkanes** have a branching connection of atoms somewhere along their chain; and **cycloalkanes** have a ring of carbon atoms. Alkanes contain no functional groups, are not soluble in water, do not conduct electricity, and are chemically inert to most common reagents. **Isomerism** is possible in alkanes having four or more carbons. **Isomers** are compounds that have the same molecular formula but have different structures and different physical properties because of different connections between atoms.

Alkanes are named in the **IUPAC system** by applying a set of **nomenclature** rules. With the exception of the four simplest alkanes—methane, ethane, propane, and butane—straight-chain alkanes are named by adding the family ending *-ane* to the Greek number that tells how many carbon atoms are present. Thus, pentane has a five-carbon chain, hexane has a six-carbon chain, heptane has a seven-carbon chain, and so on. Branched-chain alkanes are named by using the name of the longest continuous chain of carbon atoms as the parent name, and then telling what alkyl groups are present as branches off the main chain. An **alkyl group** is the part of an alkane that remains when one hydrogen is removed. Thus, —CH_3 is a **methyl group**, —CH_2CH_3 is an **ethyl group**, —$CH_2CH_2CH_3$ is a **propyl group**, and so on. Isomerism is possible in alkyl groups just as in alkanes themselves.

REVIEW PROBLEMS

Organic Molecules and Functional Groups

8.14 What characteristic do all organic compounds share?

8.15 What kind of bond predominates in organic compounds?

8.16 What special characteristic of carbon makes possible the existence of so many different organic compounds?

8.17 What are functional groups, and why are they important?

8.18 Why are most organic compounds insoluble in water and nonconducting?

8.19 If you were given two unlabeled bottles, one containing hexane and one containing water, how could you tell them apart?

8.20 What is meant by the term *polar covalent bond*? Give an example of such a bond.

8.21 Give examples of compounds that are members of the following families:

(a) alcohol   (b) amine   (c) carboxylic acid
(d) ether

8.22 Locate and identify the functional groups in these molecules:

(a)

Menthol

(b)

Aspirin (acetylsalicylic acid)

8.23 Propose structures for molecules that meet these descriptions:
(a) a ketone with the formula $C_5H_{10}O$
(b) an ester with the formula $C_6H_{12}O_2$
(c) compound with formula $C_2H_5NO_2$ that is both an amine and a carboxylic acid

8.24 Propane, more commonly known as LP gas, burns in air to yield CO_2 and H_2O. Write a balanced equation for the reaction.

8.25 Identify the functional groups in these molecules:

(a)

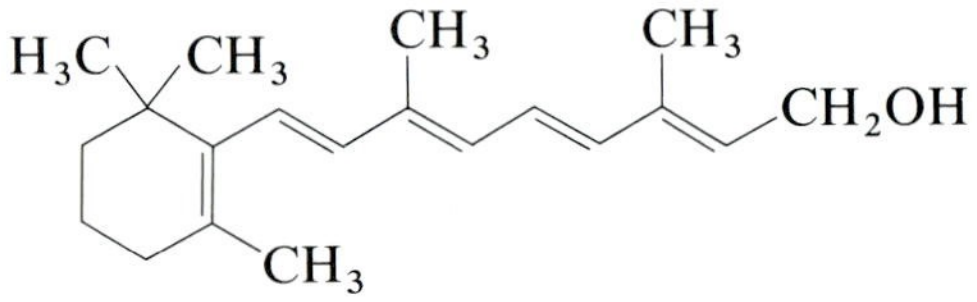

Vitamin A

(b)

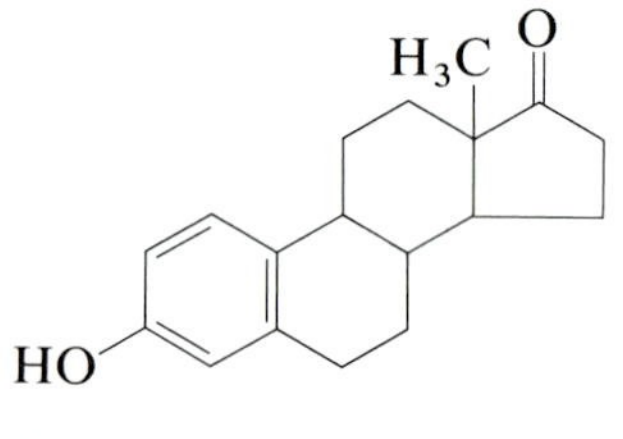

Estrone, a female sex hormone

Alkanes and Isomers

8.26 What structural feature distinguishes a straight-chain alkane from a branched-chain alkane?

8.27 What requirement must be met in order for two compounds to be isomers?

8.28 If one compound has the formula C_5H_{10} and another has the formula C_4H_{10}, are the two compounds isomers? Explain.

8.29 What is the difference between a secondary carbon and a tertiary carbon? Between a primary carbon and a quaternary carbon?

8.30 Why can't a compound have a *quintary* carbon (five R groups attached to C)?

8.31 Give examples of compounds that meet these descriptions:
(a) an alkane that has two tertiary carbons
(b) a cycloalkane that has only secondary carbons

8.32 There are three isomers with the formula C_3H_8O. Draw their structures.

8.33 Write condensed structures for each of the following molecular formulas. You may have to use rings and/or multiple bonds in some instances.
(a) C_2H_7N (b) C_4H_8 (c) C_2H_4O
(d) CH_2O_2

8.34 If someone reported the preparation of a compound with the formula C_3H_9, most chemists would be skeptical. Why?

8.35 How many isomers can you write that fit these descriptions?
(a) alcohols with formula $C_4H_{10}O$
(b) amines with formula C_3H_9N
(c) ketones with formula $C_5H_{10}O$
(d) aldehydes with formula $C_5H_{10}O$

8.36 Which of the following pairs of structures are identical, which are isomers, and which are unrelated?

(a) $CH_3CH_2CH_3$ and

$$CH_3 - CH_2CH_3$$

(b) $CH_3{-}N{-}CH_3$ and $CH_3CH_2{-}N{-}H$ (with H below each N)

(c) $CH_3CH_2CH_2{-}O{-}CH_3$ and

$$CH_3CH_2CH_2 - \overset{O}{\overset{\|}{C}} - CH_3$$

(d) $CH_3 - \overset{O}{\overset{\|}{C}} - CH_2CH_2CH(CH_3)_2$ and

$$CH_3CH_2 - \overset{O}{\overset{\|}{C}} - CH_2CH_2CH_2CH_3$$

(e) $CH_3CH{=}CHCH_2CH_2{-}O{-}H$ and

$$CH_3CH_2\overset{}{\underset{\overset{|}{CH_3}}{C}}H - \overset{O}{\overset{\|}{C}} - H$$

8.37 What is wrong with each of the following structures?

(a) $CH_3{=}CHCH_2CH_2OH$

(b)

$$CH_3CH_2CH{=}\overset{O}{\overset{\|}{C}}{-}CH_3$$

(c)

$$CH_2CH_2CH_2C{\equiv}\overset{\overset{\textstyle CH_3}{|}}{C}CH_3$$

8.38 Which of these structures represent the same compound, and which represent isomers?

(a)

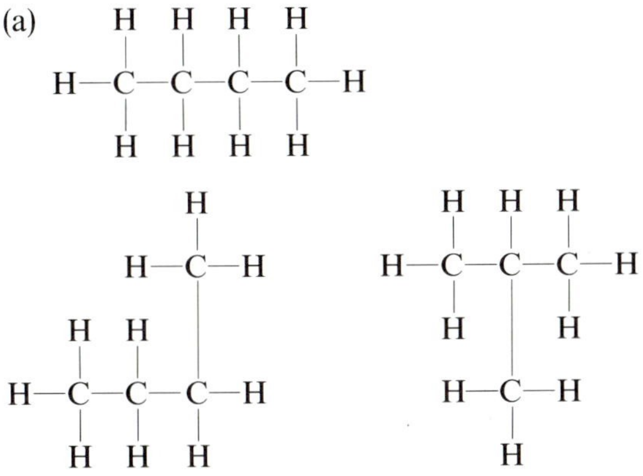

(b)

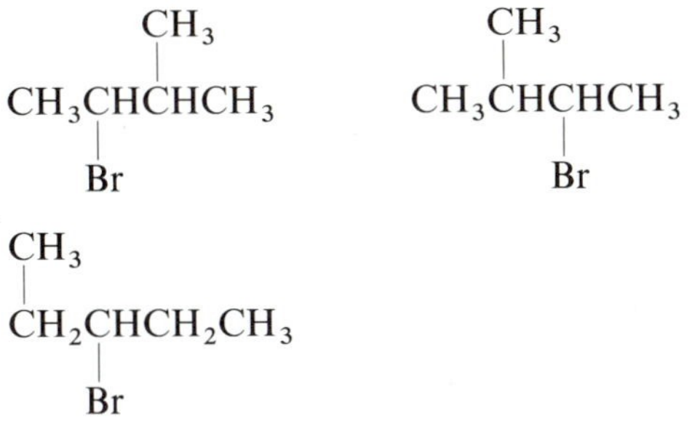

Alkane Nomenclature

8.39 What are the IUPAC names of these alkanes?

(a)

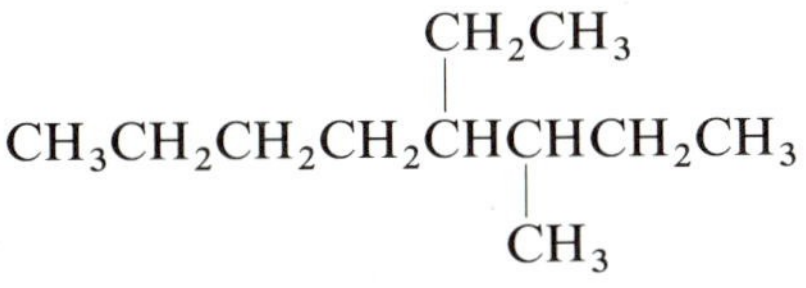

(b)

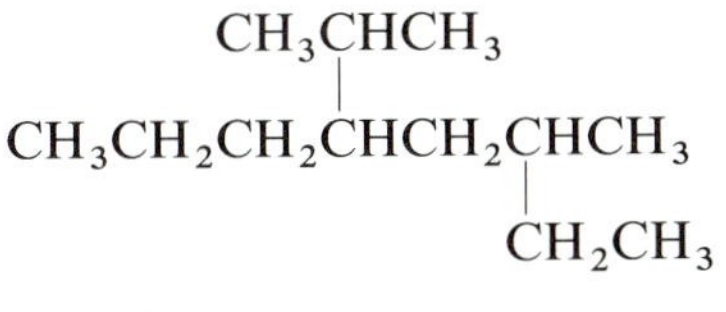

(c)

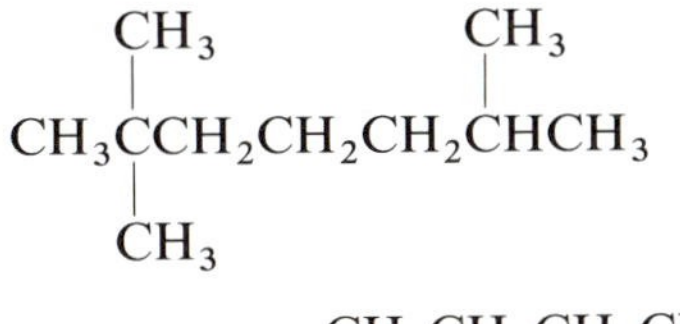

(d)

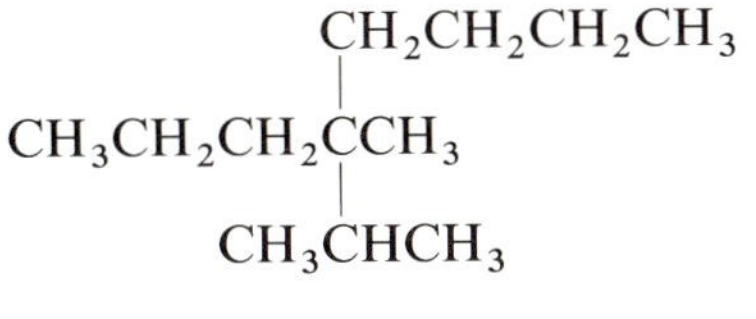

8.40 Write condensed structures for each of these compounds:
(a) 3-ethylhexane (b) 2,2,3-trimethylpentane
(c) 3-ethyl-3,4-dimethylheptane
(d) 5-isopropyl-2-methyloctane

8.41 The following compound, known trivially as isooctane, is important as a reference substance for determining the "octane" rating of gasoline. What is the proper IUPAC name of isooctane:

Isooctane

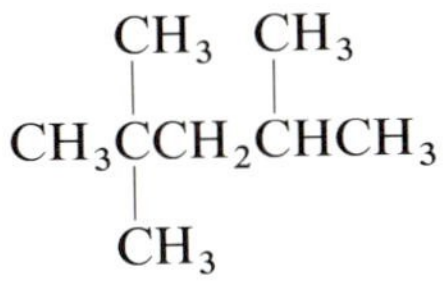

8.42 Provide IUPAC names for each of the five isomers with the formula C_6H_{14}.

8.43 Draw structures corresponding to these IUPAC names:
(a) cyclooctane (b) 1,1-dimethylcyclopentane
(c) 1,2,3,4-tetramethylcyclobutane
(d) 4-ethyl-1,1-dimethylcyclohexane

8.44 Provide IUPAC names for these cycloalkanes:

(a)

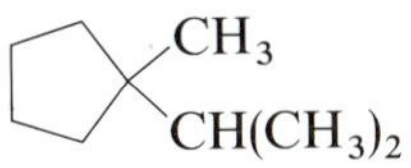

(b)

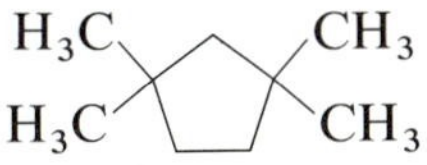

(c)

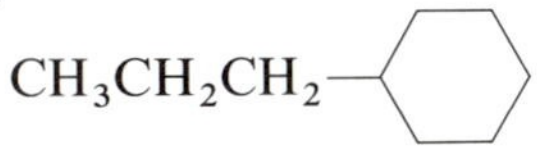

(d)

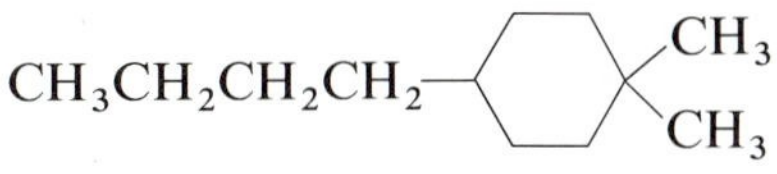

8.45 The following names are incorrect. Tell what is wrong with each and provide the correct names.

(a)

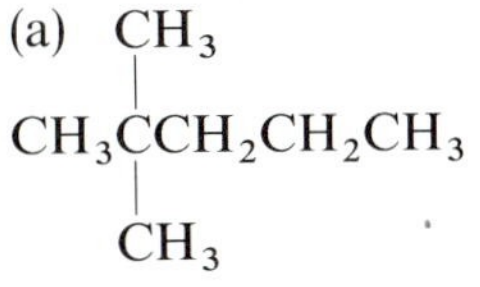

2,2-Methylpentane

(b)

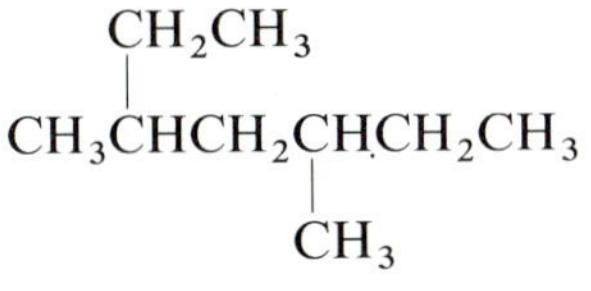

5-Ethyl-3-methylhexane

(c)

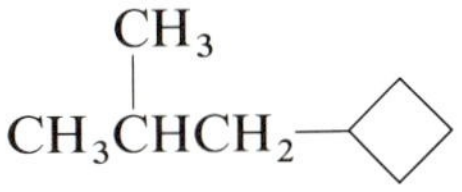

1-Cyclobutyl-2-methylpropane

8.46 Draw structures and give IUPAC names for the nine isomers of C_7H_{16}.

CHAPTER 9

Alkenes, Alkynes, and Aromatic Compounds

Why are these flamingos pink? Could it be something they ate?
In fact, the normal flamingo diet is rich in alkene pigments closely related to the
β-carotene that plays a key role in human vision (see the Application on p. 192).
These pigments are concentrated in the birds' feathers; if fed a diet from which
such pigments are absent, the flamingos eventually turn white.

The compounds in the alkane family that we dealt with in the last chapter contain only single bonds. By contrast, compounds in the three families we'll study in this chapter all contain carbon-carbon *multiple* bonds. *Alkenes* contain a carbon-carbon double bond; *alkynes* contain a carbon-carbon triple bond; and *aromatic compounds* contain a six-membered ring of carbon atoms with three double bonds. Compounds with double or triple bonds are referred to as *unsaturated*, because they have fewer hydrogens per carbon than the related alkanes, which have only single bonds and are *saturated*.

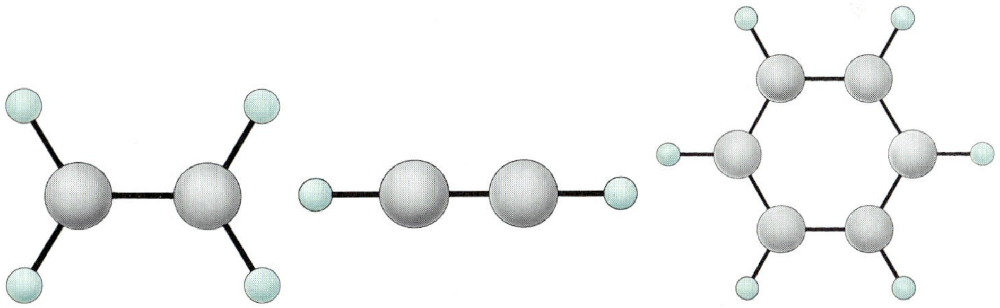

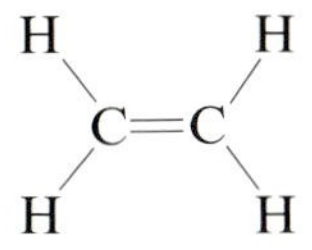

Ethylene
(an *alkene*—has
a C—C double bond)

Acetylene
(an *alkyne*—has
a C—C triple bond)

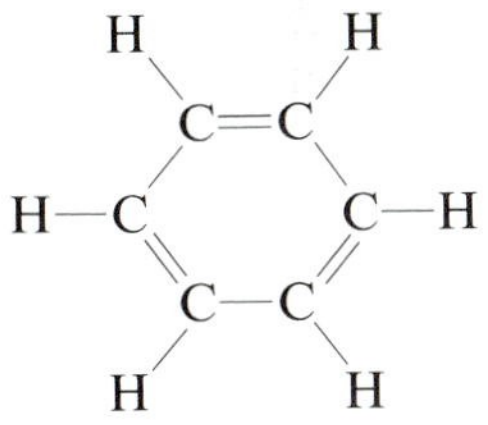

Benzene
(an *aromatic compound*—
has a six-membered ring
and three double bonds)

We'll look at the answers to these questions in this chapter:

1. **What are alkenes, alkynes, and aromatic compounds?** The goal: you should learn the structures of the different functional groups that have carbon-carbon multiple bonds.

2. **What are cis-trans alkene isomers?** The goal: you should learn how to recognize cis-trans alkene isomers and how to predict their occurrence.

3. **How are alkenes, alkynes, and aromatic compounds named?** The goal: you should learn how to name compounds with unsaturated functional groups.

185

4. **What reactions do alkenes, alkynes, and aromatic compounds undergo?**   The goal: you should learn the kinds of reactions that these compounds undergo and how to predict the products in specific cases.

5. **How do organic reactions take place?**   The goal: you should learn how alkenes and alkynes undergo addition reactions and how aromatic compounds undergo substitution reactions.

9.1   NAMING ALKENES AND ALKYNES

■ **Hydrocarbon**   A compound containing only carbon and hydrogen.

■ **Alkene**   A hydrocarbon that has a carbon-carbon double bond.

■ **Alkyne**   A hydrocarbon that has a carbon-carbon triple bond.

Any compound that contains only carbon and hydrogen is called a **hydrocarbon**. **Alkenes** are hydrocarbons that contain a carbon-carbon double bond; **alkynes** are hydrocarbons that contain a carbon-carbon triple bond. Members of both families are named by a series of three rules similar to those used for alkanes. The *-ene* ending is used in place of *-ane* for an alkene, and the *-yne* ending is used for an alkyne.

Step 1. Name the parent compound.   Find the longest chain *containing the double or triple bond*, and name that parent chain using the suffix *-ene* or *-yne*:

$$CH_3CH_2CH_2CH{=}CH_2$$

Name as a *pentene*—a five-carbon chain containing a double bond.

$$CH_3CH_2CH_2C{\equiv}CCH_3$$

Name as a *hexyne*—a six-carbon chain containing a triple bond.

$$CH_3CH_2CH_2$$
$$\qquad\qquad C{=}CHCH_3$$
$$CH_3CH_2CH_2$$

Name as a *hexene*—a six-carbon chain containing a double bond—

$$CH_3CH_2CH_2$$
$$\qquad\qquad C{=}CHCH_3$$
$$CH_3CH_2CH_2$$

not

as a heptene, since the double bond is not contained in the seven-carbon chain.

Step 2. Number the carbon atoms in the main chain.   Beginning at the end nearer the multiple bond, number each carbon atom in the chain. If the multiple bond is an equal distance from both ends, as in the second part of the following example, begin numbering at the end nearer the first branch point.

$$\underset{6\quad 5\quad 4\quad 3\quad 2\quad 1}{CH_3CH_2CH_2CH{=}CHCH_3}$$

Begin at this end because it's nearer the double bond.

$$\overset{CH_3}{\underset{1\quad 2\quad 3\quad 4\quad 5\quad 6}{CH_3CHCH{=}CHCH_2CH_3}}$$

Begin at this end because it's nearer the first branch point.

$$\overset{\qquad\qquad\qquad\qquad CH_3}{\underset{1\quad 2\quad 3\;4\quad 5\quad 6\quad 7\quad 8}{CH_3C{\equiv}CCH_2CH_2CH_2CHCH_3}}$$

Begin at this end because it's nearer the triple bond.

Step 3. Write out the full name.   Assign numbers to the branching substituents on the chain, and list them alphabetically; use commas to separate numbers, and use hyphens to separate words from numbers. Indicate the position of the multiple bond in the chain by giving the number of the *first* multiple-bonded carbon. If more than one double bond is present, identify the position of each, and use the name ending -*diene*, -*triene*, -*tetraene*, and so forth to tell the exact number of double bonds. For example:

$$\underset{5}{CH_3}\underset{4}{CH_2}\underset{3}{CH_2}\underset{2}{CH}=\underset{1}{CH_2}$$

1-Pentene

$$\underset{6}{CH_3}\underset{5}{CH_2}\underset{4}{CH_2}\underset{3}{C}\equiv\underset{2\ \ 1}{CCH_3}$$

2-Hexyne

$$\overset{\overset{\displaystyle \underset{6}{CH_3}\underset{5}{CH_2}\underset{4}{CH_2}}{\diagdown}}{\underset{\underset{\displaystyle CH_3CH_2CH_2}{\diagup}}{\underset{3}{C}=\underset{2}{CH}\underset{1}{CH_3}}}$$

3-Propyl-2-hexene

$$\underset{1}{CH_3}\underset{2}{C}\equiv\underset{3}{C}\underset{4}{CH_2}\underset{5}{CH_2}\underset{6}{CH_2}\underset{7}{\overset{\overset{\displaystyle CH_3}{|}}{CH}}\underset{8}{CH_3}$$

7-Methyl-2-octyne

$$\underset{1}{H_2C}=\underset{2}{\overset{\overset{\displaystyle CH_2CH_3}{|}}{C}}-\underset{3}{CH}=\underset{4}{CH_2}$$

2-Ethyl-1,3-butadiene

For historical reasons, there are a small number of alkenes whose names don't conform strictly to the rules. The two-carbon alkene $H_2C=CH_2$ should properly be called *ethene*, but the name *ethylene* has been used for so long that it's now accepted by IUPAC. Similarly, the three-carbon alkene *propene* ($CH_3CH=CH_2$) is often called *propylene*.

Solved Problem 9.1   What is the IUPAC name of this alkene?

$$CH_3CH_2CH_2-\overset{\overset{\displaystyle H_3C}{|}}{C}=\overset{\overset{\displaystyle CH_2CH_3}{|}}{C}-CH_3$$

Solution   First, find and circle the longest chain containing the double bond. In this case, we have to turn a corner to find it's a heptene:

$$CH_3CH_2CH_2-\overset{\overset{\displaystyle H_3C}{|}}{C}=\overset{\overset{\displaystyle CH_2CH_3}{|}}{C}-CH_3 \qquad \text{Name as a } \textit{heptene}$$

Next, number the chain from the end nearer the double bond. The first double-bond carbon is at C4 starting from the left-hand end but at C3 starting from the right:

$$\underset{7}{CH_3}\underset{6}{CH_2}\underset{5}{CH_2}-\underset{4}{\overset{\overset{\displaystyle H_3C}{|}}{C}}=\underset{3}{\overset{\overset{\displaystyle \underset{2}{CH_2}\underset{1}{CH_3}}{|}}{C}}-CH_3 \qquad \text{Name as a substituted } \textit{3-heptene}$$

Finally, identify the substituents, and write the name:

$$\underset{7}{CH_3}\underset{6}{CH_2}\underset{5}{CH_2}-\underset{4}{\overset{\overset{\displaystyle H_3C}{|}}{C}}=\underset{3}{\overset{\overset{\displaystyle \underset{2}{CH_2}\underset{1}{CH_3}}{|}}{C}}-CH_3$$

Substituents: 3-methyl, 4-methyl

Name: *3,4-dimethyl-3-heptene*

Practice Problems   **9.1**   What are the IUPAC names of these compounds?

(a) $CH_3CH_2CH_2CH{=}CHCH(CH_3)_2$

(b) $H_2C{=}CHCH_2CH_2C{=}CH_2$
$\phantom{(b) H_2C{=}CHCH_2CH_2C{=}CH_2}|$
$\phantom{(b) H_2C{=}CHCH_2CH_2}CH_3$

9.2   Draw structures corresponding to these IUPAC names:
(a) 3-methyl-1-heptene   (b) 4,4-dimethyl-2-pentyne
(c) 2-methyl-2-hexene   (d) 3-ethyl-2,2-dimethyl-3-hexene

9.2   THE STRUCTURE OF ALKENES: CIS-TRANS ISOMERISM

Alkenes and alkynes have different shapes from alkanes because of their different kinds of bonds. Whereas methane is tetrahedral, ethylene is flat (planar), and acetylene is linear (straight).

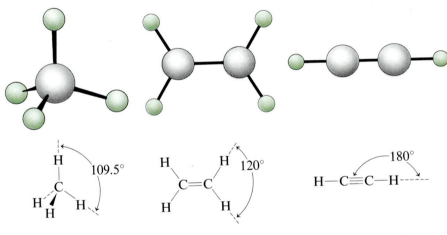

Methane—a tetrahedral molecule with bond angles of 109.5°.

Ethylene—a flat molecule with bond angles of 120°.

Acetylene—a straight molecule with bond angles of 180°.

The two carbons and four attached atoms that make up the double-bond functional group always lie in a plane. Unlike the situation in alkanes where free rotation around the C—C single bond occurs (Section 8.5), there is no free rotation around double bonds. As a consequence, a new kind of isomerism is possible for alkenes.

To see this new kind of isomerism, look at the following list of C_4H_8 compounds. When written as condensed structures, there appear to be three alkene isomers of formula C_4H_8: 1-butene, 2-butene, and 2-methylpropene. In fact, though, there are *four*. 1-Butene and 2-butene are isomers because their double bonds occur at different positions along the chain, while 2-methylpropene is isomeric with both because it has a different connection of carbon atoms. But because rotation can't occur around carbon-carbon double bonds, *there are two different 2-butenes.* In one isomer, the two methyl groups are close together on the same side of the double bond, but in the other isomer, the two methyl groups are far apart on opposite sides of the double bond.

1-Butene $H_2C{=}CHCH_2CH_3$ =

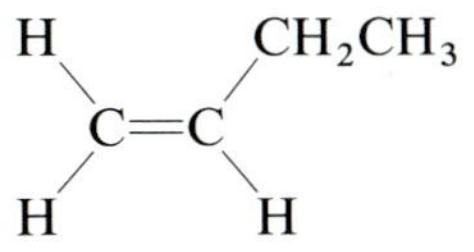

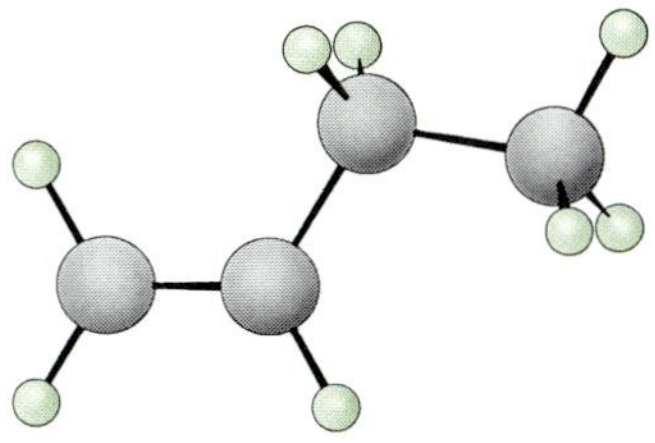

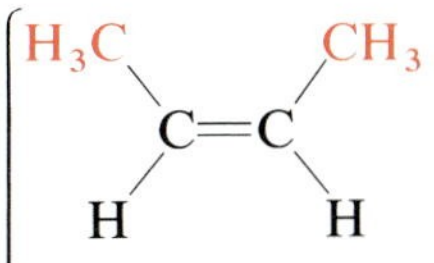

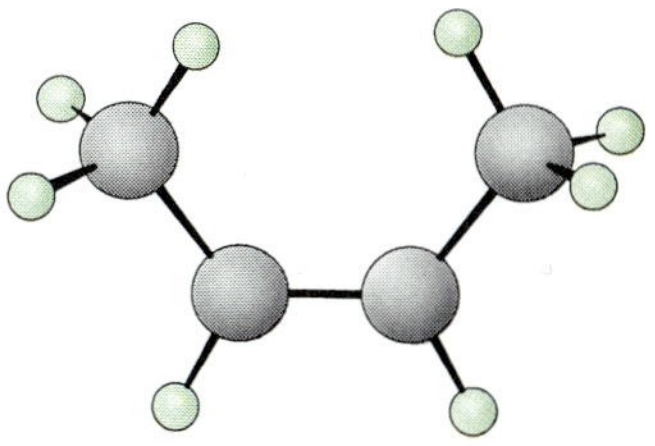

$$\underset{H}{\overset{H}{\diagdown}}C{=}C\underset{H}{\overset{CH_2CH_3}{\diagup}}$$

2-Butene $CH_3CH{=}CHCH_3$ =

$$\underset{H}{\overset{H_3C}{\diagdown}}C{=}C\underset{H}{\overset{CH_3}{\diagup}}$$

cis

and

$$\underset{H_3C}{\overset{H}{\diagdown}}C{=}C\underset{H}{\overset{CH_3}{\diagup}}$$

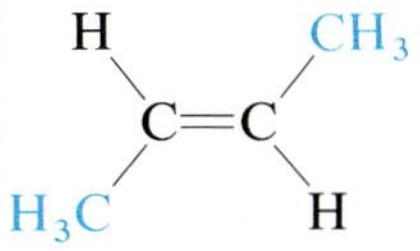

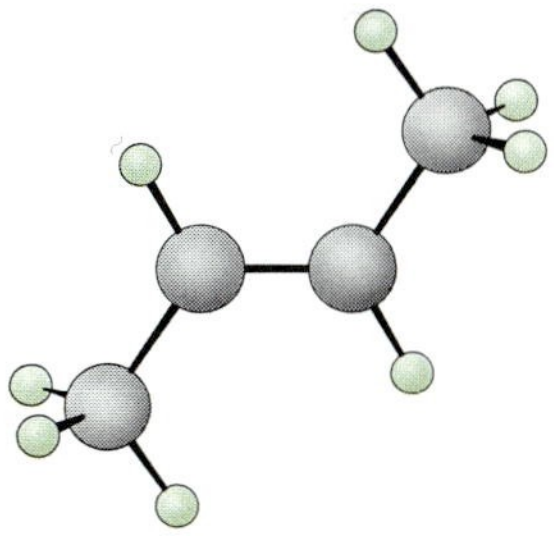

trans

2-Methylpropene $H_2C{=}C(CH_3)_2$ =

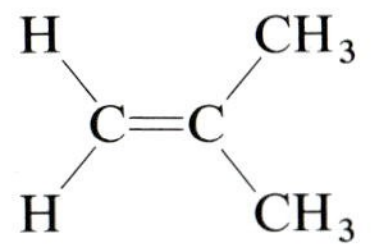

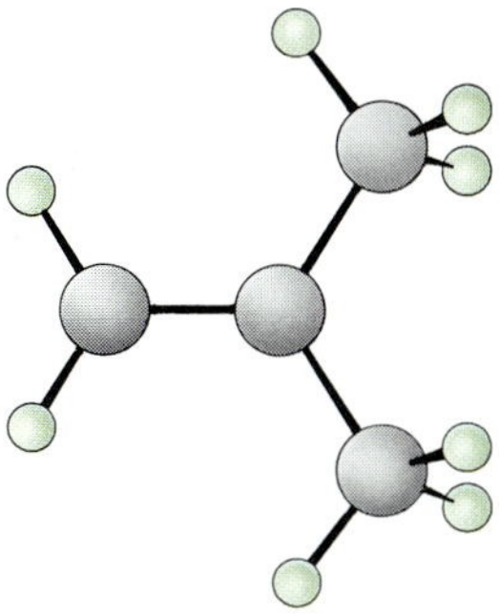

$$\underset{H}{\overset{H}{\diagdown}}C{=}C\underset{CH_3}{\overset{CH_3}{\diagup}}$$

■ **cis-trans isomers** Alkenes with the same formula and connections between atoms but different structures because of the way that groups are attached to different sides of the double bond.

The two 2-butenes are called **cis-trans isomers**; they have the same formula and connections between atoms but differ in their structure because of the way that groups are attached to different sides of the double bond. The isomer with its methyl groups on the same side of the double bond is named *cis*-2-butene. The isomer with its methyl groups on opposite sides of the double bond is named *trans*-2-butene.

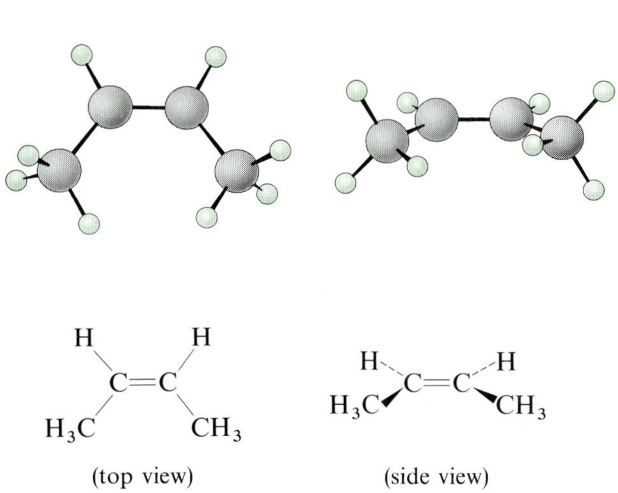

(top view)	(side view)	(top view)	(side view)

cis-2-Butene 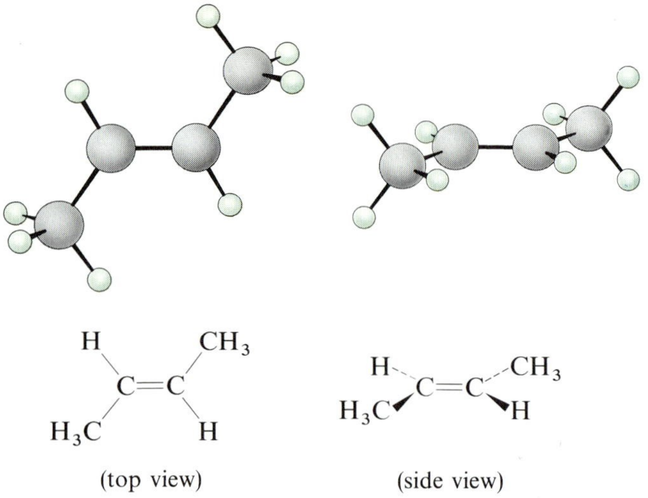   *trans*-2-Butene

Cis-trans isomerism occurs in any alkene when each double-bond carbon is bonded to two different substituent groups. If either double-bond carbon is attached to two identical groups, however, then cis-trans isomerism is not possible. In Figure 9.1, for example, the top two compounds are identical, and cis-trans isomerism is impossible because one of the double-bond carbons has two identical substituents (hydrogens) attached to it. You can convince yourself of this by mentally flipping one of the structures and seeing that it becomes identical to the other structure. In the bottom two compounds, however, the structures do not become identical when flipped.

Solved Problem 9.2   Draw structures for the cis-trans isomers of 2-hexene.

Solution   First, draw a condensed structure of 2-hexene to see what groups are attached to the double-bond carbons:

$$CH_3CH{=}CHCH_2CH_2CH_3$$
   1   2   3   4   5   6

Attached to C2: —H and —CH₃
Attached to C3: —H and —CH₂CH₂CH₃

Next, draw two double bonds. Choose one end of each double bond, and attach its groups in the same way to generate two identical part-structures:

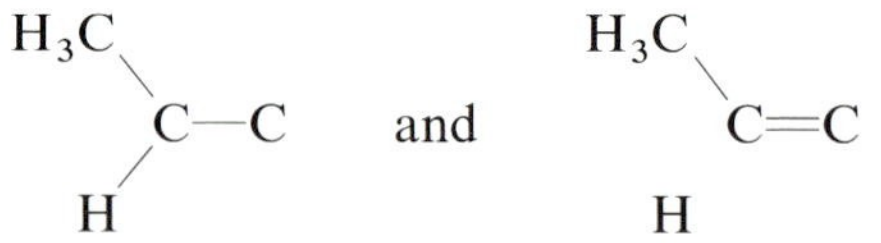

Finally, attach the proper groups to the other end in the two possible ways:

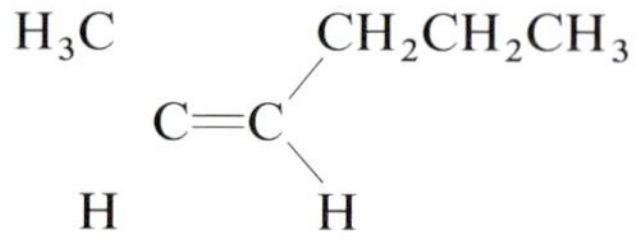

cis-2-Hexene   *trans*-2-Hexene

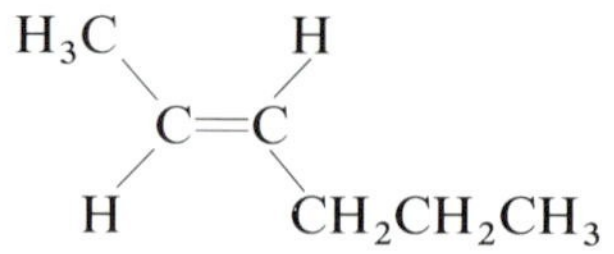

Practice Problems   **9.3**   Which of the following substances exist as cis-trans isomers?
(a) 3-heptene   (b) 2-methyl-2-hexene   (c) 5-methyl-2-hexene

9.4   Draw the cis-trans isomers of 3,4-dimethyl-3-hexene.

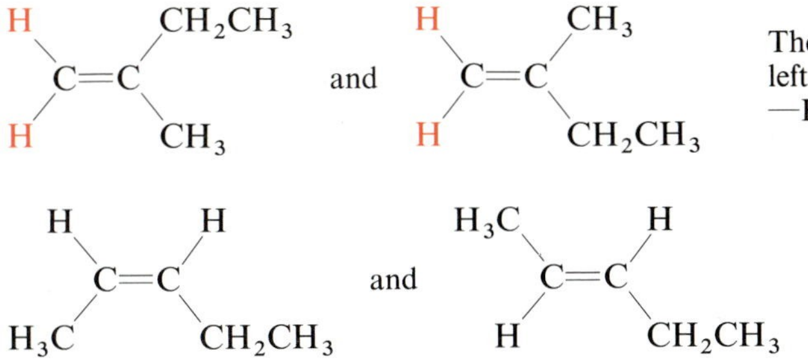

cis isomer trans isomer

These compounds are identical. Because the left-hand carbon of the double bond has two —H's attached, cis-trans isomerism is impossible.

These compounds are not identical. Neither carbon of the double bond has two identical groups attached to it.

Figure 9.1
The requirement for cis-trans isomerism. Each double-bond carbon must be bonded to two different groups in order to generate cis-trans isomers.

9.3 CHEMICAL REACTIONS OF ALKENES AND ALKYNES

■ **Addition reaction** A chemical reaction in which a reactant of the general form X—Y adds to the multiple bond in an unsaturated compound to yield a saturated product.

Most of the reactions that carbon-carbon multiple bonds undergo can be grouped under the category of **addition reactions**. That is, a reagent we might write in a general way as X-Y adds to the multiple bond in the unsaturated starting material to yield a product that has only single bonds. Alkenes and alkynes react similarly in many ways, but we'll look only at alkenes in this chapter because they are more common and more important.

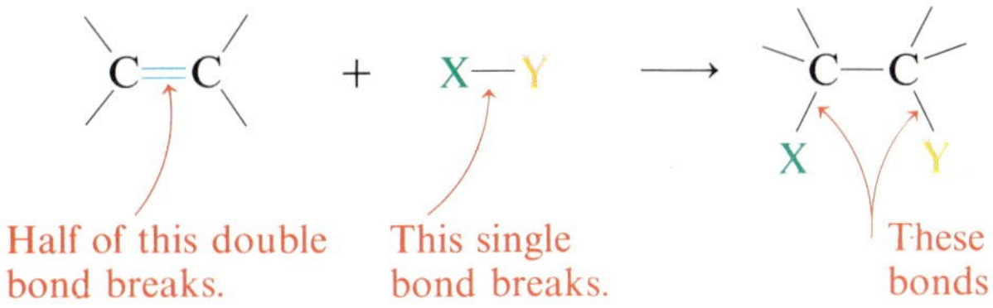

Half of this double bond breaks.
This single bond breaks.
These two single bonds form.

An addition reaction

Addition of Hydrogen to Alkenes and Alkynes: Hydrogenation Alkenes and alkynes react with hydrogen in the presence of a platinum or palladium catalyst to yield the corresponding alkane product:

(an alkene) 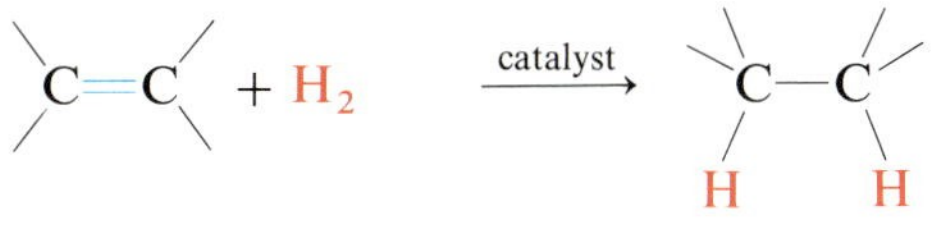 (an alkane)

(an alkyne) 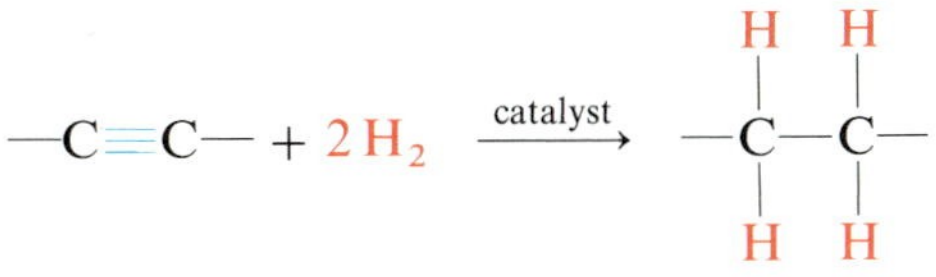 (an alkane)

For example:

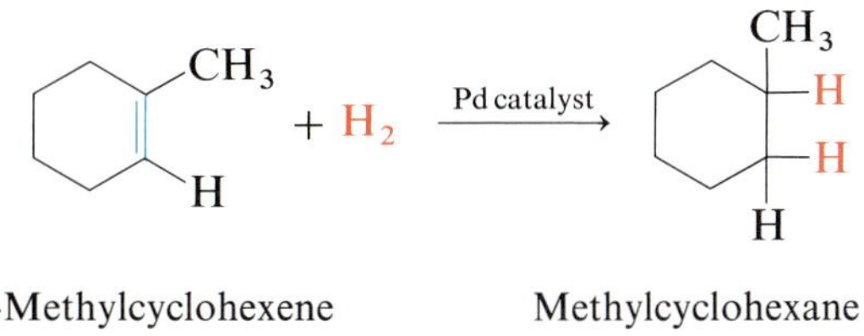

Pd catalyst

1-Methylcyclohexene Methylcyclohexane (85%)

AN APPLICATION: THE CHEMISTRY OF VISION

Does eating carrots really improve your vision? Although carrots probably don't do much to help someone who's already on a proper diet, it's nevertheless true that the chemistry of carrots and the chemistry of vision are related.

Carrots, peaches, sweet potatoes, and other yellow vegetables are rich in β-carotene, an orange-colored alkene that provides our main dietary source of vitamin A. The conversion of β-carotene to vitamin A takes place in the liver, where enzymes first cut the molecule in half and then change the geometry of the C11-C12 double bond to produce a compound named 11-*cis*-retinal. After transport from the liver to the eye, 11-*cis*-retinal reacts with

the protein *opsin* to produce the light-sensitive substance, *rhodopsin*.

The human eye has two kinds of light-sensitive cells: *rod cells*, which are responsible for seeing in dim light, and *cone cells*, which are responsible for color vision. When light strikes the rod cells of the eye, cis-trans isomerization of the C11-C12 double bond occurs and 11-*trans*-rhodopsin, also called *metarhodopsin II*, is produced. This cis-trans isomerization is accompanied by a change in molecular geometry, which in turn causes a nerve impulse to be sent to the brain, where it is perceived as vision. Metarhodopsin II is then changed back to 11-*cis*-retinal for use in another visual cycle.

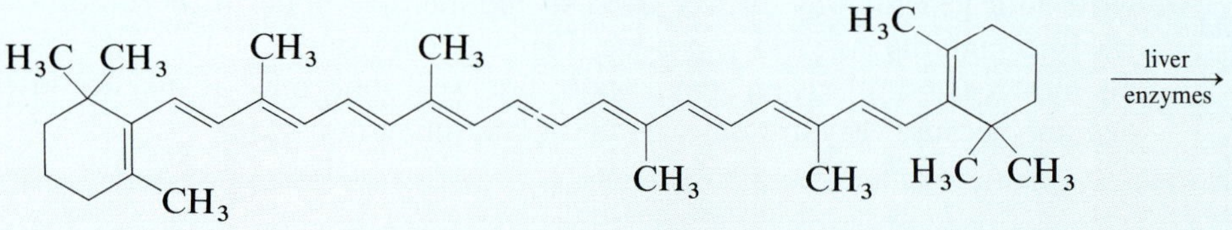

β-Carotene (a poly-alkene)

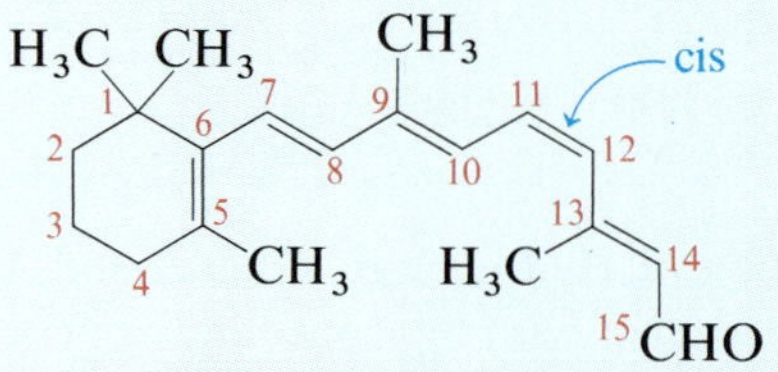

11-*cis*-Retinal

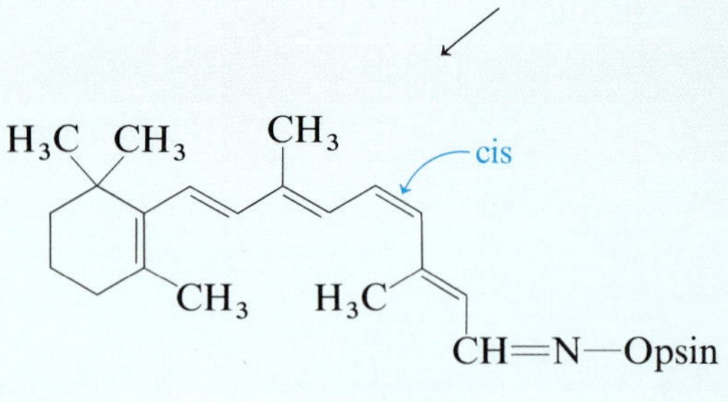

Rhodopsin

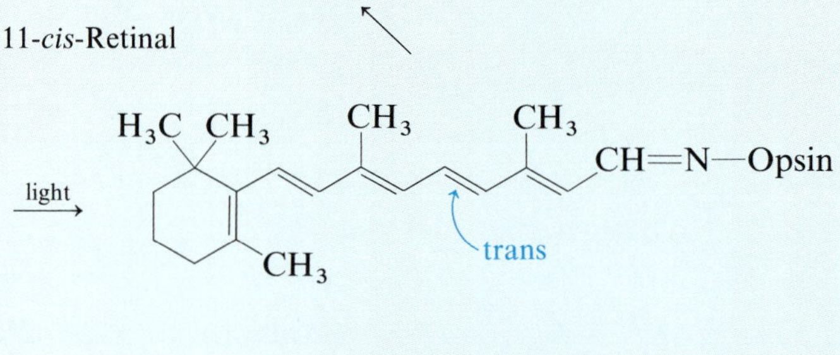

Metarhodopsin II

■ **Hydrogenation** The reaction of an alkene (or alkyne) with H₂ to yield an alkane product.

The addition of hydrogen to an alkene, often called **hydrogenation**, is used commercially to convert unsaturated vegetable oils to the saturated fats used in margarine and cooking fats. We'll see exact structures for these fats and oils in Chapter 14.

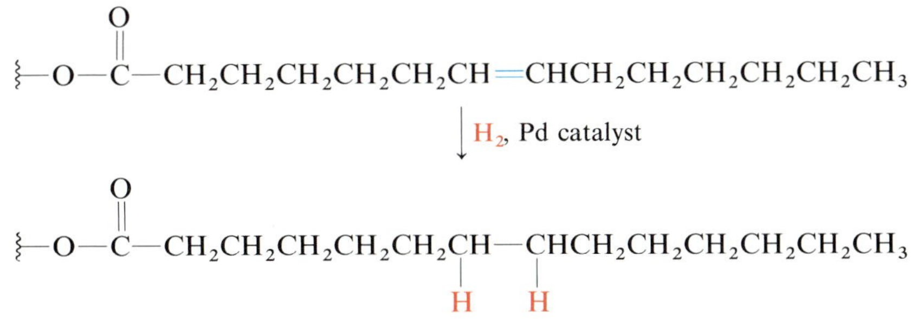

$$\overset{\text{O}}{\underset{\|}{}}$$

$$\text{{\cdot}-O-\overset{\text{O}}{\overset{\|}{C}}-CH_2CH_2CH_2CH_2CH_2CH=CHCH_2CH_2CH_2CH_2CH_2CH_3}$$ Partial structure of a vegetable oil

$$\Big\downarrow \text{H}_2, \text{Pd catalyst}$$

$$\text{{\cdot}-O-\overset{\text{O}}{\overset{\|}{C}}-CH_2CH_2CH_2CH_2CH_2\underset{\text{H}}{\overset{|}{CH}}-\underset{\text{H}}{\overset{|}{CH}}CH_2CH_2CH_2CH_2CH_2CH_3}$$ Partial structure of a saturated cooking fat

---

Solved Problem 9.3   What product would you obtain from the following reaction:

$$CH_3CH_2CH_2CH=CHCH_3 + H_2 \xrightarrow{\text{Pd}} ?$$

Solution   First rewrite the starting material, showing a single bond and two partial bonds in place of the double bond:

$$CH_3CH_2CH_2\overset{|}{C}H-\overset{|}{C}HCH_3$$

Next, add one hydrogen to each carbon atom of the double bond, and rewrite the product in condensed form:

$$CH_3CH_2CH_2\underset{\text{H}}{\overset{|}{CH}}-\underset{\text{H}}{\overset{|}{CH}}CH_3 = CH_3CH_2CH_2CH_2CH_2CH_3 \quad \text{(Hexane)}$$

The final reaction:

$$CH_3CH_2CH_2CH=CHCH_3 + H_2 \xrightarrow{\text{Pd}} CH_3CH_2CH_2CH_2CH_2CH_3$$

Practice Problem   **9.5**   Write the structures of products of these hydrogenation reactions:

(a) $CH_3CH_2CH=CH_2 + H_2 \xrightarrow{\text{Pd}} ?$ (b) *cis*-2-butene + $H_2 \xrightarrow{\text{Pd}} ?$

(c) *trans*-2-butene + $H_2 \xrightarrow{\text{Pd}} ?$

(d) 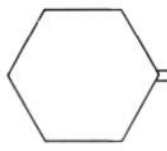 ⬡$=CH_2 + H_2 \xrightarrow{\text{Pd}} ?$

■ **Halogenation** The reaction of an alkene with a halogen (Cl_2 or Br_2) to yield a 1,2-dihaloalkane.

Addition of Cl_2 and Br_2 to Alkenes: Halogenation   Alkenes react with the halogens Br_2 and Cl_2 to give 1,2-dihaloalkane addition products, a process called **halogenation**:

$$\text{(an alkene)} \quad \overset{\diagdown}{\underset{\diagup}{}}C=C\overset{\diagup}{\underset{\diagdown}{}} + X_2 \longrightarrow -\overset{|}{\underset{X}{C}}-\overset{|}{\underset{X}{C}}- \quad \begin{array}{l}\text{(a 1,2-dihaloalkane} \\ \text{where } X = Br \text{ or } Cl)\end{array}$$

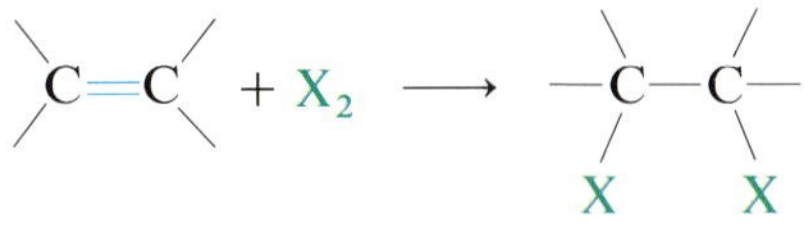

For example:

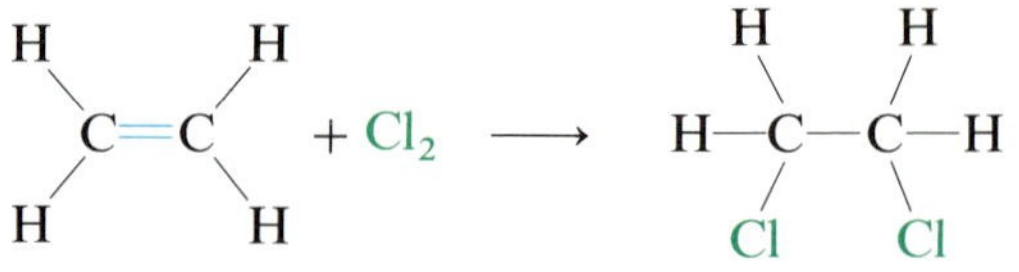

Ethylene   1,2-Dichloroethane

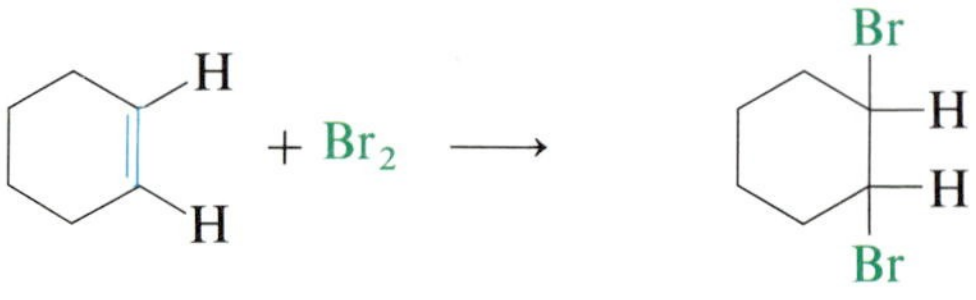

Cyclohexene   1,2-Dibromocyclohexane

The first of the previous two examples is particularly important in the chemical industry. Approximately five million tons per year of 1,2-dichloroethane are manufactured as the first step in making PVC [poly(vinyl chloride)] plastics.

Practice Problem   **9.6**   What products would you expect from these halogen addition reactions?
(a) 2-methylpropene + Br_2 $\longrightarrow$?   (b) 1-pentene + Cl_2 $\longrightarrow$?

■ **Hydrohalogenation** The reaction of an alkene with HCl or HBr to give an alkyl halide.

Addition of HBr and HCl to Alkenes   Alkenes react with hydrogen bromide (HBr) to yield *alkyl bromides* (R—Br) and with hydrogen chloride (HCl) to yield *alkyl chlorides* (R—Cl), a process called **hydrohalogenation**:

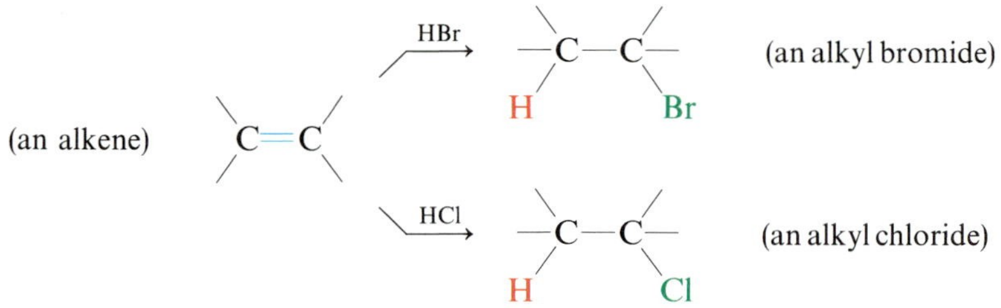

For example:

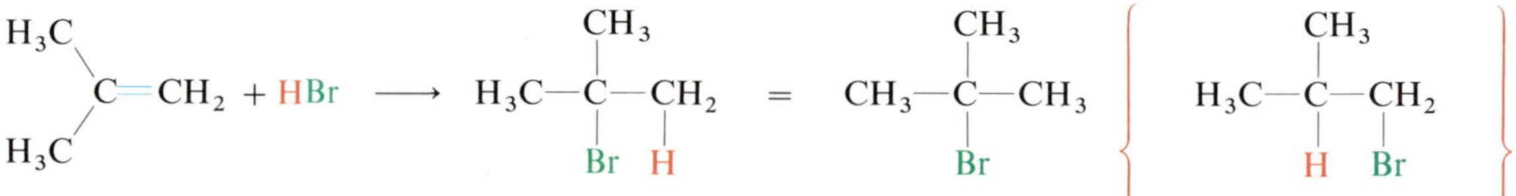

2-Methylpropene   2-Bromo-2-methylpropane   1-Bromo-2-methylpropane
                        (sole product)   (not formed)

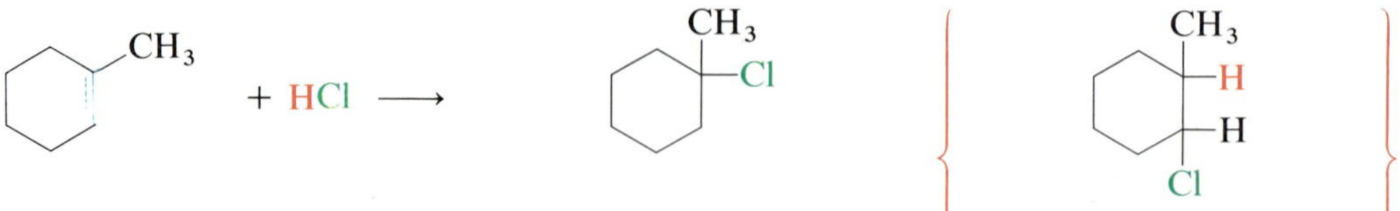

1-Methylcyclohexene   1-Chloro-1-methylcyclohexane   1-Chloro-2-methylcyclohexane
                          (sole product)   (not formed)

Look carefully at the previous two examples. In each case, only one of two possible addition products is obtained. 2-Methylpropene *could* add HBr to give 1-bromo-2-methylpropane, but it doesn't; it gives only 2-bromo-2-methylpropane. Similarly, 1-methylcyclohexene *could* add HCl to give 1-chloro-2-methylcyclohexane, but it doesn't; it gives only 1-chloro-1-methylcyclohexane. These two examples are typical of what happens when HBr and HCl add to alkenes with unsymmetrically substituted double bonds. According to **Markovnikov's rule**, formulated in 1869 by the Russian chemist Vladimir Markovnikov:

■ **Markovnikov's rule** In the addition of HX to an alkene, the H becomes attached to the carbon that already has the most H's, and the X becomes attached to the carbon that has fewer H's.

In the addition of HX to an alkene, the H becomes attached to the carbon that already has the most H's and the X becomes attached to the carbon that has fewer H's.

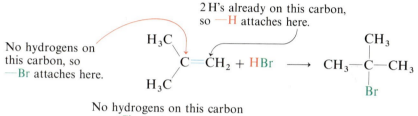

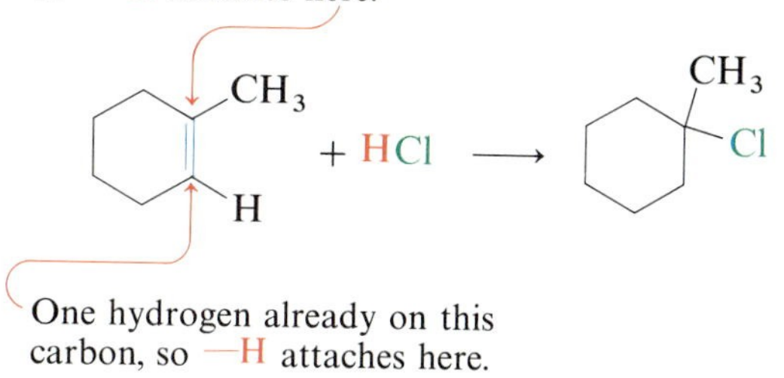

Solved Problem 9.4   What product would you expect from the following reaction?

$$CH_3CH_2\overset{\overset{\displaystyle CH_3}{|}}{C}=CHCH_3 + HCl \longrightarrow \quad ?$$

Solution   We know that reaction of an alkene with HCl leads to formation of an alkyl chloride addition product. The question here is which of two possible products will form. To make a prediction, look at the starting alkene, and count the number of hydrogens already attached to each double-bond carbon. Then write the product by attaching H to the carbon that already has more hydrogens and attaching Cl to the carbon that has fewer hydrogens.

$$CH_3CH_2\overset{\overset{\displaystyle CH_3}{|}}{C}=CHCH_3 + HCl \longrightarrow CH_3CH_2\overset{\overset{\displaystyle CH_3}{|}}{\underset{\underset{\displaystyle Cl}{|}}{C}}-CH_2CH_3$$

No hydrogens on this carbon, so —Cl attaches here.   One hydrogen already on this carbon, so —H attaches here.   3-Chloro-3-methylpentane

Practice Problems   **9.7**   What products would you expect from these following reactions?

(a) cyclohexene + HBr $\longrightarrow$?   (b) 1-hexene + HCl $\longrightarrow$?

(c) $(CH_3)_2CHCH{=}CH_2$ + HI $\longrightarrow$?   (d) 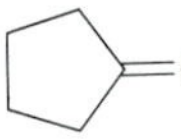 [cyclopentane]$={=}CH_2$ + HCl $\longrightarrow$?

9.8   What alkenes are the following alkyl halides likely to have been made from? (Be careful, there may be more than one answer.)
(a) 3-chloro-3-ethylpentane   (b) $(CH_3)_2CHCBr(CH_3)_2$

Addition of Water to Alkenes: Hydration   An alkene doesn't react with pure water if the two are just mixed together, but if the right experimental conditions are used, an addition reaction takes place to yield an *alcohol* (R—OH). This **hydration** reaction occurs on treatment of the alkene with water in the presence of a strong acid catalyst such as H_2SO_4. In fact, nearly 300 million gallons of ethyl alcohol (ethanol) are produced each year in the United States by this method.

■ **Hydration**   The reaction of an alkene with water to yield an alcohol.

(an alkene) $\overset{\diagdown}{\diagup}C{=}C\overset{\diagup}{\diagdown}$ + H—O—H $\xrightarrow[\text{catalyst}]{H_2SO_4}$

$$-\overset{|}{\underset{H}{C}}-\overset{|}{\underset{O-H}{C}}-$$ (An alcohol)

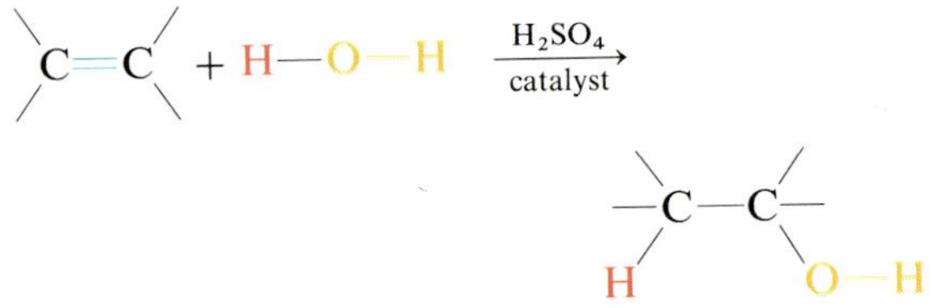

For example:

$$H_2C{=}CH_2 + H_2O \xrightarrow{H_2SO_4,\ 250°C} H_2C{-}CH_2 = CH_3CH_2OH$$
$$\qquad\qquad\qquad\qquad\qquad \underset{H}{|}\ \underset{O-H}{|}$$

Ethylene   Ethyl alcohol

As with the addition of HBr and HCl, Markovnikov's rule can be used to predict the product when water adds to an unsymmetrically substituted alkene. Thus, hydration of 2-methylpropene yields only 2-methyl-2-propanol rather than 2-methyl-1-propanol:

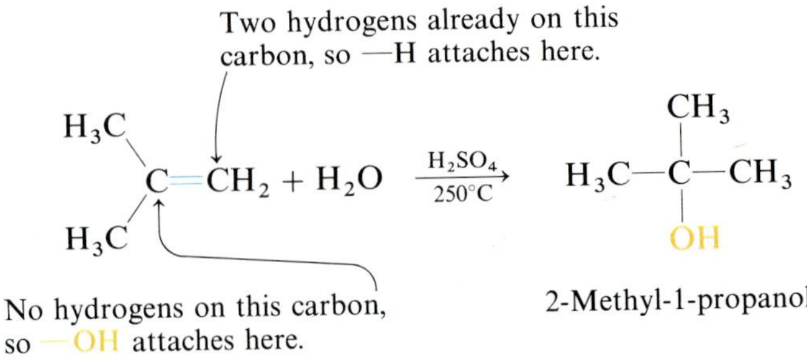

Two hydrogens already on this carbon, so —H attaches here.

$$\overset{H_3C}{\underset{H_3C}{\diagup}}C{=}CH_2 + H_2O \xrightarrow[250°C]{H_2SO_4} H_3C{-}\overset{CH_3}{\underset{OH}{\overset{|}{C}}}{-}CH_3$$

No hydrogens on this carbon, so —OH attaches here.

2-Methyl-1-propanol

Practice Problems   **9.9**   What products would you expect from these hydration reactions?

(a)

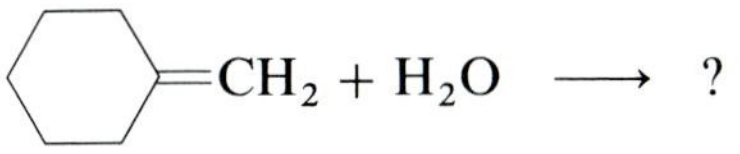

$\diagdown$=CH$_2$ + H$_2$O $\longrightarrow$?

(b)

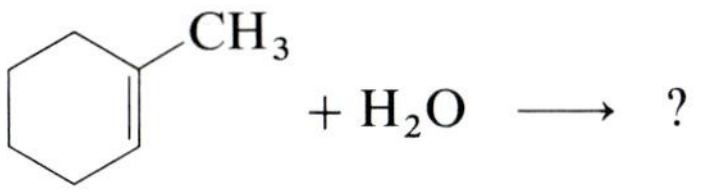

CH$_3$ + H$_2$O $\longrightarrow$?

9.10   What alkene starting material might 3-methyl-3-pentanol be made from?

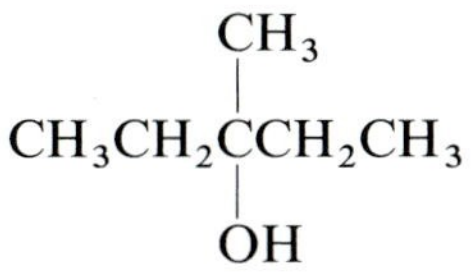

$$CH_3$$

$$CH_3CH_2CCH_2CH_3 \qquad \text{3-Methyl-3-pentanol}$$

$$OH$$

9.4   HOW AN ALKENE ADDITION REACTION OCCURS

How do addition reactions take place? Do two molecules, say ethylene and HBr, simply collide and immediately form a product molecule of bromoethane, or is the process more complex? Studies have shown that alkene addition reactions take place in two distinct steps, as illustrated in Figure 9.2 for the addition of HBr to ethylene.

Figure 9.2
How the addition reaction of HBr to an alkene occurs. The reaction takes place in two steps and involves a carbocation intermediate.

Two electrons from the C—C double bond are used to form a new single bond between the incoming hydrogen ion and one of the carbons. The other carbon now has only six electrons in its outer shell and has a positive charge.

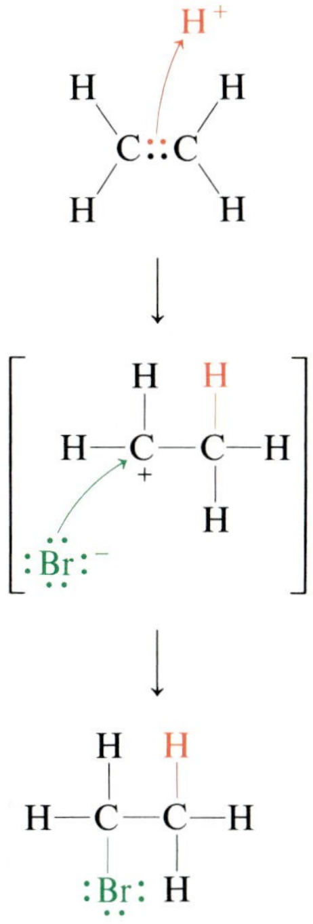

The positively charged *carbocation* then reacts with the negative bromide ion, using an electron pair from bromide to form a single bond between carbon and bromine. The second carbon thus regains an outer-shell octet.

H$^+$

H   H
  $\diagdown$ $\diagup$
    C::C
  $\diagup$ $\diagdown$
H   H

$\downarrow$

$$\left[\begin{array}{c} \text{H} \quad \text{H} \\ | \quad | \\ \text{H}-\text{C}-\text{C}-\text{H} \\ \overset{+}{} \quad | \\ \quad \text{H} \\ :\text{Br}:^- \end{array} \right]$$ *a carbocation*

$\downarrow$

H   H
| |
H—C—C—H
| |
:Br: H

In the first step, the alkene reacts with H^+ from the acid HBr. The carbon-carbon double bond partially breaks, and two electrons go to form a new covalent bond between one of the carbons and the incoming hydrogen. The other double-bond carbon, having had electrons removed from it, now has only six electrons in its outer shell and bears the positive charge. This positively charged carbon-cation, or **carbocation**, then reacts with the negatively charged bromide ion in a second step to give the neutral final product.

■ **Carbocation** A polyatomic ion with a positively charged carbon atom.

Unlike sodium ion (Na^+) and other metal cations, which are stable and easily isolated in salts like NaCl, carbocations are quite unstable and can't be isolated. As soon as it's formed by reaction of an alkene with H^+, the carbocation immediately reacts with bromide ion to form a neutral product.

Practice Problem   *9.11*   Draw the structure of the carbocation intermediate formed during the reaction of 2-methylpropene with HCl.

9.5 AROMATIC COMPOUNDS AND THE STRUCTURE OF BENZENE

■ **Aromatic compound** A compound that contains a six-membered ring of carbon atoms with three double bonds.

In the early days of organic chemistry, the word *aromatic* was used to describe certain fragrant substances from fruits, trees, and other natural sources. It was soon realized, however, that substances grouped as aromatic behaved in a chemically different manner from most other organic compounds. Today, the term **aromatic** is used to refer to the class of compounds containing a six-membered ring with three double bonds. The association with fragrance has long been lost.

Benzene is the simplest aromatic compound, but aspirin, the steroid hormone estrone, and many other important molecules also contain aromatic rings.

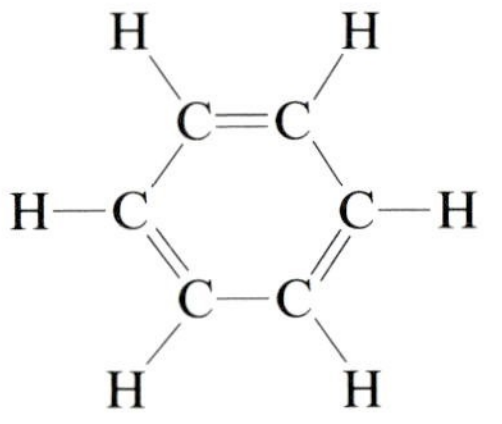

Benzene 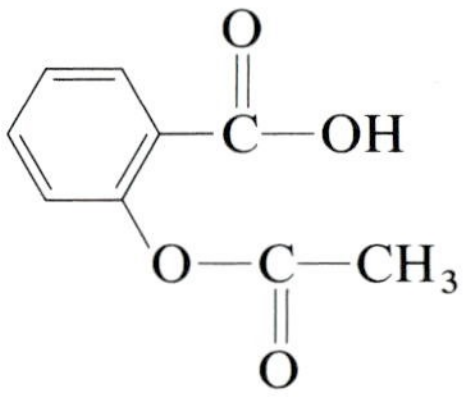   Aspirin 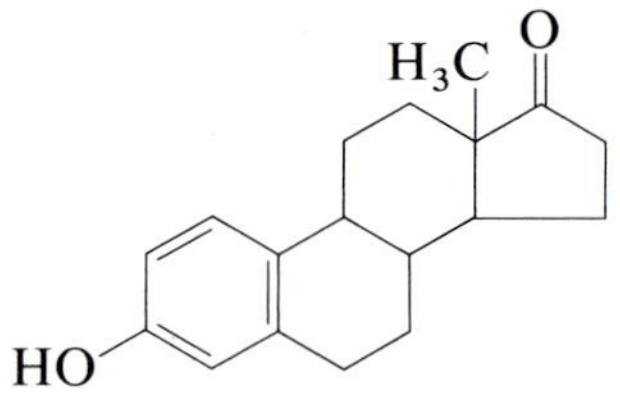   Estrone

Benzene is a flat, symmetrical molecule that is often represented as cyclohexatriene—a six-membered carbon ring with three double bonds. The problem with this representation is that it gives the wrong impression about benzene's chemical reactivity. Since benzene appears to have three double bonds, we might expect it to react with H_2, Br_2, HCl, and H_2O to give the same kinds of addition products that alkenes do. But this expectation is wrong. Benzene and other aromatic compounds are much less reactive than alkenes and don't normally undergo addition reactions.

$$\text{benzene structure} \xrightarrow{\text{HBr}} \text{No reaction}$$

$$\xrightarrow[\text{H}_2\text{SO}_4]{\text{H}_2\text{O}} \text{No reaction}$$

Benzene's remarkable stability is a consequence of its structure. If we imagine forming a ring of six carbons joined by single bonds as shown in Figure 9.3 (a), then six electrons are left over to form the necessary three double bonds. But how do these double bonds form? There are two equivalent possibilities as shown in Figure 9.3 (b).

The problem with the cyclohexatriene representation for benzene is that neither of the two equivalent structures shown in Figure 9.3 is fully correct by itself. The remarkable stability of benzene can best be explained by imagining that benzene is a completely symmetrical molecule and that the six double-bond electrons circulate around the *entire* ring. Such an idea is hard to represent by the standard conventions using lines for covalent bonds, but we sometimes indicate the electron circulation by representing the six electrons as a circle inside the six-membered ring [Figure 9.3 (c)]. Since the six electrons aren't confined to specific double bonds in the normal way, they don't react to give addition products in the normal way.

Figure 9.3
Some representations of benzene. Benzene can be represented either by the two equivalent structures in (b) or by the symmetrical structure in (c).

(a)   (b)   (c)

9.6 NAMING AROMATIC COMPOUNDS

Substituted aromatic compounds use *-benzene* as the family name. Thus, C_6H_5Br is bromobenzene, $C_6H_5CH_2CH_3$ is ethylbenzene, and so on.

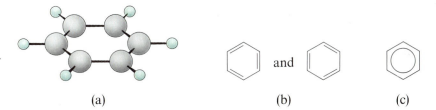

Bromobenzene   Ethylbenzene   Nitrobenzene

Disubstituted aromatic compounds are named using one of the prefixes *ortho-*, *meta-*, or *para-*. An *ortho-* or *o*-disubstituted benzene has its two substituents in a 1,2-relationship on the ring; a *meta-* or *m*-disubstituted benzene has its two substituents in a 1,3-relationship; and a *para-* or *p*-disubstituted benzene has its substituents in a 1,4-relationship.

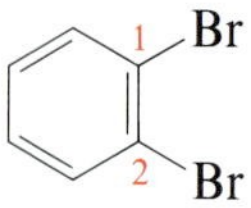

ortho-Dibromobenzene 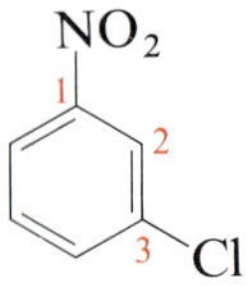   *meta*-Chloronitrobenzene 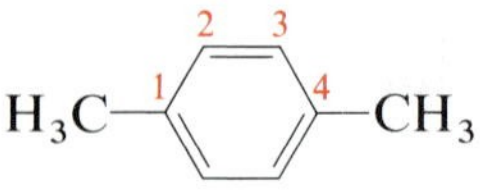   *para*-Dimethylbenzene

Although all substituted aromatic compounds can be named systematically, a number of common ones have common, nonsystematic names as well. For example, methylbenzene is familiarly known as *toluene*, hydroxybenzene as *phenol*, aminobenzene as *aniline*, and so on as shown in Table 9.1.

■ **Phenyl**   The name of the C_6H_5— unit when a benzene ring is considered as a substituent group.

Occasionally, the benzene ring itself might be considered as a substituent group attached to another parent compound. When this happens, the name **phenyl (fen**-nil) is used for the C_6H_5— unit

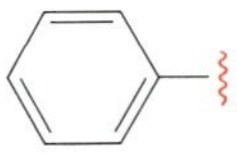

A phenyl group 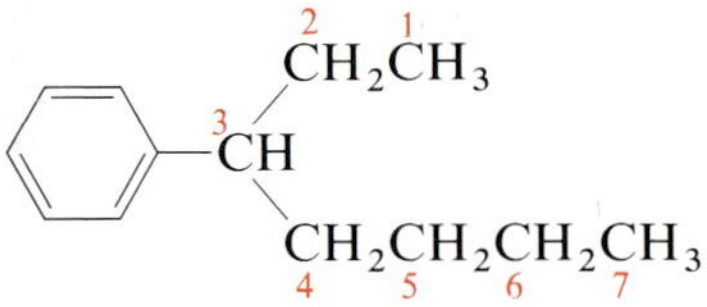   3-phenylheptane

---

Solved Problem 9.5   Draw the structure of *m*-chloroethylbenzene.

Solution   First look at the name. *m*-Chloroethylbenzene must have a benzene ring with two substituents, chloro and ethyl, in a *meta* relationship. Next, draw a benzene ring, and attach one of the substituents, say chloro, to any position:

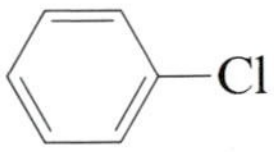

Now, go to the *meta* position two carbons away from the chloro-substituted carbon, and attach the second (ethyl) substituent:

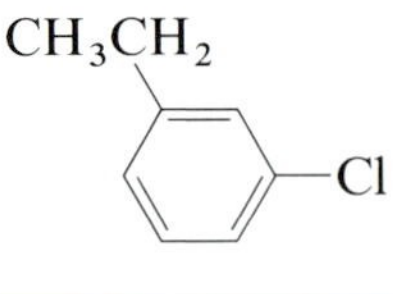

CH_3CH_2

—Cl   *m* — Chloroethylbenzene

Practice Problems   **9.12**   What are the correct IUPAC names for these compounds?

(a)   Cl
      —Br

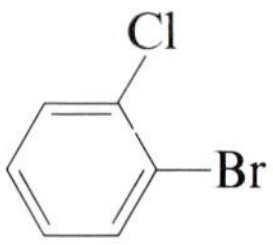

(b) $CH_2CH_2CH_2CH_3$

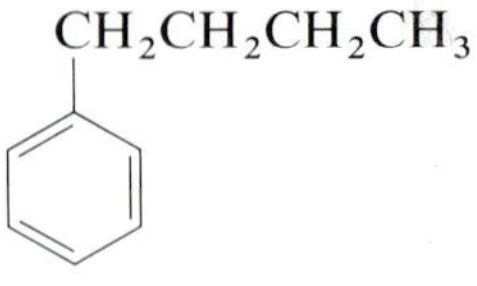

(c)   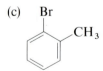 Br
      CH₃

9.13   Draw structures corresponding to these names:
(a) *o*-dibromobenzene   (b) *p*-chlorotoluene   (c) *m*-diethylbenzene

Table 9.1   Common Names of Some Aromatic Compounds

Structure	Name	Structure	Name
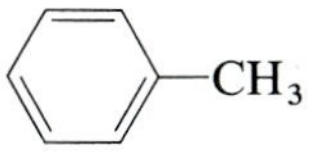 ⬡—CH$_3$	Toluene	 H$_3$C—⬡—CH$_3$	*para*-Xylene
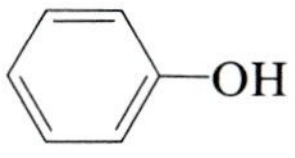 ⬡—OH	Phenol	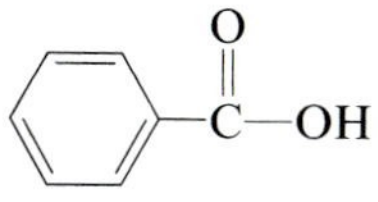 ⬡—C(=O)—OH	Benzoic acid
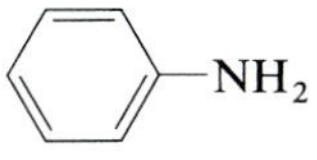 ⬡—NH$_2$	Aniline	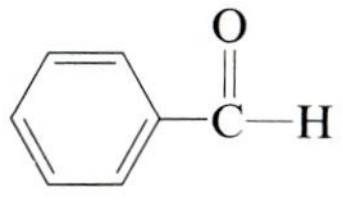 ⬡—C(=O)—H	Benzaldehyde

9.7   CHEMICAL REACTIONS OF AROMATIC COMPOUNDS

■ **Substitution reaction**   A chemical reaction in which two reactants exchange, or substitute, atoms: A + B → C + D.

Unlike alkenes, which undergo addition reactions, aromatic compounds usually undergo **substitution reactions**. That is, a group Y *substitutes* for one of the hydrogen atoms on the aromatic ring without changing the highly stable ring itself. Of course, it doesn't matter which of the six-ring hydrogens is replaced since all six are equivalent.

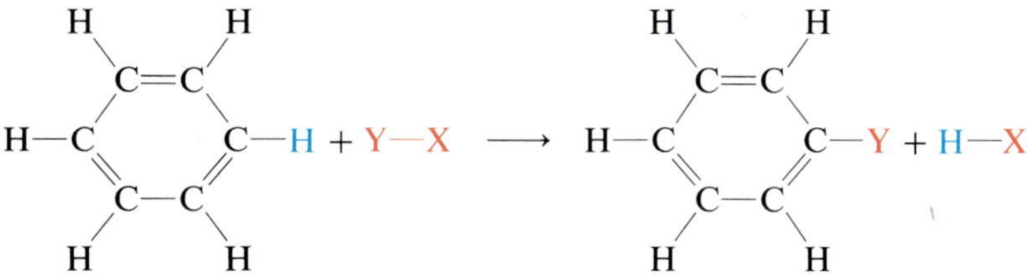

- *Nitration.* Substitution of a nitro group (—NO$_2$) for one of the ring hydrogens occurs when benzene reacts with nitric acid in the presence of sulfuric acid as catalyst:

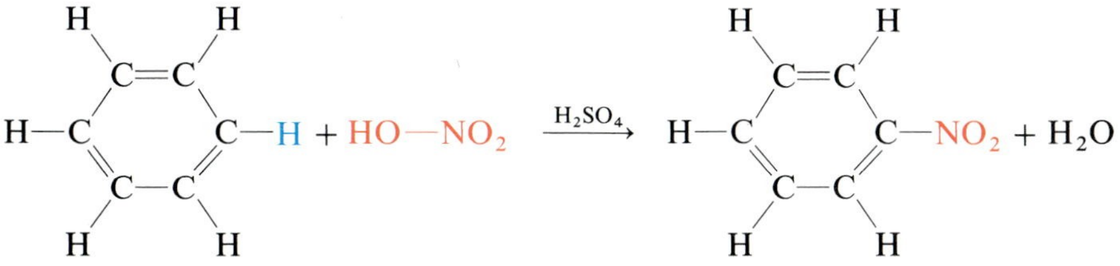

Benzene   Nitric acid   Nitrobenzene

Nitration of aromatic rings is a key step in the synthesis both of explosives like TNT (trinitrotoluene) and of many important pharmaceutical agents. Nitrobenzene itself is the industrial starting material to

Figure 9.4
A variety of dyes are derivatives of aminobenzene, or aniline. Aniline itself is made by the reduction of nitrobenzene.

prepare many of the brightly colored dyes used in clothing (Figure 9.4).

- *Halogenation.* Substitution of a bromine or chlorine for one of the ring hydrogens occurs when benzene reacts with Br_2 or Cl_2 in the presence of iron as catalyst:

Benzene   Chlorine   Chlorobenzene

The chlorination of an aromatic ring is a step used in the synthesis of numerous pharmaceutical agents such as the tranquilizer Librium.

- *Sulfonation.* Substitution of a sulfonic acid group ($-SO_3H$) for one of the ring hydrogens occurs when benzene reacts with concentrated sulfuric acid and sulfur trioxide:

Benzene   Sulfuric acid   Benzenesulfonic acid

Aromatic-ring sulfonation is a key step in the synthesis of such compounds as aspirin and the sulfa-drug family of antibiotics.

$$H_2N-\underset{\text{Sulfanilamide—a sulfa antibiotic}}{\bigcirc}-SO_3NH_2$$

Practice Problems

9.14 Write the products from reaction of these reagents with *para*-xylene (*p*-dimethylbenzene):

(a) Br_2, Fe   (b) HNO_3, H_2SO_4 (c) SO_3, H_2SO_4

9.15 Reaction of Br_2/Fe with toluene (methylbenzene) can lead to a mixture of *three* substitution products. Show the structure of each.

9.8 POLYCYCLIC AROMATIC COMPOUNDS AND CANCER

■ **Polycyclic aromatic compound** A substance that has two or more benzene-like rings fused together along their edges.

The definition of the term *aromatic* can be extended beyond simple monocyclic (one-ring) compounds to include **polycyclic aromatic compounds**—substances that have two or more benzene-like rings joined together to share a common bond. Naphthalene, known familiarly as home "moth balls," is the simplest and best known polycyclic aromatic compound.

In addition to naphthalene, there are many more complex polycyclic aromatic compounds. Benzo[a]pyrene for example, contains five benzene-like rings joined together; ordinary graphite (the "lead" in pencils) consists of enormous two-dimensional sheets of benzene rings.

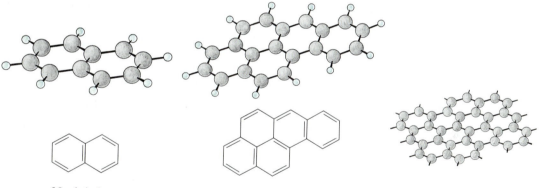

Naphthalene   Benzo[a]pyrene   A graphite segment

■ **Carcinogenic** Cancer-causing.

Benzo[a]pyrene is particularly important because it is one of the **carcinogenic** (cancer-causing) substances found in chimney soot and cigarette smoke. When taken into the body by eating or inhaling, 1,2-benzpyrene is converted by enzymes into an oxygen-containing metabolite. This metabolite is able to bind to cellular DNA, causing mutations and interfering with the normal flow of genetic information. Even benzene itself can cause certain types of cancer on prolonged exposure. Breathing the fumes of benzene and other volatile aromatic compounds in the laboratory should therefore be strictly avoided.

Benzo[a]pyrene   An oxygenated metabolite

INTERLUDE: ALKENE POLYMERS

Polymers are large molecules formed by the bonding together of many smaller molecules called *monomers*. As we'll see in later chapters, *biological polymers* occur throughout nature. Cellulose and starch are polymers built from small sugars; proteins are polymers built from amino acids; and nucleic acids are polymers built from nucleotides. Although the basic idea is the same, synthetic polymers are much simpler than biopolymers, since the starting monomer units are usually small and simple organic molecules.

Many simple alkenes (called *vinyl monomers*) undergo *polymerization* reactions when treated with the proper catalysts. Ethylene yields polyethylene on polymerization, propylene yields polypropylene, and styrene yields polystyrene. The polymer molecules that result may have anywhere from a few hundred to a few thousand monomer units incorporated into a long chain. Some of the more important alkene polymers and their uses are shown in the table.

$$\underset{\text{Vinyl monomer (where the colored}}{\overset{Z}{\underset{|}{H_2C=CH}}} \xrightarrow{\text{polymerization}}$$

Vinyl monomer (where the colored Z represents a substituent group)

$$-(-CH_2-\overset{Z}{\underset{|}{CH}}-CH_2-\overset{Z}{\underset{|}{CH}}-CH_2-\overset{Z}{\underset{|}{CH}}-)-$$

Polymer

Although we won't go into the details of how alkene polymerizations take place, the fundamental reaction involves the same sort of additions to double bonds that we saw in the previous section. The addition to an alkene of a species called an *initiator* yields a reactive intermediate, which in turn adds to a second alkene molecule to produce another reactive intermediate, which adds to a third alkene molecule, and so on.

In 1983, the artist Christo surrounded 11 islands in Biscayne Bay with 6.5 million square feet of floating pink fabric to create a striking work of environmental art. The fabric used was woven polypropylene, an alkene polymer. (© Christo, 1983.)

Some Alkene Polymers and Their Uses

Monomer Name	Structure	Polymer Name	Uses
Ethylene	$H_2C=CH_2$	Polyethylene	Packaging, bottles
Propylene	$H_2C=CH-CH_3$	Polypropylene	Bottles, rope, pails, medical tubing
Vinyl chloride	$H_2C=CH-Cl$	Poly(vinyl chloride)	Insulation, plastic pipe
Styrene	$H_2C=CH-\phenyl$	Polystyrene	Foams and molded plastics
Acrylonitrile	$H_2C=CH-C\equiv N$	Orlon, Acrilan	Fibers, outdoor carpeting
Isoprene	$H_2C=CH-\underset{\underset{CH_3}{\mid}}{C}=CH_2$	Natural Rubber	Tires, hoses

SUMMARY

Alkenes contain a carbon-carbon double bond, and **alkynes** contain a carbon-carbon triple bond. Both families are said to be **unsaturated** since they have fewer hydrogens than corresponding alkanes. Alkenes are named using the family ending -*ene*, while alkynes use the family ending -*yne*.

Alkenes and alkynes have different shapes from alkanes. Whereas alkane carbons have tetrahedral shapes, alkenes are flat (planar) and alkynes are straight (linear). The flat shape of alkenes and the lack of rotation around carbon-carbon double bonds leads to **cis-trans isomerism** for disubstituted alkenes. The cis isomer has its hydrogens on the same side of the double bond; the trans isomer has its hydrogens on opposite sides:

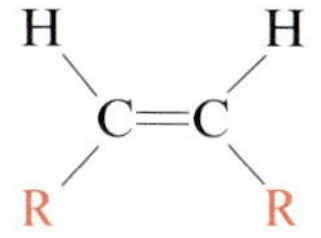

Alkenes undergo **addition reactions** readily. Addition of hydrogen to an alkene (**hydrogenation**) yields an alkane product; addition of Cl_2 or Br_2 (**halogenation**) yields a 1,2-dihaloalkane product; addition of HBr and HCl (**hydrohalogenation**) yields an alkyl halide product; and addition of water (**hydration**) yields an alcohol product. **Markovnikov's rule** lets you predict the direction of addition to an unsymmetrically substituted alkene. These addition reactions take place in two steps and involve intermediate formation of **carbocations**.

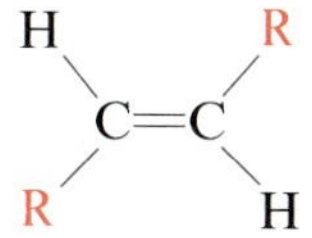

Aromatic compounds contain a six-membered ring of carbons with three double bonds. They are named using the family ending -**benzene**. Disubstituted benzenes are named with one of the prefixes *ortho* (1,2-substitution), *meta* (1,3-substitution), or *para* (1,4-substitution). Aromatic compounds are unusually stable but can be made to undergo **substitution reactions**, where one of the ring hydrogens is replaced by some other group ($C_6H_6 \rightarrow C_6H_5Y$). Among these substitutions are **nitration** (substitution of —NO_2 for —H), **halogenation** (substitution of —Br or —Cl for —H), and **sulfonation** (substitution of —SO_3H for —H).

REVIEW PROBLEMS

Naming Alkenes, Alkynes, and Aromatic Compounds

9.16 Why are alkenes, alkynes, and aromatic compounds said to be unsaturated?

9.17 Not all compounds that smell nice are called "aromatic," and not all compounds called "aromatic" smell nice. Explain.

9.18 Circle the aromatic portions of these molecules:

(a)

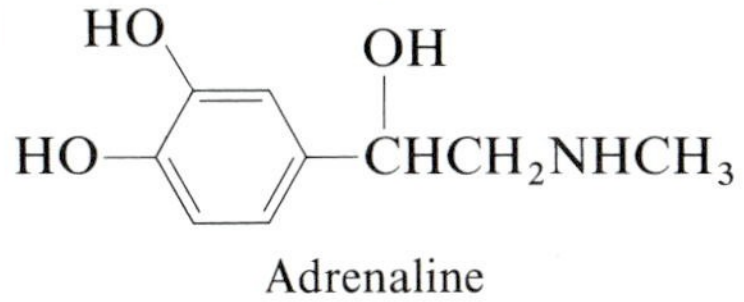

Adrenaline

(b)

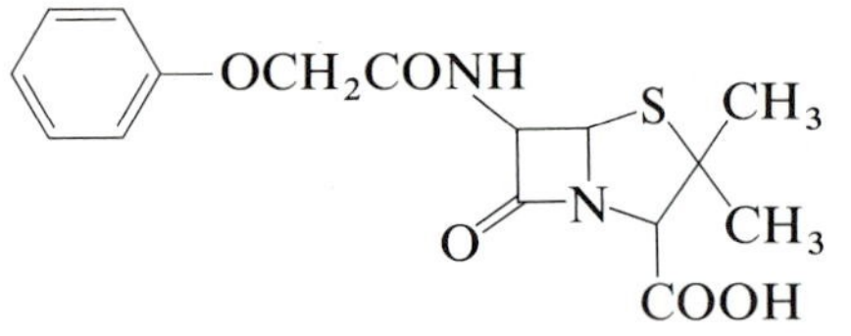

Penicillin V

9.19 What family-name endings are used for alkenes, alkynes, and aromatic compounds?

9.20 Write structural formulas for compounds that meet these descriptions:
(a) an alkene with five carbons
(b) an alkyne with four carbons
(c) a substituted aromatic compound with eight carbons

9.21 What are the IUPAC names of the following compounds?

(a) $CH_3CH_2CH_2CH{=}CH_2$

(b)
$$\underset{\qquad\;\;|}{CH_3}$$
$$CH_3CHCH_2C{\equiv}CCH_3$$

(c) $(CH_3)_2C{=}C(CH_3)_2$

9.22 Draw structures corresponding to these IUPAC names:
(a) *cis*-2-hexene   (b) 2-methyl-3-hexene
(c) 2-methyl-1,3-butadiene

9.23 There are only three alkynes with the formula C_5H_8. Draw and name them.

9.24 Provide correct IUPAC names for these aromatic compounds:

(a)

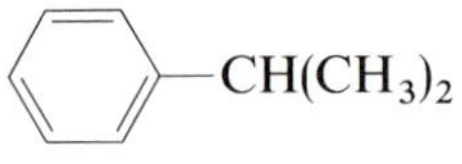

(b)

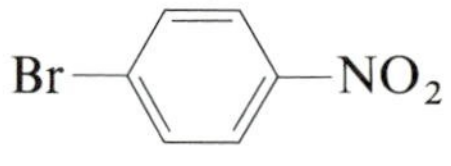

(c)

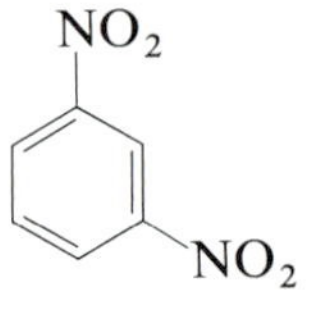

9.25 Draw structures corresponding to these names:
(a) aniline   (b) phenol   (c) *o*-xylene
(d) toluene

9.26 Draw structures corresponding to these names:
(a) *p*-nitrophenol   (b) *o*-chloroaniline
(c) *m*-bromotoluene

9.27 Draw and name all aromatic compounds with the formula C_7H_7Br.

9.28 The following names are incorrect by IUPAC rules. Draw the structures represented by these names, and write correct names.
(a) 2-methyl-4-hexene   (b) 5,5-dimethyl-3-hexyne
(c) 2-butyl-1-propene   (d) 1,5-dibromobenzene

9.29 How many dienes (compounds with two double bonds) are there with the formula C_5H_8? Draw structures of as many as you can.

Alkene cis-trans Isomers

9.30 What requirement must be met for an alkene to show cis-trans isomerism?

9.31 Why don't alkynes show cis-trans isomerism?

9.32 Excluding cis-trans isomers, there are five alkenes with the formula C_5H_{10}. Draw structures for as many as you can, and give their IUPAC names.

9.33 Which of the alkenes in Problem 9.32 can exist as cis-trans isomers?

9.34 Draw structures of the double-bond isomers of these compounds:

(a)

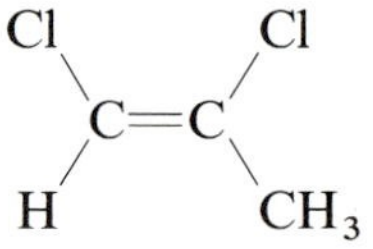

(b)

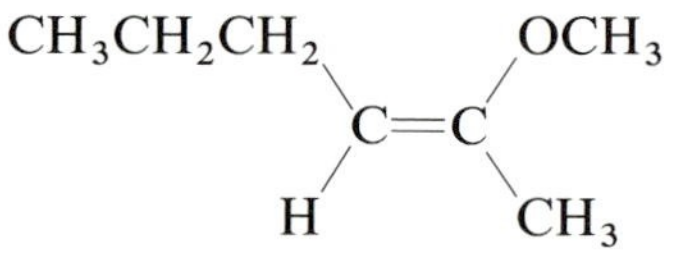

9.35 Which of the following pairs are isomers, and which are identical?

(a)

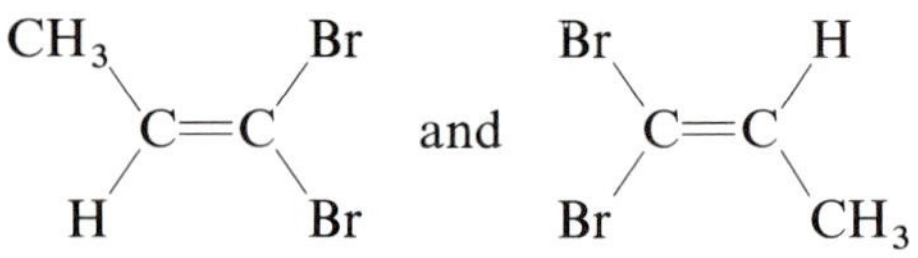

(b)

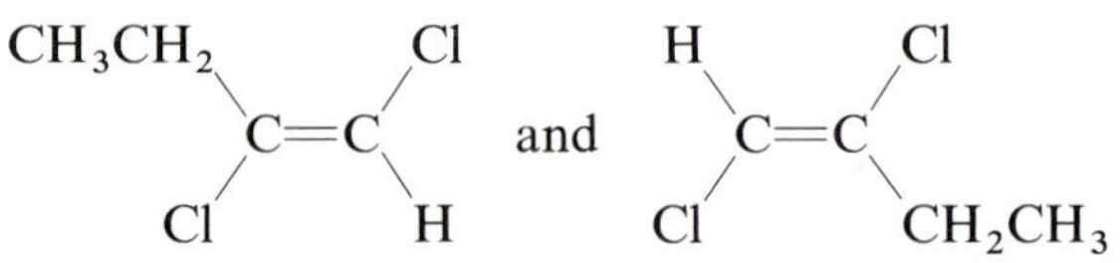

9.36 Why do you suppose small-ring cycloalkenes like cyclohexene don't exist as cis-trans isomers, whereas large-ring cycloalkenes like cyclodecene *do* show isomerism?

Reactions of Alkenes and Alkynes

9.37 What is meant by the term *addition reaction*?

9.38 Give a specific example of an alkene addition reaction.

9.39 Write equations for the reaction of 2,3-dimethyl-2-butene with these reagents:
(a) H_2 and Pd catalyst   (b) Br_2 (c) HBr
(d) H_2O and H_2SO_4 catalyst

9.40 Write equations for the reaction of 2-methyl-2-butene with the reagents shown in Problem 9.39.

Reactions of Aromatic Compounds

9.41 What is meant by the term *substitution reaction*?

9.42 Give a specific example of a substitution reaction of an aromatic compound.

9.43 Benzene normally reacts with only one of the following four reagents. Which of the four is it, and what is the structure of the product?
(a) H_2 and Pd catalyst   (b) Br_2 (c) HBr
(d) H_2O and H_2SO_4 catalyst

9.44 Write equations for the reaction of *p*-dichlorobenzene with these reagents:
(a) Br_2 and Fe catalyst
(b) HNO_3 and H_2SO_4 catalyst
(c) H_2SO_4 and SO_3
(d) Cl_2 and Fe catalyst

9.45 Benzene and other aromatic compounds don't normally react with hydrogen in the presence of a palladium catalyst. If very high pressures (200 atm) and high temperatures are used, however, benzene will add three molecules of H_2 to give an addition product. What is a likely structure for the product?

9.46 How can you account for the fact that reaction of benzene with Br_2 and an iron catalyst yields a single substitution product but the similar reaction of *o*-xylene (*o*-dimethylbenzene) with Br_2 and iron yields a mixture of two different substitution products? What are their structures?

9.47 The explosive trinitrotoluene, or TNT, is made by doing three successive nitration reactions on toluene. If these nitrations take place in the ortho and para positions relative to the methyl group, what is the structure of TNT?

Additional Problems

9.48 Salicylic acid, or *o*-hydroxybenzoic acid, is used as starting material for the industrial preparation of aspirin. Draw the structure of salicylic acid.

9.49 Assume that you have two unlabeled bottles, one with cyclohexane and one with cyclohexene. How could you tell them apart by doing chemical reactions?

9.50 Assume you have two unlabeled bottles, one with cyclohexene and one with benzene. How could you tell them apart by doing chemical reactions?

9.51 *p*-Aminobenzoic acid, or PABA, is commonly used as a sunscreen agent. Draw the structure of PABA.

9.52 Menthene, a compound found in mint plants, has the formula $C_{10}H_{18}$ and the IUPAC name 1-isopropyl-4-methylcyclohexene. What is the structure of menthene?

9.53 Cinnamaldehyde, the pleasant smelling substance found in cinnamon oil, has the following structure. What product would you expect to obtain from reaction of cinnamaldehyde with hydrogen and a palladium catalyst?

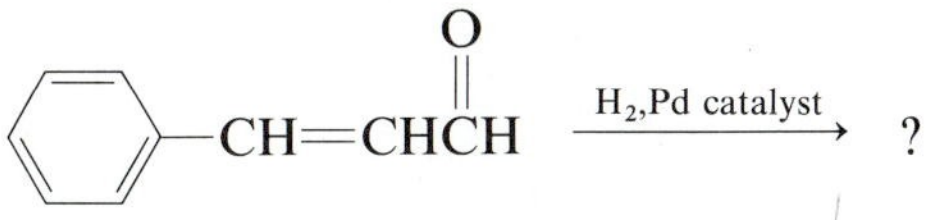

Cinnamaldehyde

9.54 Write the products of these reactions:

(a) $CH_3CH_2CH\!=\!CHCH(CH_3)_2 \xrightarrow{H_2,\ Pd} ?$

(b) 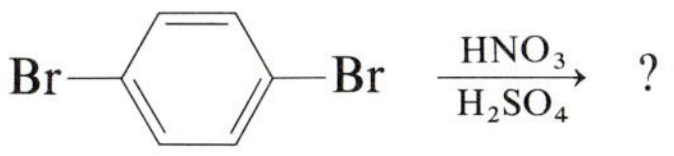
$\xrightarrow[\text{H}_2\text{SO}_4]{\text{HNO}_3} ?$

(c) 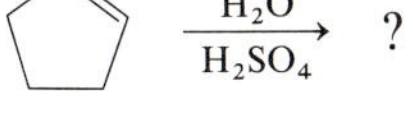
$\xrightarrow[\text{H}_2\text{SO}_4]{\text{H}_2\text{O}} ?$

(d) 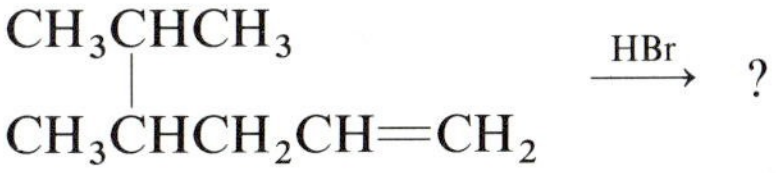 CH_3CHCH_3
     $CH_3CHCH_2CH\!=\!CH_2 \xrightarrow{HBr} ?$

CHAPTER 10

Oxygen, Nitrogen, Sulfur, and Halogen Containing Compounds

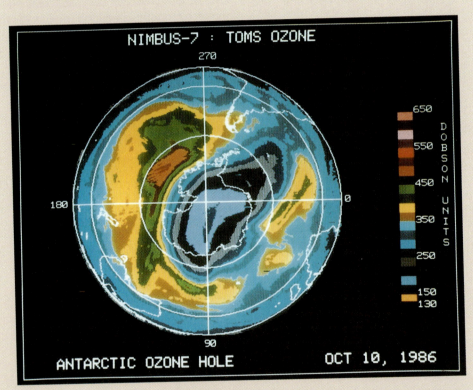

High above the Antarctic, something is eating away the protective layer of ozone that shields us from the sun's harmful ultraviolet radiation. In this chapter you'll meet the compounds believed to be responsible, as well as a variety of other substances both useful and dangerous. (This photo of the ''ozone hole'' was made from satellite data—see the Interlude.)

Unlike the alkanes, alkenes, and aromatic compounds we've seen in the previous two chapters, the families we'll study in this chapter contain other elements in addition to just carbon and hydrogen. Alcohols, phenols, and ethers contain oxygen; amines contain nitrogen; thiols and sulfides contain sulfur; and alkyl halides contain a halogen. These compounds are widely distributed in nature, and most biologically important molecules contain one or more of these functional groups. The questions we'll answer in this chapter include:

1. **What are the characteristic properties of alcohols, phenols, ethers, thiols, and amines?** The goal: you should learn to recognize these functional-group families and should learn their important uses and properties.

2. **How are these compounds named?** The goal: you should learn how to write systematic names for molecules with these functional groups.

3. **Why are alcohols and phenols weak acids?** The goal: you should learn how and why alcohols and phenols are weakly acidic.

4. **What reactions do alcohols, phenols, ethers, thiols, and amines undergo?** The goal: you should learn the characteristic chemical reactions of these functional groups.

5. **Why are amines weak bases?** The goal: you should learn how and why amines are weakly basic.

6. **What properties do alkyl halides have?** The goal: you should learn how to recognize alkyl halides and should learn their properties.

10.1 ALCOHOLS, PHENOLS, AND ETHERS

■ **Alcohol** A compound that has an —OH functional group bonded to an alkane-like carbon atom, R—OH.

■ **Hydroxyl** A name for the —OH group in an organic compound.

■ **Phenol** A compound that has an —OH functional group bonded directly to an aromatic, benzene-like ring.

■ **Ether** A compound that has an oxygen atom bonded to two carbon atoms, R—O—R.

Alcohols are compounds that have an —OH group (**hydroxyl**) bonded to a saturated, alkane-like carbon atom; **phenols** have an —OH group bonded directly to an aromatic, benzene-like ring; and **ethers** have an oxygen atom bonded to two organic groups. Compounds in all three families can be thought of as organic derivatives of water in which one or both of the water hydrogens are replaced by an organic substituent.

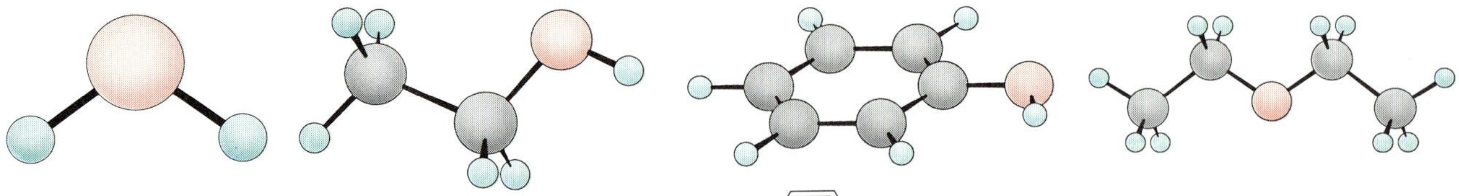

$$H-O-H$$

Water

$$CH_3CH_2-O-H$$

Ethyl *alcohol*

⟨phenol ring⟩$-O-H$

Phenol

$$CH_3CH_2-O-CH_2CH_3$$

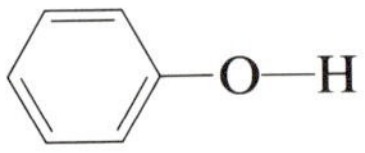

Diethyl *ether*

Practice Problem **10.1** Identify each of these compounds as an alcohol, a phenol, or an ether:

(a) $CH_3CH_2CHCH_3$
        $|$
        OH

(b) 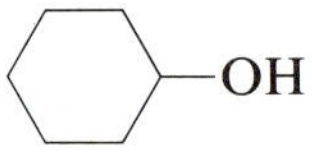 ⟨cyclohexane ring⟩$-OH$

(c) 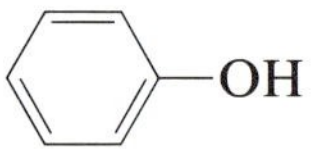 ⟨benzene ring⟩$-OH$

(d) 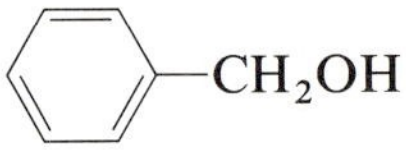 ⟨benzene ring⟩$-CH_2OH$

(e) 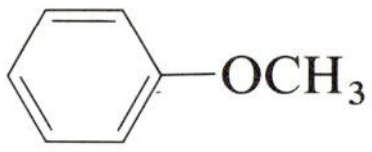 ⟨benzene ring⟩$-OCH_3$

(f) $CH_3CHOCH_2CH_3$
        $|$
        CH_3

10.2   OCCURRENCE AND USES OF ALCOHOLS, PHENOLS, AND ETHERS

■ **Wood alcohol** A common name for methyl alcohol, CH_3OH.

■ **Grain alcohol** A common name for ethyl alcohol, CH_3CH_2OH.

Simple alcohols are among the most important and commonly encountered of all organic chemicals. Methyl alcohol (CH_3OH, methanol), for example, is known to everyone as **wood alcohol** because it was once prepared by heating wood in the absence of air. Methanol is toxic to humans, causing blindness in low doses and death in larger amounts.

Ethyl alcohol (CH_3CH_2OH, ethanol) is one of the oldest known pure organic chemicals; its production by fermentation of grain and sugar goes back for many thousands of years. Sometimes called **grain alcohol**, ethanol is the "alcohol" present in all wines (10 to 13 percent), beers (3 to 5 percent), and distilled liquors (35 to 90 percent).

Isopropyl alcohol, commonly called *rubbing alcohol*, is used for rubdowns because it cools the skin through evaporation and causes pores to close. It is also used as a solvent for medicines and as a sterilant for instruments.

$$CH_3OH$$

Methyl alcohol

$$CH_3CH_2OH$$

Ethyl alcohol

$$CH_3-\overset{\overset{\displaystyle OH}{|}}{CH}-CH_3$$

Isopropyl alcohol

The word *phenol* is the name both of a specific compound (hydroxybenzene) and of a family of compounds. Phenol itself, also called *carbolic acid*, is a medical antiseptic first used by Joseph Lister in 1867. Most mouthwashes and throat lozenges, in fact, contain alkyl-substituted phenols as their active ingredients. In addition to their uses in medicine, phenols have a variety of other applications. Pentachlorophenol, for example, is a wood preservative, and BHT (butylated hydroxytoluene) is a common food preservative.

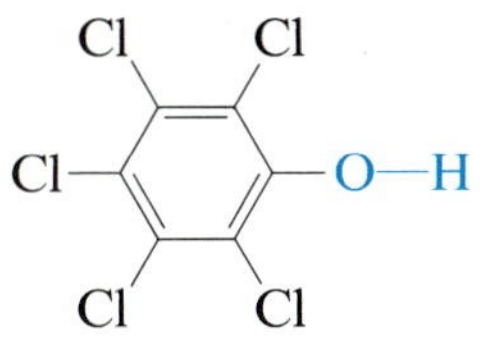

Pentachlorophenol 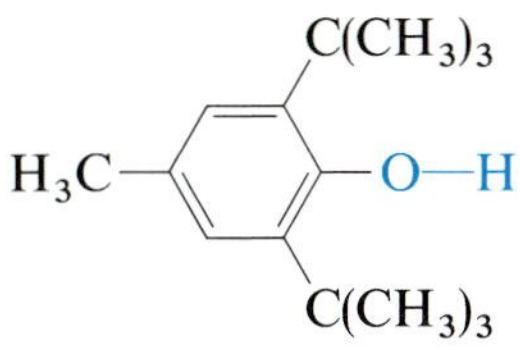   BHT

Ethers are less frequently encountered than alcohols or phenols but are widely used in the chemical industry as solvents. Diethyl ether, $CH_3CH_2OCH_2CH_3$, was also used at one time as a surgical anesthetic, although it has now been replaced by safer, nonflammable substitutes.

10.3   NAMING ALCOHOLS, PHENOLS, AND ETHERS

■ **Primary alcohol** An alcohol in which the —OH-bearing carbon atom is bonded to one other carbon (and two hydrogens), RCH_2OH.

■ **Secondary alcohol** An alcohol in which the —OH-bearing carbon atom is bonded to two other carbons (and one hydrogen), R_2CHOH.

■ **Tertiary alcohol** An alcohol in which the —OH-bearing carbon atom is bonded to three other carbons (and no hydrogens), R_3COH.

Alcohols   Alcohols are classified as either **primary** (1°), **secondary** (2°), or **tertiary** (3°), depending on the number of carbon substituents bonded to the hydroxyl-bearing carbon. Of course, the substituent groups don't have to be the same, so we'll use the representations R, R′, and R″, to indicate different groups.

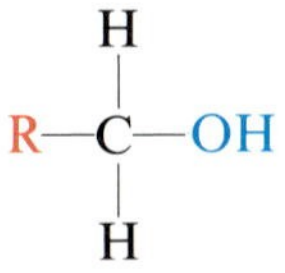

A *primary* alcohol
(one R group on
OH-bearing carbon)

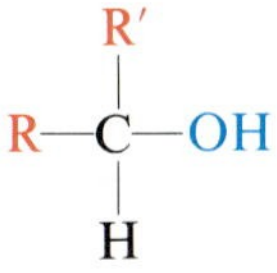

A *secondary* alcohol
(two R groups on
OH-bearing carbon)

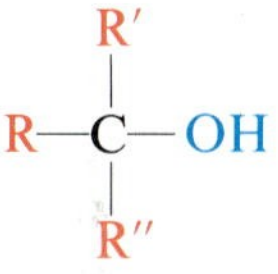

A *tertiary* alcohol
(three R groups on
OH-bearing carbon)

Alcohols are named as derivatives of the parent alkane, using the *-ol* ending, according to the following steps.

Step 1. Name the parent compound. Find the longest chain *containing the hydroxyl group*, and name the chain by replacing the *-e* ending of the corresponding alkane with *-ol*:

$$CH_3CH_2CH_2CH_2OH$$ Name as a *butanol*—a four-carbon chain containing a hydroxyl group.

Step 2. Number the carbon atoms in the main chain. Begin at the end nearer the hydroxyl group:

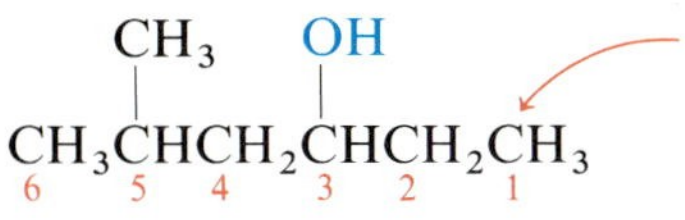

Begin at this end because it's nearer the OH group.

Step 3. Write the full name. Number all substituents according to their position on the main chain, and list them alphabetically:

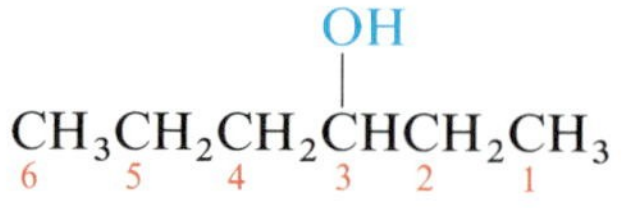

$$\underset{6\quad\ 5\quad\ 4\quad\ 3\quad\ 2\quad\ 1}{CH_3CH_2CH_2\overset{\overset{\displaystyle OH}{|}}{C}HCH_2CH_3}$$

3-Hexanol

$$\underset{1\quad\ 2\quad\ 3\quad\ 4\quad\ 5\quad\ 6\quad\ 7}{CH_3\overset{\overset{\displaystyle OH}{|}}{C}HCH_2CH_2CH_2\overset{\overset{\displaystyle CH_3}{|}}{C}HCH_3}$$

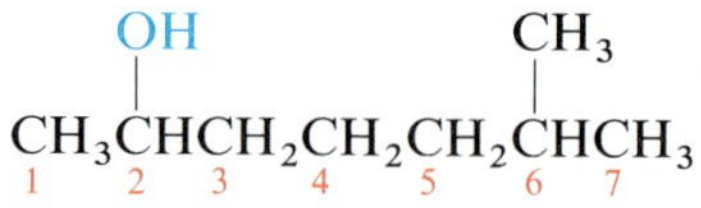

6-Methyl-2-heptanol

In addition to their systematic names, some of the simplest alcohols also have common names (Table 10.1). For instance, methanol (CH_3OH) is often called *methyl alcohol,* ethanol (CH_3CH_2OH) is often called *ethyl alcohol,* and so on.

■ **Glycol** A dialcohol, or compound that contains two —OH groups.

Notice in Table 10.1 that *dialcohols* are often called **glycols**. Ethylene glycol ($HOCH_2CH_2OH$), commonly used as automobile antifreeze, is the simplest glycol. Propylene glycol is often used as a solvent for medicines that need to be inhaled or rubbed onto the skin.

Phenols   Phenols are usually named with the family ending *-phenol* rather than *-benzene.* For example:

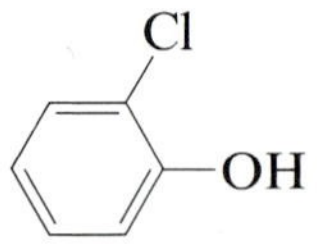

o-Chlorophenol    *p*-Methylphenol

Ethers   Simple ethers are named just by identifying the two alkyl groups bonded to oxygen and adding the word *ether.* Thus, $CH_3—O—CH_3$ is dimethyl ether, $CH_3—O—CH_2CH_3$ is ethyl methyl ether, and $CH_3CH_2—O—CH_2CH_3$ is diethyl ether.

Table 10.1   Common Names of Some Alcohols

Structure	Name	Structure	Name
CH_3OH	Methyl alcohol	$CH_2—CH_2$ $\ \ \|\quad\quad\ \|$ $OH\quad OH$	Ethylene glycol
CH_3CH_2OH	Ethyl alcohol		
$CH_3\overset{\overset{\displaystyle OH}{\|}}{C}HCH_3$	Isopropyl alcohol	$CH_2—CH—CH_3$ $\ \ \|\quad\quad\ \|$ $OH\quad OH$	Propylene glycol
$CH_3—\overset{\overset{\displaystyle CH_3}{\|}}{\underset{\underset{\displaystyle CH_3}{\|}}{C}}—OH$	*tert*-Butyl alcohol	$CH_2—CH—CH_2$ $\ \ \|\quad\quad\ \|\quad\quad\ \|$ $OH\quad OH\quad OH$	Glycerol (glycerin)

Practice Problems   **10.2**   Draw structures corresponding to these names:
(a) 3-methyl-3-pentanol   (b) isopropyl methyl ether   (c) *m*-bromophenol
(d) cyclohexanol   (e) 2-methyl-4-heptanol

10.3   Give systematic names for these compounds:

(a)
$$\underset{\displaystyle CH_3CH_2CHCH_2CH_3}{\overset{\displaystyle OH}{\vert}}$$

(b)
$$\underset{\displaystyle CH_3CH_2CHCH_2CH_2CH_3}{\overset{\displaystyle CH_2OH}{\vert}}$$

(c)
Cl—⬡—OCH$_3$

(d) Br
—⬡—OH

(e) H$_3$C
⬡—OH
H$_3$C

10.4   PROPERTIES OF ALCOHOLS, PHENOLS, AND ETHERS

As mentioned earlier, alcohols can be thought of as organic derivatives of water in which one hydrogen of H_2O has been replaced by an organic substituent. This similarity between alcohols and water extends to many chemical and physical properties as well. For example, compare the boiling points of ethanol, dimethyl ether, propane, and water:

$$\underset{\substack{\text{Ethanol}\\(\text{mol wt}=46;\\ \text{bp}=78.5°C)}}{\overset{\displaystyle H\!-\!\overset{\overset{\displaystyle H}{\vert}}{\underset{\underset{\displaystyle H}{\vert}}{C}}\!-\!\overset{\overset{\displaystyle H}{\vert}}{\underset{\underset{\displaystyle H}{\vert}}{C}}\!-\!O\!-\!H}{}} \qquad \underset{\substack{\text{Dimethyl ether}\\(\text{mol wt}=46;\\ \text{bp}=-23°C)}}{\overset{\displaystyle H\!-\!\overset{\overset{\displaystyle H}{\vert}}{\underset{\underset{\displaystyle H}{\vert}}{C}}\!-\!O\!-\!\overset{\overset{\displaystyle H}{\vert}}{\underset{\underset{\displaystyle H}{\vert}}{C}}\!-\!H}{}} \qquad \underset{\substack{\text{Propane}\\(\text{mol wt}=44;\\ \text{bp}=-42°C)}}{\overset{\displaystyle H\!-\!\overset{\overset{\displaystyle H}{\vert}}{\underset{\underset{\displaystyle H}{\vert}}{C}}\!-\!\overset{\overset{\displaystyle H}{\vert}}{\underset{\underset{\displaystyle H}{\vert}}{C}}\!-\!\overset{\overset{\displaystyle H}{\vert}}{\underset{\underset{\displaystyle H}{\vert}}{C}}\!-\!H}{}} \qquad \underset{\substack{\text{Water}\\(\text{mol wt}=18;\\ \text{bp}=100°C)}}{\overset{\displaystyle H\!-\!O\!-\!H}{}}$$

Ethanol, dimethyl ether, and propane have similar molecular weights, yet ethanol boils more than 100° higher than the other two. In fact, the boiling point of ethanol is close to that of water. Why should this be?

We saw in Section 6.6 that water has a high boiling point because of the formation of hydrogen bonds. These hydrogen bonds cause an attraction between individual molecules that prevents their easy escape into the vapor phase. In a similar manner, hydrogen bonds form between alcohol (and phenol) molecules when the positively polarized OH-group hydrogen of one molecule orients toward the negatively polarized oxygen of another molecule (Figure 10.1). Since alkanes and ethers don't have hydroxyl groups, they can't form hydrogen bonds and therefore have much lower boiling points.

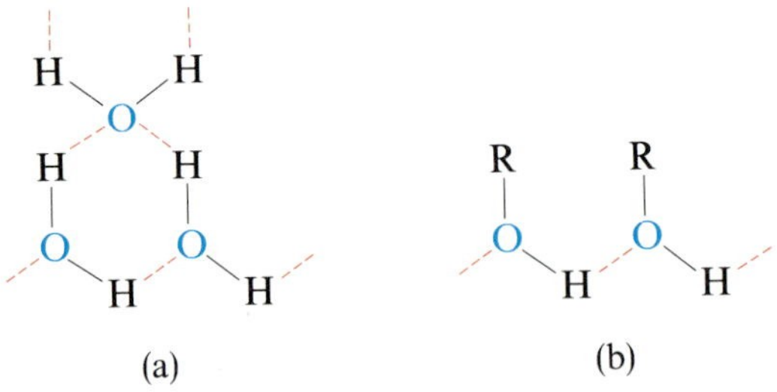

Figure 10.1
The formation of hydrogen bonds in water (a) and in alcohols (b) causes attractive forces that prevent easy escape of the molecules into the vapor phase and result in high boiling points.

(a)   (b)

Solubility behavior is another property shared by water and alcohols. Because of its polar bonds and its ability to solvate ions, water is an excellent solvent for ionic compounds but a poor solvent for nonpolar organic molecules (Section 6.7). Similarly, low molecular weight alcohols such as methanol and ethanol are quite water-like in their solubility behavior. Thus, methanol and ethanol are both infinitely soluble in water and can dissolve small amounts of many salts. Higher molecular weight alcohols such as 1-heptanol, however, are much more alkane-like and less water-like. Thus, 1-heptanol is nearly insoluble in water and can't dissolve salts but does dissolve alkanes.

$$CH_3\text{—}OH \qquad\qquad CH_3CH_2CH_2CH_2CH_2CH_2CH_2\text{—}OH$$

Methanol—has a small organic part and is therefore water-like.

1-Heptanol—has a large organic part and is therefore alkane-like.

Ethers, since they lack the hydroxyl group of water and alcohols, are alkane-like in their solubility behavior and are excellent solvents for most organic molecules. They are also alkane-like in their chemical properties and do not react with most acids, bases, or other chemical reagents.

10.5 ACIDITY OF ALCOHOLS AND PHENOLS

Like water, alcohols and phenols are weakly acidic. That is, both alcohols and phenols can donate a hydrogen ion, H^+, to a base in the same way that water can (Section 7.6).

Water: $H\text{—}O\text{—}H \rightleftharpoons H\text{—}O^- + H^+$

An alcohol: $CH_3\text{—}O\text{—}H \rightleftharpoons CH_3\text{—}O^- + H^+$

A phenol: ⟨phenyl⟩$\text{—}O\text{—}H \rightleftharpoons$ ⟨phenyl⟩$\text{—}O^- + H^+$

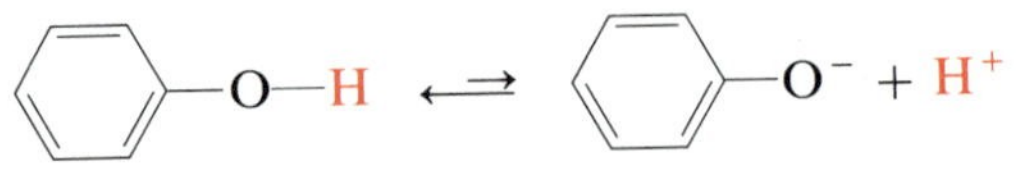

Alcohols are about as acidic as water itself, and the pH of their aqueous solutions is therefore neutral (pH = 7). Thus, alcohols don't react with sodium hydroxide and aren't soluble in NaOH solutions. Phenols, however, are considerably more acidic than water. They are readily deprotonated by NaOH and are soluble in dilute aqueous sodium hydroxide.

An alcohol: $CH_3-O-H + Na^+OH^- \longrightarrow$ No reaction

A phenol:

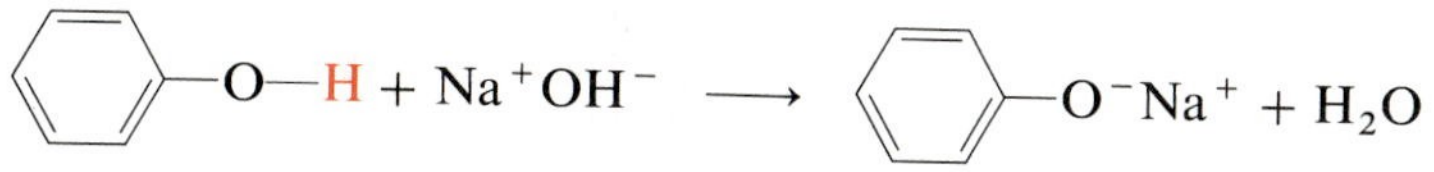

$-O-H + Na^+OH^- \longrightarrow$ $-O^-Na^+ + H_2O$

10.6   CHEMICAL REACTIONS OF ALCOHOLS

■ **Dehydration** Loss of water from an alcohol to yield an alkene.

Dehydration of Alcohols   Alcohols undergo loss of water (**dehydration**) on treatment with a strong acid catalyst. The —OH group is lost from one carbon, and an —H is lost from an adjacent carbon to yield an alkene product:

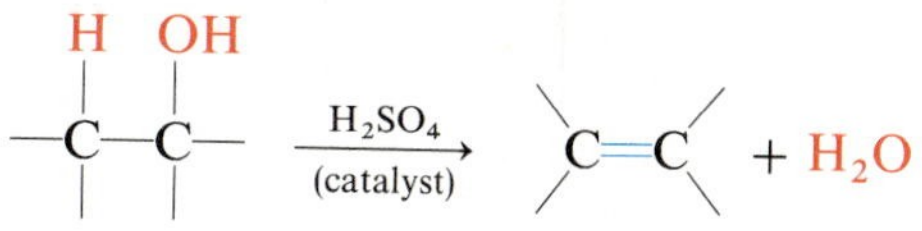

For example:

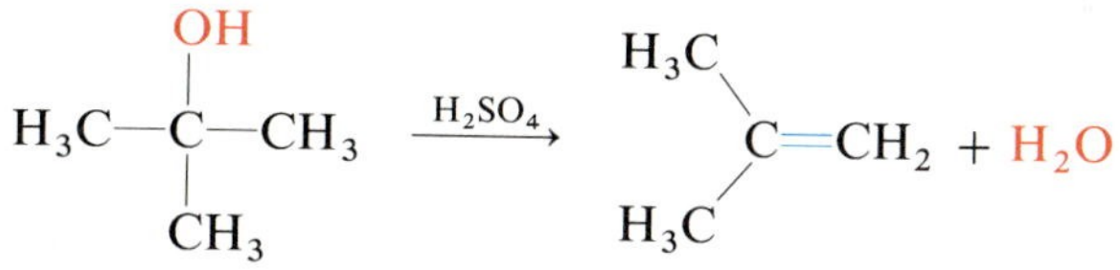

In cases where more than one alkene can result from dehydration, a mixture of products is usually formed. A good rule of thumb is that the *major* product is the one that has the greater number of alkyl groups attached to the double-bond carbons. For example, the major product obtained from dehydration of 2-butanol is 2-butene rather than 1-butene:

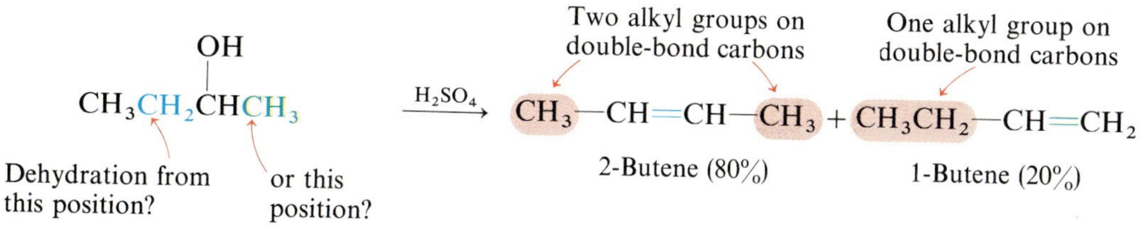

Solved Problem 10.1 What products would you expect from the following dehydration reaction? Which product will be major, and which minor?

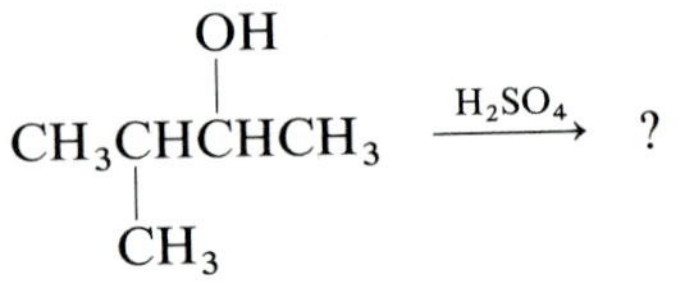

$$CH_3CHCHCH_3 \xrightarrow{H_2SO_4} ?$$

with OH on second carbon and CH$_3$ below third carbon

Solution First find the hydrogens on carbons next to the OH-bearing carbon, and rewrite the structure to emphasize these hydrogens:

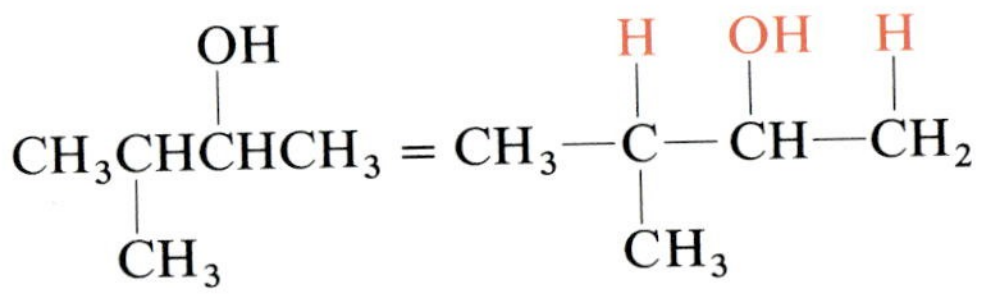

$$CH_3CHCHCH_3 = CH_3-\overset{\text{H}}{\underset{\text{CH}_3}{C}}-\overset{\text{OH}}{CH}-CH_2$$

Next, circle and remove the possible combinations of H and OH, drawing a double bond where H and OH have come from:

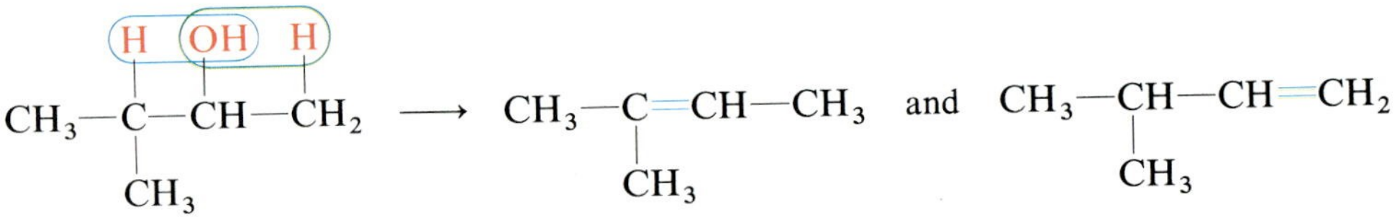

$$CH_3-\overset{\text{H}}{\underset{\text{CH}_3}{C}}-\overset{\text{OH}}{CH}-CH_2 \longrightarrow CH_3-\underset{\text{CH}_3}{C}=CH-CH_3 \text{ and } CH_3-\underset{\text{CH}_3}{CH}-CH=CH_2$$

Finally, determine which of the alkenes has the larger number of alkyl substituents on its double-bond carbons and is therefore the major product:

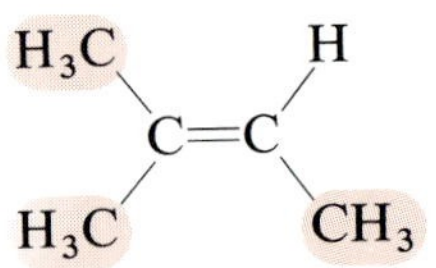

$$\begin{matrix} H_3C \\ \\ H_3C \end{matrix} C=C \begin{matrix} H \\ \\ CH_3 \end{matrix}$$

2-Methyl-2-butene
major product (three alkyl groups)

and

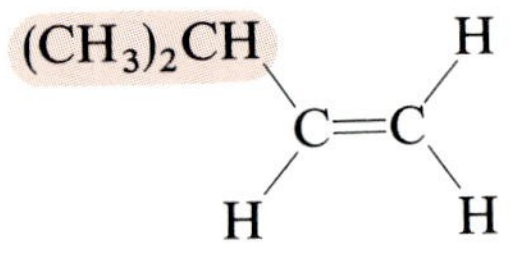

$$\begin{matrix} (CH_3)_2CH \\ \\ H \end{matrix} C=C \begin{matrix} H \\ \\ H \end{matrix}$$

3-Methyl-1-butene
minor product (one alkyl group)

Practice Problems

10.4 What alkenes might be formed by dehydration of the following alcohols? If more than one product is possible, indicate which is major.

(a) $CH_3CH_2CH_2OH$ (b) 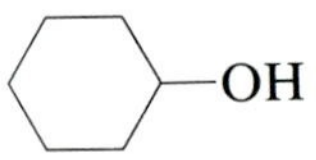 ⬡—OH (c) 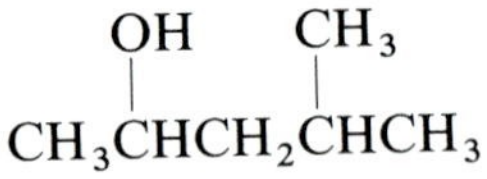 $\underset{}{CH_3CHCH_2CHCH_3}$ with OH and CH$_3$ substituents

10.5 Dehydration of what alcohols would yield these alkenes?
(a) $(CH_3)_2C=C(CH_3)_2$ (b) $CH_3CH_2CH=CH_2$

■ **Carbonyl group** A functional group consisting of a carbon atom joined to an oxygen atom by a double bond, C=O.

Oxidation of Alcohols Primary and secondary alcohols are converted into carbonyl-containing compounds on treatment with an oxidizing agent. A **carbonyl group**—pronounced car-bo-**neel**—is a functional group that has a carbon atom joined to an oxygen atom by a double bond, C=O. Many different oxidizing agents can be used—$KMnO_4$, $Na_2Cr_2O_7$, and HNO_3, for example—and it usually doesn't matter which specific reagent is chosen. Thus, we'll simply use the symbol [O] to indicate a generalized reagent.

■ **Oxidation** In organic chemistry, the removal hydrogen from a molecule or addition of oxygen to a molecule.

The net effect of an alcohol **oxidation** is the removal of two hydrogen atom joined to an oxygen atom by a double bond, C=O. Many different the carbon atom next to the —OH group. These hydrogens are converted into water during the reaction by the oxidizing agent [O].

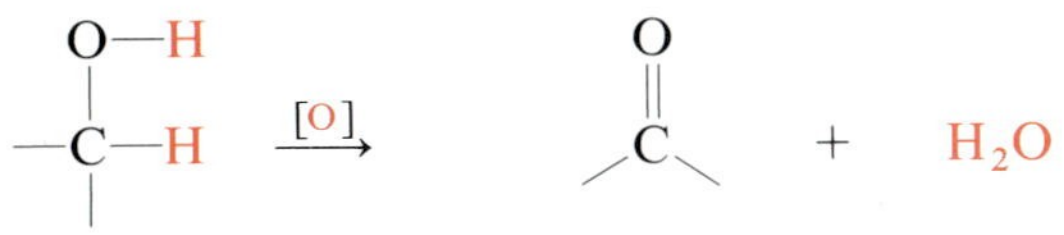

An alcohol   A carbonyl compound

■ **Aldehyde** A compound containing the —CHO functional group, RCHO.

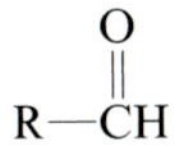

$$R-\overset{O}{\underset{}{\overset{\|}{C}}}H$$

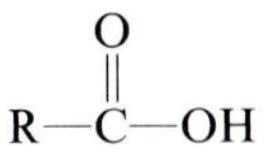

Different kinds of carbonyl-containing products are formed depending on the structure of the starting alcohol and on the exact reaction conditions. Thus, primary alcohols (RCH_2OH) are converted either into **aldehydes** (RCHO) if carefully controlled conditions are used or into **carboxylic acids** (RCOOH) if an excess of oxidant is used:

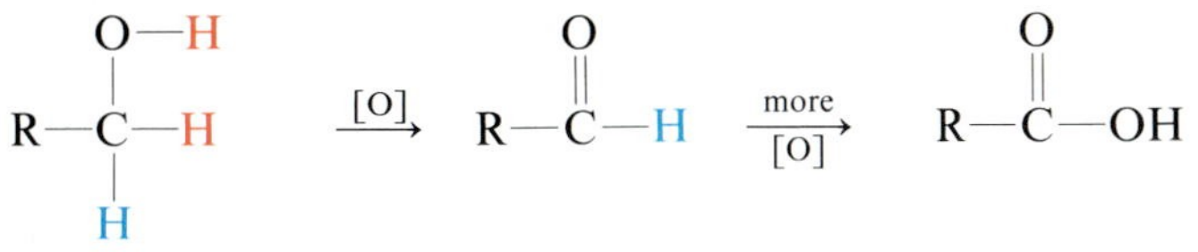

■ **Carboxylic acid** A compound containing the —COOH functional group, RCOOH.

$$R-\overset{O}{\underset{}{\overset{\|}{C}}}OH$$

A primary alcohol   An aldehyde   A carboxylic acid

For example:

1-Butanol   Butanal   Butanoic acid

■ **Ketone** A compound that has a carbonyl group bonded to two carbon atoms, R_2C=O.

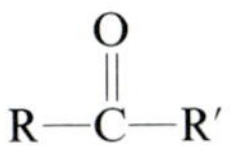

$$R-\overset{O}{\underset{}{\overset{\|}{C}}}R'$$

Secondary alcohols (R_2CHOH) are converted into **ketones** (R_2C=O) on treatment with oxidizing agents:

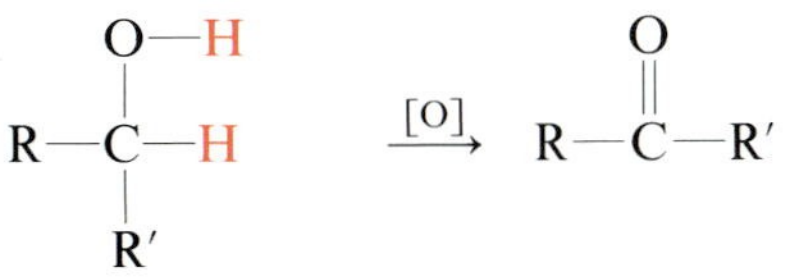

A secondary alcohol   A ketone

For example:

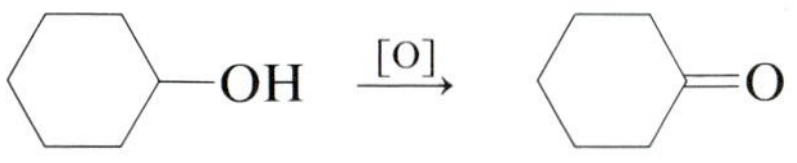

Cyclohexanol   Cyclohexanone

Tertiary alcohols don't normally react with oxidizing agents because they don't have a hydrogen on the carbon atom next to the —OH group:

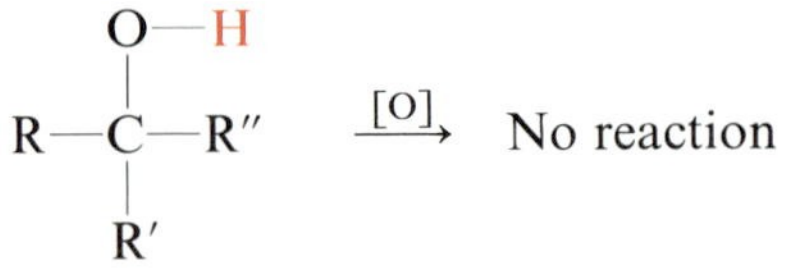

$$\underset{\text{A tertiary alcohol}}{R-\overset{\overset{\displaystyle O-H}{|}}{\underset{\underset{\displaystyle R'}{|}}{C}}-R''} \xrightarrow{\text{[O]}} \text{No reaction}$$

Alcohol oxidations are critically important steps in many biological processes. For example, when lactic acid builds up in tired, overworked muscles, the liver removes it by oxidizing it to pyruvic acid. Our bodies, of course, don't use $Na_2Cr_2O_7$ or $KMnO_4$ for the oxidation; they use specialized and highly selective enzymes to carry out their chemistry. Regardless of the details, though, the net chemical transformation is the same whether carried out in a laboratory flask or in a living cell.

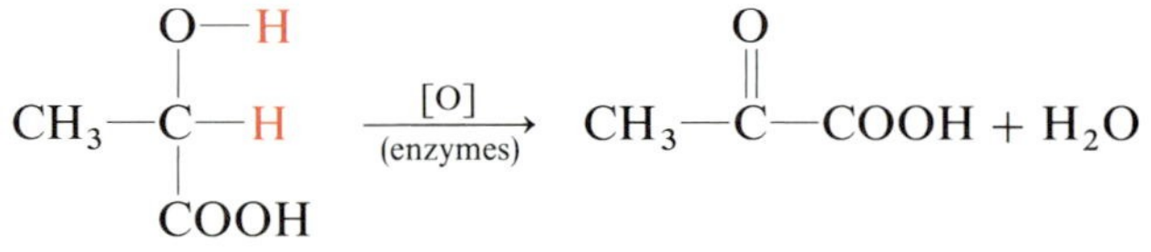

$$\underset{\substack{\text{Lactic acid}\\\text{(an alcohol-acid)}}}{CH_3-\overset{\overset{\displaystyle O-H}{|}}{\underset{\underset{\displaystyle COOH}{|}}{C}}-H} \xrightarrow[\text{(enzymes)}]{\text{[O]}} \underset{\text{Pyruvic acid}}{CH_3-\overset{\overset{\displaystyle O}{||}}{C}-COOH} + H_2O$$

Solved Problem 10.2   What are the products of the following reaction?

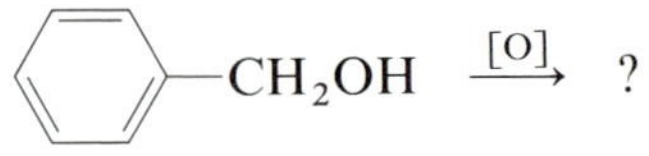

$$\underset{\text{Benzyl alcohol}}{\text{⟨benzene⟩}-CH_2OH} \xrightarrow{\text{[O]}} \text{?}$$

Solution   Since the starting material is a primary alcohol, it will be converted first to an aldehyde and then to a carboxylic acid. To find the structures of these products, first rewrite the starting alcohol to emphasize the hydrogen atoms on the OH—bearing carbon:

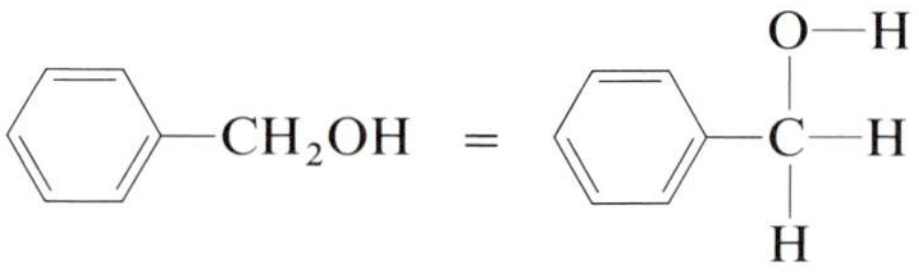

$$\text{⟨benzene⟩}-CH_2OH \quad = \quad \text{⟨benzene⟩}-\overset{\overset{\displaystyle O-H}{|}}{\underset{\underset{\displaystyle H}{|}}{C}}-H$$

Next, circle and remove two hydrogens, one from the —OH group and one from the neighboring carbon. In their place, make a C=O double bond. This is the aldehyde product that will form initially:

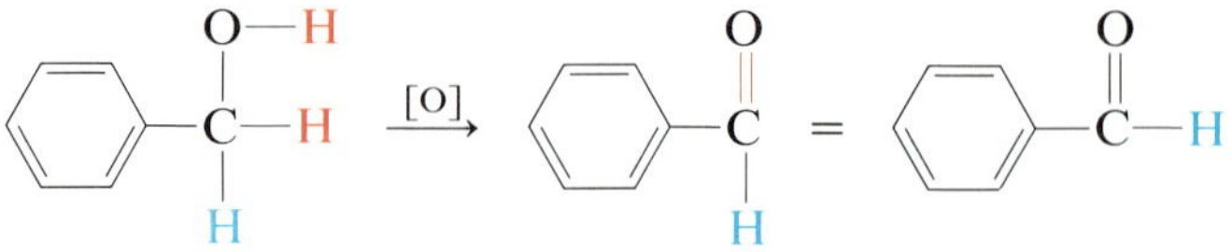

$$\text{⟨benzene⟩}-\overset{\overset{\displaystyle O-H}{|}}{\underset{\underset{\displaystyle H}{|}}{C}}-H \xrightarrow{\text{[O]}} \text{⟨benzene⟩}-\overset{\overset{\displaystyle O}{||}}{\underset{\underset{\displaystyle H}{|}}{C}} \quad = \quad \text{⟨benzene⟩}-\overset{\overset{\displaystyle O}{||}}{C}-H$$

Finally, convert the aldehyde to a carboxylic acid by removing the —CH=O hydrogen and replacing it with an —OH group. This is the final product:

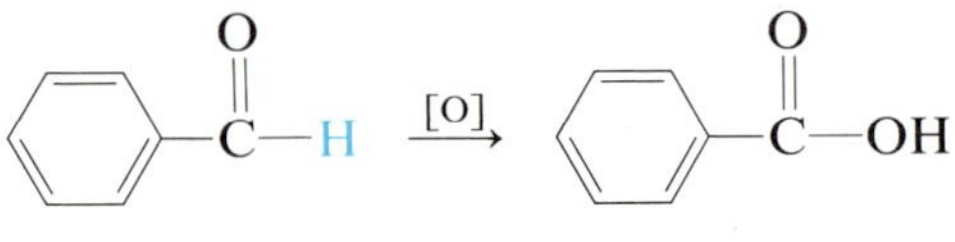

$$\text{⟨benzene⟩}-\overset{\overset{\displaystyle O}{||}}{C}-H \xrightarrow{\text{[O]}} \text{⟨benzene⟩}-\overset{\overset{\displaystyle O}{||}}{C}-OH$$

Practice Problems   **10.6**   What products would you expect from oxidation of these alcohols?

(a) $CH_3CH_2CH_2OH$

(b)
$$\overset{\displaystyle OH}{\underset{\displaystyle}{CH_3CHCH_2CH_2CH_3}}$$

(c)  ⬠—$CH(OH)CH_3$

10.7   What alcohols might these carbonyl products have come from?

(a)
$$\overset{\displaystyle O}{\underset{\displaystyle CH_3CCH_3}{\|}}$$

(b) ⬡=O

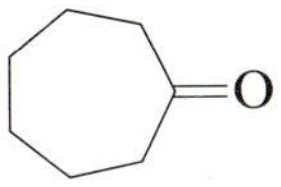

(c)
$$\overset{\displaystyle CH_3}{\underset{\displaystyle CH_3CHCH_2COOH}{|}}$$

10.7   SULFUR-CONTAINING COMPOUNDS: THIOLS AND DISULFIDES

■ **Thiol**   A compound containing the —SH functional group, R—SH.

■ **Mercaptan**   An alternate name for a thiol, R—SH.

Sulfur is the element just below oxygen in the periodic table, and many oxygen-containing compounds have sulfur analogs. For example, **thiols**, R—SH, are sulfur analogs of alcohols. Thiols, also called **mercaptans**, are named in the same way as alcohols, with the family ending *-thiol* used in place of *-ol*.

$$CH_3CH_2SH$$

$$\overset{\displaystyle CH_3}{\underset{\displaystyle CH_3CHCH_2CH_2SH}{|}}$$

$$CH_3CH=CHCH_2SH$$

Ethanethiol   3-Methyl-1-butanethiol   2-Butene-1-thiol

The most outstanding characteristic of thiols is their appalling odor. Skunk scent is caused by two of the simple thiols shown above, 3-methyl-1-butanethiol and 2-butene-1-thiol.

Thiols react with mild oxidizing agents to yield **disulfides**, R—S—S—R. Two thiols join together in this reaction, the SH hydrogen from each is lost, and the two sulfurs bond together.

■ **Disulfide**   A compound containing a sulfur-sulfur single bond, R—S—S—R.

$$H_3C—S—H + H—S—CH_3 \quad \xrightarrow[\text{(oxidizing agent)}]{[O]} \quad CH_3—S—S—CH_3 + H_2O$$

Methanethiol   Dimethyl disulfide

Thiols are important biologically because they occur as a functional group in the amino acid cysteine and in many proteins. Hair, for example, is unusually rich in —SH groups. When hair is "permed," a mild oxidizing agent causes disulfide bonds to form between —SH groups, resulting in the introduction of bends and kinks into the hair (Figure 10.2, page 220).

Practice Problem   **10.8**   What disulfides would you obtain from oxidation of these thiols?
(a) $CH_3CH_2CH_2SH$ (b) 3-methyl-1-butanethiol (skunk scent)

Figure 10.2
Giving hair a permanent wave involves forming disulfide "bridges" between—SH groups in protein molecules.

10.8 NITROGEN-CONTAINING COMPOUNDS: AMINES

■ **Amine** A compound that has one or more organic groups bonded to nitrogen, RNH_2, R_2NH, or R_3N.

■ **Primary amine** An amine that has one organic group bonded to nitrogen, RNH_2.

■ **Secondary amine** An amine that has two organic groups bonded to nitrogen, R_2NH.

■ **Tertiary amine** An amine that has three organic groups bonded to nitrogen, R_3N.

Amines are compounds that contain one or more organic groups bonded to nitrogen: RNH_2, R_2NH, or R_3N. Thus, they are organic derivatives of ammonia in the same way that alcohols and ethers are organic derivatives of water. Amines are classified either as **primary**, **secondary**, or **tertiary**, depending on how many organic substituents are bonded to the nitrogen atom.

H—N—H
|
H

Ammonia

R—N—H
|
H

A *primary* amine
(one R group on nitrogen)

R—N—H
|
R′

A *secondary* amine
(two R groups on nitrogen)

R—N—R″
|
R′

A *tertiary* amine
(three R groups on nitrogen)

Primary amines, RNH_2, are named simply by identifying the alkyl group attached to nitrogen and then adding the suffix "-*amine*." The —NH_2 group itself is referred to as an *amino* group.

$CH_3CH_2NH_2$

Ethylamine

$$CH_3\overset{\overset{\displaystyle CH_3}{|}}{C}HNH_2$$

Isopropylamine

—NH_2

Cyclohexylamine

Secondary and tertiary amines are named as *N*-substituted derivatives of a primary amine. The largest of the organic groups bonded to nitrogen is chosen as the parent, and the other groups are considered as *N*-substituents (*N* because they are attached directly to nitrogen). If two or more of the alkyl groups bonded to nitrogen are identical, the appropriate prefix *di*-, or *tri*- is used.

$$CH_3CH_2\overset{\overset{\displaystyle }{|}}{\underset{\underset{\displaystyle CH_2CH_3}{|}}{N}}CH_2CH_3$$

Triethylamine

$$H_3C\overset{\overset{\displaystyle }{|}}{\underset{\underset{\displaystyle CH_3}{|}}{N}}CH_2CH_2CH_3$$

N,N-Dimethylpropylamine

(Propylamine is the parent name; the two methyl groups are substituents on nitrogen.)

Practice Problems

10.9   Identify these compounds as primary, secondary, or tertiary amines:

(a) $CH_3CH_2CH_2NH_2$ (b) $CH_3CH_2NHCH_2CH_3$

(c)

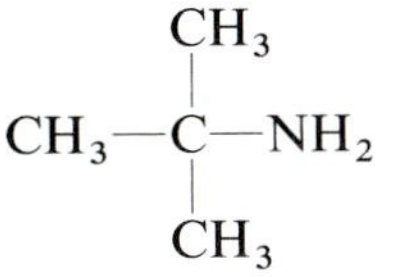

$$CH_3-\underset{\underset{CH_3}{|}}{\overset{\overset{CH_3}{|}}{C}}-NH_2$$

(d)

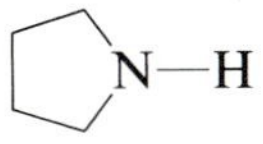

$\text{(ring)}\ N-H$

(e)

$\text{(ring)}\ N-CH_3$

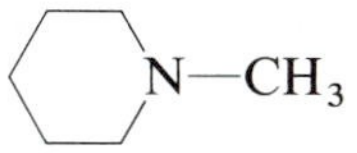

10.10   What are the names of these amines?

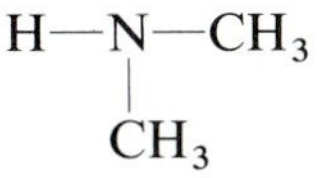

(a) $CH_3CH_2CH_2NH_2$ (b) $H-\underset{\underset{CH_3}{|}}{N}-CH_3$ (c) $\text{(benzene ring)}-NHCH_2CH_3$

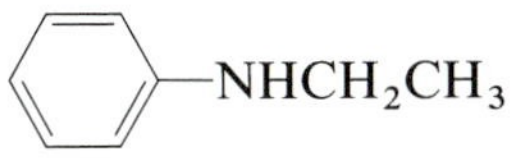

10.11   Draw structures corresponding to these names:
(a) butylamine   (b) *N*-methylethylamine   (c) *N,N*-dimethylaniline

---

10.9   SOME BIOLOGICALLY IMPORTANT AMINES

Historically, amines were among the first organic compounds isolated in pure form, and a variety of amines are widely distributed among plants and animals. Trimethylamine is responsible for the distinctive odor of fish; nicotine is a cyclic amine found in tobacco; and quinine is an antimalarial drug isolated from the bark of the South American *Cinchona* tree.

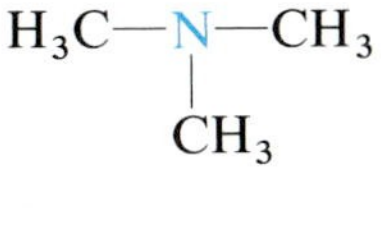

Trimethylamine 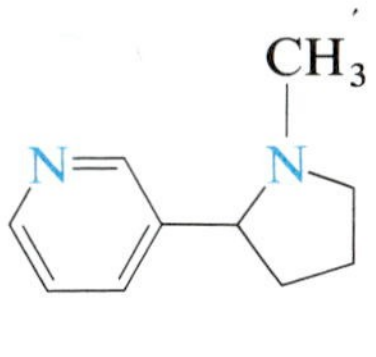   Nicotine 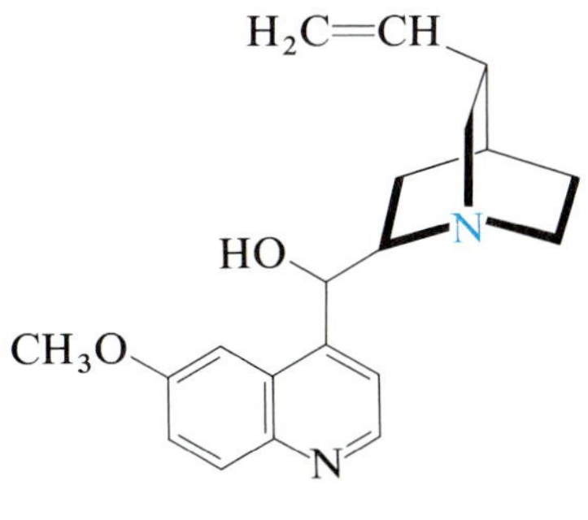   Quinine

Neurotransmitter A substance that transmits a nerve impulse in the body.

A number of naturally occurring amines are also important in human health. For example, nerve impulses are transmitted through the body by a group of **neurotransmitters** such as epinephrine (adrenaline), norepinephrine, and dopamine. These compounds have a variety of physiological effects including the ability to dilate small airways in the lung (*bronchodilation*), to contract capillaries, and to increase blood pressure. Related synthetic compounds like ephedrine and isoproterenol are widely used in the treatment of asthma and hay fever. Still other related substances such as amphetamines are strong stimulants of the central nervous system.

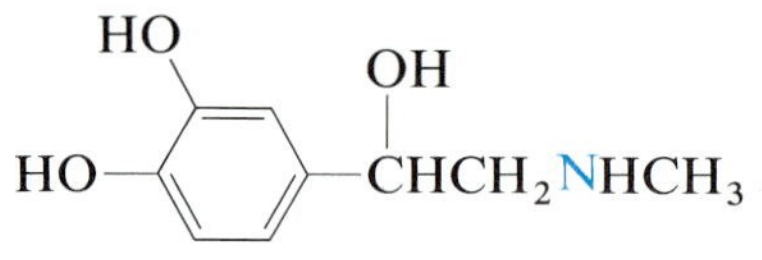

Adrenaline

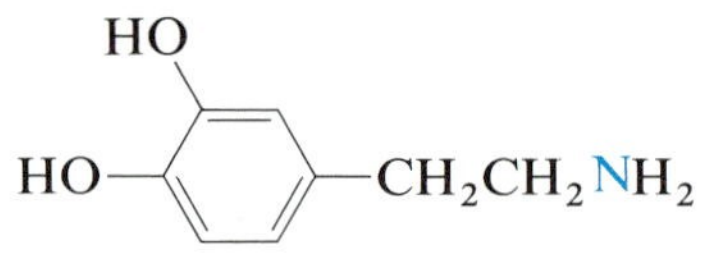

Dopamine

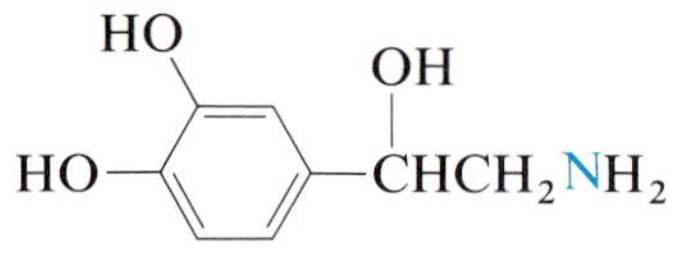

Norepinephrine

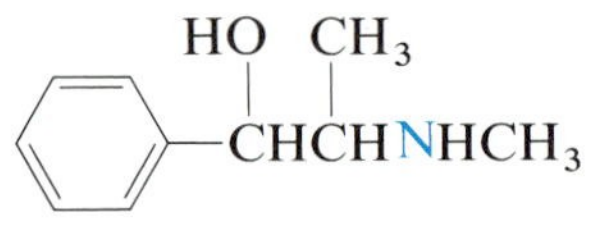

Ephedrine

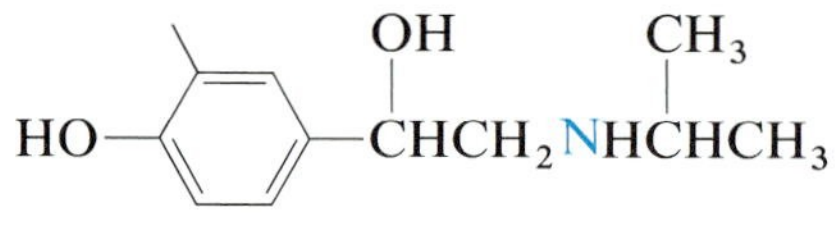

Isoproterenol

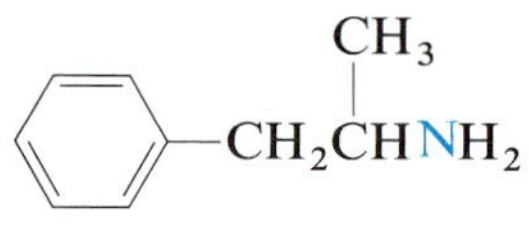

Amphetamine

10.10 PROPERTIES OF AMINES: BASICITY

Amines are analogous to ammonia in many of their physical properties, just as alcohols are analogous to water. Like ammonia, amines have polar bonds. Thus, amines form hydrogen bonds in the same way that ammonia does (Figure 10.3) and are higher boiling than alkanes of similar size. Small amines such as methylamine are soluble in water, although larger ones are more alkane-like in their behavior and are generally water-insoluble.

The most important similarity between amines and ammonia is their basicity. Recall from Section 7.4 that ammonia is a base. That is, ammonia can use its lone pair of electrons to accept a hydrogen ion from an acid, giving an ammonium salt. When dissolved in water, for example, ammonia can accept a proton from water yielding ammonium hydroxide. When reacted with hydrochloric acid, ammonia is protonated to yield ammonium chloride.

$$\text{H—O—H} + \;:\text{NH}_3 \;\;\rightleftharpoons\;\; \text{NH}_4{}^+ \text{OH}$$

Ammonia   Ammonium hydroxide

$$\text{HCl}(aq) + \;:\text{NH}_3 \;\;\longrightarrow\;\; \text{NH}_4{}^+ \;{}^-\text{Cl}(aq)$$

Ammonia   Ammonium chloride

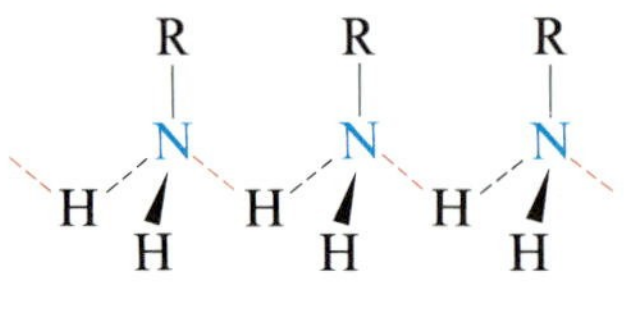

Figure 10.3
The formation of hydrogen bonds in amines.

AN APPLICATION: MORPHINE ALKALOIDS

The pain-relieving ability of the opium poppy, *Papaver somniferum*, has been known at least since the seventeenth century. Morphine was the first pure compound to be isolated from the poppy, but several close relatives including codeine also occur naturally. Heroin, another close relative of morphine, does not occur naturally but is easily synthesized in chemical laboratories. Once called "vegetable alkali" because their water solutions are basic, morphine and other naturally occurring amines are now referred to as *alkaloids*.

The opium poppy, source of morphine and other additive alkaloids.

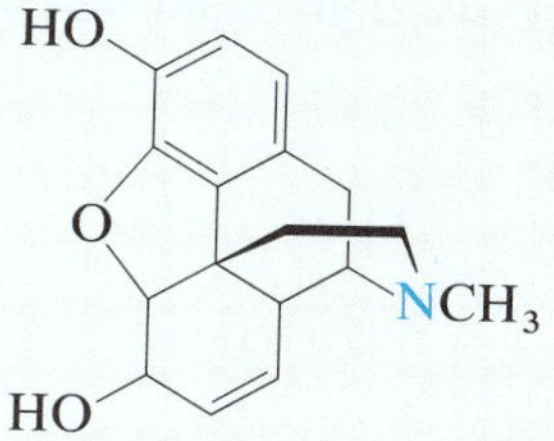

Morphine

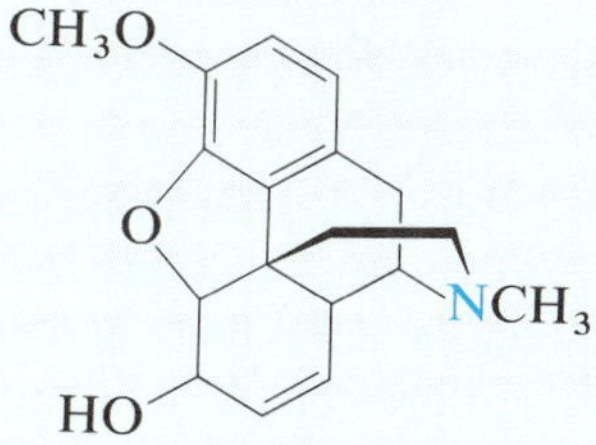

Codeine

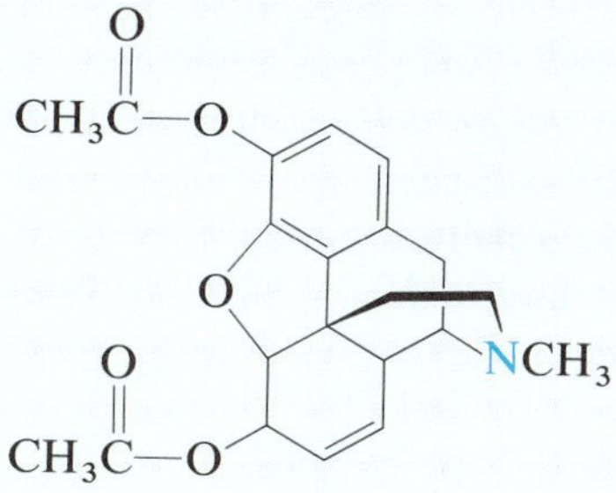

Heroin

Although enormously important as medicines, morphine and its relatives also pose a great problem because of their addictive properties. Much effort has therefore been devoted to understanding how morphine works and to developing modified morphine derivatives that retain the desired pain-killing activity but that don't cause addiction. Many nonnaturally occurring compounds, including meperidine (Demerol) and methadone, have been synthesized, but no fully satisfactory morphine substitutes have yet been prepared. Demerol is widely used as a painkiller, and methadone is used in the treatment of heroin addiction.

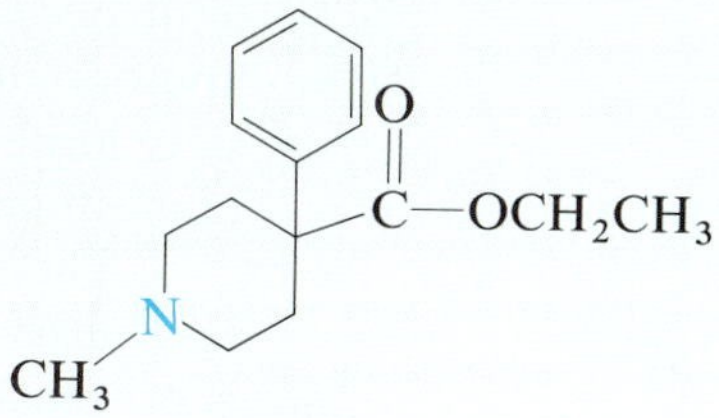

Demerol

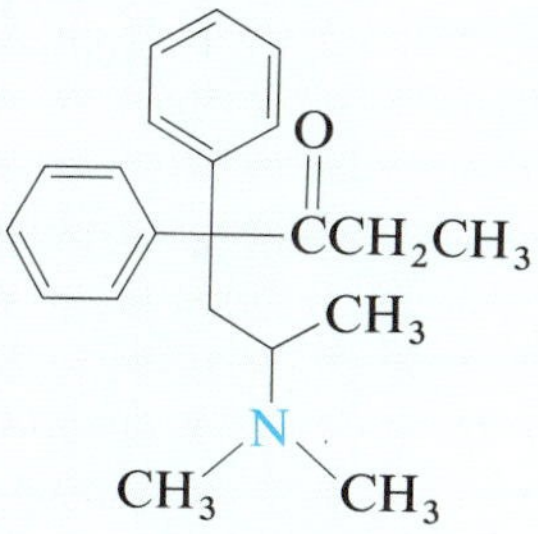

Methadone

Amines react similarly, regardless of whether they're primary, secondary, or tertiary. When treated with strong acids such as HCl, they are protonated to yield ammonium salts:

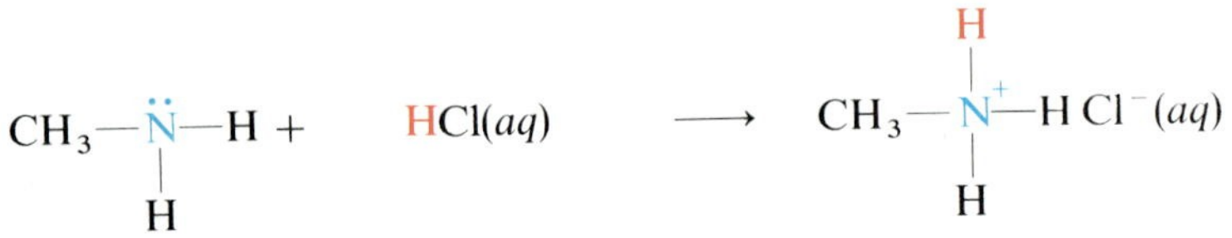

$$CH_3-\ddot{N}-H + \quad HCl(aq) \quad \longrightarrow \quad CH_3-\overset{+}{N}-H\ Cl^-\ (aq)$$

| Methylamine | Hydrochloric acid | Methylammonium chloride |

Protonated amines are much more soluble in water than neutral amines because they're ionic. Thus, a water-insoluble amine such as triethylamine dissolves readily in water when converted into its ammonium salt by reaction with HCl.

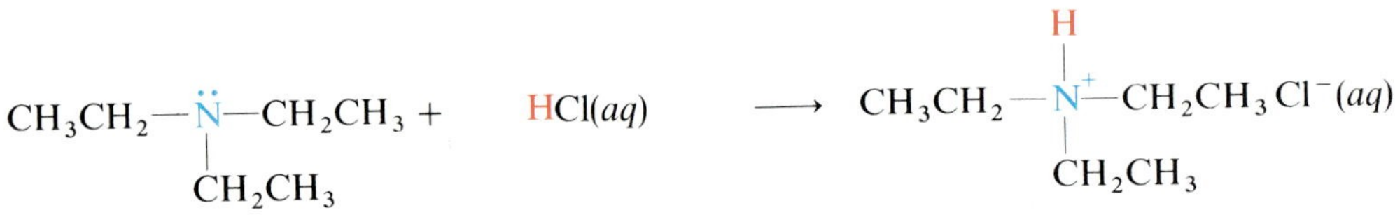

$$CH_3CH_2-\ddot{N}-CH_2CH_3 + \quad HCl(aq) \quad \longrightarrow \quad CH_3CH_2-\overset{+}{N}-CH_2CH_3\ Cl^-(aq)$$

| Triethylamine (water-insoluble) | Hydrochloric acid | Triethylammonium chloride (water-soluble) |

This increase of water solubility on conversion of an amine to its protonated salt has enormous practical consequences in drug delivery. Many important amine-containing drugs, such as morphine (a painkiller) and tetracycline (an antibiotic) are insoluble in body fluids and are thus difficult to deliver to the appropriate site within the body. By converting these drugs to their ammonium salts, however, their solubility increases to the point where delivery through the bloodstream becomes possible.

The reaction between amines and acids is reversible and can be made to take place in either direction, depending on the reaction conditions. An amine is protonated to yield an ammonium salt when treated with acid, and an ammonium salt is deprotonated to yield a neutral amine when treated with base. Thus, the exact state of the amine depends on the pH of the medium. (Remember: Acidic = pH below 7; basic = pH above 7.)

Acidic conditions:

(pH < 7)

$$R-NH_2 + HCl(aq) \longrightarrow R-NH_3^+ Cl\ (aq)$$

| An amine | An acid | An ammonium salt |

Basic conditions:

(pH > 7)

$$R-NH_3^+ Cl^-(aq) + NaOH(aq) \longrightarrow R-NH_2(aq) + NaCl(aq) + H_2O$$

| An ammonium salt | A base | An amine |

Practice Problems

10.12 Write the structures of the ammonium salts produced by reaction of these amines with HCl:

(a)

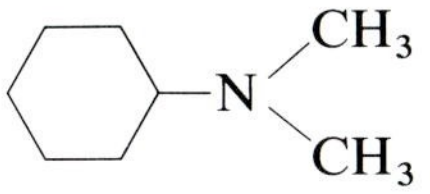

(b) diethylamine

(c)

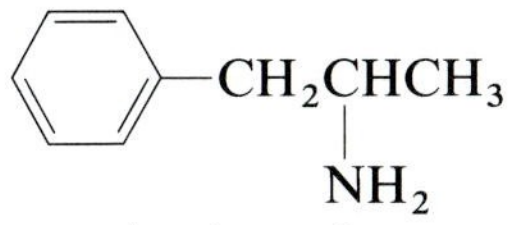

Amphetamine

10.13 Write the structures of the amines produced by reaction of these ammonium salts with NaOH:

(a) $(CH_3)_3NH^+ Cl^-$ (b) *N*-methylcyclohexylammonium chloride

10.11 HALOGEN-CONTAINING COMPOUNDS

Halogen-containing organic compounds are of less importance in health and medicine than alcohols and amines, although numerous examples of their use can be found. For example, chloroethane (ethyl chloride) is used as a topical anesthetic because of the way it cools the skin through rapid evaporation; halothane (1-bromo-1-chloro-2,2,2-trifluoroethane) has largely replaced ether as the anesthetic of choice during surgery because it is nonflammable and causes little discomfort; and thyroxine is an important iodine-containing hormone secreted by the thyroid gland. A deficiency of iodine in the human diet leads to a low thyroxine level, which causes a condition called *goiter*.

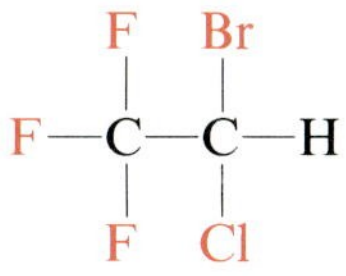

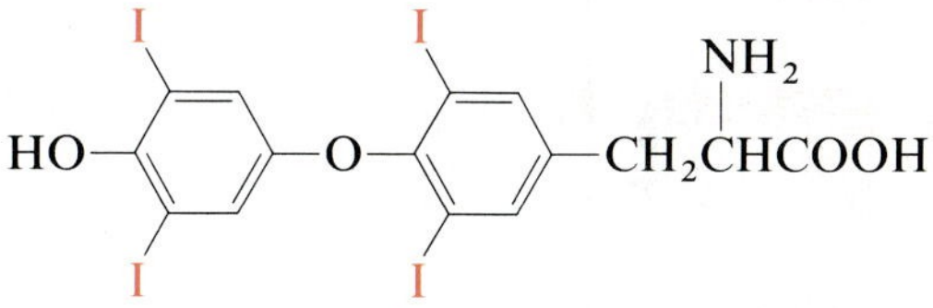

Chloroethane   Halothane   Thyroxine

In contrast to their limited use in medicine, halogenated compounds are of great importance in industry and agriculture. Thus, dichloromethane (CH_2Cl_2, methylene chloride) and trichloromethane ($CHCl_3$, chloroform) are common laboratory solvents. Chloroform was once employed as an anesthetic and as a solvent for cough syrups and other medicines but is now known to be too toxic for such uses.

Agricultural use of herbicides such as 2,4-D and fungicides such as Captan has resulted in vastly increased crop yields in recent decades, and the widespread application of chlorinated insecticides such as DDT is largely responsible for the progress made toward the worldwide control of malaria and typhus. Despite their enormous benefits, many chlorinated insecticides are so toxic to animals that their use has been greatly curtailed in recent years.

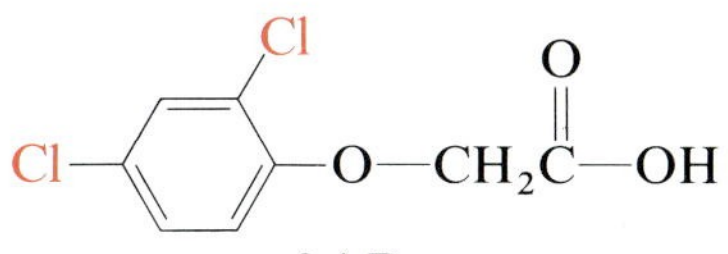

2,4-D 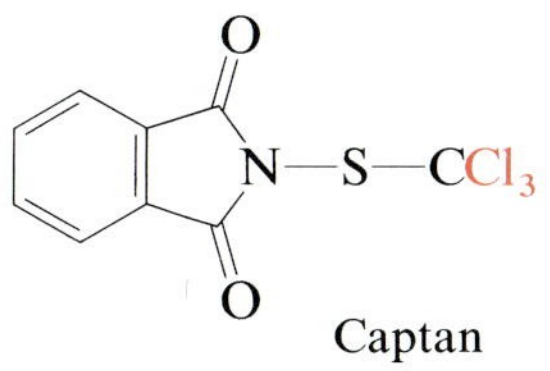   Captan 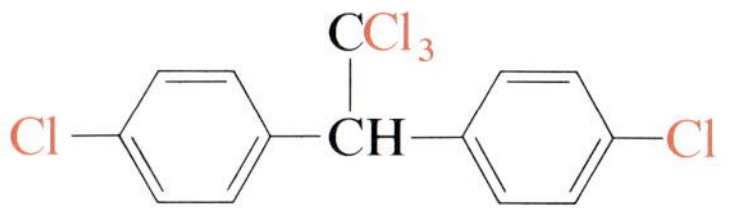   DDT

INTERLUDE: CHLOROFLUOROCARBONS AND THE OZONE HOLE

Newspapers have been full of stories recently about a "hole" in the ozone layer, the region of the atmosphere extending from about 20 to 40 km above the earth's surface. Although toxic to all life forms at high concentrations, ozone (O_3) is nevertheless of critical importance in the upper atmosphere because it acts as a shield to protect the earth from intense solar radiation. If the ozone layer were depleted, a great deal more solar radiation would reach the earth, causing an increase in the incidence of skin cancer and eye cataracts. Beginning around 1976, such an ozone depletion began showing up in the atmosphere over the South Pole—the so-called *ozone hole* (see figure).

The causes of ozone depletion, although not fully understood, almost certainly involve a group of halogen-substituted alkanes called *Freons*. The Freons are *chlorofluorocarbons*—simple alkanes in which all of the hydrogens have been replaced by either chlorine or fluorine. Fluorotrichloromethane (CCl_3F, Freon 11) and dichlorodifluoromethane (CCl_2F_2, Freon 12) are two of the most common Freons in industrial use.

For many years, Freons have been used as refrigerants in air conditioners and as propellants in aerosol cans to spray out deodorants, insecticides, and a host of other products. Unfortunately, the Freons thus released slowly find their way into the upper atmosphere, where they undergo a complex series of reactions that ultimately result in ozone destruction. Although estimates differ, a five to nine percent depletion can be expected over the next 50 to 100 years.

Ozone destruction occurs when ultraviolet (UV) light strikes a Freon molecule, breaking a carbon-chlorine bond and producing a chlorine atom.

The chlorine atom then reacts with ozone to yield oxygen and ClO:

$$CCl_2F_2 \xrightarrow{\text{UV light}} \cdot CClF_2 + Cl\cdot$$

$$Cl\cdot + O_3 \longrightarrow O_2 + ClO$$

Awareness of the problem led the U.S. government in 1980 to ban the use of Freon aerosol propellants, but other nations have not yet acted similarly. Worldwide action to reduce chlorofluorocarbon use finally began in September, 1987, however, when an international agreement was reached by the European Community and 24 other nations.

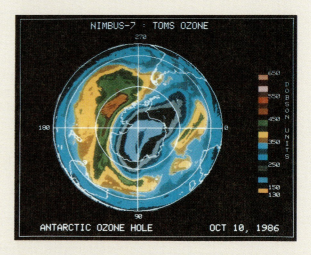

The antarctic ozone hole in October, 1986, as measured by instruments aboard the Nimbus 7 satellite. The violet area below and to the right of the South Pole (center) represents a region roughly the size of the United States where atmospheric ozone concentration is markedly reduced.

SUMMARY

Alcohols have an —OH group (**hydroxyl**) bonded to a saturated alkane-like carbon atom; **phenols** have an —OH group bonded directly to an aromatic ring; and **ethers** have an oxygen atom bonded to two organic groups. Both alcohols and phenols are "water-like" in their ability to form hydrogen bonds. Also like water, alcohols and phenols are acids that can donate their —OH protons to strong bases. Phenols are acidic enough to dissolve in aqueous NaOH.

Alcohols undergo loss of water (**dehydration**) to yield alkenes when treated with a strong acid, and undergo **oxidation** to yield carbon-containing compounds when treated with sodium dichromate or some other oxidizing agent. **Primary alcohols** (RCH_2OH) are oxidized to yield either aldehydes ($RCH{=}O$) or carboxylic acids ($RCOOH$); **secondary alcohols** (R_2CHOH) are oxidized to yield ketones ($R_2C{=}O$); and **tertiary alcohols** are not oxidized.

Thiols, or **mercaptans** (RSH), are sulfur analogs of alcohols. They react with mild oxidizing agents to yield **disulfides** (RSSR), a reaction of importance in protein chemistry.

Amines are organic derivatives of ammonia in the same sense that alcohols and ethers are organic derivatives of water. A **primary amine** has one organic group bonded to nitrogen ($R{-}NH_2$), a **secondary amine** has two organic groups bonded to nitrogen (R_2NH), and a **tertiary amine** has three organic groups bonded to nitrogen (R_3N). The dominant property of amines is their basicity. Like ammonia, amines can act as bases to accept protons from acids. Protonation of an amine yields an **ammonium salt** ($R{-}NH_2 + HCl \rightarrow R{-}NH_3{}^+Cl^-$). The protonation reaction is reversible, and ammonium salts can be reconverted to amines by treatment with NaOH.

Alkyl halides are less important than alcohols and amines in medicine but are widely used in agriculture as fumigating agents, insecticides, and herbicides.

REVIEW PROBLEMS

Alcohols, Ethers, and Phenols

10.14 How do alcohols, ethers, and phenols differ structurally?

10.15 What is the structural difference between primary, secondary, and tertiary alcohols?

10.16 Why are alcohols higher boiling than ethers of the same formula weight?

10.17 What family-name endings are used for alcohols and ethers?

10.18 Which is the stronger acid, ethanol or phenol?

10.19 The steroidal compound prednisone is often used to treat poison ivy or poison oak inflammations. Identify the functional groups present in prednisone.

Prednisone

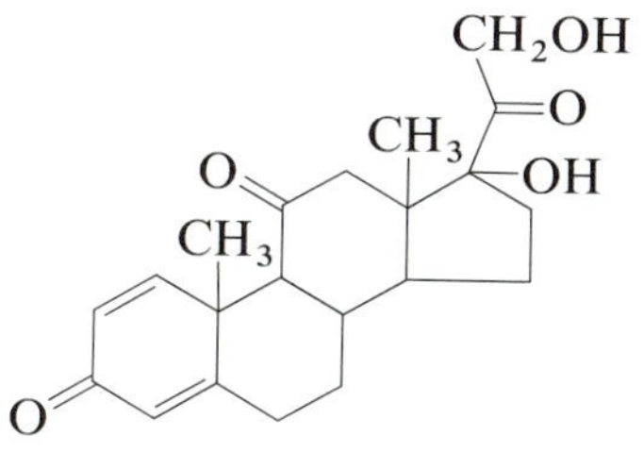

10.20 Vitamin E has the following structure. Identify the functional-group class of which each oxygen is a member.

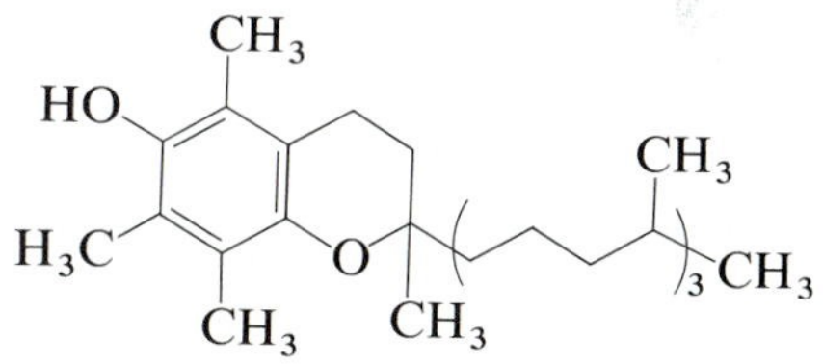

10.21 Give systematic names for these alcohols:

(a)
$$CH_3CH_2\underset{\underset{CH_2OH}{|}}{CH}CH_2CH_2CH_3$$

(b) $(CH_3)_2CHCH_2CH_2OH$

(c) $HOCH_2CH_2\underset{\underset{OH}{|}}{CH}CH_2OH$

10.22 Draw structures corresponding to these names:
(a) 2,4-dimethyl-2-pentanol
(b) 2,2-dimethylcyclohexanol
(c) 5,5-diethyl-1-heptanol
(d) 3-ethyl-3-hexanol

10.23 Identify the alcohols named in Problem 10.22 as primary, secondary, or tertiary.

10.24 Give systematic names for these compounds:

(a) [structure: benzene ring with CH₂CH₃ and OH substituents]

(b) $CH_3-CH-O-CH_3$ with CH_3

(c) O_2N—[benzene ring]—$O-CH_3$

10.25 Draw structures corresponding to these names:
(a) ethyl phenyl ether
(b) o-dihydroxybenzene (Catechol)
(c) p-bromophenyl ethyl ether

10.26 Arrange the following six-carbon compounds in order of their expected boiling points, and explain your ranking:
(a) hexane   (b) 1-hexanol   (c) dipropyl ether

Reactions of Alcohols

10.27 Give a specific example of an alcohol dehydration reaction.

10.28 What product is formed on oxidation of a secondary alcohol?

10.29 What structural feature prevents tertiary alcohols from undergoing oxidation reactions?

10.30 What product(s) can form on oxidation of a primary alcohol?

10.31 Which of these three compounds would you expect to be most soluble in water, and which least soluble? Explain.
(a) ethyl propyl ether   (b) 1-pentanol
(c) 1,2,3-propanetriol

10.32 Assume that you had samples of the following two compounds, both with formula C_7H_8O. Both compounds dissolve in ether, but only one of the two dissolves in aqueous NaOH. How could you use this information to distinguish between them?

H_3C—[benzene ring]—$O-H$ and   [benzene ring]—CH_2OH

10.33 Assume that you had samples of the following two compounds, both with formula $C_6H_{12}O$. What simple chemical reaction would allow you to distinguish between them? Explain.

[structure: cyclohexane with CH₃ and OH substituents]   and   [structure: cyclohexane with CH₃ and OH on same carbon]

10.34 The following alkenes can be prepared by dehydration of an appropriate alcohol. Show the structure of the alcohol in each case. If the alkene can arise from dehydration of more than one alcohol, show all possibilities.

(a) [cyclopentane ring]$=CHCH_3$

(b) CH_3CH_2 and CH_3CH_2 with $C=CH_2$

(c) 3-hexene

10.35 Phenols undergo the same kind of substitution reactions that other aromatic compounds do (Section 9.7). Formulate the reaction of p-methylphenol with bromine to give a mixture of two substitution products.

10.36 What carbonyl-containing products would you obtain from oxidation of the following alcohols? If no reaction occurs, write "NR."

(a) [benzene ring]—$CHCH_3$ with OH

(b) CH_3CHCH_2OH with CH_3

(c) 3-methyl-3-pentanol

10.37 What alkenes might be formed by dehydration of these alcohols? If more than one product is possible, indicate which you expect to be major.

(a) [cyclopentane with CH₃ and OH substituents]

(b) $CH_3CH_2CHCHCH_3$ with HO and CH₃

(c) [cyclopentane with CH₃ and OH on same carbon]

10.38 Thyroxine (Section 10.11) is synthesized in the body by reaction of thyronine with iodine. Formulate the reaction, and tell what kind of process is occurring.

HO—[benzene ring]—O—[benzene ring]—$CH_2CH(NH_2)\overset{\displaystyle O}{\overset{\|}{C}}-OH$

Thyronine

Amines

10.39 What family name ending is used for amines?

10.40 What is the structural difference between primary, secondary, and tertiary amines?

10.41 Draw structures corresponding to these names:
(a) propylamine (b) diethylamine
(c) *N*-methylpropylamine

10.42 Provide names for these compounds:

(a) $CH_3CH_2CH_2CH_2NH_2$ (b)

$$CH_3\overset{\overset{\displaystyle NH_2}{|}}{C}HCH_3$$

(c) 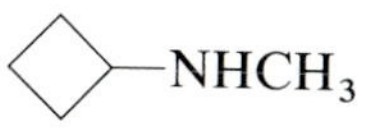 —$NHCH_3$

10.43 Identify the amines in Problems 10.41 and 10.42 as either primary, secondary, or tertiary.

10.44 Propose structures for amines that fit these descriptions:
(a) a secondary amine with formula C_5H_9N
(b) a tertiary amine with formula $C_6H_{13}N$

10.45 Cocaine has the structure indicated. Is cocaine a primary, secondary, or tertiary amine?

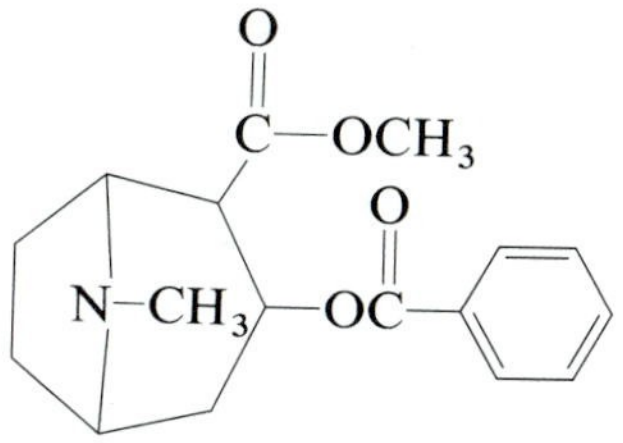

Cocaine

10.46 Most illicit cocaine is actually cocaine hydrochloride—the product from reaction of cocaine (Problem 10.45) with HCl. Show the structure of cocaine hydrochloride.

10.47 Assume that you had samples of quinine, an amine, and menthol, an alcohol. What simple chemical test could you do to distinguish between them?

10.48 The compound L-dopa [2-amino-3-(3,4-dihydroxyphenyl)propanoic acid] is used medically for its potent activity against Parkinson's disease, a chronic disease of the central nervous system. Identify the functional groups present in L-dopa.

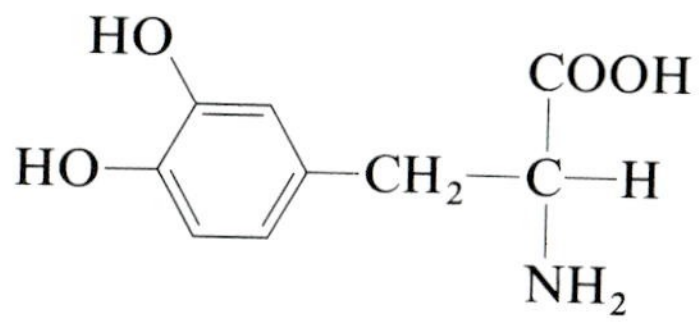

L-Dopa

10.49 Which is the stronger base, diethyl ether or diethylamine?

Thiols and Disulfides

10.50 What is the structural relationship between an alcohol and a thiol?

10.51 The amino acid cysteine forms a disulfide when oxidized. What is the structure of this disulfide?

$$HSCH_2\overset{\overset{\displaystyle NH_2}{|}}{C}HCOOH \qquad \text{Cysteine}$$

Additional Problems

10.52 Give specific examples of the following reactions:
(a) formation of an ammonium salt from a tertiary amine
(b) formation of a ketone by oxidation of an alcohol
(c) formation of a disulfide by oxidation of a thiol

10.53 Write structural formulas for compounds that meet these descriptions:
(a) a tertiary alcohol with six carbons
(b) a tertiary amine with four carbons
(c) a compound with formula $C_5H_{13}NO$ that is both an alcohol and an amine

CHAPTER 11

Aldehydes and Ketones

This bombardier beetle is spraying a potential predator with boiling-hot benzo-quinones, produced in a fraction of a second by a chemical reaction that the insect carries out within its body. Benzoquinones are ketones, members of a class of compounds you'll learn about in this chapter. (For another example of insect chemical warfare, see the Application.)

In this and the next chapter, we'll discuss the most important families of compounds in organic and biological chemistry—those that have a carbonyl group. As we saw in the last chapter, a carbonyl group contains a carbon-*oxygen* double bond in the same way that an alkene grouping contains a carbon-*carbon* double bond.

A carbonyl group

Carbonyl-containing compounds occur everywhere: Carbohydrates, fats, proteins, and nucleic acids all contain carbonyl groups; most pharmaceutical agents contain carbonyl groups; and many of the synthetic polymers used for clothing and other applications contain carbonyl groups.

We'll answer the following questions in this chapter:

1. **What are the kinds of carbonyl groups, and how do they differ?** The goal: you should learn to recognize and describe the most important families of carbonyl compounds.

2. **Why are ketones and aldehydes important?** The goal: you should learn the uses of some important ketones and aldehydes.

3. **How are ketones and aldehydes named?** The goal: you should learn the rules of nomenclature for simple ketones and aldehydes.

4. **What reactions do ketones and aldehydes undergo?** The goal: you should learn the most important reactions of ketones and aldehydes and how to predict the products in specific cases.

5. **What is the aldol reaction and why is it important in biochemistry?** The goal: you should learn what an aldol reaction is and why it is important in biochemistry.

11.1   KINDS OF CARBONYL COMPOUNDS

Since oxygen attracts electrons more strongly than carbon, carbonyl groups are strongly polarized, with a partial positive charge on carbon and a partial negative charge on oxygen. As we'll see in subsequent sections, this polarity of the carbonyl group helps to explain how many of its reactions take place.

Table 11.1   Some Kinds of Carbonyl Compounds

Name	Structure	Example	
Aldehyde	R—C(=O)—H	H_3C—C(=O)—H	Acetaldehyde
Ketone	R—C(=O)—R′	H_3C—C(=O)—CH_3	Acetone
Carboxylic acid	R—C(=O)—O—H	H_3C—C(=O)—O—H	Acetic acid
Ester	R—C(=O)—O—R′	H_3C—C(=O)—O—CH_3	Methyl acetate
Amide	R—C(=O)—N<	H_3C—C(=O)—NH_2	Acetamide

There are many different kinds of carbonyl compounds depending on what other substituents are bonded to the carbonyl-group carbon. As indicated in Table 11.1, *aldehydes* (often written as RCHO) have a carbon substituent and a hydrogen bonded to the carbonyl group; *ketones* (RCOR′) have two carbon substituents; *carboxylic acids* (RCOOH) have a carbon and a hydroxyl (—OH); *esters* (RCOOR′) have a carbon and an alkoxyl (—OR); and *amides* (RCONH₂) have a carbon and a nitrogen (—NH_2, NHR, or NR₂).

It's useful to classify carbonyl compounds into two groups based on their chemical properties. In one group are aldehydes and ketones. Since the carbonyl group in these compounds is bonded to atoms (H and C) that don't attract electrons strongly, the bonds to the carbonyl-group carbon of aldehydes and ketones are nonpolar. Thus, the compounds behave similarly.

In the second group are carboxylic acids, esters, and amides. Since the carbonyl-group carbon in these compounds is bonded to an atom (O or N) that *does* attract electrons strongly, the bonds are strongly polar. Thus, carboxylic acids, esters, and amides behave similarly. (These compounds are discussed in Chapter 12.)

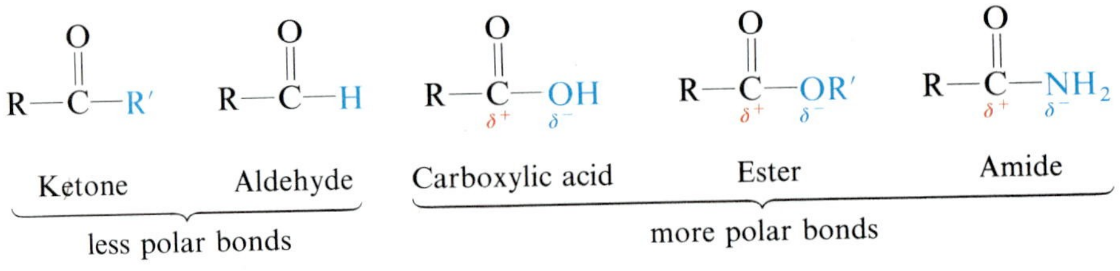

R—C(=O)—R′   R—C(=O)—H   R—C(=O)—OH   R—C(=O)—OR′   R—C(=O)—NH_2

   Ketone   Aldehyde   Carboxylic acid   Ester   Amide

        less polar bonds   more polar bonds

Practice Problem   *11.1*   Identify the kinds of carbonyl groups in these molecules:

(a)

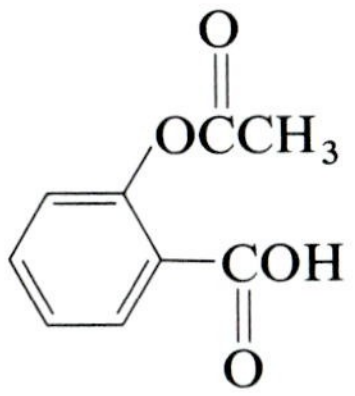

Aspirin

(b)

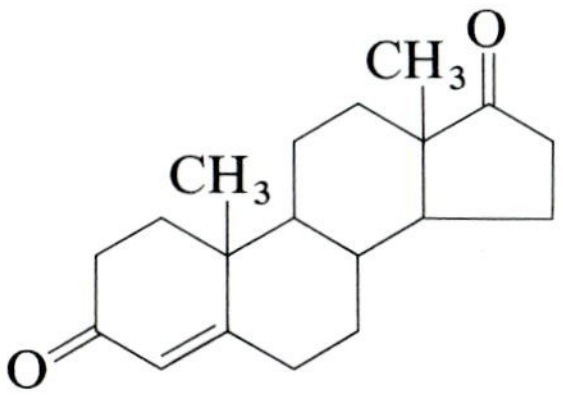

Testosterone
(a male hormone)

(c)

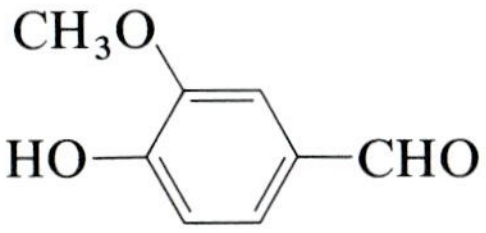

Vanillin
(a flavoring agent)

11.2   USES OF ALDEHYDES AND KETONES

Simple aldehydes and ketones are used for a multitude of purposes in chemistry and biology. For example, formaldehyde (HCHO), although a gas in the pure state, is sold as a 37% (w/w) aqueous solution under the name *formalin* for use as a biological sterilant and preservative. Formaldehyde is also used in the chemical industry as a starting material for the manufacture of the plastics *Bakelite* and *melamine*, as a component of the adhesives used to bind plywood, and as a part of the foam insulation used in houses. Although formaldehyde is an extremely useful chemical, widespread concern over possible carcinogenicity and toxicity has caused the removal of many formaldehyde-based products from the market.

Acetone (CH_3COCH_3), a liquid at room temperature, is perhaps the most widely used of all organic solvents; you may have seen cans of acetone sold in paint stores for general-purpose cleanup work. Not only can acetone dissolve most organic compounds, it is also miscible with water.

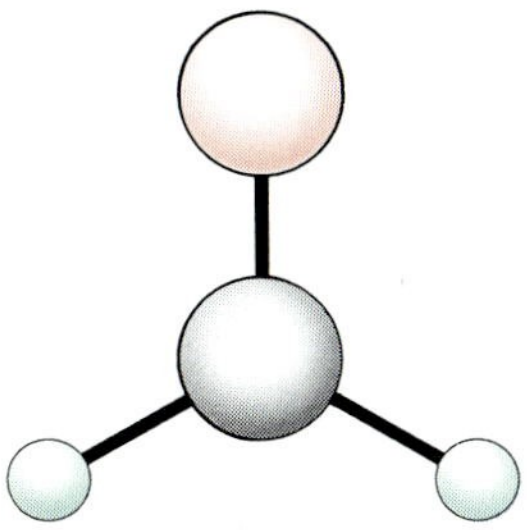

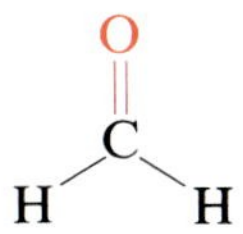

Formaldehyde

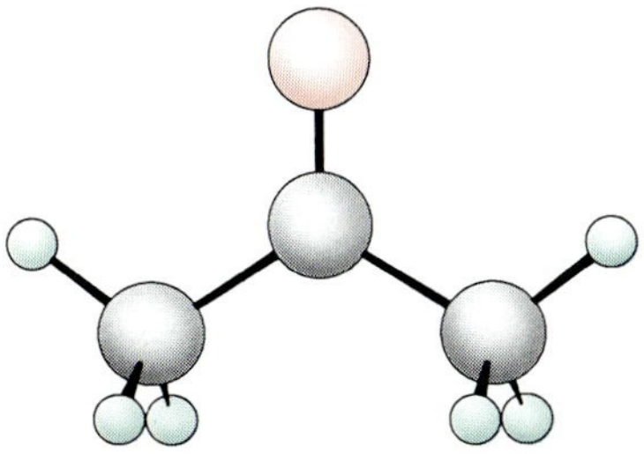

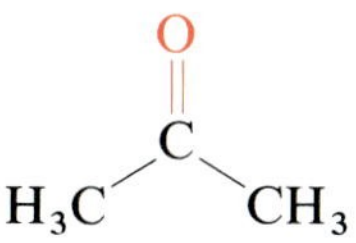

Acetone

Figure 11.1
Chemists perform "sniff tests" on potential perfume ingredients. Many aldehydes and ketones occur in nature and are used as fragrances or flavorings. The odor of violets, for example, owes much to a ketone, and the taste of vanilla derives largely from an aldehyde.

Aldehyde and ketone functional groupings are also present in a great many biologically important compounds. For example, glucose and most other sugars contain —CHO groups. Testosterone and many other steroid hormones contain keto groups. Other naturally occurring aldehydes and ketones are responsible for the odors or flavors of foods, spices, and flowers. (Figure 11.1).

Glucose—a pentahydroxyhexanal

Testosterone—a male hormone

11.3 NAMING ALDEHYDES AND KETONES

Aldehydes are named by replacing the final -*e* of the corresponding alkane name with -*al*. Thus, the three-carbon aldehyde derived from propane is propanal, the four-carbon aldehyde is butanal, and so on. When substituents are present, the chain is numbered beginning at the —CHO end. In addition to these systematic names, some simple aldehydes also have common names. The one-carbon aldehyde derived from methane (systematic name *methanal*) is always called *formaldehyde*, the two-carbon aldehyde derived from ethane (systematic name *ethanal*) is always called *acetaldehyde*, and the simplest aromatic aldehyde is *benzaldehyde*.

Formaldehyde (methanal)

Acetaldehyde (ethanal)

3-Methylbutanal

Benzaldehyde

Ketones are named by replacing the final -*e* of the corresponding alkane name with -*one* (pronounced own). The numbering of the chain begins at the end nearer the carbonyl group, as shown below for 2-pentanone, a five-carbon ketone derived from pentane. As with aldehydes, some simple ketones such as acetone and acetophenone also have common names.

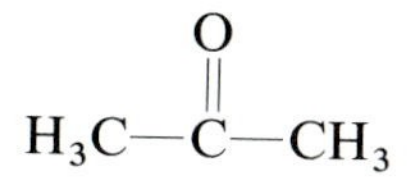

Acetone
(propanone)

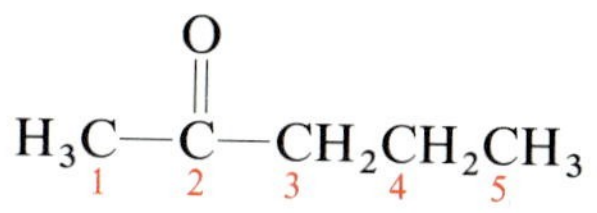

2-Pentanone

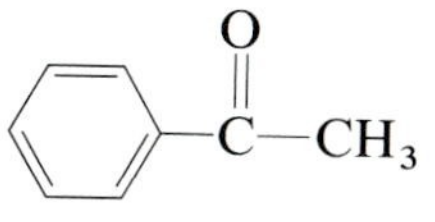

Acetophenone

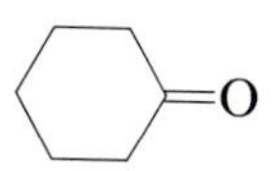

Cyclohexanone

Practice Problems

11.2   Draw structures corresponding to these names:
(a) hexanal   (b) *p*-bromoacetophenone
(c) 4-methyl-2-hexanone

11.3   Give systematic names for these compounds:

(a)
$$CH_3CH_2CH_2CH_2CH$$ with O double-bonded

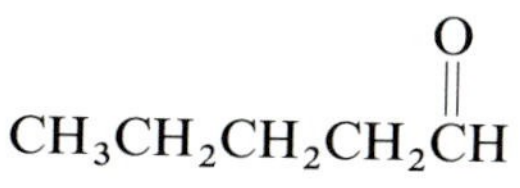

(b)
$$CH_3CH_2CCH_2CH_3$$ with O double-bonded

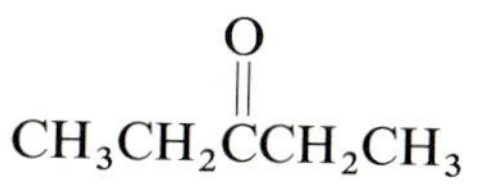

(c)
$$CH_3CH_2CHCH_2CH_2CH$$ with CH$_3$ branch and O double-bonded

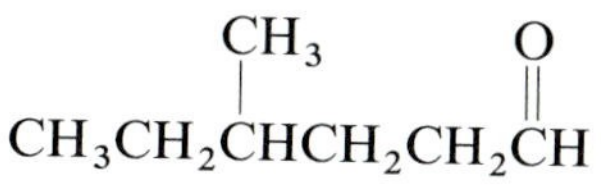

11.4   OXIDATION OF ALDEHYDES

We've already seen one important reaction of aldehydes—their oxidation to yield carboxylic acids (Section 10.6). In this reaction, the —CHO hydrogen attached to the carbonyl carbon is removed and replaced by —OH. Ketones don't have this hydrogen, however, and thus don't react with most oxidizing agents.

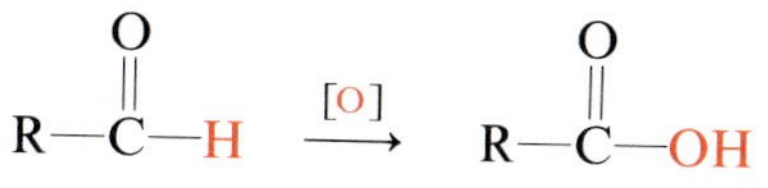

Aldehyde   Carboxylic acid 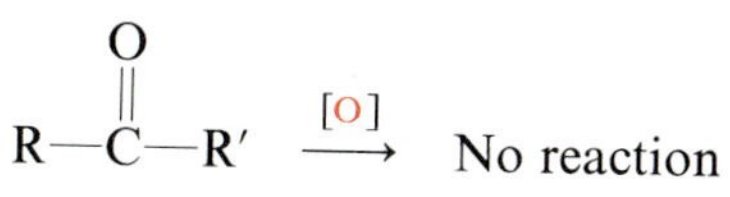   Ketone

Among the many oxidizing agents that can be used to effect the aldehyde → carboxylic acid conversion, Tollens' reagent is the most visually appealing. Treatment of an aldehyde with **Tollens' reagent**, a dilute solution of silver nitrate in aqueous ammonia, rapidly yields the ammonium salt of the carboxylic acid and silver metal as by-product. Addition of an acid such as HCl then yields the neutral carboxylic acid. If the reaction is done in a clean flask or test tube, metallic silver deposits on the inner walls of the vessel, producing a beautiful shiny mirror. In fact, a slight modification of this reaction is used in the industrial manufacture of glass mirrors.

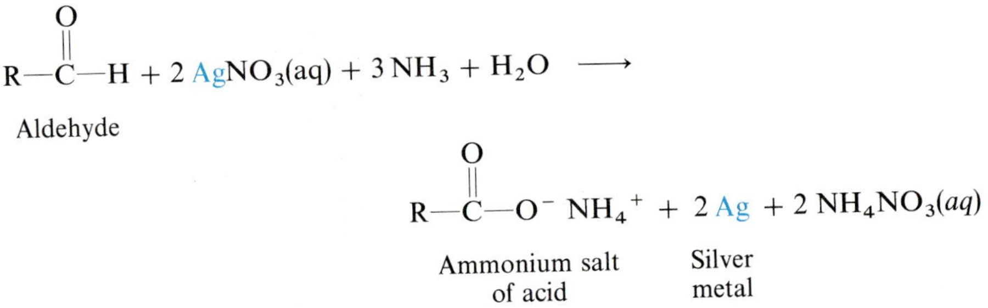

■ **Tollens' reagent** A reagent, $AgNO_3$ in aqueous NH_3, that converts an aldehyde into a carboxylic acid and deposits a silver mirror on the reaction flask.

$$R-\overset{\overset{\displaystyle O}{\|}}{C}-H + 2\,AgNO_3(aq) + 3\,NH_3 + H_2O \longrightarrow$$

Aldehyde

$$R-\overset{\overset{\displaystyle O}{\|}}{C}-O^-\,NH_4^+ + 2\,Ag + 2\,NH_4NO_3(aq)$$

Ammonium salt   Silver
of acid   metal

For example:

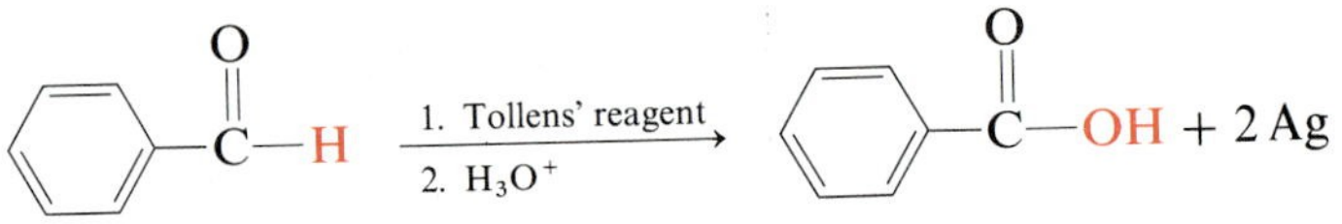

Benzaldehyde   Benzoic acid

Practice Problem   **11.4**   Draw structures of the products you would obtain by treating the following compounds with Tollens' reagent. If no reaction occurs, write "N.R."

(a) CH_3
$\quad\quad\quad |$
$CH_3CHCH_2CH_2CH_2CHO$

(b) 2,2-dimethylpentanal

(c) 2-methyl-3-pentanone

11.5   REDUCTION OF ALDEHYDES AND KETONES

Just as an alcohol can be converted into a carbonyl compound by *removing* hydrogen (an *oxidation*, Section 10.6), a carbonyl compound can be converted back into an alcohol by *adding* hydrogen to the C=O double bond (a *reduction*). **Reduction** in organic chemistry usually refers to the addition of hydrogen to a molecule.

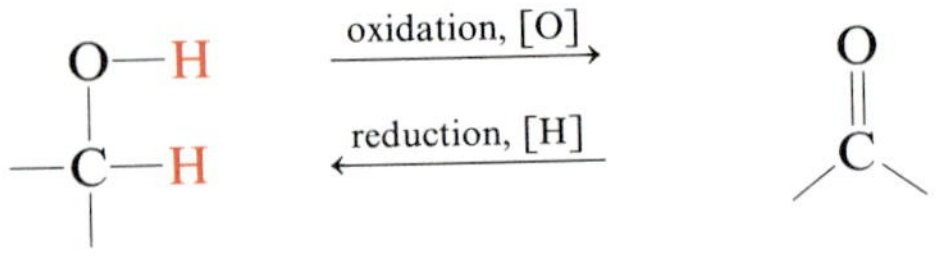

■ **Reduction** In organic chemistry, the addition of hydrogen to a carbon-oxygen double bond or carbon-carbon double bond functional group.

1° or 2° Alcohol   Aldehyde or ketone

The reduction of aldehydes and ketones is usually accomplished by treatment with sodium borohydride, $NaBH_4$. This reagent, a stable white powder, is soluble both in water and in ethanol. When mixed with a ketone or aldehyde, sodium borohydride transfers one of its hydrogens as a *hydride ion* ($:H^-$) to the carbonyl carbon, forming an anion. Aqueous acid is then added in a second step at the end of the reaction, and a neutral alcohol product results. Aldehydes are converted into primary alcohols ($RCHO \rightarrow RCH_2OH$), and ketones are converted into secondary alcohols ($R_2CO \rightarrow R_2CHOH$).

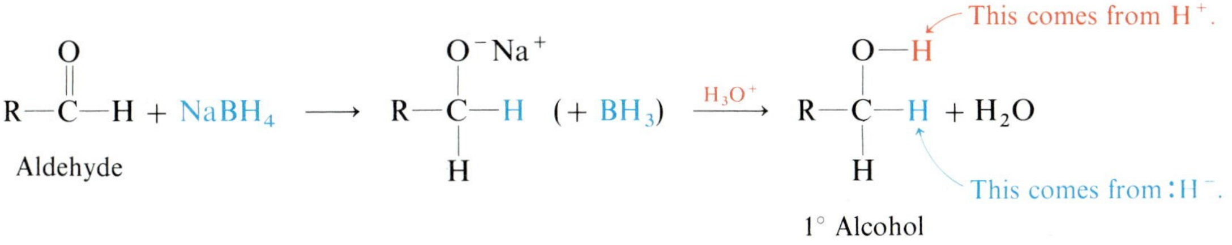

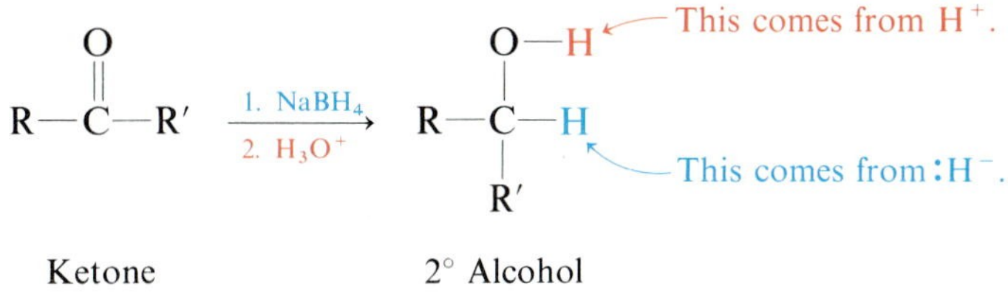

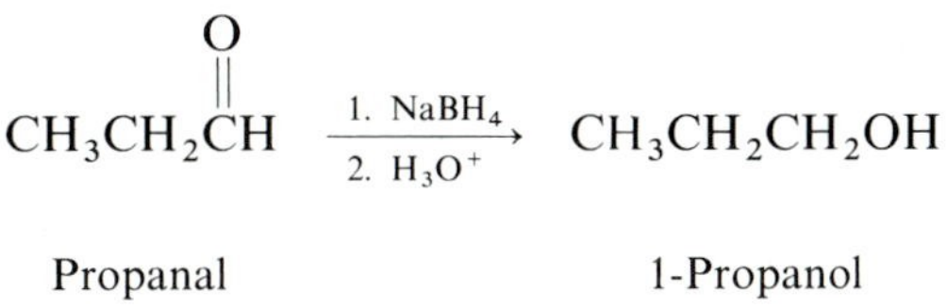

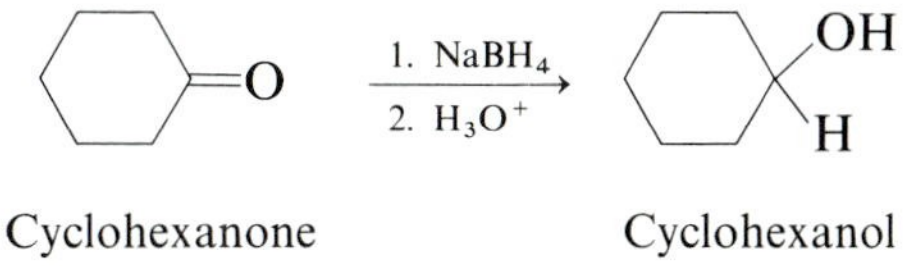

Note that the two new hydrogen atoms in the alcohol product come from different sources. The hydrogen atom attached to carbon comes from reaction with $:H^-$ (hydride ion) in the first step, whereas that attached to oxygen comes from reaction with H^+ (acid) in the second step.

The fact that the reduction of aldehydes and ketones takes two steps, addition of $:H^-$ and addition of H^+, makes good sense when you think about the polarity of the carbonyl group. Since the carbonyl-group carbon is polarized δ^+, the negatively charged hydride ion is naturally drawn to it. Similarly, a positively charged hydrogen ion is drawn to the negatively polarized carbonyl-group oxygen.

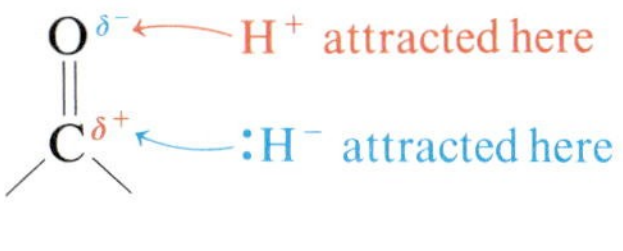

The reduction of aldehydes and ketones to yield alcohols occurs in living cells as well as in the laboratory. Although the body doesn't use $NaBH_4$ as its "reagent," there's nevertheless a remarkable similarity in the way the two kinds of reductions are done. Living organisms use a complex organic molecule called *nicotinamide adenine dinucleotide*, abbreviated NADH, as a biological source of hydride ion. NADH serves exactly the same role in the cell that $NaBH_4$ serves in the laboratory: It donates $:H^-$ to an aldehyde or ketone to yield an anion that picks up H^+ from surrounding aqueous fluids. The net result is reduction.

An example of biochemical reduction occurs when pyruvic acid, an intermediate involved in glucose metabolism, is converted into lactic acid by active skeletal muscles. Vigorous exercise causes a buildup of lactic acid in muscles, leading to a tired, flat feeling.

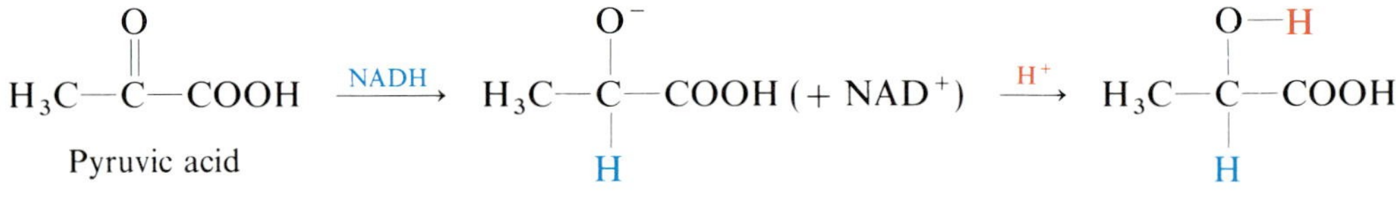

$$H_3C - \overset{\overset{O}{\|}}{C} - COOH \xrightarrow{\text{NADH}} H_3C - \overset{\overset{O^-}{|}}{\underset{H}{C}} - COOH \;(+\; NAD^+) \xrightarrow{H^+} H_3C - \overset{\overset{O-H}{|}}{\underset{H}{C}} - COOH$$

Pyruvic acid

Lactic acid

Solved Problem 11.1   What product would you obtain by reduction of benzaldehyde?

Solution   First, draw the structure of the starting material, showing the double bond in the carbonyl group. Then rewrite the structure showing only a *single* bond between C and O, along with *partial* bonds to both C and O:

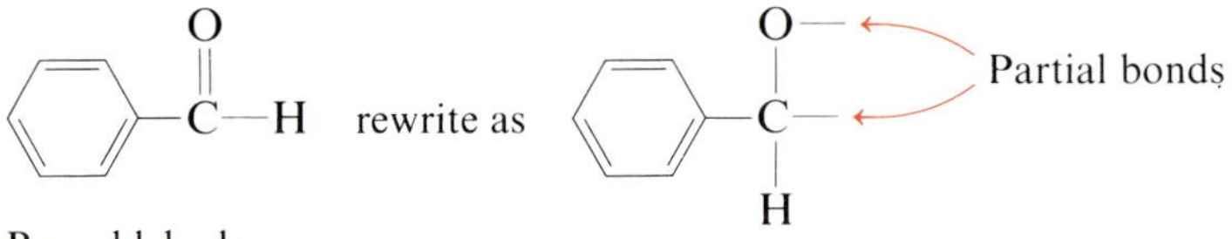

Benzaldehyde

Finally, attach hydrogen atoms to the two part-bonds, and rewrite the product:

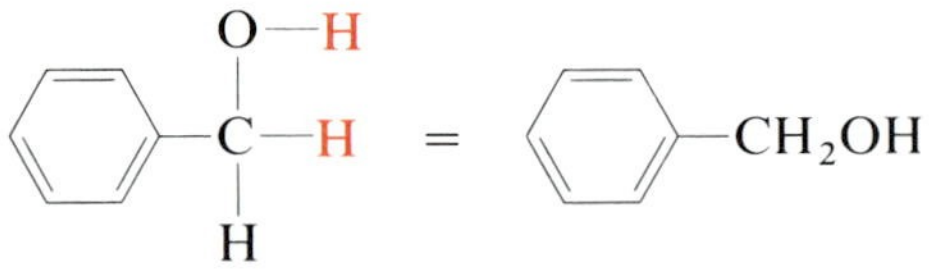

Practice Problems

11.5   What product would you obtain from reduction of the following ketones and aldehydes?
(a) $(CH_3)_2CHCHO$ (b) *m*-chlorobenzaldehyde   (c) cyclopentanone

11.6   What ketones or aldehydes might be reduced to yield the following alcohols?

(a)

(b)  $HOCH_2CH_2CH_2CHCH_3$
             $|$
             CH_3

(c) 2-methyl-1-pentanol

11.6   REACTION WITH ALCOHOLS: ACETAL FORMATION

■ **Hemiacetal**   A compound that has both an alcohol-like —OH group and an ether-like —OR group bonded to the same carbon.

Aldehydes and ketones react with alcohols to give addition products called **hemiacetals**, compounds that have both an alcohol-like —OH group and an ether-like —OR group bonded to the same carbon. During the reaction, the hydrogen from the alcohol bonds to the carbonyl-group oxygen, and the oxygen from the alcohol bonds to the carbonyl-group carbon.

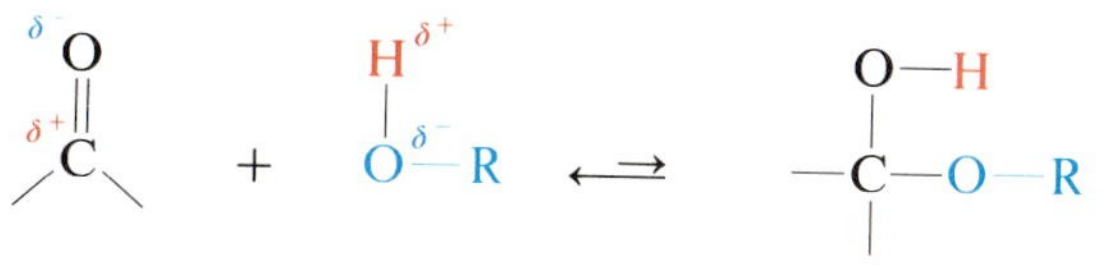

Aldehyde or ketone   Alcohol   Hemiacetal (unstable)

Two important features of this reaction need to be pointed out. First, note that it is the *negatively* polarized alcohol oxygen atom that adds to the *positively* polarized carbonyl carbon, just as a negatively charged hydride ion adds to the positively polarized carbonyl oxygen during reduction (Section 11.5). Almost all carbonyl-group reactions follow this same polarity pattern. Second, note that the reaction is reversible. Hemiacetals rapidly revert back to aldehydes or ketones by loss of alcohol.

In practice, we often find that hemiacetals are too unstable to be isolated. When the equilibrium position for the reaction is reached, very little hemiacetal is present. The one major exception to this rule occurs in the case of sugars such as glucose. As we'll see in the next chapter, most simple sugars contain a stable hemiacetal link.

■ **Acetal**   A compound that has two ether-like —OR groups bonded to the same carbon atom.

If a small amount of acid catalyst is added to the reaction of an alcohol and a carbonyl compound, the initially formed hemiacetal is converted into an acetal. An **acetal** is a compound that has two ether-like —OR groups bonded to the same carbon atom:

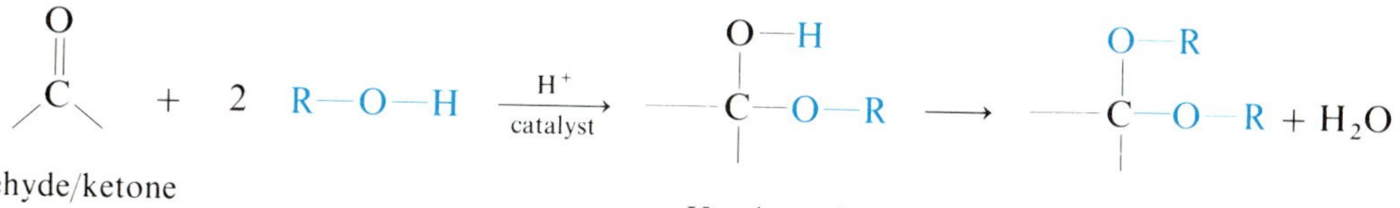

Aldehyde/ketone   Hemiacetal   Acetal

For example:

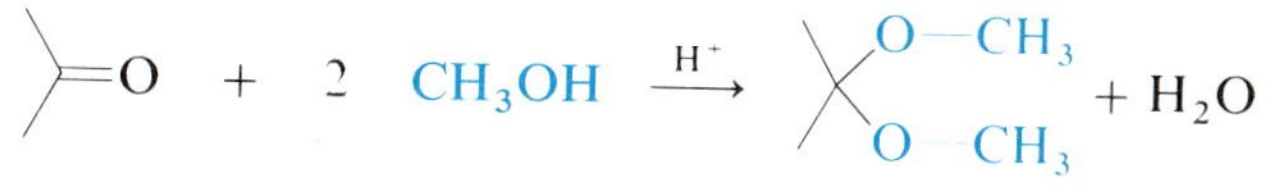

A ketone   An acetal

Unlike hemiacetals, acetals are quite stable and are easily isolated. They are so stable, in fact, that complex carbohydrates like cellulose and starch are held together simply by acetal groupings between individual sugar units. We'll learn more about carbohydrate acetals in Chapter 13.

Although stable to most reagents, acetals react with aqueous acid to regenerate the original ketone or aldehyde plus two molecules of alcohol. This reaction, called a **hydrolysis** because water is used to break down the starting material, is exactly what happens in the stomach when carbohydrates are digested.

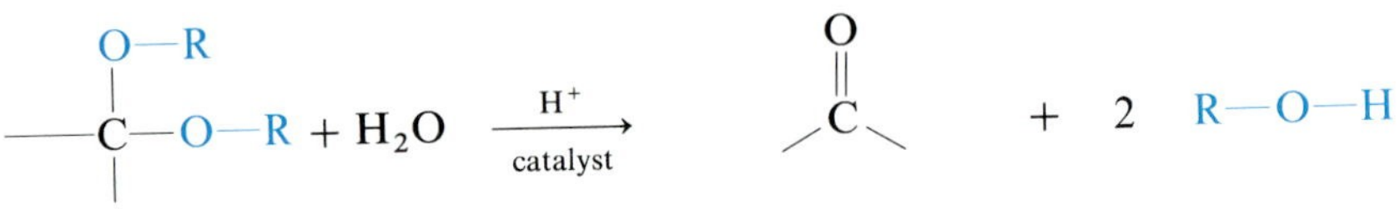

■ **Hydrolysis** The breakdown of a compound, such as an acetal, by reaction with water.

$$\underset{\text{Acetal}}{\overset{\displaystyle O-R}{\underset{|}{\overset{|}{-}C-O-R}} + H_2O} \quad \xrightarrow[\text{catalyst}]{H^+} \quad \underset{\text{Aldehyde or ketone}}{\overset{\displaystyle O}{\underset{}{\overset{\|}{C}}}} \quad + \quad 2 \quad \underset{\text{Alcohol}}{R-O-H}$$

Solved Problem 11.2 Write the structure of the intermediate hemiacetal and the acetal final product formed in the following reaction:

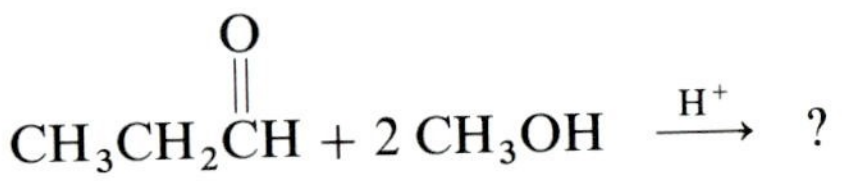

$$\underset{}{\overset{\displaystyle O}{\overset{\|}{CH_3CH_2CH}}} + 2\ CH_3OH \xrightarrow{H^+} ?$$

Solution First, rewrite the structure showing only a single bond between C and O, along with partial bonds to both C and O:

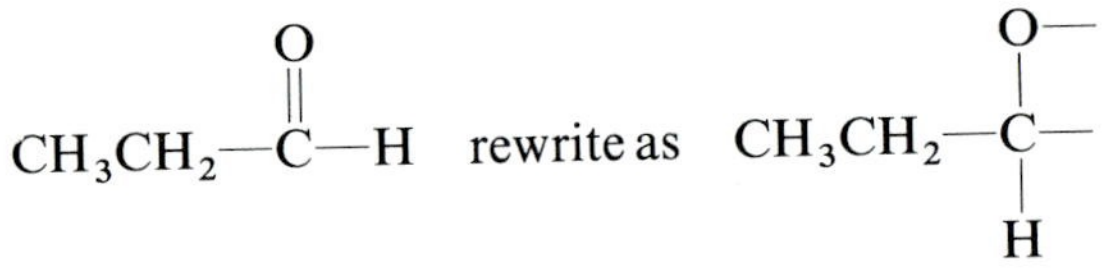

$$\underset{}{\overset{\displaystyle O}{\overset{\|}{CH_3CH_2-C-H}}} \quad \text{rewrite as} \quad CH_3CH_2-\overset{\displaystyle O-}{\underset{\displaystyle H}{\overset{|}{C}}}-$$

Next, add one molecule of the appropriate alcohol (CH_3OH in this case) by attaching —H to the oxygen part-bond and —OCH_3 to the carbon part-bond. This yields the hemiacetal intermediate:

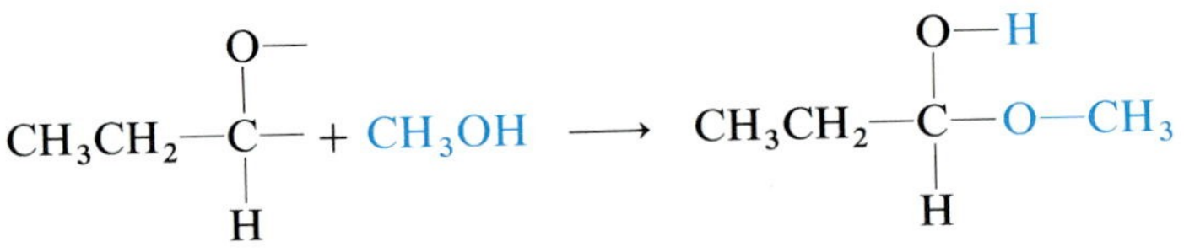

$$CH_3CH_2-\overset{\displaystyle O-}{\underset{\displaystyle H}{\overset{|}{C}}}- \ + \ CH_3OH \quad \longrightarrow \quad CH_3CH_2-\overset{\displaystyle O-H}{\underset{\displaystyle H}{\overset{|}{C}}}-O-CH_3 \qquad \text{Hemiacetal}$$

Finally, replace the —OH group of the hemiacetal with an —OCH_3 from a second molecule of alcohol. This yields the acetal product and water.

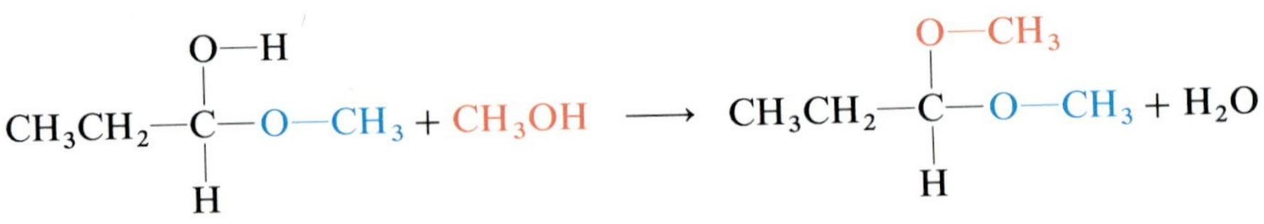

$$CH_3CH_2-\overset{\displaystyle O-H}{\underset{\displaystyle H}{\overset{|}{C}}}-O-CH_3 + CH_3OH \quad \longrightarrow \quad CH_3CH_2-\overset{\displaystyle O-CH_3}{\underset{\displaystyle H}{\overset{|}{C}}}-O-CH_3 + H_2O$$

Acetal product

Solved Problem 11.3   Write the structure of the aldehyde or ketone that would be formed by hydrolysis of the following acetal:

$$(CH_3)_2CHCH_2CH(OCH_2CH_3)_2 \xrightarrow{H_3O^+} ?$$

Solution   First, rewrite the starting acetal so that the two ether-like acetal bonds are more evident:

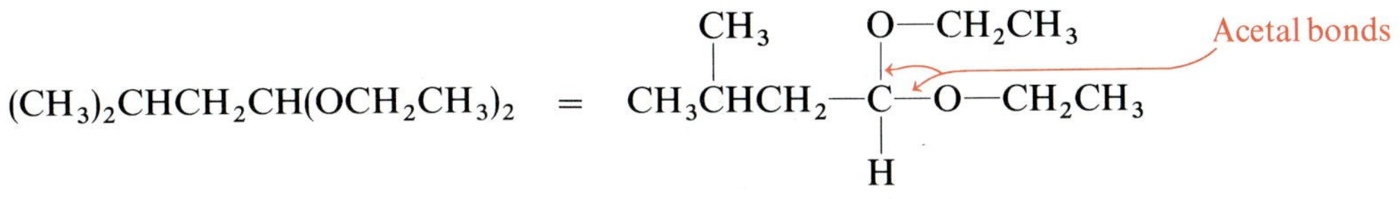

$(CH_3)_2CHCH_2CH(OCH_2CH_3)_2$ =

Acetal bonds

Next, remove the two —OR groups from the acetal carbon, converting each into a molecule of alcohol and leaving two part-bonds on carbon:

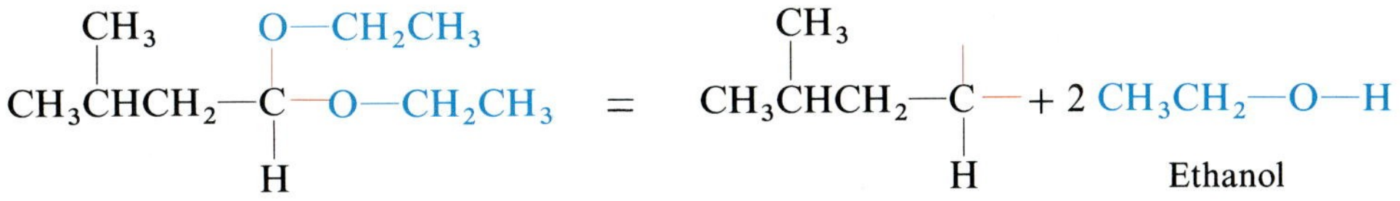

=

+ 2 CH$_3$CH$_2$—O—H

Ethanol

Finally, add an oxygen to the two part-bonds on the acetal carbon to form a carbonyl group, and rewrite the structure of the product. In this example, the product is an aldehyde.

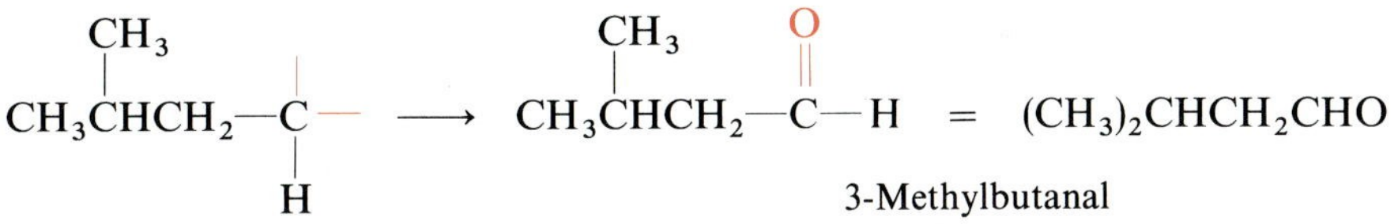

$\longrightarrow$ = $(CH_3)_2CHCH_2CHO$

3-Methylbutanal

Practice Problems   **11.7**   Draw the structures of the hemiacetals formed in these reactions:

(a) $CH_3CH_2CH_2CHO + CH_3CH_2OH \longrightarrow ?$

(b)

$$\underset{\displaystyle \overset{\displaystyle O}{\parallel}}{CH_3CH_2CCH_2CH(CH_3)_2} + CH_3OH \longrightarrow ?$$

11.8   Draw the structure of each acetal final product formed in the reactions in Problem 11.7.

11.9   What aldehydes or ketones would result from the following acetal hydrolysis reactions? What alcohol is formed in each case?

(a) 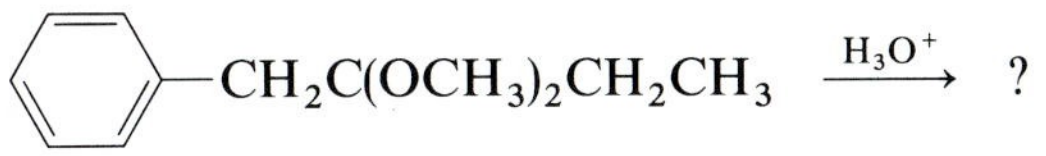 —$CH_2C(OCH_3)_2CH_2CH_3 \xrightarrow{H_3O^+} ?$

(b) $CH_3CH_2CH_2OCH_2OCH_2CH_2CH_3 \xrightarrow{H_3O^+} ?$

AN APPLICATION: CHEMICAL WARFARE AMONG THE INSECTS

Life in the insect world is a jungle. Predators abound, just waiting to make a meal of any insect that happens along. Without missiles to protect themselves, many insects have evolved extraordinarily effective means of *chemical* protection. Take the humble millipede *Apheloria corrugata*, for example. When attacked by ants, the millipede protects itself by discharging a compound called *benzaldehyde cyanohydrin*.

Cyanohydrins [RCH(OH)C≡N] are compounds formed by the addition of the toxic gas HCN (hydrogen cyanide) to ketones or aldehydes. Like the reaction of a ketone or aldehyde with an alcohol to yield a hemiacetal (Section 11.6), the reaction with HCN to yield a cyanohydrin is reversible. Thus, the benzaldehyde cyanohydrin secreted by the millipede can decompose to yield benzaldehyde and HCN. The millipede actually defends itself by discharging poisonous hydrogen cyanide at would-be attackers—a remarkably clever and very effective kind of chemical warfare.

Benzaldehyde

Using a strategy similar to that of the millipedes, apricots and peaches protect their seeds with a group of substances called *cyanogenic glycosides*. These compounds, of which amygdalin, or Laetrile, is best known because of its supposed anticancer activity, consist of benzaldehyde cyanohydrin bonded to glucose or another sugar. When eaten, the sugar unit is cleaved off by enzymes, and HCN is released.

Amygdalin (Laetrile)

Benzaldehyde cyanohydrin

11.7 ALDOL REACTION OF ALDEHYDES AND KETONES

All the reactions we've covered up to this point have involved interconversions of one functional group with another. For example, alcohols can be oxidized to yield aldehydes and ketones, which in turn can be reduced to give back alcohols. Although important, these functional-group interconversion reactions are of limited use because they don't change the size or carbon framework of molecules.

In contrast to functional-group interconversions, certain other reactions involve formation of new carbon-carbon bonds. By using such reactions, it's possible to take two small pieces, join them together, and thereby make a larger molecule. Many of the biochemical processes in living cells do exactly this.

The **aldol reaction** is a process that occurs when an aldehyde or ketone is treated with a base. In the reaction, a bond forms between the carbon atom

■ **Aldol reaction** The reaction of a ketone or aldehyde to form a hydroxy ketone product on treatment with base catalyst.

next to the carbonyl group of one molecule and the carbonyl-group carbon of the second molecule. The product, formed by joining together two molecules of starting material, is a hydroxy ketone or aldehyde.

This oxygen and this hydrogen form a bond.

This new C—C bond is formed.

$$
\underset{\substack{\text{Two ketones or}\\\text{aldehydes}}}{-\overset{|}{\underset{|}{C}}-\overset{O}{\overset{||}{C}}-} + -\overset{H}{\underset{|}{C}}-\overset{O}{\overset{||}{C}}- \xrightarrow[\text{catalyst}]{\text{NaOH}} \underset{\substack{\text{A hydroxy ketone or}\\\text{aldehyde}}}{-\overset{|}{\underset{|}{C}}-\overset{OH}{\underset{|}{C}}-\overset{|}{\underset{|}{C}}-\overset{O}{\overset{||}{C}}-}
$$

This carbon and this carbon form a bond.

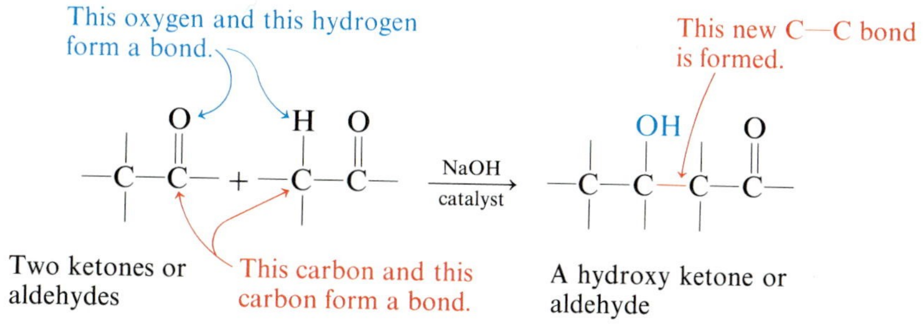

For example:

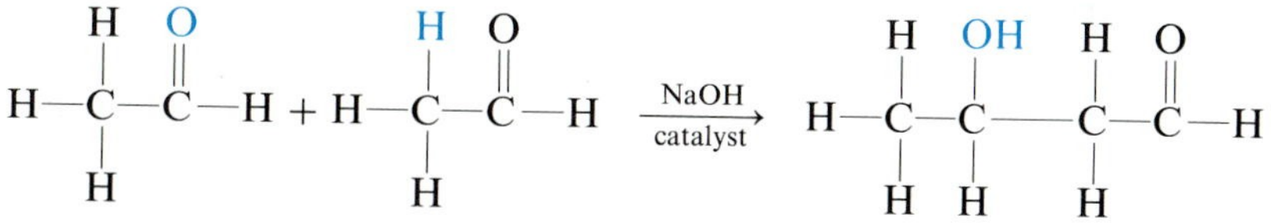

$$
\underset{2\,\text{Acetaldehyde}}{H-\overset{H}{\underset{H}{\overset{|}{C}}}-\overset{O}{\overset{||}{C}}-H + H-\overset{H}{\underset{H}{\overset{|}{C}}}-\overset{O}{\overset{||}{C}}-H} \xrightarrow[\text{catalyst}]{\text{NaOH}} \underset{3\text{-Hydroxybutanal}}{H-\overset{H}{\underset{H}{\overset{|}{C}}}-\overset{OH}{\underset{H}{\overset{|}{C}}}-\overset{H}{\underset{H}{\overset{|}{C}}}-\overset{O}{\overset{||}{C}}-H}
$$

One limitation of the aldol reaction is that the aldehyde or ketone starting material must have a hydrogen atom on the carbon next to the carbonyl group. If there is no such hydrogen present, an aldol reaction can't take place. For example, acetone can easily undergo an aldol reaction since it has six available hydrogens next to its carbonyl group, but benzaldehyde can't react in this way because it has no appropriately positioned hydrogen.

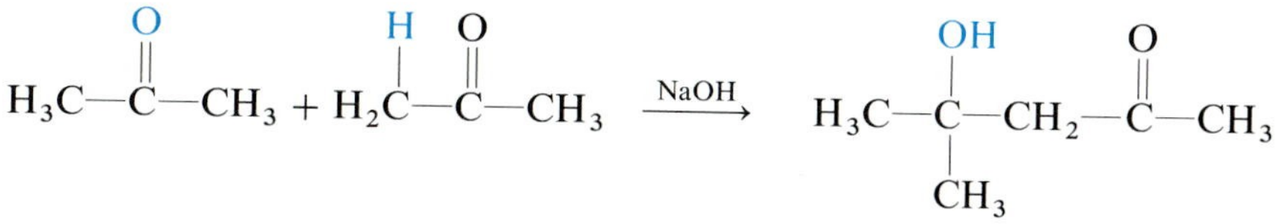

$$
\underset{\substack{\text{Two acetone molecules—}\\\text{each with six hydrogens}\\\text{next to carbonyl}}}{H_3C-\overset{O}{\overset{||}{C}}-CH_3 + H_2\overset{H}{\underset{}{C}}-\overset{O}{\overset{||}{C}}-CH_3} \xrightarrow{\text{NaOH}} \underset{\text{4-Hydroxy-4-methyl-2-pentanone}}{H_3C-\overset{OH}{\underset{CH_3}{\overset{|}{C}}}-CH_2-\overset{O}{\overset{||}{C}}-CH_3}
$$

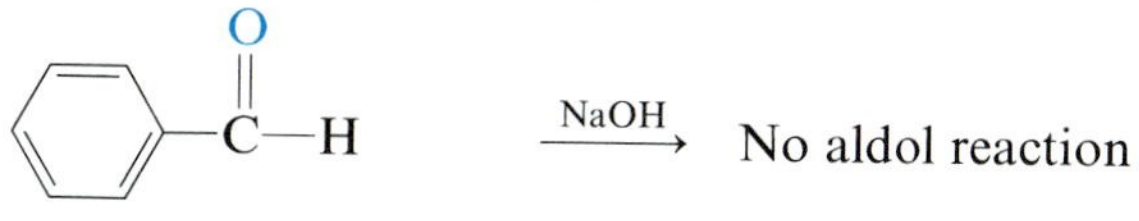

Benzaldehyde
(has no hydrogens on carbon next to carbonyl group $\xrightarrow{\text{NaOH}}$ No aldol reaction

Solved Problem 11.4   What aldol product would you obtain from this reaction?

$$CH_3CH_2CHO \xrightarrow{\text{NaOH}} ?$$

Solution   First, rewrite the reaction to emphasize the carbonyl group of one molecule and the C—H bond next to the carbonyl group of a second molecule:

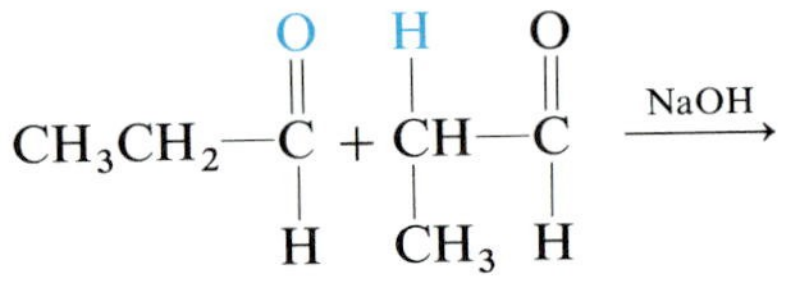

Next, draw the carbonyl group of the first molecule showing a C—O single bond and two part-bonds. Now remove the appropriate hydrogen from the second molecule, and connect it to the oxygen part-bond of the first molecule; then connect the appropriate carbon from the second molecule to the carbon part-bond of the first molecule. The structure that results is the final product.

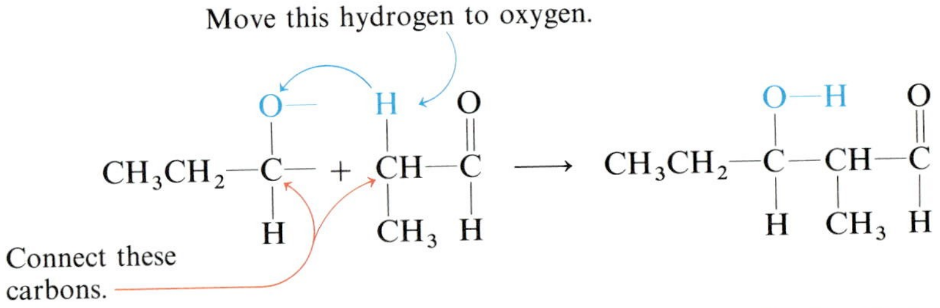

Move this hydrogen to oxygen.

Connect these carbons.

Practice Problems   **11.10**   Draw the aldol products from treatment of these aldehydes or ketones with base:

(a)

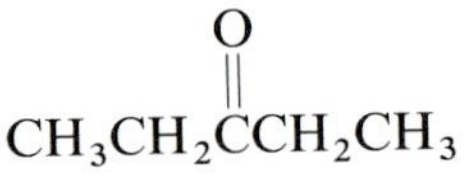

$$CH_3CH_2\overset{\overset{\displaystyle O}{\|}}{C}CH_2CH_3$$

(b)

[benzene ring]—CH_2CHO

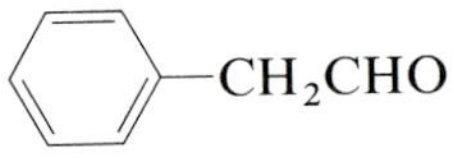

11.11   Which of the following compounds *can't* undergo aldol reactions?
(a) $CH_2\!=\!O$ (b) $(CH_3)_3CCHO$ (c) cyclopentanone

SUMMARY

Carbonyl compounds contain the carbon-oxygen double bond, C=O. Different kinds of carbonyl compounds exist, depending on what other groups are bonded to the carbonyl carbon. It's useful to classify carbonyl compounds into two groups based on their general properties: **Aldehydes (RCHO)** and **ketones (RCOR′)** behave similarly because the atoms (—H and —C) bonded to the carbonyl-group carbon don't attract electrons strongly; **carboxylic acids (RCOOH)**, **esters (RCOOR′)** and **amides (RCONH₂)** also behave similarly because the atoms (—O and —N) bound to the carbonyl-group carbon *do* attract electrons strongly.

INTERLUDE: A BIOLOGICAL ALDOL REACTION

Aldol reactions are routinely used by living organisms as a key step in the biological synthesis of many different molecules. One particularly important example takes place in green leaves during the photosynthesis of carbohydrates when two three-carbon molecules are joined to form the six-carbon sugar, fructose, which is in turn converted into glucose, sucrose, and all other sugars. Although the details of the biochemical transformation are more complex than shown here, the reaction of glyceraldehyde with dihydroxyacetone to yield fructose is clearly an aldol reaction. A hydrogen next to the dihydroxyacetone carbonyl group bonds to the carbonyl-group oxygen of glyceraldehyde, and a carbon-carbon bond forms between the two partners:

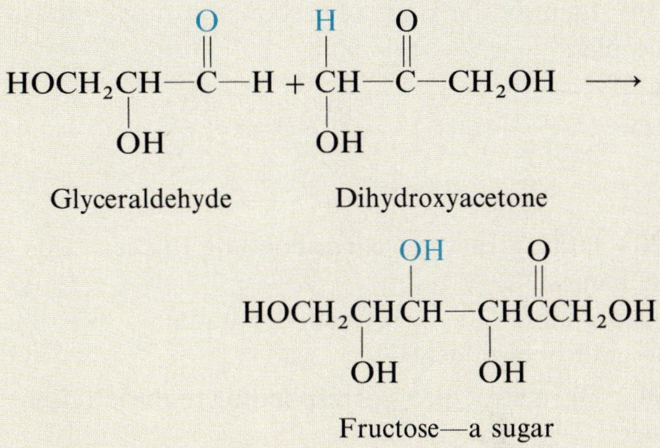

$$HOCH_2\underset{\underset{\textstyle OH}{|}}{CH}-\overset{\overset{\textstyle O}{\|}}{C}-H + \overset{\overset{\textstyle H}{|}}{\underset{\underset{\textstyle OH}{|}}{CH}}-\overset{\overset{\textstyle O}{\|}}{C}-CH_2OH \longrightarrow$$

Glyceraldehyde   Dihydroxyacetone

$$HOCH_2\overset{\overset{\textstyle OH}{|}}{CH}CH-\underset{\underset{\textstyle OH}{|}}{CH}\overset{\overset{\textstyle O}{\|}}{C}CH_2OH$$

Fructose—a sugar

The major reaction of aldehydes and ketones is the **addition** of various reagents to the carbon-oxygen double bond. For example, aldehydes and ketones both add hydrogen to yield alcohol products. This **reduction** is carried out by treating the carbonyl compound first with $NaBH_4$ to add $:H^-$ and then with aqueous acid to add H^+. Aldehydes and ketones also add alcohols to yield **hemiacetals** or **acetals** depending on the reaction conditions. A hemiacetal has an alcohol-like —OH group and an ether-like —OR group bonded to the same carbon; an acetal has two ether-like —OR groups bonded to the same carbon. Acetals react with aqueous acid (**hydrolysis**) to regenerate carbonyl compounds.

Aldehydes and ketones that have a hydrogen atom at the position next to the carbonyl group can undergo the **aldol reaction**. Because this reaction results in formation of a carbon-carbon bond between two molecules of starting material, it is useful for building larger molecules from smaller pieces.

REVIEW PROBLEMS

Aldehydes and Ketones

11.12 What is the structural difference between an aldehyde and a ketone?

11.13 How do aldehydes and ketones differ from carboxylic acids?

11.14 What family-name endings are used for aldehydes and ketones?

11.15 Use δ^+ and δ^- to show how the carbonyl group is polarized.

11.16 Draw structures for compounds that meet these descriptions:

(a) a ketone, C_5H_8O
(b) an aldehyde with eight carbons
(c) a keto-aldehyde, $C_6H_{10}O_2$
(d) a hydroxy-ketone, $C_5H_8O_2$

11.17 Which of these molecules contain carbonyl groups?

(a) 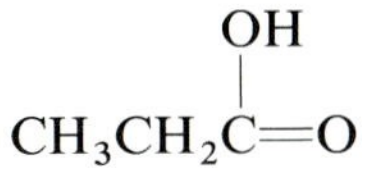
$$\underset{CH_3CH_2\overset{\underset{\textstyle |}{\textstyle OH}}{C}=O}{}$$

(b) 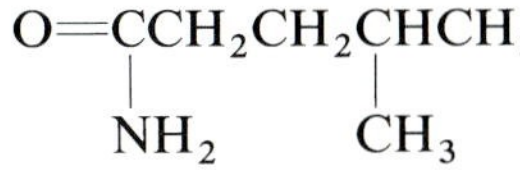 $O=CCH_2CH_2\underset{\underset{\textstyle CH_3}{|}}{CH}CH_3$
  with NH_2

(c) $CH_3CH_2-O-CH=CH_2$

11.18 Which of these molecules contain carbonyl groups?
(a) CH_3CH_2CHO (b) $CH_3CH_2COCH_2CH_3$
(c) $(CH_3)_2C(OH)CH_2CH_2CH_3$

11.19 Identify the kinds of carbonyl groups in these molecules:

(a)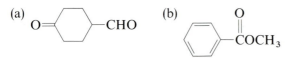
(b)

11.20 Draw structures corresponding to these aldehyde names:
(a) heptanal (b) 4,4-dimethylpentanal
(c) *o*-chlorobenzaldehyde

11.21 Draw structures corresponding to these ketone names:
(a) 3-heptanone (b) 2,4-dimethyl-3-pentanone
(c) *m*-nitroacetophenone

11.22 Give systematic names for these aldehydes:

(a) $CH_3CH_2\overset{\displaystyle |}{\underset{\displaystyle CH_3}{C}}HCHO$ (b) $CH_3CH_2CH_2\overset{\displaystyle CHO}{\underset{\displaystyle |}{C}}HCH_3$

(c) $(CH_3)_3CCHO$

11.23 Give systematic names for these ketones:

(a) $CH_3\overset{\displaystyle O}{\overset{\displaystyle \|}{C}}CH_2CH_3$ (b) $CH_3\overset{\displaystyle O}{\overset{\displaystyle \|}{C}}CH_2CH_2\overset{\displaystyle CH_3}{\underset{\displaystyle |}{C}}HCH_3$

(c) $(CH_3)_3C\overset{\displaystyle O}{\overset{\displaystyle \|}{C}}C(CH_3)_3$

11.24 The following names are incorrect. What is wrong with each?
(a) 1-butanone (b) 4-butanone (e) 2-butanal
(c) 3-butanone (d) cyclohexanal

Reactions of Aldehydes and Ketones

11.25 What kind of compound is produced when an aldehyde reacts with one equivalent of an alcohol?

11.26 What kind of compound is produced when an aldehyde reacts with two equivalents of an alcohol in the presence of acid catalyst?

11.27 What reaction does Tollens' reagent carry out?

11.28 Why does reduction of an aldehyde give a primary alcohol whereas reduction of a ketone gives a secondary alcohol?

11.29 Give specific examples of these reactions:
(a) reduction of a ketone with $NaBH_4$
(b) oxidation of an aldehyde with Tollens' reagent
(c) formation of an acetal from an aldehyde and an alcohol

11.30 Which of the following compounds would react with Tollens' reagent? Draw structures of the reaction products.

(a) cyclopentanone (b) hexanal

(c) $CH_3CH_2CH_2\overset{\displaystyle CHO}{\underset{\displaystyle |}{C}}HCH_2CH_3$

11.31 Draw structures of the products obtained by reaction of the compounds in Problem 11.30 with $NaBH_4$ followed by acid treatment.

11.32 What is the difference between an acetal and a hemiacetal?

11.33 Assume that you were given two unlabeled bottles, one containing pentanal and the other containing 2-pentanone. What simple chemical test would allow you to distinguish between the contents of the two bottles?

11.34 Draw structures of the aldehydes that might be oxidized to yield these carboxylic acids:

(a) H_3C—⬡—$COOH$

(b) $CH_3CH_2\overset{\displaystyle COOH}{\underset{\displaystyle |}{C}}HCH_2\overset{\displaystyle CH_3}{\underset{\displaystyle |}{C}}HCH_3$

11.35 Draw structures of the primary alcohols that might be oxidized to yield the carboxylic acids shown in Problem 11.34.

11.36 What ketones or aldehydes might be reduced to yield these alcohols?

(a) $CH_3CH_2CH_2\overset{\displaystyle CH_2OH}{\underset{\displaystyle |}{C}}HCH_3$

(b) 2,2-dimethyl-1-hexanol (c) HO—⬡—OH

11.37 Write the structures of the hemiacetals and acetals that result from these reactions:
(a) 2-butanone + 1-propanol
(b) butanal + isopropyl alcohol

11.38 What products would result from hydrolysis of these acetals?

(a)

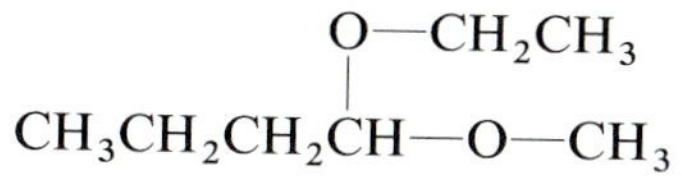

$$CH_3CH_2CH_2CH \underset{\underset{O-CH_3}{|}}{\overset{O-CH_2CH_3}{}}$$

(b)

$$\underset{H_3C}{\overset{H_3C}{}}C\underset{O-CH_2}{\overset{O-CH_2}{}}$$

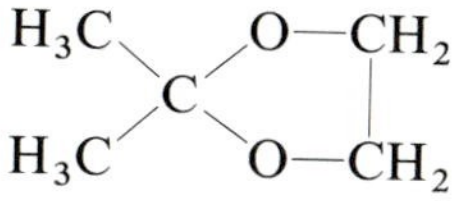

11.39 Acetals are usually made by reaction of an aldehyde or ketone with two molecules of a monoalcohol. If an aldehyde or ketone reacts with *one* molecule of a dialcohol, however, a *cyclic* acetal results. Draw the structure of the cyclic acetal formed in the following reaction:

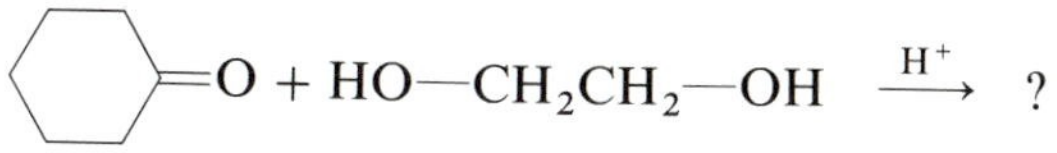

$$\bigcirc\!\!=\!\!O + HO-CH_2CH_2-OH \xrightarrow{H^+} ?$$

11.40 Cyclic hemiacetals sometimes form if an alcohol group in one part of a molecule adds to a carbonyl group elsewhere in the same molecule. What is the structure of the open-chain hydroxy aldehyde from which the following hemiacetal might form?

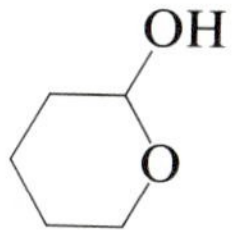

OH

A cyclic hemiacetal

11.41 Like the cyclic hemiacetal in Problem 11.40, cyclic acetals are also known. What products would you expect on hydrolysis of the following cyclic acetal?

O—CH₃

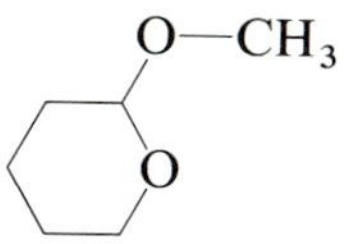

A cyclic acetal

11.42 Glucose exists largely in the cyclic hemiacetal form shown. Draw the structure of glucose in its open-chain hydroxy aldehyde form.

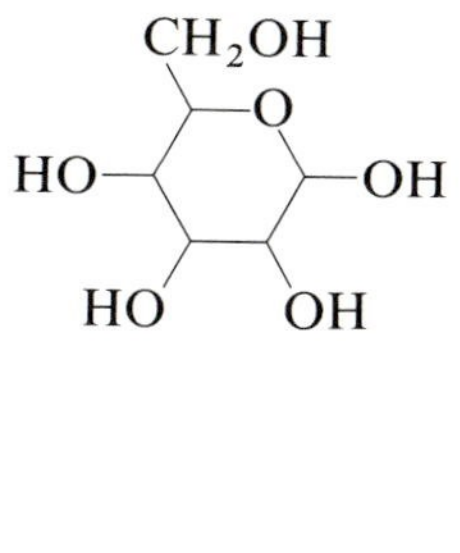

CH₂OH

HO— —OH   Glucose

HO   OH

11.43 In many respects, glucose acts as if it were an aldehyde, even though the structure shown in Problem 11.42 contains no —CHO group. For example, glucose reacts with Tollens' reagent to produce a silver mirror. Explain.

11.44 What products would result on hydrolysis of the following cyclic acetal?

$$H_2C\underset{H_2C-O}{\overset{H_2C-O}{}}\underset{}{\overset{}{\diagdown}}CH_2$$

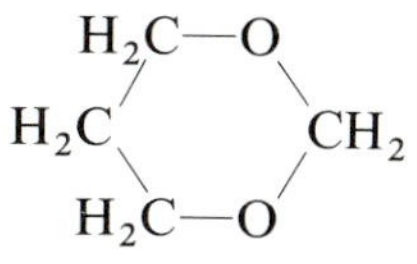

11.45 Aldosterone is a key steroid involved in controlling sodium/potassium salt balance in the body. Identify the carbonyl groups in aldosterone.

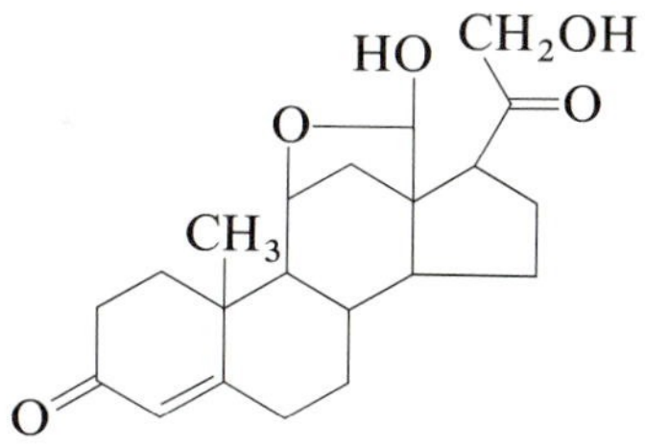

Aldosterone

11.46 Aldosterone (Problem 11.45) also contains a cyclic hemiacetal linkage. Identify the linkage, and tell whether it's derived from an aldehyde or a ketone.

11.47 3-Cyclohexenone is an example of a *difunctional* molecule, a compound that has two different functional groups. What products would you expect from treatment of 3-cyclohexenone with these reagents?
(a) H₂ and a Pd catalyst
(b) NaBH₄, then H₃O⁺
(c) Br₂

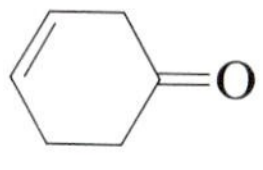

3-Cyclohexenone

Aldol Reactions

11.48 What structural requirement must be met in order for an aldehyde or ketone to undergo an aldol reaction?

11.49 Write the structures of the aldol product that

would result by base treatment of these ketones or aldehydes:

(a) CH₃CHCH₂CHO   (b) cyclohexanone
      |
      CH₃

(c) 3-pentanone

11.50 When 2-butanone is treated with base, a mixture of two different aldol products results. What are their structures?

11.51 The aldol reaction is reversible; that is, aldol products can sometimes break apart to yield aldehydes or ketones. What products would result if the following hydroxy aldehyde broke apart in a reverse aldol reaction?

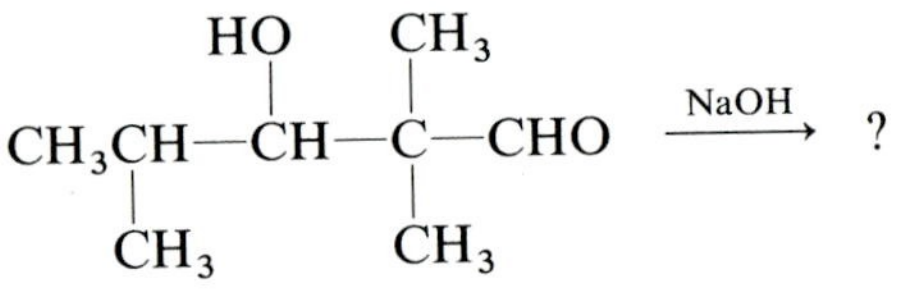

$$\underset{\underset{CH_3}{|}}{\overset{\overset{HO}{|}}{CH_3CH}}-\underset{\underset{CH_3}{|}}{\overset{\overset{CH_3}{|}}{CH}}-\overset{}{C}-CHO \xrightarrow{NaOH} \ ?$$

11.52 The following compound can be made by a mixed aldol reaction between two different carbonyl compounds. What two aldehyde or ketone starting materials would you use?

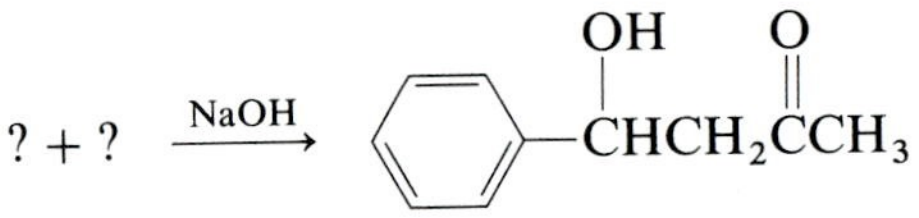

? + ? $\xrightarrow{NaOH}$ [benzene ring]—$\overset{\overset{OH}{|}}{CH}CH_2\overset{\overset{O}{\|}}{C}CH_3$

CHAPTER 12

Carboxylic Acids, Esters, and Amides

Like the flamingos in Chapter 9, we are what we eat. Among the compounds that you'll study in this chapter are the esters, which are responsible for the characteristic odors of many fruits and flowers. (Painting: "Summer," by Giuseppe Arcimboldo, 1573.)

We said in the last chapter that carbonyl compounds can be classified into two groups, based on their structural and chemical similarities. In one group are aldehydes and ketones, and in the other group are carboxylic acids, esters, and amides.

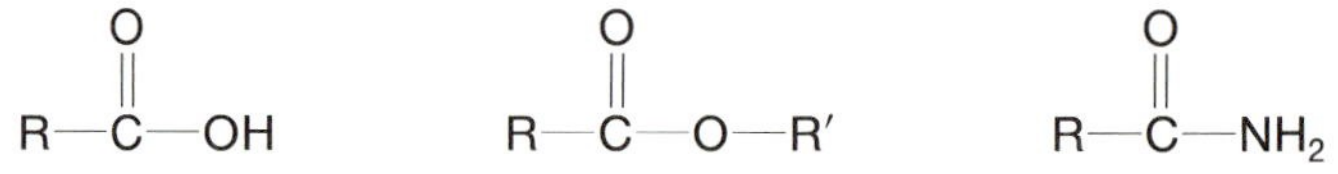

$$R-\overset{\displaystyle O}{\overset{\|}{C}}-OH \qquad R-\overset{\displaystyle O}{\overset{\|}{C}}-O-R' \qquad R-\overset{\displaystyle O}{\overset{\|}{C}}-NH_2$$

Carboxylic acid   Ester   Amide

We'll look at the chemistry of this second group of carbonyl compounds in this chapter, and answer the following questions:

1. **How are carboxylic acids, esters, and amides named?** The goal: you should learn the rules of nomenclature for these groups of compounds.

2. **What are the properties and reactions of carboxylic acids?** The goal: you should learn the important properties and reactions of carboxylic acids and how to predict the reaction products in specific cases.

3. **What are the properties and reactions of esters?** The goal: you should learn the important reactions of esters and how to predict the reaction products in specific cases.

4. **What is the Claisen condensation reaction, and why is it important in biochemistry?** The goal: you should learn how the Claisen condensation reaction occurs and why it is important in biochemistry.

5. **What are the properties and reactions of amides?** The goal: you should learn the important reactions of amides and how to predict the reaction products in specific cases.

6. **What are phosphate esters, and why are they important?** The goal: you should learn what phosphate esters are and how to recognize them.

12.1 AN OVERVIEW OF CARBOXYLIC ACID, ESTER, AND AMIDE REACTIONS

Unlike the aldehydes and ketones we studied in the last chapter, carboxylic acids, esters, and amides all have their carbonyl groups bonded to an atom (O or N) that strongly attracts electrons. In carboxylic acids, the carbonyl group

■ **Ester** A carbonyl compound that has one organic group and one —OR group bonded to the carbonyl carbon, RCOOR'.

■ **Amide** A carbonyl compound that has one organic group and one —NH$_2$, —NHR, or —NR$_2$ group bonded to the carbonyl carbon, RCONH$_2$.

is bonded to a carbon substituent and a hydroxyl (—OH); in **esters**, to a carbon substituent and an alkoxyl (—OR); in **amides**, to a carbon substituent and a nitrogen (—NH$_2$, —NHR, or —NR$_2$). Thus, the carbonyl groups are more strongly polarized in these three families of compounds than in ketones and aldehydes.

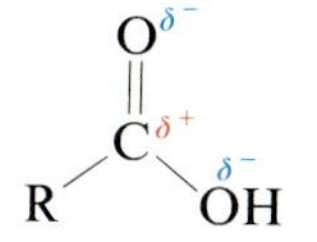

Carboxylic acid 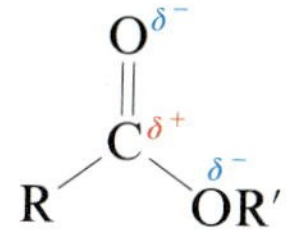   Ester 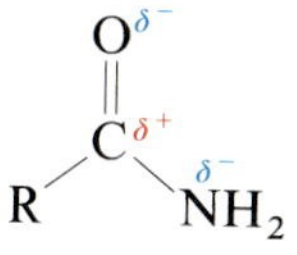   Amide

■ **Carbonyl-group substitution reaction** A reaction in which a new group X replaces (substitutes for) a group Y attached to a carbonyl-group carbon.

The structural similarity of these three families of compounds also leads to similarities in their chemistry. All three undergo **carbonyl-group substitution reactions**, in which a group we can represent as —X replaces (substitutes for) the —OH, —OR, or —NH$_2$ group of the starting material.

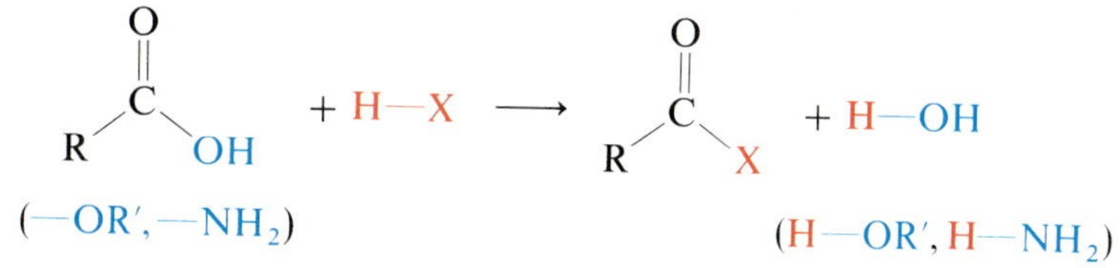

A carbonyl-group substitution reaction

We'll see numerous examples of these carbonyl-group substitution reactions in the sections that follow.

12.2 NAMING CARBOXYLIC ACIDS, ESTERS, AND AMIDES

Carboxylic acids (RCOOH) are named by replacing the final -*e* of the corresponding alkane name with -*oic acid*. The three-carbon acid is propanoic acid; the straight-chain four-carbon acid is butanoic acid; and so on. If substituents are present, the chain is numbered beginning at the —COOH end as in 3-methylbutanoic acid.

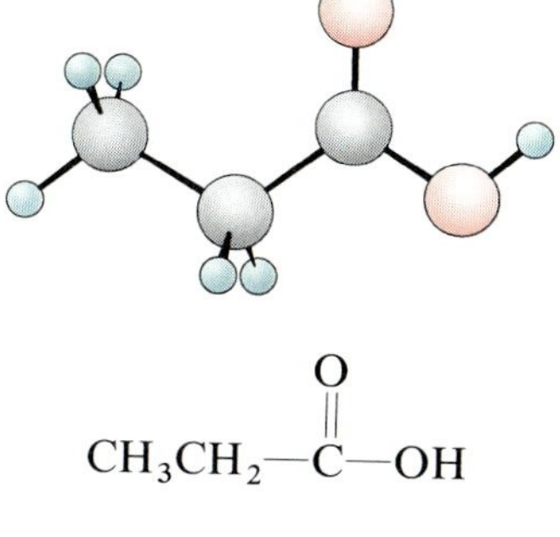

$$CH_3CH_2-\overset{\overset{\displaystyle O}{\|}}{C}-OH$$

Propanoic acid

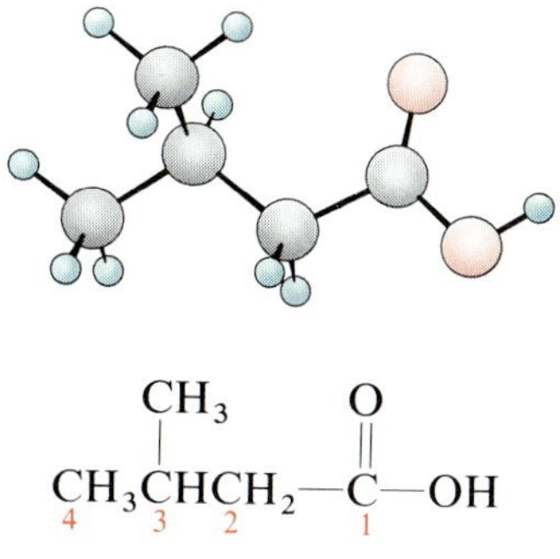

$$CH_3\overset{\overset{\displaystyle CH_3}{|}}{\underset{4}{C}}\underset{3}{H}\underset{2}{C}H_2-\overset{\overset{\displaystyle O}{\|}}{\underset{1}{C}}-OH$$

3-Methylbutanoic acid

$$\overset{\overset{\displaystyle O}{\|}}{C}-OH$$

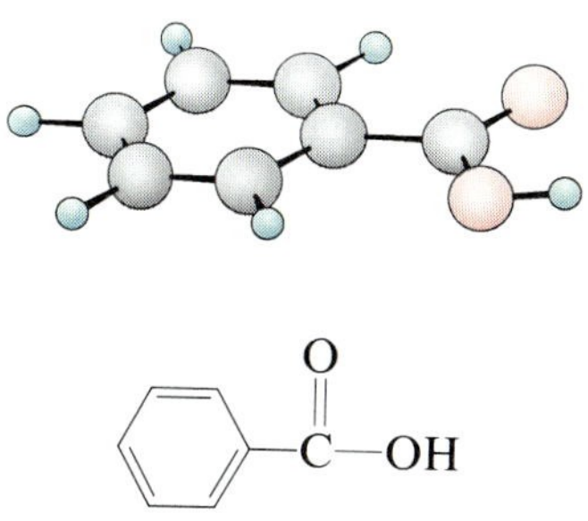

Benzoic acid

Because many simple carboxylic acids were among the first organic compounds to be isolated and purified, a number of them also have common names. For example, the one-carbon compound (HCOOH) is always called *formic acid* rather than methanoic acid; the two-carbon compound (CH₃COOH) is always called *acetic acid* rather than ethanoic acid; and the three-carbon compound is sometimes called propionic acid rather than propanoic acid. (The name *formic* is derived from the Latin, *formica*, meaning ant, since formic acid was first isolated from ants. The name *acetic* comes from the Latin, *acetum*, meaning vinegar, since vinegar is an aqueous solution of acetic acid.)

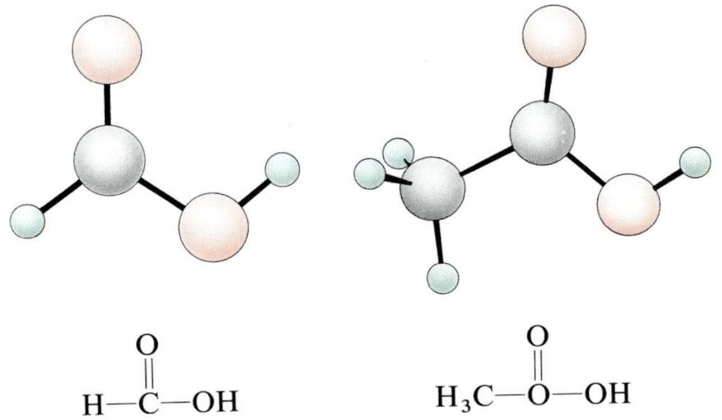

$$\overset{\displaystyle O}{\underset{}{\overset{\|}{H-C-OH}}}$$

Formic acid
(methanoic acid)

$$\overset{\displaystyle O}{\underset{}{\overset{\|}{H_3C-O-OH}}}$$

Acetic acid
(ethanoic acid)

Esters (RCOOR′) are named by first specifying the alkyl group attached to the ether-like oxygen (—O—R′) and then identifying the related carboxylic acid. The family-name ending *-ic acid* is replaced by *-ate*.

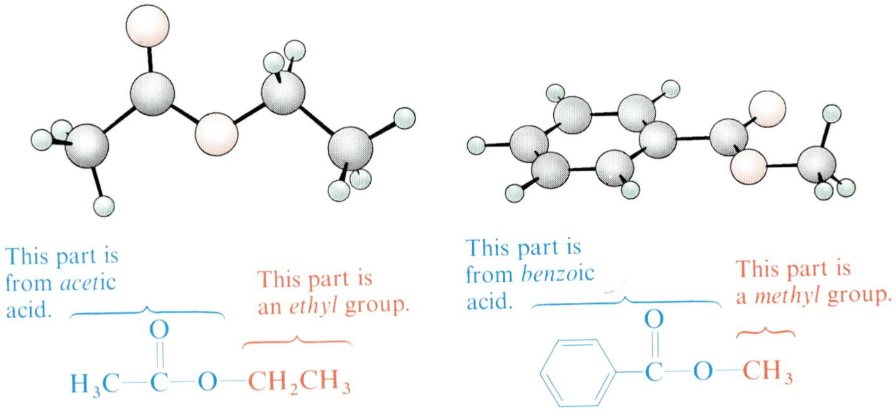

This part is from *acetic* acid. This part is an *ethyl* group.

$$H_3C-\overset{\displaystyle O}{\overset{\|}{C}}-O-CH_2CH_3$$

Ethyl acetate

This part is from *benzoic* acid. This part is a *methyl* group.

$$\text{(benzene ring)}-\overset{\displaystyle O}{\overset{\|}{C}}-O-CH_3$$

Methyl benzoate

Amides (RCONH₂) with an unsubstituted —NH₂ group are named by replacing the *-oic acid* of the corresponding carboxylic acid name with *-amide*. For example, the two-carbon amide derived from acetic acid is called *acetamide*. If the nitrogen atom of the amide has alkyl substituents on it, the compound is named by first specifying the alkyl group and then identifying the amide name. The alkyl substituents are preceded by the letter *N* to identify them as being attached directly to nitrogen.

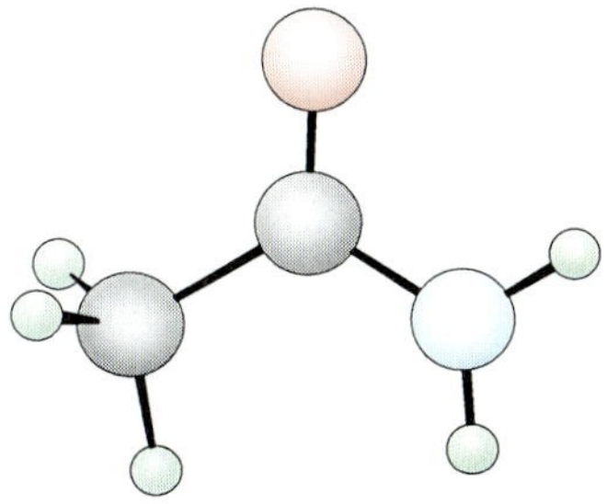

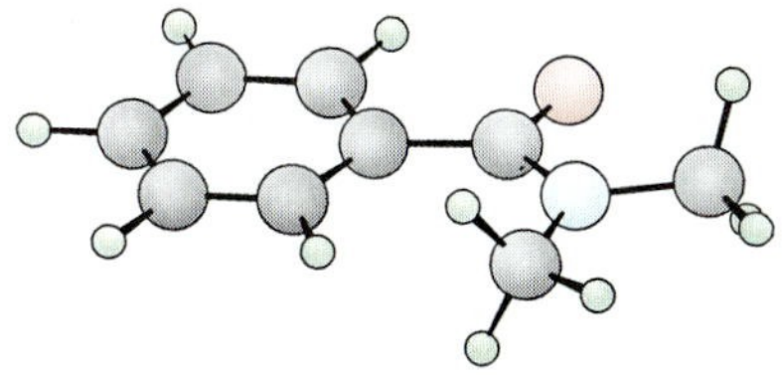

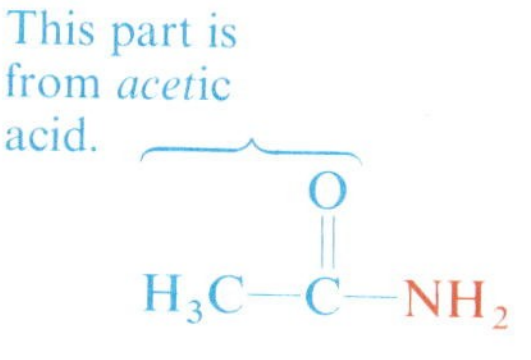

This part is from *acetic* acid.

These two *methyl* groups are attached to *Nitrogen*.

This part is from *benzoic* acid.

$$H_3C-\overset{\overset{\displaystyle O}{\|}}{C}-NH_2$$

Acetamide

$$\overset{\overset{\displaystyle O}{\|}}{C}-\overset{\overset{\displaystyle}{N}}{\underset{\overset{\displaystyle}{CH_3}}{}}-CH_3$$

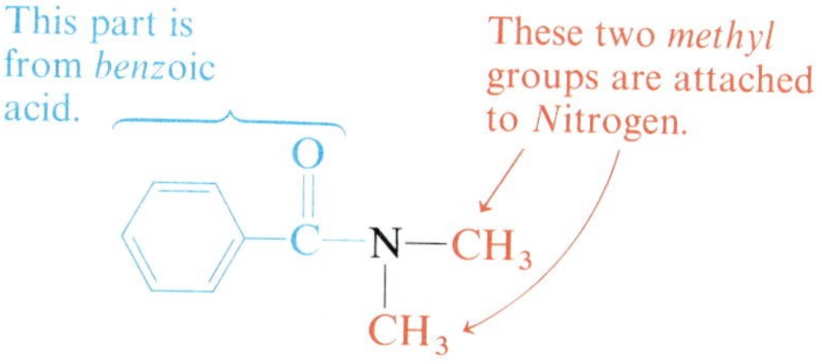

N,N-Dimethylbenzamide

Practice Problems

12.1 What are the names of these compounds?

(a)
$$\overset{\overset{\displaystyle CH_3}{|}}{CH_3CHCH_2CH_2\overset{\overset{\displaystyle O}{\|}}{C}-OH}$$

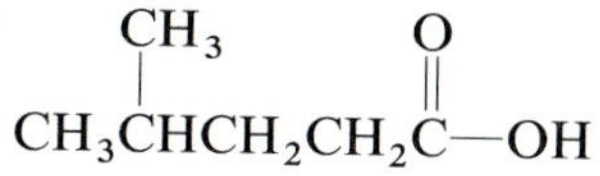

(b)
$$CH_3CH_2CH_2\overset{\overset{\displaystyle O}{\|}}{C}-O-\overset{\overset{\displaystyle CH_3}{|}}{CHCH_3}$$

$$Cl-\overset{\overset{\displaystyle O}{\|}}{C}-NHCH_3$$

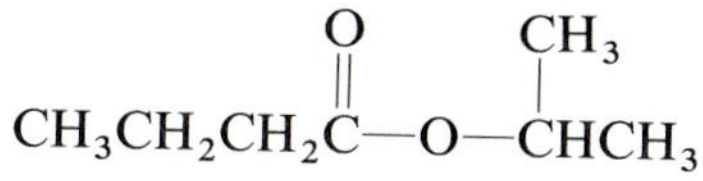

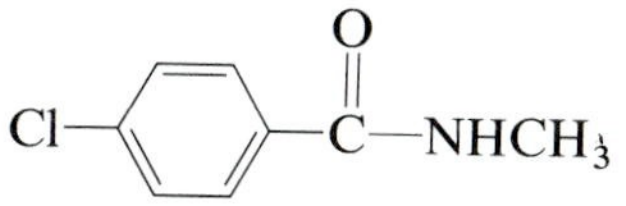

12.2 Draw structures corresponding to these names:
(a) 3-methylhexanoic acid   (b) 4-methylpentanamide   (c) propyl benzoate
(d) *o*-nitrobenzoic acid   (e) *N*-methylbutanamide   (f) ethyl propanoate

12.3 OCCURRENCE AND PROPERTIES OF CARBOXYLIC ACIDS

Carboxylic acids occur widely throughout the plant and animal kingdoms. For example, butanoic acid (from the Latin, *butyrum*, butter) is responsible for the odor of rancid butter, and cinnamic acid is partially responsible for the odor of cinnamon. We'll also see in Chapter 15 that long-chain carboxylic acids such as stearic acid are components of all animal fats and vegetable oils.

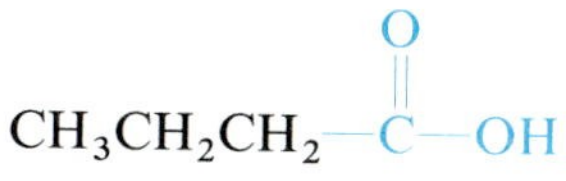

$$CH_3CH_2CH_2-\overset{\overset{\displaystyle O}{\|}}{C}-OH$$

Butanoic acid
(from butter)

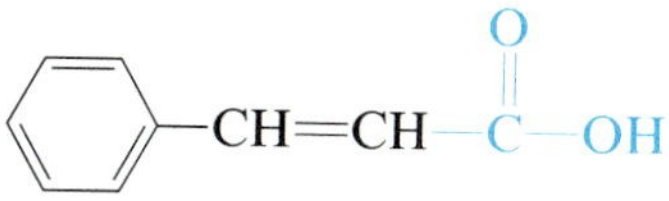

$$\overset{\overset{\displaystyle O}{\|}}{CH=CH-C-OH}$$

Cinnamic acid
(from oil of cinnamon)

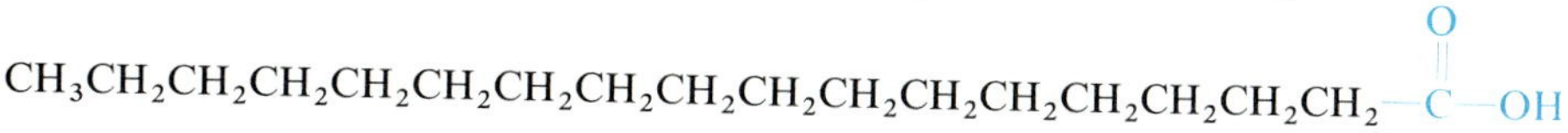

$$CH_3CH_2CH_2CH_2CH_2CH_2CH_2CH_2CH_2CH_2CH_2CH_2CH_2CH_2CH_2CH_2-\overset{\overset{\displaystyle O}{\|}}{C}-OH$$

Stearic acid ($C_{18}H_{36}O_2$, from animal fat and vegetable oil)

Figure 12.1
Hydrogen bonding between (a) pairs of carboxylic acid molecules and (b) a carboxylic acid molecule in water. The ability to form these hydrogen bonds makes simple carboxylic acids high boiling and soluble in water.

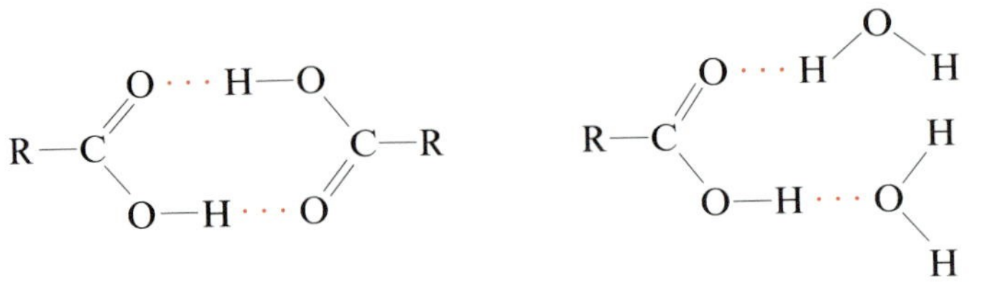

■ **Dimer** A unit formed by the joining together of two identical molecules.

Like the alcohols we saw in Section 10.4, carboxylic acids can form hydrogen bonds (Figure 12.1). Molecules pair off to form **dimers** that are twice the size of an individual molecule, and carboxylic acids thus have higher boiling points than alkanes of similar formula weight. Even formic acid, the lightest and simplest carboxylic acid, is a liquid at room temperature with a boiling point of 101°C.

Another consequence of their ability to form hydrogen bonds is that carboxylic acids with four or fewer carbons are soluble in water (Figure 12.1). As the size of the alkane-like portion increases relative to the size of the —COOH portion, however, water solubility falls off.

12.4 ACIDITY OF CARBOXYLIC ACIDS

■ **Carboxylate anion** The anion that results from dissociation of a carboxylic acid, $RCOO^-$.

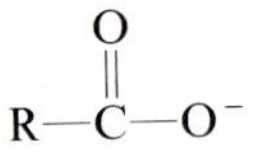

As their name implies, carboxylic acids are *acidic*. Like other acids discussed earlier in Section 7.3, carboxylic acids dissociate slightly in aqueous solution to give H_3O^+ and a **carboxylate anion**, $RCOO^-$. Carboxylic acids are much weaker than inorganic acids like HCl or H_2SO_4, however; only about one percent of acetic acid molecules dissociate in aqueous solution.

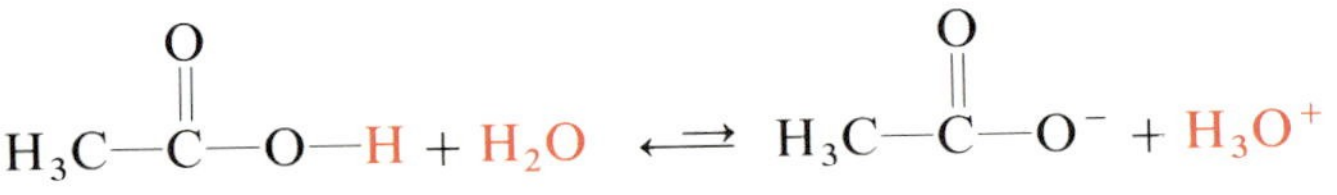

Acetic acid

Acetate ion

99% of acetic acid molecules are undissociated in a 0.10-*M* solution.

1% of acetic acid molecules are dissociated in a 0.10-*M* solution.

Although weaker than common inorganic acids, carboxylic acids nevertheless react readily with strong bases such as sodium hydroxide to give water and a salt. For example, acetic acid reacts with NaOH to give sodium acetate:

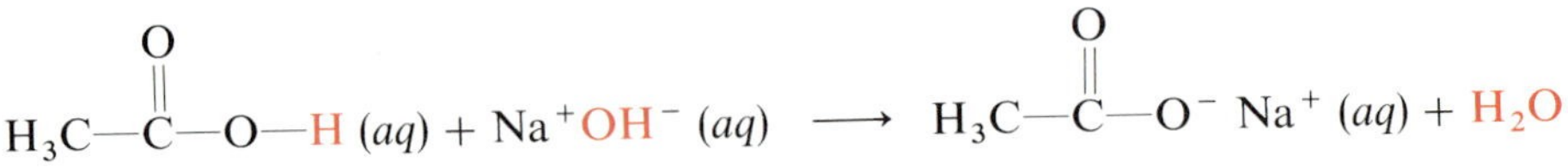

Acetic acid
(a weak acid)

Sodium hydroxide

Sodium acetate

Acetate ion, as the anion of a weak acid, is itself a base that can accept a proton from HCl or some other strong acid (Section 7.5). Thus, carboxylate anions can be converted back into free carboxylic acids by reaction with HCl or H_2SO_4.

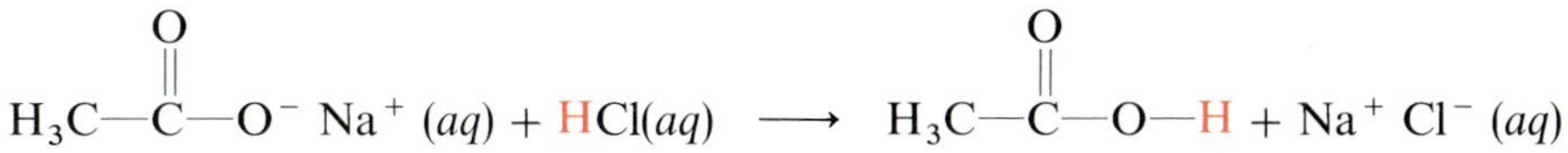

$$H_3C\overset{\displaystyle O}{\overset{\|}{-C}}-O^-\;Na^+\;(aq) + HCl(aq) \longrightarrow H_3C\overset{\displaystyle O}{\overset{\|}{-C}}-O-H + Na^+\;Cl^-\;(aq)$$

Sodium acetate   Acetic acid
(a base)

The interconversion between a carboxylic acid and its carboxylate ion is so easy that we can choose which form is present in solution merely by adjusting the pH of the medium (Section 7.7). At high pH (basic solution), carboxylate ion is present; at low pH (acidic solution), carboxylic acid is present.

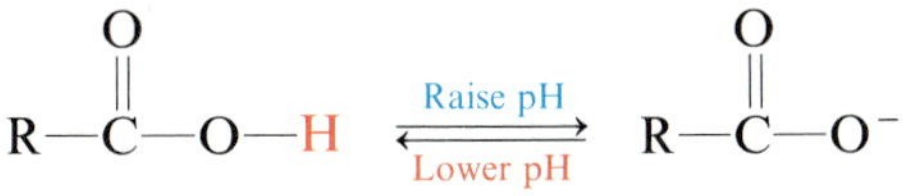

$$R\overset{\displaystyle O}{\overset{\|}{-C}}-O-H \underset{\text{Lower pH}}{\overset{\text{Raise pH}}{\rightleftharpoons}} R\overset{\displaystyle O}{\overset{\|}{-C}}-O^-$$

At low pH   At high pH

Sodium or potassium salts of carboxylic acids are ionic solids that are usually far more soluble in water than the carboxylic acids themselves. For example, benzoic acid has a water solubility of only 3.4 g/L at 25°C, whereas sodium benzoate has a water solubility of 550 g/L. (The name *sodium benzoate* might have a familiar sound: You've probably seen it listed on the labels of soft drinks and jellies as a food preservative.)

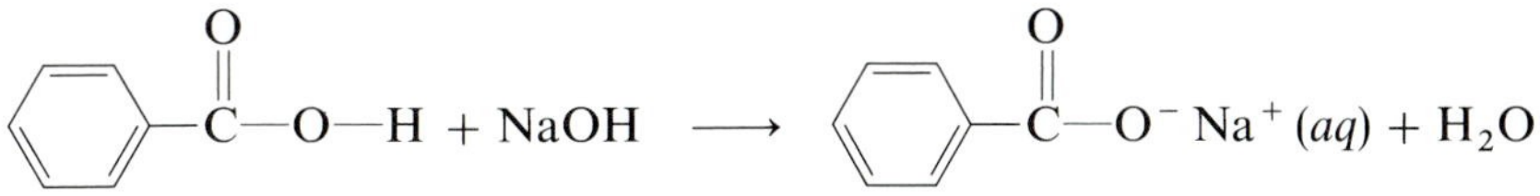

Benzoic acid   Sodium benzoate
(slightly water-soluble)   (highly water-soluble)

The conversion of an insoluble carboxylic acid into a soluble carboxylate anion is an important trick used in medicine where drugs must be able to circulate through aqueous body fluids. Penicillin G, for example, is usually administered as its sodium or potassium salt to increase its water solubility and rate of absorption by the body.

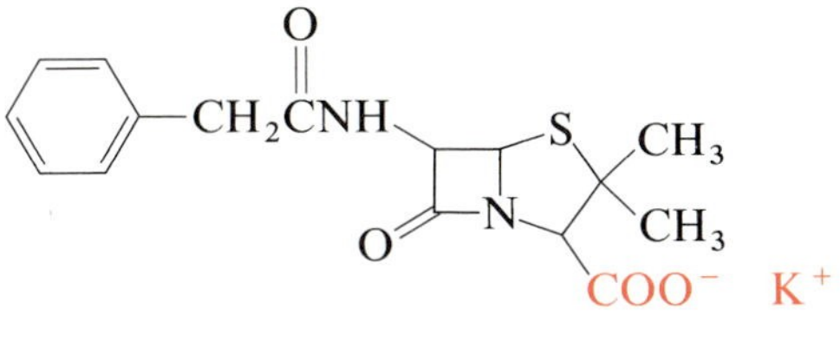

Penicillin G Potassium

Practice Problems

12.3 Write the products of these reactions:
(a) $CH_3CH_2CH_2COOH + KOH \longrightarrow$?
(b) 2-methylpentanoic acid + $Ba(OH)_2 \longrightarrow$?

12.4 How many grams of NaOH would it take to neutralize 5.0 grams of benzoic acid?

12.5 REACTIONS OF CARBOXYLIC ACIDS: ESTER FORMATION

The most important reaction of carboxylic acids, both in chemical laboratories and in living organisms, is their conversion into esters. In the laboratory, an **esterification reaction** is carried out by warming a carboxylic acid with an alcohol in the presence of a strong acid catalyst such as H_2SO_4. In so doing, an —H is lost from the alcohol, an —OH is lost from the acid, and water is formed as a by-product. The net effect is a substitution of —OH by —OR':

■ **Esterification reaction** The reaction between an alcohol and a carboxylic acid to yield an ester plus water.

This —OH group is replaced by this —OR' group.

An esterification reaction:

$$R-\overset{\displaystyle O}{\overset{\|}{C}}-OH + H-OR' \xrightarrow{H^+ \text{ catalyst}} R-\overset{\displaystyle O}{\overset{\|}{C}}-OR' + H_2O$$

A carboxylic acid   An alcohol   An ester

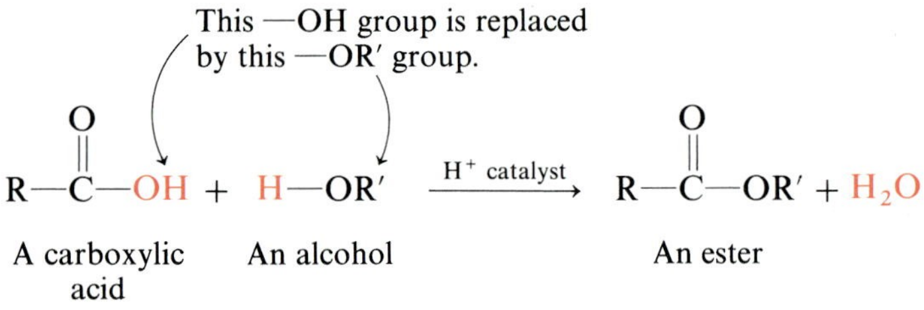

For example:

$$H_3C-\overset{\displaystyle O}{\overset{\|}{C}}-OH + H-OCH_2CH_3 \xrightarrow{H^+} H_3C-\overset{\displaystyle O}{\overset{\|}{C}}-OCH_2CH_3 + H_2O$$

Acetic acid   Ethanol   Ethyl acetate

$$CH_3CH_2CH_2-\overset{\displaystyle O}{\overset{\|}{C}}-OH + H-OCH_3 \xrightarrow{H^+} CH_3CH_2CH_2-\overset{\displaystyle O}{\overset{\|}{C}}-OCH_3 + H_2O$$

Butanoic acid   Methanol   Methyl butanoate
(in pineapple oil)

Solved Problem 12.1   The flavor ingredient in oil of wintergreen can be made by reaction of *o*-hydroxybenzoic acid with methanol. What is its structure?

$$\text{(o-hydroxybenzoic acid structure)}\ \begin{array}{c} O \\ \| \\ C-OH \end{array} + CH_3OH \xrightarrow{\ H^+\ } \ ?$$

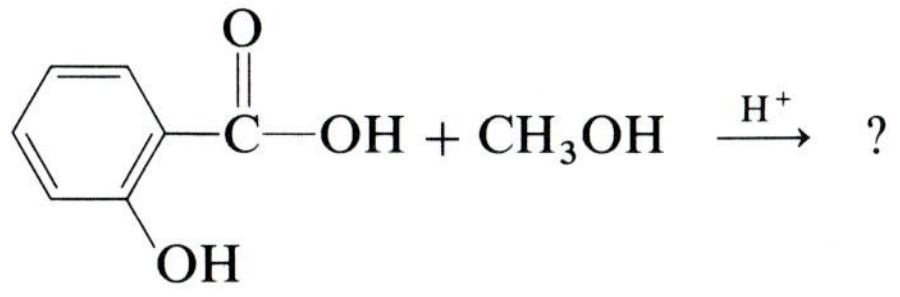

Solution   First, write the two reaction partners so that the —COOH group of the acid and the —OH group of the alcohol are facing each other:

(*o*-Hydroxybenzoic acid)
$$\begin{array}{c} O \\ \| \\ C-OH \end{array} + H-OCH_3 \qquad \text{(Methanol)} \longrightarrow H_2O$$

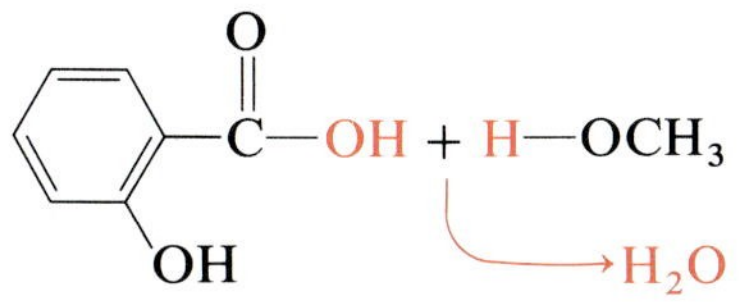

Next, remove —OH from the acid and —H from the alcohol to form water, and then join the two resulting organic fragments together. The product is the ester.

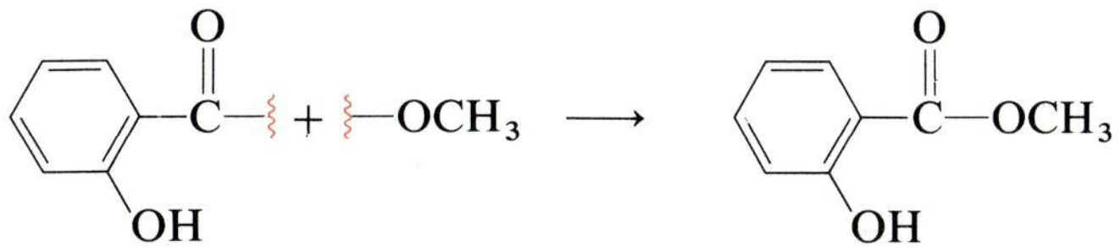

$$\begin{array}{c} O \\ \| \\ C-\!\!\!\{ \end{array} + \{-OCH_3 \longrightarrow \begin{array}{c} O \\ \| \\ C-OCH_3 \end{array}$$

Methyl *o*-hydroxybenzoate
(in oil of wintergreen)

Practice Problems

12.5   Raspberry oil contains an ester that can be made by reaction of formic acid with 2-methyl-1-propanol. What is its structure?

$$HCOOH + (CH_3)_2CHCH_2OH \longrightarrow \ ?$$

12.6   What carboxylic acid and what alcohol are needed to make each of the following esters?

(a)
$$\begin{array}{c} O \\ \| \\ \text{(cyclohexyl)}-O-CCH_2CH_2CH(CH_3)_2 \end{array}$$

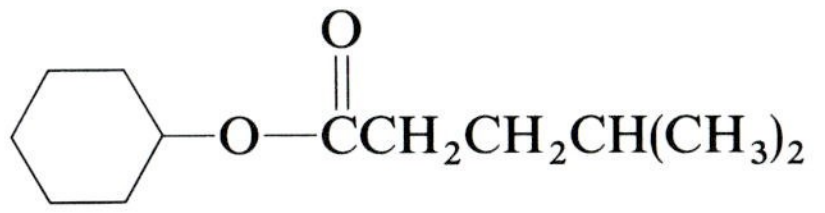

(b)
$$\begin{array}{c} O \\ \| \\ CH_3CH_2CH_2CH_2C-O-CH(CH_3)_2 \end{array}$$

12.6 OCCURRENCE AND PROPERTIES OF ESTERS

Esters have many uses in medicine, in industry, and in living systems. In medicine, a number of important pharmaceutical agents, including aspirin and the local anesthetic benzocaine, are esters. In industry, polyesters such as Dacron and Mylar are used to make synthetic fibers and films. In nature, many simple esters are responsible for the fragrant odors of fruits and flowers. For example, isopentyl acetate is found in bananas, and octyl acetate is found in oranges.

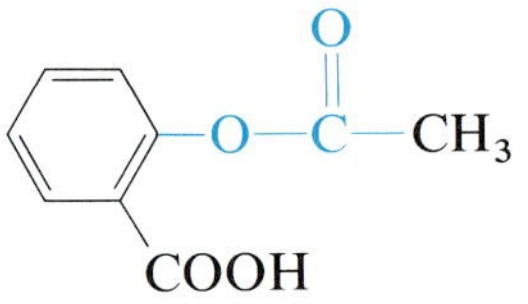

Aspirin

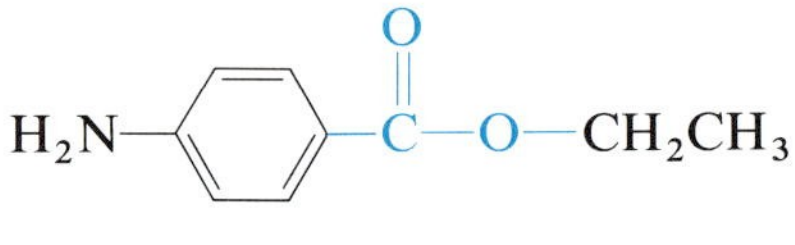

Benzocaine
(a local anesthetic)

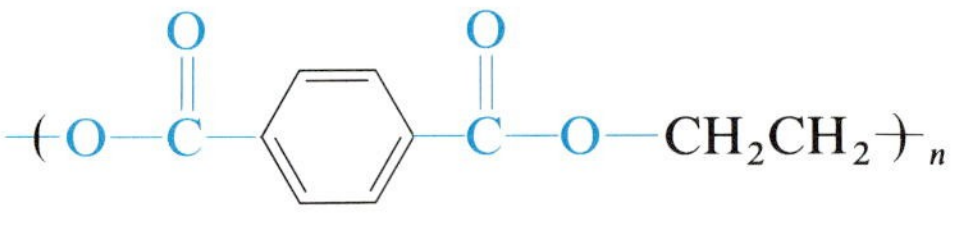

Dacron
(a polyester)

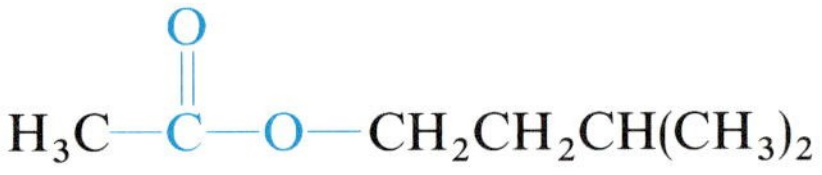

Isopentyl Acetate
(from bananas)

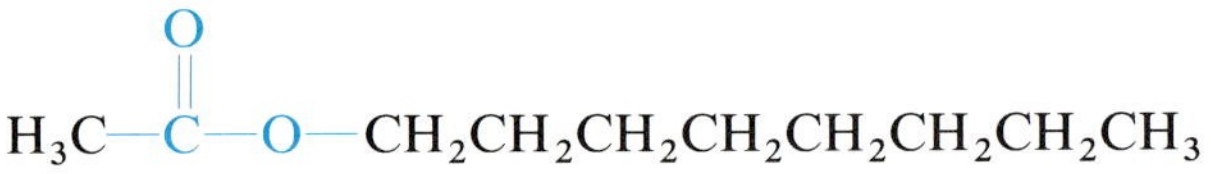

Octyl acetate
(from oranges)

Unlike alcohols and carboxylic acids, esters don't have an —OH group and can't form hydrogen bonds. Thus, esters are generally lower boiling than the acids from which they are derived and are generally insoluble in water. Ethyl acetate, for example, has a boiling point of 77°C (versus 118°C for acetic acid).

12.7 REACTIONS OF ESTERS: HYDROLYSIS

The most important reaction of esters is their conversion into carboxylic acids. Both in the laboratory and in the body, esters undergo a hydrolysis reaction with water that splits the ester molecule into a carboxylic acid and an alcohol. The net effect of the hydrolysis is a substitution of —OH for OR'.

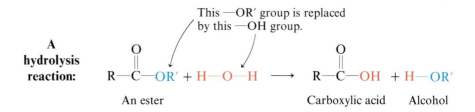

A hydrolysis reaction:

This —OR′ group is replaced by this —OH group.

An ester → Carboxylic acid   Alcohol

Ester hydrolysis is catalyzed both by acids and by bases. Acid-catalyzed hydrolysis is simply the reverse of the esterification reaction discussed in Section 12.5. An ester is treated with water in the presence of a strong acid such as H_2SO_4, and hydrolysis takes place. For example:

Ethyl benzoate + H_2O $\underset{\text{catalyst}}{\overset{H_2SO_4}{\rightleftharpoons}}$ Benzoic acid + Ethanol

■ **Saponification reaction** The reaction of an ester with aqueous hydroxide ion to yield an alcohol and the metal salt of a carboxylic acid.

Base-catalyzed hydrolysis, often called a **saponification** (sa-pon-if-i-**ca**-shun) **reaction**, after the Latin word *sapo*, "soap," takes place when an ester is treated with water in the presence of a base such as NaOH or KOH. The main difference between the acid- and base-catalyzed hydrolysis methods is that the base-catalyzed reaction yields a carboxylate anion as product rather than a free carboxylic acid. In order to isolate the acid, the anion has to be protonated by treatment with HCl or H_2SO_4.

Ester + NaOH(*aq*) ⟶ Carboxylate salt + Alcohol

Carboxylate salt $\xrightarrow{\text{HCl}(aq)}$ Carboxylic acid + NaCl(*aq*)

For example:

Methyl butanoate + NaOH(*aq*) ⟶ Sodium butanoate + Methanol

$CH_3CH_2CH_2$—C—OCH_3 + NaOH(*aq*) ⟶ $CH_3CH_2CH_2$—C—O^- Na$^+$ + CH_3OH

Sodium butanoate $\xrightarrow{\text{HCl}(aq)}$ Butanoic acid + NaCl(*aq*)

$CH_3CH_2CH_2$—C—O^- Na$^+$ $\xrightarrow{\text{HCl}(aq)}$ $CH_3CH_2CH_2$—C—OH + NaCl(*aq*)

Solved Problem 12.2 What product would you obtain from hydrolysis of ethyl formate, a flavor constituent of rum?

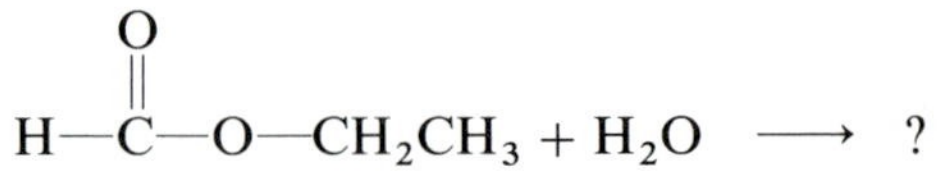

$$\overset{O}{\overset{\|}{H-C}}-O-CH_2CH_3 + H_2O \longrightarrow ?$$

Solution First, look at the name of the starting ester. Usually, the name of the ester gives a good indication of the names of the two products. Thus, *ethyl formate* yields *ethyl* alcohol and *form*ic acid. To find the product structures in a more systematic way, write the structure of the ester, and locate the bond between the carbonyl-group carbon and the —OR′ group.

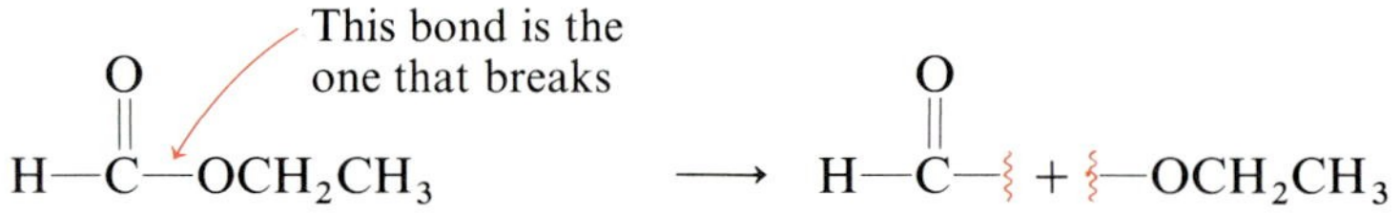

This bond is the one that breaks

$$\overset{O}{\overset{\|}{H-C}}-OCH_2CH_3 \longrightarrow \overset{O}{\overset{\|}{H-C}}- + -OCH_2CH_3$$

Next, carry out a substitution reaction. First form the carboxylic-acid product by connecting an —OH to the carbonyl-group carbon. Then add an —H to the —OCH₂CH₃ group to form the alcohol product.

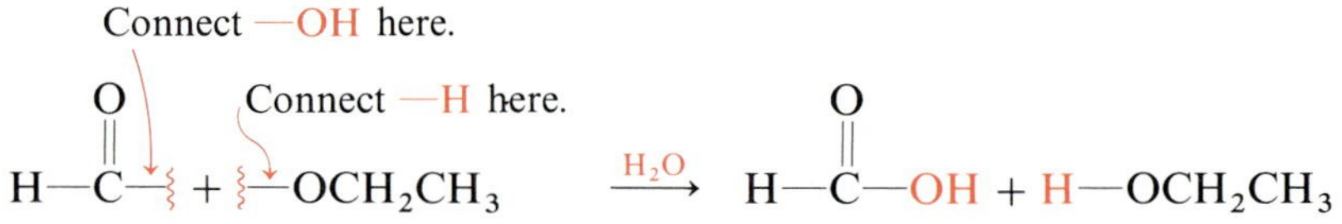

Connect —OH here.

Connect —H here.

$$\overset{O}{\overset{\|}{H-C}}- + -OCH_2CH_3 \xrightarrow{H_2O} \overset{O}{\overset{\|}{H-C}}-OH + H-OCH_2CH_3$$

Practice Problem 12.7 What products would you obtain from hydrolysis of these esters?

(a)
$$\overset{H_3C \quad\quad O \quad\quad CH_3}{\underset{CH_3CH-\overset{\|}{C}-O-CHCH_3}{\mid \quad\quad\quad\quad \mid}}$$

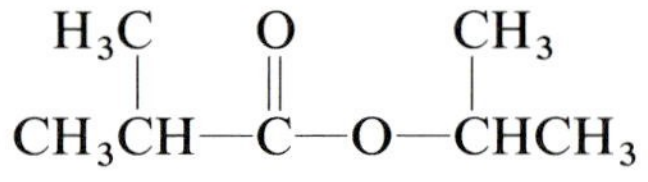

(b) $CH_3CH{=}CHCOOCH_2CH_3$

(c) propyl *p*-bromobenzoate

12.8 REACTIONS OF ESTERS: CLAISEN CONDENSATION

■ **Claisen condensation reaction** A reaction that joins two ester molecules together to yield a keto ester product.

Just as the aldol reaction (Section 11.7) joins two aldehyde or ketone molecules together, the **Claisen condensation reaction** joins two ester molecules together. A Claisen condensation takes place when an ester is treated with a strong base such as sodium methoxide, Na⁺ ⁻OCH₃, the sodium salt of methanol. In the reaction, the —OR′ group is lost from the carbonyl-group carbon of one ester molecule, and a bond forms between that carbon and the carbon atom next to the carbonyl group of the second molecule. The product is a ketone-ester, or *keto* ester.

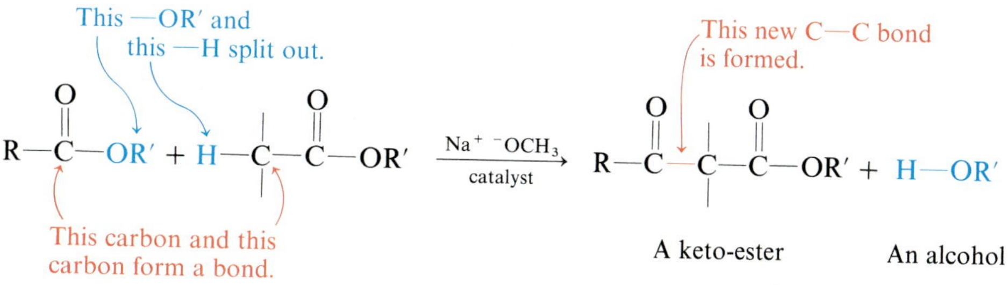

This —OR′ and
this —H split out.

This new C—C bond
is formed.

$$R-\overset{\overset{\displaystyle O}{\|}}{C}-OR' + H-\overset{|}{\underset{|}{C}}-\overset{\overset{\displaystyle O}{\|}}{C}-OR' \xrightarrow[\text{catalyst}]{Na^+\ ^-OCH_3} R-\overset{\overset{\displaystyle O}{\|}}{C}-\overset{|}{\underset{|}{C}}-\overset{\overset{\displaystyle O}{\|}}{C}-OR' + H-OR'$$

This carbon and this
carbon form a bond.

A keto-ester   An alcohol

For example:

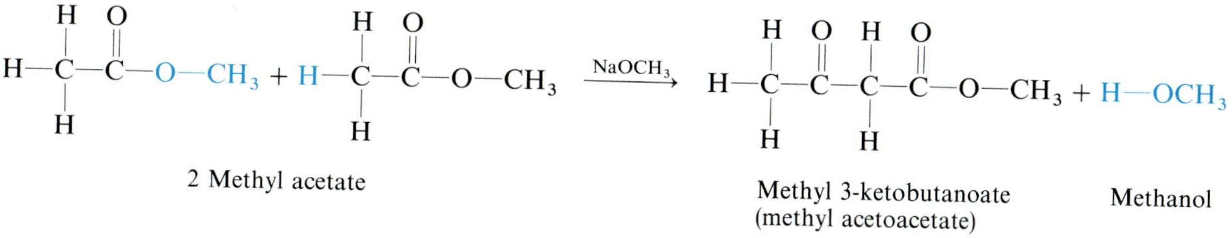

$$H-\overset{\overset{\displaystyle H}{|}}{\underset{\underset{\displaystyle H}{|}}{C}}-\overset{\overset{\displaystyle O}{\|}}{C}-O-CH_3 + H-\overset{\overset{\displaystyle H}{|}}{\underset{\underset{\displaystyle H}{|}}{C}}-\overset{\overset{\displaystyle O}{\|}}{C}-O-CH_3 \xrightarrow{NaOCH_3} H-\overset{\overset{\displaystyle H}{|}}{\underset{\underset{\displaystyle H}{|}}{C}}-\overset{\overset{\displaystyle O}{\|}}{C}-\overset{\overset{\displaystyle H}{|}}{\underset{\underset{\displaystyle H}{|}}{C}}-\overset{\overset{\displaystyle O}{\|}}{C}-O-CH_3 + H-OCH_3$$

2 Methyl acetate

Methyl 3-ketobutanoate
(methyl acetoacetate)

Methanol

Claisen condensation reactions are used by living organisms for the biological synthesis of many different kinds of molecules. Fats, sugars, steroid hormones, and many other classes of compounds are synthesized in the body by joining together small ester molecules using Claisen condensations.

Solved Problem 12.3   What product would you obtain from the following Claisen condensation reaction?

$$CH_3CH_2\overset{\overset{\displaystyle O}{\|}}{C}OCH_3 \xrightarrow{NaOCH_3} ?$$

Solution   First, rewrite the reaction to emphasize the ester group of one molecule and a C—H bond next to the ester group of the second molecule:

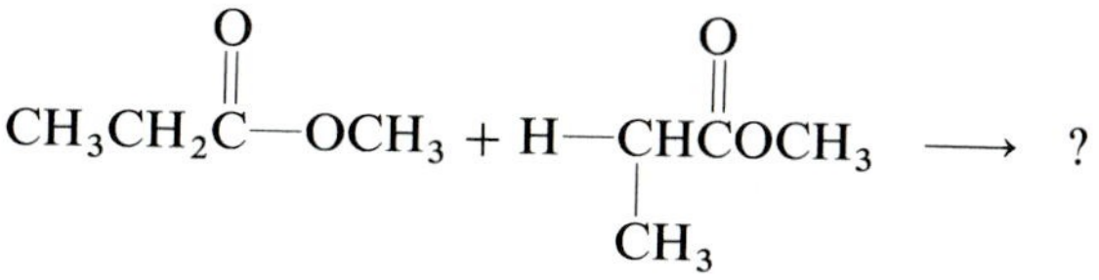

$$CH_3CH_2\overset{\overset{\displaystyle O}{\|}}{C}-OCH_3 + H-\underset{\underset{\displaystyle CH_3}{|}}{CH}\overset{\overset{\displaystyle O}{\|}}{C}OCH_3 \longrightarrow ?$$

Next, remove the —OCH₃ group from the first ester and the appropriate —H from the second ester to yield methanol. Then connect the remaining fragments to yield the keto ester product.

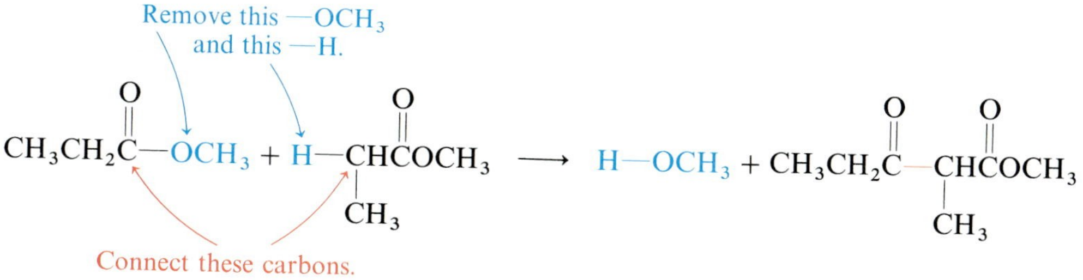

Remove this —OCH₃
and this —H.

$$CH_3CH_2\overset{\overset{\displaystyle O}{\|}}{C}-OCH_3 + H-\underset{\underset{\displaystyle CH_3}{|}}{CH}\overset{\overset{\displaystyle O}{\|}}{C}OCH_3 \longrightarrow H-OCH_3 + CH_3CH_2\overset{\overset{\displaystyle O}{\|}}{C}-\underset{\underset{\displaystyle CH_3}{|}}{CH}\overset{\overset{\displaystyle O}{\|}}{C}OCH_3$$

Connect these carbons.

Practice Problems

12.8 Draw the products from Claisen condensation of these esters:

(a)

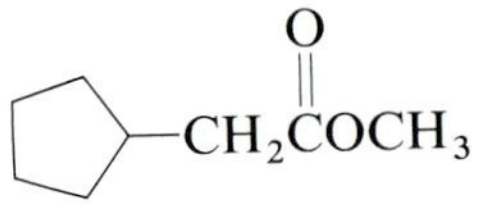

(b) methyl butanoate

12.9 Why do you suppose no Claisen reaction takes place when methyl benzoate is treated with sodium methoxide?

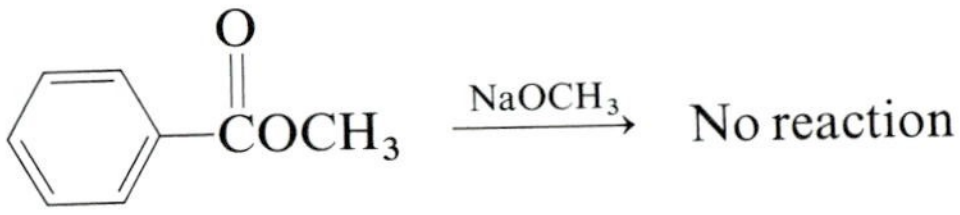

AN APPLICATION: THIOL ESTERS—BIOLOGICAL CARBOXYLIC ACID DERIVATIVES

Although the principles remain the same, many of the carbonyl-group substitution reactions that take place in living organisms use *thiol esters* in place of normal esters. A thiol ester, which is simply a sulfur-containing analog of a normal ester, has a carbonyl group bonded to one —SR group and to one carbon substituent (RCOSR′). As we'll see in Chapters 18 and 19, many of the reactions used in the body to extract energy from food involve thiol esters.

Acetyl coenzyme A, usually abbreviated as acetyl CoA, is the most common thiol ester in nature. Although it has a much more complex structure than ethyl acetate, it reacts in almost exactly the same way that ethyl acetate does. As just one example of its use by living organisms, *N*-acetylglucosamine, an important constituent of cell-surface membranes in mammals, is synthesized in nature by reaction of acetyl CoA with glucosamine.

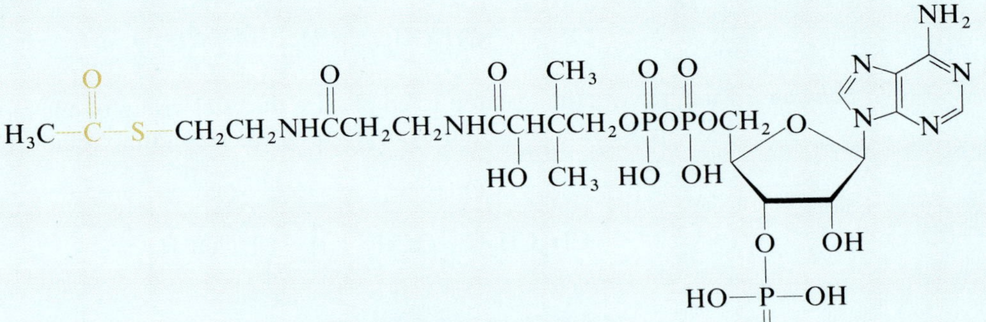

Acetyl coenzyme A—a thiol ester

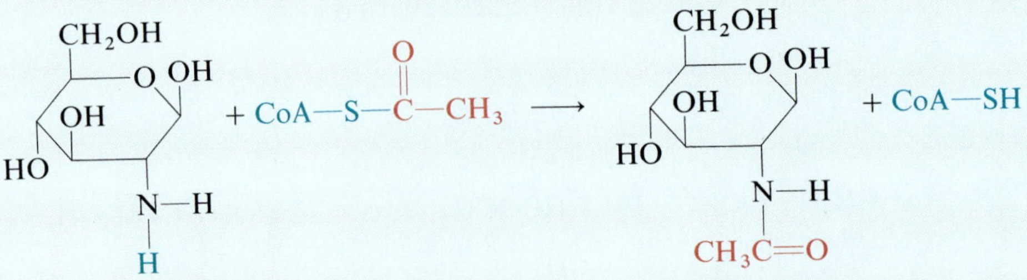

Glucosamine   Acetyl coenzyme A   *N*-Acetylglucosamine

12.9   OCCURRENCE AND PROPERTIES OF AMIDES

Without amides, there would be no life. As we'll see in Chapter 16, the amide bond between nitrogen and a carbonyl-group carbon is the fundamental link used by organisms for forming proteins. In addition, some synthetic polymers such as nylon contain amide groups, and important pharmaceutical agents such as acetaminophen, the aspirin substitute found in Tylenol and Excedrin, are amides.

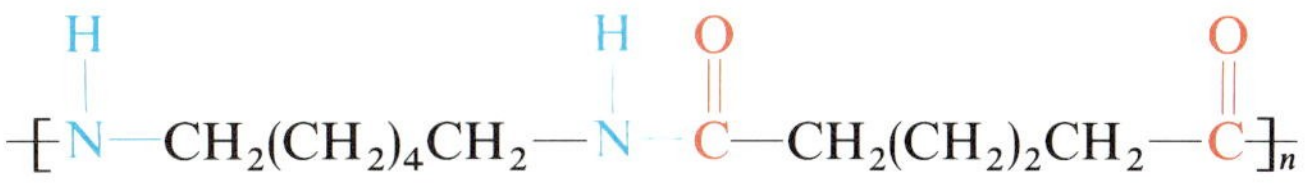

$$\left[\overset{H}{N} - CH_2(CH_2)_4CH_2 - \overset{H}{N} - \overset{O}{C} - CH_2(CH_2)_2CH_2 - \overset{O}{C} \right]_n$$

Nylon—a polyamide

$$HO - \left\langle \bigcirc \right\rangle - \overset{H}{N} - \overset{O}{C} - CH_3$$

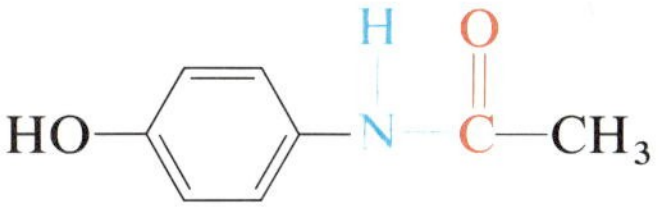

Acetaminophen

Unlike amines, which also contain nitrogen (Section 10.8), amides are neutral rather than basic. Amides do not act as proton acceptors and do not form ammonium salts when treated with acid:

$$R - \ddot{N}H_2 + HCl \longrightarrow R - NH_3^+Cl^-$$

An amine   An ammonium salt
(basic)

$$R - \overset{O^{\delta-}}{\underset{}{C}} \overset{}{\underset{\delta+}{}} \ddot{N}H_2 + HCl \longrightarrow \text{No reaction}$$

An amide
(nonbasic)

The main reason for the lack of basicity of amides is that the electron-withdrawing ability of the nearby carbonyl group causes the unshared pair of electrons on nitrogen to be held tightly and prevents it from bonding to H^+.

12.10   PREPARATION OF AMIDES FROM CARBOXYLIC ACIDS

The reaction of a carboxylic acid with an amine yields an amide, just as the reaction of a carboxylic acid with an alcohol yields an ester (Section 12.4). In both cases, water is formed as by-product, and the —OH part of the carboxylic acid is replaced.

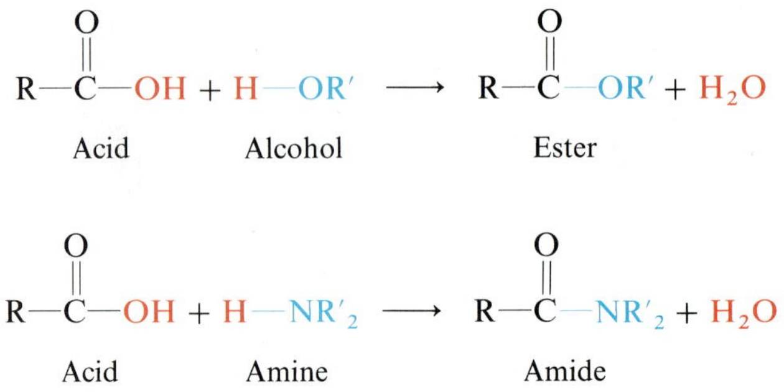

$$R-\overset{\overset{\displaystyle O}{\|}}{C}-OH + H-OR' \longrightarrow R-\overset{\overset{\displaystyle O}{\|}}{C}-OR' + H_2O$$

Acid   Alcohol   Ester

$$R-\overset{\overset{\displaystyle O}{\|}}{C}-OH + H-NR'_2 \longrightarrow R-\overset{\overset{\displaystyle O}{\|}}{C}-NR'_2 + H_2O$$

Acid   Amine   Amide

The reaction of a carboxylic acid with an amine to form an amide doesn't take place spontaneously unless another reagent is present to speed up the process. In the laboratory, a compound called *DCC* (dicyclohexylcarbodiimide) is often used. Although the exact way in which DCC works is a bit too complex to discuss in detail, its function is to activate the carboxylic acid by making it much more reactive. Once activated, the carboxylic acid then reacts rapidly with an amine to generate an amide.

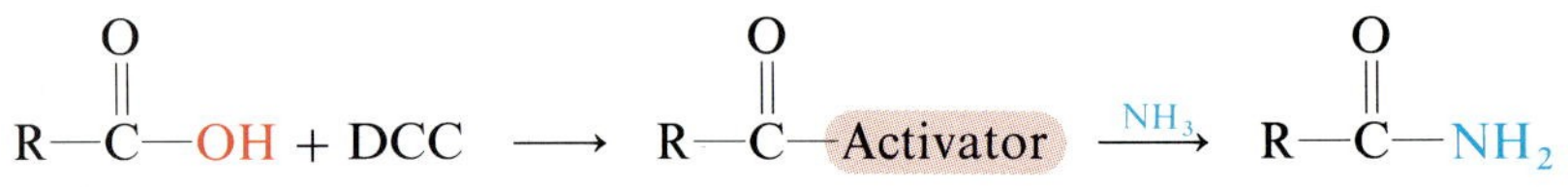

$$R-\overset{\overset{\displaystyle O}{\|}}{C}-OH + DCC \longrightarrow R-\overset{\overset{\displaystyle O}{\|}}{C}-\boxed{Activator} \xrightarrow{NH_3} R-\overset{\overset{\displaystyle O}{\|}}{C}-NH_2$$

where $DCC =$ ⬡—N=C=N—⬡

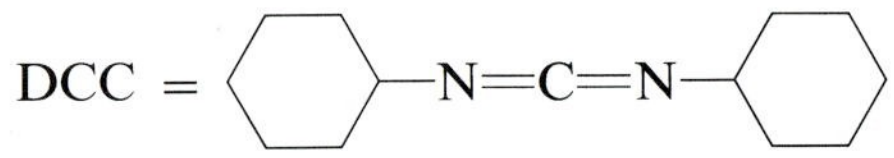

For example:

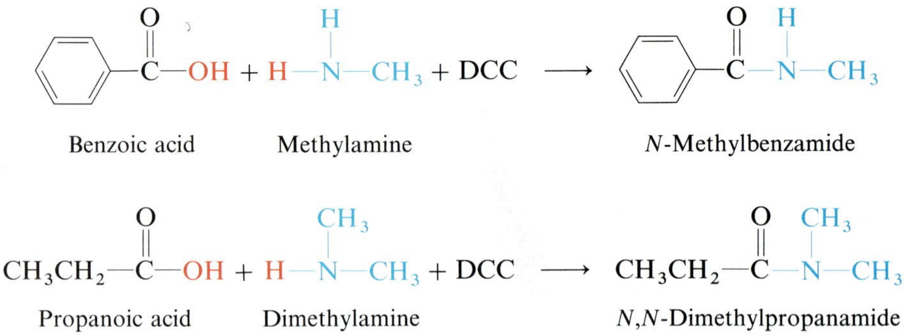

Benzoic acid + Methylamine + DCC ⟶ *N*-Methylbenzamide

$$CH_3CH_2-\overset{\overset{\displaystyle O}{\|}}{C}-OH + H-\overset{\overset{\displaystyle CH_3}{|}}{N}-CH_3 + DCC \longrightarrow CH_3CH_2-\overset{\overset{\displaystyle O}{\|}}{C}-\overset{\overset{\displaystyle CH_3}{|}}{N}-CH_3$$

Propanoic acid   Dimethylamine   *N,N*-Dimethylpropanamide

In living organisms, complex biomolecules function as activating reagents to allow amide formation from a carboxylic acid and an amine. Nevertheless, the principle behind a chemist's use of DCC in a laboratory and an organism's use of complex molecules in a cell is exactly the same.

Solved Problem 12.4   The mosquito repellent DEET (diethyltoluamide) can be prepared by reaction of diethylamine with *p*-methylbenzoic acid (toluic acid) in the presence of DCC. What is the structure of DEET?

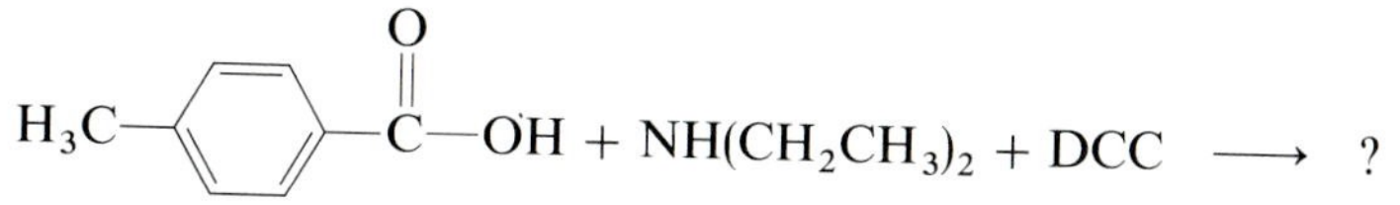

$$H_3C\text{—}\text{—}\overset{\displaystyle O}{\overset{\|}{C}}\text{—OH} + NH(CH_2CH_3)_2 + DCC \longrightarrow ?$$

Solution   First, rewrite the equation so that the —OH of the acid and the —H of the amine are facing each other:

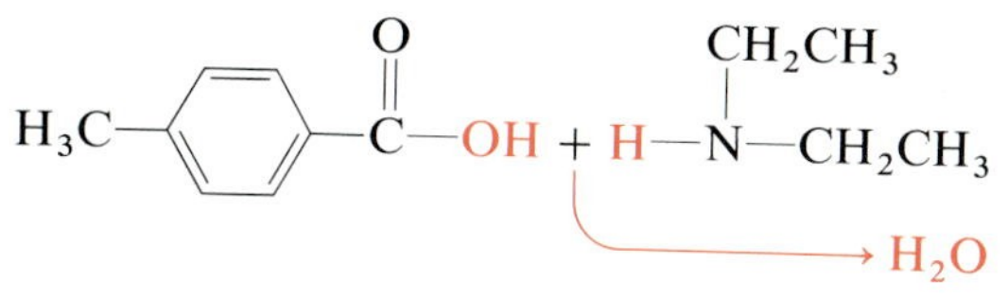

$$H_3C\text{—}\text{—}\overset{\displaystyle O}{\overset{\|}{C}}\text{—OH} + H\text{—}\overset{\displaystyle CH_2CH_3}{\underset{}{N}}\text{—}CH_2CH_3$$

$$\longrightarrow H_2O$$

Next, remove the —OH from the acid and the —H from the nitrogen atom of the amine to form water. Then join the two resulting fragments together to form the amide product.

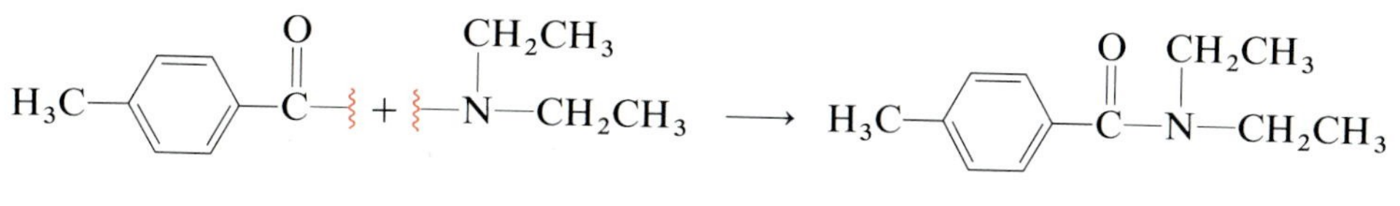

$$H_3C\text{—}\text{—}\overset{\displaystyle O}{\overset{\|}{C}}\text{—} + \text{—}\overset{\displaystyle CH_2CH_3}{\underset{}{N}}\text{—}CH_2CH_3 \longrightarrow H_3C\text{—}\text{—}\overset{\displaystyle O}{\overset{\|}{C}}\text{—}\overset{\displaystyle CH_2CH_3}{\underset{}{N}}\text{—}CH_2CH_3$$

N,N-Diethyltoluamide (DEET)

Practice Problems

12.10   Draw structures of the amides formed in these reactions:

(a) $CH_3NH_2 + (CH_3)_2CHCOOH + DCC \longrightarrow ?$

(b) ⬡—NH_2 + ⬠—$COOH$ + DCC $\longrightarrow$?

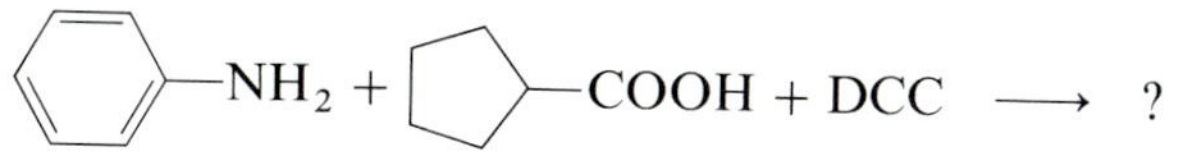

12.11   What carboxylic acid and what amine would you use if you wanted to prepare the headache remedy phenacetin?

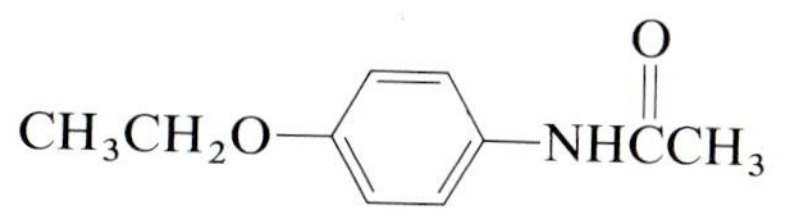

$$CH_3CH_2O\text{—}\text{—}NH\overset{\displaystyle O}{\overset{\|}{C}}CH_3 \qquad \text{Phenacetin}$$

12.11 REACTIONS OF AMIDES: HYDROLYSIS

Amides undergo a hydrolysis reaction with water in the same way that esters do. Just as an ester yields a carboxylic acid and an alcohol when hydrolyzed (Section 12.6), an amide yields a carboxylic acid and an amine. The net effect of amide hydrolysis is a substitution of NH_2 (or NR_2) by —OH. As we'll see in Chapter 16, the cleavage of amide bonds by hydrolysis is the key process that occurs in the stomach during digestion of proteins.

This—NR′R″ group is replaced
by this—OH group from water

A hydrolysis reaction:

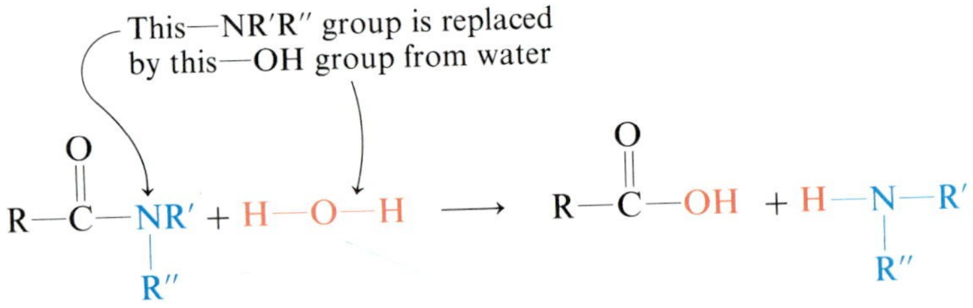

$$R-\overset{\overset{\displaystyle O}{\|}}{C}-NR' + H-O-H \longrightarrow R-\overset{\overset{\displaystyle O}{\|}}{C}-OH + H-\underset{\underset{\displaystyle R''}{|}}{N}-R'$$

        |
        R″

An amide   Carboxylic acid   Amine

Amide hydrolysis is catalyzed both by acids and by bases. If an acid catalyst such as HCl is used, the amine product is converted into its ammonium salt as soon as it's formed. If a basic catalyst such as NaOH is used, the carboxylic acid product is converted into its carboxylate ion. For the sake of simplicity, however, it's easiest to write the hydrolysis products as free carboxylic acid and free amine. For example:

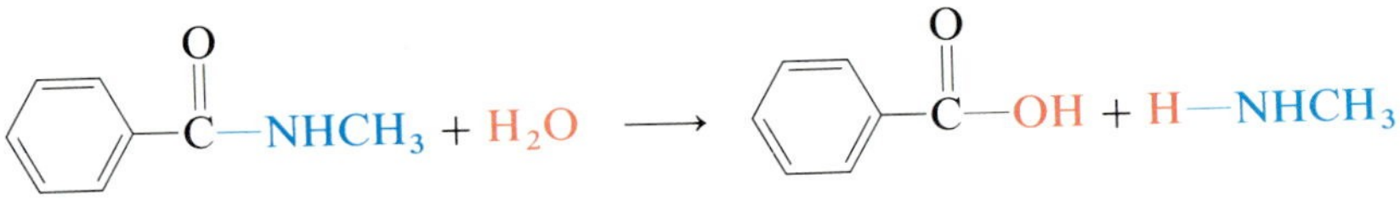

$$\underset{\text{N-Methylbenzamide}}{\text{(ring)}-\overset{\overset{\displaystyle O}{\|}}{C}-NHCH_3} + H_2O \longrightarrow \underset{\text{Benzoic acid}}{\text{(ring)}-\overset{\overset{\displaystyle O}{\|}}{C}-OH} + \underset{\text{Methylamine}}{H-NHCH_3}$$

Solved Problem 12.5   What products result from hydrolysis of *N*-ethylbutanamide?

$$CH_3CH_2CH_2-\overset{\overset{\displaystyle O}{\|}}{C}-NHCH_2CH_3 + H_2O \longrightarrow \; ?$$

Solution   First, look at the name of the starting amide. Often, the name of the amide indicates the names of the two products. Thus, *N*-ethylbutanamide will yield ethylamine and butanoic acid. To be more systematic about finding the product structures, write the amide and locate the bond between the carbonyl-group carbon and the nitrogen. Then break this amide bond, and write the two fragments:

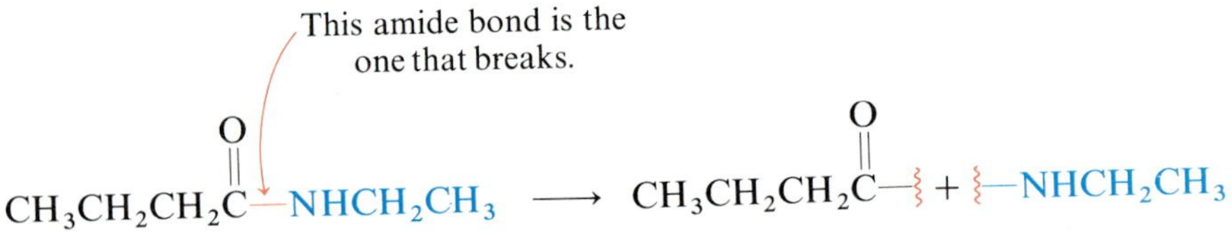

This amide bond is the
one that breaks.

$$CH_3CH_2CH_2\overset{\overset{\displaystyle O}{\|}}{C}-NHCH_2CH_3 \longrightarrow CH_3CH_2CH_2\overset{\overset{\displaystyle O}{\|}}{C}-\!\!\! + -NHCH_2CH_3$$

Next, carry out a hydrolysis reaction and form the products by connecting an —OH to the carbonyl-group carbon and an —H to the nitrogen:

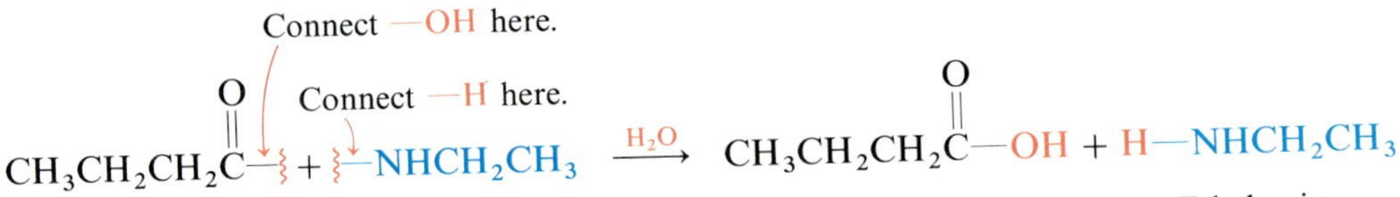

Connect —OH here.

Connect —H here.

$$CH_3CH_2CH_2\overset{\overset{\displaystyle O}{\|}}{C}-\!\!\! + -NHCH_2CH_3 \overset{H_2O}{\longrightarrow} \underset{\text{Butanoic acid}}{CH_3CH_2CH_2\overset{\overset{\displaystyle O}{\|}}{C}-OH} + \underset{\text{Ethylamine}}{H-NHCH_2CH_3}$$

12.12 What products result from hydrolysis of these amides?

$$\underset{\text{(a) } CH_3CH=CHCNHCH_3}{\overset{\overset{\textstyle O}{\|}}{}}$$

(b) *N,N*-diethyl *p*-chlorobenzamide

12.12 PHOSPHATE ESTERS

■ **Phosphate ester** A compound formed by reaction of an alcohol with phosphoric acid.

■ **Nitrate ester** A compound formed by reaction of an alcohol with nitric acid.

Certain inorganic acids such as phosphoric acid, H_3PO_4, and nitric acid, HNO_3, have structures that are similar in many respects to the structures of carboxylic acids. All three contain an atom (C, P, or N) that is singly bonded to an —OH group and doubly bonded to another oxygen. Thus, it's not surprising to find that both phosphoric acid and nitric acid react with alcohols to form **phosphate esters** and **nitrate esters**.

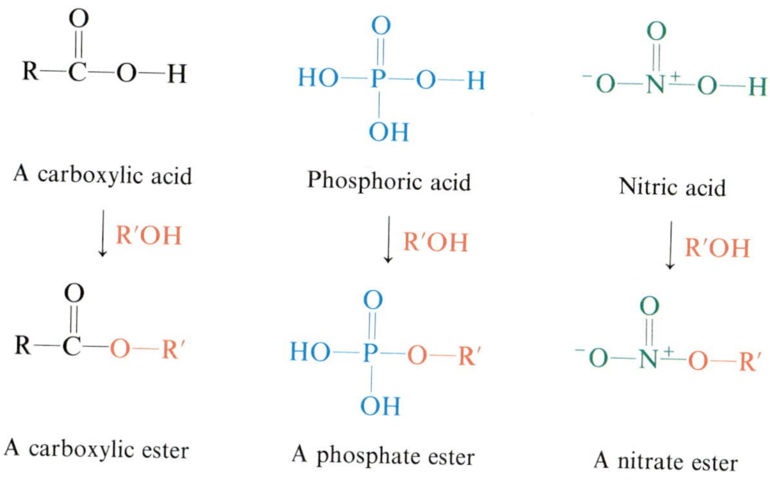

$$\underset{\text{A carboxylic acid}}{\overset{\overset{\textstyle O}{\|}}{R-C-O-H}} \qquad \underset{\underset{\text{Phosphoric acid}}{\overset{\textstyle}{OH}}}{\overset{\overset{\textstyle O}{\|}}{HO-P-O-H}} \qquad \underset{\text{Nitric acid}}{\overset{\overset{\textstyle O}{\|}}{^-O-N^+-O-H}}$$

| R′OH | R′OH | R′OH |

$$\underset{\text{A carboxylic ester}}{\overset{\overset{\textstyle O}{\|}}{R-C-O-R'}} \qquad \underset{\underset{\text{A phosphate ester}}{\overset{\textstyle}{OH}}}{\overset{\overset{\textstyle O}{\|}}{HO-P-O-R'}} \qquad \underset{\text{A nitrate ester}}{\overset{\overset{\textstyle O}{\|}}{^-O-N^+-O-R'}}$$

Nitrate esters do not occur naturally, although nitroglycerin, a triester between glycerin and three molecules of nitric acid, is well known for its use in the treatment of heart disease.

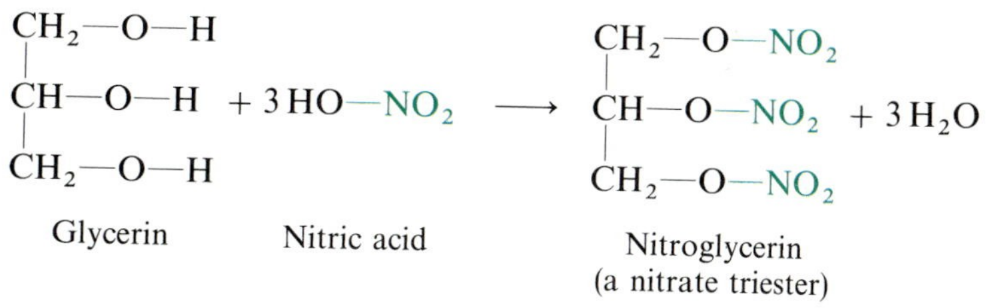

$$\begin{array}{c} CH_2-O-H \\ | \\ CH-O-H \\ | \\ CH_2-O-H \end{array} + 3\,HO-NO_2 \longrightarrow \begin{array}{c} CH_2-O-NO_2 \\ | \\ CH-O-NO_2 \\ | \\ CH_2-O-NO_2 \end{array} + 3\,H_2O$$

Glycerin   Nitric acid   Nitroglycerin
(a nitrate triester)

Phosphate esters are widespread throughout all living organisms and are key substances in nearly all metabolic pathways. Thus, adenosine monophosphate (AMP) is a constituent of DNA (deoxyribonucleic acid), and glyceraldehyde-3-phosphate is involved in carbohydrate metabolism. Note that both of these phosphate esters exist as anions at body pH.

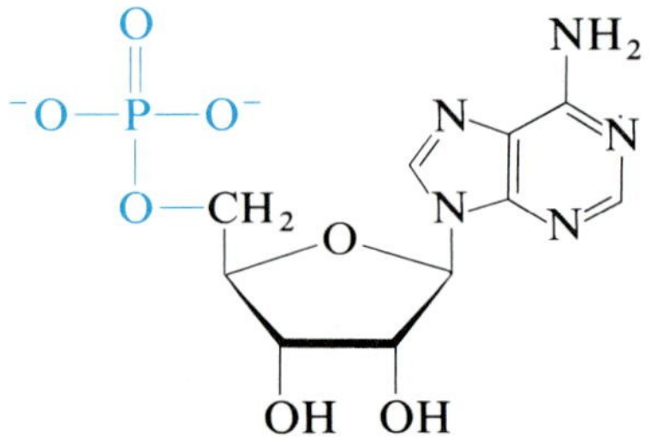

Adenosine monophosphate (AMP) 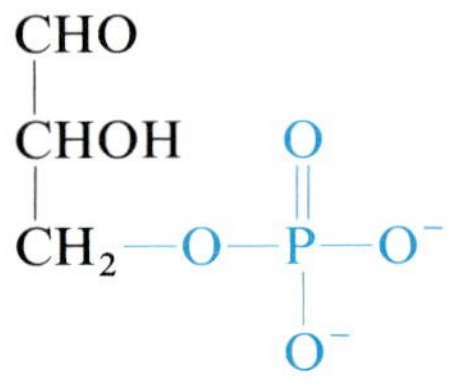   Glyceraldehyde-3-phosphate

Still other inorganic acids that can form esters include pyrophosphoric acid (sometimes called diphosphoric acid) and triphosphoric acid. Diphosphates such as adenosine diphosphate (ADP), and triphosphates such as adenosine triphosphate (ATP), are critical intermediates in all metabolic processes. We'll study them in more detail in Chapters 18 and 19.

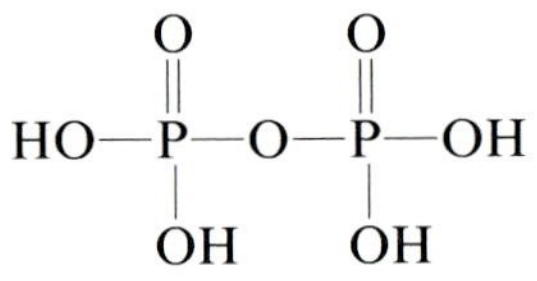

Pyrophosphoric acid 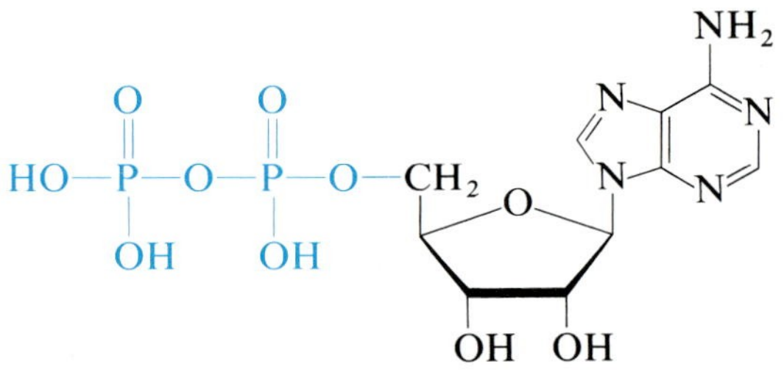   Adenosine diphosphate (ADP)

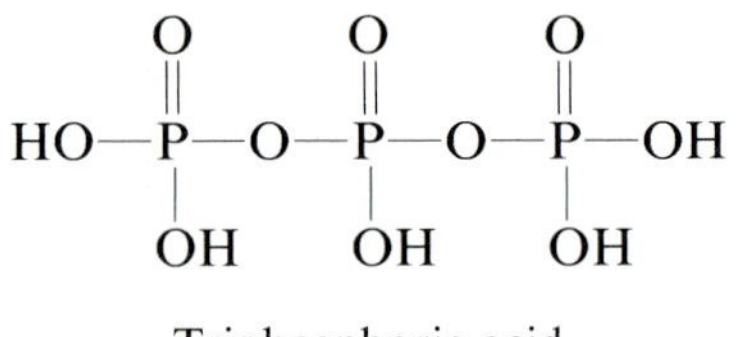

Triphosphoric acid

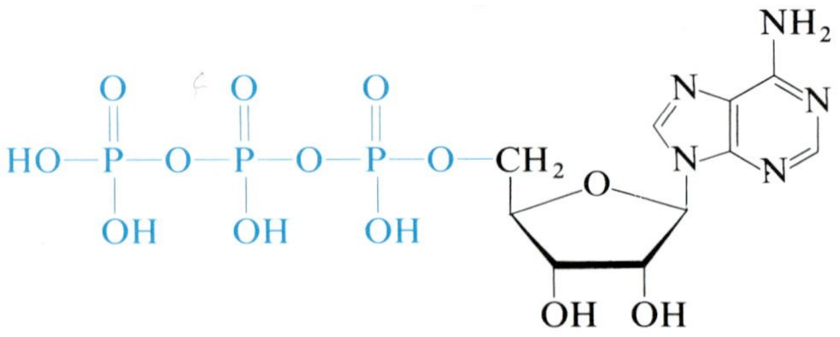

Adenosine triphosphate (ATP)

INTERLUDE: SYNTHETIC POLYMERS: POLYAMIDES AND POLYESTERS

When a reaction takes place between a carboxylic acid and an amine, the two molecules link together. Imagine what would happen, though, if a molecule with *two* carboxylic acid groups were to react with a molecule having *two* amino groups. An initial reaction would join two molecules together, but further reactions would then link more and more molecules together until a giant chain resulted. This is exactly what happens when certain kinds of synthetic polymers are made.

Nylons are *polyamides* that are prepared by reaction of a diamine with a diacid. For example, nylon 66 is prepared by heating adipic acid (hexanedioic acid) with hexamethylenediamine (1,6-hexanediamine) at 280°C.

strength and abrasion resistance make nylon an excellent material for bearings and gears, and high tensile strength makes it suitable as fibers for a range of applications from clothing to mountaineering ropes to carpets.

Just as diacids and diamines react to yield polyamides, diacids and dialcohols react to yield *polyesters*. The most industrially important polyester, made from reaction of terephthalic acid (1,4-benzenedicarboxylic acid) with ethylene glycol, is used under the trade name Dacron to make clothing fiber and under the name Mylar to make plastic film and recording tape.

$$HOOC\text{—}(CH_2)_4\text{—}COOH$$

Adipic acid
+
$$H_2N\text{—}(CH_2)_6\text{—}NH_2$$

Hexamethylenediamine

$\xrightarrow{280°C}$

$$\left[\!\!\left[\overset{O}{\overset{\|}{C}}\text{—}(CH_2)_4\text{—}\overset{O}{\overset{\|}{C}}\text{—}NH\text{—}(CH_2)_6\text{—}NH \right]\!\!\right]_n + n\,H_2O$$

Nylon 66

Nylons have a great many uses, both in engineering applications and in fibers. High impact

$$HO\text{—}\overset{O}{\overset{\|}{C}}\text{—}\!\!\bigcirc\!\!\text{—}\overset{O}{\overset{\|}{C}}\text{—}OH$$

Terephthalic acid
+
$$H\text{—}OCH_2CH_2O\text{—}H$$

Ethylene glycol

$\longrightarrow$

$$\left[\!\!\left[\overset{O}{\overset{\|}{C}}\!\!\bigcirc\!\!\overset{O}{\overset{\|}{C}}\text{—}O\text{—}CH_2CH_2\text{—}O \right]\!\!\right]_n + n\,H_2O$$

A polyester

SUMMARY

Carboxylic acids (RCOOH), **esters** (RCOOR′), and **amides** (RCONH$_2$) occur widely throughout all living organisms. Structurally, these three families are related in that all have a carbonyl group bonded to a strongly electron-attracting atom (O or N). Compounds in all three families undergo **substitution reactions** in which a group we can represent by —X substitutes for (replaces) the —OH, —OR′, or —NH$_2$ group of the starting material:

$$R\text{—}\overset{O}{\overset{\|}{C}}\text{—}OH(\text{—}OR',\,\text{—}NH_2) + H\text{—}X \longrightarrow R\text{—}\overset{O}{\overset{\|}{C}}\text{—}X + H\text{—}OH\ (H\text{—}OR',\,H\text{—}NH_2)$$

A carbonyl-group substitution reaction

As their name implies, carboxylic acids are acidic. They therefore react with bases like NaOH to form water-soluble **carboxylate anions** (RCOO⁻). The most important reaction of carboxylic acids is their conversion into esters by reaction with an alcohol in the presence of a strong-acid catalyst. The net effect of the **esterification reaction** is a substitution of —OH by —OR' (RCOOH → RCOOR').

Esters also undergo substitution reactions. For example, they undergo a **hydrolysis reaction** with water to yield a carboxylic acid and an alcohol. The reaction is catalyzed by both acids and bases, but the base-catalyzed hydrolysis (**saponification**) generally gives higher yields of products. Esters also undergo the **Claisen condensation reaction**, which joins two ester molecules together and forms a keto ester product.

Amides are usually prepared by reaction of a carboxylic acid with an amine in the presence of DCC (dicyclohexylcarbodiimide). Like esters, amides undergo acid- and base-catalyzed hydrolysis, yielding carboxylic acid and amine products.

Certain inorganic acids such as nitric acid and phosphoric acid are analogous to organic carboxylic acids in that they react with alcohols to form esters. **Phosphate esters, diphosphate esters,** and **triphosphate esters** are particularly important in many biological processes.

REVIEW PROBLEMS

Carboxylic Acids

12.13 What are the structural differences between carboxylic acids, esters, and amides?

12.14 In what general way do carboxylic acids, esters, and amides differ from ketones and aldehydes?

12.15 Write the equation for the dissociation of benzoic acid in water.

12.16 Show how two molecules of a carboxylic acid can hydrogen bond to each other.

12.17 There are four carboxylic acids with the formula $C_5H_{10}O_2$. Draw and name each one.

12.18 Assume that you have a sample of propanoic acid dissolved in water:
(a) Draw the structure of the major species present in the water solution.
(b) Now assume that aqueous HCl is added to the propanoic acid solution until pH 2 is reached. Draw the structure of the major species present.
(c) Finally, assume that aqueous NaOH is added to the propanoic acid solution until pH 12 is reached. Draw the structure of the major species present.

12.19 Give IUPAC names for these carboxylic acids:

(a) $CH_3CH_2CH_2CH_2CH_2COOH$

(b) $CH_3CH_2CH_2CHCH_3$
  |
  COOH

(c)
      COOH
      |
  $CH_3CH_2CHCH_2CH_3$

(d) ▷—CH_2CH_2COOH

(e) $BrCH_2CH_2CHCOOH$
          |
         CH_3

(f)
      CH_3
  ⬡—COOH

12.20 Give IUPAC names for these carboxylic acid salts:

(a) $CH_3CH_2CHCH_2COO^-K^+$
  |
  CH_2CH_3

(b)

  —$COO^-NH_4^+$

(c) $[CH_3CH_2COO^-]_2Ca^2$

12.21 Draw structures corresponding to these names:
(a) 3,4-dimethylpentanoic acid
(b) triphenylacetic acid
(c) *m*-ethylbenzoic acid
(d) methylammonium butanoate

12.22 Draw and name three different carboxylic acids with the formula $C_7H_{14}O_2$.

12.23 Malic acid, a dicarboxylic acid found in apples, has the IUPAC name *hydroxybutanedioic acid*. Draw its structure.

12.24 Aluminum acetate is used as an antiseptic ingredient in some skin-rash ointments. Draw its structure. (*Hint:* Review Section 3.3.)

12.25 How many grams of KOH would it take to neutralize these acids?
(a) 10.0 grams of acetic acid
(b) 250 mL of 2.0 M propanoic acid

Esters and Amides

12.26 Draw and name compounds that meet these descriptions:
(a) three different amides with the formula $C_5H_{11}NO$
(b) three different esters with the formula $C_6H_{12}O_2$

12.27 Give IUPAC names for these esters:
(a) $CH_3COOCH_2CH_2CH(CH_3)_2$
(b) $(CH_3)_2CHCH_2CH_2COOCH_3$
(c) $(CH_3)_3CCOOCH_2CH_3$

(d)

—$COOCH_2CH_3$

(e) 
CH_3CH_2COO—

12.28 Draw structures corresponding to these IUPAC names:
(a) methyl pentanoate
(b) isopropyl 2-methylbutanoate

(c) cyclohexyl acetate
(d) phenyl *o*-hydroxybenzoate

12.29 Show the structures of the carboxylic acids and alcohols you would use to prepare each of the esters in Problem 12.28.

12.30 Provide IUPAC names for these compounds:

(a) $CH_3CH_2CHCH_2CH_3$
  |
  $CONH_2$

(b)

—$CONH$—

(c) $HCON(CH_3)_2$

12.31 Show how you could prepare each of the amides in Problem 12.30 from an appropriate carboxylic acid and amine.

12.32 Draw structures corresponding to these IUPAC names:
(a) 3-methylpentanamide
(b) *N*-phenylacetamide
(c) *N*-ethyl-*N*-methylbenzamide
(d) 2,3-dibromohexanamide

12.33 What compounds would result from hydrolysis of each of the amides listed in Problem 12.32?

Reactions of Carboxylic Acids, Esters, and Amides

12.34 What general kind of reaction do carboxylic acids, esters, and amides undergo?

12.35 Methyl *o*-aminobenzoate, commonly called *methyl anthranilate*, is used as a flavoring agent in grape drinks. Write an equation for the preparation of methyl anthranilate from the appropriate alcohol and carboxylic acid.

12.36 Novocaine, a local anesthetic, has the following structure. Identify the functional groups present in novocaine, and show the structures of the alcohol and carboxylic acids you would use to prepare it.

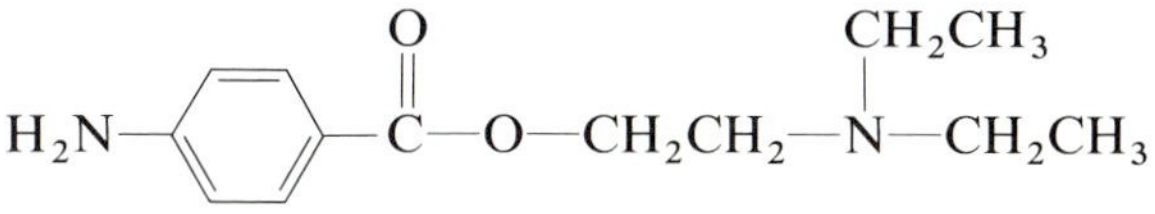

Novocaine

12.37 *Lactones* are cyclic esters in which the carboxylic acid part and the alcohol part are connected together to form a ring. What product(s) would you expect to obtain from hydrolysis of butyrolactone?

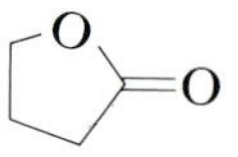

Butyrolactone (a cyclic ester)

12.38 Lidocaine (Xylocaine) is a local anesthetic closely related to novocaine. Identify the functional groups present in lidocaine, and show how you might prepare it from a carboxylic acid and an amine.

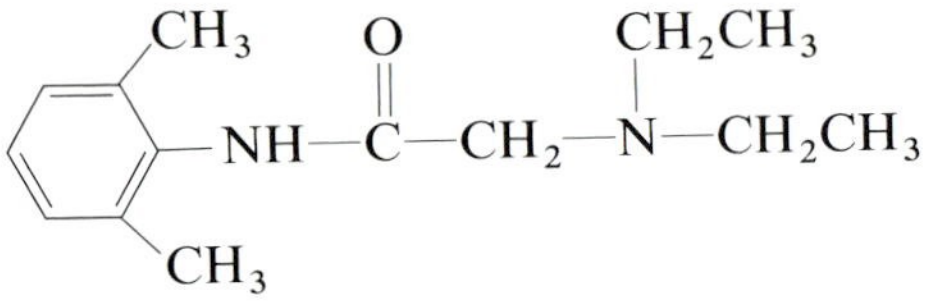

Lidocaine

12.39 Cocaine, an alkaloid isolated from the leaves of the South American coca plant, *erythroxylon coca*, has the structure indicated. Identify the functional groups present, and give the structures of the products you would obtain from hydrolysis of cocaine.

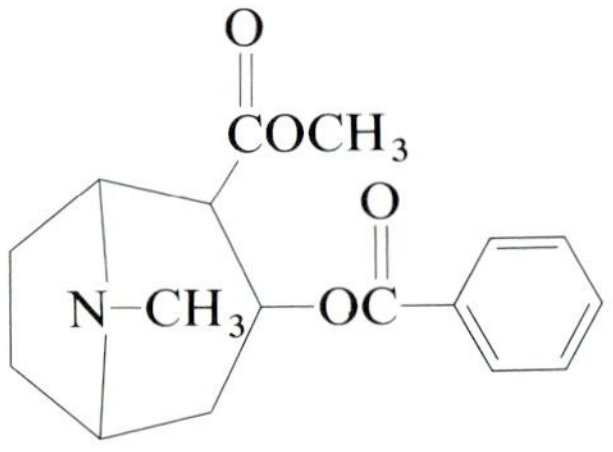

Cocaine

12.40 Household soap is a mixture of the sodium or potassium salts of long-chain carboxylic acids that arise from saponification of animal fat. Draw the structures of soap molecules produced in the following:

$$\begin{array}{ll} & CH_2-O-CO(CH_2)_{14}CH_3 \\ \text{(A fat)} & CH-O-CO(CH_2)_7CH{=}CH(CH_2)_7CH_3 \\ & CH_2-O-CO(CH_2)_{16}CH_3 \end{array}$$

$$\xrightarrow{3\,KOH} \; ?$$

12.41 A *lactam* is a cyclic amide in which the carboxylic acid part and the amine part are connected together. Draw the structure of the product(s) from hydrolysis of caprolactam, an industrial precursor of nylon.

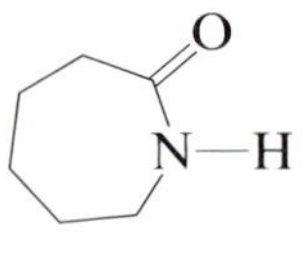

Caprolactam

12.42 Assume that you're given samples of pentanoic acid and methyl butanoate, both of which have the formula $C_5H_{10}O_2$. Describe how you can tell them apart.

12.43 What carboxylic acid and what alcohol components would you start with to prepare aspirin (acetylsalicylic acid)?

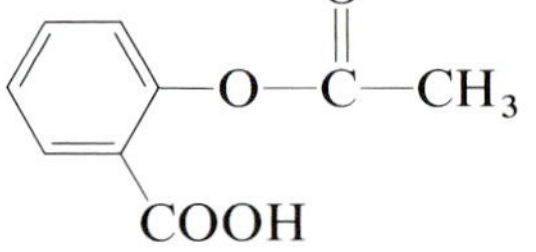

Aspirin

12.44 The following phosphate ester is an important intermediate in carbohydrate metabolism. What products would result from hydrolysis of this ester?

$$HO-CH_2-\overset{\overset{\large O}{\|}}{C}-CH_2-O-PO_3^{2-}$$

Claisen Condensation Reaction

12.45 What structural feature must an ester have in order to undergo a Claisen condensation reaction?

12.46 Which of the following esters can't undergo Claisen condensation reactions? Explain.

(a) $HCOOCH_3$ (b) $(CH_3)_3CCOOCH_2CH_3$
(c) $CH_3CH_2COOC(CH_3)_3$

12.47 Draw the Claisen condensation products of the following esters:

(a) $CH_3\underset{\underset{\large CH_3}{|}}{C}HCH_2COOCH_3$ (b) isopropyl acetate

12.48 When a mixture of methyl acetate and methyl propanoate are treated with sodium methoxide in a Claisen condensation reaction, four keto ester products are formed. What are their structures?

CHAPTER 13

The Molecules of Life: Carbohydrates

Few things are so similar, and yet so different, as a left and a right hand. Your left hand is a mirror image of your right hand—no matter how you turn them, they cannot be superimposed. You'll see in this chapter that the same is true for certain molecules, and with important consequences. (Brush drawing: "Hands of an Apostle," by Albrecht Dürer, 1508.)

Carbohydrates occur in every living organism. The starch in food and the cellulose in grass are nearly pure carbohydrate; modified carbohydrates form part of the coating around all living cells; other carbohydrates are found in the DNA that carries genetic information from one generation to the next; and still other carbohydrates such as streptomycin are valuable as medicines.

The word *carbohydrate* was used originally to describe glucose, the simplest and most readily available sugar. Because glucose has the formula $C_6H_{12}O_6$, it was once thought to be a "hydrate of carbon"—$C_6(H_2O)_6$. Although this view was soon abandoned, the name *carbohydrate* persisted until it is now used to refer to a large class of polyhydroxylated aldehydes and ketones. Glucose, for example, is a six-carbon aldehyde with five hydroxyl groups.

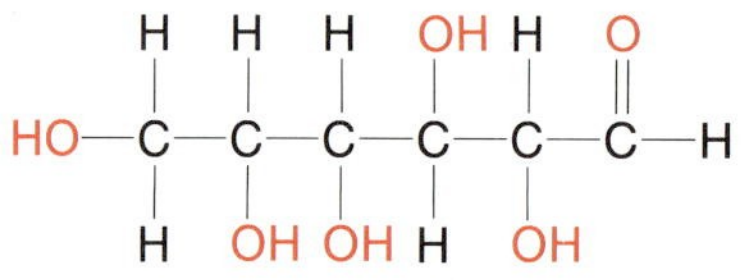

Glucose (a pentahydroxyhexanal)

Carbohydrates are synthesized in green leaves by the conversion of carbon dioxide into glucose during photosynthesis. Many molecules of glucose are then linked together to form either cellulose or starch. When eaten and later broken down in cells, glucose provides the major source of energy required by living organisms. Thus, carbohydrates act as the intermediaries by which energy from the sun is converted into chemical energy.

In this chapter, we'll look for answers to these questions:

1. **What are the different kinds of carbohydrates?** The goal: you should learn the main classifications of carbohydrates and how to classify specific examples.

2. **Why do carbohydrates have a "handedness"?** The goal: you should learn what is responsible for handedness in certain molecules.

3. **What are the structures of glucose and other important simple sugars?** The goal: you should learn how glucose differs structurally from fructose and other simple sugars.

4. **What reactions do simple sugars undergo?** The goal: you should learn the most important reactions of sugars.

5. **What are the structures of some important disaccharides?** The goal: you should learn how simple sugars join together to make disaccharides like sucrose and lactose.

6. What are the structures of some important polysaccharides? The goal: you should learn how cellulose and starch are constructed and how they differ.

13.1 CLASSIFICATION OF CARBOHYDRATES

■ **Carbohydrate** A member of a large class of naturally occurring polyhydroxy aldehydes and ketones.

■ **Simple sugar** A carbohydrate that can't be broken down into a smaller sugar by hydrolysis with acid.

■ **Monosaccharide** An alternative name for a simple sugar.

■ **Complex carbohydrate** A carbohydrate that breaks down into simple sugars when hydrolyzed with aqueous acid.

■ **Polysaccharide** An alternate name for a complex carbohydrate.

■ **Aldose** A simple sugar that contains an aldehyde carbonyl group.

■ **Ketose** A simple sugar that contains a ketone carbonyl group.

As mentioned above, **carbohydrates** are a large class of naturally occurring polyhydroxylated aldehydes and ketones. Carbohydrates are usually classified as either simple or complex. **Simple sugars**, or **monosaccharides** (mah-no-**sack**-uh-rides), are carbohydrates that can't be broken down into smaller molecules by hydrolysis with aqueous acid. Glucose and galactose are examples. **Complex carbohydrates**, or **polysaccharides**, are compounds that are made of many simple sugars linked together. On hydrolysis, polysaccharides such as cellulose and starch are cleaved to yield many molecules of simple sugars.

$$\text{Cellulose or starch} \xrightarrow{\text{H}_3\text{O}^+} \approx 1000 \text{ glucose molecules}$$

Monosaccharides can be further classified as either aldoses or ketoses. An **aldose** contains an aldehyde carbonyl group; a **ketose** contains a ketone carbonyl group; and the family-name ending *-ose* indicates a sugar. The number of carbon atoms in the sugar is specified by using one of the prefixes *tri-*, *tetr-*, *pent-*, or *hex-*. Thus, glucose is an aldohexose (a six-carbon aldehyde sugar); fructose is a ketohexose (a six-carbon ketone sugar); and ribose is an aldopentose (a five-carbon aldehyde sugar). Most commonly occurring sugars are either aldopentoses or aldohexoses.

Glucose
(an aldohexose)

Fructose
(a ketohexose)

Ribose
(an aldopentose)

Practice Problem **13.1** Classify each of these monosaccharides:

(a)

$$\underset{}{\text{HOCH}_2}-\overset{\text{OH}}{\underset{}{\text{CH}}}-\overset{\text{OH}}{\underset{}{\text{CH}}}-\overset{\text{OH}}{\underset{}{\text{CH}}}-\overset{\text{O}}{\underset{}{\text{C}}}-\text{H}$$

(b)

$$\text{HOCH}_2-\overset{\text{O}}{\underset{}{\text{C}}}-\text{CH}_2\text{OH}$$

(c)

$$\text{HOCH}_2-\overset{\text{OH}}{\underset{}{\text{CH}}}-\overset{\text{OH}}{\underset{}{\text{CH}}}-\overset{\text{O}}{\underset{}{\text{C}}}-\text{H}$$

13.2 HANDEDNESS

Are you right-handed or left-handed? Although you may not often think about it, handedness affects almost everything you do. Anyone who has played much softball knows that the last available glove always fits the wrong hand; any left-handed person sitting next to a right-handed person in a lecture knows that taking notes always involves bumping elbows. The reason for these difficulties is that your hands aren't identical. Rather, they're **mirror images**. When you hold your left hand up to a mirror, the image you see looks like your right hand (Figure 13.1). Try it.

Not all objects are handed, of course. There's no such thing as a "right-handed" tennis ball or a "left-handed" coffee mug. When a tennis ball or a coffee mug is held up to a mirror, the image reflected is identical to the ball or mug itself. Objects that have a handedness to them are said to be **chiral** (pronounced **ky**-ral, from the Greek *cheir*, meaning hand), and objects like the coffee mug that lack handedness are said to be nonchiral, or **achiral**.

Why is it that some objects are chiral (handed) but others aren't? In general, an object is not chiral if it's symmetrical. Conversely, an object *is* chiral if it's not symmetrical. Symmetrical objects are those like the coffee mug that have an imaginary plane (a **symmetry plane**) cutting through their middle so that one half of the object is an exact mirror image of the other half. If you were to cut the mug in half, one half of the mug would be the mirror image of the other half. A hand, however, has no symmetry plane and is therefore chiral. If you were to cut a hand in two, one "half" of the hand would not be a mirror image of the other half (Figure 13.2).

■ **Mirror image** The reverse image produced when an object is reflected in a mirror.

■ **Chiral** Having right or left handedness.

■ **Achiral** The opposite of chiral; not having right or left handedness.

■ **Symmetry plane** An imaginary plane cutting through the middle of an object so that one half of the object is a mirror image of the other half.

Practice Problems

13.2 Which of the following objects are handed?
(a) a glove   (b) a baseball   (c) a screw   (d) a nail

13.3 List three common objects that are handed and another three that aren't.

Figure 13.1
The meaning of *mirror image*: If you hold your left hand up to a mirror, the image you see looks like your right hand.

Left hand

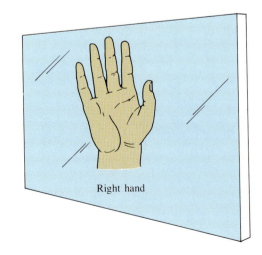

Right hand

Figure 13.2
The meaning of symmetry plane: An achiral object like the coffee mug has a symmetry plane passing through it, making the two halves mirror images. A chiral object like the hand has no symmetry plane because the two "halves" of the hand are not mirror images.

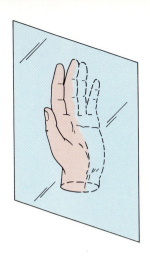

13.3 MOLECULAR HANDEDNESS: D AND L FAMILIES OF SUGARS

Just as certain objects like a hand are chiral, certain *molecules* are also chiral. For example, compare propane and glyceraldehyde, a simple aldotriose. A glyceraldehyde molecule has no symmetry plane because its two "halves" aren't mirror images. Like a hand, glyceraldehyde is chiral and can exist in two forms— a "right-handed" form and a "left-handed" form (Figure 13.3). The two forms are not identical; they are related in the same way that your left and right hands are related. Propane, however, is achiral. It has a symmetry plane cutting through the three carbons such that one half of the molecule is a mirror image of the other half. Thus, propane exists in a single form.

Figure 13.3
Glyceraldehyde (2,3-dihydroxy-propanal) has no symmetry plane; the two "halves" of the molecule are not mirror images. Thus, glyceraldehyde can exist in two forms—a "right-handed" form, referred to as D-glyceraldehyde, and a "left-handed" form, referred to as L-glyceraldehyde. Propane, however, has a symmetry plane and is achiral.

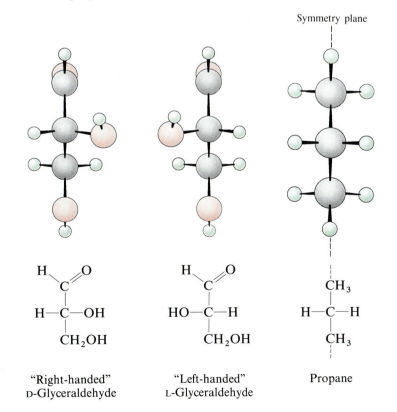

"Right-handed"
D-Glyceraldehyde

"Left-handed"
L-Glyceraldehyde

Propane

Why are some molecules chiral but others aren't? The answer has to do with the three-dimensional nature of organic compounds. As we saw in Section 3.9, carbon forms four bonds that are oriented to the four corners of an imaginary tetrahedron. Whenever a carbon atom is bonded to four *different* groups, chirality results. If a carbon is bonded to two or more of the same groups, however, no chirality is present. In glyceraldehyde, for example, carbon 2 is bonded to four different groups—a —CHO group, an —H atom, an —OH group, and a —CH$_2$OH group. Thus, glyceraldehyde is chiral. In propane, however, each of the three carbons is bonded to at least two groups—the —H atoms—that are identical. Thus, propane is achiral.

$$
\begin{array}{c}
\text{CHO} \\
| \\
\text{H—C—OH} \\
| \\
\text{CH}_2\text{OH}
\end{array}
$$

Glyceraldehyde
(chiral)

Groups attached to carbon 2

1. —CHO
2. —H
3. —OH
4. —CH$_2$OH
} different

$$
\begin{array}{c}
\text{CH}_3 \\
| \\
\text{H—C—H} \\
| \\
\text{CH}_3
\end{array}
$$

Propane
(achiral)

Groups attached to carbon 2

1. —CH$_3$
2. —CH$_3$ } identical
3. —H
4. —H } identical

The easiest way to see how tetrahedral geometry leads to chirality is to make paper models of the sort shown in Figure 13.4. Cut two large equilateral triangles out of stiff paper, fold each one so that its three corners come together to form a tetrahedron, and then color each one as indicated. When four different colors (groups) are used for the four corners of the tetrahedra, the two models are not identical but have a right-hand/left-hand relationship.

We call the two mirror-image forms of a chiral molecule like glyceraldehyde **optical isomers**, or **enantiomers** (e-**nan**-tee-o-mers). Like other isomers we've seen (Section 8.3), optical isomers have the same formula but have different structures. Although optical isomers are very closely related—the two forms of glyceraldehyde have the same physical properties such as melting point (145°C)—they are nevertheless different compounds.

Both optical isomers of glyceraldehyde are known, but only the "right-handed" or D form (from *dextro*, Latin for "right"), shown in Figure 13.3, occurs commonly in nature. The "left-handed" or L form (from *levo*, Latin for "left") does not occur naturally and must be made in the laboratory. Note how the D and L forms are shown in Figure 13.3: In the D form, the —OH group on carbon 2 comes out of the plane of the paper and points to the right when the aldehyde —CHO group is placed at the top; in the L form, the —OH group at carbon 2 comes out of the plane of the paper and points to the left when the —CHO is at the top.

■ **Enantiomer** One of the two mirror-image forms of a chiral molecule.

■ **Optical isomer** An alternative term for enantiomer.

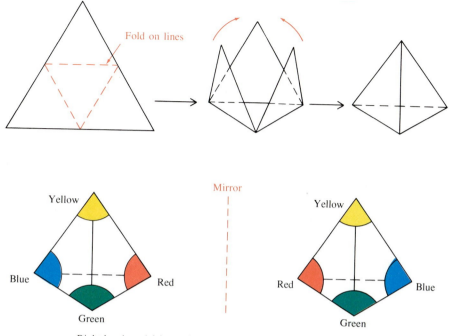

Right-hand model is a mirror image of left-hand model — *chiral*.

Figure 13.4
Paper molecular models. The two tetrahedra are mirror images (that is, are chiral) when the four corners have four different colors.

In general, *any* sugar can exist as either a right-handed D form or as a mirror-image, left-handed L form, but only the D forms are naturally occurring. When the structures are written vertically, the D form always has the hydroxyl group on the chiral carbon atom farthest from the carbonyl group pointing toward the right, while the mirror-image L form has the hydroxyl group on the chiral carbon farthest from the carbonyl group pointing toward the left. D-Glucose and D-ribose are examples.

D-Glucose

D-Ribose

This hydroxyl group points to the right.

This hydroxyl group points to the right.

Solved Problem 13.1   Lactic acid can be isolated from sour milk. Is lactic acid chiral?

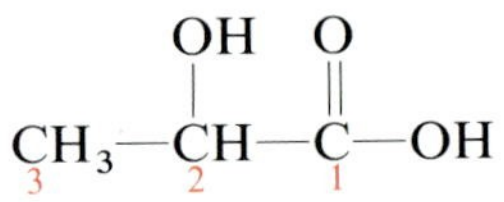

$$CH_3\!-\!\underset{3}{CH}\!-\!\underset{2}{\overset{OH}{CH}}\!-\!\underset{1}{\overset{O}{\overset{\|}{C}}}\!-\!OH \qquad \text{Lactic acid}$$

Solution   To find if lactic acid is chiral, list the groups attached to each carbon:

Groups on carbon 1	Groups on carbon 2	Groups on carbon 3
1. —OH	1. —COOH	1. —CH(OH)COOH
2. =O	2. —OH	2. —H
3. —CH(OH)CH$_3$	3. —H	3. —H
	4. —CH$_3$	4. —H

Next, look at the lists to see if any carbon is attached to four *different* groups. Of the three carbons, carbon 2 has four different groups, and lactic acid is therefore chiral.

Practice Problems

13.4   2-Propanol is an achiral molecule, but 2-butanol is chiral. Explain.

13.5   Make a rough ball-and-stick sketch of the sort shown in Figure 13.3 to show (a) the presence of a symmetry plane in 2-propanol and (b) the absence of a symmetry plane in 2-butanol.

13.6   Which of these molecules are chiral?

(a) 3-chloropentane   (b) 2-chloropentane

(c) 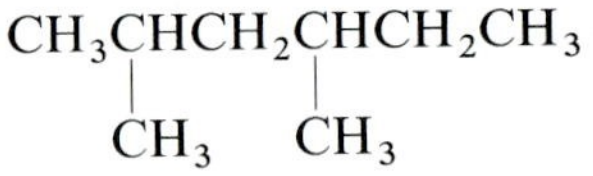 $CH_3CHCH_2CHCH_2CH_3$
    $\quad\ \ |\qquad\quad\ |$
    $\quad\ CH_3\ \ \ \ CH_3$

13.4   THE STRUCTURE OF GLUCOSE

D-Glucose, sometimes called *dextrose*, is the most widely occurring of all monosaccharides. It is found in nearly all foods, particularly fruits, and in all living organisms, where it serves as a source of energy to fuel biochemical reactions.

Before looking at the chemical structure of glucose, glance back briefly at Section 11.6. We said there that aldehydes and ketones react reversibly with alcohols to yield hemiacetal addition products:

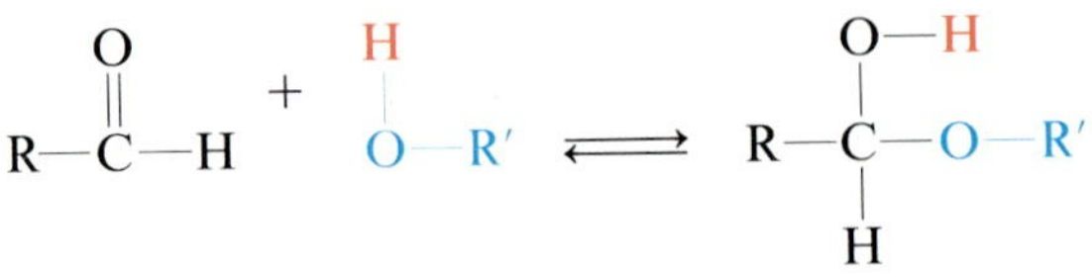

$$\underset{\text{An aldehyde}}{R\!-\!\overset{O}{\overset{\|}{C}}\!-\!H} \ +\ \underset{\text{An alcohol}}{\overset{H}{\overset{|}{O}}\!-\!R'} \ \rightleftharpoons\ \underset{\text{A hemiacetal}}{R\!-\!\overset{O-H}{\underset{H}{\overset{|}{\underset{|}{C}}}}\!-\!O\!-\!R'}$$

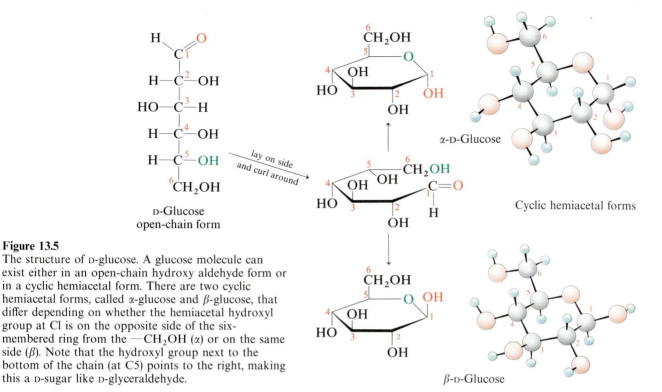

Figure 13.5

The structure of D-glucose. A glucose molecule can exist either in an open-chain hydroxy aldehyde form or in a cyclic hemiacetal form. There are two cyclic hemiacetal forms, called α-glucose and β-glucose, that differ depending on whether the hemiacetal hydroxyl group at C1 is on the opposite side of the six-membered ring from the —CH₂OH (α) or on the same side (β). Note that the hydroxyl group next to the bottom of the chain (at C5) points to the right, making this a D-sugar like D-glyceraldehyde.

Now look at the structure of D-glucose shown in Figure 13.5. Since glucose is an aldohexose—a pentahydroxy aldehyde—it has both alcohol hydroxyl groups and an aldehyde carbonyl group *in the same molecule*. Thus, an *internal* addition reaction can take place between the aldehyde carbonyl group at carbon 1 and one of the hydroxyl groups in the chain to yield a *cyclic* hemiacetal. In fact, it's the hydroxyl group at carbon 5 that reacts, leading to the six-membered-ring hemiacetal shown in Figure 13.5.

As shown in Figure 13.5, there are actually *three* forms of D-glucose—an open-chain form, a cyclic α-form, and a cyclic β-form. The cyclic α- and β-forms differ only in the side of the ring that the hemiacetal hydroxyl at C1 is on. In the β-form, the hydroxyl at C1 points upward on the same side of the ring as the —CH₂OH group connected to C5; in the α-form, the hydroxyl at C1 points downward on the opposite side of the ring from the —CH₂OH group at C5. Although this structural difference might seem small, it has enormous biological consequences. We'll see in Section 13.8, for instance, that this one small change in structure accounts for the vast difference between starch and cellulose.

Although the ordinary crystalline glucose you might take from a bottle is entirely the cyclic α-form, a water solution of glucose contains all three forms undergoing rapid interconversion. At room temperature in water, the hemiacetal ring opens and closes rapidly leading to an equilibrium mixture of the three forms in the proportion 0.02 percent open-chain, 36 percent α-form, and 64 percent β-form.

$$\text{α-D-Glucose} \rightleftharpoons \text{open-chain D-glucose} \rightleftharpoons \text{β-D-Glucose}$$

(36%)   (0.02%)   (64%)

Solved Problem 13.2   D-Altrose, an aldohexose isomer of glucose, has the following structure in its open-chain form. Draw D-altrose in its cyclic hemiacetal form.

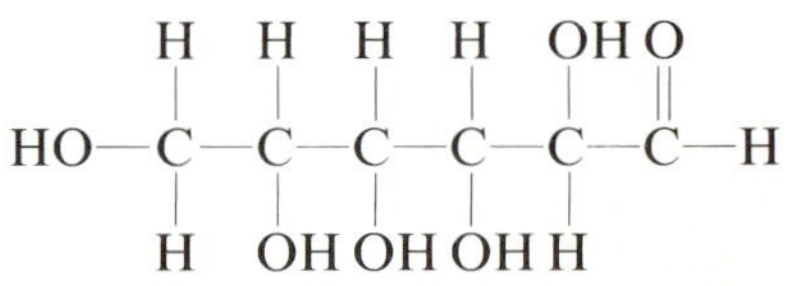

Solution   First, coil D-altrose into a circular shape by grasping the end farthest from the carbonyl group and bending it backwards into the plane of the paper:

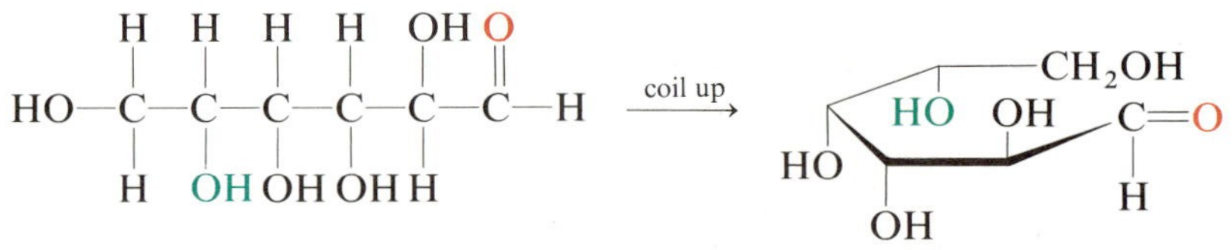

Next, rotate around the single bond between C4 and C5 so that the —CH$_2$OH group at the end of the chain is pointing up and the —OH group on C5 is pointing toward the aldehyde carbonyl group on the right:

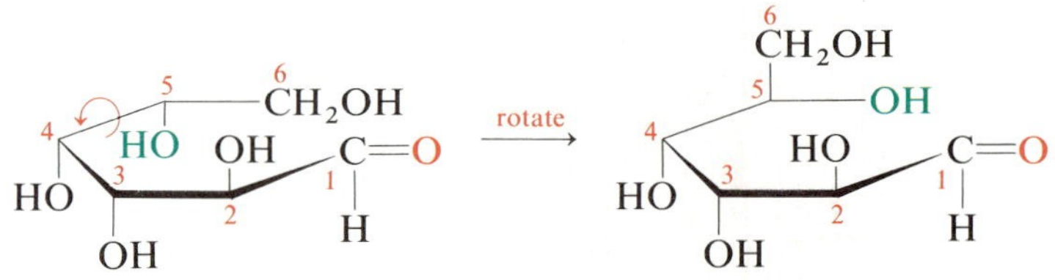

Finally, carry out an addition reaction of the —OH group at C5 to the carbonyl C=O to form a hemiacetal ring. The new —OH group formed on C1 can be either up (β) or down (α).

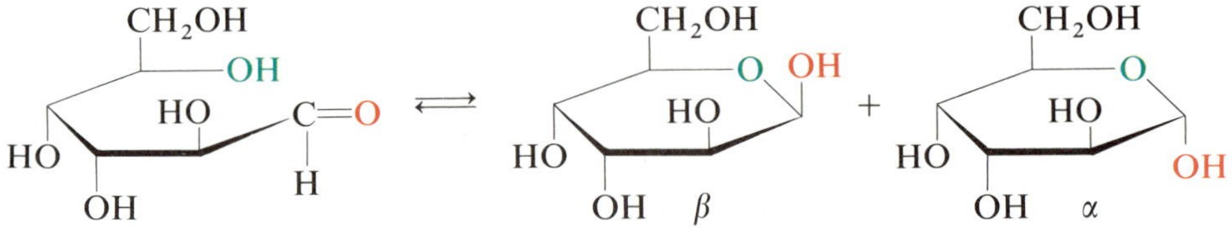

Practice Problem   **13.7**   D-Talose, a constituent of certain antibiotics, has the following open-chain structure. Draw D-talose in its cyclic hemiacetal form.

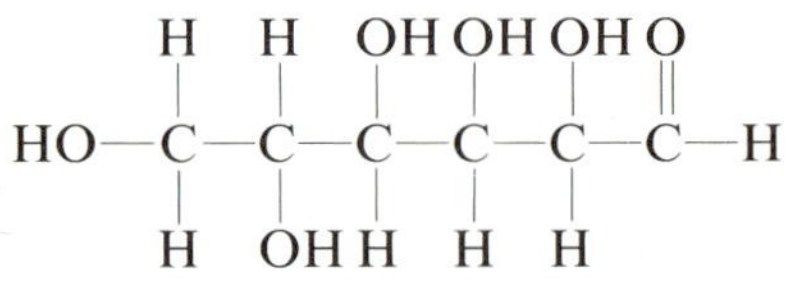

13.5 SOME IMPORTANT MONOSACCHARIDES

Galactose   D-Galactose, sometimes called "brain sugar" because of its presence in brain tissue, is widely distributed as a constituent of many plant gums and pectins. It's also a component of lactose ("milk sugar"), from which it can be formed on hydrolysis. Like glucose, galactose is an aldohexose; it differs from glucose, however, in the orientation of its hydroxyl group at C4. Whereas the hydroxyl at C4 in D-glucose points downward on the bottom side of the hemiacetal ring, the same hydroxyl in D-galactose points upward on the top side. Also like glucose, galactose exists in solution as a mixture of cyclic α- and β-forms along with a small amount of open-chain form.

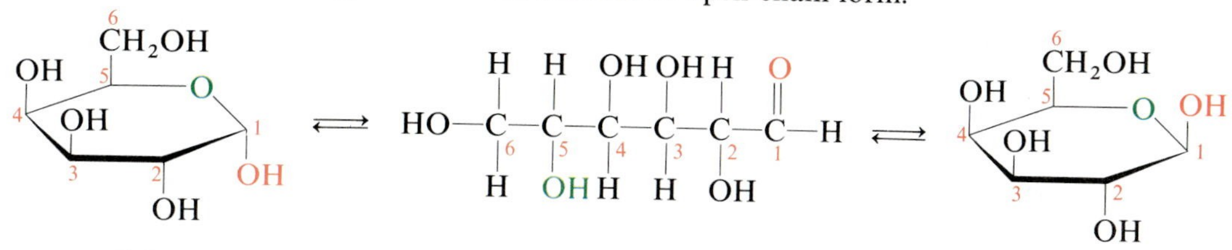

α-D-Galactose   Open-chain D-galactose   β-D-Galactose

Fructose   D-Fructose, often called *levulose* or *fruit sugar*, occurs in honey and in a large number of fruits. Unlike glucose and galactose, fructose is a *keto*hexose—a six-carbon ketone sugar. Like glucose and galactose, however, fructose can exist in solution both in open-chain form and in cyclic forms. The cyclic hemiacetal form is a five-membered ring rather than a six-membered ring.

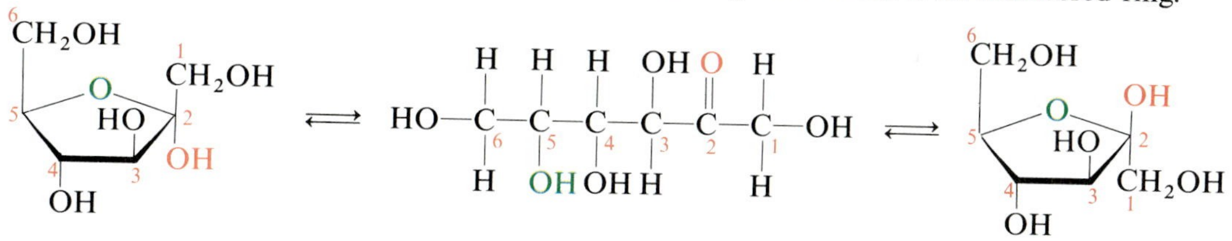

α-D-Fructose   Open-chain D-Fructose   β-D-Fructose

Ribose and Deoxyribose   Ribose and its relative deoxyribose are both aldopentoses—five-carbon aldehyde sugars. Both occur in the cells of all living organisms as constituents either of RNA (ribonucleic acid) or DNA (deoxyribonucleic acid).

As its name implies, *deoxy*ribose differs from ribose in that it is missing one oxygen atom. Since it's the hydroxyl at C2 that is missing, it's more accurate to name the compound 2-deoxyribose. Both ribose and 2-deoxyribose exist in the usual mixture of open-chain and cyclic hemiacetal forms.

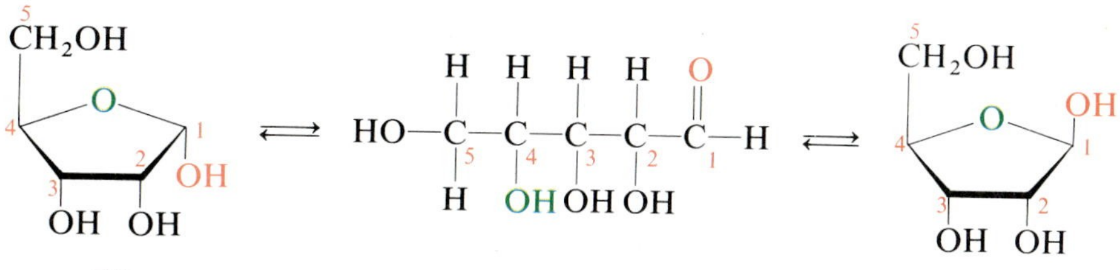

α-D-Ribose   Open-chain D-ribose   β-D-Ribose

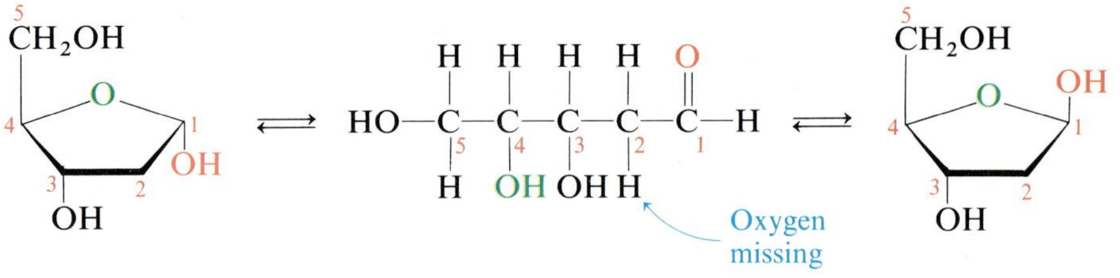

α-D-2-Deoxyribose   Open-chain D-2-deoxyribose   β-D-2-Deoxyribose

13.6 REACTIONS OF MONOSACCHARIDES

Reaction with Oxidizing Agents: Reducing Sugars   We saw in Section 11.4 that aldehydes can be oxidized to yield carboxylic acids (RCHO → RCOOH). Because they too are aldehydes, aldoses like glucose undergo exactly the same oxidation reaction. Even though only a small amount of open-chain aldehyde form is present at any one time, the hemiacetal ring opening takes place so rapidly that eventually the entire sample is oxidized. Carbohydrates that react in this manner with oxidizing agents are called **reducing sugars**.

The diabetes self-test kits sold in drugstores use this oxidation reaction to indicate abnormal glucose levels in urine. Although normal urine contains little if any glucose, the urine of persons with untreated diabetes can contain up to several percent glucose. Application of a small sample of urine to test strips impregnated with **Benedict's reagent** (Cu^{2+} and sodium citrate) yields a yellowish or reddish color if glucose is present, because the blue Cu^{2+} ion is reduced to red Cu_2O by the reducing sugar.

■ **Reducing sugar** A carbohydrate that reacts with an oxidizing agent like Benedict's reagent.

■ **Benedict's reagent** A reagent that yields a reddish or yellowish color when it reacts with a reducing sugar.

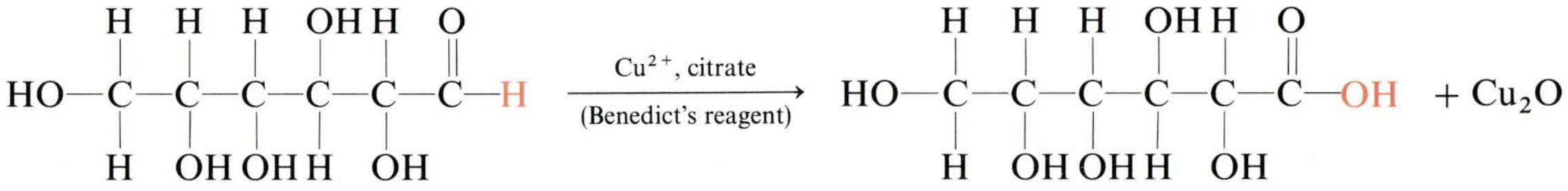

Glucose (open-chain)   A carboxylic acid

Reaction with Alcohols: Glycoside Formation   We saw in Section 11.6 that hemiacetals can react with alcohols to yield acetals, compounds that have two ether-like —OR groups bonded to the same carbon:

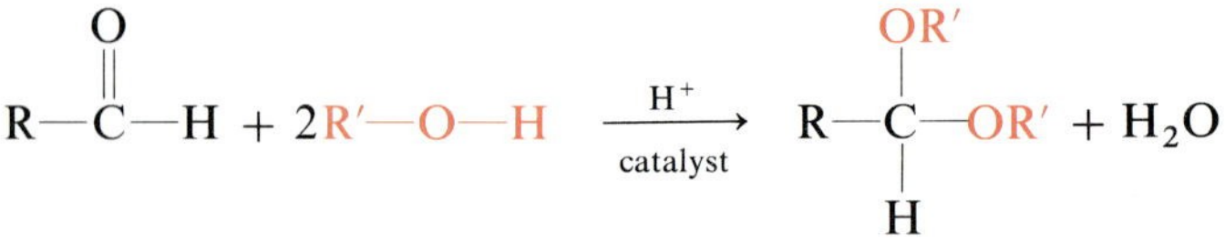

Aldehyde   Alcohol   Acetal

■ **Glycoside** A cyclic acetal formed by reaction of a simple sugar with an alcohol.

Because glucose and other simple sugars are cyclic hemiacetals, they react with alcohols to yield cyclic acetals, called **glycosides**. For example, glucose reacts with methanol to produce methyl glucoside. We'll see shortly that acetal formation can also occur between the hemiacetal group of one sugar with the alcohol group of a second sugar. The net result is a linking together of two sugars via an acetal bond.

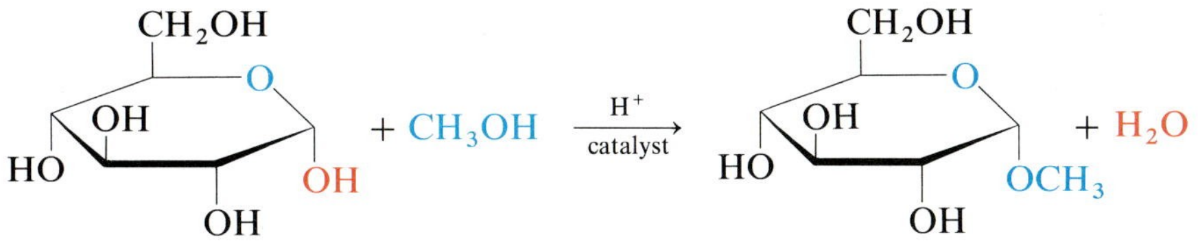

α-D-Glucose   Methyl α-D-glucoside

Practice Problem   **13.8**   Look back at the structure of α-D-galactose (Section 13.5), and write the glycoside product you would expect to obtain by reaction of galactose with methanol.

13.7   SOME IMPORTANT DISACCHARIDES

■ **Disaccharide** A complex carbohydrate formed by the bonding together of two simple sugars.

■ **1,4-Link** An acetal link between the hydroxyl group at C1 of one sugar and the hydroxyl group at C4 of another sugar.

Maltose   Maltose, often called *malt sugar*, is present in fermenting grains and can be prepared in about 80 percent yield by enzyme-catalyzed degradation of starch. Since each molecule of maltose yields two molecules of glucose on hydrolysis, maltose is a **disaccharide**. Two α-D-glucose units are bonded together by what is called a **1,4-link**—an acetal link between the hydroxyl group at C1 of one unit and the hydroxyl group at C4 of the second unit.

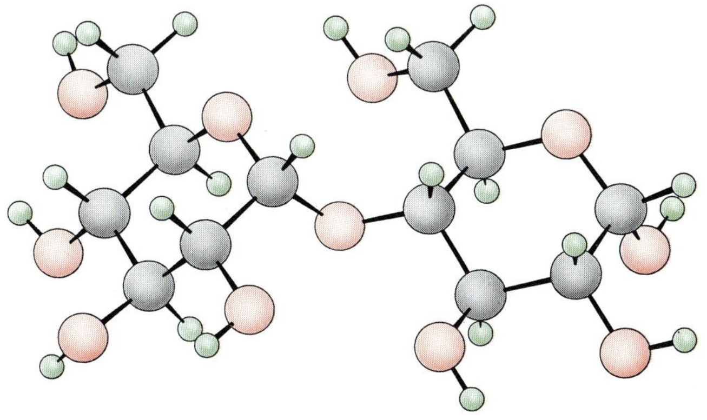

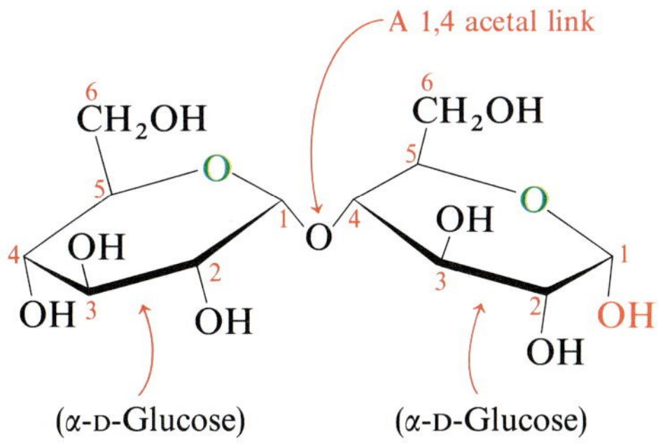

(α-D-Glucose)   (α-D-Glucose)

Maltose

A careful look at maltose shows that it's both an acetal and a hemi The α-D-glucose unit on the left contains an acetal grouping, while th the right contains a hemiacetal. Unlike hemiacetals, acetal links are they don't open and close spontaneously, and they can be hyd treatment with strong aqueous acid (Section 11.6). Thus, th two glucose units in maltose is not easily cleaved. The he glucose unit on the right opens and closes normall aldehyde group and making maltose a reducing sugar.

Practice Problem   **13.9** Draw maltose in the form in which the glucose unit on the right is an open-chain aldehyde.

Lactose   Lactose, or milk sugar, is the major carbohydrate present in mammalian milk. Human milk, for example, is about 7 percent lactose. Structurally, lactose is a disaccharide whose hydrolysis yields one molecule of glucose and one molecule of galactose. The two sugars are joined in lactose by a 1,4 acetal link between C1 of β-D-galactose and C4 of β-D-glucose. Like maltose, lactose is a reducing sugar because opening of the hemiacetal grouping in the glucose ring leads to an aldehyde that can be oxidized.

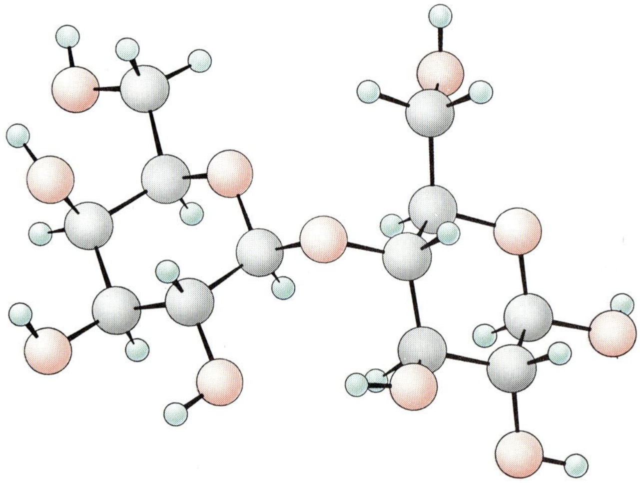

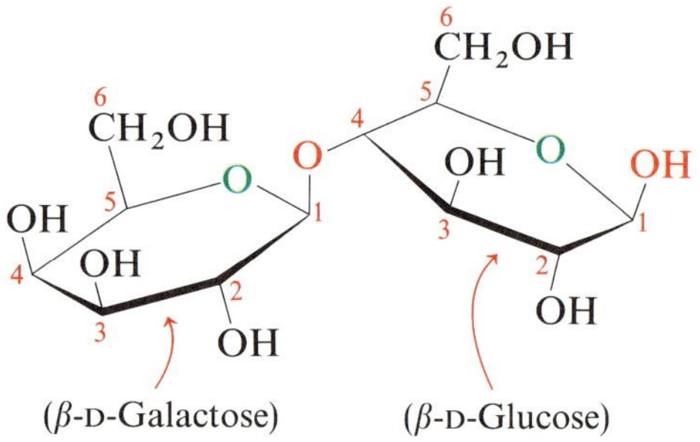

(β-D-Galactose)   (β-D-Glucose)

Lactose

Sucrose   Sucrose—plain table sugar—is probably the most common pure organic chemical in the world. Although found in many plants, sugar beets (20 percent by weight) and sugar cane (15 percent by weight) are the most common sources of sucrose. Hydrolysis of sucrose yields one molecule of glucose and one molecule of fructose. The 50:50 mixture of glucose and fructose that results, often referred to as *invert sugar*, is commonly used as a food additive.

Sucrose differs from other disaccharides we've seen in that it has no hemiacetal group and is thus not a reducing sugar.

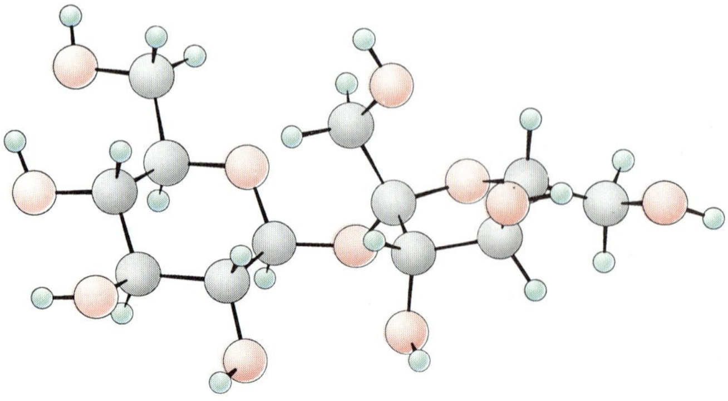

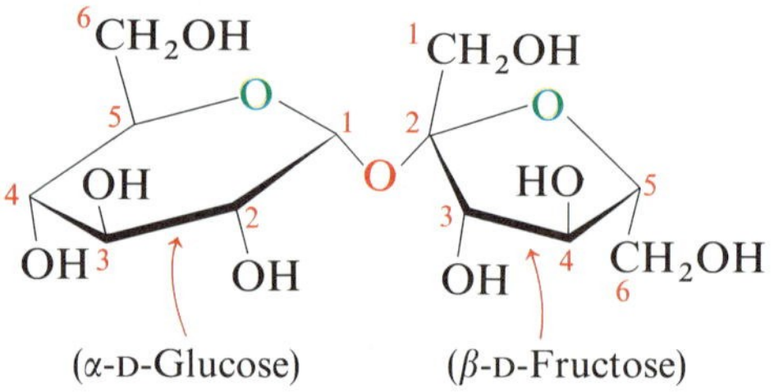

(α-D-Glucose)   (β-D-Fructose)

Sucrose

Practice Problems

13.10   The disaccharide cellobiose can be obtained by enzyme-catalyzed hydrolysis of cellulose. Would you expect cellobiose to be a reducing or nonreducing sugar? Explain.

Cellobiose

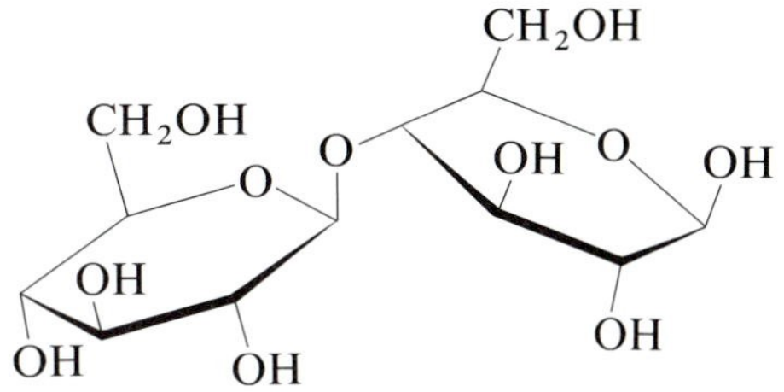

13.11   Show the structures of the two monosaccharides that would be formed on hydrolysis of cellobiose (Problem 13.10). Can you identify them?

13.8   SOME IMPORTANT POLYSACCHARIDES

Polysaccharides are complex carbohydrates that have tens, hundreds, or even thousands of simple sugars linked together through acetal bonds of the sort we've just seen in maltose and lactose. Let's look at three of the most important polysaccharides—cellulose, starch, and glycogen.

Cellulose   Cellulose—the fibrous substance used by plants as a structural material in grasses, leaves, and stems—consists entirely of several thousand β-D-glucose units linked together by 1,4 acetal bonds to form one large molecule (Figure 13.6).

Starch   Starch, like cellulose, is a polymer of glucose. There is, however, a big difference between the two because starch, unlike cellulose, is edible. Indeed, the starch in such vegetables as beans, wheat, rice, and potatoes is an essential part of the human diet.

Figure 13.6
The structure of cellulose. Several thousand β-D-glucose units are joined by 1,4 acetal links between C1 of one sugar and C4 of its neighbor.

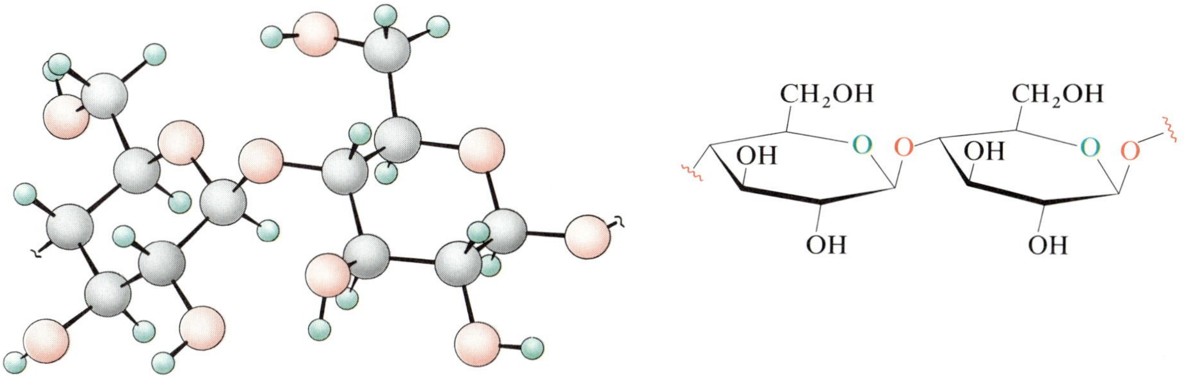

AN APPLICATION: CELL-SURFACE CARBOHYDRATES

More than 80 years ago, it was discovered that human blood can be classified into four blood-group types, called A, B, AB, and O. If a transfusion becomes necessary, blood from a donor of one type cannot be given to a recipient with blood of another type unless the two types are compatible. If an incompatible mix is made, the red blood cells clump together, or *agglutinate*, and death can result (see the table below). Agglutination indicates that the recipient's immune system has recognized foreign cells in the body and has formed antibodies to them.

Cells of types A, B, and O each have characteristic structural features called *antigenic determinants*, which can provoke an immune response

resulting in the production of antibodies. Cells of type AB have both A and B markers, however. As shown in the figure opposite, the structures of the three blood-group determinants are known. The marker for blood-group O is a trisaccharide, whereas the markers for blood-groups A and B are tetrasaccharides.

Blood to be used for transfusion must be labeled by type so that compatibility with the recipient's blood type can be assured.

Human Blood-Group Compatibilities

Donor Blood Type	Acceptor Blood Type			
	A	**B**	**AB**	**O**
A	o	x	o	x
B	x	o	o	x
AB	x	x	o	o
O	o	o	o	o

o = compatible; x = incompatible

Structurally, starch differs from cellulose in that its individual glucose units are each in the α-form, as in maltose, rather than in the β-form. Starch is also more structurally complex than cellulose. Whereas all cellulose molecules are pretty much alike, there are two kinds of starch, called amylose and amylopectin. Amylose, which accounts for about 20 percent of starch, consists of several hundred to one thousand α-D-glucose units linked together in a long chain by 1,4 acetal bonds (Figure 13.7). Amylopectin, which accounts for about 80 percent of starch, is similar to amylose but is much larger (up to 100,000 glucose units per molecule) and has branches approximately every 25 units along its chain. A glucose molecule at one of these branch points uses *two* of its hydroxyl groups (those at C4 and C6) to form acetal links to two other sugars (Figure 13.7).

When eaten, starch molecules are digested in the stomach by enzymes called **glycosidases**. These enzymes catalyze the hydrolysis of acetal bonds so that the starch chain is broken down and individual glucose molecules are freed. As is usually the case in enzyme-catalyzed reactions, glycosidases are highly specific in their action. They hydrolyze only the acetal links to an α-D sugar (as in starch) while leaving acetal links to a β-D sugar (as in cellulose) untouched. Thus, starch is easily digested, but cellulose is unaffected by digestive enzymes.

■ **Glycosidase**   An enzyme that is able to digest complex carbohydrates.

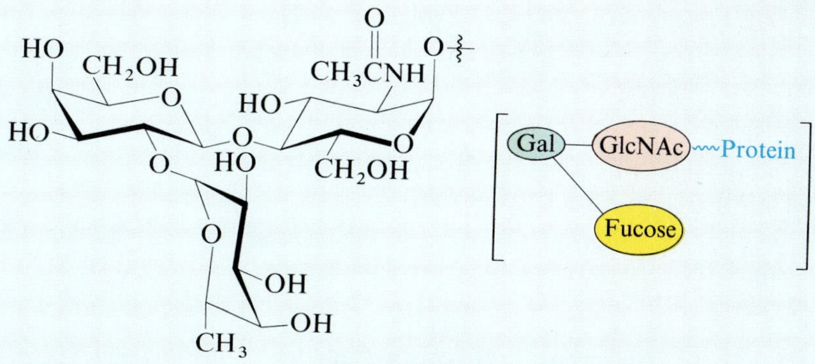

Blood group O

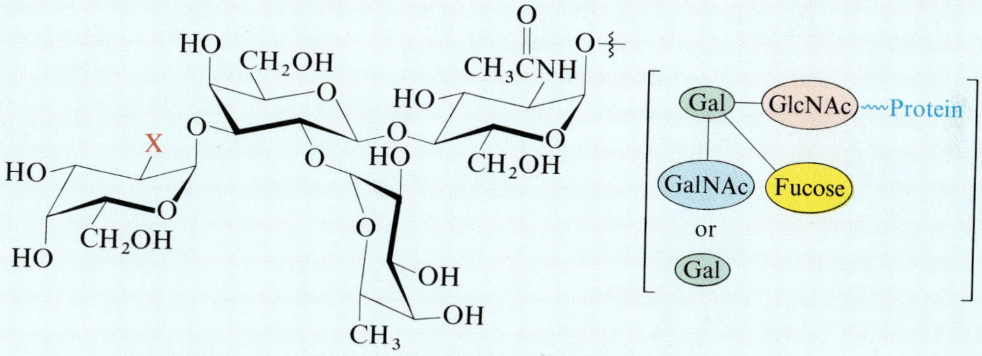

Blood group A: X = NHCOCH$_3$
Blood group B: X = OH

Structures of the A, B, and O blood-group antigenic determinants.
(Gal = D-galactose; GlcNAc = N-acetylglucosamine; GalNAc = N-acetylgalactosamine)

Practice Problem   **13.12**   According to Figure 13.7, an individual starch molecule contains thousands of glucose units but has only a single hemiacetal group at the end of the long polymer chain. Would you expect starch to be a reducing carbohydrate? Explain.

Figure 13.7
A glucose polymer in starch. Amylose consists only of linear chains of α-D-glucose units linked by 1,4 acetal bonds, whereas amylopectin has branch points about every 25 sugars in the chain. A glucose unit at a branch point uses two of its hydroxyls (at C4 and C6) to form 1,4 and 1,6 acetal links to two other sugars.

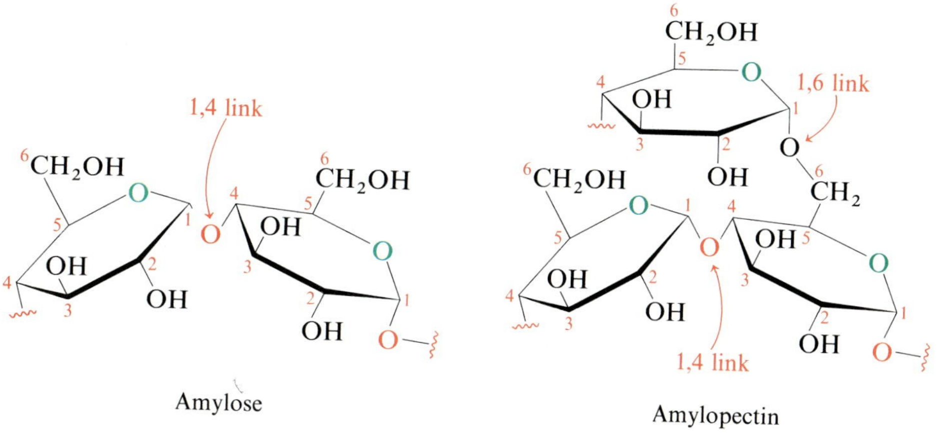

Amylose

Amylopectin

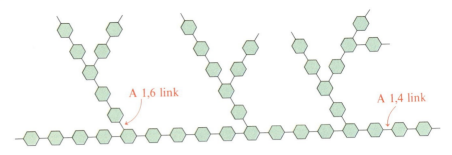

Figure 13.8
The structure of glycogen. The hexagons represent α-D-glucose units linked by 1,4 acetal bonds and (at branch points) 1,6 acetal bonds.

A 1,6 link

A 1,4 link

Glycogen   Glycogen, sometimes called *animal starch*, serves the same food-storage role in animals that starch serves in plants. After we eat starch and the body breaks it down into simple glucose units, some of the glucose is used immediately as fuel, and some is stored in the body as glycogen for later use.

Structurally, glycogen is similar to amylopectin in being a long polymer of α-D-glucose with branch points in its chain. Glycogen has many more branches than amylopectin, however, and is much larger—up to one million glucose units per molecule (Figure 13.8).

SUMMARY

Carbohydrates are polyhydroxy aldehydes and ketones. They are classified according to the number of carbon atoms and the kind of carbonyl group they contain. Glucose, for example, is an aldohexose—a six-carbon aldehyde sugar. **Simple carbohydrates** such as glucose contain only a single sugar and cannot be hydrolyzed to smaller molecules. **Complex carbohydrates** such as starch and cellulose contain many simple sugars linked together.

Certain organic molecules have a handedness to them and are said to be **chiral**; others have no handedness and are **achiral**. The difference between the two is one of symmetry. Achiral molecules have an imaginary **symmetry plane** cutting through the middle so that one half of the molecule is an exact **mirror image** of the other half. Chiral molecules lack this symmetry plane. The two right- and left-handed forms of a chiral substance are called **optical isomers, or enantiomers**. All sugars are chiral, and the naturally occurring optical isomers belong to the right-handed or (D) family. Left-handed or (L) optical isomers of sugars occur much more rarely in nature.

Monosaccharides such as **glucose, galactose, fructose, ribose**, and **2-deoxyribose** exist as a mixture of open-chain and cyclic hemiacetal forms. There are two cyclic hemiacetal forms, called the **α-form** and **β-form**, which differ in the orientation of the hemiacetal hydroxyl group. Monosaccharides undergo many of the same reactions that other aldehydes do. Thus, they react with alcohols to yield cyclic hemiacetals, called **glycosides**, and they react with oxidizing agents to yield carboxylic acids. Sugars that can be oxidized in this way are called **reducing sugars**.

Important **disaccharides** such as **maltose, lactose**, and **sucrose** all contain two simple sugars joined by an acetal link. Important **polysaccharides** include **cellulose, starch**, and **glycogen**. Cellulose is a linear polymer of up to a thousand

INTERLUDE: SWEETNESS

Mention the word *sugar* to most people and they'll automatically think of sweet-tasting foods or candies. In fact, most simple mono- and disaccharides *do* taste sweet, although the degree of sweetness varies from one compound to another. Using sucrose (table sugar) as a reference point, fructose is nearly twice as sweet, whereas glucose, galactose, lactose, and others are much less so. Exact comparisons between sugars are impossible, because sweetness is not a physical property and can't be accurately measured. Sweetness is simply a taste perception, and the ranking of different sugars is a matter of personal opinion. Nevertheless, the ordering shown in the table below is generally agreed on.

Dietary concerns and the desire of many people to cut their caloric intake have led to the widespread use of the artificial sweeteners aspartame, saccharin, and cyclamate. All are far sweeter than natural sugars, but doubts have been raised as to the long-term safety of all three. Cyclamates have been banned in the United States (but not in Canada), and saccharin has been banned in Canada (but not in the United States). As their structures indicate, none of the three bears the slightest chemical resemblance to carbohydrates.

Aspartame

Saccharin

Sodium cyclamate

Relative Sweetness of Some Sugars and Sugar Substitutes

Name	Type	Sweetness
Lactose	Disaccharide	16
Galactose	Monosaccharide	30
Maltose	Disaccharide	33
Glucose	Monosaccharide	75
Sucrose	Disaccharide	**100**
Fructose	Monosaccharide	175
Cyclamate	Artificial	3,000
Aspartame	Artificial	15,000
Saccharin	Artificial	35,000

Sugar cane contains a high concentration of the disaccharide sucrose. The stems of the plant derive much of their strength and stiffness from another carbohydrate that is not digestible by humans: the polysaccharide cellulose.

glucose units bonded together by β-acetal links between C1 of one unit and a hydroxyl at C4 of another unit. Starch and glycogen are glucose polymers in which the individual sugar units are connected by α-acetal links. The fraction of starch called **amylose** consists of linear chains, and the fraction called **amylopectin** has branched chains.

REVIEW PROBLEMS

Classification and Structure of Carbohydrates

13.13 What is a carbohydrate?

13.14 What is the difference between a simple and a complex carbohydrate?

13.15 What is the family-name ending for a sugar?

13.16 What is the structural difference between an aldose and a ketose?

13.17 Classify each of the following carbohydrates by indicating the nature of its carbonyl group and the number of carbon atoms present. For example, glucose is an aldohexose.

(a)

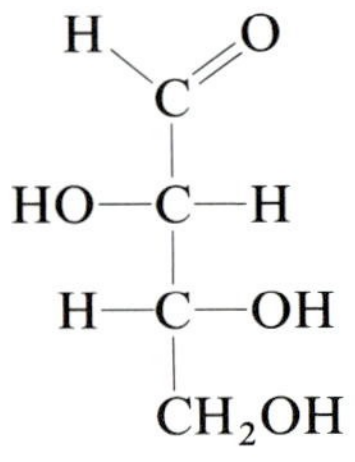

$$
\begin{array}{c}
H\diagdown \,\, O \\
C \\
HO-C-H \\
H-C-OH \\
CH_2OH
\end{array}
$$

Threose

(b)

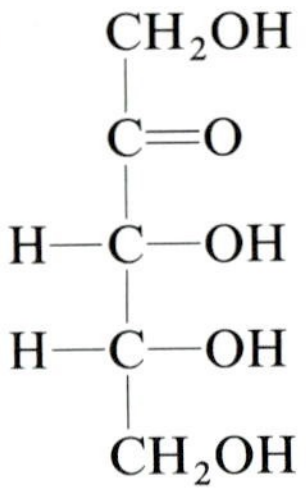

$$
\begin{array}{c}
CH_2OH \\
C=O \\
H-C-OH \\
H-C-OH \\
CH_2OH
\end{array}
$$

Ribulose

(c)

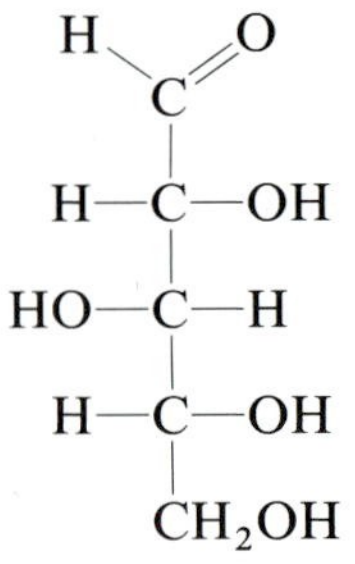

$$
\begin{array}{c}
H\diagdown \,\, O \\
C \\
H-C-OH \\
HO-C-H \\
H-C-OH \\
CH_2OH
\end{array}
$$

Xylose

(d)

$$
\begin{array}{c}
CH_2OH \\
C=O \\
HO-C-H \\
HO-C-H \\
H-C-OH \\
CH_2OH
\end{array}
$$

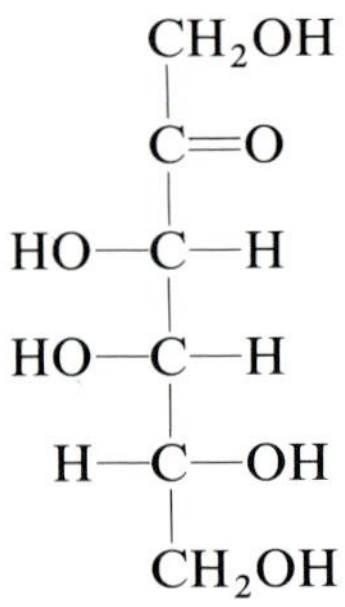

Tagatose

13.18 Write the open-chain structure of a ketotetrose.

13.19 Write the open-chain structure of a four-carbon deoxy sugar.

13.20 Give the names of three important monosaccharides, and tell where each occurs in nature.

13.21 "Dextrose" is an alternative name for what sugar?

13.22 Both galactose and glucose are aldohexoses. What is the structural difference between them?

13.23 D-Gulose, an aldohexose isomer of glucose, has the following cyclic structure. Which is shown, the α-form or the β-form?

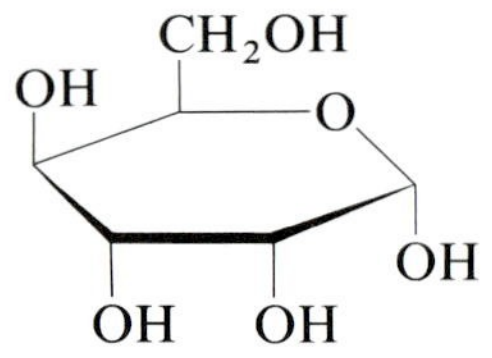

D-Gulose

13.24 Draw D-gulose (Problem 13.23) in its open-chain aldehyde form, both coiled and uncoiled.

13.25 D-Mannose, an aldohexose found in orange peels, has the following structure in open-chain form. Coil mannose around, and draw it in cyclic hemiacetal α- and β-forms.

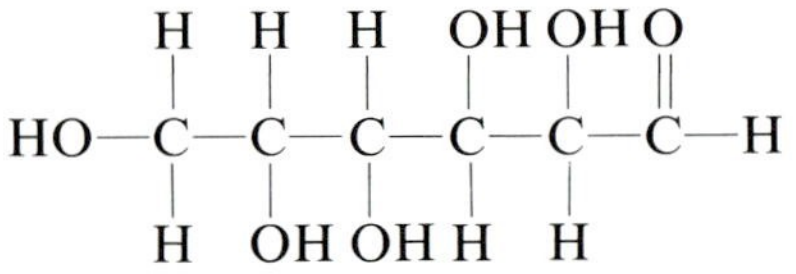

$$
\begin{array}{c}
HHHOHOHO \\
HO-C-C-C-C-C-C-H \\
HOHOHHH
\end{array}
$$ D-Mannose

13.26 D-Ribulose, a ketopentose related to ribose, has the following structure in open-chain form. Coil ribulose around, and draw it in its five-membered cyclic β-hemiacetal form.

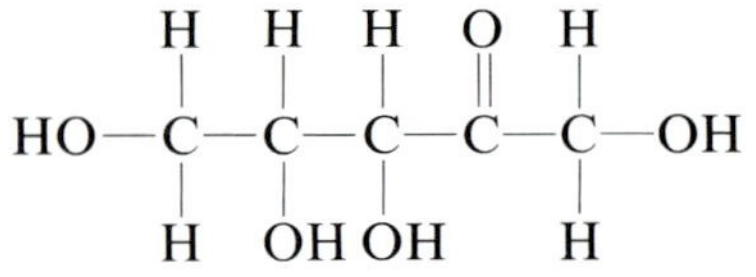

$$
\begin{array}{c}
HHHOH \\
HO-C-C-C-C-C-OH \\
HOHOHH
\end{array}
$$ D-Ribulose

13.27 D-Allose, an aldohexose, is identical with D-glucose except that the hydroxyl group at C3 points down rather than up in the cyclic hemiacetal form. Draw this cyclic form of D-allose.

13.28 Draw D-allose (Problem 13.27) in its open-chain form.

Handedness in Molecules

13.29 What do the terms *chiral* and *achiral* mean?

13.30 Give two examples of chiral objects and two examples of achiral objects.

13.31 Which of the following objects are chiral?
(a) a shoe   (b) a bed   (c) a light bulb
(d) a flower pot

13.32 What is the structural relationship of L-glucose to D-glucose?

13.33 Would you expect L-glucose to be a good food source in the human diet in the same way that D-glucose is?

13.34 2-Bromo-2-chloropropane is an achiral molecule, but 2-bromo-2-chlorobutane is chiral. Explain.

13.35 There are two D-aldotetroses, whose structures are shown. One of the two reacts with $NaBH_4$ to yield a chiral product, but the other yields an achiral product. Explain.

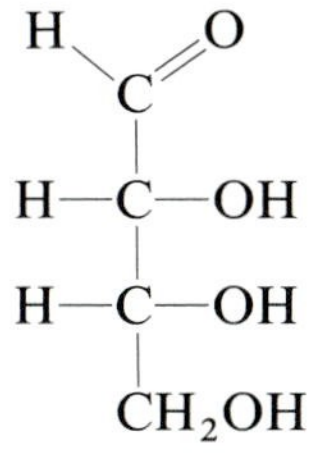

D-Erythrose

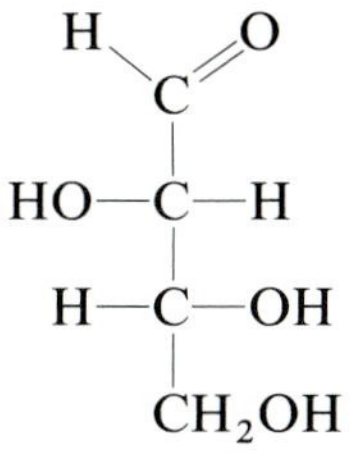

D-Threose

Reactions of Carbohydrates

13.36 What does the term *reducing sugar* mean?

13.37 What is the structural difference between a hemiacetal and an acetal?

13.38 What is the structural difference between the α-hemiacetal form of a carbohydrate and the β-form?

13.39 What are glycosides, and how can they be formed?

13.40 We saw in Section 11.5 that aldehydes react with reducing agents like $NaBH_4$ to yield primary alcohols ($RCH{=}O \rightarrow RCH_2OH$). Treatment of D-glucose with $NaBH_4$ yields *sorbitol*, a substance used as a sugar substitute by diabetics. Draw the structure of sorbitol.

13.41 Reduction of D-fructose with $NaBH_4$ (Problem 13.40) yields a mixture of D-sorbitol along with a sec-

ond, isomeric product. What is the structure of the second product?

13.42 Treatment of an aldose with an oxidizing agent like Tollens' reagent (Section 11.4) yields a carboxylic acid. Gluconic acid, the product of glucose oxidation, is used as its magnesium salt for the treatment of magnesium deficiency. Draw the structure of gluconic acid.

Disaccharides and Polysaccharides

13.43 Give the names of three important disaccharides, and tell where each occurs in nature.

13.44 Starch and cellulose are both polymers of glucose. What is the main structural difference between them, and what different roles do they serve in nature?

13.45 Starch and glycogen are both α-linked polymers of glucose. What is the structural difference between them, and what different roles do they serve in nature?

13.46 Lactose and maltose are reducing disaccharides, but sucrose is a nonreducing disaccharide. Explain.

13.47 Gentiobiose, a rare disaccharide found in saffron, has the following structure. What simple sugars would you obtain on hydrolysis of gentiobiose?

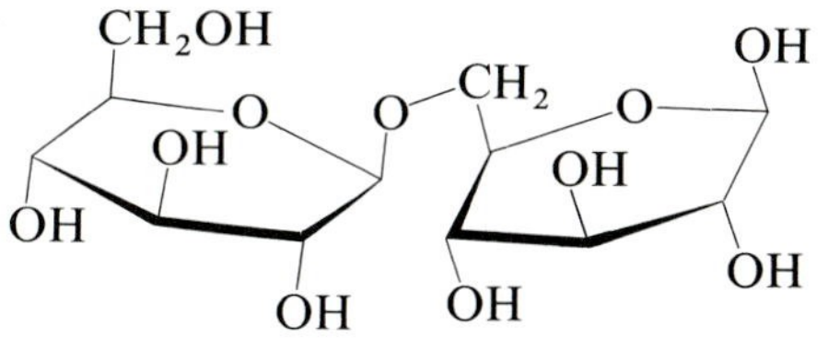

Gentiobiose

13.48 Look carefully at the structure of gentiobiose (Problem 13.47). Does gentiobiose have an acetal grouping? a hemiacetal grouping? Would you expect gentiobiose to be a reducing or nonreducing sugar?

13.49 Trehalose, a disaccharide found in the blood of insects, has the following structure. What simple sugars would you obtain on hydrolysis of trehalose?

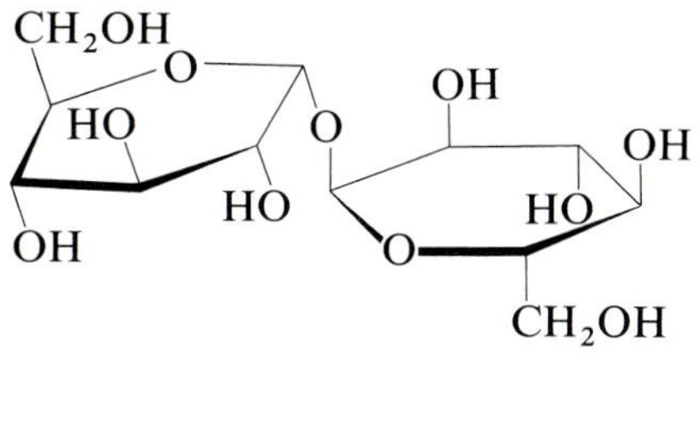

Trehalose

13.50 Does trehalose (Problem 13.49) have an acetal grouping? a hemiacetal grouping? Would you expect trehalose to be a reducing or nonreducing sugar?

13.51 Amygdalin, or Laetrile, is a glycoside isolated in 1830 from almond and apricot seeds. It is called a *cyanogenic glycoside* because hydrolysis with aqueous acid liberates hydrogen cyanide (HCN) along with benzaldehyde and two molecules of glucose. Structurally, amygdalin is a glycoside between the hemiacetal, gentiobiose (Problem 13.47), and the alcohol, mandelonitrile. Draw the structure of amygdalin.

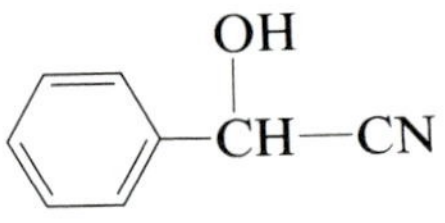

Mandelonitrile

CHAPTER 14

The Molecules of Life: Lipids

The beeswax from which this honeycomb is made is a lipid. So are a great variety of other compounds that you'll learn about in this chapter.

Lipids are less well known than carbohydrates or proteins to most people, yet they are just as essential to our diets and our well-being. Lipids have many important biological roles, serving as sources of fuel storage, as a protective coat around many plants and insects, and as a major component of the membranes that surround every living cell.

Chemically, lipids are the naturally occurring organic molecules that dissolve in nonpolar organic solvents when a sample of plant or animal tissue is crushed or ground. For example, if a plant is placed in a kitchen blender and finely ground in the presence of ether, anything that dissolves in the ether is called a lipid, while anything that remains insoluble (including carbohydrates, proteins, and inorganic salts) is not a lipid.

In this chapter, we'll answer the following questions about lipids:

1. **How are lipids classified?** The goal: you should learn how the different kinds of lipids are grouped.

2. **What are fats, waxes, and oils?** The goal: you should learn the general structures of fats, waxes, and oils.

3. **What reactions do fats, waxes, and oils undergo?** The goal: you should learn the most important reactions of these lipids.

4. **What are glycolipids and phospholipids?** The goal: you should learn the general structures and biological roles of glycolipids and phospholipids.

5. **What are cell membranes made of, and what lipids do they contain?** The goal: you should learn the general structure of cell membranes.

6. **What are steroids and prostaglandins?** The goal: you should learn the general structures of steroids and prostaglandins and how to recognize specific examples.

14.1 STRUCTURE AND CLASSIFICATION OF LIPIDS

■ **Lipid** A naturally occurring molecule that is soluble in nonpolar organic solvents.

Look at the structures of some representative lipids shown in Figure 14.1. Since **lipids** are defined by their solubility in nonpolar organic solvents (a physical property) rather than by chemical structure, it's not surprising that there are a great many different kinds. Note in all of the structures shown that the molecules contain large hydrocarbon portions, which accounts for their solubility behavior.

$$CH_3(CH_2)_{28}CH_2-O-\overset{\overset{\displaystyle O}{\|}}{C}-(CH_2)_{14}CH_3$$

A wax

$$CH_2-O-\overset{\overset{\displaystyle O}{\|}}{C}-(CH_2)_{14}CH_3$$
$$CH-O-\overset{\overset{\displaystyle O}{\|}}{C}-(CH_2)_7CH=CH(CH_2)_7CH_3$$
$$CH_2-O-\overset{\overset{\displaystyle O}{\|}}{C}-(CH_2)_{16}CH_3$$

A fat or oil

Cholesterol, a steroid

A prostaglandin

Figure 14.1
Structures of some representative lipids, isolated from plant and animal tissue by extraction with nonpolar organic solvents.

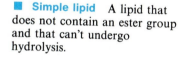

■ **Simple lipid** A lipid that does not contain an ester group and that can't undergo hydrolysis.

■ **Complex lipid** A lipid that contains an ester group and that can be hydrolyzed to yield an alcohol and a carboxylic acid.

Figure 14.2
Some important families of lipids.

Lipids are generally classified into two main groups: simple and complex. **Simple lipids**, such as cholesterol and prostaglandin (Figure 14.1), have no ester linkages and can't be hydrolyzed (saponified) by reaction with aqueous NaOH to yield simpler molecules. **Complex lipids**, such as fats, oils, and waxes, contain ester linkages that can be saponified to yield carboxylic acid and alcohol portions on reaction with aqueous NaOH (Section 12.6).

Figure 14.2 summarizes the important classes of lipids that we'll be looking at in this chapter. Most of the names are probably unfamiliar to you at this point, so you might want to check back occasionally as you study.

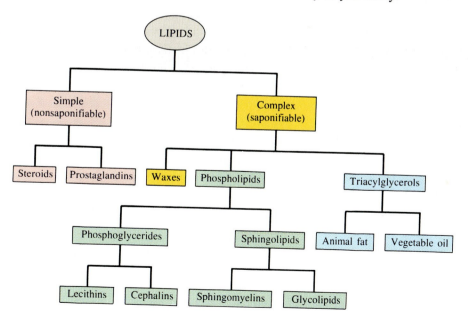

14.2 WAXES, FATS, AND OILS

■ **Wax** A mixture of complex lipids consisting of esters of long-chain carboxylic acids with long-chain alcohols.

Waxes Plant and animal waxes are among the most widely occurring kinds of lipids. Most **waxes** are mixtures of esters, formed in nature by reaction of a long-chain carboxylic acid with a long-chain alcohol. The carboxylic acid portion usually has an even number of carbons from 16 through 36, while the alcohol portion contains an even number of carbons in the range 24 through 36. For example, one of the major components in beeswax is triacontyl hexadecanoate, the ester formed from the 30-carbon alcohol (triacontanol) and the 16-carbon acid (hexadecanoic acid). The waxy protective coatings on most fruits, berries, leaves, and animal furs have similar structures.

A wax:

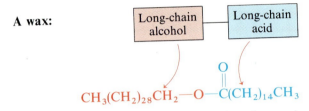

Triacontyl hexadecanoate (from beeswax)

■ **Triacylglycerol** A fat or vegetable oil, consisting of triesters of glycerol with three long-chain carboxylic acids.

■ **Fatty acid** A long-chain carboxylic acid formed by hydrolysis of animal fats and vegetable oils.

Fats and Oils Animal fats and vegetable oils are the most plentiful lipids in nature. Although they appear different—animal fats like butter and lard are usually solid, while vegetable oils like corn, olive, soybean, and peanut oil are liquid—their structures are closely related. All fats and oils are **triacylglycerols** or triglycerides—that is, triesters of glycerol (1,2,3-propanetriol, also called *glycerin*) with three long-chain carboxylic acids called **fatty acids**. The fatty-acid components of fats and oils are usually unbranched and have an even number of carbon atoms in the range 12 through 22. If double bonds are present, they are usually cis rather than trans (Section 9.2).

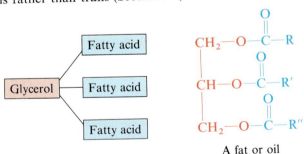

A fat or oil

For example, a typical triglyceride might have the following structure:

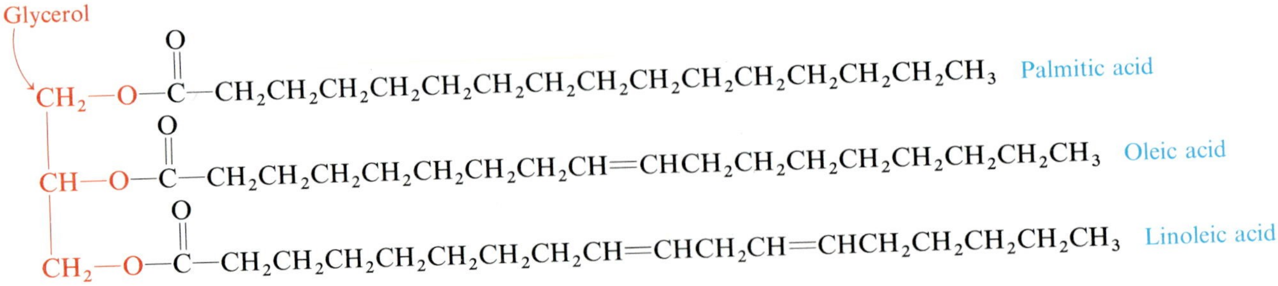

Glycerol

$$CH_2-O-\overset{\displaystyle O}{\overset{\|}{C}}-CH_2CH_2CH_2CH_2CH_2CH_2CH_2CH_2CH_2CH_2CH_2CH_2CH_2CH_2CH_3 \quad \text{Palmitic acid}$$

$$CH-O-\overset{\displaystyle O}{\overset{\|}{C}}-CH_2CH_2CH_2CH_2CH_2CH_2CH_2CH=CHCH_2CH_2CH_2CH_2CH_2CH_2CH_2CH_3 \quad \text{Oleic acid}$$

$$CH_2-O-\overset{\displaystyle O}{\overset{\|}{C}}-CH_2CH_2CH_2CH_2CH_2CH_2CH_2CH=CHCH_2CH=CHCH_2CH_2CH_2CH_2CH_3 \quad \text{Linoleic acid}$$

■ **Unsaturated** Containing one or more double or triple bonds between carbon atoms, and thus fewer than the maximum number of hydrogen atoms per carbon.

■ **Saturated** Containing only single bonds between carbon atoms, and thus unable to accommodate additional hydrogens.

As shown in the triglyceride structure above, the three fatty acids of any specific fat molecule are not necessarily the same. Furthermore, the fat or oil from a given source is always a complex mixture of many different triglycerides. Table 14.1 shows the structures of some commonly occurring fatty acids, and Table 14.2 lists the average composition of fats and oils from several different sources. Note particularly in Table 14.2 that the fatty acids in vegetable oils are largely **unsaturated**—that is, they contain only single bonds. Animal fats, by contrast, contain a much higher percentage of **saturated** fatty acids, which contain one or more double bonds. We'll see in the next section that this difference in composition is the primary reason for the different melting points of fats and oils.

Table 14.1   Structures of Some Common Fatty Acids

Name	Number of Carbons	Number of double Bonds	Structure	Melting Point (°C)
Saturated				
Lauric	12	0	$CH_3(CH_2)_{10}COOH$	44
Myristic	14	0	$CH_3(CH_2)_{12}COOH$	58
Palmitic	16	0	$CH_3(CH_2)_{14}COOH$	63
Stearic	18	0	$CH_3(CH_2)_{16}COOH$	70
Unsaturated				
Oleic	18	1	$CH_3(CH_2)_7CH{=}CH(CH_2)_7COOH$ (cis)	4
Linoleic	18	2	$CH_3(CH_2)_4CH{=}CHCH_2CH{=}CH(CH_2)_7COOH$ (all cis)	−5
Linolenic	18	3	$CH_3CH_2CH{=}CHCH_2CH{=}CHCH_2CH{=}CH(CH_2)_7COOH$ (all cis)	−11
Arachidonic	20	4	$CH_3(CH_2)_4(CH{=}CHCH_2)_4CH_2CH_2COOH$ (all cis)	−50

Table 14.2   Approximate Composition of Some Common Fats and Oils

Source	Saturated Fatty Acids (%)				Unsaturated Fatty Acids (%)	
	C_{12} Lauric	C_{14} Myristic	C_{16} Palmitic	C_{18} Stearic	C_{18} Oleic	C_{18} Linoleic
Animal Fat						
Lard	—	1	25	15	50	6
Butter	2	10	25	10	25	5
Human fat	1	3	25	8	46	10
Whale blubber	—	8	12	3	35	10
Vegetable Oil						
Corn	—	1	8	4	46	42
Olive	—	1	5	5	83	7
Peanut	—	—	7	5	60	20
Soybean	—	—	7	4	34	53

About 40 different fatty acids are known to occur naturally. Palmitic acid (16 carbons) and stearic acid (18 carbons) are the most abundant saturated acids; oleic and linoleic acids (both with 18 carbons) are the most abundant unsaturated ones. Oleic acid is *monounsaturated*, since it has only the one double bond in the C9-C10 position, but linoleic, linolenic, and arachidonic acids are **polyunsaturated fatty acids** (called **PUFAs**), because they have more than one carbon-carbon double bond.

■ **Polyunsaturated fatty acid (PUFA)** A long-chain fatty acid that has two or more carbon-carbon double bonds.

Two of the polyunsaturated fatty acids, linoleic and linolenic, are essential in the human diet. Infants grow poorly and develop severe skin lesions if fed a diet of nonfat milk for a prolonged period, although adults usually have sufficient reserves of body fat to avoid such problems. Arachidonic acid is also an essential nutrient, but the body is able to synthesize it from linolenic acid.

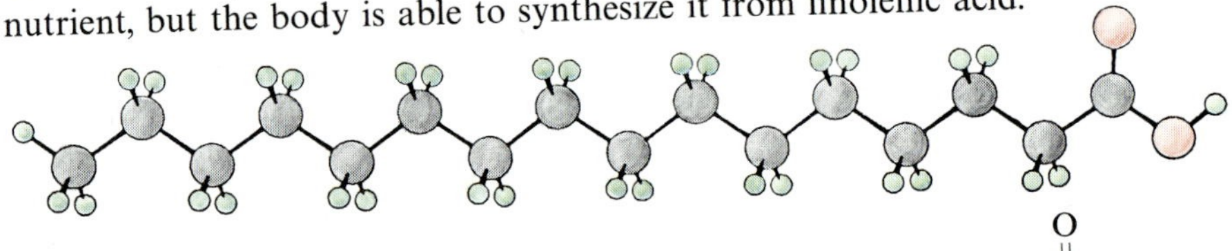

$$CH_3CH_2CH_2CH_2CH_2CH_2CH_2CH_2CH_2CH_2CH_2CH_2CH_2CH_2CH_2\overset{\displaystyle O}{\overset{\|}{C}}{-}OH$$

Palmitic acid—a saturated fatty acid

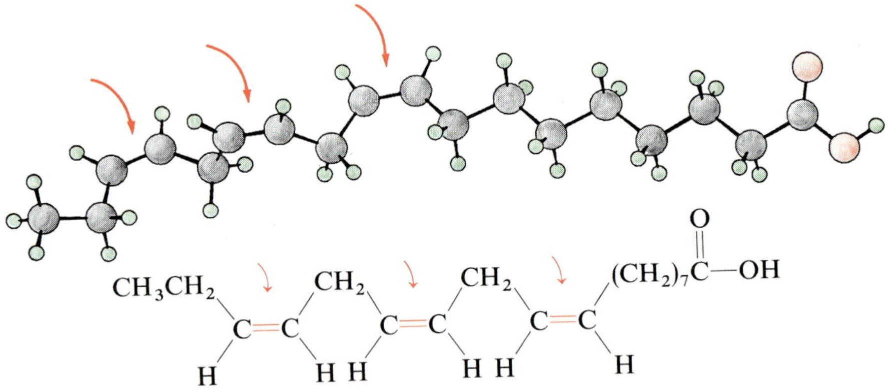

Linolenic acid—a polyunsaturated fatty acid (PUFA)

Practice Problems

14.1 Draw the full structure of arachidonic acid (Table 14.1) in a way that shows the cis stereochemistry of its four double bonds.

14.2 One of the constituents of the carnauba wax used in floor and furniture polish is an ester of a C_{32} straight-chain alcohol with a C_{20} straight-chain carboxylic acid. Draw the structure of this ester.

14.3 Show the structure of glyceryl trioleate, a fat molecule whose components are glycerol and three oleic acid units.

14.3 HYDROGENATION OF FATS AND OILS

Look at the melting points of the various fatty acids listed in Table 14.1. As a general rule, the more double bonds a fatty acid has, the lower its melting point. Thus, the saturated C_{18} acid (stearic) melts at 70°C, the monounsaturated C_{18}

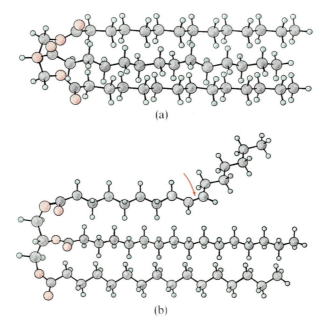

(a)

(b)

Figure 14.3
Molecular models of (a) a saturated and (b) an unsaturated triglyceride. The double bond (arrow) in (b) prevents the molecule from adopting a regular shape and crystallizing easily.

acid (oleic) melts at $4°C$, and the diunsaturated C_{18} acid (linoleic) melts at $-5°C$. This same trend also holds true for triglycerides: The more highly unsaturated a triglyceride is, the lower its melting point.

The difference in melting points between fats and oils is a consequence of the difference in their chemical structures. As shown in Table 14.2, vegetable oils generally have a higher proportion of unsaturated to saturated fatty acids than animal fats. Thus, vegetable oils are lower melting than fats and appear as liquids rather than solids. This behavior is due to the fact that saturated fats have a fairly uniform shape that allows the various fatty-acid chains to nestle together easily in a crystal. The presence of carbon-carbon double bonds in unsaturated oils causes kinks and bends in the fatty-acid chains that make it difficult for chains to fit next to each other and make crystal formation difficult. The more double bonds there are, the harder it is for a triglyceride to crystallize, an effect illustrated in Figure 14.3 with molecular models.

The carbon-carbon double bonds in vegetable oils can be hydrogenated to yield saturated fats in the same way that any alkene can react with hydrogen to yield an alkane (Section 9.3). Solid cooking fats like Crisco and Spry are produced commercially by hydrogenation of vegetable oils to give a product chemically similar to that found in animal fats.

Part-structure of an unsaturated vegetable oil

$$-O-\overset{\overset{\textstyle O}{\|}}{C}-CH_2CH_2CH_2CH_2CH_2CH_2CH_2CH=CHCH_2CH=CHCH_2CH_2CH_2CH_2CH_3$$

2 H₂, Pd catalyst

Part-structure of a saturated fat

$$-O-\overset{\overset{\textstyle O}{\|}}{C}-CH_2CH_2CH_2CH_2CH_2CH_2CH_2CH-CHCH_2CH-CHCH_2CH_2CH_2CH_2CH_3$$
$$\quad\quad\quad\quad\quad\quad\quad\quad\quad\quad\quad\quad\quad\quad\quad H\quad H\quad\quad H\quad H$$

By carefully controlling the exact extent of hydrogenation, the final product can be made to have any desired consistency. Margarine, for example, is carefully prepared so that only about two-thirds of the double bonds present in the starting vegetable oil are hydrogenated. The remaining double bonds are left unhydrogenated, so the margarine has exactly the right consistency to keep soft in the refrigerator and to melt on warm toast.

Practice Problem   14.4   Write an equation showing the complete hydrogenation of glyceryl trioleate (Problem 14.3) to yield glyceryl tristearate.

14.4   HYDROLYSIS OF FATS AND OILS: SOAP

■ **Saponification**   Reaction of a fat or vegetable oil with aqueous hydroxide ion to yield glycerol and carboxylate salts of fatty acids.

Fats and oils, like all esters, can be hydrolyzed to yield carboxylic acid and alcohol components. In the body, this hydrolysis is carried out by enzymes as the first step in lipid metabolism. In the laboratory, the same reaction is usually carried out with aqueous NaOH or KOH, where it is called **saponification** (Section 12.6). The initial products of fat (or oil) saponification are one molecule of glycerol and three molecules of fatty-acid carboxylate salts.

$$\underset{\text{A fat or oil}}{\begin{array}{l} CH_2-O-\overset{\displaystyle O}{\overset{\|}{C}}-R \\[4pt] CH-O-\overset{\displaystyle O}{\overset{\|}{C}}-R' \\[4pt] CH_2-O-\overset{\displaystyle O}{\overset{\|}{C}}-R'' \end{array}} \xrightarrow[\text{H}_2\text{O}]{\text{NaOH}} \underset{\text{Glycerol}}{\begin{array}{l} CH_2-OH \\[4pt] CH-OH \\[4pt] CH_2-OH \end{array}} + \underset{\substack{\text{Fatty-acid salts}\\(\text{soap})}}{3\,R-\overset{\displaystyle O}{\overset{\|}{C}}-O^-\,Na^+}$$

■ **Soap**   The mixture of carboxylate salts of fatty acids formed on saponification of animal fat.

The complex mixture of fatty-acid salts produced by saponification of animal fat with NaOH or KOH is known as **soap**. Crude soap curds, which contain glycerol and excess alkali as well as soap, are first purified by boiling with water and adding NaCl to precipitate the pure sodium carboxylate salts. The smooth soap that precipitates is then dried, perfumed, and pressed into bars for household use. Dyes are added for colored soaps; antiseptics are added for medicated soaps; pumice is added for scouring soaps; and air is blown in for soaps that float.

■ **Hydrophilic**   Water-loving; a hydrophilic substance dissolves in water.

■ **Hydrophobic**   Water-hating; a hydrophobic substance does not dissolve in water.

Soaps act as cleaning agents because the two ends of a soap molecule are so different. The sodium-salt end of the long-chain molecule is ionic and therefore **hydrophilic** (water-loving); it tends to dissolve in water. The long organic portion of the molecule, however, is nonpolar and therefore **hydrophobic** (water-hating); it tends to avoid water and to dissolve in grease. Because of these two opposing tendencies, soaps are attracted to *both* grease and water.

When soap molecules are dispersed in water, very few of them actually dissolve. Instead, large numbers of soap molecules aggregate together, with their

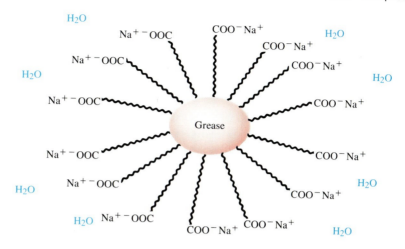

Figure 14.4
A soap micelle solubilizing a grease particle in water.

■ **Micelle** A spherical cluster formed by the aggregation of soap molecules in water.

long hydrophobic hydrocarbon tails clustering in a ball to exclude water. At the same time, their hydrophilic ionic heads on the surface of the cluster stick out into the water. These clusters, called **micelles**, are shown in Figure 14.4. Grease and dirt are made soluble in water when they are coated by the nonpolar tails of the soap molecules in the center of micelles. Once solubilized, the grease and dirt can be washed away.

Practice Problem **14.5** In hard water, which contains iron, magnesium, and calcium ions, soap often leaves a scummy residue. Draw the structure of calcium oleate, one of the components of bathtub scum.

14.5 PHOSPHOLIPIDS

■ **Phospholipids** Lipids that have an ester link between phosphoric acid and an alcohol.

Just as waxes, fats, and oils are esters that have links between an alcohol and a carboxylic acid, **phospholipids** are esters that have links between an alcohol and phosphoric acid, H_3PO_4. There are two main types of phospholipids: phosphoglycerides and sphingolipids.

■ **Phosphoglyceride** A phospholipid in which glycerin is linked by ester bonds to two fatty acids and one phosphoric acid.

Phosphoglycerides *Phosphoglycerides* are closely related to fats in that they have a glycerol unit linked by ester bonds to two fatty acids and one phosphoric acid. (Recall from Section 12.11 how phosphoric acid can form phosphate esters with alcohols.) Although the fatty acids may be any of the C_{12}–C_{20} units normally present in fats, the unit bonded to C1 of glycerol is usually saturated, while that at C2 is usually unsaturated. The phosphate group, which is always bonded to C3, is also connected by a separate ester link to a small amino alcohol like choline [$HOCH_2CH_2\overset{+}{N}(CH_3)_3$], ethanolamine [$HOCH_2CH_2NH_2$], or serine [$HOCH_2CH(NH_2)COOH$]. (See structures, page 304.)

Note that the phosphorylcholine or phosphorylethanolamine part of a phosphoglyceride is ionic. The nitrogen atom is positively charged and the phosphate oxygen atom is negatively charged under physiological conditions.

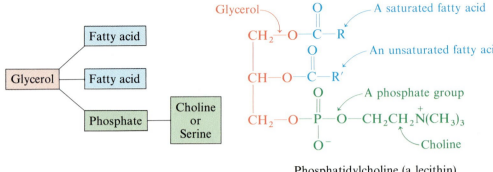

Phosphatidylcholine (a lecithin)

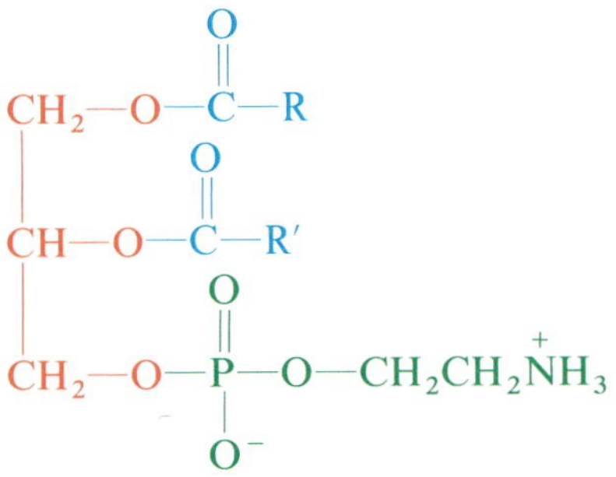

Phosphatidylethanolamine

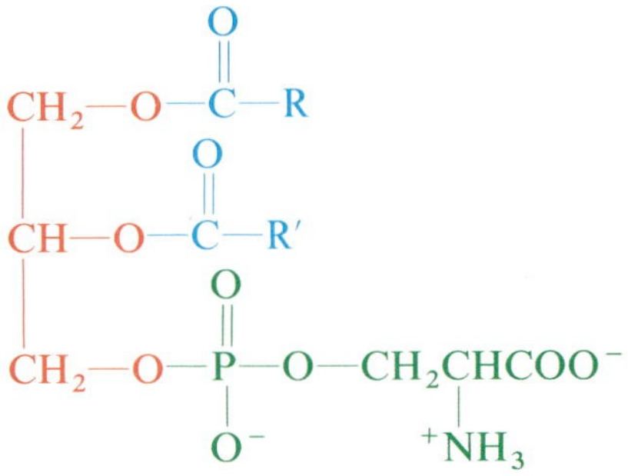

Phosphatidylserine

(Cephalins)

The most important kinds of phosphoglycerides are the *lecithins* and the *cephalins*. Lecithins, which have their phosphate group bound to choline, are major components of cell membranes in all higher organisms. Cephalins, which have their phosphate group bound to ethanolamine or serine, are particularly abundant in brain tissue.

■ **Sphingolipid** A phospho-lipid based on the amino alcohol sphingosine rather than on glycerol.

Sphingolipids   *Sphingolipids*, the second main class of phospholipids, are based on the amino alcohol *sphingosine* rather than on glycerol. A long-chain fatty acid is bonded to the C2 —NH$_2$ group by an amide link (Section 12.9), and a phosphoric acid unit is bonded to the C1 —OH group. As in the lecithins, the phosphate group in sphingolipids is also bonded to an amino alcohol. In the *sphingomyelins*, which are a major constituent of the coating around nerve fibers, the phosphate group is bonded to choline.

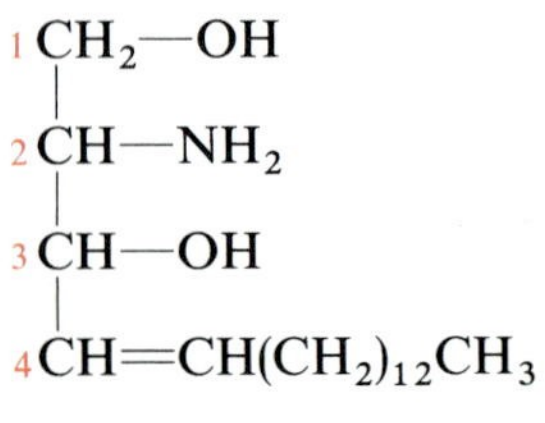

$_1$CH$_2$—OH
$_2$CH—NH$_2$
$_3$CH—OH
$_4$CH=CH(CH$_2$)$_{12}$CH$_3$

Sphingosine

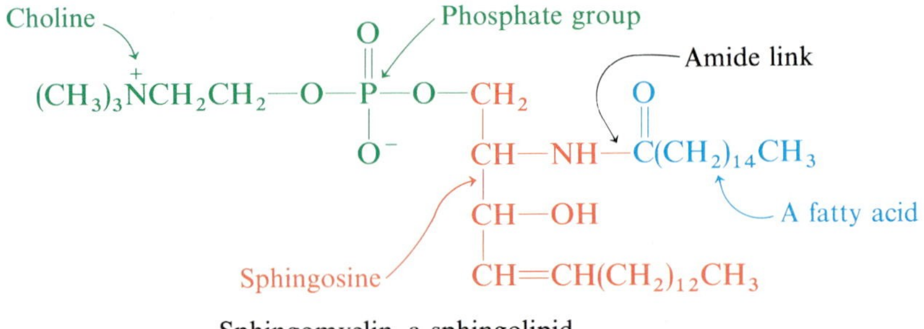

Sphingomyelin, a sphingolipid

Practice Problems   **14.6**   Show the structure of the cephalin that contains a stearic acid unit, an oleic acid unit, and a phosphorylethanolamine.

14.7   Show the structure of the sphingomyelin that contains a myristic acid unit.

14.6   GLYCOLIPIDS

■ **Glycolipid** A substance that has a lipid-like portion bonded to a carbohydrate.

Yet another class of sphingosine-based lipids are the **glycolipids**—compounds that contain both carbohydrate and lipid parts. Glycolipids differ from sphingomyelin in having the phosphorylcholine portion at C1 replaced by a sugar. In the *cerebrosides*, for example, sphingosine is connected by an amide link between a fatty acid and the —NH$_2$ group at C2, and by a glycoside link (Section 13.6) between a sugar and the —OH group at C1. Cerebrosides are particularly abundant in brain tissue, where the sugar unit is D-galactose. They are also found in nonneural tissue, where the sugar unit is D-glucose.

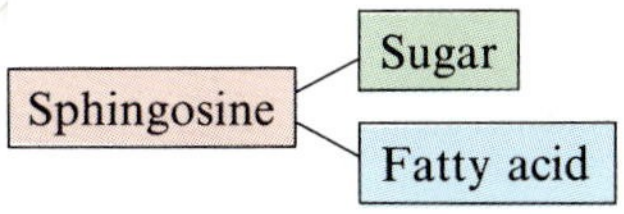

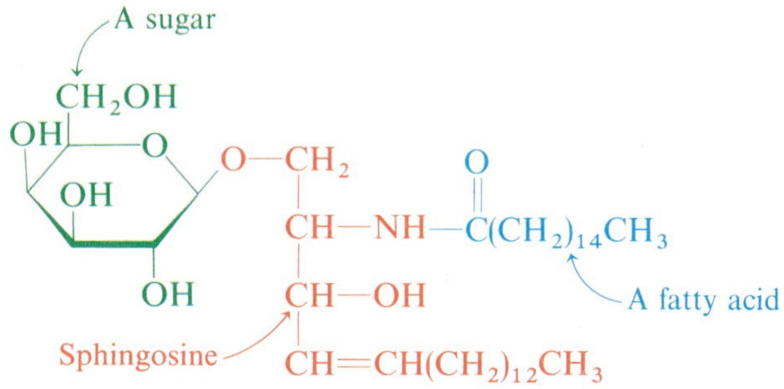

A cerebroside (a glycolipid)

Closely related to the cerebrosides are the *gangliosides*—glycolipids in which the carbohydrate component linked to sphingosine is a *polysaccharide* (Section 13.8) rather than a monosaccharide. Tay-Sachs disease, a fatal genetic disorder, is due to a defect in lipid metabolism that causes a greatly elevated concentration of gangliosides in the brain.

Practice Problem   **14.8**   Look up the structure of cellobiose shown in Problem 13.10, and draw the structure of a ganglioside that contains this disaccharide.

14.7   CELL MEMBRANES

Phospholipids form a major part of the membranes that surround all living cells. The lipid membranes serve not only to separate the interior of cells from their outside environment; they also act as a kind of semipermeable barrier to allow the selective passage of nutrients and other components into and out of cells while blocking the passage of ions.

■ **Lipid bilayer**   The basic structural unit of cell membranes; composed of two parallel sheets of lipid molecules arranged tail to tail.

Biological membranes are structurally similar in many respects to the soap micelles discussed in Section 14.4 (see Figure 14.4). Like soaps, phospholipids contain both a hydrophilic ionic *head* (the phosphate ester) and a long nonpolar hydrophobic *tail*. Unlike soaps, however, phospholipids aggregate in a closed, sheet-like membrane structure called a **lipid bilayer** rather than in globular micelles. As shown schematically in Figure 14.5, a lipid bilayer has two parallel layers of lipids oriented so that ionic head groups protrude into the aqueous environment on the inner and outer faces of the bilayer. The nonpolar tails cluster together in the middle of the bilayer where they can avoid water.

Cell membranes are a good deal more complex than the simple bilayer picture indicates. For example, the inner and outer faces of cell membranes are not identical. In human red blood cells (erythrocytes), the outer layer consists primarily of sphingomyelin and phosphatidylcholine (lecithin), while the inner layer is mainly phosphatidylethanolamine and phosphatidylserine (both cephalins).

Another complexity is that cell membranes contain many different kinds of molecules embedded in the bilayer other than just phosphoglycerides. Glycolipids and cholesterol are also present, and at least 20 percent of the weight of a membrane consists of *protein* globules (Chapter 15). Some proteins are only partially embedded, but others extend completely through the membrane, as shown in Figure 14.6. Lipid components of the membrane serve a primarily structural purpose, while proteins generally mediate the interactions of the cell

Figure 14.5
Aggregation of phospholipids into the lipid bilayer structure of a cell membrane.

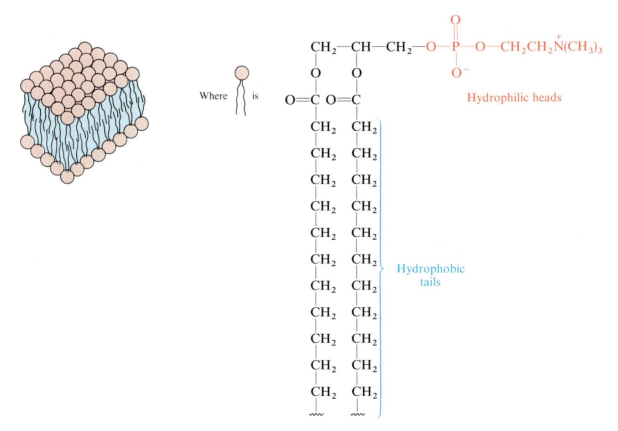

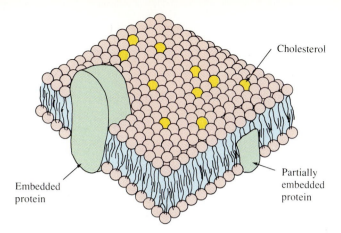

Cholesterol

Embedded
protein

Partially
embedded
protein

Figure 14.6
The fluid-mosaic model of a cell
membrane, showing how protein
globules might be embedded.

with its environment. Some proteins, for example, act as channels to allow specific molecules or ions to enter or leave the cell.

One of the central features of the *fluid-mosaic* cell membrane model shown in Figure 14.6 is that the membrane is *fluid* rather than rigid. As a consequence, the membrane is not easily ruptured, because the lipids in the bilayer can simply flow back together to repair any small hole or puncture. The effect is similar to what's often observed in cooking when a thin film of oil or melted butter lies on top of water. The film can be punctured and broken, but it immediately flows back together when left alone. Still other consequences of bilayer fluidity are that small nonpolar molecules can move easily through the membrane and that individual lipid or protein molecules can diffuse rapidly from one part of the membrane to another.

14.8 STEROIDS: CHOLESTEROL

■ **Steroid** A simple lipid
whose structure is based on a
tetracyclic (four-ring) carbon
skeleton.

In addition to triglycerides and phospholipids, plant and animal cells also contain steroids. A **steroid** is a simple lipid whose structure is based on the tetracyclic (four-ring) system shown in Figure 14.7. Three of the rings are six-membered, while the fourth is five-membered. The rings are designated A, B, C, and D, beginning at the lower left-hand corner, and the carbon atoms are numbered from 1 to 19, beginning in ring A.

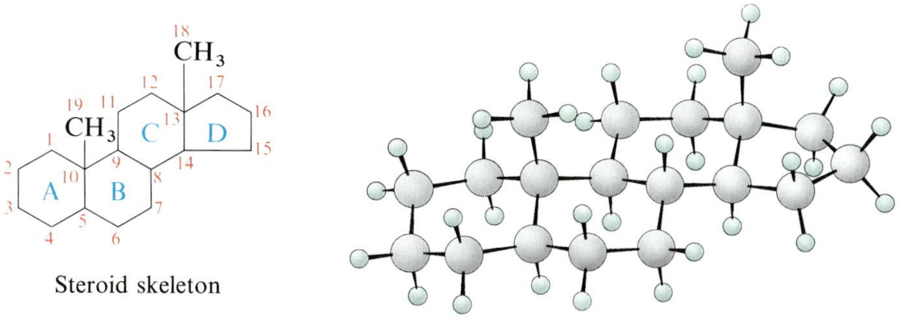

Steroid skeleton

Figure 14.7
The structure of steroids.

Steroids have many diverse roles throughout both the plant and animal kingdoms. Some, such as the plant steroid digitoxigenin isolated from the purple foxglove (*Digitalis purpurea*), are used in medicine as heart stimulants (Figure 14.8). Others, such as cortisone, are hormones; and still others, such as cholic acid found in liver bile, act as emulsifying agents to aid in the digestion of fats.

Digitoxigenin
(from purple foxglove)

Cortisone
(a human hormone)

Cholic acid
(from liver bile)

Cholesterol   Cholesterol, an unsaturated alcohol, is the most abundant animal steroid. In fact, it's been estimated that a 130-lb person has a total of about 175 g cholesterol distributed throughout the body. Much of this cholesterol is esterified with fatty acids, but some is also found as the free alcohol. Gallstones, for example, are nearly pure cholesterol.

Cholesterol, an unsaturated
steroidal alcohol

Figure 14.8
The purple foxglove, source of a steroid used as a heart stimulant. Physicians discovered the beneficial effects of this plant two centuries ago, but it was used in folk medicine long before that.

Cholesterol serves two important functions in the body. First, it is a minor component of cell membranes, where it acts as a regulator of membrane fluidity. Cholesterol is able to fit in between the fatty-acid chains of the lipid bilayer, restricting their motion and making the bilayer more rigid. Second, cholesterol serves as the body's starting material for the synthesis of all other steroids, including all the sex hormones and bile acids. Although news reports sometimes make cholesterol sound as if it were dangerous, there would be no life without it.

The human body obtains its cholesterol both by synthesis in the liver and by ingestion of food. Even on a strict no-cholesterol diet, the body of an adult is still able to synthesize approximately 800 mg per day.

14.9 STEROID HORMONES

■ **Hormone** Chemical messengers secreted by special glands for mediating biochemical events in target tissues.

As we'll see in Chapter 16, **hormones** are chemical messengers that are secreted by special glands and are circulated through the bloodstream to mediate biochemical events in target tissues. There are two major classes of steroid hormones: the *sex hormones*, which control maturation and reproduction, and the *adrenocortical hormones*, which regulate a variety of metabolic processes.

Sex Hormones. *Testosterone* and *androsterone* are the two most important male sex hormones, or *androgens*. Androgens are responsible both for the development of male secondary sex characteristics during puberty and for promoting tissue and muscle growth. Both are synthesized in the testes from cholesterol.

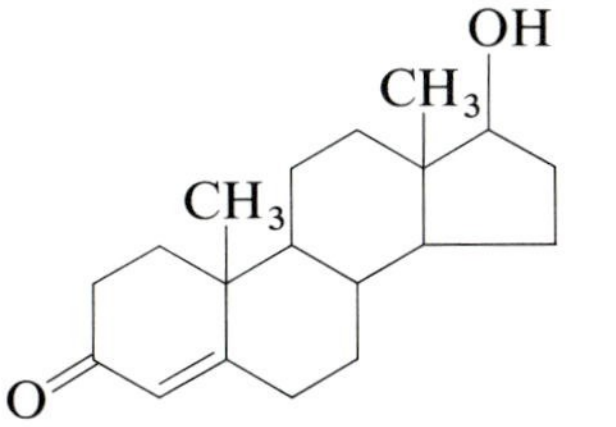

Testosterone

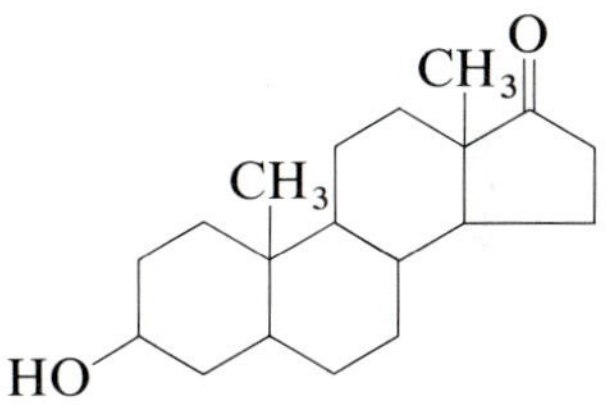

Androsterone

(androgens)

Estrone and *estradiol* are the two most important female sex hormones, or *estrogens*. These substances, which are synthesized in the ovaries from testosterone, are responsible for the development of female secondary sex characteristics and for regulation of the menstrual cycle. Note that both have a benzene-like aromatic A ring. In addition, another kind of sex hormone called a *progestin* is essential for preparing the uterus for implantation of a fertilized ovum during pregnancy. *Progesterone* is the most important progestin.

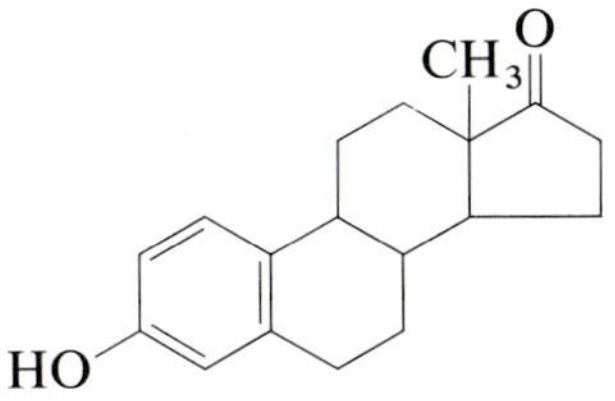

Estrone

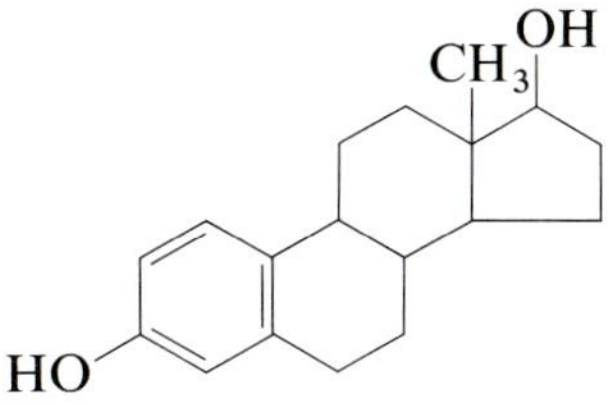

Estradiol

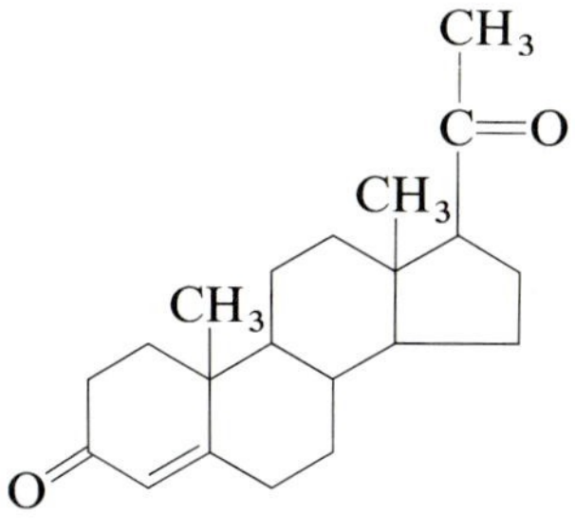

Progesterone (a progestin)

(estrogens)

AN APPLICATION: CHOLESTEROL AND HEART DISEASE

A debate about the relationship between cholesterol and heart disease has been going on for over two decades. Newspaper and television ads trumpet the idea that cholesterol is a killer and that we'd all live much healthier lives if only we switched our brand of cooking oil.

What are the facts? Several points seem clear. The first is that a diet rich in saturated animal fats generally leads to an increase in blood-serum cholesterol, at least in sedentary, overweight people. The second is that a diet lower in saturated fat and higher in PUFAs leads to a lowering of the serum cholesterol level. Since the body manufactures much of its own cholesterol, however, the serum level can't be controlled entirely by diet. (Remember also that since cholesterol is an essential component of cell membranes and the precursor of all other steroids in the body, a certain amount of serum cholesterol is necessary for life.)

The third point is that a level of serum cholesterol greater than 300 mg/dL (a normal value is 150–250 mg/dL) is weakly correlated with an increased incidence of *atherosclerosis*, a form of heart disease in which cholesterol deposits build up on the inner walls of coronary arteries, blocking the flow of blood to the heart muscles (see photo on page 311). The correlation does not hold for everyone, though, and is especially poor for people over 50.

Recent research has shown that a better indication of a person's risk of heart disease comes from a measurement of blood lipoprotein levels. People who have a high serum level of what are called *high-density lipoproteins*, or *HDLs*, appear to have a decreased risk of heart disease. A rule of thumb is that a person's risk drops about 25 percent for each increase of 5 mg/dL in HDL concentration. (Normal values are about 45 mg/dL for men and 55 mg/dL for women, perhaps helping to explain

Serum Lipoproteins

Name	Density (g/mL)	Lipid (%)	Protein (%)
Chylomicrons	< 0.94	98	2
VLDL (very-low-density lipoproteins)	0.940–1.006	90	10
LDL (low-density lipoproteins)	1.006–1.063	75	25
HDL (high-density lipoproteins)	1.063–1.210	60	40

Adrenocortical Hormones. Adrenocortical steroids are secreted from the cortex (outer part) of the adrenal glands, small organs about one-half the size of a thumb, located near the upper end of each kidney. These steroids are of two types, called *mineralocorticoids* and *glucocorticoids*. Mineralocorticoids, such as *aldosterone*, function to control tissue swelling by regulating the delicate cellular salt balance between Na^+ and K^+. Glucocorticoids, such as *hydrocortisone* and its close relative *cortisone*, are involved in the regulation of glucose metabolism and in the control of inflammation. You may well have used a glucocorticoid ointment to bring down the swelling if you've ever been exposed to poison oak or poison ivy.

Aldosterone
(a mineralocorticoid)

Hydrocortisone
(a glucocorticoid)

why women are generally less susceptible than men to heart disease.)

Lipoproteins are complex assemblages of lipids and proteins that serve to transport water-insoluble lipids throughout the body. They can be somewhat arbitrarily divided into four major types, distinguishable by their density (see the table on page 310). Since lipids are generally less dense than proteins, the chylomicrons and the very-low-density lipoproteins (VLDLs) are richer in lipid and lower in protein than the low-density lipoproteins (LDLs) and high-density lipoproteins (HDLs).

The four lipoprotein fractions have different roles. Chylomicrons and VLDL act primarily as carriers of triglycerides from where they are secreted in the intestines to peripheral tissues; LDL and HDL act as carriers of cholesterol to and from the liver. Present evidence suggests that LDL transports cholesterol as its linolenate ester *to* peripheral tissues where it is deesterified and used in membranes, while HDL *removes* cholesterol as its stearate ester from dead or dying cells back to the liver. If LDL delivers more cholesterol than is needed, and if not enough HDL is present to remove it, the excess is deposited in cells and arteries. The higher the HDL level, the less the likelihood of deposits and the lower the risk of heart disease.

If high serum levels of HDL are good, then how do we get them? The answer is simple enough:

As common sense tells you, the most important factor is a generally healthy lifestyle. Obesity, smoking, lack of exercise, and heavy drinking appear to lead to low HDL levels, while regular exercise, a prudent diet, and moderate alcohol consumption leads to high HDL levels. Runners and other endurance athletes, in particular, have HDL levels nearly 50 percent higher than the general population.

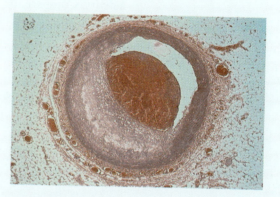

This coronary artery is severely narrowed by atherosclerosis, the build-up of cholesterol deposits in vessel walls (pink). The resulting roughening of artery walls also promotes the formation of clots (dark red) that can cut off blood flow completely, causing a heart attack.

Synthetic Steroids.   In addition to the several hundred known steroids isolated from plants and animals, a great many more have been synthesized in the laboratory in the search for new drugs. The general idea is to start with a natural hormone, carry out chemical modifications of the structure, and then see what biological properties the modified steroid might have.

Among the best-known of the synthetic, nonnaturally occurring steroids are oral contraceptive agents. Most birth-control pills are a mixture of two compounds, a synthetic estrogen such as *ethynylestradiol* and a synthetic progestin such as *norethindrone*. These synthetic steroids appear to function by tricking the body into thinking it's pregnant and therefore temporarily infertile.

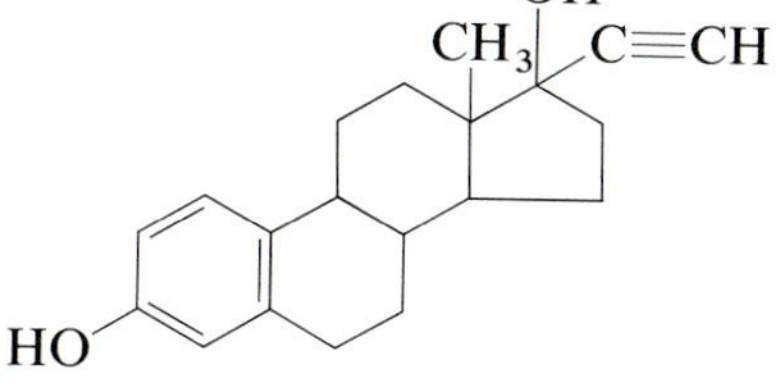

Ethynylestradiol
(a synthetic estrogen)

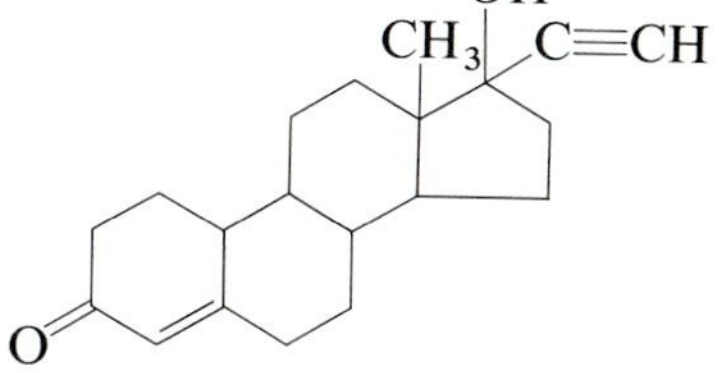

Norethindrone
(a synthetic progestin)

Practice Problems

14.9 Look at the structure of progesterone, and identify all of the functional groups in the molecule.

14.10 Look carefully at the structures of estradiol and ethynylestradiol, and point out the differences. What common structural feature do they share that makes both estrogens?

14.11 *Nandrolone* is an anabolic, or tissue-building, steroid sometimes taken by athletes seeking to build muscle mass. Among its side effects is a high level of androgenic activity. Compare the structures of nandrolone and testosterone, and point out their structural similarities.

Nandrolone, an anabolic steroid

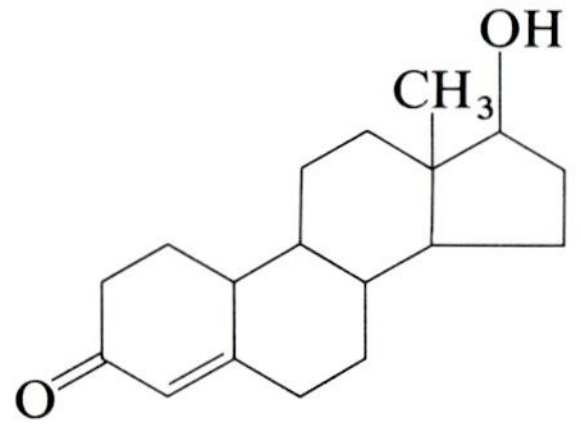

14.10 PROSTAGLANDINS

Few groups of compounds have caused as much excitement among medical researchers in the past 15 years as the prostaglandins. Although the name *prostaglandin* comes from the fact that these compounds were first isolated from the male prostate gland, they have subsequently been shown to be present in small amounts in all body tissues and fluids.

■ **Prostaglandin** A simple lipid containing a cyclopentance ring with two long side chains.

Chemically, **prostaglandins** are C_{20} carboxylic acids that contain a five-membered ring with two long side chains. They differ only in the number of oxygen atoms they contain and in the number of double bonds present. Prostaglandin E_1 (abbreviated PGE_1) and prostaglandin $F_{2\alpha}$ are examples:

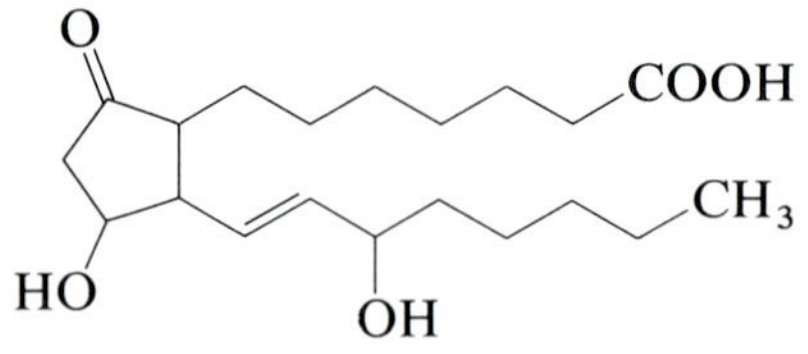

Prostaglandin E_1 (PGE_1)

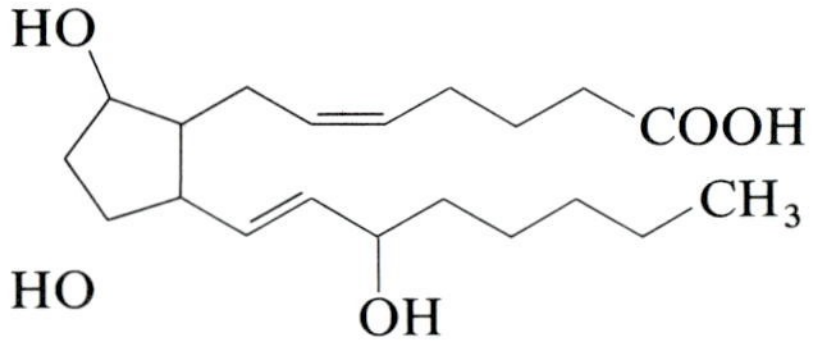

Prostaglandin $F_{2\alpha}$ ($PGF_{2\alpha}$)

The several dozen known prostaglandins have an extraordinary range of biological effects. They can lower blood pressure, influence platelet aggregation during blood clotting, stimulate uterine contractions, and lower the extent of gastric secretions. In addition, they are responsible for some of the pain and swelling that accompanies inflammation. Although the details are not fully understood, much of aspirin's anti-inflammatory action is due to its ability to inhibit prostaglandin production in the body.

In addition to the prostaglandins themselves, closely related compounds have still other important physiological actions. Interest has centered particularly on the leukotrienes (shown below), whose release in the body has been found to trigger the asthmatic response.

Prostaglandins and leukotrienes are synthesized in the body from the 20-carbon unsaturated fatty acid, arachidonic acid. Arachidonic acid, in turn, is synthesized in the body from linolenic acid, helping to explain why linolenic is one of the two essential fatty acids. The simplest way to see the relationships between these different compounds is to rewrite arachidonic acid so that it's "bent" into the proper shape.

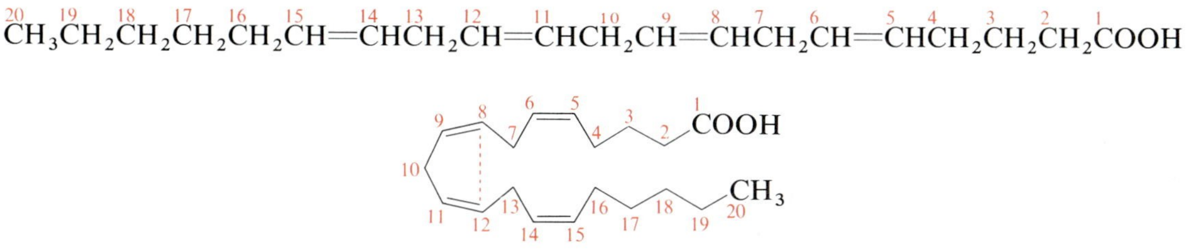

$$\overset{20}{C}H_3\overset{19}{C}H_2\overset{18}{C}H_2\overset{17}{C}H_2\overset{16}{C}H_2\overset{15}{C}H=\overset{14}{C}H\overset{13}{C}H_2\overset{12}{C}H=\overset{11}{C}H\overset{10}{C}H_2\overset{9}{C}H=\overset{8}{C}H\overset{7}{C}H_2\overset{6}{C}H=\overset{5}{C}H\overset{4}{C}H_2\overset{3}{C}H_2\overset{2}{C}H_2\overset{1}{C}OOH$$

Arachidonic acid

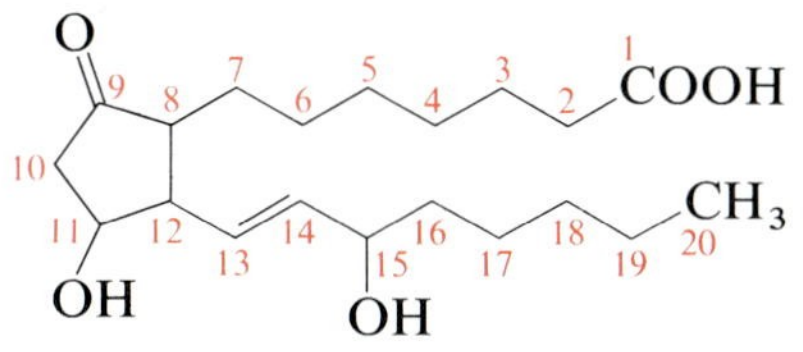

PGE$_1$

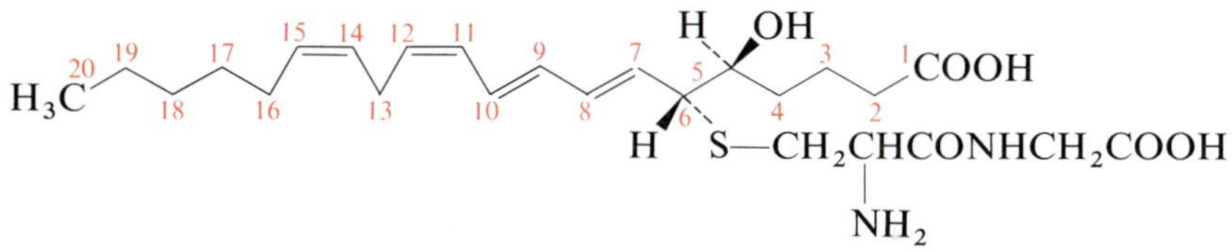

Leukotriene D$_4$

SUMMARY

Lipids are the naturally occurring organic molecules that dissolve when a plant or animal sample is extracted with a nonpolar solvent. Because they are defined by physical property rather than by structure, there are a great many different kinds of lipids.

Waxes, animal fats, and **vegetable oils** are **complex lipids**—compounds that contain ester linkages and that can be hydrolyzed to yield carboxylic acid and alcohol portions. Waxes are esters between long-chain alcohols and long-chain carboxylic acids, whereas fats and oils are **triacylglycerols** or **triglycerides**—that is, triesters of glycerol with three long-chain **fatty acids**. The fatty acids are unbranched, have an even number of carbon atoms, and may be either saturated or unsaturated. Vegetable oils contain a higher proportion of unsaturated fatty acids than animal fats and consequently have lower melting points.

Fats and oils can be **saponified** on treatment with aqueous NaOH or KOH to yield **soap**, a mixture of long-chain fatty acid carboxylate salts. Soaps act as cleansers because their two ends are so different. The negatively charged carboxylate end is ionic and **hydrophilic**, while the hydrocarbon chain end is nonpolar and **hydrophobic**. Thus, a soap molecule is attracted to both grease and water.

INTERLUDE: CHEMICAL COMMUNICATION

To be honest about it, a lot of scientific work is done for the fun of it, without great concern for immediate practical application. Practical applications often *do* follow, but the applications are not themselves the initial goal. Take the study of insects, for example. Have you ever wondered how ants are able to follow a precise path from their nests to a food source, or how male and female of an insect species find each other for mating? Such questions are interesting to the curious, but it's not clear what value the answers might have.

Insects and many other organisms deal with each other and with their surroundings primarily by sending or receiving chemical messages using substances called *semiochemicals*. Among the most important semiochemicals are *pheromones*, chemicals that are released by one member of a species to evoke a specific response in another member of the same species. Ants and termites, for example, lay down trail pheromones that mark a path from nest to food. Similarly, butterflies, moths, and other flying insects release sex pheromones to indicate their location to other interested parties.

A good semiochemical must be volatile, so that it disappears after its job has been done, and must be structurally unique, so that it is species-specific. After all, release of a sex pheromone wouldn't be of much use to a butterfly if it attracted moths, bees, houseflies, and every other flying insect within range. Low-molecular-weight lipids are ideally suited for use as insect semiochemicals, as shown by the following examples:

Gypsy moth sex pheromone

Boll weevil sex pheromone

Termite trail pheromone

Ant trail pheromone

$$CH_3\overset{O}{\underset{}{C}}-OCH_2CH_2\overset{CH_3}{\underset{}{C}HCH_3}$$
Honeybee alarm pheromone

What about practical applications? An understanding of the chemical messages used by insects offers hope for developing environmentally safe, species-specific means of insect control. For example, gypsy moth infestations in hardwood forests of the northeast United States are now being fought with *pheromone traps* rather than with broad-range insecticides. A tiny amount of the gypsy moth sex pheromone is used to lure the moths into a trap, without harming insects of any other species.

Finally, what about *human* sex pheromones? That's another story, but one that is actively being worked on.

These mating gypsy moths probably didn't meet just by chance. Volatile lipids used as pheremones enable insects to attract members of the opposite sex, often from great distances.

Phospholipids are compounds that have an ester link between an alcohol and phosphoric acid. **Phosphoglycerides** contain a glycerol portion bonded to two fatty acids and to one phosphoric acid. The phosphate group, in turn, is also bonded to a small amino alcohol like choline, ethanolamine, or serine. **Sphingolipids**, the second main class of phospholipids, are based on the alcohol sphingosine rather than on glycerol. Phospholipids oriented into a **lipid bilayer** form a major part of the membranes that surround living cells. The ionic phosphate head groups orient on the outside of the bilayer toward the aqueous environment, while the nonpolar hydrocarbon chains cluster in the middle of the bilayer to avoid water.

Steroids and **prostaglandins** are two of the major classes of **simple lipids**, compounds that do not contain ester links and that can't be hydrolyzed. Steroids have a structure with four rings joined together, whereas prostaglandins are C_{20} carboxylic acids that have a single five-membered ring connected to two long side chains.

REVIEW PROBLEMS

Fats, Waxes, and Oils

14.12 What is a lipid?

14.13 Why are there so many different structural kinds of lipids?

14.14 What is the chemical difference between a simple lipid and a complex lipid?

14.15 What is a fatty acid?

14.16 What does it mean to say that fats and oils are triacylglycerols?

14.17 Draw the structure of a typical fat molecule.

14.18 How does animal fat differ chemically from vegetable oil?

14.19 Spermaceti, a fragrant substance isolated from sperm whales, was commonly found in cosmetics until its use was banned in 1976 to protect the whales from extinction. Chemically, spermaceti is cetyl palmitate, the ester of palmitic acid with cetyl alcohol (the straight-chain C_{16} alcohol). Show the structure of spermaceti.

14.20 What kind of lipid is spermaceti (Problem 14.19)—a fat, wax, or steroid?

14.21 Which of the following are complex lipids, and which are simple lipids?
(a) prostaglandin E_1 (b) a lecithin
(c) progesterone   (d) a sphingomyelin
(e) a cerebroside   (f) glyceryl trioleate

14.22 Identify the component parts of each complex lipid listed in Problem 14.21.

14.23 There are two isomeric fat molecules whose components are glycerol, one palmitic acid, and two stearic acid units. Draw the structures of both, and explain how they differ.

14.24 One of the two molecules in Problem 14.23 is *chiral* (Section 13.3). Which molecule is chiral, and why?

14.25 Write the structures of these molecules:
(a) sodium palmitate   (b) decyl oleate
(c) glyceryl palmitodioleate

Chemical Reactions of Lipids

14.26 How would you convert a vegetable oil like soybean oil into a solid cooking fat?

14.27 Show the structures of all products you would obtain by saponification of the following lipid with aqueous KOH. What are the names of the products?

$$CH_2OC(CH_2)_{16}CH_3$$

$$CHOC(CH_2)_7CH=CH(CH_2)_7CH_3$$

$$CH_2OC(CH_2)_7CH=CHCH_2CH=CHCH_2CH=CHCH_2CH_3$$

14.28 If the average molecular weight of a sample of soybean oil is 1500, how many grams of NaOH are needed to saponify 5.0 g of the oil?

14.29 What products would you obtain by treatment of oleic acid with these reagents?
(a) Br_2 (b) H_2, Pd catalyst
(c) CH_3OH, HCl catalyst

Phospholipids, Glycolipids, and Cell Membranes

14.30 What is the difference between a fat and a phospholipid?

14.31 How do sphingomyelins and cerebrosides differ structurally?

14.32 Why is it that phosphoglycerides rather than triglycerides are found in cell membranes?

14.33 Why are phosphoglycerides more soluble in water than triglycerides?

14.34 How does a soap micelle differ from a membrane bilayer?

14.35 What constituents besides phospholipids are present in a cell membrane?

14.36 What are the names of the two different kinds of sphingosine-based lipids?

14.37 Show the structure of a cerebroside made up of D-galactose, sphingosine, and myristic acid.

14.38 The *plasmalogens* are a group of lipids found in nerve and muscle cells. Look carefully at the following plasmalogen structure, and tell how it differs from lecithins and cephalins.

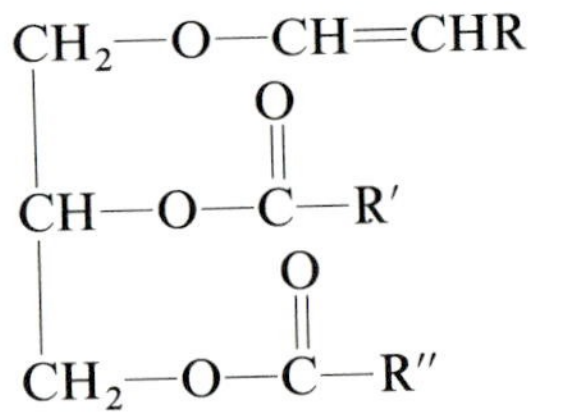

$$CH_2-O-CH=CHR$$
$$\begin{array}{c}O\\ \|\\ CH-O-C-R'\end{array}$$
$$\begin{array}{c}O\\ \|\\ CH_2-O-C-R''\end{array}$$

A plasmalogen

14.39 *Cardiolipin*, a compound found in heart muscle has the following structure:

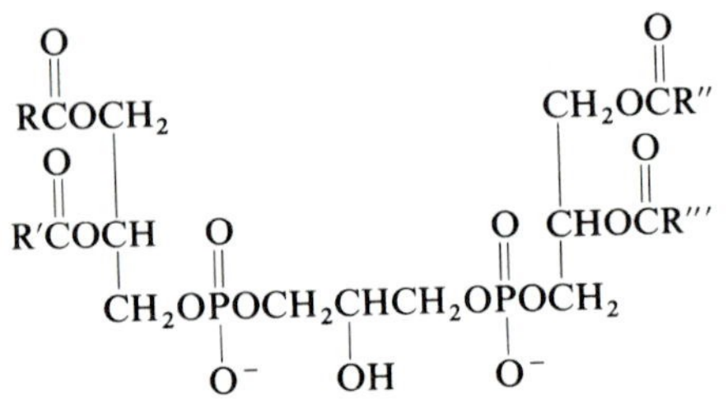

Cardiolipin

What products would be formed if all ester bonds in the molecule were saponified by treatment with aqueous NaOH?

Steroids and Prostaglandins

14.40 What functional groups are present in the two sex hormones estradiol and testosterone?

14.41 How do the sex hormones estradiol and testosterone differ structurally?

14.42 The female sex hormone estrone has four chiral centers. Identify them.

14.43 The concentration of cholesterol in the blood serum of a normal adult is approximately 200 mg/dL. How much total blood cholesterol does a person with a blood volume of 5.75 L have?

14.44 What function does cholesterol serve in cell membranes?

14.45 Draw the products you would expect from treatment of cholesterol with these reagents:
(a) Br_2 (b) H_2, Pd catalyst
(c) [O], an oxidizing agent

14.46 Diethylstilbestrol (DES) exhibits estradiol-like hormonal activity even though it is not itself a steroid. Once used widely as an animal food additive, DES has been implicated as a causative agent in several types of cancer. Show how DES can be drawn so that it is structurally similar to estradiol.

Diethylstilbestrol

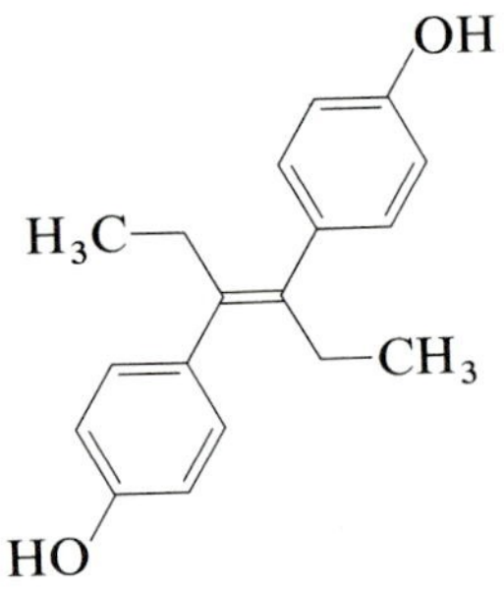

14.47 Thromboxane A_2 is a lipid involved in the blood-clotting process. To what category of lipids does thromboxane A_2 belong?

Thromboxane A_2

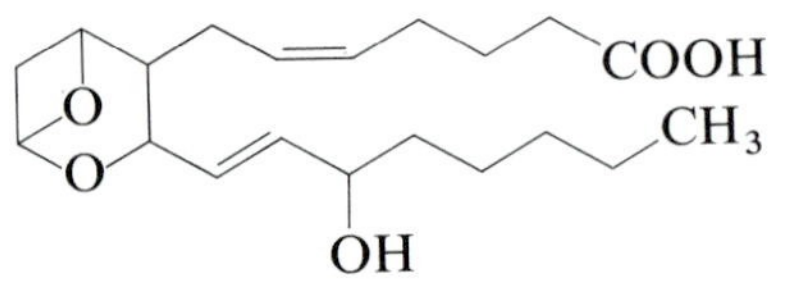

14.48 What fatty acid do you think serves as a biological precursor to thromboxane A_2 (Problem 14.47)?

CHAPTER 15

The Molecules of Life: Amino Acids and Proteins

The horns of this Alaskan Dall ram are made of a tough, insoluble fibrous protein. So are his hooves, fur, and much of his skin. Beneath the skin, his muscles, tendons, and ligaments are also largely protein. The hemoglobin that carries oxygen in his blood, the antibody molecules that protect him from disease, and the enzymes that carry out the chemical reactions needed to keep him alive are all proteins too. You'll meet many of these versatile molecules in this chapter.

The word *protein* is familiar to everyone. Taken from the Greek *proteios*, meaning "primary," the name protein is an apt description for a group of biological molecules that are of primary importance to all living organisms. Without proteins, there could be no life. Approximately 50 percent of your body's dry weight is protein; virtually every molecule in your body was made by proteins; and every reaction that occurs in your body is catalyzed by proteins. In fact, a human body contains well over *100,000* different kinds of proteins.

Proteins are of many different types and have many different biological functions. Some proteins, such as the *keratin* in skin, hair, and fingernails, and the *collagen* in connective tissue, serve a structural purpose. Other proteins, such as the *insulin* that controls glucose metabolism, serve as hormones to regulate specific body processes. And still other proteins, such as *DNA polymerase*, serve as enzymes, the biological catalysts that carry out all body chemistry.

In this chapter, we'll look at the following questions about proteins:

1. **What are the structures of amino acids?** The goal: you should learn what amino acids are and the structures of some simple ones.

2. **How are proteins classified?** The goal: you should learn the differences among major classes of proteins.

3. **What are the primary structures of proteins?** The goal: you should learn how amino acids join together to make proteins.

4. **What are the secondary, tertiary, and quaternary structures of proteins?** The goal: you should learn the meaning of these terms and the most common kinds of secondary structure for proteins.

5. **What chemical properties do proteins have?** The goal: you should learn the simple chemical behavior of proteins.

15.1 AN OVERVIEW OF PROTEIN STRUCTURE

■ **Protein** A large biological molecule made of many amino acids linked by amide bonds.

■ **Amino acids** Molecules that contain both an amino group and a carboxyl group; the building blocks of proteins.

Regardless of their differing biological functions, all proteins are chemically similar. **Proteins** are made up of many amino acids, linked together like building blocks to form a long chain. As their name implies, **amino acids** are molecules that contain two functional groups, an acidic carboxyl group ($-COOH$) and a basic amino group ($-NH_2$). Glycine is the simplest example of the amino acid structure.

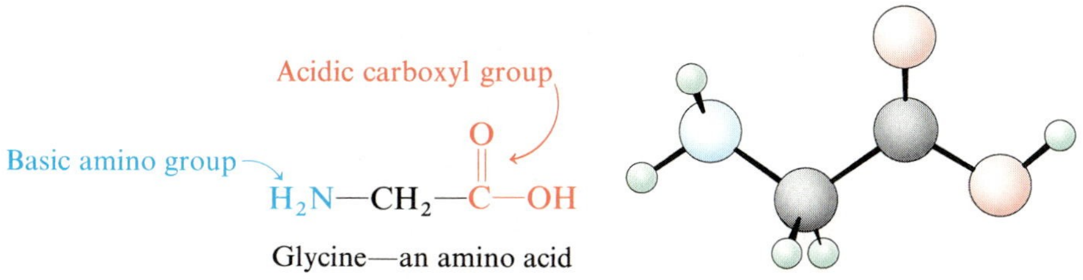

Acidic carboxyl group

Basic amino group

$$H_2N—CH_2—\overset{\displaystyle O}{\overset{\displaystyle \|}{C}}—OH$$

Glycine—an amino acid

Peptide bond An amide bond that links two amino acids together.

Two or more amino acids can link together by forming amide bonds (Section 12.9), or **peptide bonds**. A *dipeptide* results from the linking together of two amino acids by formation of a peptide bond between the —NH$_2$ group of one and the —COOH group of the second; a *tripeptide* results from linkage of three amino acids via two peptide bonds; and so on. Any number of amino acids can link together to form a long chain. For classification purposes, chains with fewer than 50 amino acids are called *polypeptides*, while the term *protein* is usually reserved for larger chains.

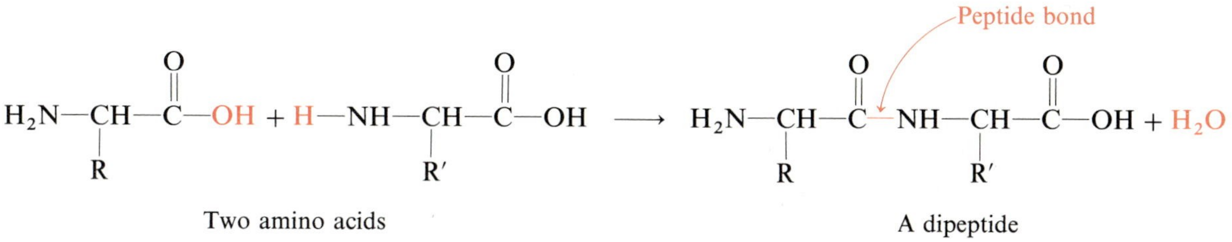

Peptide bond

$$H_2N—\underset{R}{CH}—\overset{O}{\overset{\|}{C}}—OH + H—NH—\underset{R'}{CH}—\overset{O}{\overset{\|}{C}}—OH \longrightarrow H_2N—\underset{R}{CH}—\overset{O}{\overset{\|}{C}}—NH—\underset{R'}{CH}—\overset{O}{\overset{\|}{C}}—OH + H_2O$$

Two amino acids   A dipeptide

(R and R' may be the same or different)

15.2 STRUCTURES OF AMINO ACIDS

Alpha (α) amino acid An amino acid in which the amino group is bonded to the carbon atom next to the —COOH group.

There are 20 different amino acids commonly found in proteins. As shown in Table 15.1, all 20 are **alpha (α) amino acids**; that is, the amino group in each is connected to the carbon atom *alpha to* (next to) the carboxylic acid group. The 20 amino acids differ only in the nature of the R group (called the *side chain*) attached to the α carbon.

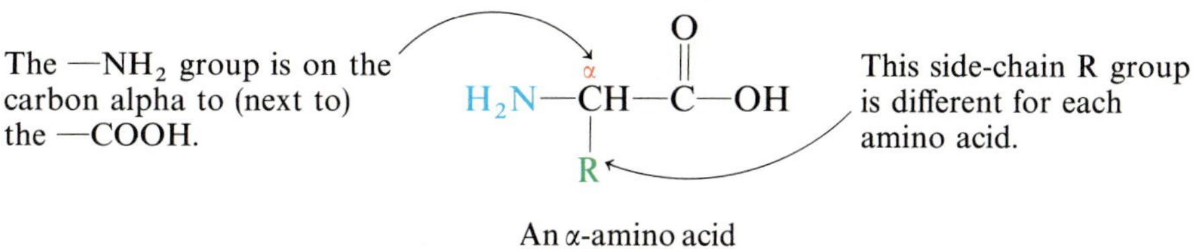

The —NH$_2$ group is on the carbon alpha to (next to) the —COOH.

$$H_2N—\underset{R}{\overset{\alpha}{CH}}—\overset{O}{\overset{\|}{C}}—OH$$

This side-chain R group is different for each amino acid.

An α-amino acid

Nineteen of the twenty common amino acids are primary amines, R—NH$_2$, but the remaining one (proline) is a secondary amine whose nitrogen and α-carbon atoms are joined together in a five-membered ring. Each amino acid is usually referred to by a three-letter shorthand code, such as Ala (alanine), Gly (glycine), Pro (proline), and so on. In addition, a new space-saving one-letter code is currently gaining popularity. These codes are shown in Table 15.1.

Table 15.1   Structures of the 20 Common Amino Acids[a]

Name	Abbreviations	Molecular Weight	Structure	Isoelectric Point
Neutral Amino Acids				
Alanine	Ala (A)	89	$CH_3-\overset{\overset{\displaystyle NH_2}{|}}{\underset{\underset{\displaystyle H}{|}}{C}}-COOH$	6.0
Asparagine	Asn (N)	132	$H_2N-\overset{\overset{\displaystyle O}{\|}}{C}CH_2-\overset{\overset{\displaystyle NH_2}{|}}{\underset{\underset{\displaystyle H}{|}}{C}}-COOH$	5.4
Cysteine	Cys (C)	121	$HSCH_2-\overset{\overset{\displaystyle NH_2}{|}}{\underset{\underset{\displaystyle H}{|}}{C}}-COOH$	5.0
Glutamine	Gln (Q)	146	$H_2N-\overset{\overset{\displaystyle O}{\|}}{C}CH_2CH_2-\overset{\overset{\displaystyle NH_2}{|}}{\underset{\underset{\displaystyle H}{|}}{C}}-COOH$	5.7
Glycine	Gly (G)	75	$H-\overset{\overset{\displaystyle NH_2}{|}}{\underset{\underset{\displaystyle H}{|}}{C}}-COOH$	6.0
Isoleucine	Ile (I)	131	$CH_3CH_2\overset{\overset{\displaystyle CH_3}{|}}{CH}-\overset{\overset{\displaystyle NH_2}{|}}{\underset{\underset{\displaystyle H}{|}}{C}}-COOH$	6.0
Leucine	Leu (L)	131	$CH_3\overset{\overset{\displaystyle CH_3}{|}}{CH}CH_2-\overset{\overset{\displaystyle NH_2}{|}}{\underset{\underset{\displaystyle H}{|}}{C}}-COOH$	6.0
Methionine	Met (M)	149	$CH_3SCH_2CH_2-\overset{\overset{\displaystyle NH_2}{|}}{\underset{\underset{\displaystyle H}{|}}{C}}-COOH$	5.7
Phenylalanine	Phe (F)	165	$C_6H_5-CH_2-\overset{\overset{\displaystyle NH_2}{|}}{\underset{\underset{\displaystyle H}{|}}{C}}-COOH$	5.5
Proline	Pro (P)	115	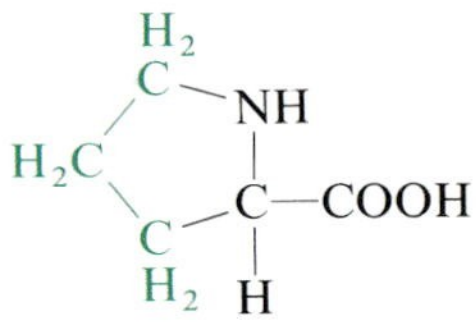 (pyrrolidine ring structure) $-COOH$	6.3

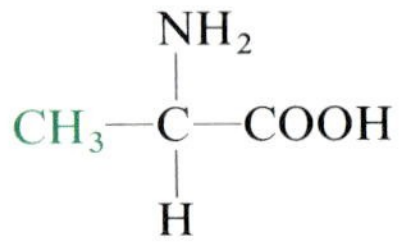

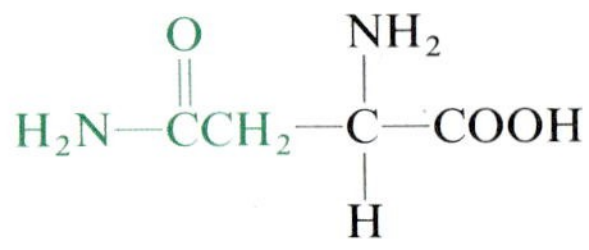

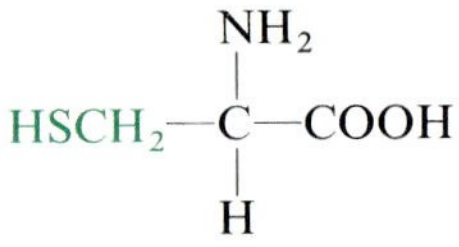

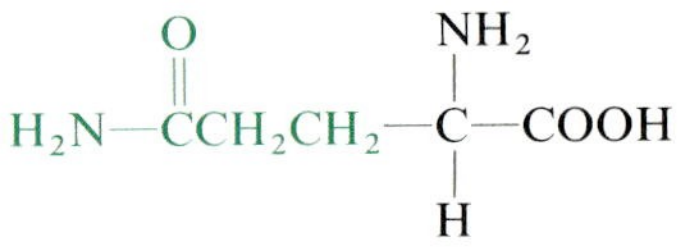

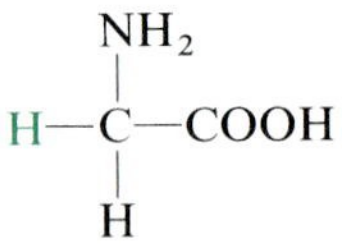

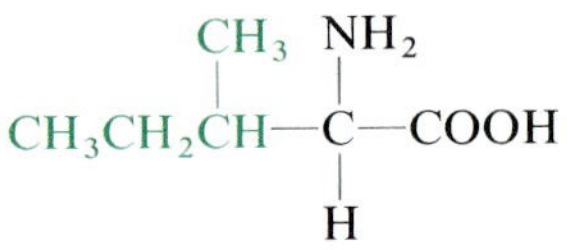

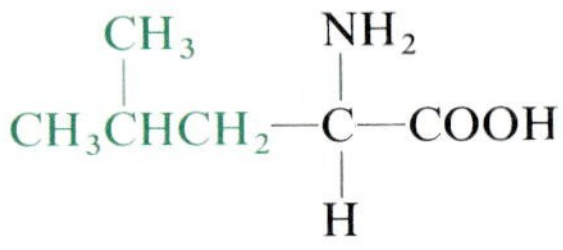

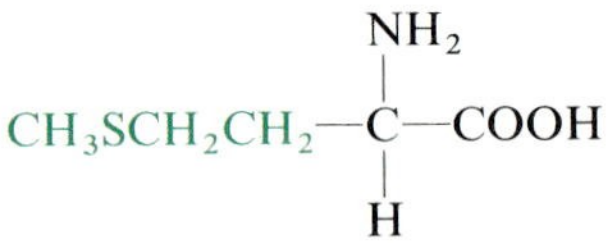

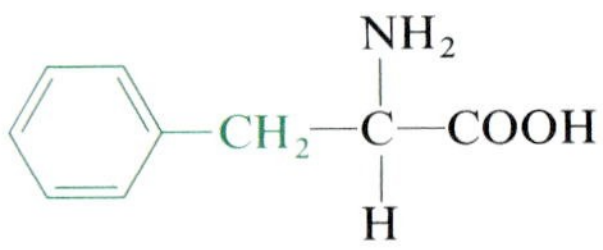

Table 15.1 *(Continued)*

Name	Abbreviations	Molecular Weight	Structure	Isoelectric Point
Serine	Ser (S)	105	$\underset{\underset{H}{\mid}}{\overset{\overset{NH_2}{\mid}}{HOCH_2-C-COOH}}$	5.7
Threonine	Thr (T)	119	$\underset{\underset{H}{\mid}}{\overset{\overset{OH\ \ NH_2}{\mid\ \ \ \ \mid}}{CH_3CH-C-COOH}}$	5.6
Tryptophan	Trp (W)	204	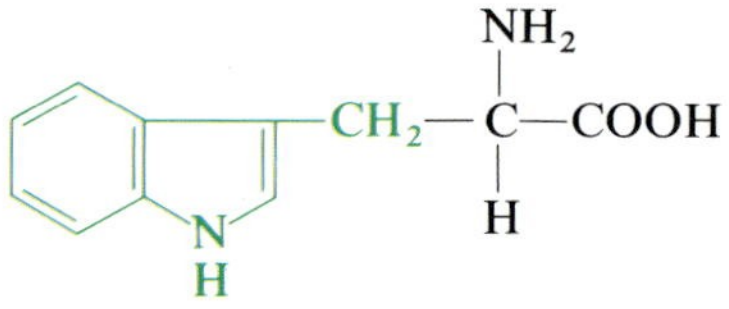 $\underset{\underset{H}{\mid}}{\overset{\overset{NH_2}{\mid}}{CH_2-C-COOH}}$ (indole ring)	5.9
Tyrosine	Tyr (Y)	181	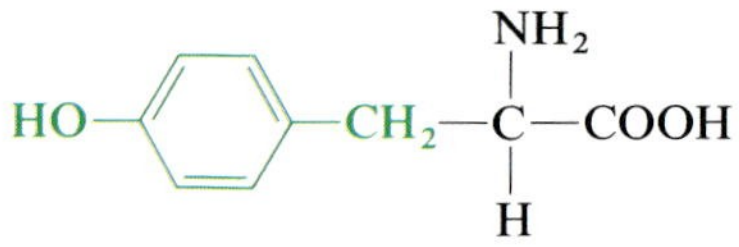 $HO-$(benzene ring)$-\underset{\underset{H}{\mid}}{\overset{\overset{NH_2}{\mid}}{CH_2-C-COOH}}$	5.7
Valine	Val (V)	117	$\underset{\underset{H}{\mid}}{\overset{\overset{CH_3\ \ NH_2}{\mid\ \ \ \ \mid}}{CH_3CH-C-COOH}}$	6.0
Acidic Amino Acids				
Aspartic acid	Asp (D)	133	$\underset{}{\overset{\overset{O}{\parallel}}{HO-C}}CH_2-\underset{\underset{H}{\mid}}{\overset{\overset{NH_2}{\mid}}{C}}-COOH$	3.0
Glutamic acid	Glu (E)	147	$\underset{}{\overset{\overset{O}{\parallel}}{HO-C}}CH_2CH_2-\underset{\underset{H}{\mid}}{\overset{\overset{NH_2}{\mid}}{C}}-COOH$	3.2
Basic amino acids				
Arginine	Arg (R)	174	$\underset{}{\overset{\overset{N}{\parallel}}{H_2N-C}}-NHCH_2CH_2CH_2-\underset{\underset{H}{\mid}}{\overset{\overset{NH_2}{\mid}}{C}}-COOH$	10.8
Histidine	His (H)	155	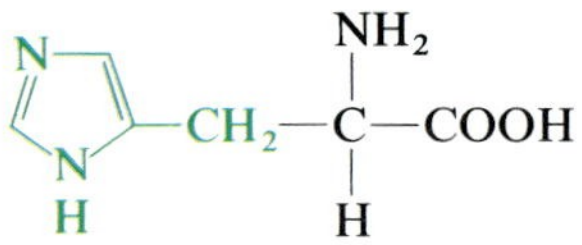 (imidazole ring)$-CH_2-\underset{\underset{H}{\mid}}{\overset{\overset{NH_2}{\mid}}{C}}-COOH$	7.6
Lysine	Lys (K)	146	$H_2NCH_2CH_2CH_2CH_2-\underset{\underset{H}{\mid}}{\overset{\overset{NH_2}{\mid}}{C}}-COOH$	9.7

[a] Amino acids essential to the human diet are shown in red.

The 20 common amino acids can be classified as either *neutral*, *basic*, or *acidic*, depending on the structure of their side chains. Fifteen of the twenty have neutral side chains, two (aspartic acid and glutamic acid) have an extra carboxylic acid group in their side chains, and three (lysine, arginine, and histidine) have basic amine groups in their side chains.

■ **Essential amino acid** One of ten amino acids that cannot be synthesized by the body and so must be obtained in the diet.

All 20 amino acids are needed to make proteins, but our bodies can synthesize only 10. The remaining 10 are called **essential amino acids** because they must be obtained from dietary sources. Failure to receive an adequate dietary supply of the essential amino acids leads to retarded growth and development in children and to disease and body deterioration in adults.

Practice Problems

15.1 How many of the 20 common amino acids contain an aromatic ring? How many contain sulfur? How many are alcohols? How many have alkyl-group side chains?

15.2 Draw alanine, and show the tetrahedral geometry of its α carbon.

15.3 Is alanine chiral (Section 13.3)? Explain.

15.3 HANDEDNESS OF AMINO ACIDS

Look at the amino-acid structures in Table 15.1. Are any of them chiral? We saw in Section 13.3 that chiral compounds are those that have a carbon atom bonded to four different atoms or groups of atoms. Such compounds lack a plane of symmetry and can exist in either a right-handed D form or a left-handed L form. Of the 20 common amino acids, 19 are chiral because they have four different groups bonded to their α carbons, —H, —NH_2, —COOH, and —R (the side chain). Only glycine, H_2NCH_2COOH, is achiral. Because of their structural similarity to L sugars (Section 13.3), the naturally occurring α-amino acids are often referred to as L-*amino acids*. D-amino acids occur very rarely in nature.

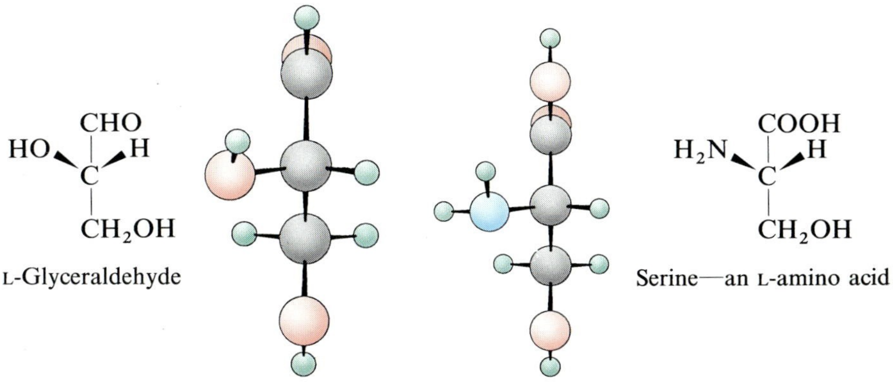

$$\underset{\text{L-Glyceraldehyde}}{\overset{\displaystyle \text{CHO}}{\text{HO}\diagdown \underset{\displaystyle \text{CH}_2\text{OH}}{\overset{|}{\text{C}}} \diagup\text{H}}}$$

$$\underset{\text{Serine—an L-amino acid}}{\overset{\displaystyle \text{COOH}}{\text{H}_2\text{N}\diagdown \underset{\displaystyle \text{CH}_2\text{OH}}{\overset{|}{\text{C}}} \diagup\text{H}}}$$

Practice Problem

15.4 Two of the twenty common amino acids have *two* chiral carbon atoms in their structures. Identify them.

15.4 DIPOLAR STRUCTURE OF AMINO ACIDS

■ **Zwitterion** A neutral compound that contains both + and − charges in its structure.

Amino acids contain both acidic and basic groups in the same molecule. They therefore undergo an *internal* acid–base reaction and exist primarily as dipolar ions, called **zwitterions** (German *zwitter*, "hybrid").

$$H_2N-\underset{R}{CH}-\overset{O}{\overset{\|}{C}}-OH \; \rightleftharpoons \; H_3\overset{+}{N}-\underset{R}{CH}-\overset{O}{\overset{\|}{C}}-O^-$$

Amino acid—nonpolar form　　　　Amino acid—zwitterion form

Amino-acid zwitterions have many of the physical properties we associate with salts (Section 7.4). Thus, amino acids are crystalline, have high melting points, and are soluble in water but not in hydrocarbon solvents. In addition, amino acids can react either as acids or as bases depending on the circumstances. In acid solution at low pH, amino acids *accept* a proton on the —COO^- group to yield a cation; in base solution at high pH, amino acids *lose* a proton from the —NH_3^+ group to yield an anion.

■ **Isoelectric point** The pH at which a large sample of amino acid molecules has equal numbers of + and − charges.

The exact structure of an amino acid at any given time depends on the particular amino acid and on the pH of the medium. The pH at which the numbers of positive and negative charges in a large sample are equal and an amino acid exists predominantly in its neutral dipolar form is called the amino acid's **isoelectric point**. As indicated in Table 15.1, the 15 amino acids with neutral side chains have isoelectric points near neutrality in the pH range 5.0–6.5; the two amino acids with acidic side chains have isoelectric points at more acidic (lower) pH values; and the three amino acids with basic side chains have isoelectric points at more basic (higher) pH values.

$$H_3\overset{+}{N}-\underset{R}{CH}-\overset{O}{\overset{\|}{C}}-O-H \qquad H_3\overset{+}{N}-\underset{R}{CH}-\overset{O}{\overset{\|}{C}}-O^- \qquad H_2N-\underset{R}{CH}-\overset{O}{\overset{\|}{C}}-O^-$$

Predominant form　　　　Predominant form　　　　Predominant form
at low pH (acidic)　　　　at isoelectric point　　　　at high pH (basic)

Practice Problem　　　**15.5**　　Look up the structure of valine in Table 15.1, and draw it at low pH, neutral pH, and high pH.

15.5 STRUCTURE OF PEPTIDES AND PROTEINS

■ **Residue** An alternative name for an amino-acid unit in a polypeptide or protein.

Peptides and proteins are amino-acid polymers in which the individual amino acids, called **residues**, are linked together by peptide (amide) bonds. An amino group from one residue forms a peptide bond with the carboxylic acid group of a second residue; the amino group of the second forms a peptide bond with the carboxylic acid of a third; and so on. The repeating chain of amide linkages that the side chains are attached to is called the *backbone*.

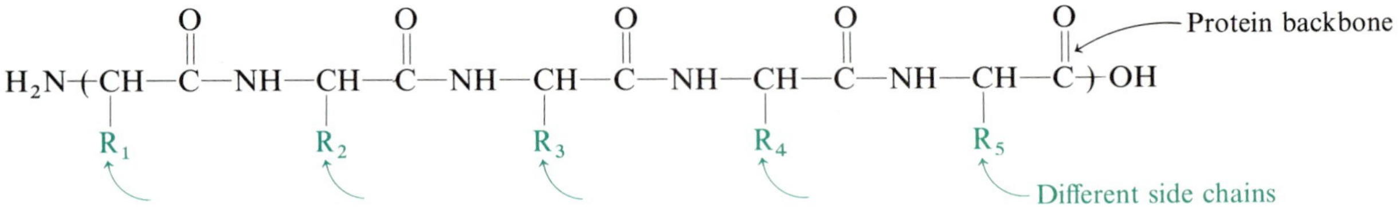

As a simple example, the dipeptide alanylserine results when a peptide bond is made between the alanine —COOH and the serine —NH$_2$:

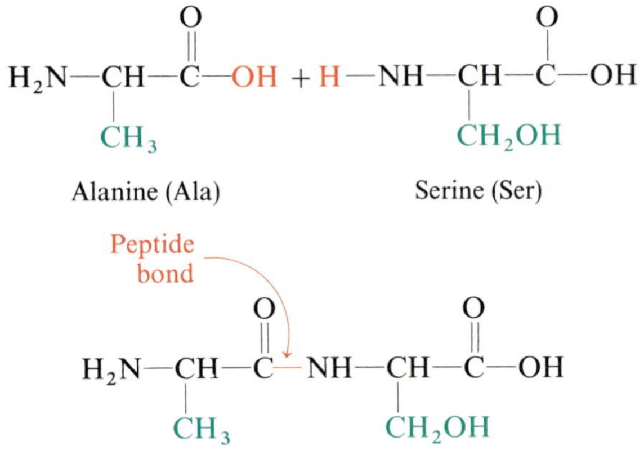

Alanine (Ala)   Serine (Ser)

Peptide
bond

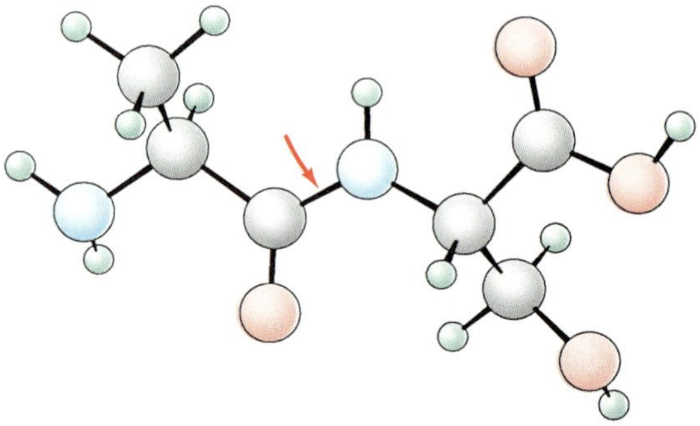

Alanylserine (H$_2$N-Ala-Ser-COOH)

Note that two *different* dipeptides can result from reaction of alanine with serine depending on which —COOH group reacts with which —NH$_2$ group. If the alanine —NH$_2$ reacts with the serine —COOH, serylalanine results.

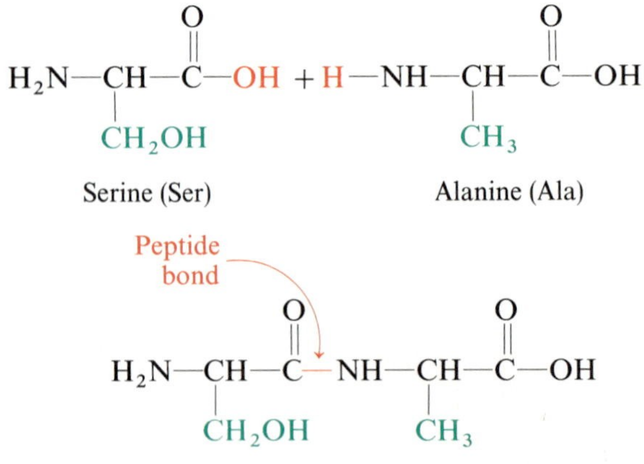

Serine (Ser)   Alanine (Ala)

Peptide
bond

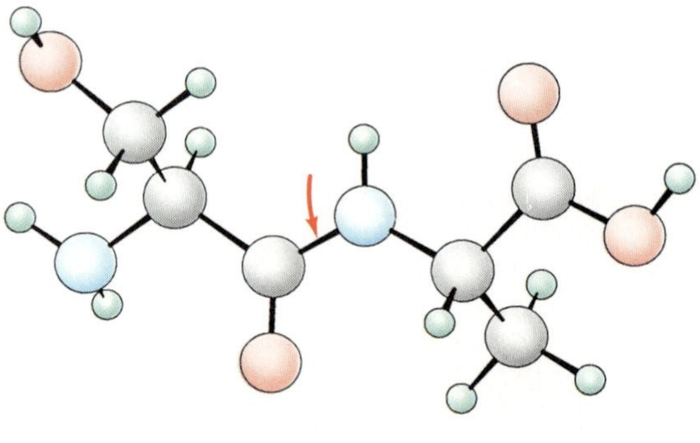

Serylalanine (H$_2$N-Ser-Ala-COOH)

By convention, peptides and proteins are always written with the *N-terminal amino acid* (the one with the free —NH$_2$ group) on the left, and the *C-terminal amino acid* (the one with the free —COOH group) on the right. The name of the peptide is then indicated by using the three-letter abbreviations listed in Table 15.1. An H$_2$N— is sometimes added to the abbreviation of the leftmost amino acid, and a —COOH is sometimes added to the abbreviation of the rightmost amino acid. Thus, alanylserine can be abbreviated H$_2$N—Ala—Ser—COOH, and serylalanine can be similarly abbreviated H$_2$N—Ser— Ala—COOH.

The number of possible isomeric peptides increases rapidly as the number of amino acid residues increases. There are six ways in which three amino acids can be joined, more than 40,000 ways in which the eight amino acids of the blood-pressure–regulating hormone angiotensin II can be joined (Figure 15.1, page 326), and an inconceivably vast number of ways in which the *1800* amino acids in myosin, the major component of muscle filaments, can be arranged (approximately 10^{1800}!—a far larger number than there are atoms in the universe).

Solved Problem 15.1   Draw the structure of the dipeptide H$_2$N—Ala—Gly—COOH.

Solution   First look up the names and structures of the two amino acids, Ala (alanine) and Gly (glycine).

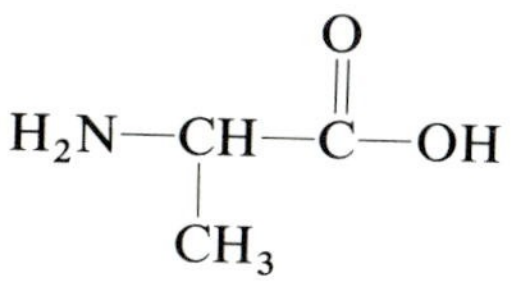

$$H_2N-CH-\overset{\displaystyle O}{\overset{\displaystyle \|}{C}}-OH \qquad\qquad H_2N-CH_2-\overset{\displaystyle O}{\overset{\displaystyle \|}{C}}-OH$$
$$\underset{\displaystyle CH_3}{|}$$

Alanine (Ala) 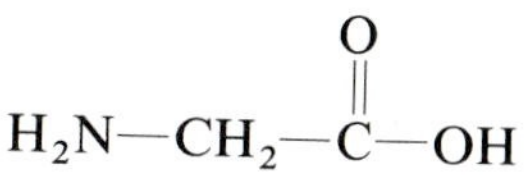   Glycine (Gly)

Since alanine is N-terminal and glycine is C-terminal, H$_2$N—Ala—Gly—COOH must have a peptide bond between the alanine —COOH and the glycine —NH$_2$.

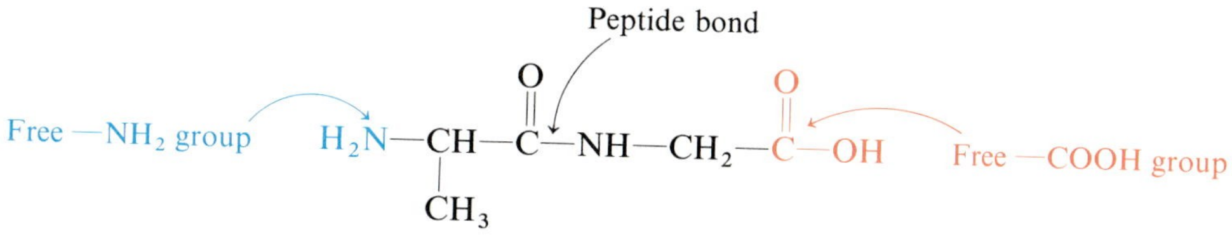

Peptide bond

Free —NH$_2$ group $H_2N-CH-\overset{\displaystyle O}{\overset{\displaystyle \|}{C}}-NH-CH_2-\overset{\displaystyle O}{\overset{\displaystyle \|}{C}}-OH$ Free —COOH group
$\qquad\qquad\qquad\qquad\quad \underset{\displaystyle CH_3}{|}$

Alanine   H$_2$N-Ala-Gly-COOH   Glycine

Practice Problems   **15.6**   Use the three-letter shorthand notations to name the two isomeric dipeptides that could be made from valine and cysteine. Draw the structure of each.

15.7   Name the six tripeptides that contain valine, tyrosine, and glycine.

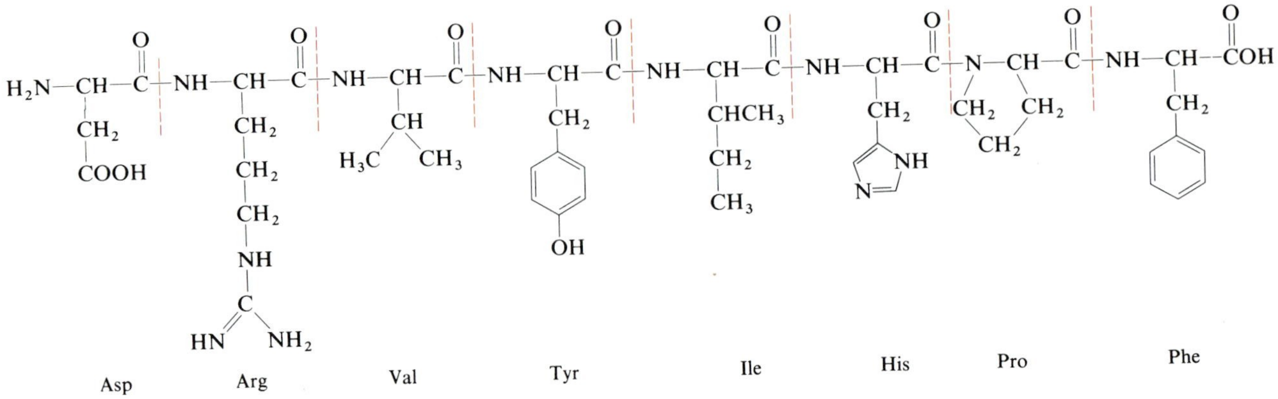

Figure 15.1
The structure of angiotensin II, a blood-pressure–regulating hormone present in blood plasma.

15.6 DISULFIDE BRIDGES IN PROTEINS

■ **Disulfide bridge** An S—S bond formed between two cysteine residues that can join two peptide chains together or cause a loop in a peptide chain.

Although the peptide bond is the fundamental link between amino-acid residues, a second kind of covalent bonding sometimes occurs when a **disulfide bridge** forms between two cysteine residues. Recall from Section 10.7 that thiols, R—S—H, can react with mild oxidizing agents to yield disulfides, R—S—S—R. If the thiol groups are on the side chains of two cysteine amino acids, then disulfide bond formation will link the cysteine residues together. The linkage is sometimes indicated by writing CyS with a capital "S" (for sulfur) and then drawing a line from one CyS to the other: CyS CyS.

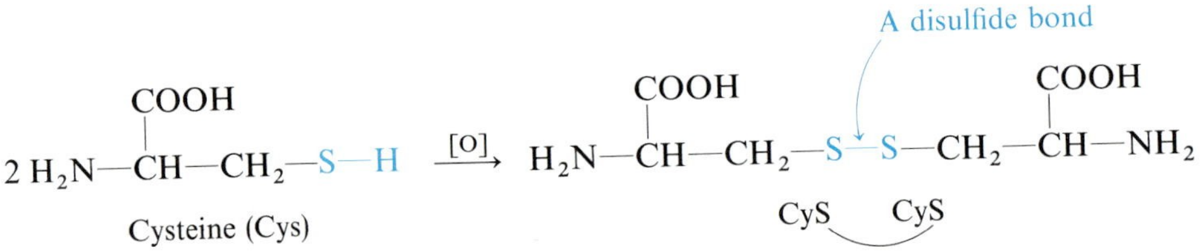

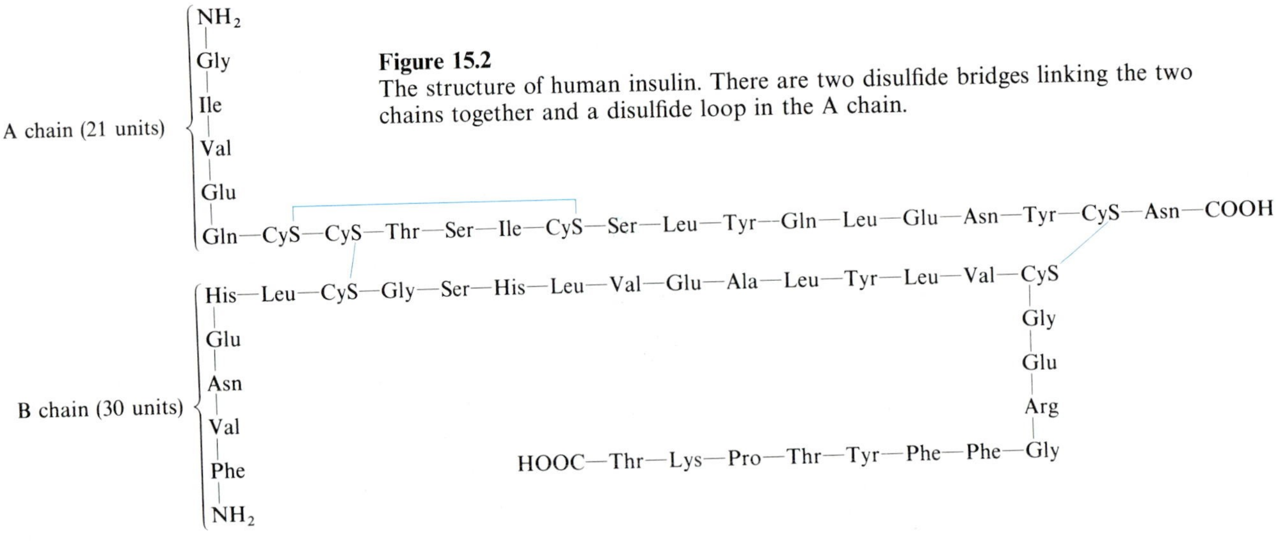

Figure 15.2
The structure of human insulin. There are two disulfide bridges linking the two chains together and a disulfide loop in the A chain.

AN APPLICATION: PROTEIN AND NUTRITION

Protein is often thought of as "muscle food" because of its popularity with weightlifters and other athletes. Although it's true that protein is needed for growing muscles, it's also true that *everyone* needs substantial amounts of protein in their diet. Children need large amounts of protein for proper growth, and adults need protein to replace what is lost each day by normal biochemical reactions in the body.

Our requirement for dietary protein is due to the fact that our bodies can synthesize only 10 of the 20 common amino acids from simple precursor molecules; the remaining 10 amino acids must be obtained from the diet by digestion of edible proteins. The first table shows the estimated amino-acid requirements of an infant and an adult.

Not all foods are equally good sources of protein. A high-quality food source is one that provides the 10 essential amino acids in sufficient amount to meet our minimum daily needs. Most meat and dairy products meet this requirement, but many vegetable sources such as wheat and corn don't. The term *incomplete protein* refers to protein in which one or more of the 10 essential amino acids is present in too low a quantity to sustain the growth of laboratory animals. For example, wheat is low in lysine, corn low in both lysine and tryptophan.

Using incomplete food as the sole source of protein can cause nutritional deficiencies, particularly in growing children. Vegetarians must therefore be certain to adopt a varied diet that provides proteins from several different sources. Thus, legumes and nuts are particularly valuable in overcoming the deficiencies of wheat and grains. Some of the limiting amino acids found in various foods are listed in the second table.

Estimated Amino-Acid Requirements

Amino Acid	Daily Requirement (mg/kg body weight)	
	Infant	Adult
Arginine	?	?
Histidine	33	?
Isoleucine	83	12
Leucine	135	16
Lysine	99	12
Cysteine + Methionine	49	10
Tyrosine + Phenylalanine	141	16
Threonine	68	8
Tryptophan	21	3
Valine	92	14

Limiting Amino Acids in Some Foods

Food Category	Limiting Amino Acid
Wheat, other grains	Lysine, threonine
Peas, beans, other legumes	Methionine, tryptophan
Nuts and seeds	Lysine
Leafy green vegetables	Methionine

If a disulfide bond forms between two cysteine residues in different peptide chains, the otherwise separate chains are linked together. Alternatively, if a disulfide bond forms between two cysteines in the same chain, a loop is formed in the chain. Insulin, for example, consists of two polypeptide chains linked together by disulfide bridges in two places. One of the chains also has a loop in it caused by a third disulfide bridge (Figure 15.2).

■ **Simple protein**   A protein that yields only amino acids when hydrolyzed.

■ **Conjugated protein**   A protein that yields one or more other compounds in addition to amino acids when hydrolyzed.

15.7 CLASSIFICATION OF PROTEINS

Proteins can be classified in several different ways. The simplest way is to group them according to composition. **Simple proteins**, such as blood-serum albumin, are those that yield only amino acids and no other compounds on hydrolysis. **Conjugated proteins**, which are far more common than simple proteins, yield

Table 15.2 Some Different Kinds of Conjugated Proteins

Name	Composition
Glycoproteins	Proteins bonded to a carbohydrate. Cell membranes have a glycoprotein coating.
Lipoproteins	Proteins bonded to fats and oils (lipids). These proteins transport cholesterol and other fats through the body.
Metalloproteins	Proteins bonded to a metal ion. The enzyme cytochrome oxidase, necessary for biological energy production, is an example.
Phosphoproteins	Proteins bonded to a phosphate group. Milk casein, which serves to store nutrients for a growing embryo, is an example.
Nucleoproteins	Proteins bonded to RNA (ribonucleic acid). These are found in cell ribosomes.

other compounds in addition to amino acids on hydrolysis. Table 15.2 lists some examples of conjugated proteins.

Another way to classify proteins is according to their three-dimensional shape as either *fibrous* or *globular*. **Fibrous proteins**, such as collagen and the keratins, consist of polypeptide chains arranged side by side in long filaments. Because these proteins are tough and insoluble in water, nature uses them for structural materials like tendons, hair, ligaments, and muscle. **Globular proteins**, by contrast, are usually coiled into compact, nearly spherical shapes, as indicated by the computer-generated picture of an immunoglobulin molecule in Figure 15.3. These proteins, which might have anywhere from 100 to well over 1000 amino acids in their chains, include most of the 2000 or so known enzymes. They are generally soluble in water and are mobile within cells. Table 15.3 lists some common examples of both fibrous and globular proteins.

Yet a third way to classify proteins is to group them according to biological function. As indicated in Table 15.4, proteins have an extraordinary diversity of roles in the body.

■ **Fibrous protein** A tough, insoluble protein whose peptide chains are arranged in long filaments.

■ **Globular protein** A water-soluble protein that adopts a compact, coiled-up shape.

Figure 15.3
Computer-generated structure of an immunoglobulin. Proteins of this type, also known as antibodies, help protect us against harmful invaders. They bind to the surfaces of viruses, bacteria, and similar microorganisms, immobilizing them and tagging them for disposal by other body defensive systems.

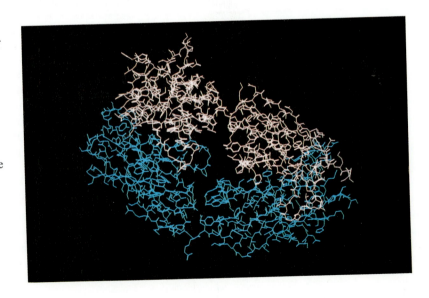

Table 15.3   Some Common Fibrous and Globular Proteins

Name	Occurrence and use
Fibrous proteins (*insoluble*)	
Keratins	Found in skin, wool, feathers, hooves, silk, fingernails
Collagens	Found in animal hide, tendons, bone, eye cornea, and other connective tissue
Elastins	Found in blood vessels and ligaments, where ability of the tissue to stretch is important
Myosins	Found in muscle tissue
Fibrinogen	Found in blood; necessary for blood clotting
Globular proteins (*soluble*)	
Insulin	Regulatory hormone for controlling glucose metabolism
Ribonuclease	Enzyme controlling RNA synthesis
Immunoglobulins	Proteins involved in immune response
Hemoglobin	Protein involved in oxygen transport

Table 15.4   Some Biological Functions of Proteins

Type	Function and example
Enzymes	Proteins such as alcohol dehydrogenase that act as biological catalysts
Hormones	Proteins such as insulin and growth hormone that regulate body processes
Storage proteins	Proteins such as ferritin that store nutrients
Transport proteins	Proteins such as hemoglobin that transport oxygen and other substances through the body
Structural proteins	Proteins such as keratin, elastin, and collagen that form an organism's structure
Protective proteins	Proteins such as the antibodies that help fight infection
Contractile proteins	Proteins such as actin and myosin found in muscles
Toxic proteins	Proteins such as the snake venoms that play a defensive role for the plant or animal

15.8   PRIMARY STRUCTURES OF PEPTIDES AND PROTEINS

■ **Primary protein structure**
The sequence in which amino acids are linked together.

■ **Secondary protein structure**
The way in which segments of a protein chain are oriented into a regular pattern.

■ **Tertiary protein structure**
The way in which an entire protein molecule is coiled and folded into its specific three-dimensional shape.

■ **Quaternary protein structure**
The way in which protein chains aggregate to form large, ordered structures.

With molecular weights of up to one-half *million*, many proteins are so large that the word *structure* takes on a broader meaning when applied to these immense molecules than it does with simple organic molecules. In fact, chemists usually speak about four levels of structure when describing proteins. The **primary structure** of a protein specifies the sequence in which the various amino acids are linked together. **Secondary structure** refers to how segments of the protein chain are oriented into a regular pattern; **tertiary structure** refers to how the entire protein chain is coiled and folded into a specific three-dimensional shape; and **quaternary structure** refers to how several protein chains can aggregate together to form a larger structure.

Primary structure is the most important of the four structural levels because it is the protein's amino-acid sequence that determines its overall shape, function, and properties. So crucial is primary structure to function that the change of only one amino acid out of several hundred can drastically alter biological properties. Sickle-cell anemia, for example, is caused by a genetic defect in blood hemoglobin whereby valine is substituted for glutamic acid at only one position in a chain of 146 amino acids.

15.9   SECONDARY STRUCTURES OF PROTEINS

When looking at the lengthy primary structure of a protein like insulin (Figure 15.2), you might get the idea that the molecule is simply a long thread, stretching from the N-terminal amino acid at one end to the C-terminal amino acid at the other end. The fact is, though, that proteins are not thread-like. Most proteins fold themselves in such a way that segments of the protein chain orient into regular patterns. Called *secondary structures*, there are two common kinds of patterns: the α-helix and the β-pleated sheet. There is in addition a third, less common kind: the triple helix.

The α-Helix: Secondary Structure of α-Keratin   α-Keratin is a fibrous structural protein found in wool, hair, fingernails, and feathers. Studies have shown that α-keratin wraps around into a helical coil, much like the cord on a telephone (Figure 15.4). Called an **α-helix**, this structure is stabilized by the formation of hydrogen bonds (Section 6.6) between the N—H group of one amino acid and

■ **α-Helix**   A common secondary protein structure in which a protein chain wraps into a coil stabilized by hydrogen bonds.

Figure 15.4
The helical secondary structure of α-keratin. The amino-acid backbone winds in a spiral, much like that of a telephone cord.

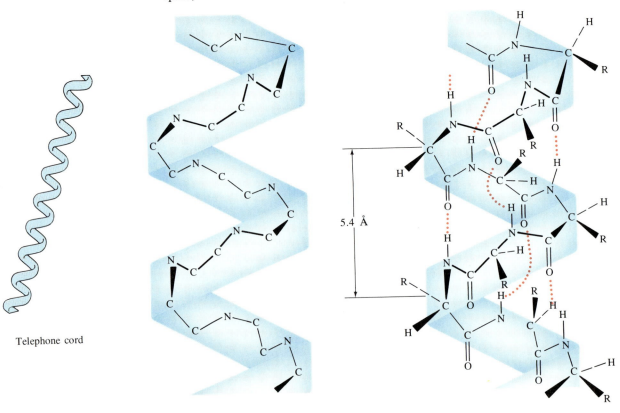

Telephone cord

5.4 Å

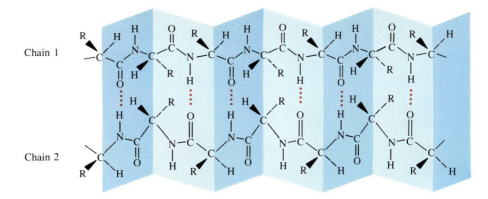

Figure 15.5
The β-pleated sheet, secondary structure of silk fibroin.

the carbonyl group of another amino acid four residues away. Although the strength of each individual hydrogen bond is small, the large number present in the helix results in an extremely stable secondary structure. Each coil of the helix contains 3.6 amino-acid residues, with a distance between coils of 0.54 nanometers (5.4 Å).

Although α-keratin is the best example of an almost entirely helical protein, a great many globular proteins contain α-helical segments. The chains of both hemoglobin and myoglobin, for example, contain many short helical sections.

The β-Pleated Sheet: Secondary Structure of Fibroin   Fibroin, the fibrous protein found in silk, has a secondary structure called a **β-pleated sheet**. In this pleated-sheet structure, polypeptide chains line up in a parallel arrangement held together by hydrogen bonds (Figure 15.5). Although not as common as the α-helix, small β-pleated sheet regions are often found in proteins where sections of peptide chains double back on themselves.

■ **β-Pleated sheet** A common secondary protein structure in which segments of a protein chain fold back on themselves to form parallel strands held together by hydrogen bonds.

The Triple Helix: Secondary Structure of Tropocollagen   Collagen is the most abundant of all proteins in mammals, making up 30 percent or more of the total protein. A fibrous protein, collagen is the major constituent of skin, tendons, bones, blood vessels, and connective tissues. The basic structural unit of collagen is *tropocollagen*, a large protein that consists of three chains of about 1000 amino acids each, wrapped around each other to form a stiff, rod-like **triple helix** (Figure 15.6) held together by hydrogen bonds. Each tropocollagen rod has a length of about 300 nm and a width of 1.5 nm. (For comparison, a simple organic molecule like propane is only about 0.5 nm long, and a typical globular protein like myoglobin is about 5.0 nm in diameter.)

■ **Triple helix** The secondary protein structure of tropocollagen in which three protein chains coil around each other to form a long fiber.

Figure 15.6
The triple helix in tropocollagen. Three individual protein strands wrap around each other to form a stiff helical cable.

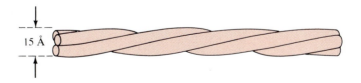

15 Å

15.10   TERTIARY AND QUATERNARY STRUCTURES OF PROTEINS

The secondary protein structures that we saw in the previous section result primarily from hydrogen-bonding interactions between different amide linkages along the protein backbone. By contrast, higher levels of structure result primarily from interactions of side-chain R groups in the protein. Let's look at two examples.

Tertiary Protein Structure: Myoglobin   Myoglobin is a globular protein with a single chain of 153 amino acid residues. A relative of hemoglobin, myoglobin is found in the skeletal muscles of sea mammals, where it stores oxygen needed to sustain the animals during long dives. Structurally, myoglobin consists of eight straight segments, each of which adopts an α-helical secondary structure. These helical sections then fold up further to form a compact, nearly spherical, tertiary structure (Figure 15.7).

Why does myoglobin adopt the shape it does? The forces that determine the tertiary structure of myoglobin and other globular proteins are the same simple forces that act on all molecules, regardless of size. By bending and twisting in a precise way, myoglobin can achieve maximum stability. Although the bends appear to be irregular, and the three-dimensional structure appears to be random, this is not the case. All myoglobin molecules adopt this same shape because it is more stable than any other.

The single most important force stabilizing a protein's tertiary structure results from the hydrophobic (water-hating) interactions of hydrocarbon side chains on amino acids. In a manner much like that of micelle formation in soaps (Section 14.4), those amino acids with neutral, nonpolar side chains have a strong tendency to congregate together on the hydrocarbon-like interior of a protein molecule, away from the aqueous medium. Those acidic or basic amino acids with charged side chains, by contrast, are usually found on the exterior of the protein where they can be solvated by water.

Also important for stabilizing a protein's tertiary structure are the formation of disulfide bridges between cysteine residues, the formation of hydrogen bonds between nearby amino acids, and the formation of ionic attractions, called *salt bridges*, between positively and negatively charged sites on the protein.

Figure 15.7
Secondary and tertiary structures of myoglobin.

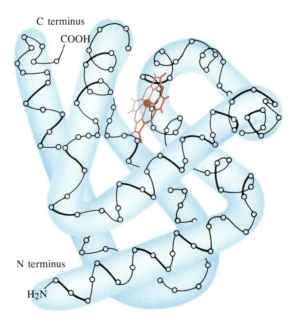

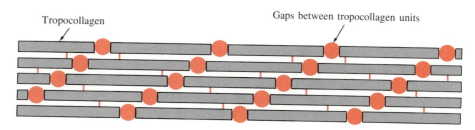

Tropocollagen

Gaps between tropocollagen units

Figure 15.8
The quarter-stagger arrangement of tropocollagen triple helix units aggregating into a collagen fiber.

Quaternary Protein Structure: Collagen   We saw in the previous section that tropocollagen is a stiff, rod-like protein with three chains coiled into a triple helix. Collagen itself, the actual protein present in skin, teeth, bones, and connective tissues, has a complex quaternary structure formed when a great many tropocollagen strands aggregate together by overlapping lengthwise in a *quarter-stagger arrangement* (Figure 15.8).

Depending on the exact purpose that the collagen serves in the body, further structural modifications also occur. In connective tissue like tendons, chemical bonds form between strands to give collagen fibers with a rigid, cross-linked structure. In teeth and bones, a mineral called *calcium hydroxyapatite*, $Ca_5(PO_4)_3OH$, deposits in the gaps between chains to further harden the overall assembly.

Figure 15.9 shows another example of quaternary structure: hemoglobin, which carries oxygen in the blood. A hemoglobin molecule consists of four polypeptide chains, two of 141 amino acids each and two of 146 amino acids each. The chains are held together by the same kinds of forces that create tertiary structure.

15.11   CHEMICAL PROPERTIES OF PROTEINS

Protein Hydrolysis   Just as a simple amide can be hydrolyzed to yield an amine and a carboxylic acid (Section 12.10), a protein can be hydrolyzed to yield many amino acids. In fact, digestion of proteins involves nothing more

Figure 15.9
Computer-generated model of the oxygen-carrying protein hemoglobin. The hemoglobin molecule consists of four polypeptide chains (red), each of which is linked to a non-protein group called heme (green). In the center of each heme is an iron atom (yellow), which actually binds the oxygen molecule.

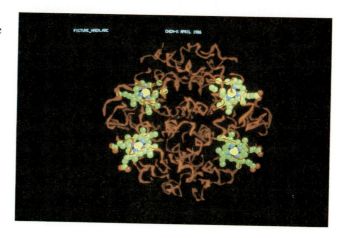

than breaking the numerous peptide bonds that link amino acids together. For example:

$$-NH-CH(CH_3)-C(=O)\;|\;NH-CH_2-C(=O)\;|\;NH-CH(CH_2SH)-C(=O)\;|\;NH-CH(CH_2COOH)-C(=O)-$$

$$\xrightarrow[\text{or digestive enzymes}]{H_3O^+}$$

$$H_2N-CH(CH_3)-C(=O)-OH + H_2N-CH_2-C(=O)-OH + H_2N-CH(CH_2SH)-C(=O)-OH + H_2N-CH(CH_2COOH)-C(=O)-OH$$

Alanine   Glycine   Cysteine   Aspartic acid

Although a chemist in the laboratory might choose to hydrolyze a protein by heating it with a solution of concentrated hydrochloric acid, most digestion of proteins in the body takes place in the small intestine, where the process is catalyzed by enzymes. Once formed, individual amino acids are absorbed through the wall of the intestine.

Practice Problem   **15.8**   Look up the structure of angiotensin II in Figure 15.1, and draw the products that would be formed during digestion.

■ **Denaturation** The disruption of tertiary protein structure brought on by heat or a change in pH.

Protein Denaturation   We've seen that the secondary and tertiary structures of globular proteins are delicately held together by a combination of weak forces such as those resulting from formation of hydrogen bonds. Often, a modest change in temperature or pH will disrupt the tertiary structure and cause the protein to become *denatured*. **Denaturation** occurs under such mild conditions that peptide and disulfide bonds aren't affected. The protein's primary structure remains intact, but its tertiary structure unfolds from a well-defined globular shape to a randomly looped chain (Figure 15.10). In addition, parts of the α-helical or β-pleated sheet regions making up the protein's secondary structure are also disrupted.

Denaturation is accompanied by changes in both physical and biological properties. Solubility is often decreased by denaturation, as occurs when egg white is cooked and the albumins coagulate into an insoluble mass. Moreover, enzymes lose all their catalytic activity, and other proteins can no longer perform their biological functions when their shapes are altered by denaturation.

Most denaturation is irreversible—for instance, hard-boiled eggs don't soften when their temperature is lowered. Many cases are known, however, in which spontaneous renaturation of an unfolded protein occurs. Renaturation is accompanied by a full recovery of biological activity, indicating that the protein has completely returned to its stable secondary and tertiary structure.

Figure 15.10
Denaturation of a protein occurs when its stable secondary and tertiary structure is disrupted.

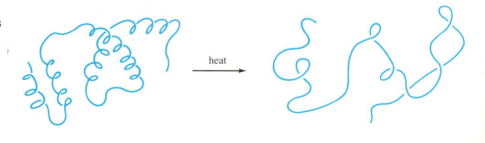

INTERLUDE: DETERMINING PROTEIN STRUCTURE

Determining the primary structure of a peptide or protein requires answers to three questions: What amino acids are present? How many of each are present? Where does each occur in the peptide chain? The answers to these questions are provided by two remarkable instruments, the amino-acid analyzer and the protein sequenator.

An *amino-acid analyzer* is an automated instrument for determining the identity and amount of each amino acid in a protein. The protein is first broken down into its constituent amino acids by reducing all disulfide bonds and hydrolyzing all amide bonds. The amino-acid mixture that results is then separated by placing it at the top of a glass column filled with a special adsorbent material and pumping a series of aqueous buffer solutions through the column. Different amino acids pass down the column at different rates, depending on their structures, and are thus separated.

As each different amino acid passes from the end of the glass column, it is mixed with a solution of *ninhydrin*, a reagent that forms a deep purple color on reaction with α-amino acids. The purple color is detected by an electronic sensor, and its intensity is measured. Since the amount of time required for a given amino acid to pass through a standard column is reproducible, the identity of all amino acids present in the sample can be determined simply by noting the time at which each comes off. The amount of each amino acid can be measured by determining the intensity of the purple color resulting from the ninhydrin reaction. The first figure shows the results of amino acid analysis of a standard equimolar mixture of 17 amino acids.

With the identity and amount of each amino acid known, the final task is to *sequence* the peptide—to find out in what order the amino acids are linked together. As shown in the second figure, the general idea of peptide sequencing is to cleave selectively one amino-acid residue at a time from the end of the peptide chain, separate and identify that amino acid, and then repeat the process on the chain-shortened peptide until the entire structure is known. Although the chemistry of the cleavage process is complex, automated protein sequenators are available that allow a series of 30 or more repetitive sequencing steps to be carried out.

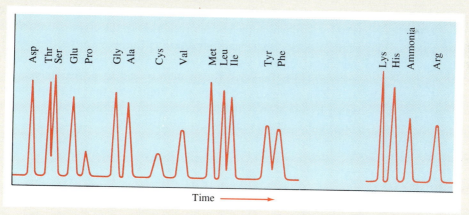

Amino-acid analysis of an equimolar amino-acid mixture. Each peak represents a different amino acid passing through the analysis column. The identities of the amino acids are determined by noting the times at which the peaks appear (horizontal axis), and the amounts are determined by measuring the area of each peak.

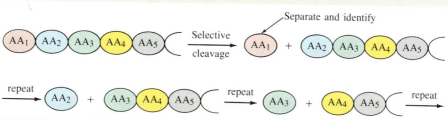

Protein sequencing. One amino acid at a time is selectively cleaved from the end of the protein chain, then separated and identified, and the process is repeated.

SUMMARY

Proteins are large biomolecules consisting of **α-amino acid residues** linked together by amide bonds. Twenty amino acids are commonly found in proteins. All are α-amino acids, and all except glycine have stereochemistry similar to that of L-sugars.

Structure determination of a large polypeptide or protein is carried out in several steps. The identity and amount of each amino acid present in a peptide can be determined by **amino-acid analysis**. The peptide is first hydrolyzed to its constituent α-amino acids, which are separated and identified. The peptide is then **sequenced** by selective cleavage of one residue at a time from the end of the peptide chain. Identification of the cleaved residue and repetition of the process on the chain-shortened peptide gives the entire structure.

Proteins can be classified by either composition, shape, or biological function. By composition, proteins are either simple or complex. **Simple proteins** yield only amino acids on hydrolysis; **conjugated proteins** yield other compounds in addition to amino acids. By shape, proteins are either fibrous or globular. **Fibrous proteins** such as α-keratin are tough and water-insoluble; **globular proteins** such as myoglobin are water-soluble and mobile within cells. By biological function, proteins have an enormous diversity of roles: Some are enzymes, some are hormones, and some serve to store nutrients. Others act as structural, protective, or transport agents.

Proteins are so large that the word *structure* has several meanings. A protein's **primary structure** is its amino-acid sequence. Its **secondary structure** is the way in which segments of the protein chain are oriented into a regular pattern, such as an α-helix, a β-pleated sheet, or a triple helix. Its **tertiary structure** is the way in which the entire protein molecule is coiled into a three-dimensional shape, and its **quaternary structure** is the way in which several protein chains aggregate to form a larger structure. When the tertiary structure of a protein is disrupted by heating, the protein is said to be **denatured**.

The chemistry of proteins is similar to that of simple amides. Hydrolysis, either by reaction with aqueous acid or by digestive enzymes, breaks a protein into its individual amino-acid constituents.

REVIEW PROBLEMS

Amino Acids

15.9 What is an amino acid?

15.10 What does the prefix "α" mean when referring to α-amino acids?

15.11 Why are the naturally occurring amino acids referred to as L-amino acids?

15.12 What amino acids do these abbreviations stand for?
(a) Ser (b) Thr (c) Pro (d) Phe (e) Cys

15.13 Draw the structures of the amino acids listed in Problem 15.12.

15.14 Name and draw the structures of amino acids that fit these descriptions:
(a) contains an isopropyl group
(b) contains a secondary alcohol group
(c) contains a thiol group
(d) contains a phenol group

15.15 What do the following terms mean as they apply to amino acids?
(a) zwitterion   (b) essential amino acid
(c) isoelectric point

15.16 What is the structural relationship of L-alanine to D-alanine?

15.17 L-Alanine has the three-dimensional structure indicated. Draw the structure of D-alanine, a rare amino acid found in earthworm larvae.

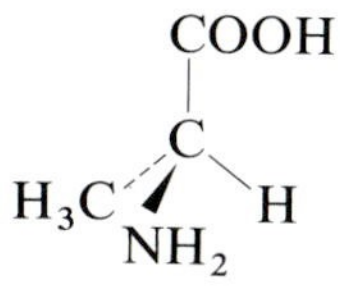

L-Alanine

15.18 Classify these amino acids as neutral, basic, or acidic:
(a) lysine   (b) phenylalanine
(c) glutamic acid   (d) proline

15.19 At what pH would you expect aspartic acid to have the following structures, pH 3, pH 7, or pH 13?

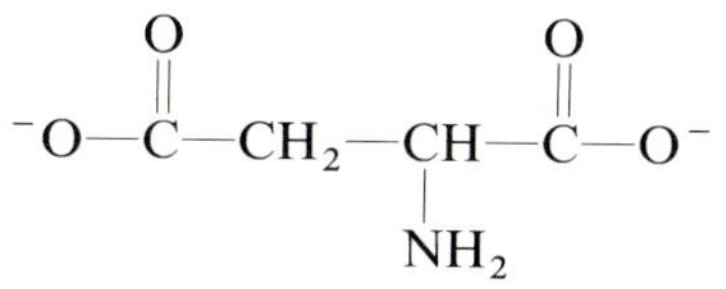

15.20 Draw the structures of valine at pH 1.0, 6.3, and 9.7.

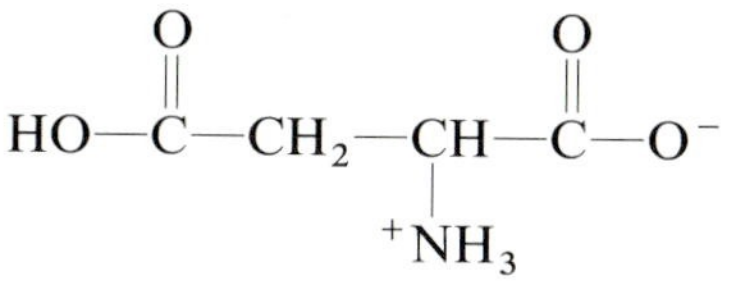

15.21 Draw the structures of lysine at pH 1.0, 6.3, and 9.7.

Peptides and Proteins

15.22 What is a protein?

15.23 What is the difference between a peptide and a protein?

15.24 What is the difference between a simple protein and a conjugated protein?

15.25 What kind of change takes place in a protein when it is denatured?

15.26 How can cross-links form between cysteine residues in a protein?

15.27 What kinds of molecules are found in these kinds of conjugated proteins in addition to the protein part?
(a) nucleoproteins   (b) lipoproteins
(c) glycoproteins

15.28 What is the difference between fibrous and globular proteins?

15.29 Name three different biological functions that proteins have in the body.

15.30 What is meant by the following terms as they apply to proteins?
(a) primary structure   (b) secondary structure
(c) tertiary structure   (d) quaternary structure

15.31 What kind of bonding is it that stabilizes helical and β-pleated-sheet secondary protein structures?

15.32 Why is cysteine such an important amino acid for defining the tertiary structure of proteins?

15.33 Look at the structure of angiotensin II in Figure 15.1, and then identify both the N-terminal and C-terminal amino acids.

15.34 Use the three-letter abbreviations to name all tripeptides containing methionine, isoleucine, and lysine.

15.35 Write structural formulas for the two dipeptides containing phenylalanine and glutamic acid.

15.36 Which of the following amino acids are most likely to be found on the outside of a globular protein, and which on the inside? Explain.
(a) valine   (b) leucine   (c) aspartic acid
(d) asparagine

15.37 Why do you suppose diabetics must receive insulin subcutaneously by injection rather than orally?

15.38 The *endorphins* are a group of naturally occurring neuroproteins that act in a manner similar to morphine to control pain. Research has shown that the biologically active part of the endorphin molecule is a pentapeptide called an *enkephalin*, with structure H_2N—Tyr—Gly—Gly—Phe—Met— COOH. Draw the complete structure of this enkephalin.

15.39 Identify the N-terminal and C-terminal amino acids in enkephalin (Problem 15.38).

Properties and Reactions of Amino Acids and Proteins

15.40 Much of the chemistry of amino acids is the familiar chemistry of carboxylic acid and amine functional groups. What products would you expect to obtain from these reactions of glycine?

(Section 12.4)

(b)
$$H_2N-CH_2-\overset{\overset{\displaystyle O}{\|}}{C}-OH + HCl \longrightarrow \text{ ?}$$

(Section 10.10)

15.41 How can you account for the fact that glycine is a solid that decomposes without melting when heated to 262°C, whereas the ethyl ester of glycine is a liquid at room temperature?

15.42 We saw in Section 12.9 that amides can be prepared by reaction between a carboxylic acid and an amine in the presence of DCC. What problem would you expect to encounter if you tried to prepare the simple dipeptide glycylglycine by treatment of glycine with DCC?

$$2\,H_2N-CH_2-\overset{\overset{\displaystyle O}{\|}}{C}-OH + DCC \longrightarrow$$

$$H_2N-CH_2-\overset{\overset{\displaystyle O}{\|}}{C}-NH-CH_2-\overset{\overset{\displaystyle O}{\|}}{C}-OH$$

15.43(a) Identify the amino acids present in the following hexapeptide:

$$H_2N-\underset{\underset{\displaystyle (CH_3)_2CH}{|}}{CH}-\overset{\overset{\displaystyle O}{\|}}{C}-NH-\underset{\underset{\displaystyle H}{|}}{CH}-\overset{\overset{\displaystyle O}{\|}}{C}-NH-\underset{\underset{\displaystyle HOCH_2}{|}}{CH}-\overset{\overset{\displaystyle O}{\|}}{C}-NH-\underset{\underset{\displaystyle CH_3SCH_2}{|}}{CH}-\overset{\overset{\displaystyle O}{\|}}{C}-NH-\underset{\underset{\displaystyle CH_3}{|}}{CH}-\overset{\overset{\displaystyle O}{\|}}{C}-NH-\underset{\underset{\displaystyle CH_2COOH}{|}}{CH}-\overset{\overset{\displaystyle O}{\|}}{C}-OH$$

(b) Identify the N-terminal and C-terminal amino acids of this hexapeptide.

(c) Show the structures of the products that would be obtained on digestion of this hexapeptide.

15.44 Which would you expect to be more soluble in water, a peptide rich in aspartic acid and lysine residues, or a peptide rich in valine and alanine residues? Explain.

15.45 Proteins are generally least soluble in water at their isoelectric points. Explain.

15.46 *Aspartame*, marketed under the trade name Nutra-Sweet for use as a nonnutritive sweetener, is the methyl ester of a simple dipeptide.

$$H_2N-\underset{\underset{\displaystyle HOOCCH_2}{|}}{CH}-\overset{\overset{\displaystyle O}{\|}}{C}-NH-\underset{\underset{\displaystyle CH_2-}{|}}{CH}-\overset{\overset{\displaystyle O}{\|}}{C}-O-CH_3$$

Aspartame (Nutra-Sweet)

Identify the two amino acids present in aspartame, and show *all* the products of digestion, assuming both amide and ester bonds are hydrolyzed in the stomach.

15.47 Both of the amino acids in aspartame (Problem 15.46) have the normal L geometry. If aspartame is made from amino acids having D geometry, the product tastes bitter rather than sweet. Explain.

CHAPTER **16**

The Molecules of Life: Enzymes, Vitamins, and Hormones

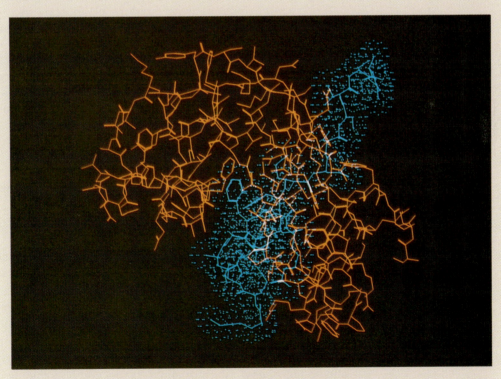

The blue chain in this computer-generated picture is ribonucleic acid, or RNA (Chapter 17). It is about to be cut in two by a molecule of the enzyme ribonuclease (orange). Snipping molecules is only one of the many things enzymes do. You'll learn more about these remarkable molecular machines in this chapter.

Think of your body as a walking chemical laboratory. Although the analogy isn't perfect, there's a good deal of truth to it nevertheless. In a laboratory, chemical reactions are carried out one at a time by mixing pure chemicals in test tubes or flasks. In your body, chemical reactions take place in cells rather than in test tubes, and many thousands of reactions take place simultaneously.

The main difference between chemistry in a laboratory and chemistry in a living organism is *control*. In a laboratory, the speed of a reaction is controlled by adjusting experimental conditions such as temperature, solvent, reagent concentrations, and pH. In an organism, however, these conditions can't be adjusted. After all, the human body must maintain a temperature of $37.0°C$, the "solvent" must be water, and the pH must be nearly neutral.

How then does an organism control the tens of thousands of different reactions taking place so that all occur to the proper extent? How are the speeds of the many individual reactions regulated so that the entire organism functions smoothly? The answer is that all reactions in living organisms are governed by biological catalysts called *enzymes*. It has been estimated that the body has at least 50,000 enzymes for regulating the multitude of reactions that take place inside us.

In this chapter, we'll learn how it is that enzymes control biological reactions, and we'll answer these questions:

1. **What kinds of enzymes are there, and how are they classified?** The goal: you should learn the main classes of enzymes.
2. **How do enzymes work, and why are they so specific?** The goal: you should learn how enzymes act as biological catalysts.
3. **How does the body regulate enzymes?** The goal: you should learn some of the ways that the body turns enzyme activity on and off.
4. **What are vitamins, and how do they function?** The goal: you should learn what vitamins are and how they interact with enzymes.
5. **What are hormones, and how do they function?** The goal: you should learn what hormones are and the biochemical roles they play.
6. **What are neurotransmitters, and how do they function?** The goal: you should learn what neurotransmitters are and how they work.

16.1   ENZYMES

■ **Enzyme** A protein that acts as a catalyst for biological reactions.

Enzymes are large proteins that act as catalysts for biological reactions. A *catalyst*, as we saw in Section 4.8, is something that speeds up the rate of a chemical reaction without itself undergoing change. For example, palladium metal catalyzes the reaction of an alkene with hydrogen gas to yield an alkane (Section 9.3), and sulfuric acid catalyzes the reaction of a carboxylic acid with an alcohol to yield an ester (Section 12.5). Neither reaction would occur very rapidly if the catalyst were not present.

Reactions would occur very slowly without catalyst

$$H_2C{=}CH_2 + H_2 \xrightarrow{\text{Pd catalyst}} CH_3{-}CH_3$$

$$CH_3{-}\overset{\overset{\displaystyle O}{\|}}{C}{-}OH + HOCH_3 \xrightarrow[\text{catalyst}]{H_2SO_4} CH_3{-}\overset{\overset{\displaystyle O}{\|}}{C}{-}OCH_3 + H_2O$$

Enzymes are similar to palladium and sulfuric acid in that they catalyze reactions that might otherwise occur very slowly, but they differ in two important respects. First, enzymes are far larger, more complicated molecules than simple inorganic catalysts. Second, enzymes are far more specific in their action. Whereas sulfuric acid catalyzes the reaction of nearly *every* carboxylic acid with nearly *every* alcohol, enzymes often catalyze only a *single* reaction of a single compound, called the enzyme's **substrate**. For example, the enzyme *amylase* found in human digestive systems is able to catalyze the breakdown of starch to yield glucose but has no effect on cellulose, even though the two polysaccharides are structurally similar (Section 13.8). Thus, humans can digest potatoes (starch) but not grass (cellulose).

■ **Substrate** A molecule acted upon by an enzyme.

$$\text{Starch} + H_2O \xrightarrow{\text{amylase}} \text{Many glucose molecules}$$

$$\text{Cellulose} + H_2O \xrightarrow{\text{amylase}} \text{No reaction}$$

Different enzymes differ greatly in their substrate specificity. Although amylase is specific for starch and has no effect on other polysaccharides, many other enzymes catalyze reactions of a range of substrates. For example, the enzyme *papain*, a globular protein of 212 amino acids isolated from papaya fruit, catalyzes the hydrolysis of a wide variety of different peptide bonds. It is this ability to hydrolyze peptides, in fact, that accounts for the use of papain in meat tenderizers and in contact-lens cleaners.

$$\text{---NH--CH--}\overset{\overset{\displaystyle O}{\|}}{C}\text{--NH--CH--}\overset{\overset{\displaystyle O}{\|}}{C}\text{---} \xrightarrow[\text{papain}]{H_2O} \text{---NH--CH--}\overset{\overset{\displaystyle O}{\|}}{C}\text{--OH} + H_2N\text{--CH--}\overset{\overset{\displaystyle O}{\|}}{C}\text{---}$$
$$\quad\quad\quad\;\; R \quad\quad\quad\quad R' \quad\quad\quad\quad\quad\quad\quad\quad\quad R \quad\quad\quad\quad\quad R'$$

Note that enzymes don't affect the equilibrium point of a reaction and can't bring about a reaction that is energetically unfavorable. In terms of the energy diagrams we used in Section 4.7 to depict the changes that take place

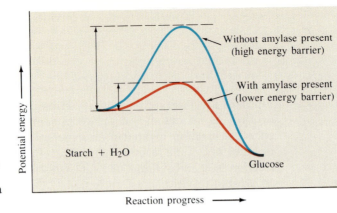

Figure 16.1
A reaction energy diagram for the hydrolysis of starch in the presence and absence of amylase. The enzyme-catalyzed reaction takes place faster because it has a lower energy barrier.

during reactions, the difference between a catalyzed and an uncatalyzed process is shown in Figure 16.1. Starch and water react very slowly under normal conditions because the energy barrier is too high. When amylase is present, the energy barrier is lowered, and starch is rapidly converted into glucose. The energy difference between starting material (starch) and product (glucose) remains exactly the same, though, whether the enzyme is present or not.

The catalytic activity of an enzyme is measured by its **turnover number**, which is defined as the number of substrate molecules acted on by one molecule of enzyme per unit time. As indicated in Table 16.1, enzymes vary greatly in their turnover number. Although most enzymes have values in the 1 to 10,000 range, carbonic anhydrase is able to catalyze the reaction of *600,000* substrate molecules per second.

■ **Turnover number** The number of substrate molecules acted on by one molecule of enzyme per unit time.

Table 16.1 Turnover Numbers of Some Enzymes

Enzyme	Turnover Number (per second)
Carbonic anhydrase	600,000
Acetylcholinesterase	25,000
β-Amylase	18,000
Penicillinase	2,000
DNA Polymerase I	15

16.2 ENZYME STRUCTURE

Structurally, enzymes are globular proteins. Thus, they have a primary structure defined by their amino acid sequence, a secondary structure defined by α-helical or β-pleated sheet regions, and a tertiary structure defined by the exact nature in which the overall molecule coils into a three-dimensional shape (Sections 15.7 through 15.9).

In addition to their protein part, many enzymes contain small, nonprotein portions called **cofactors**. In such enzymes, the protein part is called an **apoenzyme**, while the entire assembly of apoenzyme plus cofactor is called a **holoenzyme**. Only holoenzymes are active as catalysts; neither apoenzyme nor cofactor alone can catalyze a reaction.

■ **Cofactor** A small, nonprotein part of an enzyme that is essential to the enzyme's catalytic activity.

■ **Apoenzyme** The protein portion of an enzyme.

■ **Holoenzyme** The combination of apoenzyme and cofactor that is active as a biological catalyst.

$$\text{Holoenzyme} = \text{Apoenzyme} + \text{Cofactor}$$

■ **Coenzyme** A small organic molecule that acts as an enzyme cofactor.

An enzyme cofactor can be either an inorganic ion (usually a metal) or a small organic molecule called a **coenzyme**. The requirement that many enzymes have for metal-ion cofactors is the main reason behind our dietary need for trace minerals (Section 2.2). Iron, zinc, copper, manganese, molybdenum, cobalt, nickel, and selenium are all known to be essential trace elements that function as enzyme cofactors. In addition, vanadium, chromium, cadmium, lead, tin, lithium, and arsenic may also be enzyme cofactors, although the exact biological role of these elements is not yet known.

A large number of different organic molecules serve as coenzymes. Often, though not always, the coenzyme is a vitamin, which we'll discuss later in this chapter. Thiamin (vitamin B_1), for example, is the precursor of thiamin pyrophosphate, a coenzyme of the pyruvate decarboxylase required in the metabolism of carbohydrates.

Thiamin pyrophosphate

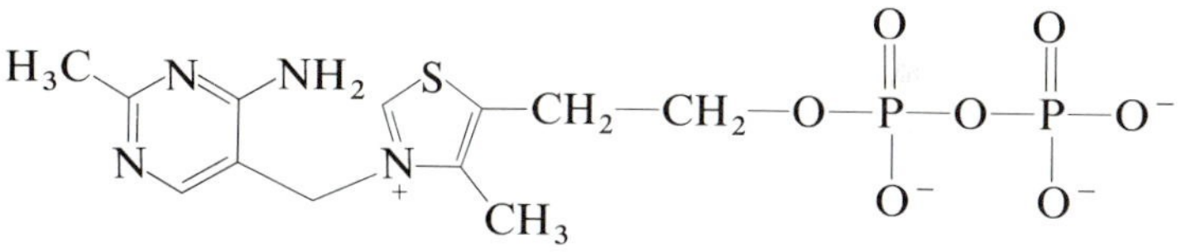

Pyruvic acid   Acetaldehyde

Practice Problem   **16.1** Check the label on a bottle of vitamin/mineral supplements, and look at the correlation between the metals listed and the metals known to be present in the body as enzyme cofactors.

16.3 ENZYME CLASSIFICATION

Enzymes are arranged into six main classes according to the general kind of reaction they catalyze, and each main class is further subdivided (Table 16.2). Most of the class names are self-explanatory: thus, *oxidoreductases* catalyze oxidations and reductions of substrate molecules; *transferases* catalyze the transfer of a group from one substrate to another; *hydrolases* catalyze the hydrolysis of substrates; and *isomerases* catalyze the isomerization of substrates. In addition, *lyases* (from the Greek, *lein*, meaning "to break") catalyze the breaking away of a small molecule like H_2O from a substrate, and *ligases* (from the Latin, *ligare*, meaning "to tie together") catalyze the bonding together of two substrates.

Although many enzymes like papain and trypsin were historically given uninformative common names, the systematic name of a specific enzyme has two parts: The first part identifies the substrate molecule on which the enzyme operates, and the second part indicates the enzyme class. For example, *alcohol*

Table 16.2   Classification of Enzymes

Main Class	Some Subclasses	Type of Reaction Catalyzed
Oxidoreductases	Oxidases	Oxidation of a substrate
	Reductases	Reduction of a substrate
	Dehydrogenases	Introduction of double bond (oxidation) by formal removal of H_2 from substrate
Transferases	Transaminases	Transfer of an amino group to substrate
	Kinases	Transfer of a phosphate group to substrate
Hydrolases	Lipases	Hydrolysis of ester groups in lipids
	Proteases	Hydrolysis of amide groups in proteins
	Nucleases	Hydrolysis of phosphate groups in nucleic acids
Lyases	Dehydrases	Loss of H_2O from substrate
	Decarboxylases	Loss of CO_2 from substrate
Isomerases	Epimerases	Isomerization of chiral center in substrate
Ligases	Synthetases	Forms new bond between two substrates
	Carboxylases	Forms new bond between substrate and CO_2

dehydrogenase is an oxidoreductase enzyme that oxidizes an alcohol to yield an aldehyde by removal of two hydrogens:

Ethanol   Acetaldehyde

Note that all enzymes have the family name ending *-ase*.

Practice Problems   **16.2**   What reactions would you expect these enzymes to catalyze?
(a) fumaric acid hydrase   (b) squalene oxidase
(c) glucose kinase   (d) cellulose hydrolase

16.3   To what class of enzymes does pyruvate decarboxylase belong?

16.4   ENZYME SPECIFICITY: THE LOCK-AND-KEY MODEL

■ **Lock-and-key model**   A model for enzyme specificity that pictures an enzyme as having an irregular cleft into which only specific substrate molecules fit.

■ **Active site**   A small three-dimensional portion of the enzyme with the specific shape and structure necessary to bind a substrate.

Any theory of how enzymes work must explain two facts: It must explain why enzymes are so specific, and it must explain how enzymes speed up reactions. Our current picture of enzyme action explains the first of these facts by postulating what's called the **lock-and-key model**. In this picture, an enzyme is a large, irregularly shaped molecule with a cleft or crevice in its middle. Inside the crevice is an **active site**, a small three-dimensional region of the enzyme with the specific shape necessary to bind the proper substrate and catalyze the appropriate reaction. In other words, the shape of the active site acts like a lock into which only a specific key (substrate) can fit (Figure 16.2). Note that the

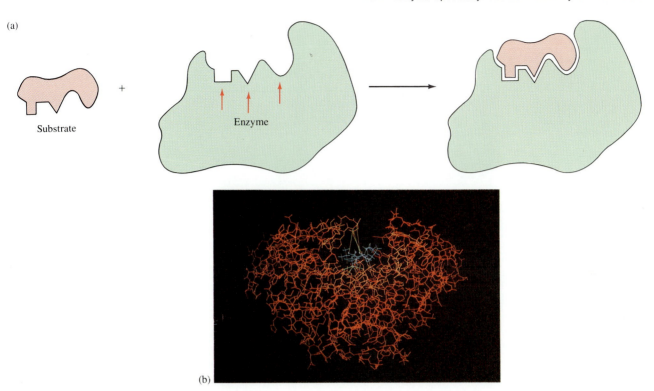

Figure 16.2
(a) The lock-and-key model of enzymes. Enzymes are large, three-dimensional molecules containing a crevice with a well-defined active site. Only a substrate whose shape and chemical nature is complementary to that of the active site can fit into the enzyme. (b) The active site of the enzyme (yellow) is clearly visible in this computer-generated structure of penicillopepsin, as is the fit of the substrate (blue) in the active site. The weak bonds that bind the substrate to groups in the active site are shown in green.

binding of a substrate by an enzyme involves only weak electrical attractions of the sort that stabilize a protein's tertiary structure (Section 15.10). No covalent bonds are formed.

An enzyme's active site is formed by various amino acid residues in the protein backbone. As we've seen, the 20 different amino acids that make up proteins have different kinds of side chains—some are hydrocarbons, some are hydroxylic (—OH containing), some are acidic, and some are basic. The active site of a given enzyme usually has a number of different kinds of amino acids, with the various acidic, basic, and neutral side chains properly positioned for maximum interaction with the substrate.

Although some enzymes are so specific that they operate on only a single substrate molecule, other enzymes are less selective. For example, the lipase enzymes secreted by the pancreas will hydrolyze all triglycerides, regardless of their exact structure. Presumably, these less selective enzymes are conformationally flexible enough so that they can change the shapes and sizes of their active sites to fit the spatial requirements of different substrates, as a pair of stretch socks adjusts to different sizes of feet. Called the **induced-fit model**, this picture of enzyme-substrate interaction is illustrated in Figure 16.3.

■ **Induced-fit model**   A model that pictures an enzyme with a conformationally flexible active site that can change shape to accommodate a range of different substrate molecules.

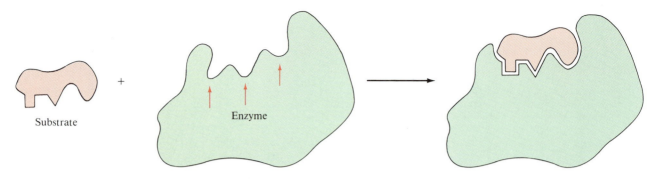

Figure 16.3
The induced-fit model of enzyme action. Some enzymes have a conformation that is flexible enough to allow them to adapt the size and shape of their active site to the shapes of several different substrates.

16.5   HOW ENZYMES WORK

■ **Enzyme-substrate complex**
A complex of enzyme and substrate in which the two are not linked by covalent bonds.

Now that we know why enzymes are so specific, we can get an idea of how they speed up reactions. Enzyme-catalyzed reactions begin with migration of substrate into the active-site to form an **enzyme-substrate complex**. No covalent bonds are formed; the enzyme and substrate are held together by hydrogen bonds and by weak electrical attractions between functional groups. With enzyme and substrate now held together in a precisely defined arrangement, the appropriately positioned functional groups in the active site bring about a chemical reaction in the substrate molecule, and enzyme plus product then separate.

E	+	S	$\rightleftharpoons$	E—S	$\longrightarrow$	E	+	P
Enzyme		Substrate		Complex		Enzyme		Product

An example of enzyme action is that of the enzyme hexose kinase, which catalyzes the reaction of adenosine triphosphate (ATP) with glucose to yield glucose 6-phosphate and adenosine diphosphate (ADP). As illustrated schematically in Figure 16.4, the enzyme first binds a molecule of ATP cofactor at

Figure 16.4
The hexose-kinase-catalyzed reaction of glucose with ATP. Glucose enters the cleft of the enzyme and binds to the active site, where it reacts with a molecule of ATP cofactor already bound nearby. Glucose 6-phosphate and ADP are then released from the enzyme.

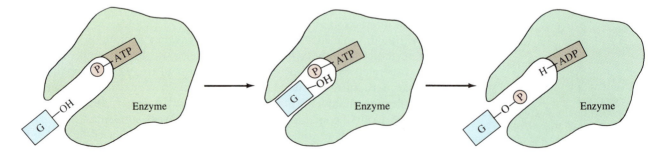

a position near its active site. Glucose then bonds to the enzyme's active site with its C6 hydroxyl group held rigidly in position right next to the ATP molecule. A reaction ensues, and the two products are then released from the enzyme.

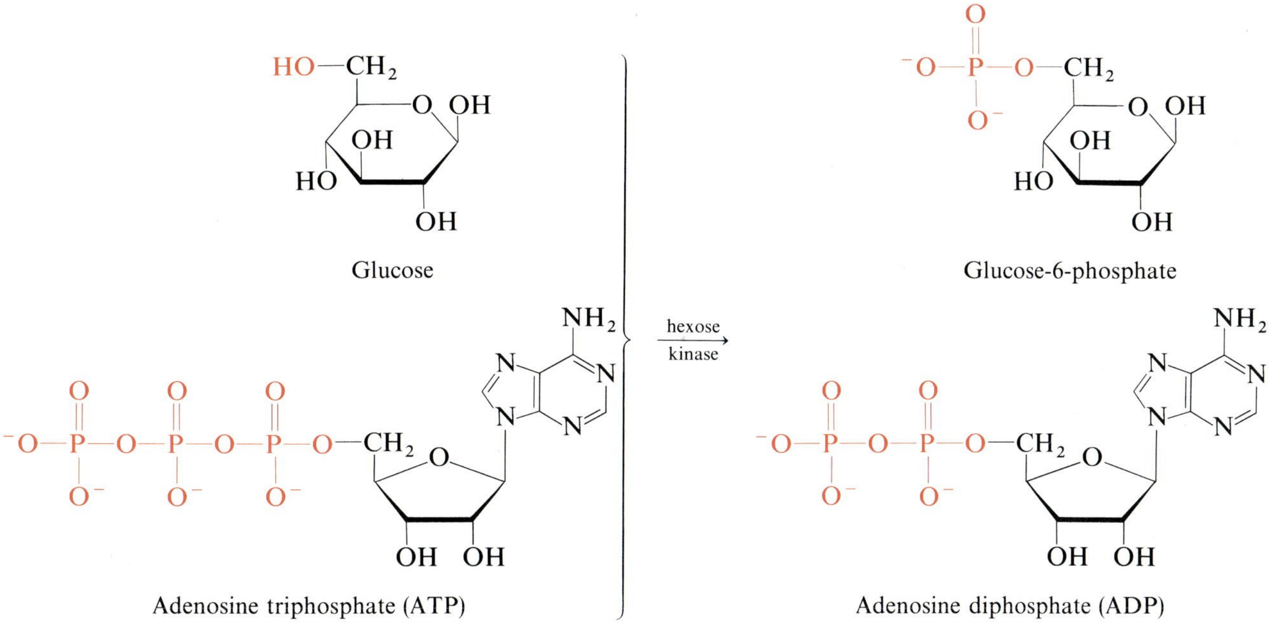

Glucose

Glucose-6-phosphate

Adenosine triphosphate (ATP)

Adenosine diphosphate (ADP)

hexose kinase

Compare the enzyme-catalyzed reaction shown in Figure 16.4 with the same reaction in the absence of enzyme. Without enzyme present, the two reagents would spend most of their time surrounded by solvent molecules, far away from each other and only occasionally bumping together. With an enzyme present, however, the two reagents are forced into close contact. Enzymes act as catalysts because of their ability to bring reagents together, to hold them at the exact distance and with the exact orientation necessary for reaction, and to provide acidic or basic sites as required.

16.6   ENZYME REGULATION AND INHIBITION

The control of biological reactions by enzymes is only half the overall picture. Equally important is that the enzymes themselves must somehow be regulated. After all, if an enzyme were continually functioning at full speed, it would soon run out of substrate, and the body would soon have a huge oversupply of the enzyme's product. Enzyme activity is regulated in the body by several types of inhibition processes.

Competitive Inhibition   What would happen if an enzyme were to encounter a molecule with a shape and size similar to that of its normal substrate? The imposter molecule could bind to the enzyme's active site and thereby prevent a normal substrate molecule from binding to the same site. Thus, the enzyme would be inactivated (Figure 16.5). Of course, if the inhibitor later happened to

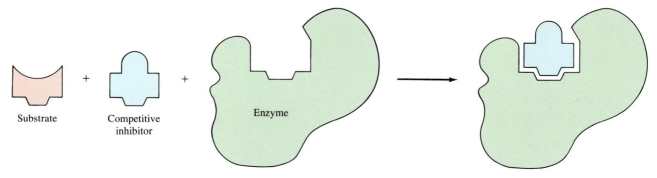

Figure 16.5
A competitive inhibitor blocks an enzyme's active site by mimicking the shape and size of the normal substrate, thereby preventing the enzyme from functioning.

■ **Competitive enzyme inhibition** A mechanism of enzyme regulation in which an inhibitor competes with a similarly shaped substrate for binding to the enzyme active site.

migrate *out* of the active site, the enzyme would once again be able to bind with substrate and would be fully active. Thus, inhibition of this sort is reversible.

The kind of inhibition illustrated in Figure 16.5 is called **competitive inhibition**, because the inhibitor competes with substrate for binding to the active site. The extent of inhibition depends on how good a mimic the inhibitor is, on what its concentration is, and on how tightly it binds compared to the substrate.

A good example of competitive inhibition is involved in the treatment of methanol poisoning. Although not harmful itself, methanol (wood alcohol) is oxidized in the body to formaldehyde, which is highly toxic ($CH_3OH \rightarrow H_2C{=}O$). Because of its similarity to methanol, ethanol is able to act as a competitive inhibitor of the methanol oxidase enzyme, thereby blocking oxidation and allowing methanol to be excreted harmlessly. Thus, the medical treatment of methanol poisoning involves administering high levels of ethanol.

Competitive inhibition is also a common pathway used by organisms for natural enzyme regulation. Since the products of many enzyme catalyzed reactions are structurally similar to their precursor substrates, these products often act as competitive inhibitors. Thus, much enzyme activity naturally slows down as product concentration builds up.

Noncompetitive Inhibition   A second kind of reversible inhibition occurs when an inhibitor binds to an enzyme at some place other than the active site. Such binding often leads to a change in the shape of the enzyme, with resultant change in the shape and binding ability of the active site. Binding of substrate thus becomes more difficult, and enzyme activity diminishes. Called **noncompetitive inhibition** because the inhibitor does not compete with substrate for the active site, this kind of effect is illustrated in Figure 16.6.

■ **Noncompetitive enzyme inhibition** A mechanism of enzyme regulation in which an inhibitor binds to an enzyme, thereby changing the shape of the enzyme's active site.

The high toxicity of lead, mercury, and other heavy metals is due to their ability to act as noncompetitive inhibitors of a variety of different enzymes. These metals bind strongly to thiol groups (—SH) on cysteine units in enzymes, thus altering the enzymes' shapes.

Irreversible Inhibition   Yet a third type of inhibition occurs when a molecule enters an enzyme's active site and forms a covalent bond to the enzyme. Since the inhibitor is firmly bonded in place, rather than just loosely held by electrical

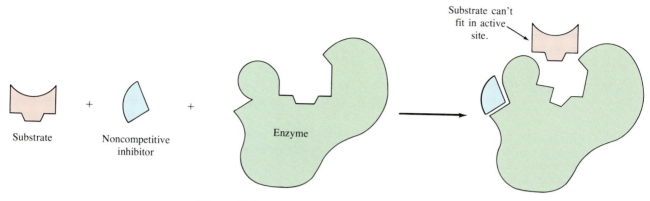

Figure 16.6
By binding elsewhere on the enzyme, a noncompetitive inhibitor changes the shape of the active site and decreases the ability of the enzyme to bind with substrate.

attractions, the active site is permanently blocked, and the enzyme is irreversibly inactivated.

■ **Irreversible enzyme inhibition**   A mechanism of deactivation regulation in which an inhibitor forms covalent bonds to the active site.

Irreversible inhibition accounts for the toxicity of nerve gases like diisopropylphosphofluoridate, which reacts with the enzyme acetylcholinesterase to block nerve transmission. Acetylcholinesterase has at its active site a serine residue that covalently bonds to the inhibitor.

Serine amino acid at active site of acetylcholinesterase   Diisopropylphosphofluoridate

Covalent bond that irreversibly binds the inhibitor to the enzyme

16.7   INFLUENCE OF pH AND TEMPERATURE ON ENZYMES

Enzymes have been finely tuned through evolution so that their maximum catalytic ability is highly dependent on pH and temperature. As you might expect, optimum conditions vary slightly for each enzyme but are generally near neutral pH and body temperature.

Effect of Temperature   We saw in Section 4.8 that an increase in temperature leads to an increase in rate for most chemical reactions. Thus, enzyme-catalyzed reactions often show a doubling in rate for each 10°C rise in temperature. Unlike many simple reactions, however, enzyme-catalyzed processes always reach an optimum temperature, after which their rate again falls (Figure 16.7). Often, this temperature optimum is just above body temperature.

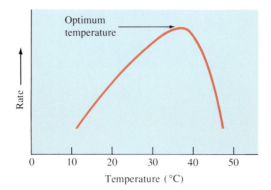

Figure 16.7
The effect of temperature on the rate of an enzyme-catalyzed reaction. An optimum temperature is always present.

AN APPLICATION: MEDICAL USES OF ENZYMES: ISOENZYMES

In a healthy person, enzymes are found almost entirely within cells. When some diseases occur, however, enzymes are released from dying cells into the blood, where their increased levels can be measured by sensitive clinical instruments. The table below lists some of the presently available enzyme tests, along with the medical conditions indicated by abnormal blood levels.

Perhaps the most useful enzyme assays are those done for diagnosis of heart disease. Three enzymes, creatine kinase (CK), glutamic-oxaloacetic transaminase (GOT), and lactate dehydrogenase (LD), are found in the muscle cells of a healthy heart. When a heart attack, or *myocardial infarction* (MI), occurs, some cells are damaged and their enzymes leak into the bloodstream. As shown in the figure below, the blood levels of CK, GOT, and LD all increase markedly in the hours immediately following a heart attack. The CK level rises almost immediately following an MI, reaching a sixfold increase over normal values after about 30 hours; the GOT level triples after about 40 hours; and the LD level doubles after about 4 days.

Some Enzyme Assays in Body Fluids

Enzyme	Condition Indicated by Abnormal Level
Lactate dehydrogenase (LD)	Heart disease, liver diseases
Creatine kinase (CK)	Heart disease
Glutamic-oxaloacetic transaminase (GOT)	Heart disease, liver diseases
Glutamic-pyruvic transaminase (GPT)	Heart disease, liver diseases
Glutamyl transferase (GMT)	Liver diseases
Alkaline phosphatase (ALP)	Bone disease, liver diseases
Amylase	Pancreatic diseases
Lipase (LPS)	Pancreatic diseases
Acid phosphatase (ACP)	Prostate cancer
Renin	Hypertension
Glucose-6-phosphate dehydrogenase (GPD)	Hemolytic anemia

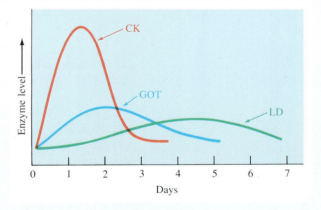

Blood levels of creatine kinase (CK), glutamic-oxaloacetic transaminase (GOT), and lactate dehydrogenase (LD) in the days following a heart attack (myocardial infarction).

The falloff in reaction rate after the optimum temperature is due to the fact that most enzymes are so sensitive they begin to denature (Section 15.10) when heated too strongly. Their delicately maintained tertiary structure begins to come apart, with a resultant change in shape and loss of the active site.

Most enzymes begin to denature and lose their catalytic activity above 50–60°C, a fact that explains why medical instruments and laboratory glassware can be sterilized by heating with steam in an autoclave. The high temperature of the autoclave denatures the enzymes of any bacteria present, thereby killing the organisms.

Effect of pH The catalytic activity of many enzymes depends on the pH of the surroundings and often has a well-defined optimum point. For example, trypsin, a protease enzyme secreted by the pancreas to aid digestion of proteins

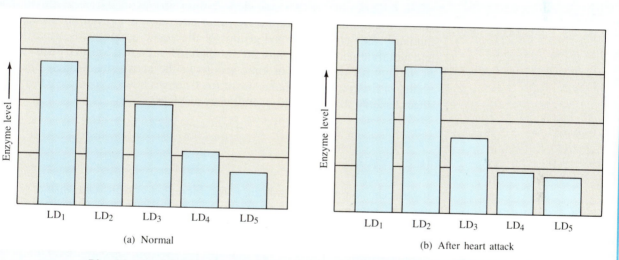

(a) Normal (b) After heart attack

Blood levels of the five lactate dehydrogenase (LD) isoenzymes in a normal person and in a heart attack victim after 48 hours. Note the flip of LD_1 and LD_2 levels following a heart attack.

Confirmation that a heart attack has occurred can be gained by a careful analysis of the individual enzyme levels. Recent work has shown that a number of enzymes, including CK and LD, are actually mixtures of several closely related compounds called *isoenzymes* that have slightly different structures but that catalyze the same reaction. For example, creatine kinase is a mixture of three isoenzymes, denoted CK(MM), CK(MB) and CK(BB). Brain tissue is rich in CK(BB), skeletal muscles are rich in CK(MM), and heart tissue is rich in CK(MB). Similarly, lactate dehydrogenase is a mixture of five isoenzymes, denoted LD_1, LD_2, LD_3, LD_4, and LD_5. Of the five, heart tissue contains primarily LD_1.

Since heart tissue contains primarily the CK(MB) and LD_1 isoenzymes, these levels increase the most following a heart attack. Thus, separation and analysis of the five individual LD isoenzymes shows that the LD_2 level is higher than LD_1 in a normal profile but that the two levels "flip" following a heart attack (see figure above). Similarly, an analysis of the three CK isoenzymes shows that the CK(MB) level increases to a much greater extent than the other two following a heart attack.

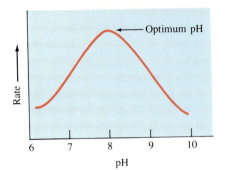

Figure 16.8
The effect of pH on the catalytic activity of trypsin, a protease digestive enzyme of the small intestine.

in the small intestine, has optimum activity at pH = 8.0. Enzyme activity falls dramatically at a pH either higher or lower than 8.0 (Figure 16.8).

The reasons for the effect of pH on enzyme activity involve changes both in enzyme structure and in the ability of the active site to bind a substrate. One effect, for example, is that side-chain amino groups in the basic amino acids lysine, guanidine, and histidine are protonated at low pH but not at high pH. Similarly, side-chain carboxyl groups of the acidic amino acids, aspartic acid and glutamic acid, are deprotonated at high pH but not at low pH. Such changes in the number and kinds of ionic groups at the active site strongly affect the ability of the enzyme to bind a substrate. Changes elsewhere in the enzyme can affect tertiary structure and overall enzyme shape.

16.8 VITAMINS

■ **Vitamin** A small organic molecule that must be obtained in the diet and that is essential in trace amounts for proper biological functioning.

It has been known since antiquity that there is a critical relationship between diet and health. Lime and other citrus juices are known to cure scurvy; meat and milk cure pellagra; and cod-liver oil cures rickets. The active substances in all three cases are vitamins. **Vitamins** are small organic molecules that must be obtained through the diet and that are required in trace amounts for proper growth and biological function. As the following representative structures show, vitamins differ widely in structure.

Vitamin C
(ascorbic acid)

Niacin

Vitamin A
(retinol)

Biochemically, the function of most vitamins is to act as enzyme cofactors. Table 16.3 lists the 13 known vitamins required in the human diet and, if known, lists the enzyme function they are needed for.

Table 16.3 Vitamins and Their Enzyme Functions

Vitamin	Enzyme Function	Deficiency Symptom
Water-Soluble Vitamins		
Ascorbic acid (Vitamin C)	Hydroxylases	Scurvy, bleeding gums, bruising
Thiamin (Vitamin B_1)	Reductases	Beriberi, fatigue, depression
Riboflavin (Vitamin B_2)	Reductases	Cracked lips, scaly skin, sore tongue
Pyridoxine (Vitamin B_6)	Aminotransferases	Anemia, irritability, skin lesions
Niacin	Reductases	Pellagra: dermatitis, dementia
Folic acid (Vitamin M)	Methyltransferases	Megaloblastic anemia
Vitamin B_{12}	Isomerases	Megaloblastic anemia, neurodegeneration
Pantothenic acid	Acyltransferases	Weight loss, irritability
Biotin (Vitamin H)	Carboxylases	Dermatitis, anorexia, depression
Fat-Soluble Vitamins		
Vitamin A	Visual system	Night blindness, dry skin
Vitamin D	Calcium metabolism	Rickets, osteomalacia
Vitamin E	Antioxidant	Hemolysis of red blood cells
Vitamin K	Blood clotting	Hemorrhage, delayed blood clotting

Vitamins are grouped by solubility into two classes: water-soluble and fat-soluble. Water-soluble vitamins are found in the aqueous environment inside cells, where they serve as precursors for specific coenzymes. The fat-soluble vitamins A, D, E, and K are lipids (Chapter 14) and are stored in the body's fat deposits. As indicated in Table 16.3, less is known about the specific enzyme functions of the fat-soluble vitamins than about their water-soluble counterparts.

Practice Problems

16.4 Vitamin D, a substance necessary for proper growth of bones and teeth, has the following structure. To what general class of compounds does vitamin D belong?

Vitamin D

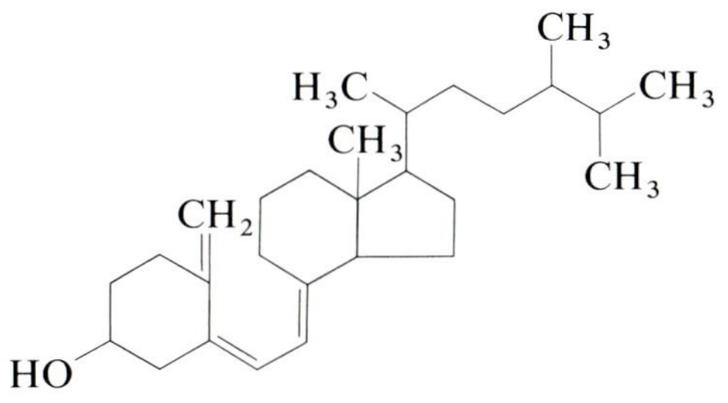

16.5 The structures of vitamin A and vitamin C are shown in the text. What structural features does each have that make one water-soluble but the other fat-soluble?

16.9 HORMONES

At this point, we've seen some of the many kinds of enzyme-catalyzed reactions that take place in cells—oxidations, reductions, bond formations, and so forth. What we've not seen is how the individual reactions are tied together. Clearly, the many thousands of separate reactions that take place in cells throughout the body do not occur randomly; there must be an overall control mechanism that coordinates the individual activities in different cells and keeps the entire organism in chemical balance.

The body's control mechanism for regulating and coordinating cellular activity is centered in the glands of the **endocrine system**, shown in Figure 16.9. Different endocrine glands located in various parts of the body secrete chemical messengers called **hormones** that are transported through the bloodstream to their target tissues. Once at the target tissue, hormones interact through a lock-and-key-like fit with specific receptors on the cell surface.

Hormone binding to a receptor on the cell's *outer* surface leads to activation of an enzyme on the cell's *inner* surface, which in turn carries the message into the cell, where it elicits a biological response. This response might involve activation of an enzyme, it might involve gene regulation, or it might involve changing membrane permeability to aid the passage of molecules into and out of cells. In all cases, however, hormones do not themselves carry out any chemical reactions; they act solely by turning existing biological mechanisms on or off.

The endocrine system is organized into successive stages under the overall control of the *hypothalamus*, a thumbnail-sized gland located at the base of the brain. The hypothalamus secretes tiny amounts of peptide hormones called *releasing factors* that pass through the blood stream to the pituitary gland, where they in turn stimulate the release of other hormones. These hormones then activate other endocrine glands to produce yet a third round of hormones, which ultimately act on their target tissues, as shown in Figure 16.10.

■ **Endocrine system**   A combination of different glands that regulate and control cellular activity in the body.

■ **Hormone**   A chemical messenger, secreted by an endocrine gland and transported through the bloodstream to elicit response from a specific target tissue.

Figure 16.9
Location of the endocrine glands.

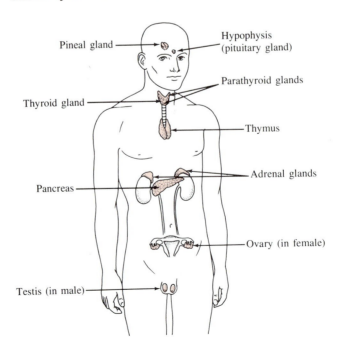

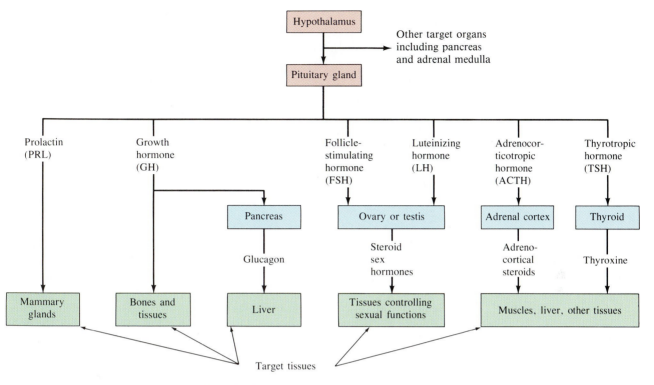

Figure 16.10
The primary organization of the endocrine system, together with the names and functions of some major hormones.

Hormones vary greatly in structure. Some, such as insulin (which regulates blood-glucose levels) and the six releasing factors secreted by the hypothalamus, are polypeptides. Others, such as thyroxine, are small organic molecules; and still others, such as estrone, are steroids.

Thyroxine
(a thyroid hormone)

Estrone
(a sex hormone)

16.10   HOW HORMONES WORK: THE ADRENALINE RESPONSE

Let's look at a specific example of hormone action to see how the overall process works. *Epinephrine*, better known as *adrenaline*, is often called the "fight or flight" hormone because its release prepares the body for instant response to

Figure 16.11
A grizzly bear exhibits the "fight or flight" reaction to a threatening situation. The hormone adrenaline produces a variety of physiological responses (such as increased heart rate and blood sugar level) and behavioral responses (such as these bared fangs).

danger (Figure 16.11). We've all felt the rush of adrenaline that accompanies a near-miss accident or a sudden loud noise. Chemically, epinephrine is a simple aromatic amine that is secreted into the bloodstream by the adrenal medulla, small glands located just above each kidney. Although it produces a variety of responses from numerous target tissues, epinephrine's main function is to raise the body's blood-sugar level by increasing the rate of glycogen breakdown in the liver. (Recall from Section 13.8 that glycogen is the polysaccharide the body uses to store glucose.)

$$HO \overset{OH}{\underset{HO}{\diagdown}} \diagdown CH-CH_2-N \overset{CH_3}{\underset{H}{\diagdown}} \qquad \text{Epinephrine}$$

Binding of epinephrine to a receptor site on the outside of a liver cell stimulates an enzyme (adenylate cyclase) inside the cell to begin production of what is called a *second messenger*. This second messenger, cyclic adenosine monophosphate (cyclic AMP or cAMP) in the present case, is released into the cell's interior. There it interacts in a series of steps with the phosphorylase enzymes necessary for the production of free glucose from glycogen (Figure 16.12). When enough glucose has been produced, an enzyme called *phosphodiesterase* catalyzes the hydrolysis of cyclic AMP, and glucose production ends. The entire series of events from initial stimulus to blood-glucose production takes only seconds.

Figure 16.12
The series of events by which release of the hormone epinephrine raises blood-glucose level. Interaction of the hormone with a cell-membrane receptor causes the second messenger cyclic AMP to be produced inside the target cell, in turn activating a series of steps that results in phosphorylation of glycogen and release of glucose to the bloodstream.

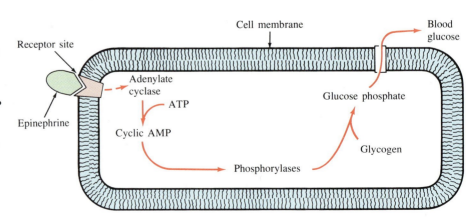

As we'll see in the next chapter, cyclic AMP is closely related to adenosine monophosphate (AMP), one of the four nucleotide building blocks that make up ribonucleic acid (RNA).

16.11   NEUROTRANSMITTERS

The hormones we discussed in the previous two sections are closely related to another group of biologically important chemicals called *neurotransmitters*. Whereas hormones are substances that carry a chemical message from a gland to a target tissue some distance away in the body, **neurotransmitters** are substances that mediate the flow of nerve impulses by transmitting a signal between neighboring nerve cells, or *neurons* (Figure 16.13).

■ **Neurotransmitter**   A chemical messenger that transmits a nerve impulse between neighboring nerve cells.

Structurally, nerve cells have a bulb-like body connected to a long, thin stem called an *axon*. Short, tentacle-like appendages called *dendrites* protrude from the bulbous end of the neuron, while numerous feathery filaments protrude from the axon at the opposite end. At the site where an impulse must pass between two neutrons, the axon of one neuron butts up against the dendrites or cell body of the next neuron, separated only by a narrow gap called a *synapse*.

Transmission of a nerve impulse between cells occurs when neurotransmitter molecules are released from filaments in the axon of a *presynaptic* neuron, cross the synaptic cleft, and fit into appropriately shaped receptor sites on the dendrites of the second, *postsynaptic* neuron (Figure 16.14, page 358). This stimulates the receiving cell to transmit a signal down its own axon and further pass along the nerve impulse.

The human nervous system has two main parts: the *central nervous system* (CNS), which consists of nerves in the brain and spinal cord, and the *peripheral nervous system* (PNS), which consists of sensory and motor nerves. The PNS,

Figure 16.13
Nerve cells in the spinal cord. Each neuron can receive impulses from, and transmit impulses to, many other neurons.

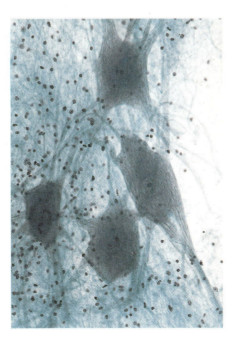

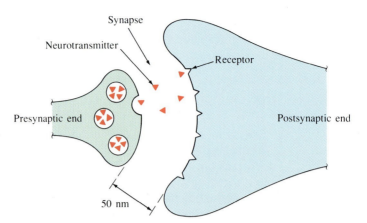

Figure 16.14
Transmission of a nerve signal between neurons occurs when a neurotransmitter molecule is released by the presynaptic neuron, crosses the synaptic cleft, and fits into a receptor site on the postsynaptic neuron.

in turn, is divided into *autonomic* and *somatic* parts. The autonomic part deals with unconsciously controlled functions like blood circulation and digestion, while the somatic part deals with consciously controlled functions like movement.

Neurons in the PNS can be divided into *cholinergic* and *adrenergic* types, depending on what kind of neurotransmitter molecules they release. Cholinergic neurons, which include all of those in the somatic part and many in the autonomic part, act by releasing *acetylcholine*. Adrenergic neurons, which are found only in the autonomic part, act by releasing *norepinephrine*.

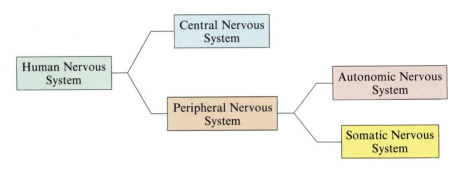

Acetylcholine   Norepinephrine

Unlike the PNS, the CNS uses a surprising variety of chemicals as neurotransmitters. Among the most interesting are dopamine and serotonin, which appear to be vital to such higher brain functions as perception and emotion. Both of these compounds are structurally related to norepinephrine. In addition, several amino acids and small peptides have been found to act as neurotransmitters in certain regions of the brain.

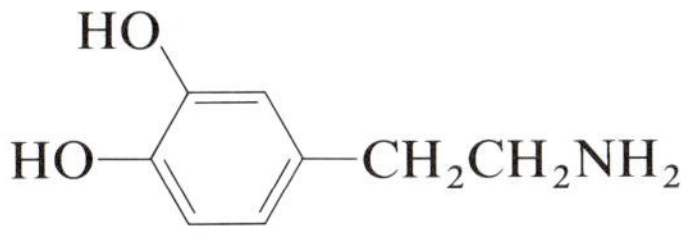

Dopamine

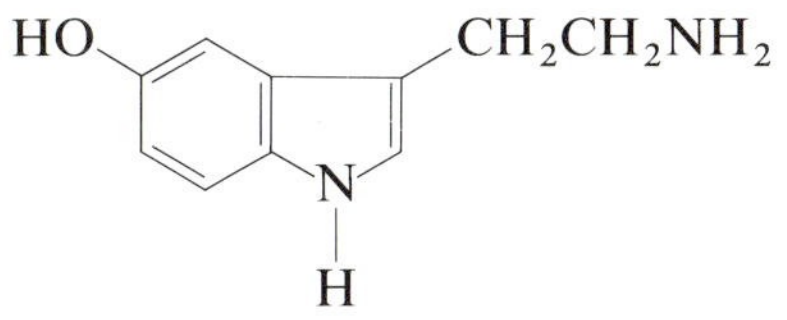

Serotonin

Once a neurotransmitter has done its job, it must be rapidly removed from the receptor site so that the postsynaptic neuron is ready to receive another impulse. In cholinergic neurons, for example, this removal is accomplished by the enzyme *acetylcholinesterase*. Nerve gases and certain insecticides act as cholinesterase inhibitors to block the activity of the enzyme, thereby allowing acetylcholine to accumulate at the receptor sites. Convulsions, paralysis of respiratory muscles, and death can result.

Malfunctions in the nervous system are involved either as symptoms or causes in a number of serious diseases. Parkinson's disease and schizophrenia, for example, have both been linked to defects in the brain's use of dopamine. Alzheimer's disease, a progressive disorder involving memory loss and deterioration of mental function, appears to involve a deficiency of the enzyme that synthesizes acetylcholine.

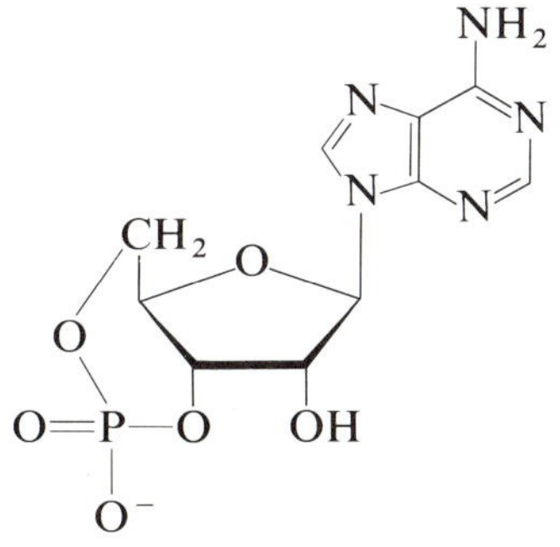

Cyclic AMP (cAMP)

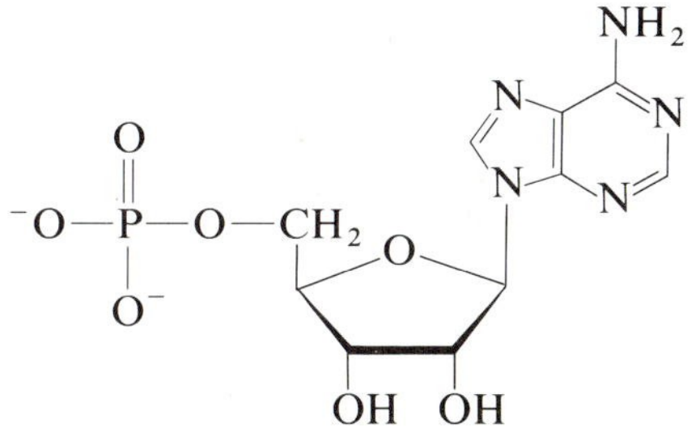

Adenosine monophosphate
(a ribonucleotide)

SUMMARY

Enzymes are large proteins that function as biological catalysts. Like all catalysts, enzymes speed up the rate of a reaction without themselves being changed. Unlike typical inorganic catalysts, however, enzymes are large, complex molecules that are quite specific in their action. They are classified into six groups according to the kind of reaction they catalyze: **oxidoreductases** catalyze oxidations and reductions; **transferases** catalyze transfers of groups; **hydrolases** catalyze hydrolysis; **isomerases** catalyze isomerizations; **lyases** catalyze bond breakages; and **ligases** catalyze bond formations.

Scarlet fever, diphtheria, bubonic plague, and rheumatic fever are all easily controlled diseases that are now almost unknown in modern industrialized countries. But if you had the bad fortune to contract one of these diseases before 1940, you might well not have survived. The difference between then and now is due to the discovery in 1928 and the subsequent isolation in 1940 of penicillin, the first antibiotic.

Antibiotics are substances that are produced by one microorganism and are toxic to other microorganisms. Penicillin, the first antibiotic, was discovered by the British bacteriologist Alexander Fleming as the result of a chance observation made when several culture plates of staphylococcus bacteria he was growing became contaminated. Fleming noticed that the contaminant, a mold subsequently identified as *Penicillium notatum*, produced a substance that appeared to kill the staphylococcus.

Chemically, the penicillins are known as *beta-lactam* antibiotics because they contain a four-membered amide ring. More than 30 naturally occurring penicillins with closely related structures are known, and several thousand synthetic derivatives have been prepared in the laboratories of pharmaceutical companies.

Acyl side chain

β-lactam ring

Different penicillins have different acyl side chains (in blue) attached to nitrogen. Penicillin G, the most common one, is shown.

The penicillins kill bacteria by acting as irreversible inhibitors of a critical transpeptidase enzyme that is involved in the synthesis of bacterial cell walls. Bacterial cells, unlike those of higher organisms, are enveloped by a protective coating called a *cell wall*. In the final stages of cell-wall synthesis, strands of glycoprotein known as peptidoglycans are cross-linked together to form the final three-dimensional web. The crucial cross-linking step involves enzyme-catalyzed reaction of a D-Ala-D-Ala terminus on one strand with a Gly terminus on another strand.

Peptidoglycan—Gly—NH$_2$ D-Ala—Peptidoglycan
                              |
                              D-Ala

transpeptidase

Peptidoglycan—Gly—D-Ala—Peptidoglycan + D-Ala

Penicillin's three-dimensional shape is evidently similar enough to that of the D-Ala-D-Ala end of the peptidoglycan side chain that it is able to fit into the active site of the transpeptidase enzyme. Once in the active site, penicillin binds irreversibly to the enzyme by covalent bond formation between the enzyme and the carbonyl group of the β-lactam ring. With bacterial cell-wall synthesis thus halted, the cell contents leak out through the weakened wall, and the cell dies. Since the cells of higher organisms have no cell walls, penicillin is completely specific for bacteria and is nontoxic to all other organisms.

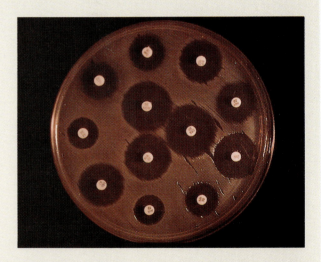

Testing the sensitivity of bacteria to different antibiotics. The clear circular areas on the culture dish are regions free of bacterial growth. The bacteria have been killed or prevented from reproducing by antibiotics diffusing outward from the smaller spots at the center of each circle.

In addition to their protein part, many enzymes contain **cofactors**, which can be either metal ions or small organic molecules. If the cofactor is an organic molecule, it is called a **coenzyme**. The combination of protein part (**apoenzyme**) plus coenzyme is called a **holoenzyme**. Often, the coenzyme is a **vitamin**, a small molecule that must be obtained in the diet and that is required in trace amounts for proper growth. There are 13 known human vitamins, 9 of which are water-soluble and 4 of which are fat-soluble.

Enzyme specificity is due to a lock-and-key fit between enzyme and substrate. Enzymes contain a crevice, inside which is an **active site**, a small three-dimensional region of the enzyme with the specific shape necessary to bind the proper substrate. No covalent bond formation occurs when a substrate binds in the active site; the **enzyme-substrate complex** is held together by hydrogen bonds and electrical attractions. The catalytic action of enzymes is due to their ability to bring reagents together and hold them in the exact position necessary for reaction.

Enzymes are highly sensitive molecules that usually function best at an optimum pH and temperature. Their action can be suppressed either by **competitive inhibitors**, which compete with substrate for binding to the active site, by **noncompetitive inhibitors**, which act by changing the shape of the active site, or by **irreversible inhibitors**, which act by covalently bonding to and blocking the active site.

The many thousands of enzyme-catalyzed reactions that take place throughout the body are coordinated by the **endocrine system**, functioning under the overall control of the hypothalamus gland. Different endocrine glands in various parts of the body secrete chemical messengers called **hormones** that are transported through the bloodstream to their target tissues. Once at the target tissue, hormones act by binding to a receptor on the outside of a cell, triggering an event inside the cell that turns a specific biological mechanism on or off. In a similar manner, **neurotransmitters** are secreted by nerve cells to transmit a nerve impulse from cell to cell.

REVIEW PROBLEMS

Structure and Classification of Enzymes

16.6 What is the family name ending for an enzyme?

16.7 What is an enzyme, and what general kind of structure does it have?

16.8 There are approximately 50,000 different enzymes in the body. Why are so many needed?

16.9 How do catalysts speed up reactions?

16.10 What are the two primary differences between an inorganic catalyst and an enzyme?

16.11 What general kinds of reactions do these classes of enzymes catalyze?
(a) hydrolases   (b) isomerases   (c) lyases

16.12 What is a holoenzyme, and what are its two parts called?

16.13 What is the difference between a cofactor and a coenzyme?

16.14 What feature of enzymes makes them so specific in their action?

16.15 Describe in general terms how enzymes act as catalysts.

16.16 What classes of enzymes would you expect to catalyze these reactions:

(a)

$$H_2N-\underset{\underset{R}{|}}{CH}-\underset{\underset{}{\overset{\overset{O}{\|}}{C}}}-NH-\underset{\underset{R'}{|}}{CH}-\underset{\underset{}{\overset{\overset{O}{\|}}{C}}}-OH + H_2O \longrightarrow$$

$$H_2N-\underset{\underset{R}{|}}{CH}-COOH + H_2N-\underset{\underset{R'}{|}}{CH}-COOH$$

(b)

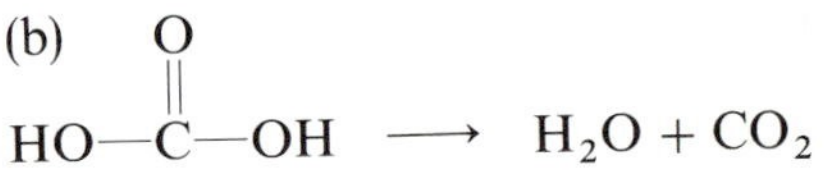

$$HO-\overset{\overset{\displaystyle O}{\|}}{C}-OH \longrightarrow H_2O + CO_2$$

(c)

$$HOOC-CH_2-CH_2-COOH \longrightarrow$$
$$HOOC-CH=CH-COOH$$

16.17 What kind of reaction does each of the following enzymes catalyze?
(a) a protease   (b) a DNA ligase
(c) a transmethylase

16.18 The following reaction is catalyzed by an enzyme called urease. To what class of enzymes does urease belong?

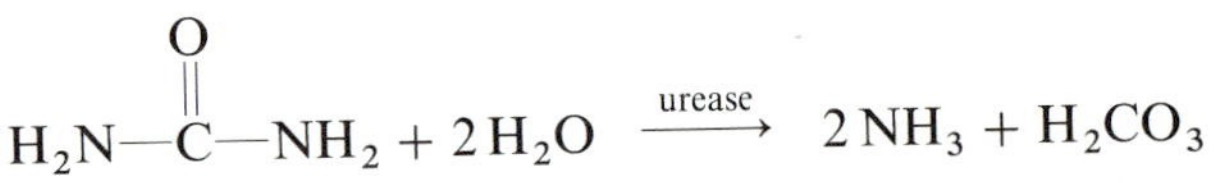

$$H_2N-\overset{\overset{\displaystyle O}{\|}}{C}-NH_2 + 2\,H_2O \xrightarrow{\text{urease}} 2\,NH_3 + H_2CO_3$$

16.19 The catalytic activity of urease (Problem 16.18) can be inhibited by adding dimethylurea to the reaction. What kind of inhibition is probably occurring?

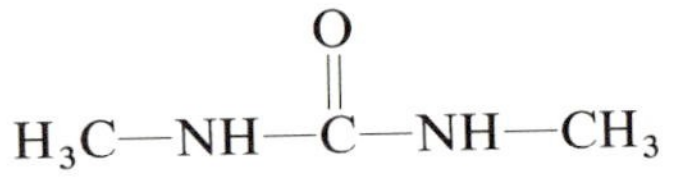

$$H_3C-NH-\overset{\overset{\displaystyle O}{\|}}{C}-NH-CH_3 \qquad \text{Dimethylurea}$$

Enzyme Function and Regulation

16.20 What is the difference between the lock-and-key model of enzyme action and the induced-fit model?

16.21 Which enzyme is more specific in its action, a lock-and-key enzyme or an induced-fit enzyme?

16.22 Honeybees produce an enzyme called *invertase* that catalyzes the conversion of sucrose to a 1:1 mixture of glucose and fructose.
(a) Formulate the reaction (see Section 13.7).
(b) To what class of enzymes does invertase belong?

16.23 Draw an energy diagram for the exothermic enzyme-catalyzed hydrolysis of sucrose (Problem 16.22). Label the energy levels of reactants and products, the activation energy, and the overall energy difference in the reaction.

16.24 Which part of a holoenzyme would you expect to be affected by denaturation, the coenzyme or the apoenzyme? Explain.

16.25 Before the development of modern antibiotics, a drop of dilute $AgNO_3$ was often placed in the eyes of newborn infants to prevent gonorrheal eye infections. How might $AgNO_3$ kill bacteria?

16.26 What general effects would you expect the following changes to have on the speed of an enzyme-catalyzed reaction?
(a) lowering the reaction temperature from 37°C to 27°C
(b) raising the pH from 7.5 to 10.5.
(c) adding a heavy-metal salt like $Hg(NO_3)_2$
(d) raising the reaction temperature from 37°C to 40°C

16.27 How can you explain the observation that pepsin, a digestive enzyme found in the stomach, has a high catalytic activity at pH = 1.5 while trypsin, a digestive enzyme of the small intestine, has no activity at the same pH?

16.28 What are the three kinds of enzyme inhibitors?

16.29 How does each of the three kinds of enzyme inhibitors work?

16.30 What kinds of bond are formed between an enzyme and each of the three kinds of enzyme inhibitors?

16.31 Poisoning by which of the three types of enzyme inhibitors is probably the most difficult to treat medically?

16.32 The meat tenderizer used in cooking is primarily *papain*, a protease enzyme isolated from the papaya tree. Why do you suppose papain is so effective at tenderizing meat?

16.33 A substantial number of calories in regular beer are due to the presence of complex carbohydrates called *amylopectins*. By adding an amyloglucosidase enzyme during brewing, the amylopectins can be hydrolyzed into glucose and then fermented. What two characteristics does the resultant light beer have? (*Hint:* The first characteristic is that it tastes great.)

Hormones and Vitamins

16.34 What is the difference between a hormone and a vitamin?

16.35 What is the difference between a hormone and a neurotransmitter?

16.36 What is the relationship between vitamins and enzymes?

16.37 The adult recommended daily allowance (RDA) of riboflavin is 1.6 mg. If one glass (100 mL) of red wine contains 0.014 mg of riboflavin, how much wine would an adult have to consume to obtain the RDA?

16.38 What is the structural difference between an enzyme and a hormone?

16.39 What are the four fat-soluble vitamins?

16.40 What are the two parts of the human nervous system?

16.41 What is a synapse, and what role does it play in nerve transmission?

16.42 Name two classes of neurotransmitters.

16.43 Describe in general terms how a nerve impulse is passed from one neuron to another.

16.44 What is the general purpose of the body's endocrine system?

16.45 What gland is primarily responsible for controlling the endocrine system?

16.46 How is a hormone transported from its secretory gland to its target tissue?

16.47 Name as many endocrine glands as you can.

16.48 Describe in general terms how a hormone works.

16.49 What is the relationship between enzyme specificity and tissue specificity of hormones?

16.50 What is the cellular role of cyclic AMP?

16.51 What biological effect does the hormone epinephrine have?

16.52 Which of the three are required in the diet: hormones, enzymes, or vitamins?

16.53 Identify the functional groups in these vitamins:

(a)

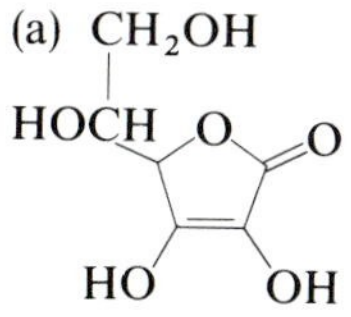

Vitamin C

(b)

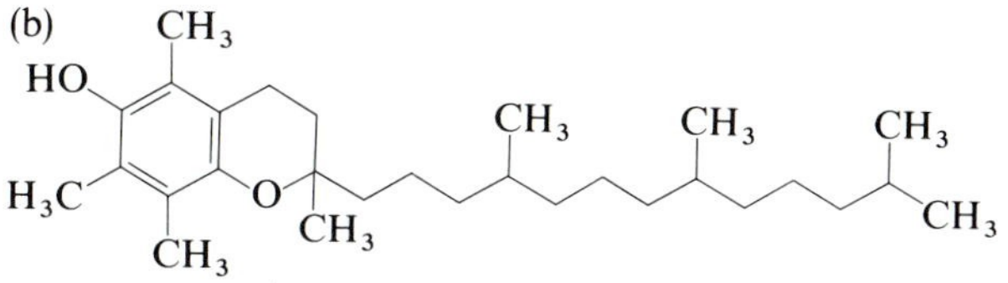

Vitamin E

16.54 Thyrotropin releasing factor, one of the primary hormones of the hypothalamus, is the C-terminal amide of the simple tripeptide PyroGlu-His-Pro, where PyroGlu is pyroglutamic acid:

PyroGlu =

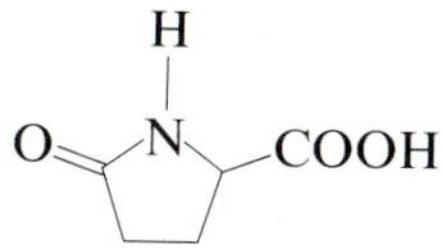

Draw the full structure of thyrotropin releasing factor.

16.55 What is the chemical relationship of pyroglutamic acid (Problem 16.54) to glutamic acid, one of the 20 common amino acids?

CHAPTER 17

The Molecules of Life: Nucleic Acids

An electron micrograph of a cell that has been invaded by the human immuno-deficiency virus (HIV-1) responsible for AIDS. A virus is simply a molecule of nucleic acid surrounded by a protein coat, "a piece of bad news wrapped up in protein," in the words of one scientist. It takes over the genetic machinery of the cell and directs it to produce new viruses. One such virus particle (orange) is seen here budding out from the surface of the infected cell (blue).

How does a seed "know" what kind of plant to become? How does a fertilized ovum know how to grow into a human being? And how does a cell know what part of the body it's in, whether brain or big toe, so that it can produce the right chemicals necessary for sustaining life? The answers to these and myriad other questions about all living organisms involve the biological molecules called nucleic (nu-**clay**-ic) acids.

The nucleic acids, *deoxyribonucleic acid (DNA)* and *ribonucleic acid (RNA),* are the chemical carriers of an organism's genetic information. Coded in an organism's DNA is all of the information that determines the nature of the organism, whether dandelion, goldfish, or human being. Coded also in DNA are all the instructions needed for cellular functioning, and all the directions needed for producing the tens or hundreds of thousands of different proteins required by the organism. In this chapter, we'll answer the following questions about nucleic acids:

1. **What are the nucleic acids, DNA and RNA?** The goal: you should learn what DNA and RNA are and what their biochemical functions are.

2. **What are the structures of DNA and RNA?** The goal: you should learn the chemical structures of DNA and RNA, and you should understand the phenomenon of base pairing in the DNA double helix.

3. **How is DNA involved in heredity, and how is it replicated?** The goal: you should learn the role of DNA in heredity and should understand how a DNA molecule is replicated to preserve hereditary information.

4. **How do organisms synthesize RNA?** The goal: you should learn how RNA is synthesized from DNA.

5. **How does RNA control protein synthesis?** The goal: you should learn how organisms use the hereditary information passed from DNA to RNA to synthesize proteins.

6. **How is the structure of DNA segments determined?** The goal: you should learn how DNA sequences are determined in the laboratory.

17.1   DNA, CHROMOSOMES, AND GENES

■ **Chromosome**   A thread-like strand of DNA in the cell nucleus that stores genetic information.

■ **Gene**   A small segment of a DNA chain where the genetic instructions for making an individual protein are encoded.

Most DNA of higher organisms, both plant and animal, is found in the nucleus of cells in the form of thread-like strands that are coated with proteins and wound into complex assemblies called **chromosomes**. Each chromosome is made up of several thousand **genes**, where a gene is a section of the DNA chain that codes for a single piece of information needed by the cell.

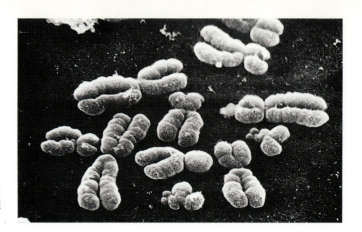

Figure 17.1
A human chromosome pictured under an electron microscope. Chromosomes are roughly X-shaped during cell division and have four banded arms connected at a centromere.

Different organisms differ in their complexity and therefore have different numbers of chromosomes and genes. A frog, for example, has 26 chromosomes (13 pairs), whereas a human has 46 (23 pairs). Each chromosome contains a single immense molecule of DNA that, in humans, has a molecular weight of up to 150 *billion* amu and a length of up to 12 *centimeters*. Chromosomes are so large that they can even be seen during cell division under powerful electron microscopes, as shown in Figure 17.1.

17.2   COMPOSITION OF NUCLEIC ACIDS

■ **Nucleic acid**   A biological polymer made by the linking together of nucleotide units.

■ **Nucleotide**   A building block for nucleic acid synthesis, consisting of a five-carbon sugar bonded to a cyclic amine base and to phosphoric acid.

Like proteins, **nucleic acids** are made up of many individual building blocks linked together to form a long chain. Unlike proteins, however, RNA and DNA are built up from **nucleotides** rather than amino acids. A nucleotide can itself be further broken down by hydrolysis to yield three components—a simple aldopentose sugar (Section 13.5), an amine base (Section 10.10), and phosphoric acid, H_3PO_4 (Section 12.12).

Nucleic acid $\xrightarrow{\text{hydrolysis}}$ **many nucleotides** $\xrightarrow{\text{hydrolysis}}$ **aldopentose** + **amine bases** + H_3PO_4

■ **RNA**   Ribonucleic acid.

■ **DNA**   Deoxyribonucleic acid.

The sugar component in **RNA** (*ribo*nucleic acid) is D-ribose, whereas the sugar in **DNA** (*deoxyribo*nucleic acid) is 2-deoxyribose. (The prefix "2-deoxy" means that an oxygen atom is missing from the C2 position of ribose.)

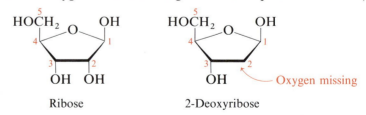

Ribose   2-Deoxyribose

■ **Heterocycle**   A ring that contains nitrogen or some other atom in addition to carbon.

Four different amine bases are found in DNA. All are **heterocycles**, meaning that the basic amine nitrogen atom occurs within a ring. Two of the heterocyclic amines are substituted *purines* (*adenine* and *guanine*), and two are substituted *pyrimidines* (*cytosine* and *thymine*). Adenine, guanine, and cytosine also occur in RNA, but thymine is replaced in RNA by a different pyrimidine base called *uracil*.

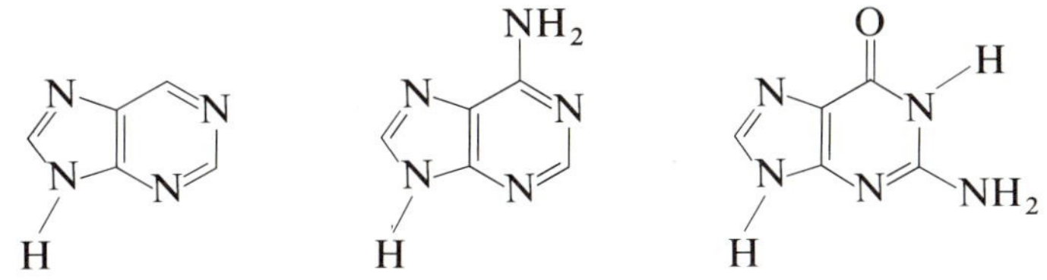

Purine   Adenine   Guanine

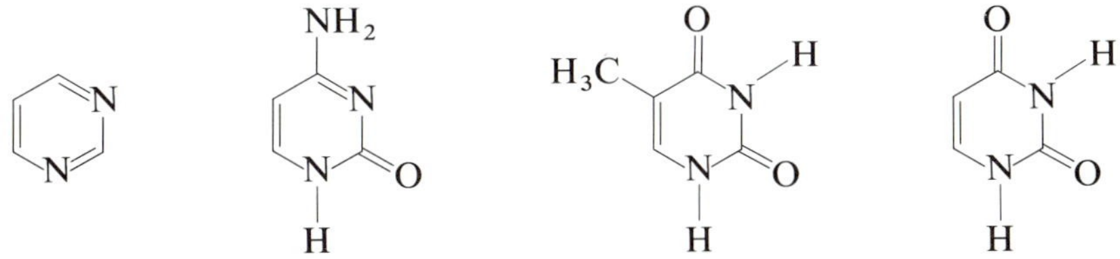

Pyrimidine   Cytosine   Thymine (DNA)   Uracil (RNA)

Structurally, nucleotides have the heterocyclic amine base bonded to carbon 1′ of the sugar, with the phosphoric acid connected by a phosphate ester bond to the hydroxyl group at carbon 5′. Thus, the four nucleotides found in DNA, called *deoxyribonucleotides*, have the structures shown in Figure 17.2. (In discussing RNA and DNA, numbers with a prime [′] superscript refer to positions on the sugar, and numbers without a prime refer to positions on the heterocyclic amine base.)

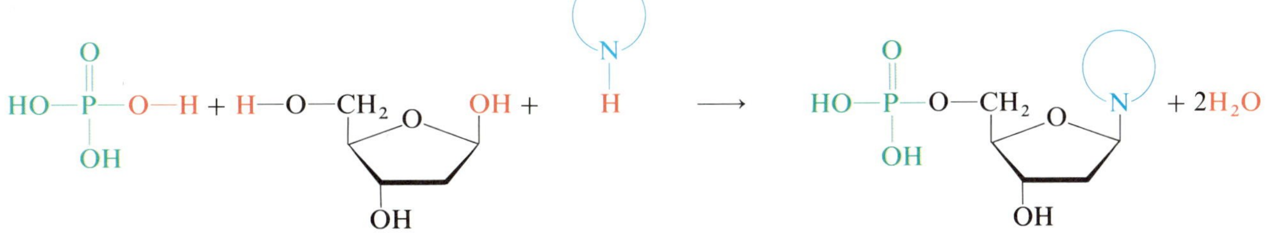

Phosphoric acid   2-Deoxyribose   Heterocyclic amine   A DNA nucleotide

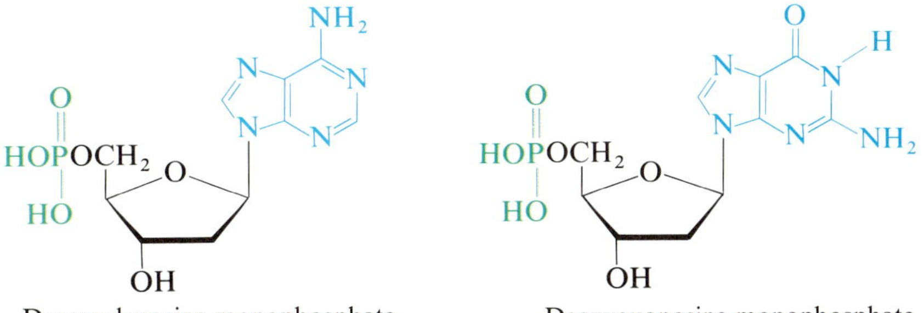

Deoxyadenosine monophosphate   Deoxyguanosine monophosphate

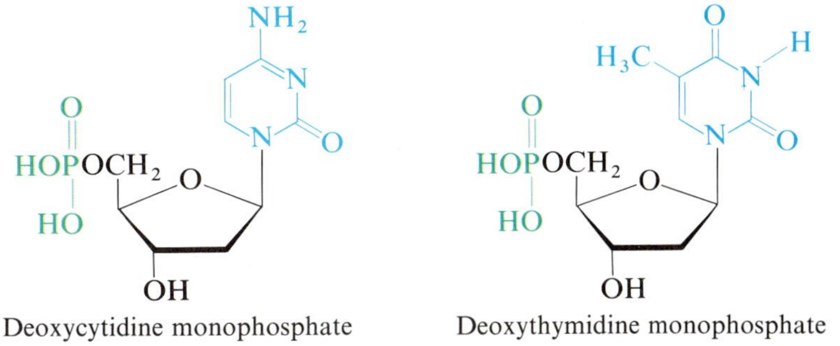

Deoxycytidine monophosphate   Deoxythymidine monophosphate

Figure 17.2
The structures of the four deoxyribonucleotides in DNA. The ribonucleotides in RNA are similar but have an additional —OH group at C2′ on the sugar.

Practice Problems   ***17.1***   Show how adenine and phosphoric acid can combine with ribose to form the RNA nucleotide, adenosine monophosphate.

17.2   Show how ribose and phosphoric acid can combine with uracil to form uridine monophosphate, the RNA nucleotide corresponding to deoxythymidine monophosphate.

17.3   THE STRUCTURE OF NUCLEIC ACID CHAINS

Two nucleotides join together in DNA and RNA by forming a phosphate ester bond between the phosphate component of one nucleotide and the carbon 3′ —OH group on the sugar component of the second nucleotide. Additional nucleotides then join by formation of still more phosphate ester bonds until ultimately an immense chain is formed (Figure 17.3). Regardless of how long the chain is, though, one end of the nucleic acid molecule always has a free —OH group at C3′ (called the *3′ end*), and the other end of the molecule always has a phosphoric acid group at C5′ (the *5′ end*).

   Just as the exact structure of a protein depends on the sequence in which the individual amino acids are connected (Section 15.4), the exact structure of a nucleic acid molecule depends on the sequence in which individual nucleotides

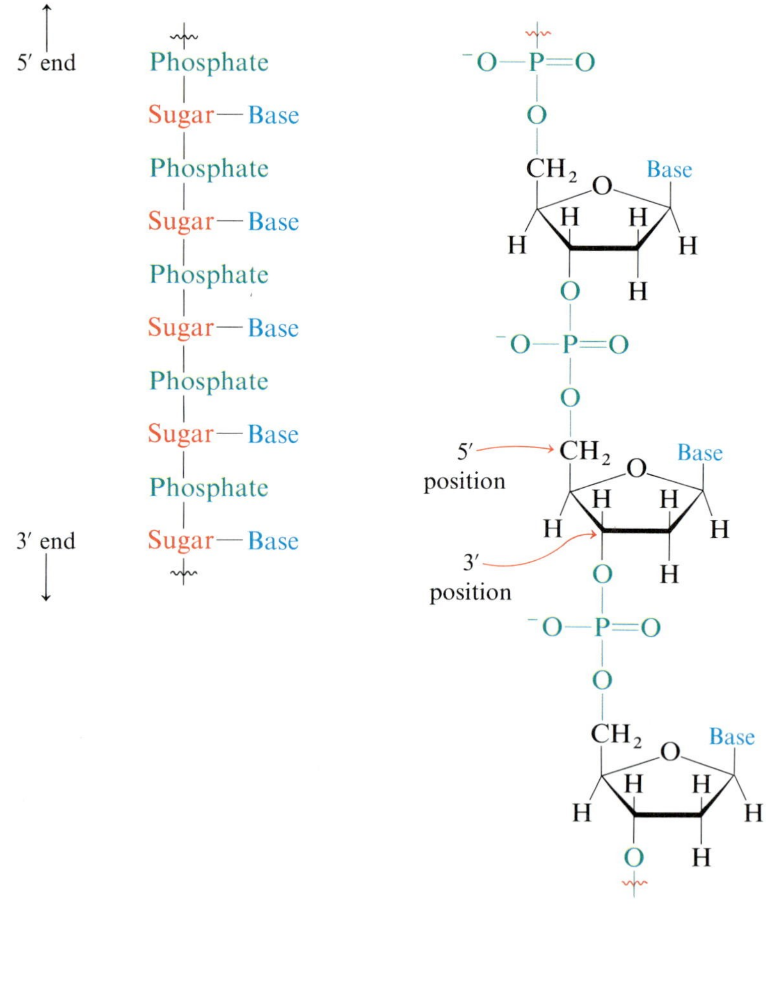

Figure 17.3
The generalized structure of a nucleic acid chain showing how phosphate ester bonds link the individual nucleotides together.

are connected. To carry the analogy even further, just as a protein has a polyamide backbone with different side chains attached to it at regular intervals, a nucleic acid has an alternating sugar-phosphate backbone with different heterocyclic amine bases attached to it at regular intervals.

A protein:

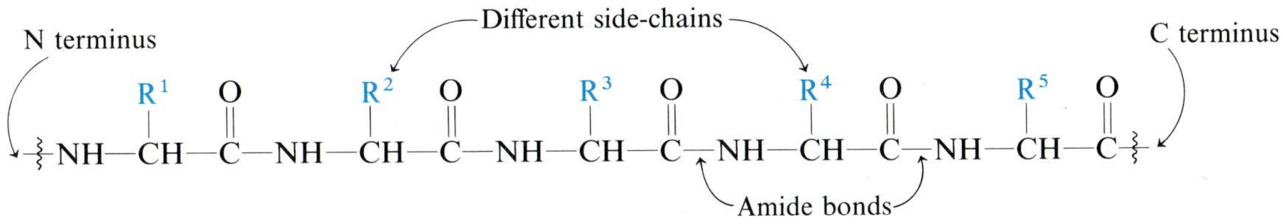

A nucleic acid:

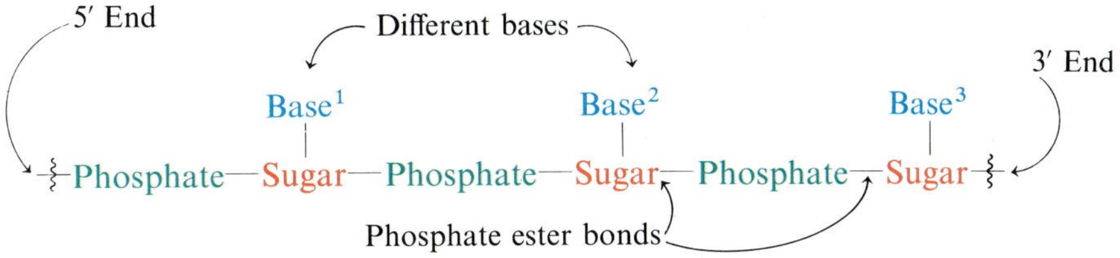

The sequence of nucleotides in a chain is described by starting at the 5' end and identifying the bases in order of occurrence. Rather than write the full name of each nucleotide, however, it's more convenient to use simple abbreviations: A for adenosine, T for thymine, G for guanosine, and C for cytidine. Thus, a typical sequence might be written as -T-A-G-G-C-T-.

Practice Problem **17.3** Write out the full structure of the DNA dinucleotide A-G.

17.4 BASE PAIRING IN DNA: THE WATSON-CRICK MODEL

DNA samples from different cells of the same species have the same proportions of the four heterocyclic bases, but samples from different species can have quite different proportions of bases. For example, human DNA contains about 30 percent each of adenine and thymine, and 20 percent each of guanine and cytosine. The bacterium *E. coli*, however, contains only about 24 percent each of adenine and thymine, and 26 percent each of guanine and cytosine. Note that in both cases, the bases occur in *pairs*. A and T are usually present in equal amounts, as are G and C. Why should this be?

In 1953, James Watson and Francis Crick proposed a structure for DNA that not only accounts for the pairing of bases but that also provides a simple method for the storage and transfer of genetic information. According to the *Watson-Crick model* (Figure 17.4), DNA consists of *two* polynucleotide strands coiled around each other in a helical, screw-like fashion. The sugar-phosphate backbone is on the outside of the so-called **double helix**, and the heterocyclic bases are on the inside so that a base on one strand points directly toward a base on the second strand (Figure 17.5).

■ **Double helix** Two nucleotide strands coiled around each other in a screw-like fashion.

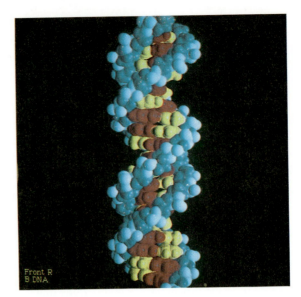

Figure 17.4
The double-helical structure of DNA. The sugar-phosphate backbone (blue) runs along the outside of the helix, while the amine bases (purines in red, pyrimidines in green) hydrogen bond to one another on the inside.

Figure 17.5
Complementary base pairing in the DNA double helix.

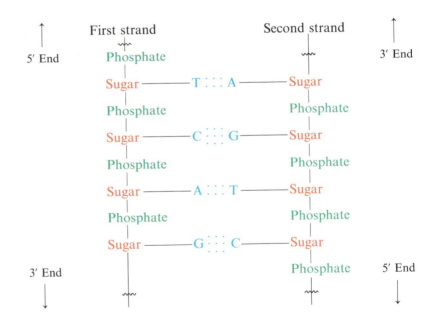

The two strands of the DNA double helix run in opposite directions and are held together by hydrogen bonds between bases. This hydrogen bonding is not random, however. Adenine and thymine form two strong hydrogen bonds to each other but not to cytosine or guanine. Similarly, cytosine and guanine form three strong hydrogen bonds to each other but not to adenine or thymine.

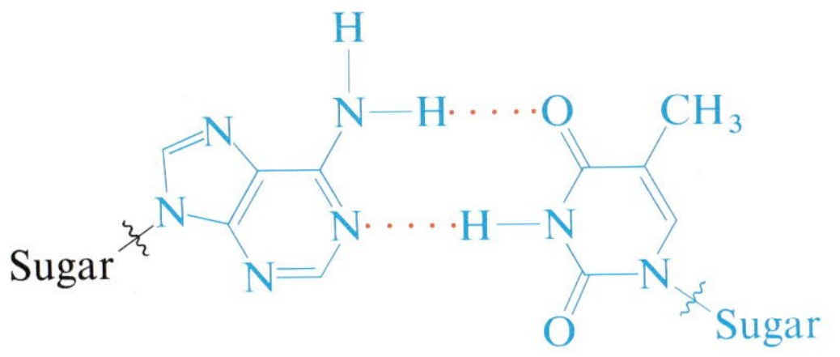

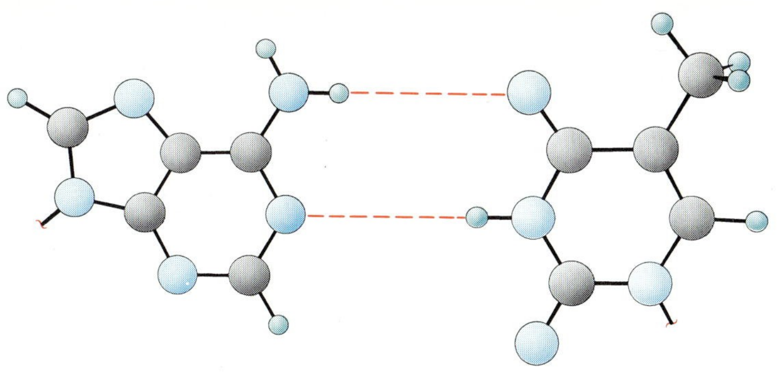

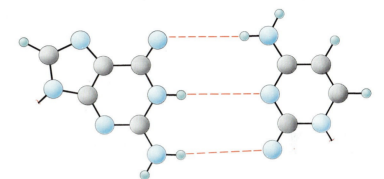

Adenine–Thymine (two hydrogen bonds)

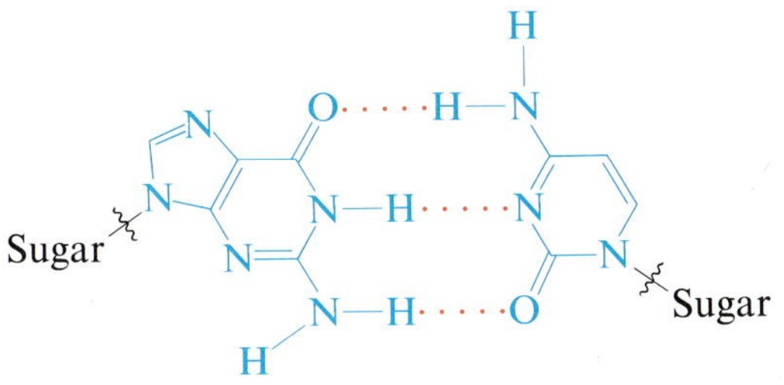

Guanine–Cytosine (three hydrogen bonds)

The two strands of the DNA double helix are *complementary* rather than identical. Whenever an A base occurs in one strand, a T base occurs opposite it in the other strand; when a C base occurs in one strand, a G base occurs in the other. This complementary *base pairing* in the two strands explains why A/T and C/G always occur in equal amounts. As indicated in Figure 17.4, the DNA double helix is like a badly twisted ladder, with the sugar-phosphate backbone making up the sides and the hydrogen-bonded base pairs the rungs. X-ray measurements show that the double helix is 2.0 nm wide, that there are exactly ten nucleotides in each full turn, and that each turn is 3.4 nm long. A helpful way to remember the base pairing in DNA is to memorize the phrase "Pure silver taxi."

Pure	Silver	Taxi
Pur	**Ag**	**TC**
The purines	A and G	pair with T and C.

Solved Problem 17.1 What sequence of bases of one strand of DNA is complementary to the sequence T-A-T-G-C-A-G on the other strand?

Solution Remembering that A and G (silver) bond to T and C (taxi) respectively, we go through the original sequence replacing each A by T, each G by C, each T by A, and each C by G.

Original: T-A-T-G-C-A-G

Complement: A-T-A-C-G-T-C

Practice Problem **17.4** What sequences of bases on one DNA strand are complementary to these sequences on another strand?
(a) G-C-C-T-A-G-T   (b) A-A-T-G-G-C-T-C-A

17.5 NUCLEIC ACIDS AND HEREDITY

How do organisms use nucleic acids to store and express genetic information? If the information is to be preserved and passed on from generation to generation, a mechanism must exist for copying DNA. If the information is to be used, mechanisms must exist for decoding the information and for carrying out the instructions they contain.

According to what has been called the *central dogma of molecular genetics*, the function of DNA is to store information and pass it on to RNA, while the function of RNA is to read, decode, and use the information received from DNA to make proteins. Each of the thousands of individual genes on each chromosome contains the instructions necessary to make a specific protein that is in turn needed for a specific biological purpose. By decoding the right genes at the right time in the right place, an organism can use genetic information to synthesize the many thousands of proteins necessary to carry out the biochemical reactions required for smooth functioning.

Three fundamental processes take place in the transfer of genetic information:

■ **Replication** The process by which copies of DNA are made in the cell.

■ **Transcription** The process by which the information in DNA is read and used to make RNA.

■ **Translation** The process by which RNA directs protein synthesis.

1. **Replication** is the process by which a replica, or identical copy, of DNA is made so that information can be preserved and handed down to offspring.
2. **Transcription** is the process by which the genetic messages contained in DNA are "read," or transcribed, and carried out of the nucleus to parts of the cell called *ribosomes* where protein synthesis occurs.
3. **Translation** is the process by which the genetic messages are decoded and used to build proteins.

Let's look at the three processes in order.

17.6 REPLICATION OF DNA

The Watson-Crick double-helix model of DNA does more than just explain base pairing; it also provides an ingenious way for DNA molecules to reproduce exact copies of themselves. DNA replication begins with a partial unwinding of the double helix. As the two DNA strands separate and the bases are exposed, new nucleotides line up by hydrogen bonding to each strand in an exactly complementary manner, A to T and G to C. When the enzyme DNA polymerase then catalyzes the bonding of these new nucleotides to their neighbors, two new strands begin to grow. Since each new strand is complementary to its old template strand, two identical new copies of the DNA double helix are produced. The process is shown schematically in Figure 17.6.

Crick probably described the DNA replication process best when he described the fitting together of two DNA strands as being like a hand in a glove.

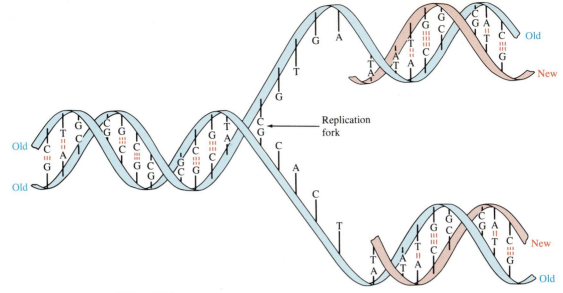

Figure 17.6
A schematic representation of DNA replication. The original DNA double helix partially unzips at a point called the *replication fork*, and complementary new nucleotides line up on each strand. When the new nucleotides are joined by DNA polymerase, two identical new DNA molecules result.

The hand and glove separate; a new hand forms inside the old glove, and a new glove forms around the old hand. Two identical copies now exist where only one existed before.

It's difficult to conceive of the magnitude of the replication process. The sum of all genes in a human cell—the **genome**—is estimated to be approximately three billion base pairs. A single DNA chain might have a length of over 12 cm and contain up to 250 million pairs of bases. Regardless of the size of these enormous molecules, their base sequence is faithfully copied during replication. The copying process takes only minutes, and an error occurs only about once each 10–100 billion bases.

■ **Genome** The sum of all genes in an organism.

17.7 STRUCTURE AND SYNTHESIS OF RNA: TRANSCRIPTION

RNA is structurally similar to DNA—both are sugar-phosphate polymers and both have heterocyclic bases attached—but there are important differences. We've seen, for example, that the sugar component in RNA is ribose rather than 2-deoxyribose and that uracil is present instead of thymine. Uracil in RNA forms strong hydrogen bonds to its complementary base, adenine, just as thymine does in DNA.

Adenine   Uracil (in RNA)   Thymine (in DNA)

■ **Messenger RNA (mRNA)**
The RNA whose function is to carry genetic messages transcribed from DNA.

■ **Ribosome** A structure in the cell where protein synthesis occurs.

■ **Ribosome RNA (rRNA)**
The structural material of ribosomes.

■ **Transfer RNA (RNA)** The RNA whose function is to transport a specific amino acid into position for protein synthesis.

Another difference between DNA and RNA is that there are three main kinds of ribonucleic acid, each of which has a specific function. **Messenger RNA's (mRNA)** are the primary information-bearing molecules that carry genetic messages from DNA to **ribosomes**, small granular structures in the cytoplasm of a cell that act as "factories" for making proteins. **Ribosomal RNA's (rRNA)** are structural molecules that bond to protein and help to provide the physical makeup of the ribosomes. **Transfer RNA's (tRNA)** are both informational and structural in that they transport specific amino acids to the ribosomes where they are joined together to make proteins. All three kinds of RNA are similar to DNA, but they are much smaller molecules than DNA, and they remain single-stranded rather than double-stranded.

The process of converting the information contained in a DNA segment into proteins begins with the synthesis of mRNA molecules that contain anywhere from several hundred to several thousand ribonucleotides, depending on the size of the protein to be made. Each of the 100,000 or so proteins in the human body is synthesized from a different mRNA that has been transcribed from a specific gene on DNA. Messenger RNA is synthesized in the cell nucleus by *transcription* of DNA, a process similar to DNA replication. As in replication, a small section of the DNA double helix unwinds, and the bases on the two strands are exposed. RNA nucleotides (*ribonucleotides*) line up in the proper order by hydrogen bonding to their complementary bases on DNA, the nucleotides are joined together by the enzyme RNA polymerase, and mRNA results.

Unlike what happens in DNA replication, where both strands are copied, only one of the two DNA strands is transcribed into mRNA. The strand that is transcribed is called the *template strand*, while its complement is called the *informational strand*. Since the template strand and the informational strand are complementary, and since the template strand and the mRNA molecule are also complementary, it follows that *the messenger RNA molecule produced during transcription is a copy of the DNA informational strand*. The only difference is that the mRNA molecule has a U-base everywhere that the DNA informational strand has a T-base. The transcription process is shown schematically in Figure 17.7, an electron micrograph of a gene being transcribed in Figure 17.8.

Transcription of DNA by the process just discussed raises as many questions as it answers. How does the DNA know where to unwind? Where along the chain does one gene stop and the next one start? How do the ribonucleotides know exactly the right place along the template strand to begin lining up and the right place to stop? Although these questions are extremely difficult to answer, our current picture is that a DNA chain contains certain base sequences called *promoter sites*, which bind to the RNA polymerase enzyme that actually carries out RNA synthesis, thus signalling the beginning of a gene. Similarly, there are other base sequences at the end of the gene that signal a stop.

Another part of the picture is the recent discovery that genes are not necessarily continuous segments of the DNA chain. Often a gene will begin in one small section of DNA, called an **exon**, then be interrupted by a seemingly nonsensical section called an **intron** that does not code for any part of the protein to be synthesized, and then take up again farther down the chain in another exon. The final mRNA molecule only results after the nonsense sections are cut out and the remaining pieces spliced together (Figure 17.9, page 376). Current evidence is that up to 90 percent of human DNA is made up of introns and only about 10 percent of DNA actually contains genetic instructions.

■ **Exon** A DNA segment in a gene that codes for part of a protein molecule.

■ **Intron** A seemingly nonsensical segment of DNA that occurs in a gene but does not code for part of the protein.

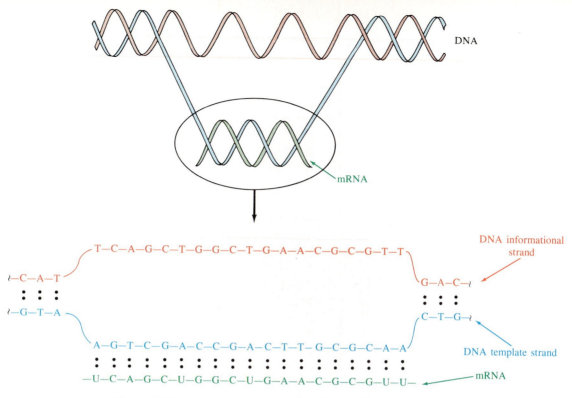

T—C—A—G—C—T—G—G—C—T—G—A—A—C—G—C—G—T—T

}—C—A—T
: : :
}—G—T—A

A—G—T—C—G—A—C—C—G—A—C—T—T—G—C—G—C—A—A
: : : : : : : : : : : : : : : : : : :
—U—C—A—G—C—U—G—G—C—U—G—A—A—C—G—C—G—U—U—

G—A—C—{   DNA informational strand
: : :
C—T—G—{   DNA template strand

mRNA

Figure 17.7
The transcription of DNA to synthesize mRNA. The mRNA molecule produced is complementary to the template strand from which it is transcribed and is identical to the informational strand except for the replacement of T by U.

Figure 17.8
Electron micrograph of several genes (long strand) undergoing transcription simultaneously by many molecules of RNA polymerase. The growing mRNA molecules extend perpendicularly from the DNA.

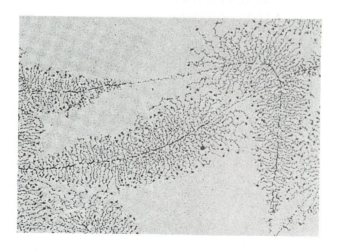

Figure 17.9
As initially transcribed from DNA, mRNA contains both informational segments (exons, in red) and nonsense segments (introns, in black). The finished mRNA molecule has the introns removed.

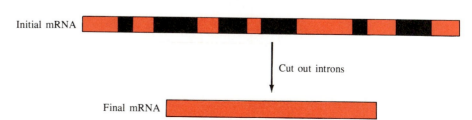

Initial mRNA

Cut out introns

Final mRNA

Practice Problem   **17.5**   What mRNA sequences are complementary to each of the following DNA template sequences?
(a) -G-A-T-T-A-C-C-G-T-A-
(b) -T-A-T-G-G-C-T-A-G-G-C-A-

17.8   THE GENETIC CODE

How does an organism use the thousands of genetic messages transcribed from DNA into mRNA? Since all biological processes are ultimately regulated by proteins, the primary function of mRNA is to direct the biosynthesis of the thousands of diverse peptides and proteins required by an organism. Protein synthesis takes place outside the cell nucleus on ribosomes, each of which consists of about 60 percent rRNA and 40 percent protein and has a total molecular weight of approximately 5,000,000 amu.

■ **Codon**   A sequence of three ribonucleotides in the RNA chain that codes for a specific amino acid.

The ribonucleotide sequence in an mRNA chain acts like a long coded sentence to specify the order in which different amino-acid residues should be joined to form a protein. Each "word" or **codon** in the mRNA sentence consists of a series of three ribonucleotides that is specific for a particular amino acid. For example, the series cytosine-uracil-guanine (C-U-G) on an mRNA chain is a codon directing incorporation of the amino acid leucine into a growing protein. Similarly, the sequence guanine-adenine-uracil (G-A-U) codes for aspartic acid. Of the 64 possible three-base combinations in RNA, 61 code for specific amino acids, and 3 code for chain termination. The meaning of each

Table 17.1   Codon Assignments of Base Triplets

First Base (5' end)	Second Base	Third Base (3' end)			
		U	C	A	G
U	U	Phe	Phe	Leu	Leu
	C	Ser	Ser	Ser	Ser
	A	Tyr	Tyr	Stop	Stop
	G	Cys	Cys	Stop	Trp
C	U	Leu	Leu	Leu	Leu
	C	Pro	Pro	Pro	Pro
	A	His	His	Gln	Gln
	G	Arg	Arg	Arg	Arg
A	U	Ile	Ile	Ile	Met
	C	Thr	Thr	Thr	Thr
	A	Asn	Asn	Lys	Lys
	G	Ser	Ser	Arg	Arg
G	U	Val	Val	Val	Val
	C	Ala	Ala	Ala	Ala
	A	Asp	Asp	Glu	Glu
	G	Gly	Gly	Gly	Gly

■ **Genetic code** The assignment of mRNA codons to specific amino acids.

codon—the **genetic code** universal to all living organisms—is shown in Table 17.1. Note that most amino acids are specified by more than one codon.

Practice Problems

17.6 List possible codon sequences for the following amino acids:
(a) Ala   (b) Phe   (c) Leu   (d) Val   (e) Tyr

17.7 What amino acids do the following sequences code for?
(a) A-U-U   (b) G-C-G   (c) C-G-A   (d) A-A-C

17.9   RNA AND PROTEIN BIOSYNTHESIS: TRANSLATION

The message contained in mRNA is decoded by tRNA. There are 61 different tRNA's, one for each "word" in the genetic code. Each different tRNA acts as a carrier to transport a specific amino acid into place on the ribosome so that it can be incorporated into the growing protein chain. A typical tRNA is shaped something like a cloverleaf, as shown in Figure 17.10. It contains about 70 to 100 ribonucleotides and is bonded to its specific amino acid by an ester linkage

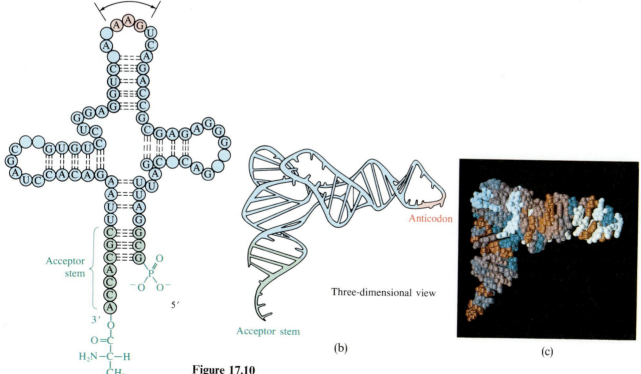

Figure 17.10
(a) Schematic, "flattened" view of the structure of a tRNA molecule. The cloverleaf-shaped tRNA contains an anticodon triplet on one "leaf" and a covalently bonded amino acid at its 3′ end. The example shown is a yeast tRNA that codes for phenylalanine. (The nucleotides not specifically identified are slightly altered analogs of the four normal ribonucleotides.) (b) The three-dimensional shape of the molecule. (c) A computer-generated space filling model.

between the —COOH of the amino acid and the free —OH group at the C3 position of ribose on the 3′ end of the tRNA chain.

Let's look at the three main steps in protein synthesis: initiation, elongation, and termination.

Initiation   Protein synthesis is initiated when an mRNA molecule approaches a ribosome. Ribosomes consist of two parts or subunits, the larger one called the *60S subunit* and the smaller one called the *40S subunit*. Although the full details are not clear, it's thought that the two subunits come apart, the mRNA attaches to the smaller subunit, and the two subunits then join. Transfer RNA molecules, each carrying its assigned amino acid, then approach the ribosome (Figure 17.11).

Each tRNA molecule contains near its halfway point a section called an **anticodon**, a sequence of three ribonucleotides complementary to a specific codon sequence on mRNA. For example, the codon sequence C-U-G on mRNA matches up with the complementary anticodon sequence G-A-C on a leucine-bearing tRNA. Similarly, the codon sequence G-A-U matches up with the anticodon sequence C-U-A on an aspartic acid-bearing tRNA. The first codon on mRNA is always A-U-G, which acts as a "start" signal for the translation machinery and codes for the introduction of a methionine unit. Thus, a methionine tRNA molecule approaches the ribosome and attaches sto the larger subunit at what is called the *P binding site*. It is at this P site that peptide bond formation occurs.

Elongation   Next to the P binding site on the ribosome is the *A binding site*, the point where the next (second) codon on mRNA is exposed. All 61 tRNA molecules can approach and try to fit, but only the one with the right anticodon sequence can bind. Let's assume that the second codon in our mRNA is C-U-G, which codes for leucine. A leucine-bearing tRNA with a G-A-C anticodon thus binds to the A site, a peptide bond is then formed between leucine and methionine, and protein synthesis is underway.

> ■ **Anticodon** A sequence of three ribonucleotides on tRNA that recognizes the complementary sequence on mRNA.

Figure 17.11
Events in the initiation of protein synthesis. An mRNA molecule joins with two ribosomal subunits and a methionine tRNA at the P binding site.

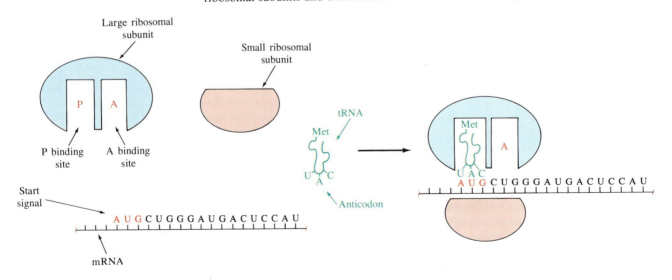

With the first peptide bond formed, the entire ribosome shifts three positions (one codon) along the mRNA chain, a process called *translocation*. At the same time, the methionine tRNA previously at the P site is freed from the ribosome, the leucine tRNA previously at the A site is moved to the P site, and the A site is opened up to accept the next tRNA carrying the next amino acid for the peptide chain.

The three elongation steps now repeat: The next appropriate tRNA binds to the A site, peptide bond formation occurs to attach the newly arrived amino acid to the growing chain, and translocation takes place to free the A site for a further elongation cycle. The entire sequence of events is shown schematically in Figure 17.12, and an electron micrograph showing ribosomes translating a strand of mRNA is shown in Figure 17.13 (page 380).

Termination   When synthesis of the proper protein is completed, a "stop" codon (U-A-A, U-G-A, or U-A-G) signals the end of the process. An enzyme called a *releasing factor* frees the polypeptide chain from the last tRNA, and the mRNA molecule is released from the ribosome. The mRNA molecule is then free to join with another ribosome and repeat the entire translation process many times to provide many copies of the protein for which it codes.

Figure 17.12
Schematic representation of protein elongation. (a) After initiation, the second tRNA approaches the ribosome and (b) binds at the A site. (c) Peptide bond formation between the first two amino acids then occurs, and translocation shifts the entire ribosome one codon further down the mRNA chain. The first tRNA is thus freed from the P site, the tRNA previously at the A site shifts to the P site, and the A site is made available for the next tRNA. (d) The cycle then repeats.

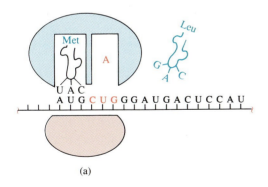

(a)

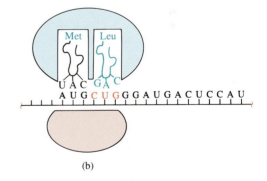

(b)

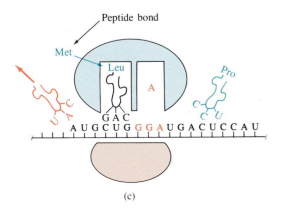

(c)

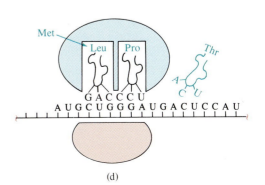

(d)

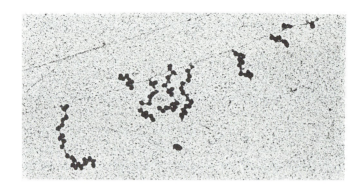

Figure 17.13
An electron micrograph of ribosomes clustering on several mRNA chains to carry out protein synthesis.

Solved Problem 17.2   What amino-acid sequence is coded for by the mRNA base sequence AUC-GGU?

Solution   Table 17.1 indicates that AUC codes for isoleucine and that GGU codes for glycine. Thus, AUC-GGU codes for H_2N-Ile-Gly-COOH.

Practice Problems   **17.8**   What amino-acid sequence is coded for by the following mRNA base sequence?

CUU-AUG-GCU-UGG-CCC-UAA

17.9   What anticodon sequences of tRNA's are coded for by the mRNA in Problem 17.8?

17.10   What was the base sequence in the original DNA template strand on which the mRNA sequence in Problem 17.8 was made?

17.10   SEQUENCING OF DNA

Newspapers and magazines are full of stories about the remarkable advances in genetic engineering that have revolutionized biology in the last decade. None of these advances would have been possible, however, were it not for the development by Allan Maxam and Walter Gilbert in 1977 of a remarkably powerful and efficient method of DNA sequencing. The **Maxam-Gilbert method** involves five steps:

■ **Maxam-Gilbert method**   A rapid and efficient method for sequencing long strands of DNA.

■ **Restriction endonuclease**   An enzyme that is able to cut a DNA strand at a point in the chain where a specific base sequence occurs.

Step 1.   A DNA molecule is first cut by enzymes called **restriction endonucleases** to yield smaller, more manageable DNA fragments containing 100 to 200 base pairs. Each different restriction enzyme, of which more than 200 are available, cleaves a DNA molecule between two nucleotides at those points along the chain where a specific base sequence occurs.

By cleavage of large DNA molecules with a given restriction enzyme, many different and well-defined segments of manageable length are produced. If the original DNA molecule is cut with another restriction enzyme having a different specificity for cleavage, still other segments are produced, whose sequences

AN APPLICATION: VIRUSES

Viruses are small structures consisting of nucleic acid wrapped in a protective coat of protein. As indicated in the table, the nucleic acid may be either DNA or RNA, it may be either single-stranded or double-stranded, and it may consist either of a single piece or of several pieces. Many hundreds of different viruses are known, each of which can infect a particular type of plant or animal cell.

Viruses occupy the grey area between living and nonliving. By itself, a virus has none of the cellular machinery necessary for replication. Once it enters a living cell, though, a virus can take over the host cell and force it to produce additional copies of the virus. Some infected cells eventually die, but others continue to produce viral copies, which then leave the host and spread the infection to other cells.

Viral infection begins when a virus particle enters a host cell, loses its protein coat, and releases its nucleic acid. What happens next depends on whether the virus is based on DNA or RNA. If the infectious agent is a DNA virus, the host cell first replicates the viral DNA and then decodes it in the normal way. The viral DNA is transcribed to produce RNA, and the RNA is translated to synthesize viral coat proteins. Copies of the viral DNA don their protein coats, and the newly formed virus particles are released from the cell. If the infectious agent is an RNA virus, however, a problem exists. Either the cell must transcribe and produce proteins directly from the viral RNA template, or else it must first produce DNA from the viral RNA by a process called *reverse transcription*. Viruses that follow the reverse transcription route are called *retroviruses*; the HIV-1 virus responsible for AIDS is an example.

Some Typical Viruses

Virus	Type of Nucleic Acid	Mol. Wt. ($\times 10^6$)	Number of Genes
Influenza	Single-stranded RNA	4*	12
Polio	Single-stranded RNA	2.5	7
Reovirus	Double-stranded RNA	15*	22
Parvovirus	Single-stranded DNA	1.5	5
Herpes simplex	Double-stranded DNA	100	150

* Virus contains several RNA strands

partially overlap those produced by the first enzyme. Sequencing of all the segments, followed by identification of the overlapping sections, then allows complete DNA structure determination.

Step 2. The various double-stranded fragments are isolated, and each is marked by incorporating a radioactively labeled phosphate group (^{32}P) onto the 5′-hydroxyl of the terminal nucleotide. The fragments are then separated into two strands by heating, and the strands are isolated. For example, imagine that we have now isolated a single-stranded DNA fragment of approximately 100 nucleotides with the following partial structure:

Radioactive phosphate

(5′ end) ^{32}P-A-G-T-A-C-C-G-A-T-T---   (3′ end)

Step 3. A radioactively labeled fragment is subjected to four parallel sets of chemical reactions whose conditions have been carefully defined so that: (a) splitting of the DNA chain next to A occurs, (b) splitting next to G occurs, (c) splitting next to C occurs, and (d) splitting next to T occurs. Mild reaction conditions are chosen so that *only a few of the many possible splittings occur in each reaction.* In our example, the pieces shown in Table 17.2 might be produced.

Table 17.2 Splitting of a DNA Fragment under Four Sets of Conditions

Cleavage Conditions	DNA Pieces Produced
Original DNA fragment	^{32}P-A-G-T-A-C-C-G-A-T-T---
A	^{32}P-A-G-T ^{32}P-A-G-T-A-C-C-G + larger pieces
G	^{32}P-A ^{32}P-A-G-T-A-C-C + larger pieces
C	^{32}P-A-G-T-A ^{32}P-A-G-T-A-C + larger pieces
T	^{32}P-A-G ^{32}P-A-G-T-A-C-C-G-A ^{32}P-A-G-T-A-C-C-G-A-T + larger pieces

Step 4. Product mixtures from the four cleavage reactions are separated by *gel electrophoresis.* When each mixture is placed at one end of a strip of buffered gelatinous polyacrylamide and a voltage is applied across the ends of the strip, electrically charged pieces migrate along the gel. Each piece moves at a rate that depends both on its size and on the number of negatively charged phosphate groups (that is, the number of nucleotides) it contains. Smaller pieces move rapidly, and larger pieces move more slowly. The technique is so sensitive that up to 250 DNA pieces, differing in size by only one nucleotide, can be separated.

With separation of the pieces accomplished, the position on the gel of each radioactive piece is determined by exposing the gel to a photographic plate. Only the pieces containing the radioactively labeled 5′ phosphate-end group show up as dark spots on the plate; unlabeled pieces from the middle of the chain don't appear. The gel electrophoresis pattern shown in Figure 17.14 would be obtained in our hypothetical example.

Step 5. The DNA sequence is read directly from the gel. The band that appears farthest from the origin is the terminal mononucleotide (the smallest piece) and can't be identified. Since the terminal mononucleotide appears in the G column, however, *it must have been produced by splitting next to a G.* Thus, the *second* nucleotide in the DNA fragment is a G.

The second farthest band from the origin is a dinucleotide that appears in the T column and is produced by splitting next to the third nucleotide, which must therefore be a T. The third farthest band appears in the A column, which means that the fourth nucleotide is an A. Continuing in this manner, the entire sequence of the DNA fragment is read from the gel simply by noting in what column the successively larger labeled polynucleotide pieces appear.

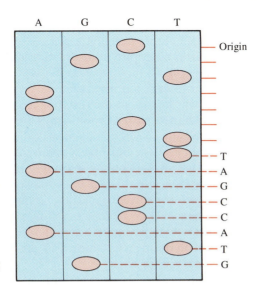

Figure 17.14
Representation of a gel electrophoresis pattern. The products of the four cleavage experiments are placed at the top of the gel, and a voltage is applied between top and bottom. Smaller products migrate down the gel at a faster rate and thus appear at the bottom. The DNA sequence can be read by exposing the gel to a photographic plate.

So powerful and efficient is the Maxam-Gilbert method that a trained person can sequence up to 2000 base pairs per day. DNA strands of up to *170,000* base pairs have already been sequenced, and a campaign is being mounted to undertake sequencing of the entire human genome containing *3,000,000,000* base pairs. Such an undertaking might take up to ten years and cost several billion dollars.

Practice Problems

17.11   Finish assigning the sequence to the gel electrophoresis pattern shown in Figure 17.14.

17.12   Show the labeled products you would expect to obtain if the following DNA segment were subjected to each of the four cleavage reactions:

$$^{32}P\text{-G-A-T-A-C-G-G-A-T-C-G-G}$$

17.13   Sketch what you would expect the gel electrophoresis pattern to look like if the DNA segment in Practice Problem 17.12 were sequenced.

17.11   GENE MUTATION AND HEREDITARY DISEASE

The base-pairing mechanism of DNA replication and RNA transcription is an extremely efficient and accurate method for preserving and using genetic information, but it's not perfect. Occasionally an error occurs, resulting in the incorporation of an incorrect base at some point.

If an occasional error occurs during the transcription of an mRNA molecule, the problem is not too serious. After all, large numbers of mRNA molecules are continually being produced, and an error that occurs only one out of a million or so times would hardly be noticed. If an occasional error occurs during

■ **Mutation**   An error in base sequence occurring during DNA replication.

■ **Mutagen**   A substance that causes mutations.

the replication of a *DNA* molecule, however, the consequences can be far more damaging. Each chromosome in a cell contains only *one* DNA molecule, and if that molecule is miscopied during replication, then the error is passed on when the cell divides.

An error in base sequence that occurs during DNA replication is called a **mutation**. Some mutations occur spontaneously, others are caused by exposure to ionizing radiation such as cosmic rays and gamma rays, and still others are caused by exposure to certain chemicals called **mutagens**. Some, perhaps most, mutations are harmless, but others can be devastating. Imagine, for example, that the sequence A-T-G on the informational strand of DNA is miscopied as A-C-G during replication. The mRNA transcribed from the corresponding template strand will then have the incorrect codon sequence A-C-G rather than the correct sequence A-U-G. But since A-C-G codes for threonine whereas A-U-G codes for methionine, an incorrect amino acid will be inserted into the corresponding protein during translation. Furthermore, *every* copy made of the protein will have the same error.

The biological effects of incorporating an incorrect amino acid into a protein can range from negligible to catastrophic. If the incorrect amino acid occurs at some unimportant site, there may be little or no change in the biological properties of the protein. If, however, the error occurs at an important point, the biological activity of the protein can be completely changed.

■ **Somatic cell**   Any cell other than a reproductive one.

■ **Germ cell**   A reproductive (sperm or egg) cell.

When mutation occurs in a **somatic cell**, meaning any cell other than sperm or egg, that cell might undergo uncontrolled growth leading to cancer. When mutation occurs in a **germ cell** (sperm or egg), then the genetic alteration is passed on to the offspring, where it might show up as a hereditary disease. There are some 2000 known hereditary diseases that affect humans, with consequences that are sometimes fatal. Some of the more common ones are listed in Table 17.3.

The biochemical nature of the genetic defects in some hereditary diseases is well understood, but the nature of many others has eluded investigation. Sickle-cell anemia, a hereditary disease that affects one of every 400 persons of African ancestry, is caused by the substitution of only one amino acid in blood hemoglobin (Figure 17.15). A valine residue is substituted for a glutamic-acid residue six amino acids in from the N-terminus of hemoglobin. Cystic fibrosis, however, a hereditary disease affecting one of every 1600 Caucasians, remains of unknown origin.

Table 17.3   Some Common Hereditary Diseases and Their Causes

Name	Nature and Cause of Defect
Phenylketonuria	Brain damage in infants caused by the defective enzyme phenylalanine hydroxylase
Albinism	Absence of skin pigment caused by the defective enzyme tyrosinase
Tay-Sachs disease	Mental retardation caused by a defect in production of the enzyme hexosaminidase A
Cystic fibrosis	Bronchopulmonary, liver, and pancreatic obstructions by thickened mucus; cause unknown
Sickle-cell anemia	Anemia and obstruction of blood flow caused by a defect in hemoglobin

Figure 17.15
Photograph of sickled red blood cells of a patient with sickle-cell anemia. (Compare the normal cells shown in Figures 1.5 and 6.7.)

Practice Problems

17.14   If the sequence T-G-G in a gene on a DNA informational strand mutated and became T-G-A, the protein encoded by that gene would stop short of completion. Explain.

17.15   What changes would result in the proteins made from genes in which the following mutations had occurred?
(a) A-T-C becomes A-T-G   (b) C-C-T becomes C-G-T

SUMMARY

Deoxyribonucleic acid (DNA) and **ribonucleic acid (RNA)** are the chemical carriers of an organism's genetic information. Most DNA in higher organisms is found in the nucleus of cells in the form of thread-like strands called **chromosomes**. Each chromosome consists of several thousand **genes**, or sections of the DNA chain, that encode instructions for constructing a single protein.

Nucleic acids are made up of many individual building blocks, called **nucleotides**, linked together to form a long chain. Each nucleotide consists of a **heterocyclic amine base** linked to C1 of an aldopentose sugar, with the sugar in turn linked through its C5 —OH group to phosphoric acid. The sugar component in RNA is ribose, while that in DNA is 2-deoxyribose. DNA contains two **purine** amine bases (**adenine** and **guanine**) and two **pyrimidine** bases (**cytosine** and **thymine**); RNA contains adenine, guanine, cytosine, and a second pyrimidine base called **uracil** in place of thymine. Nucleotides join together to yield nucleic acid chains by formation of an ester link between the phosphate component of one nucleotide and the —OH group at C3 on the sugar of the neighboring nucleotide.

Molecules of DNA consist of two polynucleotide strands, held together by hydrogen bonds between bases on the two strands and coiled into a **double helix**. Adenine (A) and thymine (T) form hydrogen bonds only to each other, as do cytosine (C) and guanine (G). Thus, the two strands are **complementary**

INTERLUDE: RECOMBINANT DNA

The revolution in molecular biology that began in the late 1970s owes its existence to the discovery of powerful techniques for manipulating the transfer and expression of genetic information. Using what has come to be called *recombinant DNA* technology, it is actually possible to cut a specific gene out of one organism and recombine it into the genetic machinery of a second organism. Normally, a specific gene from a higher organism is spliced into the DNA of a bacterium, and the bacterium then manufactures the specified protein, perhaps insulin or some other valuable material.

Bacterial cells, unlike the cells of higher organisms, contain part of their DNA in circular pieces called *plasmids*, each of which carries just a few genes. Plasmids are extremely easy to isolate. Several copies of each plasmid are present in a cell, and each plasmid replicates through the normal base-pairing pathway.

To prepare a plasmid for insertion of a foreign gene, the plasmid is first cut open by use of a restriction endonuclease—the same technique used in Maxam-Gilbert sequencing. For example, the much-used restriction endonuclease *Eco*RI is capable of cutting a plasmid between G and A in the sequence G-A-A-T-T-C. The cut is not straight, however, but is slightly offset so that both DNA strands are left with a few unpaired bases on each end—so-called *sticky ends*.

Once the plasmid has been cut, the specific gene to be inserted is then cut from its chromosome using the same restriction endonuclease so that the sticky ends on the gene fragment are complementary to the sticky ends on the opened plasmid. The gene fragment and opened plasmid are then mixed in the presence of a DNA ligase enzyme that joins them together and reconstitutes the now altered plasmid. The entire sequence of events is shown schematically in the figure.

Once the altered plasmid has been made, it can be inserted back into a bacterial cell where the normal processes of transcription and translation take place to synthesize the protein encoded by the inserted gene. Since bacteria multiply rapidly, there are soon a large number of them, all containing the altered plasmid and all manufacturing the desired protein. In a sense, the bacterium has been harnessed and put to work as a protein factory.

Among the present commercial uses of recombinant DNA technology are the syntheses of human insulin, human growth hormone, and interferon, an antiviral protein that shows promise of being able to provide some protection against the common cold. Future research may well provide the ability to apply recombinant DNA techniques to humans, perhaps providing cures for hereditary diseases.

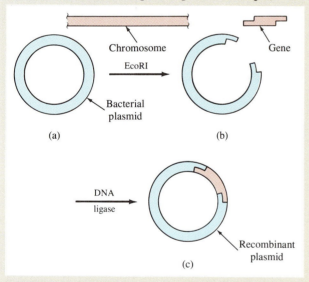

A schematic representation for preparing recombinant DNA. (a) A bacterial plasmid and a chromosome of a higher organism are cut with the same restriction enzyme, leaving "sticky ends." (b) The opened plasmid and the desired gene fragment are then joined by DNA ligase enzyme to yield (c) an altered plasmid that is reinserted into a bacterial cell.

rather than identical. Whenever an A occurs in one strand, a T occurs opposite it in the other strand; when a G occurs in one strand, a C occurs opposite it. Three main processes take place in the transfer of genetic information:

1. **Replication** of DNA is the process by which identical copies of DNA are made and genetic information is preserved. The DNA double helix partially unwinds, complementary deoxyribonucleotides line up in order, and two new DNA molecules are produced.

2. **Transcription** is the process by which RNA is produced to carry the genetic information from the cell nucleus to **ribosomes**. A segment of the DNA double helix unwinds, and complementary ribonucleotides line up on the **template strand** of DNA. The **messenger RNA (mRNA)** that is produced is an exact copy of the **informational strand** of the DNA.

3. **Translation** is the process by which mRNA directs protein synthesis. Each mRNA has three-base segments called **codons** along its chain. Each codon is recognized by a small amino-acid–carrying molecule of **transfer RNA (tRNA)** that delivers the appropriate amino acid to the ribosomes in the correct order for protein synthesis.

DNA sequencing is done by the **Maxam-Gilbert method**, in which chemical reactions are carried out to cause specific cleavages of the DNA chain, followed by separation of the fragments on a gel. The DNA sequence is then read directly from the gel.

Mutations occur when an error is made during DNA replication, an incorrect base is placed into the DNA chain, and a protein containing an incorrect amino-acid sequence is produced. If the genetic alteration occurs in a sperm or egg cell, the mutation is passed on to offspring.

REVIEW PROBLEMS

Structure and Function of Nucleic Acids

17.16 What are the full names of the two kinds of nucleic acid?

17.17 What is a nucleotide, and what three kinds of components does it contain?

17.18 What are the names of the sugars in DNA and RNA, and how do they differ?

17.19 What does the prefix *deoxy* mean when used in DNA?

17.20 What are the names of the four heterocyclic bases in DNA?

17.21 What are the names of the four heterocyclic bases in RNA?

17.22 Where in the cell is most DNA found?

17.23 What are the three main kinds of RNA, and what is the function of each?

17.24 What is meant by these terms?
(a) base pairing   (b) replication (of DNA)
(c) translation   (d) transcription

17.25 What is the difference between a gene and a chromosome?

17.26 What genetic information does a single gene contain?

17.27 Approximately how many genes are in human DNA?

17.28 What kind of bonding holds the DNA double helix together?

17.29 What does it mean to speak of bases as being "complementary"?

17.30 Rank the following kinds of nucleic acids in order of size: DNA, mRNA, tRNA.

17.31 What is the difference between a nucleotide's 3′ end and its 5′ end?

17.32 Show by drawing structures how the phosphate and sugar components of a nucleic acid are joined.

17.33 Show by drawing structures how the sugar and heterocyclic base components of a nucleic acid are joined.

17.34 Draw the complete structure of deoxycytidine 5′-phosphate, one of the four deoxyribonucleotides.

Cytosine

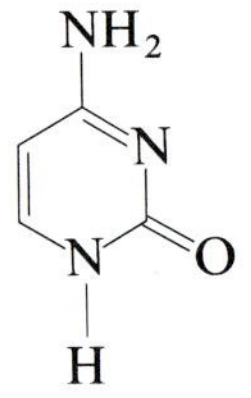

Nucleic Acids and Heredity

17.35 What are the names of the two DNA strands, and how do they differ?

17.36 What is a codon, and on what kind of nucleic acid is it found?

17.37 What is an anticodon, and on what kind of nucleic acid is it found?

17.38 What is the general shape and structure of a tRNA molecule?

17.39 What are "exons" and "introns"? Which of the two carries genetic information? Which of the two is more abundant in DNA?

17.40 How can the phrase "Pure silver taxi" help you to remember the nature of base pairing in DNA?

17.41 The DNA from sea urchins contains about 32 percent A and about 18 percent G. What percentages of T and C would you expect in sea urchin DNA? Explain.

17.42 Look at Table 17.1, and find codons for these amino acids:
(a) Pro   (b) Lys   (c) Met

17.43 What amino acids are specified by these codons?
(a) A-C-U   (b) G-G-A   (c) C-U-U

17.44 What anticodon sequences on tRNA are complementary to the codons in Problem 17.43?

17.45 If the sequence T-A-C-C-G-A appeared on the informational strand of DNA, what sequence would appear opposite it on the template strand?

17.46 What sequence would appear on the mRNA molecule transcribed from the DNA in Problem 17.45?

17.47 What dipeptide would be synthesized from the gene sequence in Problem 17.45?

17.48 What is a mutation, and how can it be caused?

17.49 Why does a mutation in RNA have a much smaller effect on an organism than a mutation in DNA?

17.50 If the gene sequence A-T-T-G-G-C-C-T-A on the informational strand of DNA mutated and became A-C-T-G-G-C-C-T-A, what effect would the mutation have on the sequence of the protein produced?

17.51 What kind of cell must undergo mutation in order for the genetic error to be passed down to future generations?

17.52 What is a restriction endonuclease, and why is it used in Maxam-Gilbert DNA sequencing?

17.53 Why is a radioactive phosphate label attached to the end of the DNA chain prior to Maxam-Gilbert DNA sequencing?

17.54 What problem would you foresee if codons were made of only two nucleotides rather than three? How many codons are possible using combinations of two bases?

17.55 In general terms, what is the cause of hereditary diseases?

17.56 A substantial percentage of chemical agents that cause mutations also cause cancer. Explain in general terms why this might be.

17.57 Metenkephalin is a small peptide having morphine-like properties and found in animal brains. Give an mRNA sequence that will code for the synthesis of metenkephalin:

H$_2$N-Tyr-Gly-Gly-Phe-Met-COOH

17.58 Give a DNA gene sequence that will code for metenkephalin (Problem 17.57).

17.59 List the steps you would go through in order to sequence the DNA gene for metenkephalin (Problem 17.58) by the Maxam-Gilbert method.

17.60 Sketch a representation of what the gel electrophoresis pattern might look like if the DNA gene in Problem 17.58 were sequenced by the Maxam-Gilbert method.

CHAPTER 18

Metabolism I:
The Generation
of Biochemical Energy

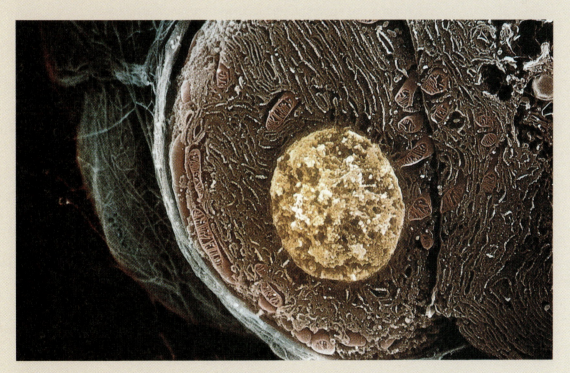

In this electron micrograph of a glandular cell, the nucleus and its contents are shown in yellow. The numerous red structures are mitochondria. All the energy that powers the life of the cell—and of the organism to which it belongs—are produced in these organelles. In this chapter you'll learn how.

Food is to humans what gasoline is to a car: It's the fuel that keeps us going. Once we've eaten, though, how do our bodies "burn" the fuel? How do we extract from food the energy that keeps us warm and active? We'll answer these questions in this and the next chapter as we look at metabolic pathways and biological energy transfer. Unlike the rapid burning of gasoline in a car, the biological breakdown of food molecules and the extraction of chemical energy take place in a carefully controlled series of chemical reactions.

In this chapter, we'll take an overview of metabolism and a close look at biochemical energy—what it is, where it comes from, and how it is stored. Among the specific questions we'll answer are these:

1. **How are cells constructed?** The goal: you should learn the general make-up of mammalian cells.

2. **How is energy released during metabolism?** The goal: you should understand the overall process by which food is converted into biochemical energy.

3. **What is the metabolic role of ATP, and how does ATP store energy?** The goal: you should understand the central role of ATP in biological energy production and use.

4. **What is the citric-acid cycle?** The goal: you should learn what occurs in the citric-acid cycle and what role the cycle plays in energy production.

5. **What is the respiratory chain?** The goal: you should learn the sequence of events in the respiratory chain leading to ATP synthesis.

6. **How much energy is released during respiration?** The goal: you should learn how the amount of energy produced by the citric-acid cycle and the respiratory chain is calculated.

18.1 CELLS AND THEIR STRUCTURE

■ **Prokaryotic cell** Cell that has no nucleus; found in bacteria and algae.

■ **Eukaryotic cell** Cell with a membrane-enclosed nucleus; found in all higher organisms.

There are two main categories of cells: **prokaryotic** (pro-**carr**-y-ah-tic) **cells**, found in bacteria and algae, and **eukaryotic** (you-**carr**-y-ah-tic) cells, found in higher organisms, both plant and animal. Prokaryotic cells are much simpler in structure: They have no nucleus, their DNA is dispersed throughout their contents, and they are quite small. Eukaryotic cells, by contrast, have a membrane-enclosed nucleus containing their chromosomal DNA and are about a thousand

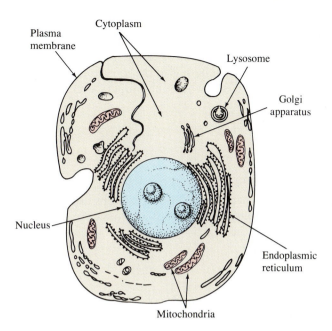

Figure 18.1
A cutaway diagram of a typical eukaryotic cell, showing some of its organelles.

■ **Plasma membrane**   The lipid bilayer membrane surrounding a eukaryotic cell.

■ **Organelle**   A small organized unit in the cell that performs a specific function.

■ **Cytoplasm**   The jelly-like fluid filling a cell.

■ **Mitochondria**   An egg-shaped organelle where small molecules are broken down to provide the energy to power an organism.

Figure 18.2
A cutaway diagram of a mitochondrion, showing the double membrane and the inner cristae.

times larger than bacterial cells. The entire cell is enclosed by a **plasma membrane** consisting primarily of a lipid bilayer (Section 14.6).

A further important difference between cell types is that eukaryotic cells contain a large number of subcellular structures called **organelles**, small functional units that perform specialized tasks within the cell. In addition to the nucleus, other organelles include ribosomes, endoplasmic reticulum, lysosomes, golgi apparatus, and mitochondria. All of these organelles are surrounded by cellular **cytoplasm**, the jelly-like material that fills the interior of the cell. A diagram of a typical eukaryotic cell is shown in Figure 18.1.

Among the most important of cellular organelles are the mitochondria. **Mitochondria** (Figure 18.2) are often called the cell's "power plants" because it is here that carbohydrates, lipids, and amino acids are chemically broken down to extract the energy that powers all cellular processes. A typical liver cell contains about 800 mitochondria, totaling about 20 percent of the cellular volume.

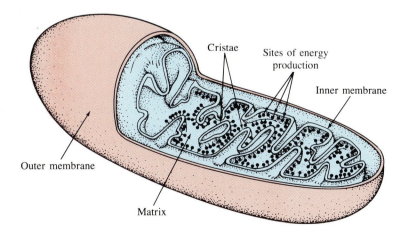

■ **Cristae**   The inner folds of a mitochondrion.

A mitochondrion is a roughly egg-shaped structure, composed of a smooth outer membrane and a folded inner membrane (Figure 18.2). The inner folds are called **cristae**, and the inner space between folds is called the mitochondrial *matrix*. Adhering to the inner folds are small knobby protuberances that make up the site where the primary events of energy production occur.

18.2   AN OVERVIEW OF METABOLISM AND ENERGY PRODUCTION

■ **Metabolism**   The overall sum of the many reactions taking place in a cell.

The sum total of the many organic reactions that go on in cells is referred to as **metabolism**. These reactions usually occur in long sequences called *metabolic pathways*, in which the product of one reaction serves as the starting material for the next:

$$A \longrightarrow B \longrightarrow C \longrightarrow D \longrightarrow \cdots$$

■ **Catabolism**   Metabolic reactions that break down molecules into smaller pieces.

■ **Anabolism**   Metabolic reactins that build larger biological molecules from smaller pieces.

Those pathways that break molecules apart are known collectively as **catabolism**, while those that put building blocks back together to assemble larger molecules are known as **anabolism**. Because they break high-energy bonds and form low-energy products, catabolic reactions release energy that is used to power living organisms. Anabolic reactions, by contrast, generally absorb energy.

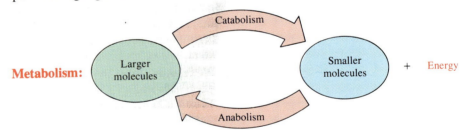

Metabolism: Larger molecules ⟶ (Catabolism) Smaller molecules + Energy ⟶ (Anabolism) Larger molecules

The overall picture of catabolism and energy production is simple: Eating provides fuel; breathing provides oxygen; and our bodies oxidize the fuel to extract energy. The process can be roughly divided into the four stages shown in Figure 18.3.

The first stage, *digestion*, takes place in the stomach and small intestine when bulk food is broken down into individual small molecules, such as simple sugars, fatty acids, and amino acids. In stage 2, these small molecules are further degraded in the cytoplasm of cells to yield 2-carbon acetyl groups (CH_3CO—) that are attached to a large carrier molecule called *coenzyme A*. The resultant compound, *acetyl coenzyme A (acetyl CoA)*, is an intermediate in the breakdown of all main classes of food molecules.

Acetyl groups are oxidized in the third stage, the *citric-acid cycle*, to yield carbon dioxide and water. This stage, which takes place inside mitochondria, releases a great deal of energy that is used in the fourth stage, the *respiratory chain*, to make molecules of *adenosine triphosphate (ATP)*. This stage also takes place in mitochondria.

In the remainder of this chapter, we'll discuss stages 3 and 4, because that's where energy production occurs. We'll then return to look at the first two stages in the next chapter.

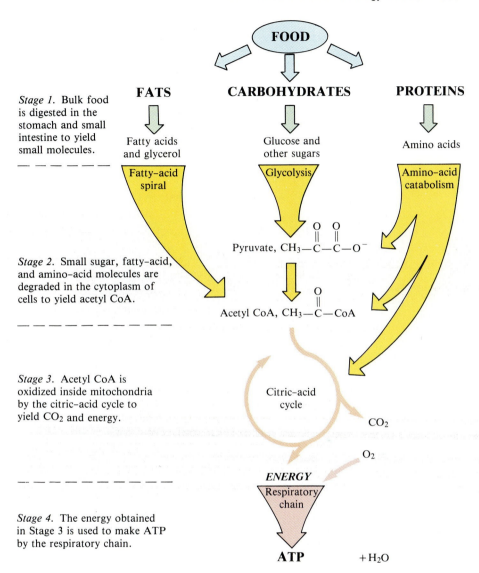

Stage 1. Bulk food is digested in the stomach and small intestine to yield small molecules.

Stage 2. Small sugar, fatty-acid, and amino-acid molecules are degraded in the cytoplasm of cells to yield acetyl CoA.

Stage 3. Acetyl CoA is oxidized inside mitochondria by the citric-acid cycle to yield CO_2 and energy.

Stage 4. The energy obtained in Stage 3 is used to make ATP by the respiratory chain.

Figure 18.3
An overview of catabolic pathways for the degradation of food and the production of biochemical energy.

18.3   ATP AND ITS ROLE IN ENERGY TRANSFER

Adenosine triphosphate (ATP), the final product of food catabolism, plays a pivotal role in the production of biological energy. As the molecule that stores the energy needed to power all life processes, it has been called the "energy currency of the living cell." Catabolic reactions "pay off" in ATP by synthesizing it from adenosine diphosphate (ADP) plus phosphate ion. Anabolic reactions "spend" ATP by transferring a phosphate group to other molecules, thereby regenerating ADP. The entire process of energy production thus revolves around the ATP $\rightleftarrows$ ADP interconversion (Figure 18.4).

Since the primary metabolic function of ATP is to store energy, we often refer to it as a "high-energy molecule" or an "energy storehouse." We don't mean by this that ATP is somehow different from other compounds; we only

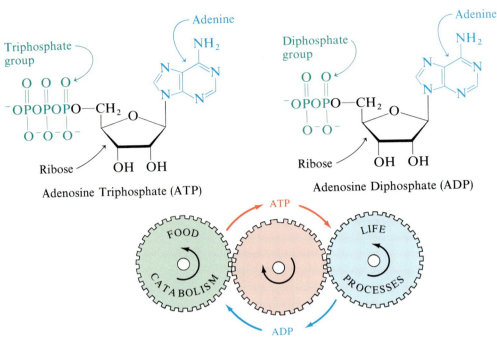

Figure 18.4
The interconversion of ATP with ADP serves as the central cog in the overall process of producing energy from food.

mean that ATP is more reactive than many other biological molecules because of the large amount of energy released when its P—O—P (*phosphoric anhydride*) bonds are broken.

Energy is stored in ATP when it is synthesized from ADP during various catabolic pathways. We'll see in Section 19.2, for example, that one of the steps in carbohydrate catabolism involves formation of ATP by reaction of ADP with a substance called *phosphoenolpyruvic acid* (*PEP*).

$$H_2C=\overset{\overset{\displaystyle O-PO_3^{2-}}{|}}{C}-COOH \quad \xrightarrow{\text{ADP} \quad \text{ATP}} \quad CH_3-\overset{\overset{\displaystyle O}{\|}}{C}-COOH$$

Phosphoenolpyruvic acid   Pyruvic acid

Notice how this reaction is written. Instead of showing ATP and ADP in their entirety, it's much more convenient to write complex biochemical reactions in a manner that highlights the important transformation without bothering to show the full structures of all reactants and products. The curved arrow intersecting the normal straight reaction arrow shows that ADP is a reactant and that ATP is a product.

Practice Problem   **18.1**   Guanosine triphosphate, GTP, is involved instead of ATP in certain biochemical reactions. Look up guanosine in Section 17.2, and then draw the structure of GTP.

18.4   ENERGY CHANGES IN BIOCHEMICAL REACTIONS: USING ATP

What does the body do with the ATP it makes? To understand the full role of ATP in biological processes, you should first go back and reread Sections 4.6–4.8 to recall some general ideas about chemical reactivity. We said in those sections that in order for any chemical reaction to occur spontaneously, energy has to be given off. In other words, the products of a reaction must have less energy and be more stable than the starting reactants. Biological reactions in living organisms are no different from laboratory reactions in test tubes; both follow the same physical laws, and both have the same kinds of energy requirements.

Reactions in which the products are *higher* in energy (less stable) than the starting reactants can also take place under the right circumstances, but such reactions can't be spontaneous. In other words, energy has to be added to the reactants in order for an energetically "uphill" change to occur. As shown by the energy diagram in Figure 18.5, the favorable reaction whose curve is on the left has its products farther downhill on the energy scale than the reactants, whereas the unfavorable reaction whose curve is on the right has its products uphill from the reactants.

What usually happens in order for an energetically unfavorable uphill reaction to occur is that it is "coupled" to an energetically favorable downhill reaction so that the *overall* energy change for the two reactions together is favorable. For example, take the reaction of glucose with hydrogen phosphate ion (HPO_4^{2-}) to yield glucose 6-phosphate plus water:

Hydrogen phosphate ion   Glucose   Glucose 6-phosphate

Figure 18.5
Energy diagrams for two reactions. In the favorable reaction, diagrammed on the left, the products have less energy than the reactants. In the unfavorable reaction, diagrammed on the right, the products have more energy than the reactants.

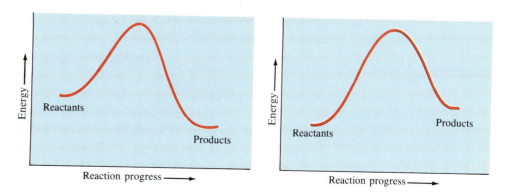

The reaction simply doesn't take place spontaneously because it is energetically unfavorable. The two products are about 3.3 kcal/mol higher in energy than the starting materials.

But now let's see what the energy picture looks like when ATP is involved. The reaction of ATP with water to yield ADP plus hydrogen phosphate ion is energetically *favorable* by about 7.3 kcal/mol. Thus, if the two reactions are coupled, the overall process for the synthesis of glucose 6-phosphate is favorable by about 4 kcal/mol. That is, the reaction of ATP with water releases more than enough energy than is needed for the unfavorable reaction of glucose with hydrogen phosphate. The net effect is that glucose 6-phosphate forms easily:

(*unfavorable*)	Glucose + HPO_4^{2-} → Glucose 6-phosphate + H_2O	Requires 3.3 kcal/mol
(*favorable*)	ATP + H_2O → ADP + HPO_4^{2-}	Releases 7.3 kcal/mol
Net Change:		
(*favorable*)	Glucose + ATP → Glucose 6-phosphate + H_2O	Releases 4.0 kcal/mol

It is this ability to "drive" otherwise unfavorable reactions that makes ATP so useful. As we'll see in the next chapter when we discuss anabolic pathways, many of the thousands of reactions going on in your body every minute are powered by energy from ATP. It's no exaggeration to say that the transfer of a phosphate group from ATP is the one chemical reaction that makes life possible.

Practice Problems

18.2 One of the steps in lipid metabolism is the reaction of glycerol ($HOCH_2$—CHOH—CH_2OH) with ATP to yield glycerol 1-phosphate. Formulate the reaction, and draw the structure of glycerol 1-phosphate.

18.3 Look at the reaction energy diagrams in Figure 18.5, and identify the parts of the curves representing the activation energies, E_{act}, and the parts representing the heats of reaction. (See Sections 4.6 and 4.7.)

18.5   THE CITRIC-ACID CYCLE

Now that we have an overview of how energy is stored and used, let's take a more detailed look at how it is produced. As noted earlier in Figure 18.3, the first two stages of food catabolism result in the conversion of fats, carbohydrates, and proteins into acetyl groups. These acetyl groups then enter the third stage, the citric-acid cycle. The cycle takes place within cell mitochondria and its reactions are directly coupled to the fourth stage of food catabolism, the respiratory chain.

■ **Citric-acid cycle**   A series of biochemical reactions used by the cell to obtain energy from food molecules.

The **citric-acid cycle**, also called the *tricarboxylic acid cycle* (*TCA*) or the *Krebs cycle* after its discoverer, is a sequence of eight enzyme-catalyzed steps that act as the body's primary energy producer. The eight steps and a brief explanation of what is happening at each point are shown in Figure 18.6. As its name implies, the citric-acid *cycle* is a closed loop of reactions; that is, the

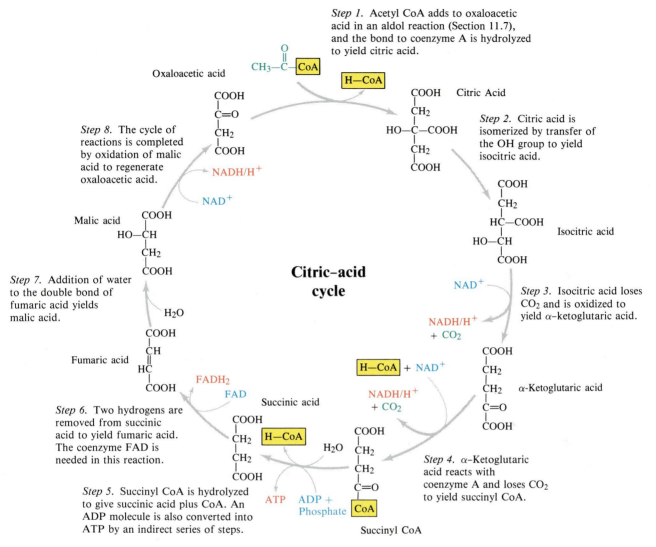

Figure 18.6
The citric-acid cycle, an eight-step series of reactions whose net effect is the metabolic breakdown of acetyl groups (from acetyl CoA) into two molecules of carbon dioxide plus energy. Here, and throughout Chapters 18 and 19, energy-rich forms (ATP, reduced coenzymes) are shown in red and their lower energy counterparts (ADP, oxidized coenzymes) are shown in blue.

product of the final step in the sequence is a *reactant* in the initial step. The intermediates are constantly regenerated, but there is a continuous flow of material through the cycle. Acetyl groups from acetyl CoA enter the cycle at step 1, and two molecules of CO_2 leave at steps 3 and 4. In the process, energy is released in the form of reduced coenzymes.

$$CH_3\overset{O}{\overset{\|}{C}}\text{—CoA} + 3\,H_2O + \text{Oxidized Coenzymes} \longrightarrow 2\,CO_2 + H\text{—CoA} + \text{Reduced Coenzymes (energy)}$$

Most of the energy released during a loop of the citric-acid cycle is not used to make ATP directly. Rather, it is used to reduce the coenzymes *nicotinamide adenine dinucleotide* (NAD^+) and *flavin adenine dinucleotide* (FAD) in Steps 3, 4, 6, and 8. These reduced coenzymes then enter the respiratory chain discussed in the next section, where they give up electrons and release their energy.

Practice Problems

18.4 Which of the substances in the citric-acid cycle are tricarboxylic acids (thus giving the cycle its alternate name)?

18.5 Write a balanced equation for the conversion of acetyl CoA into CO_2 and coenzyme A.

$$CH_3{-}\overset{\overset{\displaystyle O}{\|}}{C}{-}CoA + O_2 \longrightarrow CO_2 + H{-}CoA + H_2O$$

18.6 Write a balanced equation for the overall catabolism of glucose.

$$C_6H_{12}O_6 + O_2 \longrightarrow CO_2 + H_2O$$

18.6 THE RESPIRATORY CHAIN

■ **Respiratory chain** A series of biochemical reactions that harness the energy released in the citric-acid cycle.

The **respiratory chain**, also called the *electron transport system*, is a series of enzyme-catalyzed reactions whose overall purpose is to use the energy extracted from food to synthesize ATP. In essence, small organic molecules produced in the citric-acid cycle donate two of their hydrogen atoms (H·) to a coenzyme, either NAD^+ or FAD, generating a reduced coenzyme. Electrons from those hydrogen atoms are then passed along from enzyme to enzyme in a series of oxidation/reduction reactions. As each enzyme passes the electrons to the next carrier in the chain, the donor is oxidized (loses electrons) and the acceptor is reduced (gains electrons). Energy is captured in each step and used for the

Figure 18.7
An overview of reactions in the respiratory chain. A complex series of coupled reactions leads to the formation of water, the generation of 52.7 kcal/mol energy, and the transport of six H^+ ions through the mitochondrial membrane. The oxidized products of each reaction in the chain are shown in blue, and the reduced products in red.

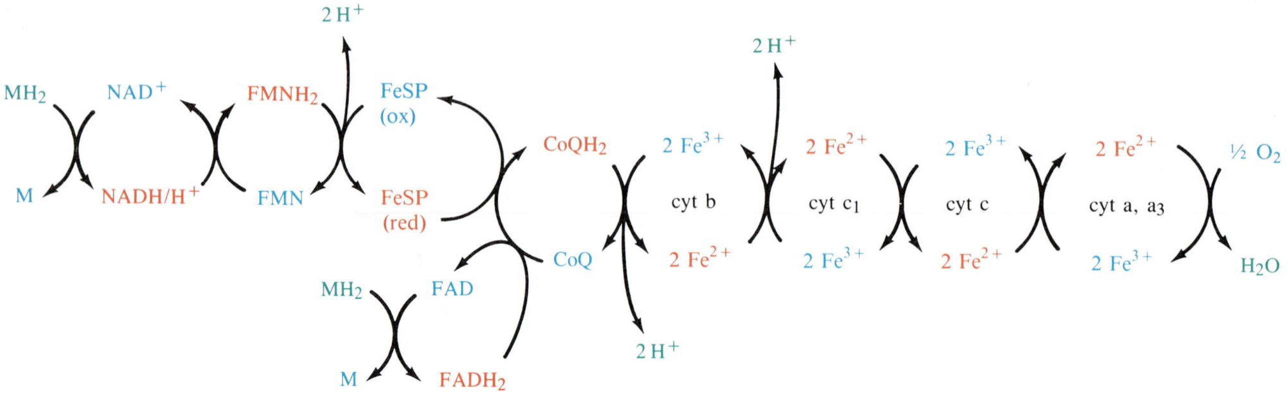

preparation of ATP from ADP and phosphate ion. Ultimately, the hydrogens and their electrons combine with oxygen obtained from breathing to produce water.

An overview of the many reactions that make up the respiratory chain is shown in Figure 18.7. We'll look at each step individually to see what is happening.

18.7 REACTIONS IN THE RESPIRATORY CHAIN

Nicotinamide Adenine Dinucleotide (NAD⁺)   The initial event in the respiratory chain involves a transfer of two hydrogen atoms from a substrate molecule (abbreviated MH_2), often a secondary alcohol, to the coenzyme *nicotinamide adenine dinucleotide* (NAD^+). (If you glance back at the citric-acid cycle in Figure 18.6, you'll see that this kind of reaction takes place in steps 3, 4, and 8.) In accepting the hydrogens, the coenzyme thereby undergoes a reduction to yield reduced NAD^+ and H^+ (abbreviated as *NADH/H⁺*). At the same time, the alcohol substrate that gives up the hydrogens is oxidized.

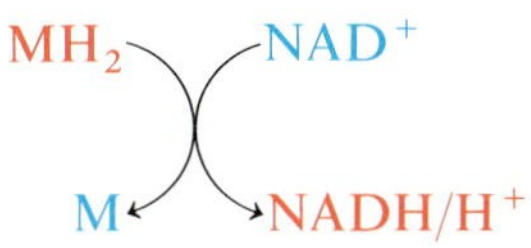

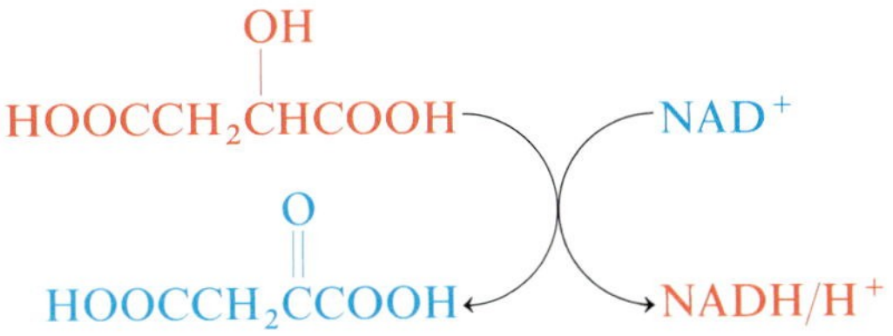

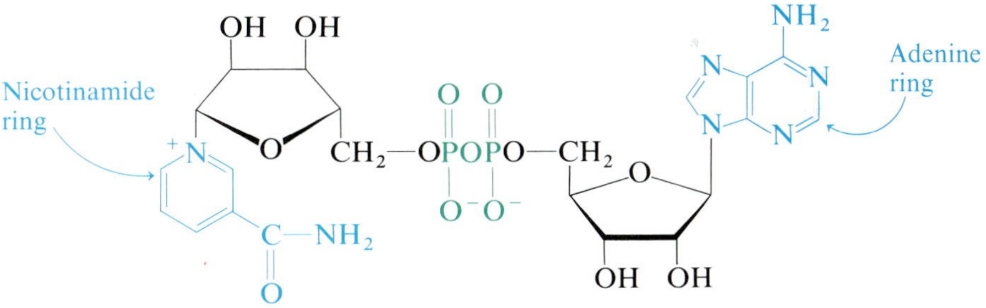

Nicotinamide adenine dinucleotide (NAD⁺)

Flavin Mononucleotide (FMN)   The reduced NADH formed in the first step is next oxidized to regenerate NAD⁺, which can thus recycle back into the respiratory chain for use over and over again. The two hydrogens that were gained by NADH/H⁺ are transferred to another coenzyme called *flavin mononucleotide* (*FMN*) to generate its reduced form, $FMNH_2$.

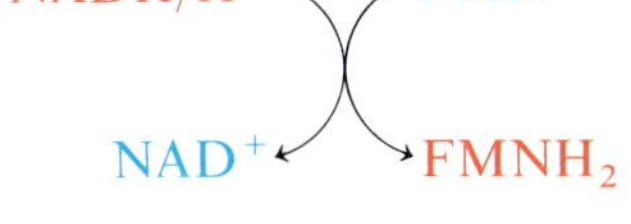

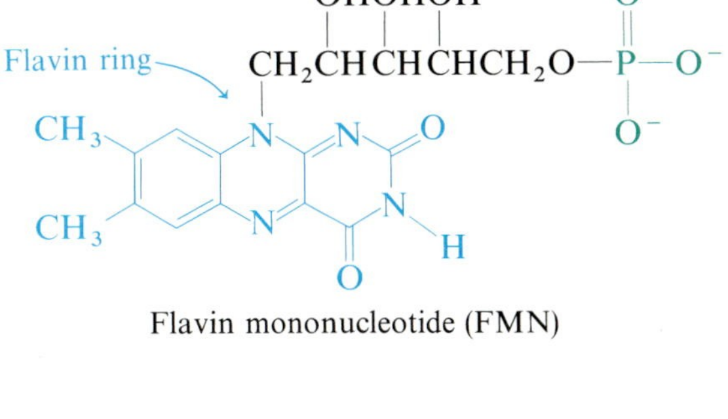

Flavin mononucleotide (FMN)

Following the formation of $FMNH_2$, the respiratory chain takes a different course. No longer are hydrogen atoms (H·) transferred with their electrons; from this point on, only the electrons are transferred. The two hydrogens, now stripped of their electrons, are released as two H^+ ions into the space between inner and outer membranes. The substance involved in transferring electrons from $FMNH_2$ is an iron/sulfur-containing protein that we'll abbreviate as FeSP.

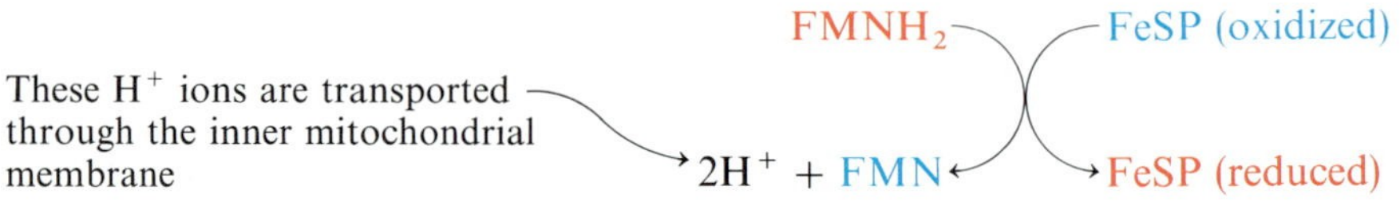

These H^+ ions are transported through the inner mitochondrial membrane

$$FMNH_2 \quad FeSP \text{ (oxidized)}$$
$$2H^+ + FMN \quad FeSP \text{ (reduced)}$$

Flavin Adenine Dinucleotide (FAD)   Although NAD^+ is the primary coenzyme that reacts directly with food-derived molecules to initiate the respiratory chain, a second coenzyme called *flavin adenine dinucleotide* (*FAD*) can react similarly. The only difference is that NAD^+ removes two hydrogens from an alcohol to yield a ketone, whereas FAD usually removes two hydrogens from a carbon chain to yield an alkene double bond. You can see an example of this kind of reaction in step 6 of the citric-acid cycle (Figure 18.6). The reduced coenzyme is called $FADH_2$.

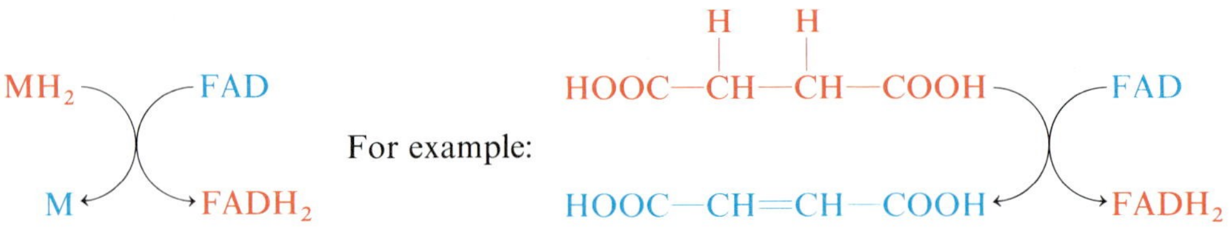

For example:

$$MH_2 \quad FAD$$
$$M \quad FADH_2$$

$$HOOC-\overset{\overset{H}{|}}{C}H-\overset{\overset{H}{|}}{C}H-COOH \quad FAD$$
$$HOOC-CH=CH-COOH \quad FADH_2$$

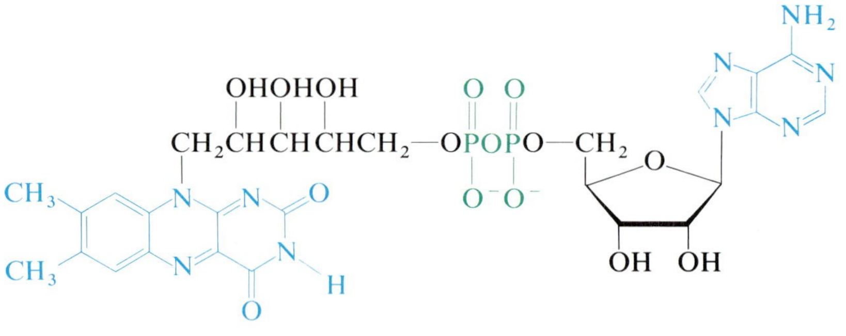

Flavin adenine dinucleotide (FAD)

Practice Problem   **18.7**   Which of these reactions probably requires NAD^+ as a coenzyme, and which requires FAD?

(a)
$$HOOC-CH_2-\overset{\overset{OH}{|}}{C}H-COOH \longrightarrow HOOC-CH_2-\overset{\overset{O}{||}}{C}-COOH$$

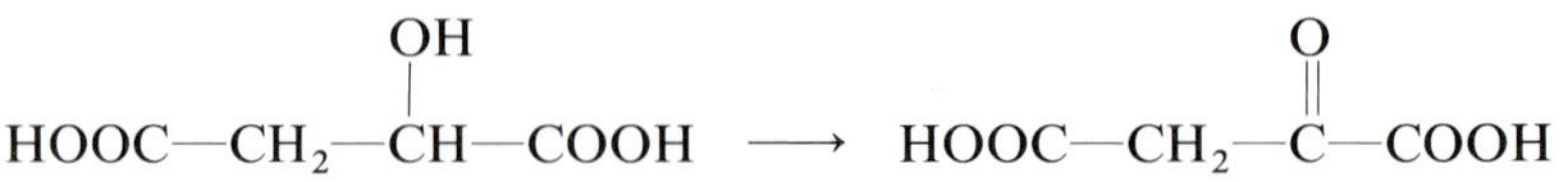

(b) $CH_3-\overset{\overset{O}{||}}{C}-CH_2-CH_2-COOH \longrightarrow CH_3-\overset{\overset{O}{||}}{C}-CH=CH-COOH$

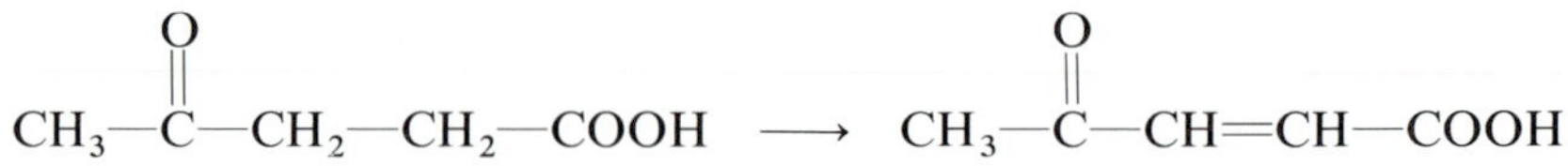

Coenzyme Q   The FAD and NAD^+/FMN/FeSP pathways converge at the next step of the respiratory chain when $FADH_2$ and the iron/sulfur protein pass their electrons on to *coenzyme Q (CoQ)*. Reduced coenzyme Q $(CoQH_2)$ is formed, and $FADH_2$ and FeSP are both converted into their oxidized forms for recycling back into the earlier steps of the respiratory chain.

FADH₂ — CoQ   FeSP (red) — CoQ
            or
FAD → CoQH₂   FeSP (ox) → CoQH₂

$$CH_3O-\underset{\underset{O}{\parallel}}{\overset{\overset{O}{\parallel}}{}} ... (CH_2CH=\overset{CH_3}{\underset{\mid}{C}}CH_2)_nH$$

Coenzyme Q (CoQ)

Cytochromes   Following the formation of $CoQH_2$, electrons are transferred to a group of enzymes called the *cytochromes*. There are a number of different cytochromes, abbreviated cyt a, cyt b, and so forth. All are closely related in structure, having the same heme coenzyme but slightly different apoenzymes. Centered in each cytochrome is a heme coenzyme (Figure 18.8) with an iron atom that can be in either the 2+ or 3+ oxidation state (Section 3.4).

Electrons are shuttled down the respiratory chain when an iron atom in one cytochrome gives up an electron to an iron atom in the next cytochrome and is oxidized from Fe^{2+} to Fe^{3+} in the process. At the same time, the iron atom in the cytochrome that accepts the electron is reduced from Fe^{3+} to Fe^{2+}. Ultimately, the electrons are donated to an oxygen atom (obtained from breathing), which combines with two H^+ to yield water.

Oxidation: $Fe^{2+} \longrightarrow Fe^{3+} + e^-$

Reduction: $Fe^{3+} + e^- \longrightarrow Fe^{2+}$

Figure 18.8
(a) The structure of the iron-containing heme coenzyme present in th cytochromes.
(b) Computer-generated structure of two molecules of cytochrome c. The heme groups are shown in light green.

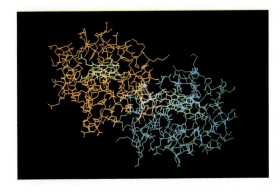

Now that we've looked at the most important parts of the respiratory chain, you should look back at the entire sequence shown in Figure 18.7. As the figure indicates, four more H^+ ions are transported through the inner mitochondrial membrane during the cytochrome series of electron-transport reactions. Thus, a total of six H^+ ions are transported out of the mitochondrial matrix during the NAD^+ pathway, and four are transported during the FAD pathway. The overall effect of the reactions in the respiratory chain is therefore to "pump" H^+ ions through the inner membrane and out of the matrix. Experiments show that the proton concentration difference between the inner and outer faces of the mitochondrial membrane is about 1.4 pH units, with the outside being more acidic. As we'll see in the next section, it is this pH difference that drives ATP formation.

Practice Problems

18.8 Without looking at Figure 18.7, identify the missing substances in the following series of coupled reactions:

$$NAD^+ \quad ? \quad FeSP\ (ox) \quad CoQH_2$$

$$? \quad FMN \quad ? \quad ?$$

18.9 Identify each of the steps in the cytochrome part of the respiratory chain (Figure 18.7) as an oxidation or a reduction.

18.8 OXIDATIVE PHOSPHORYLATION: THE SYNTHESIS OF ATP

How does the complex sequence of reactions that make up the respiratory chain lead to the synthesis of ATP? Despite many years of work, the full answer to this question is still not known. It seems clear, however, that the key point is the establishment of a proton concentration difference between the inner and outer sides of the mitochondrial membrane.

■ **Chemiosmotic hypothesis**
A theory to explain how the establishment of a pH difference across a cell membrane leads to the synthesis of ATP in the respiratory chain.

According to the **chemiosmotic hypothesis**, proposed in 1961, the protons that are pumped outside the mitochondrial membrane are unable to diffuse back inside spontaneously. In other words, membrane passage of a proton is easier in one direction than in the other, and a concentration difference is therefore maintained. By a mechanism whose details are unknown, the protons are able to cross back into the mitochondrial matrix only at certain sites. These sites, which are visible at very high magnification as knobby protrusions on the inner membrane, evidently house the ATP synthetase enzyme necessary for combining ADP with phosphate to yield ATP.

$$ADP + HPO_3{}^{2-} \xrightarrow{\text{ATP synthetase}} ATP + H_2O$$

■ **Oxidative phosphorylation**
A term for the synthesis of ATP from ADP and phosphate ion.

A schematic view of ATP synthesis—called **oxidative phosphorylation** because it is the *oxidation* of food that leads to the formation of a *phosphate* bond—is shown in Figure 18.9. It's believed that the knobby protrusions inside the mitochondrion extend completely through the inner membrane, providing a channel through which H^+ ions can flow. Exactly how this proton flow activates the ATP synthetase enzyme is not yet known.

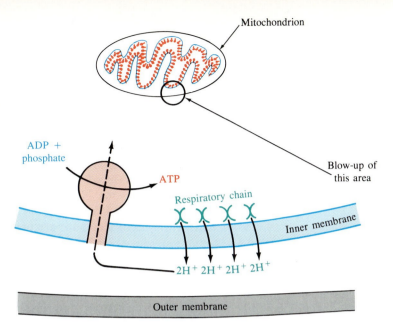

Figure 18.9
A schematic view of oxidative phosphorylation. Protons released in the respiratory chain are transported through the inner membrane of the mitochondrion, thereby establishing a concentration difference between inner and outer faces. Return flow of the protons into the matrix occurs at knobs that house the enzymes necessary for ATP synthesis.

AN APPLICATION: BARBITURATES

The barbiturates, discovered in 1864 by the German chemist Adolph von Baeyer, are among the first pharmaceutical agents to come from the chemical laboratory rather than from nature. Different barbiturates are extremely easy to synthesize, and more than 2500 analogs have been prepared. All are cyclic amides that differ only in the nature of the two groups attached to the ring. Among the more common ones are Veronal (named after the Italian city of Verona), Phenobarbital, Secobarbital (Seconal), and Amobarbital (Amytal).

Veronal

Phenobarbital

Secobarbital

Amobarbital

Barbiturates are usually classified into two groups: short-acting ones like Secobarbital, and long-acting ones like Phenobarbital. All act as tranquilizers at low doses and as sleep-inducers at somewhat higher doses. So popular are they, in fact, that barbiturates are among the most often prescribed (and most often abused) drugs in the country. At one point several years ago it was estimated that enough barbiturates were being produced in the United States to put 10 million people to sleep every night.

Although medically useful at doses in the 50- to 100-mg range, barbiturates are extremely toxic in higher amounts. At doses above 1.5 g, barbiturates depress the respiratory system, leading to severe oxygen deficiency in the brain and ultimately to death. These toxic effects appear to arise from an enzyme inhibition that blocks the NAD^+ branch of the respiratory chain. Barbiturates somehow prevent electron transfer in the FMN/FeS/CoQ part of the respiratory chain, thereby preventing the major part of the chain from functioning.

403

18.9   ENERGY OUTPUT DURING RESPIRATION

Let's take an overall look at the entire process to see exactly how much energy the citric-acid cycle and respiratory chain produce. One turn around the citric-acid cycle converts an acetyl group into two molecules of carbon dioxide, one molecule of ATP, and four molecules of reduced coenzymes (three NADH plus one $FADH_2$). ATP is produced when these reduced coenzymes are recycled back into their oxidized forms, NAD^+ and FAD, in the respiratory chain.

Three molecules of ATP are produced for each NADH oxidized (steps 3, 4, and 6), and two molecules of ATP are produced for each $FADH_2$ oxidized (step 8). In addition, one molecule of ATP is produced indirectly during step 5 of the citric-acid cycle when succinyl CoA is converted into succinic acid. Thus, a single turn of the citric-acid cycle yields 12 ATP molecules from each acetyl group. We'll see in the next chapter that there is an additional energy harvest during the production of acetyl groups in stage 2 of catabolism.

Citric-Acid Cycle Step				ATP Yield per Acetyl Group
3	NADH	$\xrightarrow{\text{Respiratory chain}}$	NAD^+	3
4	NADH	$\xrightarrow{\text{Respiratory chain}}$	NAD^+	3
5	Succinyl CoA	$\longrightarrow$	Succinic acid	1
6	NADH	$\xrightarrow{\text{Respiratory chain}}$	NAD^+	3
8	$FADH_2$	$\xrightarrow{\text{Respiratory chain}}$	FAD	2

Net yield per acetyl group 12

SUMMARY

There are two broad categories of cells: **prokaryotic** cells, found in algae and bacteria, and **eukaryotic** cells, found in higher organisms. Eukaryotic cells contain a large number of small subcellular structures called **organelles** that perform specialized tasks in the cell. Among the most important organelles are the **mitochondria**, which house the enzymes necessary for producing energy from the breakdown of food-derived molecules.

Metabolism is the sum total of all chemical reactions going on in the body. Those reactions that break down large molecules into smaller fragments are called **catabolism**, while those that build up large molecules from small pieces are called **anabolism**. Lipids, carbohydrates, and proteins are all catabolized to yield acetyl CoA, which is then further degraded in the citric-acid cycle.

The **citric-acid cycle**, also called the **tricarboxylic-acid cycle** or the **Krebs cycle**, is a series of eight enzyme-catalyzed steps that act as the body's primary energy producer. The cycle is a closed loop of reactions that convert an acetyl group into two molecules of carbon dioxide plus a large amount of energy. The energy output of the various steps in the citric-acid cycle is coupled to the **respiratory chain**, a series of enzyme-catalyzed reactions whose ultimate purpose

INTERLUDE: DIETS, BABIES, AND HIBERNATING BEARS

Most warm-blooded animals generate enough heat to maintain body temperature through normal metabolic reactions. In certain cases, however, normal metabolism is not able to satisfy the body's need for warmth, and a supplemental method of heat generation is needed. This occurs, for example, in newborn infants, and in hibernating animals that must survive for long periods without food. Nature has therefore provided babies, bears, and many other creatures with *brown fat tissue*. Brown fat tissue, so called because its color contrasts with that of normal white adipose tissue, contains high concentrations of blood vessels and mitochondria. In addition, it contains a special protein that acts as an *uncoupler* of ATP synthesis.

We saw in Section 18.6 that the electron-transport reactions of the respiratory chain are coupled to the synthesis of ATP. In other words, the purpose of the respiratory chain is to harness the energy released during catabolism to synthesize ATP. In the presence of an uncoupling agent like the protein in brown fat, however, the electron-transport reactions continue to take place, but no ATP is synthesized. As a result, the energy that would otherwise be used for ATP synthesis is released simply as heat, and the temperature of the organism rises. In essence, brown fat acts as a furnace to produce heat energy rather than stored energy.

The uncoupling of ATP synthesis from the respiratory chain can also be caused by chemical agents. It has been found, for example, that 2,4-dinitrophenol can cause uncoupling even in normal white adipose tissue of adults.

2,4-Dinitrophenol

Imagine the possibilities for a simple chemical agent that causes food to be converted directly into heat rather than ATP. The body, in critical need of ATP to drive its many reactions, would continue to burn fat at a frantic rate. Pound after pound of fat would literally melt away, giving every previously unsuccessful dieter the figure of a movie star. In fact, the idea works, and weight-control pills containing 2,4-dinitrophenol were marketed for a brief period in the 1930s. Unfortunately, the dose needed to *slow* ATP synthesis is very close to the dose that *stops* ATP synthesis, with lethal results—a fact that was discovered only after several fatalities had occurred. Thus, there's still no magic pill for dieters.

is to use the energy to synthesize **adenosine triphosphate (ATP)**. ATP is called a "high-energy" molecule and is suitable for energy storage because it contains reactive P—O—P (**phosphoric anhydride**) bonds that allow it to transfer a phosphate group.

As the respiratory chain commences, small molecules donate two of their hydrogen atoms to either a **nicotinamide adenine dinucleotide (NAD^+)** or a **flavin adenine dinucleotide (FAD)** coenzyme. Electrons from these reduced coenzymes are then passed through a series of other enzymes (**cytochromes**) and donated to oxygen to produce water. At the same time, hydrogen ions (protons, H^+) migrate from the mitochondrial matrix through the inner membrane to establish a proton concentration difference across the membrane. According to the **chemiosmotic hypothesis**, the protons migrate back into the matrix at certain knob-like sites on the membrane, thereby activating the enzymes that carry out ATP synthesis (**oxidative phosphorylation**).

REVIEW PROBLEMS

Cells and Their Structure

18.10 List several differences between prokaryotic and eukaryotic cells.

18.11 What kinds of organisms have prokaryotic cells, and what kinds have eukaryotic cells?

18.12 What is an organelle, and what is its general function?

18.13 What is the name of the substance that fills the interior of cells?

18.14 Describe in general terms the structural make-up of a mitochondrion.

18.15 What are cristae, and why are they important?

Metabolism

18.16 What is the difference between digestion and metabolism?

18.17 What is the difference between catabolism and anabolism?

18.18 What key metabolic substance is formed from the catabolism of all three major classes of foods—carbohydrates, fats, and proteins?

18.19 Put the following events in the correct order of their occurrence: respiratory chain; digestion; oxidative phosphorylation; citric-acid cycle.

18.20 What is the full name of the substance formed during catabolism to store chemical energy? How is this substance related to nucleic acids?

18.21 What is the chemical difference between ATP and ADP?

18.22 What general kind of chemical reaction does ATP carry out?

18.23 Adenosine *mono*phosphate is an important intermediate in certain biochemical pathways. Draw its structure.

Energy Changes

18.24 What does it mean when we say that two reactions are coupled?

18.25 What energy requirement must be met in order for a reaction to be spontaneous?

18.26 Why is ATP called a high-energy molecule?

18.27 What does the term *oxidative phosphorylation* mean?

18.28 Draw a reaction energy diagram for an unfavorable reaction.

18.29 Draw a reaction energy diagram for a favorable reaction.

The Citric-Acid Cycle

18.30 Where in the cell does the citric-acid cycle take place?

18.31 By what other names is the citric-acid cycle known?

18.32 What substance acts as the starting point of the citric-acid cycle, reacting with acetyl CoA in the first step and being regenerated in the last step? Draw its structure.

18.33 Look at the eight steps of the citric-acid cycle (Figure 18.6), and answer the following questions.
(a) Which steps involve oxidation reactions?
(b) Which steps involve decarboxylations (loss of CO_2)?
(c) Which step involves a hydration reaction?

18.34 The fumaric acid produced in step 6 of the citric-acid cycle must have a trans double bond in order to continue on in the cycle. Suggest a reason why the corresponding cis double-bond isomer cannot continue in the cycle.

18.35 Look up the structures of the intermediates in the citric-acid cycle, and then account for the fact that isocitrate is chiral while citrate is achiral (Section 13.3).

18.36 What other intermediate in the citric-acid cycle is chiral besides isocitrate (Problem 18.35)? Explain.

The Respiratory Chain

18.37 By what other name is the respiratory chain known?

18.38 What two coenzymes initiate the events of the respiratory chain?

18.39 What are the ultimate products of the respiratory chain?

18.40 What do the following abbreviations stand for?
(a) FAD (b) CoQ (c) NADH/H$^+$ (d) cyt

18.41 What atom in the cytochromes undergoes oxidation and reduction in the respiratory chain?

18.42 Put the following substances in the correct order of their action in the respiratory chain: iron/sulfur protein; cytochrome a; coenzyme Q; NAD$^+$

18.43 Fill in the missing substances in these coupled reactions:

$$FAD \diagdown CoQH_2 \diagdown \; ?$$
$$? \diagup \qquad ? \diagup 2Fe^{2+}$$

18.44 With what class of enzymes (Section 16.1) are the coenzymes NAD$^+$ and FAD associated?

CHAPTER 19

Metabolism II: Catabolic and Anabolic Pathways

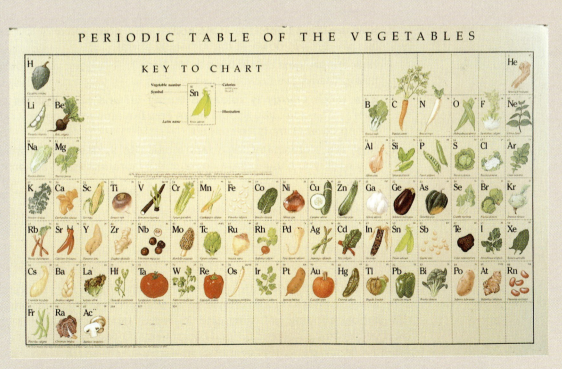

Eating your vegetables at every meal won't do you any good if your body can't utilize the chemical substances they contain. The human body uses a great variety of enzymes to break down large molecules in food, extract their energy, and recycle their parts to build the new molecules needed for tissue repair and growth. You'll learn more about these processes in this chapter.

We saw in the previous chapter how biochemical energy is extracted from food by the citric-acid cycle and the respiratory chain to fuel the synthesis of ATP from ADP. ATP molecules, in turn, provide the energy to carry out the thousands of different reactions necessary to sustain life. In the present chapter, we'll study the catabolic pathways by which fats, carbohydrates, and proteins are degraded, as well as some anabolic pathways by which important biological molecules are made in the body.

We'll look at the following questions in this chapter:

1. **What occurs during digestion?** The goal: you should learn how food is broken down to simpler molecules during digestion.

2. **How are carbohydrates catabolized in the body?** The goal: you should learn the general pathway for the metabolic breakdown of carbohydrates.

3. **How are fats catabolized in the body?** The goal: you should learn how fats are broken down in the fatty-acid spiral.

4. **How are proteins catabolized in the body?** The goal: you should learn how proteins are broken down.

5. **How much energy is released during fat and carbohydrate catabolism?** The goal: you should learn how the amount of energy released during catabolism is calculated.

6. **How are biologically important molecules synthesized in the body?** The goal: you should learn some of the more important anabolic pathways.

19.1   DIGESTION

■ **Digestion** A general term for the breakdown of food into small molecules.

The first step in the food path is **digestion**, a catch-all term used to describe the breakdown of bulk food into individual small molecules. Digestion entails both the physical grinding, softening, and mixing of food, as well as the enzyme-catalyzed hydrolysis of specific food components. Digestion begins in the mouth when foods are chewed; it continues in the stomach; and it concludes in the small intestine (Figure 19.1).

The main digestive action on starch, a carbohydrate polymer, is to hydrolyze it through the action of amylase enzymes into many glucose molecules. Fats (triglycerides) are similarly hydrolyzed by lipase enzymes into glycerol and fatty-acid molecules, and proteins are hydrolyzed by protease enzymes into amino-acid molecules.

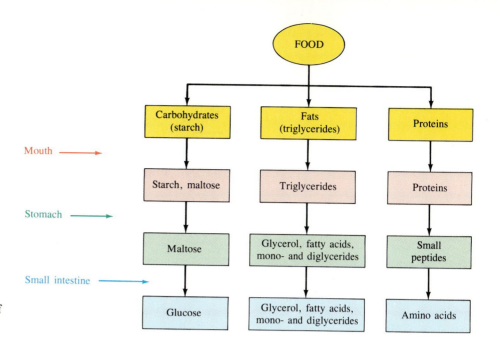

Figure 19.1
The products at several stages of food digestion.

The products of food digestion—glucose, fatty acids, and amino acids—are all relatively small molecules that are easily transported through the bloodstream and into target cells. Once inside the cells, they are further broken down to produce energy. Some of the fragments produced by cellular breakdown of food molecules are excreted, and some are used as building blocks to synthesize the many biomolecules that an organism needs to sustain life.

Practice Problems

19.1 What products would you expect from the digestion of glyceryl tristearate?

19.2 Insulin is a peptide hormone (Section 16.10) that regulates carbohydrate metabolism and that is critically important in controlling diabetes. Why do you suppose insulin is effective only when injected? Why is an insulin pill ineffective when swallowed?

19.2 CARBOHYDRATE CATABOLISM: GLYCOLYSIS

■ **Glycolysis** A series of biochemical reactions that break down a molecule of glucose into two molecules of pyruvate ion plus energy.

Glycolysis, also called the *Embden-Meyerhof pathway* after its discoverers, is a series of ten enzyme-catalyzed reactions that break down a molecule of glucose into two molecules of pyruvic acid with release of energy. Most living organisms carry out the glycolysis sequence, although the ultimate fates of the pyruvic acid molecules differ from one organism to another. In humans, glycolysis is carried out primarily in the cytoplasm of muscle, fat, and liver cells. The pathway is summarized in Figure 19.2.

Glycolysis:

Glucose $\xrightarrow{\text{10 steps}}$ $2\,CH_3\!-\!\overset{\displaystyle O}{\overset{\|}{C}}\!-\!\overset{\displaystyle O}{\overset{\|}{C}}\!-\!OH$ + energy

Glucose   Pyruvic acid

Glucose

Step 1. Glucose undergoes reaction with ATP to yield glucose 6-phosphate plus ADP.

ATP

ADP

Glucose
6-phosphate

Step 2. Isomerization of glucose 6-phosphate yields fructose 6-phosphate. The reaction is catalyzed by the mutase enzyme, phosphoglucoisomerase.

Fructose
6-phosphate

Step 3. Fructose 6-phosphate reacts with a second molecule of ATP to yield fructose 1,6-diphosphate plus ADP.

ATP

ADP

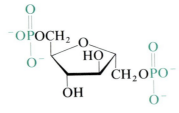

Fructose
1,6-diphosphate

Figure 19.2
The glycolysis pathway converting glucose to pyruvic acid.

Step 4. The six-carbon chain of fructose 1,6-diphosphate is cleaved into two three-carbon pieces by the enzyme aldolase
(Continued on next page.)

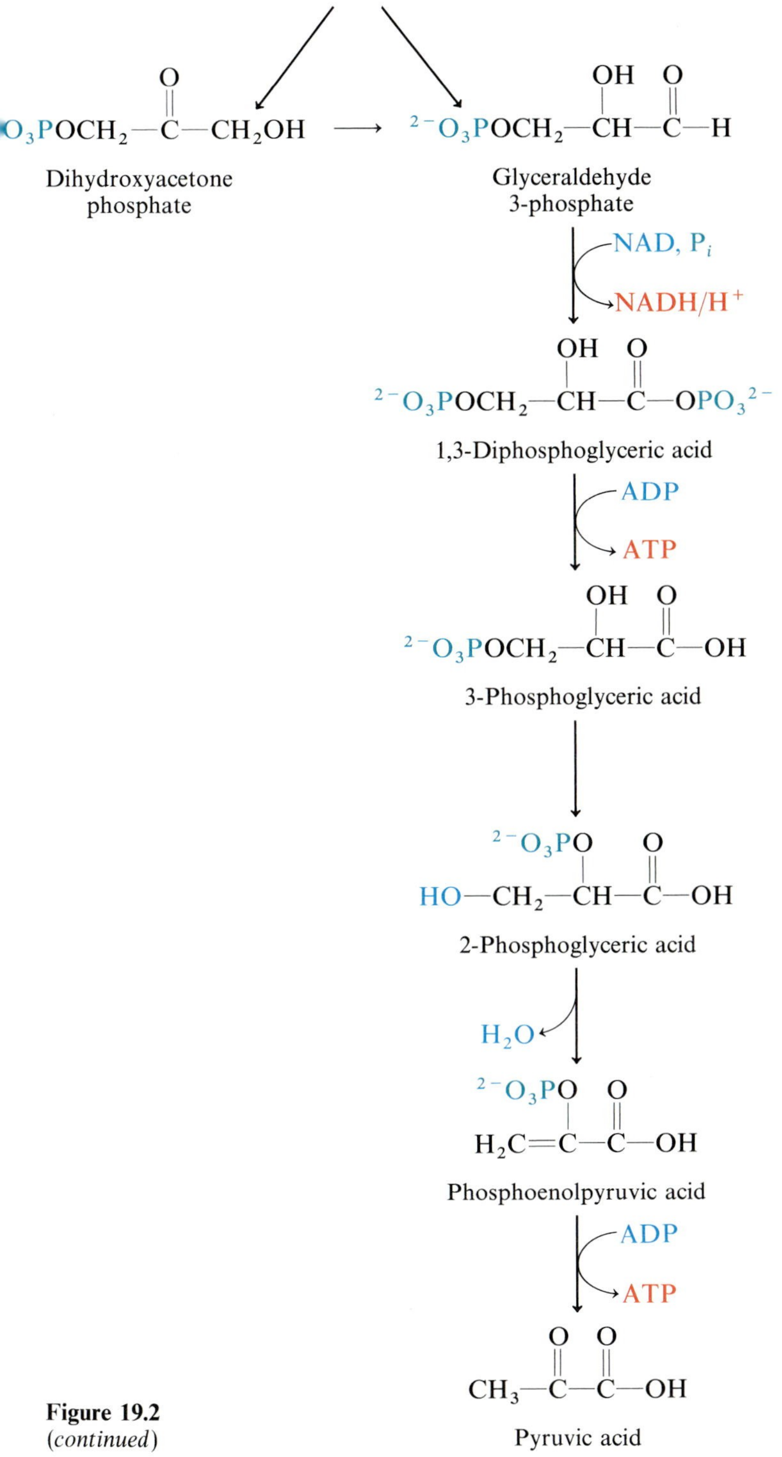

$$\text{O}_3\text{POCH}_2{-}\overset{\overset{\text{O}}{\|}}{\text{C}}{-}\text{CH}_2\text{OH} \longrightarrow \ ^{2-}\text{O}_3\text{POCH}_2{-}\overset{\overset{\text{OH}}{|}}{\text{CH}}{-}\overset{\overset{\text{O}}{\|}}{\text{C}}{-}\text{H}$$

Dihydroxyacetone phosphate   Glyceraldehyde 3-phosphate

NAD, P_i
NADH/H$^+$

$$^{2-}\text{O}_3\text{POCH}_2{-}\overset{\overset{\text{OH}}{|}}{\text{CH}}{-}\overset{\overset{\text{O}}{\|}}{\text{C}}{-}\text{OPO}_3{}^{2-}$$

1,3-Diphosphoglyceric acid

ADP
ATP

$$^{2-}\text{O}_3\text{POCH}_2{-}\overset{\overset{\text{OH}}{|}}{\text{CH}}{-}\overset{\overset{\text{O}}{\|}}{\text{C}}{-}\text{OH}$$

3-Phosphoglyceric acid

$$\text{HO}{-}\text{CH}_2{-}\overset{\overset{^{2-}\text{O}_3\text{PO}}{|}}{\text{CH}}{-}\overset{\overset{\text{O}}{\|}}{\text{C}}{-}\text{OH}$$

2-Phosphoglyceric acid

H_2O

$$\text{H}_2\text{C}{=}\overset{\overset{^{2-}\text{O}_3\text{PO}}{|}}{\text{C}}{-}\overset{\overset{\text{O}}{\|}}{\text{C}}{-}\text{OH}$$

Phosphoenolpyruvic acid

ADP
ATP

$$\text{CH}_3{-}\overset{\overset{\text{O}}{\|}}{\text{C}}{-}\overset{\overset{\text{O}}{\|}}{\text{C}}{-}\text{OH}$$

Pyruvic acid

Figure 19.2
(*continued*)

Step 5. The two products of step 4 are both three-carbon sugars, but only glyceraldehyde 3-phosphate can continue in the glycolysis pathway. Dihydroxyacetone phosphate must first be isomerized by the enzyme triose phosphate isomerase.

Step 6. Two reactions occur as glyceraldehyde 3-phosphate is first oxidized to a carboxylic acid and then phosphorylated by the enzyme glyceraldehyde 3-phosphate dehydrogenase. The coenzyme nicotinamide adenine dinucleotide (NAD) and inorganic phosphate ion (P_i) are required.

Step 7. A phosphate group from 1,3-diphosphoglyceric acid is transferred to ADP, resulting in synthesis of ATP. This is an important step in capturing some of the energy in glucose.

Step 8. A phosphate group is next transferred from carbon 3 to carbon 2 of phosphoglyceric acid in a step catalyzed by the enzyme phosphoglyceromutase.

Step 9. Loss of water from 2-phosphoglyceric acid produces phosphoenolpyruvic acid (PEP). The dehydration is catalyzed by the enzyme enolase.

Step 10. Transfer of the phosphate group from phosphoenolpyruvic acid to ADP yields pyruvic acid and generates an additional ATP.

411

As shown in Figure 19.2, glucose is activated by reaction with two molecules of ATP in steps 1 and 3, and is then split into two three-carbon molecules in step 4. The energy used in steps 1 and 3 is more than recouped when four molecules of ATP (two from each three-carbon piece) are generated in steps 7 and 10. In addition, the NADH/H$^+$ produced in step 6 can enter the respiratory chain (Section 18.9) where additional energy is generated.

Practice Problem   **19.3**   Write a balanced equation for the overall glycolysis reaction of glucose ($C_6H_{12}O_6$) to pyruvic acid ($C_3H_4O_3$). Oxygen is also needed as a reactant, and water is a product.

19.3   THREE REACTIONS OF PYRUVIC ACID

■ **Aerobic**   In the presence of oxygen.

■ **Anaerobic**   Without oxygen.

■ **Fermentation**   The breakdown of glucose to ethanol plus carbon dioxide by the action of yeast enzymes.

The breakdown of glucose to yield pyruvic acid is a central metabolic pathway in most living systems. The further reactions of pyruvic acid, however, depend on the conditions and on the nature of the organism. Under the normal oxygen-rich (**aerobic**) conditions in liver cells, pyruvic acid is converted into acetyl CoA. Under oxygen-free (**anaerobic**) conditions, such as might occur in muscle cells during heavy exercise, it is converted into lactic acid. Finally, when glucose is broken down by **fermentation** with yeast, the pyruvic acid is converted into ethanol.

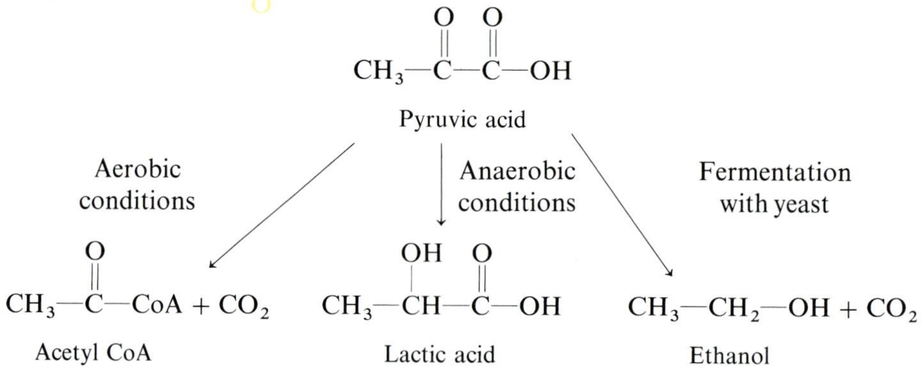

All three transformations of pyruvic acid occur commonly in nature. For example, the aerobic conversion of pyruvic acid to acetyl CoA occurs in the cytoplasm of many cells. Similarly, the anaerobic conversion of pyruvic acid to lactic acid takes place both in bacteria (as when milk turns sour) and in muscle tissue of higher organisms when the oxygen supply has been used up by intense exercise. Finally, the transformation of pyruvic acid into ethanol plus carbon dioxide occurs during the fermentation of sugar by yeasts to produce beer, wine, and other alcoholic drinks (Figure 19.3).

Of the three transformations that pyruvic acid can undergo, the aerobic conversion into acetyl CoA is by far the most important because it sets the stage for the further events of the citric-acid cycle and the respiratory chain discussed in the previous chapter. Three steps occur in the overall transformation of pyruvic acid to acetyl CoA:

Figure 19.3
Yeast cells ferment the sugar in grape juice, producing wine

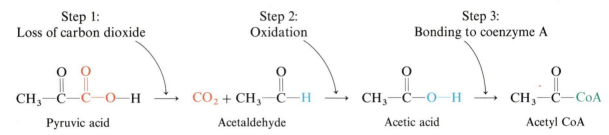

Step 1:
Loss of carbon dioxide

Step 2:
Oxidation

Step 3:
Bonding to coenzyme A

Pyruvic acid

Acetaldehyde

Acetic acid

Acetyl CoA

Three enzymes and five coenzymes are involved in the conversion of pyruvic acid to acetyl CoA. Although we won't go into the details, six molecules of ATP are produced from each glucose (three from each of two pyruvic acids) during the process.

19.4   ENERGY OUTPUT DURING CARBOHYDRATE CATABOLISM

Let's take an overall look at the entire process of carbohydrate catabolism—glycolysis, pyruvic-acid synthesis, citric-acid cycle, and respiratory chain—to see how much ATP is produced from a single glucose molecule. A look at the individual steps of the glycolysis pathway (Figure 19.2) shows that six molecules of ATP are produced for each glucose molecule converted into pyruvate. Steps 1 and 3 each use one ATP; steps 7 and 10 each generate two ATPs per starting glucose (one from each of two pyruvic acid molecules); and step 6 generates a further four molecules of ATP per glucose.

The ATP produced in glycolysis step 6 is generated indirectly when the reduced NADH coenzyme is eventually recycled back to its oxidized form, NAD$^+$. Although we saw in Section 18.9 that *three* molecules of ATP are generated from NADH when NAD$^+$ is produced inside mitochondria by the respiratory chain, only *two* molecules of ATP are generated when NAD$^+$ is produced in the cytoplasm, as occurs in glycolysis.

	Change in Number of ATPs Per Glucose
Step 1: Glucose $\longrightarrow$ Glucose 6-phosphate	-1
Step 3: Fructose 6-phosphate $\longrightarrow$ Fructose 1,6-diphosphate	-1
Step 6: 2 NADH $\longrightarrow$ 2 NAD$^+$	4
Step 7: 2 1,3-diphosphoglyceric acid $\longrightarrow$ 2 3-Phosphoglyceric acid	2
Step 10: 2 Phosphoenolpyruvic acid $\longrightarrow$ 2 Pyruvic acid	2
Net change for glycolysis:	6

In addition to the six ATPs produced during glycolysis and the six produced during the conversion of pyruvic acid to acetyl CoA, there are 24 ATPs produced from each glucose molecule by the citric-acid cycle and the respiratory chain (Section 18.9). Thus, a total of 36 molecules of ATP result from the catabolism of one glucose molecule, two-thirds of which come from the citric-acid cycle and the respiratory chain. So important is the production of ATP to all bodily processes that an active adult human synthesizes and consumes each day an amount of ATP approximately equal to his or her own body weight.

	Change in Number of ATPs Per Glucose
Glycolysis: Glucose $\longrightarrow$ Pyruvic acid	6
Pyruvic acid $\longrightarrow$ Acetyl CoA	6
Citric-acid cycle + Respiratory chain 2 Acetyl CoA $\longrightarrow$ 2 CO$_2$	24
Net change for complete glucose catabolism:	36

Solved Problem 19.1 How many moles of ATP are generated by glycolysis of 10.0 g of glucose (mol wt = 180 amu)?

Solution First determine how many moles of glucose are in 10.0 g.

$$10.0 \text{ g} \times \frac{1 \text{ mol}}{180 \text{ g}} = 0.0556 \text{ mol of glucose}$$

Next use your knowledge that glycolysis of 1 mol of glucose generates 6 moles of ATP as a conversion factor to arrive at the answer:

$$0.0556 \text{ mol glucose} \times \frac{6 \text{ mol ATP}}{1 \text{ mol glucose}} = 0.337 \text{ mol ATP}$$

Practice Problems

19.4 How many moles of ATP are generated by glycolysis of 0.5 mol glucose? By catabolism of 0.2 mol acetyl CoA?

19.5 How many grams of ATP (mol wt = 507 amu) are produced by catabolism of 1.8 g of glucose (mol wt = 180 amu)?

19.5 LIPID CATABOLISM

Lipids, like carbohydrates, undergo a series of enzyme-catalyzed metabolic reactions that result in the production of acetyl CoA. The process begins during digestion when triglycerides are hydrolyzed in the small intestine to yield a mixture of di- and monoglycerides, together with some glycerol and fatty acids. These components cross over into cells of the intestine, where they are reassembled into triglycerides and wrapped in protein to form lipoproteins called *chylomicrons*. The chylomicrons are then carried through the bloodstream either to fatty (*adipose*) tissues or to the liver, where lipid catabolism occurs. A diagram of the overall process is shown in Figure 19.4.

19.6 THE FATTY-ACID SPIRAL

Lipid catabolism begins with the hydrolysis of triglycerides to yield glycerol and three fatty-acid molecules. Glycerol is then converted into glyceraldehyde 3-phosphate, which enters the glycolysis pathway (Section 19.5) and is further

Figure 19.4
An overview of lipid catabolism.

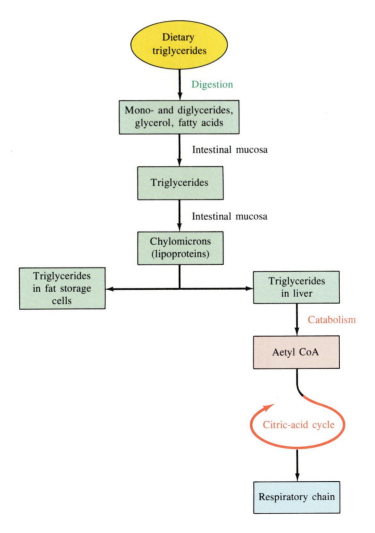

catabolized. As you might guess by looking at the structures, the transformation involves the phosphorylation of one —OH group followed by oxidation of another.

$$HO-CH_2-\overset{\overset{\displaystyle OH}{|}}{CH}-CH_2 \xrightarrow[\text{2. Oxidation}]{\text{1. Phosphorylation}} {}^{2-}O_3PO-CH_2-\overset{\overset{\displaystyle OH}{|}}{CH}-\overset{\overset{\displaystyle O}{\|}}{C}-H$$

Glycerol   Glyceraldehyde 3-phosphate

■ **Fatty-acid spiral** A repetitive series of biochemical reactions that degrade fatty acids to acetyl CoA.

Fatty-acid catabolism takes place in cell mitochondria by a repetitive four-step sequence of reactions called the **fatty-acid spiral** (Figure 19.5). A fatty acid is first bonded to coenzyme A, and each turn of the spiral then results in the cleavage of a two-carbon acetyl group from the fatty-acid chain until the entire molecule is ultimately degraded. As each acetyl group is cleaved, it enters the citric-acid cycle and is further catabolized.

To see how the fatty-acid spiral works, look at the catabolism of palmitic acid. One turn of the spiral converts the 16-carbon palmitoyl CoA into the 14-carbon myristyl CoA plus acetyl CoA; a second turn of the spiral converts myristyl CoA into the 12-carbon lauryl CoA plus acetyl CoA; a third turn converts lauryl CoA into the 12-carbon capryl CoA plus acetyl CoA; and so

Figure 19.5
The events of the fatty-acid spiral leading to the cleavage of an acetyl group from the end of the fatty-acid chain.

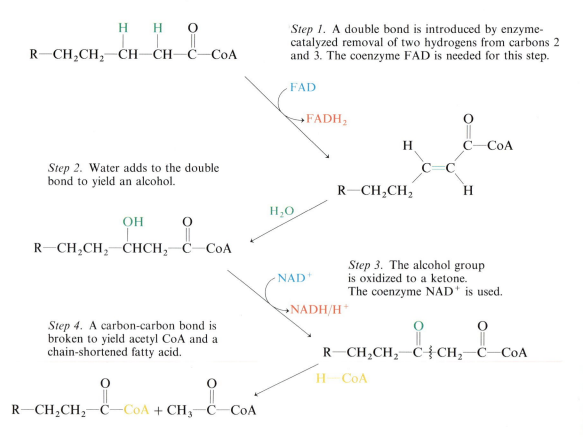

Step 1. A double bond is introduced by enzyme-catalyzed removal of two hydrogens from carbons 2 and 3. The coenzyme FAD is needed for this step.

Step 2. Water adds to the double bond to yield an alcohol.

Step 3. The alcohol group is oxidized to a ketone. The coenzyme NAD⁺ is used.

Step 4. A carbon-carbon bond is broken to yield acetyl CoA and a chain-shortened fatty acid.

on (Figure 19.6). Since fatty acids usually have an even number of carbon atoms (Section 14.2), none are left over.

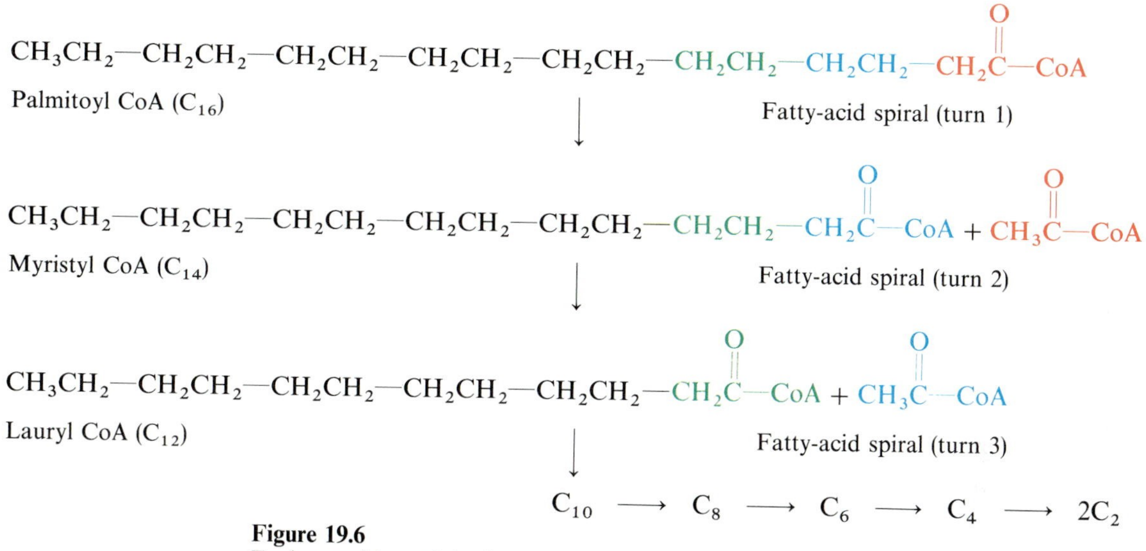

$$CH_3CH_2-CH_2CH_2-CH_2CH_2-CH_2CH_2-CH_2CH_2-CH_2CH_2-CH_2CH_2-CH_2\overset{\overset{\textstyle O}{\|}}{C}-CoA$$

Palmitoyl CoA (C_{16})   Fatty-acid spiral (turn 1)

$$CH_3CH_2-CH_2CH_2-CH_2CH_2-CH_2CH_2-CH_2CH_2-CH_2CH_2-CH_2\overset{\overset{\textstyle O}{\|}}{C}-CoA + CH_3\overset{\overset{\textstyle O}{\|}}{C}-CoA$$

Myristyl CoA (C_{14})   Fatty-acid spiral (turn 2)

$$CH_3CH_2-CH_2CH_2-CH_2CH_2-CH_2CH_2-CH_2CH_2-CH_2\overset{\overset{\textstyle O}{\|}}{C}-CoA + CH_3\overset{\overset{\textstyle O}{\|}}{C}-CoA$$

Lauryl CoA (C_{12})   Fatty-acid spiral (turn 3)

$$C_{10} \longrightarrow C_8 \longrightarrow C_6 \longrightarrow C_4 \longrightarrow 2C_2$$

Figure 19.6
Each repetition of the fatty-acid spiral cleaves two carbons from the end of the chain, yielding a molecule of acetyl CoA and a chain-shortened fatty acid.

You can predict how many molecules of acetyl CoA will be obtained from a given fatty acid simply by counting the number of carbon atoms and dividing by two. For example, the 14-carbon myristic acid yields seven molecules of acetyl CoA after six turns of the spiral, as shown in Solved Problem 19.2. (The number of turns of the spiral is always one less than the number of acetyl CoA molecules produced, because the last turn cleaves a four-carbon chain into two acetyl CoA's.)

Solved Problem 19.2   How many molecules of acetyl CoA are obtained by catabolism of myristic acid, $C_{14}H_{28}O_2$?

Solution   Draw the structure of the fatty acid starting from the carboxylic acid end and grouping the carbons into pairs:

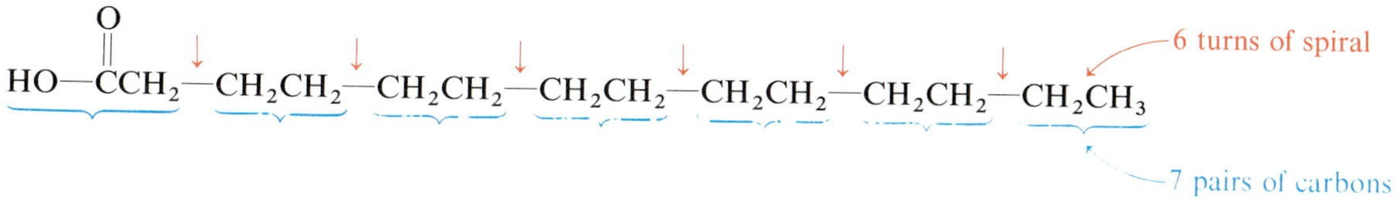

$$HO-\overset{\overset{\textstyle O}{\|}}{C}CH_2-CH_2CH_2-CH_2CH_2-CH_2CH_2-CH_2CH_2-CH_2CH_2-CH_2CH_3$$

6 turns of spiral

7 pairs of carbons

Now simply count the numbers of carbon pairs to find how many molecules of acetyl CoA are produced on catabolism, and count the number of bonds linking the pairs to find how many turns of the fatty-acid spiral are required.

Practice Problem   19.6   How many molecules of acetyl CoA are produced by catabolism of these fatty acids, and how many turns of the fatty-acid spiral are needed?
(a) lauric acid, $CH_3(CH_2)_{10}COOH$ (b) arachidic acid, $CH_3(CH_2)_{18}COOH$

19.7   ENERGY OUTPUT DURING FATTY-ACID CATABOLISM

We can calculate the amount of energy released during the catabolism of fatty acids in the same way we did for carbohydrates. The major energy source, of course, is acetyl CoA. As in glycolysis, each acetyl group produced in the fatty-acid spiral enters the citric-acid cycle and generates 12 molecules of ATP. Thus, a fatty acid that yields n molecules of acetyl CoA releases $12n$ ATPs.

In addition, each turn of the fatty-acid spiral releases five molecules of ATP indirectly. The $FADH_2$ coenzyme generated in step 1 releases two molecules of ATP when it is recycled, and the NADH coenzyme generated in step 3 releases three ATPs. To arrive at the final total, however, we have to subtract two molecules of ATP that are required for the initial bonding of the free fatty acid with coenzyme A. Thus the overall energy score card looks like this:

AN APPLICATION: DIABETES, A METABOLIC DISORDER

Diabetes mellitus, one of the most common metabolic diseases, is estimated to affect up to four percent of the population. Diabetes is not a single disease but is divided into two types, *insulin-dependent* and *noninsulin-dependent*. The insulin-dependent type, also called *juvenile-onset* diabetes, usually occurs in young people. By contrast, the less serious but more common noninsulin-dependent type, also called *adult-onset* diabetes, usually occurs in obese older people. Although usually thought of only as a disease of glucose metabolism, diabetes affects protein and fat metabolism as well.

Insulin-dependent diabetes is caused by an insufficient supply of the hormone insulin as a result of damage to secretory cells in the pancreas. Although insulin has multiple physiological effects, its primary role is to aid the passage of blood glucose into cells of the liver, skeletal muscle, and adipose tissue. When the supply of insulin is diminished and glucose is unable to enter cells, extremely high levels of blood sugar build up, leading to a condition called *hyperglycemia*. Some of this excess blood sugar spills over into the urine where it is excreted, leading to the loss of excessive amounts of water and electrolytes.

With glucose unavailable to its cells, the body responds by increasing the rate of fat metabolism as the only way of obtaining energy. Acetyl CoA molecules are manufactured at an ever-increasing rate until the citric-acid cycle is overloaded and cannot degrade them as rapidly as they are produced. As a result, acetyl CoA builds up inside cells, from which it is removed by a new series of metabolic reactions that transform it into substances called *ketone bodies*.

Ketone bodies:

$$CH_3-\overset{\overset{\displaystyle O}{\|}}{C}-CH_2-\overset{\overset{\displaystyle O}{\|}}{C}-OH$$
Acetoacetic acid

$$CH_3-\overset{\overset{\displaystyle O}{\|}}{C}-CH_3 \qquad CH_3-\overset{\overset{\displaystyle OH}{|}}{CH}-CH_2-\overset{\overset{\displaystyle O}{\|}}{C}-OH$$
Acetone   3-Hydroxybutyric acid

Ketone bodies enter the bloodstream and the urine, where they cause a condition called *ketosis*. Acetone, the most volatile of the three ketone bodies, is exhaled through the lungs and can often be smelled on the breath of patients with a severe case of diabetes. As a patient's condition worsens, the buildup of acetoacetic acid and 3-hydroxybutyric acid lowers the pH of the blood, producing *ketoacidosis*. Severe ketoacidosis may lead to coma and diminished brain function, but the effects can all be reversed by supplying insulin.

	Change in Number of ATPs per Fatty Acid
Initial bonding to coenzyme A	-2
$12 \times$ (molecules of acetyl CoA)	$12n$
$5 \times$ (number of turns of spiral)	$5(n - 1)$
Net change for complete fatty-acid catabolism:	$17n - 7$

where n = number of molecules of acetyl CoA produced
             = 1/2 number of carbons in the fatty-acid chain
and $n - 1$ = number of turns of the fatty-acid spiral

Comparing the amount of ATP produced by catabolism of a fatty acid to the amount produced by catabolism of glucose shows why our bodies use fat rather than carbohydrate for long-term energy storage. One mol of glucose (180 g) generates 36 mol of ATP, whereas one mol of lauric acid (200 g) generates 95 mol of ATP. Thus, fats yield more than twice as much energy per gram as carbohydrates (and do twice as much damage to diets).

Solved Problem 19.3   Calculate the number of molecules of ATP produced during the catabolism of lauric acid, $CH_3(CH_2)_{10}COOH$.

Solution   First, count the number of carbons in the lauric acid chain, and then divide by two to find how many molecules of acetyl CoA are produced:

$$CH_3(CH_2)_{10}COOH \Rightarrow C_{12} \Rightarrow 12/2 = 6 \text{ molecules of acetyl CoA}$$

Next, find how many turns of the fatty-acid spiral are needed to degrade lauric acid by subtracting 1 from the number of acetyl CoA molecules:

$$6 \text{ acetyl CoA} - 1 \Rightarrow 5 \text{ turns of the fatty-acid spiral}$$

Finally, do the arithmetic:

Initiation	-2 ATP
6 acetyl CoA $\times$ 12 ATP/acetyl CoA	72 ATP
5 turns $\times$ 5 ATP/turn	25
Net change	95 ATP

Practice Problems   **19.7**   Calculate the amount of ATP produced by catabolism of 1 mol of palmitic acid, $CH_3(CH_2)_{14}COOH$.

19.8   How many grams of ATP (form. wt = 507 amu) are produced by catabolism of 1.0 g of palmitic acid (form. wt = 256 amu)?

19.8 PROTEIN CATABOLISM

■ **Amino-acid pool** The sum of free amino acids available in the body.

Protein breakdown begins during digestion in the stomach and small intestine, where peptide bonds are hydrolyzed to give a mixture of the 20 amino acids. These amino acids are then absorbed through the intestinal wall, after which they enter circulation and become a part of the body's **amino-acid pool**, a name given to the entire collection of free amino acids found anywhere in the body.

Since living organisms are dynamic rather than static, all tissues and biomolecules in the body are constantly being degraded, repaired, and replaced. Thus, amino acids enter the body pool not only by digestion of proteins but also by breakdown of tissues. In fact, a normal adult breaks down approximately 400 g of protein every day. Once in the pool, amino acids meet several fates. Most are used as building blocks for the resynthesis of tissues, but some are used as a source of nitrogen in the synthesis of nucleic acid bases and other biomolecules, and some are catabolized for energy if the body needs additional fuel (Figure 19.7).

Amino-acid catabolism is quite complex since each of the 20 different amino acids is degraded through its own unique pathway. The general idea of all the pathways, however, is that amino acids are converted into one of the intermediates in the citric-acid cycle (Section 18.5). They then enter the cycle, where they are ultimately degraded to CO_2.

■ **Oxidative deamination** A reaction that converts an amino group ($-NH_2$) into a carbonyl group.

The catabolism of amino acids begins with **oxidative deamination**, a reaction that converts the $-NH_2$ group of an α-amino acid into a keto group. The process occurs by an enzyme-catalyzed reaction between the α-amino acid and α-ketoglutaric acid, yielding an α-keto acid and glutamic acid. The α-keto acid is then further converted into one of the intermediates in the citric-acid cycle, while the glutamic acid is recycled back to α-ketoglutaric acid plus ammonia.

This —NH$_2$ group and this =O group exchange places.

$$R-\underset{\underset{\text{An α-amino acid}}{}}{\overset{NH_2}{CH}}-COOH + HOOC-\underset{\underset{\text{α-Ketoglutaric acid}}{}}{\overset{O}{C}}-CH_2CH_2COOH$$

transaminase

$$R-\underset{\underset{\text{An α-keto acid}}{}}{\overset{O}{C}}-COOH + HOOC-\underset{\underset{\text{Glutamic acid}}{}}{\overset{NH_2}{CH}}-CH_2CH_2COOH$$

NAD^+
$NADH/H^+$

Citric-acid cycle

$$HOOC-\underset{\underset{\text{α-Ketoglutaric acid}}{}}{\overset{O}{C}}-CH_2CH_2COOH + NH_3$$

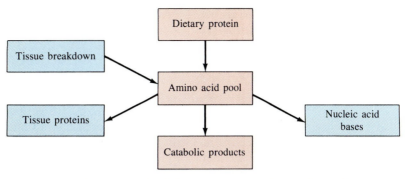

Figure 19.7
The body's amino-acid pool.

About half of the 20 different α-keto acids produced by oxidative deamination enter the citric-acid cycle through conversion to acetyl CoA, but the rest enter after being converted into succinic acid, fumaric acid, oxaloacetic acid, or α-ketoglutaric acid. The exact fate of each amino acid is shown in Figure 19.8.

Solved Problem 19.4   What α-keto acid results from oxidative deamination of leucine?

Solution   First, look up leucine in Table 15.1, and write its structure:

$$CH_3CHCH_2-CH-COOH \qquad Leucine$$

with CH_3 on the third carbon and NH_2 on the α carbon.

Next, remove the —NH_2 and —H groups bonded to the α carbon, and replace them by a =O. The product is the desired α-keto acid.

$$CH_3CHCH_2-C-COOH \qquad Leucine\ oxidative\ deamination\ product$$

with CH_3 on the third carbon and O double-bonded to the α carbon.

Practice Problems   **19.9**   What α-keto acid is produced by oxidative deamination of aspartic acid?

19.10   What further series of catabolic reactions is the α-keto acid produced from aspartic acid in Problem 19.9 likely to undergo?

Figure 19.8
Amino-acid catabolism. Different amino acids enter the citric-acid cycle at the indicated points.

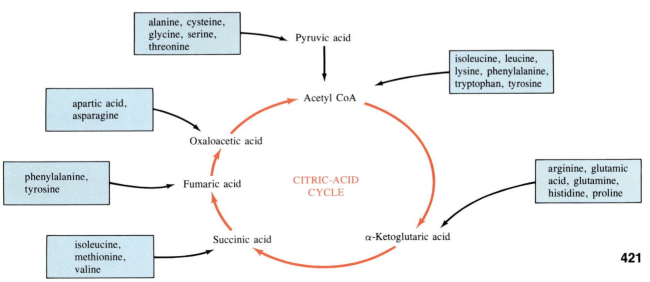

19.9 CARBOHYDRATE ANABOLISM: GLUCONEOGENESIS

We've seen in considerable detail over the last two chapters how our bodies degrade glucose to extract energy. It's perhaps surprising, therefore, to find out that our bodies also need to be able to *make* glucose from noncarbohydrate precursors. Certain tissues, particularly those in the brain, have an enormous requirement for glucose—as much as 120 g per day—but our bodies store only about 200 g. Thus, the brain would come dangerously close to running out of fuel after only one day without eating, unless the body were able to synthesize glucose.

■ **Gluconeogenesis** The biological synthesis of glucose from simpler molecules.

The overall process for **gluconeogenesis**, the biological synthesis of glucose from simpler molecules, is shown in Figure 19.9. The major precursors are pyruvic acid, formed from lactic acid and some amino acids, and oxaloacetic acid, formed from other amino acids. The reactions take place in the cytoplasm of liver cells and require a total of four ATP molecules, two GTP (guanosine triphosphate) molecules, and two NADH coenzymes to supply the necessary energy. After synthesis, the glucose migrates through the cell membrane and circulates through the blood to reach other tissues.

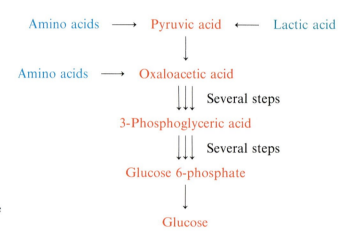

Figure 19.9
The anabolic pathway for glucose synthesis by gluconeogenesis.

19.10 LIPID ANABOLISM: LIPIGENESIS

Just as our bodies need to synthesize glucose (gluconeogenesis), they also need to synthesize fat (**lipigenesis**). Through lipigenesis, excess dietary carbohydrate is converted into fat for long-term energy storage and for the synthesis of specialized membrane lipids.

■ **Lipigenesis** The biological synthesis of fat molecules from excess dietary carbohydrate.

Fatty-acid anabolism takes place in the cytoplasm of liver cells by a spiral process that involves the sequential joining together of two-carbon acetyl groups (Figure 19.10). Two acetyl groups join to yield a four-carbon chain, a third acetyl group joins to yield a six-carbon chain, and so on until an entire fatty-acid molecule is made. Energy for each round of the process is provided by one molecule of ATP and two molecules of NADH.

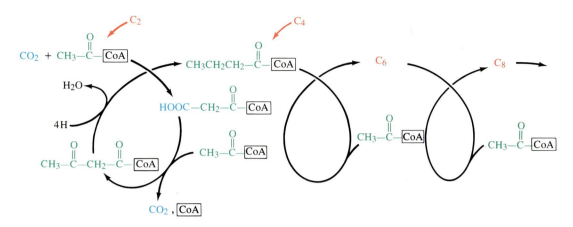

Figure 19.10
The lipigenesis of fatty acids from acetyl CoA. Each turn of the spiral adds two more carbon atoms to the growing chain.

Practice Problem   **19.11**   How many rounds of the lipigenesis spiral are necessary to synthesize palmitic acid, $C_{15}H_{31}COOH$? How many molecules of ATP are necessary?

19.11 AMINO-ACID ANABOLISM

We saw in Section 15.2 that humans are able to synthesize only 10 of the 20 different amino acids found in proteins; the remaining 10 essential amino acids must be obtained from dietary sources. Nine of the ten amino acids synthesized in the body are made from glutamic acid, itself made from ammonia and α-ketoglutaric acid. The key reaction is a *reductive amination*, which is simply the reverse of the oxidative deamination step so common in amino-acid catabolism (Section 19.8). The coenzyme necessary for this transformation is NADPH (nicotinamide adenine dinucleotide phosphate), a close relative of NAD.

$$NH_3 + \underset{\alpha\text{-Ketoglutaric acid}}{HOOCCH_2CH_2\overset{\displaystyle O}{\overset{\|}{C}}COOH} \xrightarrow[\;H^+\quad\quad H_2O\;]{NADPH,\quad NADP^+,} \underset{\text{Glutamic acid}}{HOOCCH_2CH_2\overset{\displaystyle NH_2}{\overset{|}{C}HCOOH}}$$

Glutamic acid then reacts with various other α-keto acids to transfer its —NH_2 group and generate α-amino acids.

$$\underset{\text{An }\alpha\text{-keto acid}}{R-\overset{\displaystyle O}{\overset{\|}{C}}-COOH} + \underset{\text{Glutamic acid}}{HOOCCH_2CH_2\overset{\displaystyle NH_2}{\overset{|}{C}HCOOH}} \longrightarrow \underset{\text{An }\alpha\text{-amino acid}}{R-\overset{\displaystyle NH_2}{\overset{|}{C}H}-COOH} + \underset{\alpha\text{-Ketoglutaric acid}}{HOOCCH_2CH_2\overset{\displaystyle O}{\overset{\|}{C}}COOH}$$

Practice Problem   **19.12**   What α-amino acids would you obtain from these α-keto acids?
(a) pyruvic acid   (b) oxaloacetic acid

INTERLUDE: EXERCISE AND WEIGHT

Surveys indicate that 59 percent of adult Americans claim to exercise regularly, and more than 75,000 of them have completed a marathon. Surveys also indicate, though, that nearly 60 percent of adult Americans are obese or have a problem with weight control. The relationship between exercise and body weight is by no means simple—some slender individuals consume more calories per day and get less exercise than some obese persons—but there *is* a relationship. Assuming that caloric intake remains the same, body weight *must* go down as the amount of exercise goes up, because more of the body's fuel must be metabolized to provide energy. For example, a half hour of jogging or brisk walking "burns" about 350 calories (large calories, or kcal—see Section 1.11). Since a pound of fat represents about 3500 calories, this exercise corresponds to one-tenth of a pound of weight lost or, if you're already at a desirable weight, one more ice-cream cone you can get away with.

According to recommendations of the American College of Sports Medicine, an effective exercise program should meet three criteria: (1) The exercise should be done at least three times per week; (2) each session should be intense enough to maintain a heart rate near 90 percent of the individual's maximum effort for at least 15 minutes; and (3) the exercise should involve the large muscle groups in the legs. Running, swimming, biking, aerobic dancing, and cross-country skiing are perfect choices, while tennis, basketball, and others are effective if done vigorously.

The physiological effects of such exercise extend far beyond weight control. As shown in the following table, a large number of beneficial changes in both the cardiovascular and skeletal muscle systems occur for someone who has trained to a marathon level of endurance.

Physiological Responses to Sustained Exercise

Muscle Responses	Cardiovascular Responses
Number of mitochondria up 120%	Blood volume up 20%
Glycolytic enzymes up 10–100%	Heart volume up 25%
Glycogen storage up 150%	Stroke volume of heart up 60%
Lactic acid production down 30%	Resting pulse rate down 20%
Fat catabolic enzymes up 100–300%	Blood pressure down 15%
Muscle fiber size up 30%	High-density lipoproteins up 30%

SUMMARY

The metabolic breakdown of carbohydrates begins with **digestion** in the stomach and small intestine where amylase enzymes hydrolyze starch to yield glucose. Glucose is then further degraded by **glycolysis**, also called the *Embden-Meyerhof pathway*, a series of ten enzyme-catalyzed reactions that convert glucose into two molecules of pyruvic acid. Pyruvic acid, in turn, is broken down in the cytoplasm of cells under **aerobic** (oxygen-rich) conditions to yield acetyl CoA plus CO_2. Under **anaerobic** (oxygen-free) conditions in muscle tissue, pyruvic acid is reduced to give lactic acid.

Lipids also undergo a series of enzyme-catalyzed catabolic reactions that result in the formation of acetyl CoA. Lipids are first hydrolyzed to yield glycerol plus three fatty acids, and the fatty acids are degraded by a four-step sequence of reactions called the **fatty-acid spiral**. Each turn through the sequence causes the cleavage of a two-carbon acetyl CoA fragment from the end of the fatty-acid chain until the entire molecule is ultimately degraded.

Protein catabolism begins with digestion when amide bonds are hydrolyzed and individual free amino acids circulate into the body's **amino-acid pool**. Each of the 20 different amino acids is catabolized by its own unique pathway, although acetyl CoA is once again the final product. **Oxidative deamination**, the conversion of an α-amino group into a keto group, is the opening step.

Carbohydrates, fats, and proteins can be made as well as degraded in the body. Glucose synthesis from pyruvic acid (**gluconeogenesis**) is a particularly important anabolic pathway, because it serves to keep brain tissues supplied with fuel. **Lipigenesis**, the synthesis of fatty acids from acetyl CoA, is important as a mechanism for long-term energy storage. Amino acids are made by anabolic pathways starting from α-keto acids.

REVIEW PROBLEMS

Digestion and Metabolism

19.13 Where in the body does digestion occur, and what kinds of chemical reactions does it involve?

19.14 What is meant by the words *aerobic* and *anaerobic*?

19.15 What three products are formed from pyruvic acid under aerobic, anaerobic, and fermentation conditions?

19.16 What are chylomicrons, and how are they involved in metabolism?

Carbohydrate Metabolism

19.17 What is the name and structure of the final product of carbohydrate glycolysis?

19.18 Where in the cell does glycolysis occur?

19.19 What is the name of the anabolic pathway for making glucose?

19.20 Where in the cell does glucose synthesis take place?

19.21 What two molecules serve as starting materials for glucose synthesis?

19.22 By what other name is the glycolysis pathway known?

19.23 Lactic acid can be converted into pyruvic acid by the enzyme lactic acid dehydrogenase and the coenzyme NAD^+. Write the reaction in the standard biochemical format, using a curved arrow to show the involvement of NAD^+.

19.24 Look at the ten steps in glycolysis, and then answer these questions:
(a) Which steps involve phosphorylation?

(b) Which steps involve isomerizations?
(c) Which step is an oxidation?

19.25 How many moles of ATP are produced by:
(a) glycolysis of one mole of glucose?
(b) aerobic conversion of one mole of pyruvic acid to one mole of acetyl CoA?
(c) catabolism of one mole of acetyl CoA in the citric-acid cycle?

19.26 How many grams of ATP (mol wt = 570 amu) are released by catabolism of 5.0 g of glucose (mol wt = 180 amu)?

19.27 If fructose is isomerized to yield glucose prior to glycolysis, how many moles of ATP would you expect to obtain from complete catabolism of 1.0 mol of sucrose (Problem 19.26)?

19.28 How many moles of acetyl CoA are produced by catabolism of 1.0 mol of sucrose (Problem 19.27)?

19.29 How many grams of CO_2 would you expect to obtain from catabolism of 0.10 mol of glucose?

19.30 How many moles of ATP are produced by complete catabolism of one mole of maltose to yield CO_2 (Section 13.7)?

Fat Metabolism

19.31 Where in the cell does the fatty-acid spiral occur?

19.32 What initial chemical transformation takes place on a fatty acid to activate it for catabolism?

19.33 Why do you suppose the sequence of reactions that catabolize fats is called the fatty-acid *spiral* rather than the fatty-acid *cycle*?

19.34 What is the name of the anabolic pathway for synthesizing fatty acids?

19.35 Where in the cell does fatty-acid synthesis occur?

19.36 What is the starting material for fatty-acid synthesis?

19.37 How many moles of ATP are produced by one turn of the fatty-acid spiral?

19.38 Arrange these three molecules in order of their biological energy content per mole:
(a) glucose   (b) capric acid, $CH_3(CH_2)_8COOH$
(c) alanine

19.39 We saw in Section 19.5 that the glycerol derived from digestion of dietary fat is converted into glyceraldehyde 3-phosphate, which then enters into step 6 of the glycolysis pathway. What further transformations are necessary to convert glyceraldehyde 3-phosphate into pyruvic acid?

19.40 If the conversion of glycerol to glyceraldehyde 3-phosphate releases one molecule of ATP, how many molecules of ATP are released during the conversion of glycerol to pyruvic acid? (See Problem 19.39.)

19.41 How many molecules of ATP are released in the overall catabolism of glycerol to acetyl CoA?

19.42 How many molecules of ATP are released in the complete catabolism of glycerol to CO_2 and H_2O? (See Problem 19.41.)

19.43 Show the products of each step in the following fatty-acid spiral on hexanoic acid:

(1)

$$CH_3CH_2CH_2CH_2CH_2\!-\!\overset{\displaystyle O}{\overset{\|}{C}}\!-\!CoA \xrightarrow[\substack{\text{acetyl CoA}\\\text{dehydrogenase}}]{\text{FAD}\quad\text{FADH}_2} \;?$$

(2) Product of (1) $\xrightarrow[\text{enoyl CoA hydratase}]{H_2O}$?

(3) Product of (2) $\xrightarrow[\substack{\beta\text{-hydroxyacyl-CoA}\\\text{dehydrogenase}}]{\text{NAD}^+ \quad \text{NADH/H}^+}$?

(4) Product of (3) $\xrightarrow[\substack{\text{acetyl CoA}\\\text{transferase}}]{\text{H}-\text{CoA}}$?

19.44 Write the equation for the final step in the catabolism of any fatty acid with an even number of carbons.

19.45 How many molecules of acetyl CoA result from catabolism of these compounds?
(a) caprylic acid, $CH_3(CH_2)_6COOH$
(b) myristic acid, $CH_3(CH_2)_{12}COOH$

19.46 How many turns of the fatty-acid spiral are necessary to completely catabolize caprylic and myristic acids (Problem 19.45)?

19.47 How many molecules of acetyl CoA result from catabolism of glyceryl trimyristate?

19.48 Are any of the intermediates in the fatty-acid spiral chiral (Section 13.3)? Explain.

19.49 How many grams of ATP are released by catabolism of 5.0 g of glyceryl trimyristate (mol wt = 722 amu)?

Protein Metabolism

19.50 What is the body's amino-acid pool?

19.51 What are the fates of amino acids in the amino-acid pool?

19.52 What is meant by an oxidative deamination reaction?

19.53 Write the structures of the α-keto acids produced by oxidative deamination of (a) phenylalanine, and (b) tryptophan.

19.54 What substance serves as the starting material for amino-acid anabolism?

CHAPTER 20

Nuclear Chemistry

NON - IRRADIATED -

IRRADIATED - (0.2 M RAD)

STRAWBERRIES -

15 DAYS STORAGE 38ºF (4ºC)

Both boxes of strawberries have been stored for two weeks at refrigerator temperatures. The box on the right was first subjected to low-level radiation to kill organisms that cause food spoilage (such as the molds now covering the strawberries on the left). We'll learn more about the effects and uses of radiation in this chapter.

All the chemistry we've seen in the past 19 chapters has been based on a single assumption—an assumption so basic that it's hardly been mentioned. We've simply taken it for granted that chemical changes entail only the making and breaking of *bonds*, while the identities of the *atoms* themselves remain unchanged. In other words, all the reactions we've seen have involved changes in the sharing of *electrons* between atoms, while *nuclei* have been unaffected. Carbon atoms have remained carbon, hydrogen atoms have remained hydrogen, and so on.

It's now time to examine this assumption. Although it's true that the chemistry of living organisms and the chemistry in a normal laboratory do not affect atomic nuclei, some high-energy processes such as those taking place inside a nuclear reactor *do* change the very atoms themselves. In this chapter, we'll answer the following questions:

1. **What is radioactivity?** The goal: you should learn what radioactivity is and should be familiar with the terms used to describe it.
2. **What are the different kinds of radioactivity?** The goal: you should learn the characteristics of the three fundamental kinds of radiation— α, β, and γ.
3. **How is radioactivity measured?** The goal: you should learn the common units for measuring radiation.
4. **What changes take place in atoms when they emit radiation?** The goal: you should learn how one element can change into another by emitting radiation.
5. **What is nuclear fission and how is atomic energy used?** You should learn how atomic energy can be used to generate power.

20.1 THE DISCOVERY OF RADIOACTIVITY

The discovery of radioactivity dates to the year 1896 when the French physicist Henri Becquerel made a mistake. While investigating the nature of fluorescence, Becquerel happened to place a sample of a uranium-containing mineral called *pitchblende* on top of a photographic plate that had been wrapped in black paper and placed in a drawer to protect it from sunlight. On developing the plate, Becquerel was surprised to find a silhouette of the mineral. He immediately concluded that the mineral was producing some kind of unknown radiation that passed through the paper and exposed the photographic plate.

Becquerel's student, Marie Sklodowska Curie, and her husband Pierre took up the challenge and began a series of investigations into this new phenomenon, which they termed **radioactivity**. They found that the source of the radioactivity was the element uranium (U) and that two previously unknown elements, which they named polonium (Po) and radium (Ra), were also radioactive. For these achievements, Becquerel and the Curies shared the 1903 Nobel Prize in physics.

20.2   THE NATURE OF RADIOACTIVITY: α, β, AND γ RADIATION

Further work on radioactivity by the English scientist Ernest Rutherford soon established that there were at least two types of radiation, which he named **alpha** (α) and **beta** (β) after the first two letters of the Greek alphabet. Shortly thereafter, a third type of radiation was found and named for the third Greek letter, **gamma** (γ).

Each of the three types of radiation shows behavior quite different from the other two. For example, Marie Curie was able to demonstrate by a very simple experiment that α radiation is positively charged, that β radiation is negatively charged, and that γ radiation has no charge. When the three kinds of radiation are passed between electrically charged plates, α rays (positive) bend toward the negative plate, β rays (negative) bend toward the positive plate, and γ rays continue straight ahead (Figure 20.1).

A further difference among the three kinds of radiation is that α and β radiations are actually small particles, while γ radiation consists of very high energy light waves that have no mass. Rutherford was able to show that an α particle is simply a helium nucleus and that a β particle is an electron. Recall that a helium atom (atomic number = 2 and atomic weight = 4) consists of two protons, two neutrons, and two electrons. When the two electrons are removed, the remaining helium *nucleus*, or α particle, has only the two protons and two neutrons.

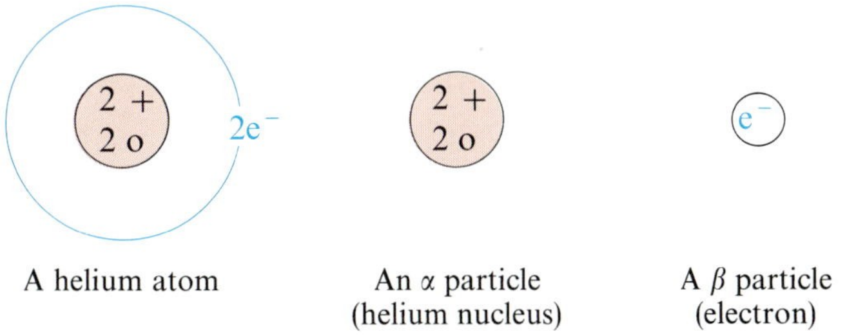

- **Radioactivity**   Spontaneous emission of radiation (alpha, beta, or gamma rays).

- **Alpha radiation**   Emission of helium nuclei, $^4_2$He.

- **Beta radiation**   Emission of electrons.

- **Gamma radiation**   Emission of high-energy electromagnetic waves.

A helium atom

An α particle
(helium nucleus)

A β particle
(electron)

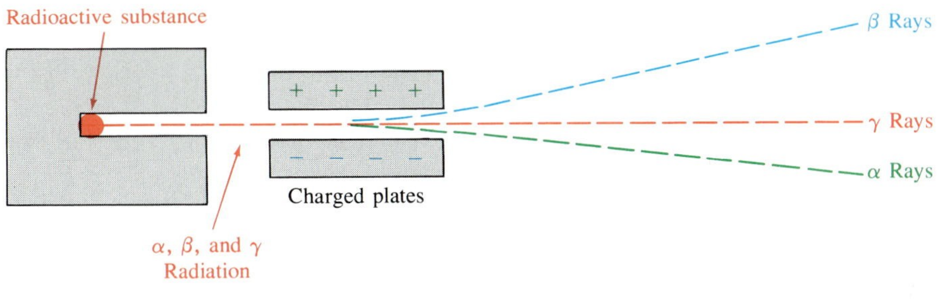

Figure 20.1
Determining the electrical charges of different kinds of radiation by passing them between charged plates. α rays, since they are attracted toward the negative plate, must be positively charged; β rays, since they are attracted toward the positive plate, must be negatively charged; and γ rays, since they are not attracted to either plate, must be neutral.

Radioactive substance

β Rays

γ Rays

α Rays

Charged plates

α, β, and γ
Radiation

Table 20.1   Characteristics of α, β, and γ Radiation

Type of Radiation	Symbol	Charge	Composition	Mass (*amu*)	Velocity	Penetrating Power
Alpha	α, ^4_2He	$+2$	Helium nucleus	4	Up to 10% light speed	Low (1)
Beta	β, $^0_{-1}\text{e}$	-1	Electron	1/1840	Up to 90% light speed	Medium (100)
Gamma	γ	0	High-energy radiation	0	Light speed	High (1000)

Yet a third difference between the three kinds of radiation is their penetrating power. Because of their relatively large mass, α particles move slowly (up to about one-tenth the speed of light) and can be stopped by a few sheets of paper or by the top layer of skin. Beta particles, because they are much lighter, move at up to nine-tenths the speed of light and have about 100 times the penetrating power of alpha particles. A block of wood or heavy protective clothing is necessary to stop beta radiation, which would otherwise penetrate the skin, causing burns. Gamma rays move at the speed of light and have about 1000 times the penetrating power of alpha particles. A lead block several inches thick is needed to stop gamma radiation, which would otherwise penetrate and damage the body's internal organs. The characteristics of the three kinds of radiation are summarized in Table 20.1.

Note that a beta particle is represented by $^0_{-1}\text{e}$ because its mass number is 0 and its charge is -1. Similarly, an alpha particle (a helium nucleus) is represented in Table 20.1 by the symbol ^4_2He. As discussed earlier in Section 2.3, atoms are represented by giving their mass number as a superscript and their atomic number as a subscript. Even though an alpha particle has a $+2$ charge, we don't indicate the charge when writing the symbol:

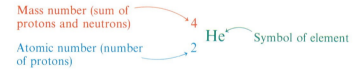

Mass number (sum of protons and neutrons) — 4

Atomic number (number of protons) — 2

He ← Symbol of element

20.3   NUCLEAR DECAY

Radioisotope A radioactive isotope.

Radioactivity is not limited to just the few elements studied by the Curies. Every element in the periodic table has at least one radioactive isotope (**radioisotope**), and there are nearly 2000 different radioisotopes known. Most do not occur naturally but are made in high-energy particle accelerators by techniques we'll mention briefly in Section 20.7. (Recall from Section 2.3 that *isotopes* are atoms that have the same number of protons in their nuclei but different numbers of neutrons. Hydrogen, for example, has three known isotopes—^1_1H, ^2_1H, and ^3_1H—but only ^3_1H is radioactive. It's not fully understood why some isotopes are radioactive and others aren't.)

Alpha Emission   Think for a minute about the consequences of what you've just learned. If radioactivity involves the spontaneous emission of a small particle

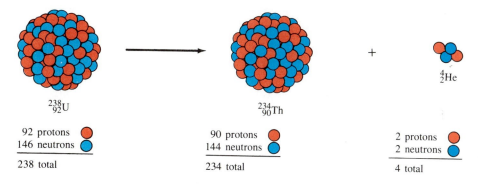

Figure 20.2
Emission of an alpha particle from an atom of uranium-238 leaves behind an atom of thorium-234.

from an unstable atomic nucleus, then the atom itself must undergo a change. For example, if an atom of uranium-238 ($^{238}_{92}U$) were to give off an alpha particle, the atom would have two fewer protons and two fewer neutrons after the emission. Since the number of protons in the nucleus has now changed from 92 to 90, the identity of the atom has changed from uranium-238 to thorium-234 ($^{234}_{90}Th$) (Figure 20.2). The process of particle emission from a nucleus is called **nuclear decay**, and the change of one element into another brought about by nuclear decay is called **transmutation**.

■ **Nuclear decay** The emission of a particle from the nucleus of an element.

■ **Transmutation** The change of one element into another brought about by nuclear decay.

Note that the nuclear reaction shown in Figure 20.2 is not balanced in the chemical sense because the kinds of atoms are not the same on both sides of the arrow. Instead, we say that a nuclear equation is balanced when the sums of the mass numbers are the same on both sides of the equation and when the sums of the atomic numbers are the same.

Solved Problem 20.1 Polonium-218 is one of the α emitters studied by Marie Curie. Write the equation for the decay, and identify the element formed.

Solution First, look up the atomic number of polonium in the periodic table, and write part of the nuclear equation using the symbol for polonium-218 in the standard format:

$$^{218}_{84}Po \longrightarrow {}^{4}_{2}He + ?$$

Next, finish the nuclear equation by calculating the mass number and atomic number of the product element. The mass number is $218 - 4 = 214$, and the atomic number is $84 - 2 = 82$. A look at the periodic table identifies the element with atomic number 82 as lead (Pb).

$$^{218}_{84}Po \longrightarrow {}^{4}_{2}He + {}^{214}_{82}Pb$$

Finally, check your answer by seeing that the mass numbers and atomic numbers of the two sides of the equation are balanced:

Mass numbers: $218 = 4 + 214$

Atomic numbers: $84 = 2 + 82$

Practice Problems

20.1 High levels of radioactive radon-222 ($^{222}_{86}Rn$) have been found in many homes built on radium-containing rock, leading to the possibility of health hazards. What product results from alpha emission by radon-222?

20.2 What isotope of radium (Ra) is converted into radon-222 ($^{222}_{86}Rn$) by alpha emission?

Beta Emission   Beta emission also leads to nuclear decay, but in a different way than alpha emission. Whereas alpha emission involves the loss of two protons and two neutrons from the nucleus, beta emission involves the *decomposition of a neutron* to yield an electron and a proton. The electron is ejected as a beta particle, and the proton is retained by the nucleus. The net result is that the atomic number of the atom increases by 1, but the mass number of the atom remains the same since the total number of protons and neutrons is not changed. For example, iodine-131 ($^{131}_{53}I$), a radioisotope used in detecting thyroid problems, undergoes nuclear decay by beta emission to yield xenon-131 ($^{131}_{54}Xe$) (Figure 20.3).

Solved Problem 20.2 What element is produced by beta emission of cobalt-60, a radiation source used in many hospitals?

Solution First, look up the atomic number of cobalt in the periodic table, and write part of the nuclear equation:

$$^{60}_{27}Co \longrightarrow {}^{0}_{-1}e + ?$$

Next, finish the nuclear equation by calculating the mass number and atomic number of the product element. The mass number remains the same (60), but the atomic number increases by 1 since it gains a proton (27 + 1 = 28). A look at the periodic table identifies the element with atomic number 28 as nickel (Ni).

$$^{60}_{27}Co \longrightarrow {}^{0}_{-1}e + {}^{60}_{28}Ni$$

Finally, check your answer by seeing that the mass numbers and atomic numbers of the two sides of the equation are balanced:

Mass numbers:	$60 = 0 + 60$
Atomic numbers:	$27 = -1 + 28$

Practice Problems

20.3 Carbon-14, a beta emitter, is a rare isotope of great use in dating archaeological artifacts. Write a nuclear equation for decay of carbon-14.

20.4 Write nuclear equations for beta emission of these isotopes:
(a) $^{3}_{1}H$ (b) $^{210}_{82}Pb$

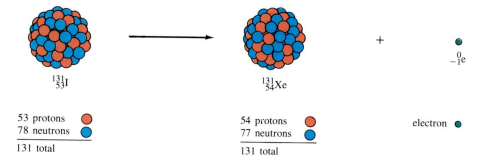

Figure 20.3
Emission of a beta particle from an atom of iodine-131 leaves behind an atom of xenon-131.

$^{131}_{53}I$

53 protons ●
78 neutrons ●
131 total

$^{131}_{54}Xe$

54 protons ●
77 neutrons ●
131 total

$+$

$^{\ 0}_{-1}e$

electron ●

Gamma Emission   Gamma rays, unlike alpha and beta particles, have no mass; they are simply very high-energy light waves. Gamma emission usually accompanies α or β emission but is often omitted when writing nuclear equations because it affects neither mass number nor atomic number. Because of their great penetrating power, however, gamma rays are by far the most dangerous kind of radiation to humans.

Table 20.2 summarizes the three different kinds of radioactive decay.

Table 20.2   Summary of Radioactive Decay

Kind of Decay	Effect on Mass Number	Effect on Atomic Number
α	Mass number decreases by 4	Atomic number decreases by 2
β	Mass number unchanged	Atomic number increases by 1
γ	Mass number unchanged	Atomic number unchanged

20.4   HALF-LIFE

The rate of radioactive decay varies greatly from one isotope to another. Some isotopes such as uranium-235 decay at a barely perceptible rate, while others decay almost instantly. Rates of decay are measured in units of **half-life ($t_{1/2}$)**, where one half-life is defined as the amount of time required for one-half of the radioactive sample to decay. For example, the half-life of iodine-131, a radioisotope used in thyroid testing, is 8 days. If today you have a certain amount of $^{131}_{53}I$, say 1.0 g, then 8 days from now you will have only 0.50 g of $^{131}_{53}I$ remaining because one-half of the sample will have decayed. After 8 more days (16 total), only 0.25 g of $^{131}_{53}I$ will remain; after a further 8 days (24 total), only 0.125 g will remain; and so on. Each passage of a half-life causes the decay of one-half of whatever sample remains (Figure 20.4, page 434).

■ **Half-life ($t_{1/2}$)** The amount of time required for 50 percent of a sample to undergo radioactive decay.

$$1.0 \text{ g } ^{131}_{53}I \quad \xrightarrow[\text{days}]{8} \quad \begin{array}{l} 0.50 \text{ g } ^{131}_{53}I \\ 0.50 \text{ g } ^{131}_{54}Xe \end{array} \quad \xrightarrow[\text{days}]{16} \quad \begin{array}{l} 0.25 \text{ g } ^{131}_{53}I \\ 0.75 \text{ g } ^{131}_{54}Xe \end{array} \quad \xrightarrow[\text{days}]{24} \quad \begin{array}{l} 0.125 \text{ g } ^{131}_{53}I \\ 0.875 \text{ g } ^{131}_{54}Xe \end{array} \quad \longrightarrow$$

One half-life   Two half-lives   Three half-lives

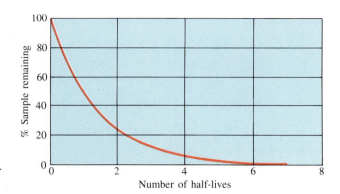

Figure 20.4
A plot showing the radioactive decay of $^{131}_{53}\text{I}$ ($t_{1/2} = 8$ days) over time.

The half-lives of some useful radioisotopes are given in Table 20.3. As you might expect, those radioisotopes that are used in medical applications have fairly short half-lives so that they decay rapidly and cause no long-term health hazards.

Table 20.3   Half-lives of Some Useful Radioisotopes

Radioisotope	Symbol	Radiation	Half-life	Use
Tritium	$^{3}_{1}\text{H}$	β	12.26 years	Biochemical tracer
Carbon-14	$^{14}_{6}\text{C}$	β	5730 years	Archaeological dating
Sodium-24	$^{24}_{11}\text{Na}$	β, γ	15.0 hours	Blood studies
Phosphorus-32	$^{32}_{15}\text{P}$	β	14.3 days	Leukemia therapy
Potassium-42	$^{42}_{19}\text{K}$	β, γ	12.4 hours	Nutrition studies
Cobalt-60	$^{60}_{27}\text{Co}$	β, γ	5.3 years	Cancer therapy
[a] Technetium-99*m*	$^{99m}_{43}\text{Tc}$	γ	6.02 hours	Brain scans
Iodine-123	$^{123}_{53}\text{I}$	γ	13.3 hours	Thyroid therapy
Iodine-131	$^{131}_{53}\text{I}$	β, γ	8.07 days	Thyroid studies
Uranium-235	$^{235}_{92}\text{U}$	α, γ	7.1×10^8 years	Nuclear reactors

[a] The *m* in technetium-99*m* stands for *metastable*, meaning that it decays (by γ decay) without changing its mass number or atomic number.

Solved Problem 20.3   Phosphorus-32 has a half-life of approximately 14 days. What percent of a sample will remain after 8 weeks?

Solution   This kind of problem is a lot easier than it looks; you just have to think it through logically. First, determine how many half-lives have elapsed. Since one half-life of $^{32}_{15}\text{P}$ is 14 days (2 weeks), 8 weeks represents four half-lives.

Next, do the calculation. Since each half-life decreases the amount of the sample by half, we have to multiply the starting amount (100 percent) by 1/2 for each of the four half-lives that has elapsed:

Four half-lives

$$\text{Final percentage} = 100\% \times (1/2) \times (1/2) \times (1/2) \times (1/2)$$
$$= 100\% \times (1/16) = 6.25\%$$

Practice Problem   **20.5**   The half-life of carbon-14 is 5730 years. What percentage of $^{14}_{6}\text{C}$ remains in a sample estimated to be about 17,000 years old?

20.5 IONIZING RADIATION

■ **Ionizing radiation** Radiation capable of dislodging an electron from (ionizing) a molecule it strikes.

■ **X rays** High-energy electromagnetic radiation.

■ **Cosmic rays** Energetic particles, primarily protons, from space.

High-energy radiation of all kinds is usually grouped together under the name **ionizing radiation**. The name comes from the fact that interaction of any of these kinds of radiation with a molecule knocks an electron from the molecule, causing the molecule to become an extremely reactive ion with an odd number of electrons. Ionizing radiation includes not only α particles, β particles, and γ rays, but also **X rays** and **cosmic rays**. X rays are like γ rays in that they are high-energy light waves rather than particles, whereas cosmic rays are energetic particles coming from interstellar space. Cosmic rays consist primarily of protons, along with some alpha and beta particles.

$$\text{Ionizing radiation} \quad \rightsquigarrow \quad \text{Molecule} \quad \longrightarrow \quad \text{Molecule}^+\cdot + e^-\cdot$$

An ion with an odd number of electrons

The effects of ionizing radiation can be lethal because the highly reactive ions formed can attack other important cellular molecules and cause serious disruptions of normal metabolism. The consequences are likely to be especially far-reaching when penetrating radiation strikes a cell nucleus and damages its genetic machinery. The resultant changes might lead to a genetic mutation, to cancer, or to cell death. (Bursts of radiation are purposely used to kill tumor cells in cancer therapy, for example.) If a sperm or egg cell is affected, a genetic mutation can be passed on to offspring.

■ **Background radiation** The sum of low-level radiation from naturally occurring sources.

No amount of radiation can be called absolutely safe. All ionizing radiation, even the small amount of **background radiation** we are all exposed to every day from naturally occurring radioisotopes and cosmic rays, has a small potential for causing damage. It makes sense, though, to try to limit exposure when possible and to use protective shields when working with known radiation sources like X-ray machines.

20.6 DETECTION AND MEASUREMENT OF RADIATION

Small amounts of naturally occurring radiation have always been present, but it's only been within the past hundred years that people have been aware of it. The problem, of course, is that radiation is invisible. We can't see, hear, smell, touch, or taste radiation, no matter how high the dose. We can, however, detect radiation by taking advantage of its ionizing properties.

The simplest device for detecting radiation is the photographic film badge, worn by all people who routinely work with radioactive materials. Any radiation striking the badge causes it to fog (remember Becquerel's mistake). Perhaps the best known way of detecting and measuring radiation is the *Geiger counter*. A Geiger counter consists of an argon-filled tube containing two electrodes (Figure 20.5). The inner walls of the tube are coated with an electrically conducting material and given a negative charge, while a wire in the center of

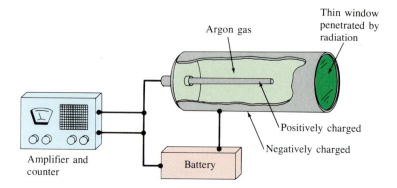

Figure 20.5
A Geiger counter. As radiation enters the tube through a thin window, it ionizes argon atoms, which conduct a tiny electric current from the negatively charged walls to the positively charged center electrode.

Argon gas

Thin window penetrated by radiation

Positively charged

Negatively charged

Amplifier and counter

Battery

the tube is given a positive charge. As radiation enters the tube through a thin window, it strikes and ionizes argon atoms, which briefly conduct a tiny electric current from the negatively charged walls to the positively charged center electrode. The passage of the current is detected, amplified, and used to produce a clicking sound in earphones worn by the operator. The more radiation that enters the tube, the more frequent the clicks.

Radiation intensity is expressed in several ways, depending on what is being measured. Some units measure the number of nuclear decay occurrences; others measure the biological consequences of radiation (Table 20.4).

■ **Curie** A unit for measuring the number of radioactive disintegrations per second.

- The **curie (Ci)** measures the number of radioactive disintegrations occurring each second in a sample. One curie is an extremely large unit, and **millicuries (mCi)** are more often used.

$$1 \text{ Ci} = 3.7 \times 10^{10} \text{ disintegrations per second}$$
$$1 \text{ mCi} = 1/1000 \text{ Ci} = 3.7 \times 10^{7} \text{ disintegrations per second}$$

■ **Roentgen** A unit for measuring the ionizing intensity of radiation.

- The **roentgen (R)** is a unit that measures the ionizing intensity of gamma or X radiation. One roentgen is an amount of radiation that produces 2.1×10^{9} units of charge in 1 cm^3 of dry air at atmospheric pressure.

■ **Rad** A unit for measuring the amount of radiation absorbed per gram of tissue.

- The **rad (D**, *radiation absorbed dose*) is a unit that measures the amount of radiation absorbed per gram of tissue. For most purposes, the roentgen and the rad are so close that they can be considered identical: 1 R = 1 D.

■ **Rem** A unit for measuring the amount of tissue damage caused by radiation.

- The **rem** (*roentgen equivalent for man*) is a unit that measures the amount of tissue damage caused by radiation. One rem is the amount of radiation that produces the same damage as 1 R of X rays.

Table 20.4 Units for Measuring Radiation

Unit	Quantity Measured	Description
Curie (Ci)	Amount of radioactivity	Amount of sample that undergoes 3.7×10^{10} disintegrations/sec
Roentgen (R)	Ionizing intensity of radiation	Amount of radiation that produces 2.1×10^{9} charges in 1 cm^3 air
Rad (D)	Amount of radiation absorbed per gram of tissue	For X- and γ-rays, 1 D = 1 R
Rem	Amount of tissue damage	Amount of radiation producing the same tissue damage as 1 R of X rays

Table 20.5   Biological Effects of Short-term Radiation on Humans

Dose (rems)	Biological Effects
0–25	No detectable effects
25–100	Temporary decrease in white blood cell count
100–200	Nausea, vomiting, longer term decrease in white blood cells
200–300	Vomiting, diarrhea, loss of appetite, listlessness
300–600	Vomiting, diarrhea, hemorrhaging, eventual death in some cases
above 600	Eventual death in nearly all cases

Rems are the preferred units for medical purposes because they measure equivalent doses of different kinds of radiation. For example, one rad of α rays and one rad of γ rays cause different amounts of tissue damage because they are different kinds of radiation, but one rem of α rays and one rem of γ rays cause the same amount of tissue damage. In effect, the rem takes *both* ionizing intensity and biological effect into account, while the rad deals only with intensity. The biological consequences of different radiation doses are given in Table 20.5.

Although the effects shown in Table 20.5 sound fearful, the average radiation dose received annually by most people is only about 0.12 rem (120 *milli*rem). About 70 percent of this radiation comes from natural sources (rocks and cosmic rays); the remaining 30 percent comes from medical procedures like X rays (Figure 20.6). The amount due to emissions from nuclear power plants and to fallout from atmospheric testing of nuclear weapons in the 1950s is barely detectable.

Practice Problem   **20.6**   Initial estimates of radiation loss from the 1986 Chernobyl nuclear power plant disaster in Russia predict a worldwide increase in background radiation of about 5 mrem. By what percentage will this amount increase the annual dose of most people?

Figure 20.6
Sources of average human exposure to radiation.

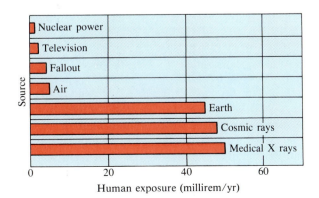

AN APPLICATION: MEDICAL USES OF RADIOACTIVITY

The origins of nuclear medicine reach back to 1901 when the French physician Henri Danlos first used radium in the treatment of a tuberculous skin lesion. Since that time, a great many techniques have been devised, and nuclear procedures have become a crucial part of modern medical care, both diagnostic and therapeutic. Currently used nuclear techniques can be grouped into four classes: (1) imaging procedures, (2) *in vivo* studies, (3) *in vitro* tests, and (4) radiation therapy.

Imaging Procedures Imaging procedures are those that give diagnostic information about the health of body organs by analyzing the distribution pattern of radioisotopes introduced into the body. In essence, a radioactive pharmaceutical agent known to concentrate in a specific tissue or organ is injected into the body, and its distribution pattern is monitored by external radiation detectors. Depending on the exact disease and organ, a diseased organ might concentrate more of the radiopharmaceutical than a normal organ and thus show up as a radioactive "hot spot" against a cold background. Alternatively, the diseased organ might concentrate less of the radiopharmaceutical than a normal organ and thus show up as a "cold spot" on a hot background.

An example of this method—used here for research rather than clinical diagnosis—is provided by the accompanying photograph, which shows a thin slice of a rat's brain. The brain tissue has been saturated with a compound in which some of the hydrogen atoms have been replaced with tritium (^{3}H), a radioactive isotope. The compound used is one that binds strongly to receptors for the neurotransmitters dopamine and serotonin (Chapter 16). When a sheet of film sensitive to the radiation from tritium is exposed to the tissue slice, a "map" of the distribution of nerve cells that respond to these particular neurotransmitters is produced. The technique of radioactively "labeling" structures and then allowing them to take their own picture in this way is called *autoradiography*.

An alternate imaging procedure makes use of a technique called *nuclear magnetic resonance*, or NMR. In spite of its name, NMR uses no radioisotopes and has no known hazards associated with it. Instead, NMR imaging makes use of an extremely powerful magnet to stimulate certain nuclei in mol-

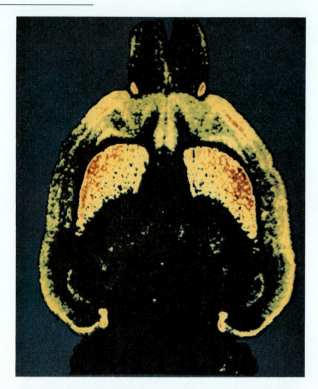

Autoradiograph of a slice of a rat's brain, showing the concentration of neurons that respond to dopamine and serotonin (see text). Red represents the highest concentration of receptors, dark blue the lowest.

ecules. Those stimulated nuclei (normally the hydrogen nuclei in H_2O molecules) then give off a signal that can be measured, interpreted, and correlated with their environment in the body. The figure on page 439, for example, shows a brain scan carried out by NMR and indicates the position of a cerebrovascular accident (CVA, commonly called a stroke).

In Vivo Procedures In vivo studies—those that take place *inside* the body—are carried out to assess the functioning of a particular organ or body system. A radiopharmaceutical agent is administered, and its path in the body—whether absorbed, excreted, diluted, or concentrated—is determined by analysis of blood or urine samples.

Among the many in vivo procedures utilizing radioactive agents is a simple method for determination of whole-blood volume by injecting a known quantity of red blood cells labeled with Cr-151. After a suitable interval to allow the labeled

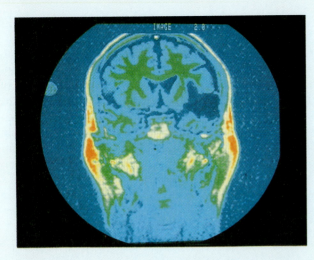

NMR image of a patient's brain. The dark blue area near the left temple (at right in this photo) is a region of dead tissue, caused by obstruction of blood flow through one of the cerebral arteries. Such an obstruction, often the result of a blood clot, produces a cerebrovascular accident, or stroke.

cells to equilibrate throughout the body, a blood sample is taken, and blood volume is calculated by measuring the amount of dilution that has taken place. A similar determination of plasma volume can be carried out by injecting a known volume of I-125 radioiodinated serum albumin (RISA) and then measuring dilution.

In Vitro Procedures In vitro studies—those that take place *outside* the body—are usually done on blood or urine samples and involve measuring small concentrations of drugs or hormones by *radioim-munoassay (RIA)* techniques. The principle of the RIA technique is quite simple: Antibodies are first obtained that bind to the substance to be measured, and the binding ability of a standard amount of radiolabeled sample is measured. By then comparing the binding ability of a biological fluid with that of the standard, the amount of sample in the fluid can be determined. In some peptide hormone analyses, such as that carried out on diabetics for insulin, as little as 10^{-14} M concentration can be measured, and the test can be carried out on a single drop of blood.

Therapeutic Procedures Therapeutic procedures (those in which radiation is purposely used as a weapon to kill diseased tissue) can involve either external or internal sources of radiation. External radiation therapy used in the treatment of cancer involves gamma rays emanating from a Co-60 source. The highly radioactive source is shielded by a thick lead container and has a small opening directed toward the site of the tumor.

Internal radiation therapy is a much more selective technique, but is largely limited to the treatment of thyroid disease because of the lack of tissue-specific drugs. The idea of internal radiation therapy is to administer a radioactive substance such as I-131 iodide ion, a powerful beta emitter that is known to localize in a specific target tissue, in this case the thyroid. Since beta particles can penetrate only several millimeters of tissue, the localized I-131 produces a high radiation dose that destroys only the surrounding diseased tissue.

Radioimmunoassay (RIA) involves measuring the binding of a substrate to an antibody against that substrate. Antibodies of known binding ability are prepared and added to a sample of biological fluid containing the substrate to be measured (●). A known amount of labeled substrate is also added (●). The unknown amount of substrate in the fluid can be determined by counting the amount of labeled substrate bound to the antibodies.

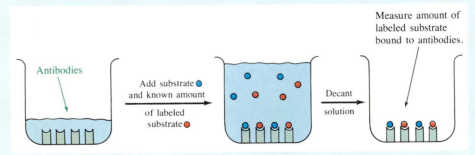

20.7 ARTIFICIAL TRANSMUTATION

■ **Artificial transmutation** The change of one element into another by bombardment with a high-energy particle.

Most of the nearly 2000 known radioisotopes, including all of those used in medicine, do not occur naturally. Rather, they are made by **artificial transmutation** of stable isotopes. Artificial transmutation involves the bombardment of an atom with a high-energy particle such as a proton, neutron, or alpha particle. In the ensuing collision between particle and atom, a nuclear change occurs, and a different element is produced. For example, Ernest Rutherford found when he allowed a beam of alpha particles to pass through a tube containing nitrogen-14 that oxygen-17 was produced, along with protons ($_1^1H$). In essence, the nitrogen and helium nuclei fuse together by collision to yield the larger nucleus of a different element.

$$_7^{14}N + {}_2^4He \longrightarrow {}_8^{17}O + {}_1^1H$$

Other particles besides alpha particles can also be used to induce nuclear transmutations. For example, bombardment of nitrogen-14 with a beam of neutrons results in the formation of carbon-14 plus a proton. This transmutation occurs naturally in the upper atmosphere when neutrons produced by bombardment of other atoms with cosmic rays collide with atmospheric nitrogen. In the collision, a neutron dislodges a proton from the nitrogen nucleus as the neutron and nucleus fuse together.

$$_7^{14}N + {}_0^1n \longrightarrow {}_6^{14}C + {}_1^1H$$

Still other artificial transmutations can lead to the synthesis of entirely new elements never before seen on earth. In fact, all of the *transuranium elements*—those elements with atomic number greater than 92—are artificially produced. For example, the element plutonium (Pu) can be made by bombardment of uranium-238 with alpha particles:

$$_{92}^{238}U + {}_2^4He \longrightarrow {}_{94}^{241}Pu + {}_0^1n$$

The plutonium-241 that results from uranium bombardment is itself radioactive with a half-life of 13.2 years, decaying by beta emission to yield americium-241. Americium-241 is also radioactive, decaying by alpha emission with a half-life of 458 years. (If the name sounds vaguely familiar, it's because americium is used commercially in making smoke detectors.)

$$_{94}^{241}Pu \longrightarrow {}_{95}^{241}Am + {}_{-1}^0e$$

Note that all of the equations just mentioned for artificial transmutations are balanced. The sum of the mass numbers and the sum of the atomic numbers are the same on both sides of each equation.

Practice Problems

20.7 What isotope of what element results from alpha decay of the americium-241 in smoke detectors?

20.8 The element berkelium, first prepared at the University of California at Berkeley in 1949, is made by alpha bombardment of americium-241 ($^{241}_{95}$Am). Two neutrons are also produced during the reaction. What isotope of berkelium results from this transmutation? Write a balanced nuclear equation.

20.9 Write a balanced nuclear equation for the reaction of argon-40 with a proton:

$$^{40}_{18}\text{Ar} + ^{1}_{1}\text{H} \longrightarrow \ ? + ^{1}_{0}n$$

20.8 NUCLEAR FISSION AND ATOMIC ENERGY

We saw in the previous section that particle bombardment of various elements causes artificial transmutation and results in the formation of new, usually heavier elements. With a very few isotopes, however, a somewhat different result occurs. In popular terms, it's called "splitting the atom"; in more scientific terms, it's called **nuclear fission**; and it arises when an atom is split apart by neutron bombardment to give small fragments.

■ **Nuclear fission** The splitting apart of an atom by neutron bombardment to give smaller atoms.

Uranium-235 is the only naturally occurring isotope that undergoes nuclear fission. When U-235 is bombarded by a stream of relatively slow neutrons, its nucleus is split apart like a glass window that's been struck by a baseball. The split can take place in any number of ways to give any of over 200 different isotope fragments. For example, barium-139 and krypton-94 are sometimes produced (Figure 20.7).

$$^{235}_{92}\text{U} + ^{1}_{0}n \longrightarrow \ ^{139}_{56}\text{Ba} + ^{94}_{36}\text{Kr} + 3\,^{1}_{0}n + \gamma + \text{heat energy}$$

As indicated by the balanced nuclear equation shown above, *one* neutron is used to initiate fission of a U-235 nucleus, but *three* neutrons are released. Thus a nuclear **chain reaction** can be started: One neutron initiates one fission that releases three neutrons; the three neutrons initiate three more fissions that release nine neutrons; the nine neutrons initiate nine fissions that release 27 neutrons; and so on at an ever faster pace as long as U-235 remains (Figure 20.8).

■ **Chain reaction** A self-sustaining reaction that continues unchecked once it has been started.

Figure 20.7
Neutron-induced fission splits a $^{235}_{92}$U nucleus into $^{139}_{56}$Ba, $^{94}_{36}$Kr, and three neutrons.

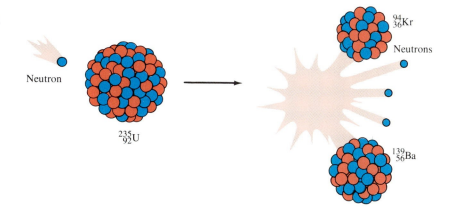

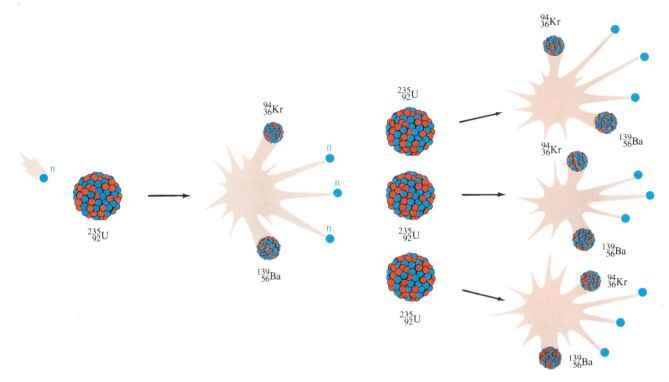

Figure 20.8
A nuclear chain reaction caused by fission of uranium-235. Each fission event releases three neutrons that can initiate further fissions at an ever increasing rate until nuclear detonation occurs.

In order for a chain reaction to occur, the fissionable sample must have a certain minimum size. Otherwise, there is a high probability that many neutrons will simply escape from the sample before they chance to encounter another uranium nucleus, and the chain reaction will stop. If, however, enough sample is present (an amount called the *critical mass*), the chain reaction is self-sustaining.

A great deal of heat is produced during nuclear fission, heat that can be used to turn water to steam, thereby turning huge generators to produce electric power. At present, about 12 percent of the electric power generated in the United States comes from nuclear plants. Although the uranium used in these nuclear power plants contains only about 3 percent of $^{235}_{92}U$ (the remainder is nonfissionable $^{238}_{92}U$), there is more than enough fissionable material present to sustain a chain reaction.

The rate of the fission reaction in power plants—and therefore the rate of heat production—is moderated by adjustable control rods made of cadmium and boron. When lowered into the reactor core, these rods absorb some of the neutrons produced, thereby slowing down the chain reaction. Should accidents make it impossible to control the reaction, however, heat continues to build in the reactor until the structural materials of the reactor melt.

The major objections causing so much public debate about nuclear power plants are twofold: safety and waste disposal. Although a nuclear explosion is impossible (the concentration of $^{235}_{92}U$ in fuel rods is not high enough to sustain such a rapid reaction), there is a serious potential radiation hazard posed by nuclear power plants. Should the containment vessel surrounding a reactor ever be breached, as occurred during the April, 1986, accident at the unit 4 reactor of the Chernobyl plant in Russia, the accidental release of radioactive fallout could be disastrous.

Perhaps even more important than reactor safety in the long term is the problem posed by disposal of radioactive wastes from nuclear plants. Many of these wastes are highly radioactive and have such long half-lives that hundreds or even thousands of years must elapse before they will be safe for humans to go near. It has been estimated, for example, that at least 600 years will be necessary for strontium-90 ($t_{1/2} = 28.8$ years) to decay to safe levels. Plutonium-239, with $t_{1/2} = 24,000$ years, will require much longer. How to dispose safely of such extraordinarily hazardous materials is an unsolved, and perhaps unsolvable, problem.

Practice Problem   **20.10**   What other isotope besides tellurium-137 is produced by nuclear fission of uranium-235 in the following equation?

$$^{235}_{92}U + {}^{1}_{0}n \longrightarrow {}^{137}_{52}Te + 2\,{}^{1}_{0}n + \,?$$

SUMMARY

Radioactivity is the spontaneous emission of radiation from the nucleus of an unstable atom. The three major kinds of radiation are called **alpha** (α), **beta** (β), and **gamma** (γ). Alpha radiation consists of helium nuclei, small particles containing two protons and two neutrons (${}^{4}_{2}He$); beta radiation consists of electrons (${}^{0}_{-1}e$); and gamma radiation consists of very high-energy light waves that have no mass. Every element in the periodic table has at least one radioactive isotope (**radioisotope**), and nearly 2000 different radioisotopes are known.

Particle emission from a radioactive nucleus leads to **nuclear decay** and to the change of one element into another (**transmutation**). Thus, loss of an alpha particle from an atom leads to a new atom whose atomic number is two less than the starting atom. Loss of a beta particle arises by decomposition of a neutron and leads to an atom whose atomic number is one greater than the starting atom:

$$\text{Alpha emission:} \quad {}^{238}_{92}U \longrightarrow {}^{234}_{90}Th + {}^{4}_{2}He$$

$$\text{Beta emission:} \quad {}^{131}_{53}I \longrightarrow {}^{131}_{54}Xe + {}^{0}_{-1}e$$

The rate of nuclear decay is expressed in units of **half-life ($t_{1/2}$)**, where one half-life is the amount of time necessary for one-half of the radioactive sample to decay.

High-energy radiation of all types—α particles, β particles, γ rays, X rays, and cosmic rays—is called **ionizing radiation**. When any of these kinds of radiation strikes a molecule, it dislodges an electron and gives a reactive ion. This ionizing effect can be lethal to cells.

Radiation intensity is expressed in different ways according to what property of radiation is being measured. The **curie (Ci)** measures the number of

INTERLUDE: ARCHAEOLOGICAL RADIOCARBON DATING

Biblical scrolls are found in a cave near the Dead Sea; are they authentic? A mummy is discovered in an Egyptian tomb; how old is it? The burned bones of a man are dug up near Lubbock, Texas; how long ago were humans on the North American continent? Using a technique called *radiocarbon dating*, archaeologists can answer these and many other questions. (The Dead Sea Scrolls are 1900 years old and authentic, the mummy is 3100 years old, and the human remains found in Texas are 9900 years old.)

Radiocarbon dating of archaeological artifacts depends on the fact that the earth's upper atmosphere is under constant bombardment by cosmic rays. When these highly energetic particles strike molecules in the upper atmosphere, they dislodge neutrons that in turn collide with other molecules. Collision of a neutron with atmospheric nitrogen-14 results in transmutation and the production of carbon-14. Unlike carbon-12, the most abundant carbon isotope, carbon-14 is radioactive, with a half-life of 5730 years.

$$^{14}_{7}N + ^{1}_{0}n \longrightarrow ^{14}_{6}C + ^{1}_{1}H$$

Carbon-14 atoms produced in the upper atmosphere combine with oxygen to yield $^{14}CO_2$, which slowly mixes with ordinary $^{12}CO_2$ and is then taken up by plants during photosynthesis. When these plants are then eaten by animals, carbon-14 enters the food chain and is ultimately distributed evenly throughout all living organisms.

As long as a plant or animal is living, a dynamic equilibrium is established. An organism excretes or exhales the same amount of ^{14}C that it takes in, and the ratio of ^{14}C to ^{12}C that the organism contains is the same as that in the atmosphere—about 1 part in 10^{12}. When the plant or animal dies, however, it no longer takes in more ^{14}C. Thus, the ^{14}C to ^{12}C ratio in the organism slowly diminishes as ^{14}C undergoes radioactive decay. At 5730 years (one ^{14}C half-life) after the death of the organism, the ^{14}C to ^{12}C ratio has decreased by a factor of two; at 11,460 years after death, the ^{14}C to ^{12}C ratio has decreased by a factor of four; and so on.

By simply measuring the amount of ^{14}C that remains in the traces of any once-living organism, archaeologists can determine how long ago the organism died. Human hair from well-preserved remains, charcoal or wood fragments from once-living trees, and cotton or linen from once-living plants, are all useful sources for radiocarbon dating. The accuracy of the technique lessens as samples get older and the amount of ^{14}C they contain diminishes, but artifacts with an age of 1000 to 20,000 years can be dated with reasonable accuracy.

Objects between 500 and 50,000 years old that contain some organic material, such as this Egyptian mummy, are ideal subjects for radiocarbon dating.

radioactive disintegrations per second in a sample; the **roentgen (R)** measures the ionizing ability of radiation; the **rad (D)** measures the amount of radiation absorbed per gram of tissue; and the **rem** measures the amount of tissue damage caused by radiation. Radiation effects become noticeable with a human exposure of 25 rem and become lethal at an exposure above 600 rem.

Most known radioisotopes do not occur naturally but are made by bombardment of an atom with a high-energy particle. In the ensuing collision between particle and atom, a nuclear change occurs, and a new element is produced by **artificial transmutation**. With a very few isotopes, including $^{235}_{92}U$, the nucleus is actually split apart by neutron bombardment to give smaller fragments. A large amount of energy is released during this **nuclear fission**, leading to the use of the reaction for generating electric power.

REVIEW PROBLEMS

Radioactivity

20.11   What does it mean to say that a substance is radioactive?

20.12   What word is used to describe the change of one element into another?

20.13   Describe how α, β, and γ radiation differ from each other.

20.14   What symbol is used for an α particle in a nuclear equation?

20.15   What symbol is used for a β particle in a nuclear equation?

20.16   Describe a simple experiment that would let you determine the kind of electrical charge on α, β, and γ radiation.

20.17   Which kind of radiation, α, β, or γ, has the highest penetrating power, and which has the lowest?

20.18   Approximately how much protective shielding do you need to protect yourself from α radiation? from β radiation? from γ radiation?

20.19   What happens when ionizing radiation strikes a molecule?

20.20   How does ionizing radiation lead to cell damage?

20.21   What are the main sources of background radiation?

20.22   How can a nucleus emit an electron during β decay when there are no electrons present in the nucleus to begin with?

20.23   What is the difference between an α particle and a helium atom?

20.24   What does it mean when we say that strontium-90, a waste product of nuclear power plants, has a half-life of 28.8 years?

20.25   What percent of the original radioactivity remains in a sample after two half-lives have passed?

20.26   Look at Figure 20.4 and estimate what percent of original radioactivity remains in a sample after 0.5 half-lives have passed.

Measuring Radioactivity

20.27   What two devices are commonly used for measuring radioactivity?

20.28   Describe in your own words how a Geiger counter works.

20.29   Why are rems the preferred unit for measuring the health effects of radiation?

20.30   Approximately what amount (in rems) of short-term exposure to radiation produces noticeable effects in humans?

20.31   Match the units in the left-hand column with the property being measured in the right-hand column:
(1) curie   (a) ionizing intensity of radiation
(2) rem   (b) amount of tissue damage
(3) rad   (c) number of disintegrations per second
(4) roentgen   (d) amount of radiation per gram of tissue

Nuclear Decay and Transmutation

20.32   What does it mean to say that a nuclear equation is balanced?

20.33 What happens to the mass number and atomic number of an atom that emits an α particle? a β particle? a γ ray?

20.34 Harmful chemical spills can often be cleaned up by treatment with another chemical. For example, a spill of H_2SO_4 might be neutralized by addition of $NaHCO_3$. Why can't the harmful radioactive wastes from nuclear power plants be cleaned up just as easily?

20.35 What is the difference between natural transmutation of a radioisotope and artificial transmutation?

20.36 What is a transuranium element, and how is it made?

20.37 How does nuclear fission differ from normal radioactive decay?

20.38 What characteristic of uranium-235 fission causes it to become a chain reaction?

20.39 Why must there be a critical mass of U-235 present before a self-sustaining chain reaction can occur?

20.40 What is the function of the cadmium/boron control rods in nuclear reactors, and how do they work?

20.41 Selenium-75, a beta emitter with a half-life of 120 days, is used medically for pancreas scans. What is the product of selenium-75 decay?

20.42 Approximately how much selenium-75 would remain from a 0.050-g sample that had been stored for 1 year? (See Problem 20.41.)

20.43 Approximately how long would it take a sample of selenium-75 to lose 99 percent of its radioactivity? (See Problem 20.41.)

20.44 What products result from radioactive decay of these beta emitters?

(a) $^{35}_{16}S$ (b) $^{24}_{10}Ne$ (c) $^{90}_{38}Sr$

20.45 The half-life of mercury-197 is 65 hours. If a patient undergoing a kidney scan were given 5.0 ng of mercury-197, how much would remain after 6 days? after 27 days?

20.46 Identify the starting radioisotopes that give these products:

(a) ? $\longrightarrow$ $^{140}_{56}Ba + ^{0}_{-1}e$

(b) ? $\longrightarrow$ $^{242}_{94}Pu + ^{4}_{2}He$

20.47 What products result from radioactive decay of these alpha emitters?

(a) $^{190}_{78}Pt$ (b) $^{208}_{87}Fr$ (c) $^{245}_{96}Cm$

20.48 What products are formed in these nuclear reactions?

(a) $^{109}_{47}Ag + ^{4}_{2}He \longrightarrow$?

(b) $^{10}_{5}B + ^{4}_{2}He \longrightarrow$? $+ ^{1}_{0}n$

20.49 For centuries alchemists dreamed of turning base metals into gold. This dream finally became reality when it was shown that mercury-198 can be converted into gold-198 upon bombardment by neutrons. What small particle is produced in addition to gold-198? Write a balanced nuclear equation for the reaction.

20.50 Balance these equations for the nuclear fission of $^{235}_{92}U$:

(a) $^{235}_{92}U + ^{1}_{0}n \longrightarrow ^{160}_{62}Sm + ^{72}_{30}Zn + ? ^{1}_{0}n$

(b) $^{235}_{92}U + ^{1}_{0}n \longrightarrow ^{87}_{35}Br + ? + 3 ^{1}_{0}n$

20.51 Element 109 ($^{266}_{109}Une$), the heaviest known element, was prepared in 1982 by bombardment of bismuth-209 atoms with iron-58. What other product must also have been formed? Write a balanced nuclear equation for the transformation.

Exponential Notation

Numbers that are either very large or very small are usually represented in *exponential notation* as a number between 1 and 10 multiplied by a power of 10. In this kind of expression, the small raised number to the right of the 10 is the *exponent*.

Number	Exponential Form	Exponent
1,000,000	1×10^6	6
100,000	1×10^5	5
10,000	1×10^4	4
1,000	1×10^3	3
100	1×10^2	2
10	1×10^1	1
1		
0.1	1×10^{-1}	-1
0.01	1×10^{-2}	-2
0.001	1×10^{-3}	-3
0.000 1	1×10^{-4}	-4
0.000 01	1×10^{-5}	-5
0.000 001	1×10^{-6}	-6
0.000 000 1	1×10^{-7}	-7

Numbers greater than 1 have *positive* exponents, which tell how many times a number must be *multiplied* by 10 to obtain the correct value. For example, the expression 5.2×10^3 means that 5.2 must be multiplied by 10 three times:

$$5.2 \times 10^3 = 5.2 \times 10 \times 10 \times 10 = 5.2 \times 1000 = 5200$$

Note that doing this means moving the decimal point three places to the right:

5 2 0 0.

1 2 3

The value of a positive exponent indicates *how many places to the right the decimal point must be moved* to give the correct number in ordinary decimal notation.

Numbers less than 1 have *negative exponents*, which tell how many times a number must be *divided* by 10 (or multiplied by one-tenth) to obtain the correct value. Thus, the expression 3.7×10^{-2} means that 3.7 must be divided by 10 two times:

$$3.7 \times 10^{-2} = \frac{3.7}{10 \times 10} = \frac{3.7}{100} = 0.037$$

Note that doing this means moving the decimal point two places to the left:

$$0.\,0\,3\,7$$
$$2\ 1$$

The value of a negative exponent indicates *how many places to the left the decimal point must be moved* to give the correct number in ordinary decimal notation.

CONVERTING DECIMAL NUMBERS TO EXPONENTIAL NOTATION

To convert a number greater than 1 from decimal notation to exponential notation, first move the decimal point to the *left* until there is only a single digit to the left of the decimal point. The *positive* exponent needed for exponential notation is the same as *the number of places the decimal point was moved.*

$$6\,3\,5\,7\,8\,1. = 6.35781 \times 10^5$$
$$5\ 4\ 3\ 2\ 1$$

To convert a number smaller than 1 from decimal notation to exponential notation, first move the decimal point to the *right* until there is *a single nonzero digit* to the left of the decimal point. The *negative* exponent needed for exponential notation is the same as *the number of places the decimal point was moved.*

$$0.\,0\,0\,0\,4\,2\,6 = 4.26 \times 10^{-4}$$
$$1\ 2\ 3\ 4$$

MULTIPLYING EXPONENTIAL NUMBERS

To multiply two numbers in exponential form, the exponents are *added*. For example:

$$(3.5 \times 10^3) \times (4.2 \times 10^4) = 3.5 \times 4.2 \times 10^{(3+4)} = 14.7 \times 10^7$$
$$= 1.47 \times 10^8 = 1.5 \times 10^8 \quad \text{(rounded off)}$$
$$(5.2 \times 10^4) \times (4.6 \times 10^{-3}) = 5.2 \times 4.6 \times 10^{[4+(-3)]} = 23.92 \times 10^1$$
$$= 2.392 \times 10^2 = 2.4 \times 10^2 \quad \text{(rounded off)}$$

DIVIDING EXPONENTIAL NUMBERS

To divide two numbers in exponential form, the exponents are *subtracted*. For example:

$$\frac{4.1 \times 10^4}{6.2 \times 10^6} = \frac{4.1}{6.2} \times 10^{(4-6)} = 0.6613 \times 10^{-2} = 6.6 \times 10^{-3}$$

$$\frac{6.6 \times 10^3}{8.4 \times 10^{-2}} = \frac{6.6}{8.4} \times 10^{[3-(-2)]} = 0.7857 \times 10^5 = 7.9 \times 10^4$$

Accuracy and Precision in Measurement

Although most of us use the words *accuracy* and *precision* interchangeably, there's actually an important distinction between the two terms. **Accuracy** refers to how close a given measurement is to the true value, whereas **precision** refers to how carefully a measurement is done with the equipment at hand.

To see the difference between accuracy and precision, look at the different dart boards in the figure below. The thrower on the left has clustered the darts close together but off target, a result that's *precise* but inaccurate. (Perhaps there's a slight breeze blowing from left to right.) The second thrower's darts are widely scattered, but their distribution is centered on the bullseye. The aim is *accurate* because the average of the five darts is centered, but the precision is low because the individual throws are inconsistent. The third thrower has missed wildly, a result that's neither accurate nor precise, and the fourth thrower has clustered all three darts dead on target, a result that's both accurate and precise.

The difference between accuracy and precision. The dart thrower in (a) is precise but not accurate; the thrower in (b) is accurate but not precise; the thrower in (c) is neither accurate nor precise; and the thrower in (d) is both accurate and precise.

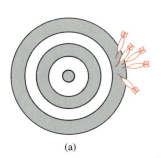

(a)

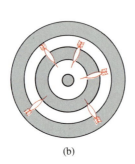

(b)

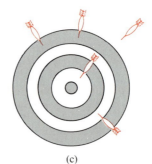

(c)

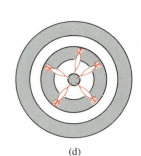

(d)

The same kinds of results illustrated by the different dart boards often occur in experimental measurements. It's not unusual to make a series of measurements that are consistent (precise) but incorrect (not accurate). You might, for example, be using excellent equipment that has been improperly calibrated. (Obviously, if your balance reads "1.0 g" when empty, all your measured masses will be 1.0 g too high, no matter how sensitive the balance is.) Alternatively, there might be some external factor that hasn't been taken into account. (Masses measured at a high altitude will be lower than those measured at sea level, for example.) Inaccuracies of this kind are said to be due to *systematic error*.

On the other hand, you might make a series of measurements that vary widely but can be averaged to give an accurate value. In such cases, the results are usually attributed to *random error*—slight differences in measurement due to carelessness or poor equipment.

For an illustration of the importance of precision in chemistry, imagine that you have an unlabeled bottle of an industrial solvent that you have to identify. You know it's either acetone or butanone, and you decide to determine which by measuring the density of the solvent and comparing your measured value with known values you look up in a handbook—0.7899 g/mL for acetone at 20°C and 0.8054 g/mL for butanone at 20°C. Unfortunately, the only equipment you have at your disposal is a small graduated cylinder and a single-pan balance.

Working as carefully as possible, you measure out 5.0 mL of the solvent and determine its weight as 4.02 g. You then calculate the density and arrive at a value of 0.804 g/mL:

$$\text{Density} = \frac{4.02 \text{ g}}{5.0 \text{ mL}} = 0.804 \text{ g/mL}$$

What can you conclude from your measurement? You might well be tempted to conclude that the solvent is butanone, since your measurement of 0.804 g/mL is much closer to the density of butanone (0.8054 g/mL) than to that of acetone (0.7899 g/mL). In fact, though, you can't properly conclude *anything*, because your measurement isn't precise enough. No matter how careful you are in using the scale and graduated cylinder, the precision of your result is limited by the precision of the equipment. Since your volume measurement of 5.0 mL is precise to only two significant figures, your density is also precise to only two significant figures and must be rounded off to 0.80. Furthermore, since the last digit is just an estimate, your actual value could be anywhere from 0.79 to 0.81, which encompasses the densities of both acetone and butanone. Only with precise equipment can you obtain a precise measurement.

APPENDIX C

Conversion Factors

Length   SI Unit; Meter (m)

    1 meter = 0.001 kilometers (km)
           = 100 centimeters (cm)
           = 1.0936 yards (yd)
    1 centimeter = 10 millimeters (mm)
                = 0.3937 inch (in.)
    1 nanometer = 1×10^{-9} meters
    1 Angstrom (Å) = 1×10^{-10} meter
    1 inch = 2.54 centimeters
    1 mile = 1.6094 kilometers

Volume   SI Unit: Cubic meter (m^3)

    1 cubic meter = 1000 liters (L)
    1 liter = 1000 cubic centimeters (cm^3)
         = 1000 milliliters (mL)
         = 1.056710 quarts (qt)
    1 cubic inch = 16.4 cubic centimeters

Temperature   SI Unit: Kelvin (K)

    0 K = $-273.15°C$
        = $-459.67°F$
    $°F = (9/5)°C + 32°$
    $°C = (5/9)(°F - 32°)$
    $K = °C + 273.15°$

Mass   SI Unit: Kilogram (kg)

    1 kilogram = 1000 grams (g)
             = 2.205 pounds (lb)
    1 gram = 1000 milligrams (mg)
         = 0.0353 ounces (oz)
    1 pound = 453.6 grams
    1 atomic mass unit = 1.66054×10^{-24} grams

Pressure   SI Unit: Pascal (Pa)

    1 pascal = 9.869×10^{-7} atmospheres
    1 atmosphere = 101,325 pascals
               = 760 mm Hg (Torr)
               = 14.70 lb/in^2.

Energy   SI Unit: Joule (J)

    1 joule = 0.23901 calorie (cal)
    1 calorie = 4.184 joule

Glossary

Acetal A compound that has two ether-like —OR groups bonded to the same carbon atom.

Achiral The opposite of chiral; not having (right or left) handedness.

Acid A substance that is able to donate a hydrogen ion, H^+.

Acid (base) equivalent The amount in grams of an acid (or base) that can donate one mole of H^+ (or OH^-) ions.

Acidosis The medical condition that results when blood pH drops below 7.35.

Actinides The series of 14 elements following actinium in the periodic table.

Activation energy (E_{act}) The amount of energy that reactants must have in order to surmount the energy barrier to reaction.

Active site A small three-dimensional portion of the enzyme with the specific shape and structure necessary to bind a substrate.

Acyclic alkane An alkane that contains no rings.

Addition reaction A chemical reaction in which two reactants add together to yield a single product: $A + B \rightarrow C$.

Aerobic Presence of oxygen.

Alcohol A compound that contains an —OH functional group bonded to an alkane-like carbon atom, R—OH.

Aldehyde A compound that has a carbonyl group bonded to one organic substituent and one hydrogen, RCHO.

Aldol reaction The reaction of a ketone or aldehyde to form a hydroxy ketone product on treatment with base catalyst.

Aldose A simple sugar that contains an aldehyde carbonyl group.

Alkali metal An element in Group 1A of the periodic table (Li, Na, K, Rb, Cs, Fr).

Alkaline earth metal An element in Group 2A of the periodic table (Be, Mg, Ca, Sr, Ba, Ra).

Alkalosis The medical condition that results when blood pH rises above 7.45.

Alkane A molecule that contains only carbon and hydrogen and that has only single bonds.

Alkene A compound that contains only carbon and hydrogen and that has a carbon-carbon double bond.

Alkyl group The part of an alkane that remains when one hydrogen atom is removed.

Alkyne A compound that contains only carbon and hydrogen and that has a carbon-carbon triple bond.

Alpha (α) amino acid An amino acid in which the amino group is bonded to the carbon atom next to the —COOH group.

Alpha (α) radiation Emission of helium nuclei, ^4_2He.

Amide A carbonyl compound that has one organic

group and one —NH$_2$,—NHR, or —NR$_2$ group bonded to the carbonyl carbon.

Amine   A compound that has one or more organic groups bonded to nitrogen, RNH$_2$, R$_2$NH, or R$_3$N.

Amino acid   A molecule that contains both an amino group and a carboxylic-acid functional group; used as a building block to construct proteins.

Amino-acid pool   The sum of free amino acids available in the body.

Ammonium salt   An ionic substance formed by reaction of ammonia with an acid.

Anabolism   Metabolic reactions that build larger biological molecules from smaller pieces.

Anaerobic   Absence of oxygen.

Angstrom (Å)   A convenient unit of length on the atomic scale; equal to 10^{-10} m.

Anion   A negatively charged ion.

Anode   The positive electrode of a battery.

Anticodon   A sequence of three ribonucleotides on tRNA that recognizes the complementary sequence on mRNA.

Apoenzyme   The protein portion of an enzyme.

Aqueous   Referring to a solution with water as the solvent.

Aromatic compound   A compound that contains a six-membered ring of carbon atoms with three double bonds.

Arrhenius acid   A substance that increases the concentration of H$_3$O$^+$ when dissolved in water.

Artificial transmutation   The change of one element into another by bombardment with a high-energy particle.

Atom   The smallest and simplest piece that an element can be broken into while still maintaining the chemical properties of the element.

Atomic mass unit (amu)   A convenient unit of mass on the atomic scale; approximately equal to the mass of a proton or neutron.

Atomic number   The primary characteristic that distinguishes atoms of different elements; equal to the number of protons in an atom's nucleus.

Atomic weight   The average mass (in amu) of a large sample of an element's atoms.

Avogadro's law   Equal volumes of gases at the same temperature and pressure contain equal numbers of molecules.

Avogadro's number   The number of particles (atoms, ions, or molecules) in one mole (6.02×10^{23}).

Background radiation   The sum of low-level radiation from naturally occurring sources.

Balanced   Describing an equation in which the numbers and kinds of atoms on both sides of the reaction arrow are the same.

Base   A substance that can accept a hydrogen ion (H$^+$) from an acid.

Benedict's reagent   A reagent that yields a reddish-yellow color when it reacts with a reducing sugar.

Beta (β) radiation   Emission of electrons.

Boiling point (bp)   The temperature at which the vapor pressure of a liquid is equal to atmospheric pressure.

Bond angle   The angle formed between any two adjacent covalent bonds.

Boyle's law   The pressure of a gas at constant temperature is inversely proportional to its volume.

Branched-chain alkane   An alkane that has a branching connection of carbon atoms along its chain.

Buffer   A combination of substances, usually a weak acid and its anion, that act together to prevent a large change in the pH of a solution.

Calorie (cal)   The amount of energy (heat) necessary to raise the temperature of one gram of water by one Celsius degree.

Carbocation   A polyatomic ion with a positively charged carbon atom.

Carbohydrate   A member of a large class of naturally occurring polyhydroxy ketones and aldehydes.

Carbonyl group   A functional group that has a carbon atom joined to an oxygen atom by a double bond, C=O.

Carbonyl-group substitution reaction   A reaction in which a new group X replaces (substitutes for) a group Y attached to a carbonyl-group carbon: RCOY + X$^-$ → RCOX + Y$^-$.

Carboxylate anion   The anion that results from dissociation of a carboxylic acid, RCOO$^-$.

Carboxylic acid   A compound that has a carbonyl group bonded to one organic substituent and one —OH group, RCOOH.

Carcinogenic   Cancer-causing.

Catabolism   Metabolic reactions that break down molecules into smaller pieces.

Catalyst   A substance that speeds up a chemical reaction without itself undergoing change.

Cathode   The negative electrode of a battery.

Cation   A positively charged ion.

Celsius degree (C°)   The metric unit of temperature; equal to 1.8 Fahrenheit degree.

Centimeter (cm)   A unit of length equal to 1/100 meter, or about 0.3937 inch.

Chain reaction   A self-sustaining reaction that continues unchecked once it has been started.

Charles' law   The volume of a gas at constant pressure is directly proportional to its kelvin temperature.

Chemical bonds   The forces that hold atoms together in chemical compounds.

Chemical compound   A chemical substance formed by joining together atoms of different elements.

Chemical equation   The written expression that describes a chemical reaction.

Chemical equilibrium   The point in a reversible reaction at which forward and reverse reactions take place at the same rate, so that the concentrations of products and reactants no longer change.

Chemical property   A property that involves a chemical reaction of a substance.

Chemical reaction   A chemical change brought about by making and breaking of bonds between atoms.

Chemiosmotic hypothesis   A theory to explain how the establishment of a pH difference across a cell membrane leads to the synthesis of ATP in the respiratory chain.

Chemistry   The study of the nature, properties, and transformations of matter.

Chiral   Having (right or left) handedness.

Chromosome   A thread-like strand of DNA in the cell nucleus that stores genetic information.

Cis-trans isomers   Alkenes that have the same formula and connections between atoms but have different structures because of the way that groups are attached to different sides of the double bond.

Citric-acid cycle   A series of biochemical reactions used by the cell to obtain energy from food molecules.

Claisen condensation reaction   A reaction that joins two ester molecules together to yield a keto ester product.

Codon   A sequence of three ribonucleotides in the RNA chain that codes for a specific amino acid.

Coefficient   A number placed before a substance in a chemical equation to tell how many units of that substance are required to balance the equation.

Coenzyme   A small organic molecule that acts as an enzyme cofactor.

Cofactor   A small, nonprotein part of an enzyme that is essential to the enzyme's catalytic activity.

Colloid   A homogeneous mixture that contains particles larger than a typical molecule.

Combustion   The burning of an organic substance with oxygen.

Competitive enzyme inhibition   A mechanism of enzyme regulation in which an inhibitor competes with a similarly shaped substrate for binding to the enzyme active site.

Complex carbohydrate   A carbohydrate that breaks down into simple sugars when hydrolyzed with aqueous acid.

Complex lipid   A lipid that contains an ester group and that can be hydrolyzed to yield an alcohol and a carboxylic acid.

Compound   A specific chemical substance.

Concentration   A measure of the amount of dissolved substance per unit volume of solvent.

Condensed structure   A shorthand way of drawing an organic structure in which carbon-carbon and carbon-hydrogen bonds are "understood" rather than shown.

Conformation   The exact three-dimensional structure of a molecule.

Conjugated protein   A protein that yields one or more other compounds in addition to amino acids when hydrolyzed.

Conversion factor   A fraction that states the relationship between different units of measure.

Cosmic rays   Energetic particles, primarily protons, from space.

Covalence   The number of bonds formed by an atom in a molecule.

Covalent bond   A bond that results when two atoms share one or more pairs of electrons.

Crenation   The shriveling of red blood cells due to a loss of water.

Cristae   The inner folds of a mitochondrion.

Cubic centimeter (cc or cm³)   An alternative name for a milliliter.

Cubic meter (m³)   The SI unit of volume; equal to 264.2 gallons.

Curie A unit for measuring the number of radioactive disintegrations per second.

Cycloalkane An alkane that contains a ring of carbon atoms.

Cytoplasm The jelly-like fluid filling a cell.

Dalton An alternate name for atomic mass unit.

Dalton's law of partial pressure The total pressure exerted by a mixture of gases is equal to the sum of the partial pressures exerted by each individual gas.

Dehydration Loss of water from an alcohol to yield an alkene.

Denaturation The disruption of tertiary protein structure brought on by heat or a change in pH.

Density The mass of an object per unit of volume.

Diatomic molecule A molecule that consists of two atoms bonded together.

Digestion A general term for the breakdown of food into small molecules.

Dilution A decrease in the concentration of a solution caused by addition of solvent.

Dilution factor The ratio of original-to-final volumes of a solution being diluted.

Dimer A unit formed by the joining together of two identical molecules.

Diprotic acid A substance that has two acidic hydrogen atoms.

Disaccharide A complex carbohydrate formed by the bonding together of two simple sugars.

Dissociation The splitting apart of a substance to yield two or more ions.

Disulfide A compound that contains a sulfur-sulfur single bond, R—S—S—R.

Disulfide bridge An S—S bond formed between two cysteine residues that can join two peptide chains together or cause a loop in a peptide chain.

DNA Deoxyribonucleic acid.

Double bond A covalent bond that results from sharing four electrons between atoms.

Double helix Two nucleotide strands coiled around each other in a screw-like fashion.

Electrolyte A substance that conducts electricity when dissolved in water.

Electron A negatively charged subatomic particle that orbits around the nucleus.

Electron configuration The specific way that an atom's electrons are distributed into shells and subshells.

Electronegativity The ability of an atom in a molecule to attract electrons.

Element A fundamental substance that cannot be chemically broken down into any simpler substance.

Enantiomer One of the two mirror-image forms of a chiral molecule.

End point The point at which a titration is complete.

Endocrine system A combination of different glands that regulate and control cellular activity in the body.

Endothermic reaction A reaction that absorbs heat from the surroundings.

Enzyme A protein that acts as a catalyst for biological reactions.

Enzyme-substrate complex A complex of enzyme and substrate not linked by covalent bonds.

Equivalent (Eq) The amount of an ion in grams that contains Avogadro's number of charges.

Essential amino acid One of ten amino acids that cannot be synthesized by the body and so must be obtained in the diet.

Ester A carbonyl compound that has one organic group and one —OR group bonded to the carbonyl carbon, RCOOR'.

Esterification reaction The reaction between an alcohol and a carboxylic acid to yield an ester plus water.

Ether A compound that has an oxygen atom bonded to two carbon atoms, R—O—R.

Ethyl group —CH_2CH_3, the alkyl group derived from ethane.

Eukaryotic cell Cell with a membrane-enclosed nucleus; found in all higher organisms.

Evaporation The spontaneous conversion of a liquid to a gas.

Exon A DNA segment in a gene that codes for part of a protein molecule.

Exothermic reaction A reaction that gives off heat to the surroundings.

Factor-label method A method of problem solving in which equations are set up so that unwanted units cancel and only the correct units remain.

Fatty acid A long-chain carboxylic acid formed by hydrolysis of animal fats and vegetable oils.

Fatty-acid spiral A repetitive series of biochemical reactions that degrade fatty acids to acetyl CoA.

Fermentation The breakdown of glucose to ethanol plus carbon dioxide by the action of yeast enzymes.

Fibrous protein A tough, insoluble protein whose peptide chains are arranged in long filaments.

Formula weight The sum of individual atomic weights for all atoms in a molecule or ion.

Functional group A part of a larger molecule composed of an atom or group of atoms that has characteristic chemical behavior.

Gamma (γ) Alternative name for a microgram.

Gamma (γ) radiation Emission of high-energy electromagnetic waves.

Gas A substance that has neither a definite volume nor a definite shape.

Gas laws A series of laws that describe the behavior of gases under conditions of differing pressure, volume, and temperature.

Gay-Lussac's law The pressure of a gas at a constant volume is directly proportional to its kelvin temperature.

Gene A small segment of a DNA chain where the genetic instructions for making an individual protein are encoded.

Genetic code The assignment of mRNA codons to specific amino acids.

Genome The sum of all genes in an organism.

Germ cell A reproductive (sperm or egg) cell.

Globular protein A water-soluble protein that adopts a compact, coiled-up shape.

Gluconeogenesis The biological synthesis of glucose from simpler molecules.

Glycol A dialcohol, or compound that contains two —OH groups.

Glycolipid A substance that has a lipid-like portion bonded to a carbohydrate.

Glycolysis A series of biochemical reactions that break down a molecule of glucose into two molecules of pyruvate ion plus energy.

Glycosidase An enzyme that is able to digest complex carbohydrates.

Glycoside A cyclic acetal formed by reaction of a simple sugar with an alcohol.

Grain alcohol A common name for ethyl alcohol, CH_3CH_2OH.

Gram (g) The metric unit of mass; equal to 1/1000 kilogram.

Group A vertical column of elements in the periodic table.

Half-life ($t_{1/2}$) The amount of time required for 50% of a sample to undergo radioactive decay.

Halogen An element in Group 7A of the periodic table (F, Cl, Br, I).

Halogenation The reaction of an alkene with a halogen (Cl_2 or Br_2) to yield a 1,2-dihaloalkane product.

Heat A form of energy transferred from a hotter object to a colder one.

Heat of fusion The amount of heat necessary to convert one gram of solid into a liquid when the solid is at its melting point.

Heat of reaction The exact amount of heat released or absorbed during a chemical reaction.

Heat of vaporization The amount of heat necessary to convert one gram of liquid into a gas when the liquid is at its boiling point.

α Helix A common secondary protein structure in which a protein chain wraps into a coil stabilized by hydrogen bonds.

Hemiacetal A compound that has both an alcohol-like —OH group and an ether-like —OR group bonded to the same carbon.

Hemolysis Bursting of red blood cells due to a buildup of pressure in the cell.

Henry's law The solubility of a gas in a liquid varies with its pressure if temperature is held constant.

Heterocycle A ring that contains nitrogen or some other atom in addition to carbon.

Heterogeneous mixture A mixture that is visually nonuniform.

Holoenzyme The combination of apoenzyme and cofactor that is active as a biological catalyst.

Homogeneous mixture A mixture that is uniform to the naked eye.

Hormone A chemical messenger, secreted by an endocrine gland and transported through the bloodstream to elicit response from a specific target tissue.

Hydrate A solid substance that has water molecules included in its crystals.

Hydration The reaction of an alkene with water to yield an alcohol.

Hydrocarbon A compound that contains only carbon and hydrogen.

Hydrogen bond A weak attraction between a hydrogen and a nearby oxygen, nitrogen, or fluorine atom.

Hydrogenation The reaction of an alkene (or alkyne) with H_2 to yield an alkane product.

Hydrohalogenation   The reaction of an alkene with HX (HCl or HBr) to yield an alkyl halide product.

Hydrolysis   The breakdown of a compound, such as an acetal, by reaction with water.

Hydrometer   A weighted bulb used to measure specific gravity of a liquid.

Hydronium ion H_3O^+, the species formed when an acid is dissolved in water.

Hydrophilic   Water-loving. A hydrophilic substance dissolves in water.

Hydrophobic   Water-hating. A hydrophobic substance does not dissolve in water.

Hydroxide ion   The OH^- ion formed by loss of H^+ from water.

Hydroxyl   A name for the —OH group in an organic compound.

Hygroscopic   Having the ability to attract water vapor from the air.

Hyperthermia   The medical condition that results from an uncontrolled rise in body temperature.

Hypertonic   Referring to a solution that has higher osmolarity than another, usually blood.

Hyperventilation   Heavy breathing that depletes carbon dioxide from the lungs and blood.

Hypothermia   The medical condition that results from uncontrolled loss of body temperature.

Hypotonic   Referring to a solution that has lower osmolarity than another, usually blood.

Indicator   A dye that changes color to indicate the pH of a solution.

Induced-fit model   A model that pictures an enzyme with a conformationally flexible active site that can change shape to accommodate a range of different substrate molecules.

Intron   A seemingly nonsensical segment of DNA that occurs in a gene but does not code for part of a protein.

Ion   An electrically charged atom or group of atoms.

Ion product constant (K_w)   The product of H_3O^+ and OH^- molar concentrations in water ($K_w = 10^{-14}$).

Ionic bond   The force of electrical attraction between an anion and a cation.

Ionic solid   A chemical compound held together by ionic bonds between anions and cations.

Ionizing radiation   Radiation capable of dislodging an electron from (ionizing) a molecule it strikes.

Irreversible enzyme inhibition   A mechanism of enzyme regulation in which an inhibitor forms covalent bonds to the active site.

Isoelectric point   The pH at which a large sample of amino-acid molecules has equal numbers of + and − charges.

Isoenzymes   Closely related enzymes that differ slightly in structure but have similar biological activity.

Isomers   Compounds with the same molecular formula but with different connections between atoms and different chemical structures.

Isopropyl group   —CH(CH$_3$)$_2$, the alkyl group derived by removing a hydrogen atom from the central carbon of propane.

Isotonic   Referring to a solution that has the same osmolarity as another, usually blood.

Isotopes   Atoms of the same element that have different numbers of neutrons in their nuclei.

Joule (J)   The SI unit of energy; equal to 4.184 calories.

Kelvin (K)   The SI unit of temperature; equal to 1.9 Fahrenheit degree.

Ketone   A compound that has a carbonyl group bonded to two carbon atoms, $R_2C{=}O$.

Ketose   A simple sugar that contains a ketone carbonyl group.

Kilocalorie (Kcal)   A unit of energy equal to 1000 calories.

Kilogram (kg)   The SI unit of mass; equal to 2.205 pounds.

Kinetic theory of gases   A set of four assumptions for explaining the general behavior of gases.

Lanthanides   The series of 14 elements following lanthanum in the periodic table.

Law of definite proportions   Every chemical compound is formed by a combination of elements in a defined proportion.

Length   The distance an object extends in a particular direction.

Lewis structure   A representation of a molecule that uses dots to represent outer-shell electrons.

Limiting reactant   The reactant present in limiting amount that restricts the extent to which a reaction can occur.

Line-bond structure   A representation of a molecule that uses lines between atoms to represent covalent bonds.

1,4-Link   An acetal link between the hydroxyl group

at C1 of one sugar and the hydroxyl group at C4 of another sugar.

Lipid   A naturally occurring molecule that is soluble in nonpolar organic solvents.

Lipid bilayer   The basic structural unit of cell membranes; composed of two parallel sheets of lipid molecules arranged tail to tail.

Lipigenesis   The biological synthesis of fat molecules from excess dietary carbohydrate.

Liquid   A substance that has a definite volume but that changes shape to fill the container it's placed in.

Liter (L)   The metric unit of volume; equal to 1.057 quarts.

Litmus   A well-known indicator used to distinguish an acid from a base.

Lock-and-key model   A model for enzyme specificity that pictures an enzyme as having a large irregular cleft into which only specific substrate molecules can fit.

Lone pair   A pair of outer-shell electrons not used by an atom for forming bonds.

Main group element   An element group on the far right (Groups 3A–8A) or far left (Groups 1A–2A) of the periodic table.

Manometer   A mercury-filled U-tube used to measure pressure.

Markovnikov's rule   In the addition of HX to an alkene, the H becomes attached to the carbon that already has the most H's, and the X becomes attached to the carbon that has fewer H's.

Mass   The amount of matter in an object.

Mass number   The sum of an atom's protons and neutrons.

Matter   The physical material that makes up the universe; anything that has mass and occupies space.

Maxam-Gilbert method   A rapid and efficient method for sequencing long strands of DNA.

Melting point (mp)   The temperature at which a solid turns into a liquid.

Mercaptan   An alternate name for a thiol, R—SH.

Mercury barometer   A mercury-filled glass tube used to measure atmospheric pressure.

Messenger RNA (mRNA)   The RNA whose function is to carry genetic messages transcribed from DNA.

Metabolism   The overall sum of the many reactions taking place in a cell.

Metal   A malleable element with a lustrous appearance that is a good conductor of heat and electricity.

Meter (m)   The SI unit of length; equal to 3.280 feet.

Methyl group   —CH_3, the alkyl group derived from methane.

Metric units   Units of a common system of measure used in commerce throughout the world.

Micelle   A spherical cluster formed by the aggregation of soap molecules in water.

Microgram (μg)   A unit of mass equal to 1/1000 milligram.

Milliequivalent (mEq)   The amount of an ion equal to one-thousandth of an equivalent.

Milligram (mg)   A unit of mass equal to 1/1000 gram.

Milliliter (mL)   A unit of volume equal to 1/1000 liter.

Millimeter (mm)   A unit of length equal to 1/1000 meter, or about the thickness of a dime.

Millimeter of mercury (mm Hg)   A unit of pressure equal to the force exerted by a 1-mm column of mercury.

Mirror image   The reverse image produced when an object is reflected in a mirror.

Miscible   Soluble in all proportions without limit.

Mitochondria   An egg-shaped organelle where small molecules are broken down to provide the energy to power an organism.

Mixture   A physical blend of two or more substances, each of which retains its chemical identity.

Molarity (M)   Concentration expressed as the number of moles of solute per liter of solution.

Mole (mol)   An amount of a substance in grams equal to the formula weight of the substance.

Molecular Weight   The sum of individual atomic weights for all atoms in a covalent molecule.

Molecule   A group of atoms bonded together in a discrete unit.

Monoprotic acid   A substance that has one acidic hydrogen atom.

Monosaccharide   An alternative name for a simple sugar.

Mutagen   A substance that causes mutations.

Mutation   An error in base sequence occurring during DNA replication.

***n*-Propyl group**   —$CH_2CH_2CH_3$, the alkyl group derived by removing a hydrogen atom from an end carbon of propane.

Neurotransmitter A chemical messenger than transmits a nerve impulse between neighboring nerve cells.

Neutralization reaction The reaction of an acid with a base to yield water and a salt.

Neutron An electrically neutral subatomic particle found in the nucleus of atoms.

Nitrate ester A compound formed by reaction of an alcohol with nitric acid.

Noble gas An element in Group 8A of the periodic table (He, Ne, Ar, Kr, Xe, Rn).

Nomenclature The system for naming molecules.

Noncompetitive enzyme inhibition A mechanism of enzyme regulation in which an inhibitor binds to an enzyme, thereby changing the shape of the enzyme's active site.

Nonelectrolyte A substance that does not conduct electricity when dissolved in water.

Nonmetal An element at the right side of the periodic table that is a poor conductor of heat and electricity.

Normality (N) A measure of acid (or base) concentration expressed as the number of acid (or base) equivalents per liter of solution.

Nuclear decay The emission of a particle from the nucleus of an element.

Nuclear fission The splitting apart of an atom by neutron bombardment to give smaller atoms.

Nucleic acid A biological polymer made by the linking together of nucleotide units.

Nucleotide A building block for nucleic-acid synthesis, consisting of a five- carbon sugar bonded to a cyclic amine base and to phosphoric acid.

Nucleus The dense, positively charged mass at the center of an atom where protons and neutrons are located.

Octet rule Atoms undergo reactions in order to attain a noble-gas electronic configuration with eight outer-shell electrons.

Optical isomer An alternative term for enantiomer.

Orbital A specifically shaped region of space around an atom, denoted s, p, d, or f, where electrons of a specific energy level are found.

Organelle A small organized unit in the cell that performs a specific function.

Organic chemistry The chemistry of carbon compounds.

Osmolarity (Osmol) The number of moles of dissolved solute particles (ions or molecules) per liter of solution.

Osmosis The passage of solvent molecules across a semipermeable membrane from a more dilute solution to a more concentrated solution.

Osmotic pressure The amount of external pressure that must be applied to halt the passage of solvent molecules across a semipermeable membrane.

Oxidation In inorganic chemistry, the loss of electrons by a reactant in a chemical reaction. In organic chemistry, the removal of hydrogen from a molecule or addition of oxygen to a molecule.

Oxidation state The charge on an ion.

Oxidative deamination A reaction that converts an amino group ($-NH_2$) into a carbonyl group.

Oxidative phosphorylation A term for the synthesis of ATP from ADP and phosphate ion.

Oxidizing agent The reactant that causes an oxidation by taking electrons.

Paraffin A mixture of waxy alkanes having 20 to 36 carbon atoms.

Partial pressure The contribution to total gas pressure caused by each individual component of a mixture of gases.

Pascal (Pa) The SI unit of pressure; equal to 0.007500 mm Hg.

Peptide bond An amide bond that links two amino acids together.

Peptide sequence The order in which individual amino acids are bonded together in the peptide chain.

Period A horizontal row of elements in the periodic table.

Periodic table The chart displaying the elements in order of increasing atomic number so that elements with similar properties fall into groups.

pH A number between 0 and 14 that describes the acidity of an aqueous solution. Mathematically, pH is equal to the negative logarithm of a solution's H_3O^+ concentration.

Phenol A compound that has an $-OH$ functional group bonded directly to an aromatic, benzene-like ring.

Phenyl The name of the C_6H_5- unit when a benzene ring is considered as a substituent group.

Phosphate ester A compound formed by reaction of an alcohol with phosphoric acid.

Phosphoglyceride A phospholipid in which glycerin is linked by ester bonds to two fatty acids and one phosphoric acid.

Phospholipid Lipids that have an ester link between phosphoric acid and an alcohol.

Physical property A property that does not involve a chemical change in a substance or object.

Physical quantity A physical property that can be measured.

Plasma The fluid surrounding blood cells.

Plasma membrane The lipid bilayer membrane surrounding a eukaryotic cell.

β-Pleated sheet A common secondary protein structure in which segments of a protein chain fold back on themselves to form parallel strands held together by hydrogen bonds.

Polar covalent bond A covalent bond in which one atom attracts bonding electrons more strongly than the other atom.

Polarized Having a partial positive or negative charge as the result of being in a polar covalent bond.

Polyatomic ion An ion that contains two or more atoms linked by covalent bonds.

Polycyclic aromatic compound A substance that has two or more benzene-like rings fused together along their edges.

Polysaccharide An alternate name for a complex carbohydrate.

Polyunsaturated fatty acid (PUFA) A long-chain fatty acid that has two or more carbon-carbon double bonds.

Potential energy Energy that is stored because of the position or composition of an object.

Pressure The force per unit area exerted on a surface.

Primary (1°) carbon A carbon atom that is bonded to one other carbon.

Primary protein structure The sequence in which amino acids are linked together.

Product A substance formed as the result of a chemical reaction.

Prokaryotic cell A cell that has no nucleus; found in bacteria and algae.

Property A characteristic or trait useful for identifying a substance or object.

Prostaglandin A simple lipid containing a cyclopentane ring with two long side chains.

Protein A large biological molecule made of many amino acids linked together through amide bonds.

Proton A positively charged subatomic particle found in the nucleus of atoms.

Quaternary (4°) carbon A carbon atom that is bonded to four other carbons.

Quaternary protein structure The way in which protein chains aggregate to form large, ordered structures.

R— The general symbol for an alkyl group.

Rad A unit for measuring the amount of radiation absorbed per gram of tissue.

Radioactivity Spontaneous emission of radiation (alpha, beta, or gamma rays).

Radioisotope A radioactive isotope.

Reactant A starting substance that undergoes change in a chemical reaction.

Reaction energy diagram A pictorial way of representing the energy changes that occur during a chemical reaction.

Redox reaction A reaction in which oxidations and reductions occur.

Reducing agent The reactant that causes a reduction by giving electrons.

Reducing sugar A carbohydrate that reacts with an oxidizing agent like Benedict's reagent.

Reduction In inorganic chemistry, the gain of electrons by a reactant in a chemical reaction. In organic chemistry, the addition of hydrogen to a carbon-oxygen double bond or carbon-carbon double bond functional group.

Rem A unit for measuring the amount of tissue damage cause by radiation.

Replication The process by which copies of DNA are made in the cell.

Residue An alternative name for an amino-acid unit in a polypeptide or protein.

Respiratory chain A series of biochemical reactions that harness the energy released in the citric-acid cycle.

Restriction endonuclease An enzyme that is able to cut a DNA strand at a point in the chain where a specific base sequence occurs.

Reversible reaction A reaction that can proceed in either the forward or the reverse direction because the reactants and products are of similar stability.

Ribosomal RNA (rRNA) The structural material of ribosomes.

Ribosome   A structure in the cell where protein synthesis occurs.

RNA   Ribonucleic acid.

Roentgen   A unit for measuring the ionizing intensity of radiation.

Rounding off   The procedure for expressing values with the correct number of significant figures.

Salt   An ionic substance composed of a positive ion other than H^+ and a negative ion other than HO^-.

Saponification   The reaction of an ester with aqueous hydroxide ion to yield an alcohol and the metal salt of a carboxylic acid, as in the making of soap from animal fat.

Saturated   Containing only single bonds between carbon atoms, and thus unable to accommodate additional hydrogens.

Saturated solution   A solution in which the solute has reached its solubility limit.

Scientific notation   A way of representing a large or small number as a power of 10.

Secondary (2°) carbon   A carbon atom that is bonded to two other carbons.

Secondary protein structure   The way in which segments of a protein chain are oriented into a regular pattern.

Semipermeable membrane   A thin membrane that allows water or other small solvent molecules to pass through but that blocks the flow of larger solute molecules or ions.

Shell   An imaginary layer surrounding an atom's nucleus where electrons are located.

SI Unit   (Système International d'Unités) An internationally agreed on unit of measure derived from the metric system.

Significant figures   In describing a quantity, the total of the number of digits whose values are known with certainty, plus one estimated digit.

Simple lipid   A lipid that does not contain an ester group and that can't undergo hydrolysis.

Simple protein   A protein that yields only amino acids when hydrolyzed.

Simple sugar   A carbohydrate that can't be chemically broken down into a smaller sugar by hydrolysis with acid.

Single bond   A covalent bond that results from sharing two electrons between atoms.

Soap   The mixture of metal salts of fatty acids formed on saponification of animal fat.

Solid   A substance that has a definite shape and volume.

Solubility   The amount of a substance that can be dissolved in a given volume of solvent.

Solute   The minor substance in a homogeneous mixture.

Solution   A homogeneous mixture containing molecule-sized particles uniformly dispersed in another material.

Solvation   The surrounding of a solute ion or molecule by solvent molecules.

Solvent   The major substance in a homogeneous mixture.

Somatic cell   Any cell other than a reproductive one.

Specific gravity   The density of a substance divided by the density of water at the same temperature.

Specific heat   The amount of heat that will raise the temperature of one gram of a substance by one Celsius degree.

Sphingolipid   A phospholipid based on the amino alcohol sphingosine, rather than on glycerin.

Standard molar volume   The volume of one mole of a gas at standard temperature and pressure (22.4 L).

Standard temperature and pressure (STP)   Standard conditions for a gas, defined as 0°C and 1 atm pressure.

State of matter   The physical state of a substance as a solid, a liquid, or a gas.

Steroid   A simple lipid whose structure is based on a tetracyclic (four-ring) carbon skeleton.

Straight-chain alkane   An alkane that has all its carbon atoms connected in a row.

Subatomic particle   An elementary particle from which atoms are made.

Subshell   A subregion of a shell where electrons of the same energy level are located.

Substitution reaction   A chemical reaction in which two reactants exchange, or substitute, atoms: $A + B \rightarrow C + D$.

Substrate   A molecule acted upon by an enzyme.

Suspension   A homogeneous mixture containing particles just large enough to be visible to the naked eye.

Symmetry plane   An imaginary plane cutting through the middle of an object so that one half of the object is a mirror image of the other half.

Temperature   The measure of how much heat energy an object contains.

Tertiary (3°) carbon   A carbon atom that is bonded to three other carbons.

Tertiary protein structure   The way in which an entire protein molecule is coiled and folded into its specific three-dimensional shape.

Tetrahedron   A geometrical figure with four identical triangular faces.

Thiol   A compound that contains the —SH functional group, R—SH.

Titration   An experimental method for determining acid (or base) concentration by neutralizing a sample with a base (or acid) of known concentration.

Tollens' reagent   A reagent ($AgNO_3$ in aqueous NH_3) that converts an aldehyde into a carboxylic acid and deposits a silver mirror on the reaction flask.

Torr   Alternate name for mm Hg.

Transcription   The process by which the information in DNA is read and used to synthesize RNA.

Transfer RNA (tRNA)   The RNA whose function is to transport a specific amino acid into position for protein synthesis.

Transition metal group   An element group in the middle of the periodic table.

Translation   The process by which RNA directs protein synthesis.

Transmutation   The change of one element into another brought about by nuclear decay.

Triacylglycerol   A fat or vegetable oil, consisting of triesters of glycerol with three long-chain carboxylic acids.

Triple bond   A covalent bond that results from sharing six electrons between atoms.

Triple helix   The secondary protein structure of tropocollagen in which three protein chains coil around each other to form a long fiber.

Triprotic acid   A substance that has three acidic hydrogen atoms.

Turnover number   The number of substrate molecules acted on by one molecule of enzyme per unit time.

Unit   A specific quantity, used for measurement.

Universal gas law   A law that relates the effects on a gas sample of temperature, pressure, volume, and molar amount ($PV = nRT$).

Unsaturated   Containing one or more double or triple bonds between carbon atoms, and thus fewer than the maximum number of hydrogen atoms per carbon.

Vapor   The gaseous state of a substance that is normally a liquid.

Vapor pressure   The pressure of a vapor at equilibrium with its liquid.

Vitamin   A small organic molecule that must be obtained in the diet and that is essential in trace amounts for proper biological functioning.

Volume   The amount of space occupied by an object.

Volume/volume percent concentration [(v/v)%]   Concentration expressed as the number of mL solute per 100 mL solution.

Volumetric flask   A flask whose volume has been precisely calibrated.

Wax   A mixture of complex lipids consisting of esters of long-chain carboxylic acids with long-chain alcohols.

Weight   The measure of the gravitational force exerted on an object by the earth, moon, or other massive body.

Weight/volume percent concentration [(w/v)%]   Concentration expressed as the number of grams of solute dissolved in 100 mL of solution.

Weight/weight percent concentration [(w/w)%]   Concentration expressed as the number of grams of solute per 100 g of solution.

Wood alcohol   A common name for methyl alcohol, CH_3OH.

X rays   High-energy electromagnetic radiation.

Zwitterion   A neutral compound that contains both + and − charges in its structure.

Answers to In-Chapter Practice Problems

Chapter 1

1.1 centiliter
1.2 (a) milliliter
   (b) kilogram
   (c) centimeter
   (d) kilometer
   (e) gram
1.3 (a) L
   (b) μL
   (c) nm
   (d) mm
1.4 (a) 3
   (b) 4
   (c) 5
   (d) exact
1.5 2.78×10^2 pm
1.6 (a) 5.8×10 g
   (b) 4.6792×10^4 m
   (c) 6.720×10^{-4} cm
   (d) 3.453×10^{22} kg
1.7 (a) 48,850 mg
   (b) 0.0000083 m
   (c) 0.0400 mL
1.8 (a) 2.30 g
   (b) 188.38 mL
   (c) 0.009 L
   (d) 1.000 kg
1.9 (a) 50.9 mL
   (b) 0.078 g
   (c) 51 mg

1.10 (a) 3.4 kg
   (b) 120 mL
1.11 (a) 10.6 mg/kg
   (b) 36 mg/kg
1.12 39.4°C
1.13 -38.9°F
1.14 7700 cal
1.15 0.21 cal/g °C
1.16 Float: ice, human fat, cork, balsa wood. Sink: gold, table sugar, earth.
1.17 8.392 mL

Chapter 2

2.1 (a) U
   (b) Ti
   (c) W
2.2 (a) sodium
   (b) calcium
   (c) palladium
   (d) potassium
   (e) strontium
   (f) tin
2.3 92 protons, 143 neutrons
2.4 ^{35}Cl: 17 protons, 18 neutrons; ^{37}Cl: 17 protons, 20 neutrons
2.5 $^{35}_{17}$Cl and $^{37}_{17}$Cl
2.6 (a) $1s^2 2s^2 2p^2$
   (b) $1s^2 2s^2 2p^6 3s^1$

   (c) $1s^2 2s^2 2p^6 3s^2 3p^5$
   (d) $1s^2 2s^2 2p^6 3s^2 3p^6 4s^2$
2.7 Metals: scandium, technetium. Nonmetal: selenium.
2.8 Group 6A
2.9 $C_6H_{12}O_6$

Chapter 3

3.1 Potassium, $1s^2 2s^2 2p^6 3s^2 3p^6 4s^1$, must lose one electron.
3.2 Magnesium, $1s^2 2s^2 2p^6 3s^2$, must lose two electrons.
3.3 Oxygen, $1s^2 2s^2 2p^4$, must gain two electrons.
3.4 Bromine is the oxidizing agent; potassium is the reducing agent. K^+ has oxidation state $+1$; Br^- has oxidation state -1.
3.5 (a) Oxidizing agent: chlorine; reducing agent: calcium. $Ca^{2+} = +2$; $Cl^- = -1$
   (b) Oxidizing agent: oxygen; reducing agent: barium. $Ba^{2+} = +2$; $O^{2-} = -2$
   (c) Oxidizing agent: iodine; reducing agent, magnesium. $Mg^{2+} = +2$; $I^- = -1$.
3.6 (a) $V = +3$
   (b) $Mg = +2$

(c) $Sn = +4$

(d) $Cr = +6$

3.7 helium, $1s^2$

3.8 **(a)** $P = 3$, $H = 1$

(b) $H = 1$, $Se = 2$

3.9 **(a)** $H:\overset{..}{\underset{..}{P}}:H = H-P-H$ with H below

(b) $H:\overset{..}{\underset{..}{Se}}:H = H-Se-H$

3.10

$H:\overset{H}{\underset{H}{\overset{..}{C}}}:\overset{..}{N}:H$ $H-\overset{H}{\underset{H}{C}}-N-H$

CH_3NH_2

3.11

$H:\overset{H}{\underset{H}{C}}:\overset{H}{\underset{H}{C}}:\overset{H}{\underset{H}{C}}:H$ $H-\overset{H}{\underset{H}{C}}-\overset{H}{\underset{H}{C}}-\overset{H}{\underset{H}{C}}-H$

$CH_3CH_2CH_3$

3.12 $H:\overset{:\overset{..}{O}:}{C}:H$ $H-\overset{\overset{O}{\|}}{C}-H$

3.13 $H:C:::N:$ $H-C\equiv N$

3.14 $\overset{..}{P}$ with H, H, H   The $H-P-H$ bond angle is approximately $109°$.

3.15 30 protons, 32 electrons

3.16 11 protons, 10 electrons. Ammonium ion is tetrahedral.

3.17 $BaSO_4$

3.18 silver(I) sulfide

3.19 **(a)** copper(II) oxide

(b) boron fluoride

(c) sodium nitrate

(d) copper(I) sulfate

(e) lithium phosphate

3.20 **(a)** $Ba(OH)_2$

(b) $CuCO_3$

(c) $Mg(HCO_3)_2$

(d) CrO_3

Chapter 4

4.1 **(a)** balanced

(b) not balanced

(c) balanced

(d) not balanced

(e) not balanced

4.2 $2\,Na + Cl_2 \longrightarrow 2\,NaCl$

4.3 $3\,O_2 \longrightarrow 2\,O_3$

4.4 **(a)** $Ca(OH)_2 + 2\,HCl \longrightarrow CaCl_2 + 2\,H_2O$

(b) $4\,Al + 3\,O_2 \longrightarrow 2\,Al_2O_3$

(c) $Ag_2O + 2\,HCl \longrightarrow 2\,AgCl + H_2O$

(d) $2\,CH_3CH_3 + 7\,O_2 \longrightarrow 4\,CO_2 + 6\,H_2O$

4.5 **(a)** 17 amu

(b) 18 amu

(c) 159.6 amu

4.6 formula wt $= 342$ amu; 1.0×10^{24} molecules

4.7 **(a)** 1.2×10^{24}

(b) 1.5×10^{23}

(c) 6.02×10^{23}

4.8 0.217 mol; 4.6 g

4.9 **(a)** 0.427 mol

(b) 0.298 mol

(c) 0.336 mol

(d) 0.139 mol

4.10 **(a)** 74.2 g

(b) 4.5 g

4.11 4.34 g C_2H_4; 5.66 g HCl

4.12 4 mol excess CH_4 remains; 1 mol CO_2 and 2 mol H_2O formed.

4.13 7.5 g CH_4 remains; 28 g H_2O and 34 g CO_2 formed.

4.14 42.6 kcal; 106 kcal

4.15 160 kcal; 12.0 g CH_4

Chapter 5

5.1 0.289 atm; 4.25 psi; 29,300 Pa

5.2 9.3 atm He; 0.2 atm O_2

5.3 Identical

5.4 450 L

5.5 1.2 atm; 62 atm

5.6 0.38 L; 0.55 L

5.7 410 K $= 137°C$

5.8 4460 mol; 7.14×10^4 g CH_4; 1.96×10^5 g CO_2

5.9 5.0 atm

5.10 1110 mol; 4440 g

Chapter 6

6.1 **(a)** colloid

(b) solution

(c) suspension

(d) solution

6.2 **(a)** 12 g

(b) 1.5 g

6.3 160 mL

6.4 Place 38 mL acetic acid in flask, and dilute to 500 mL.

6.5 **(a)** 22 mL

(b) 18 mL

6.6 **(a)** 0.025 mol

(b) 1.6 mol

6.7 **(a)** 25 g

(b) 68 g

6.8 695 mL

6.9 1.5 g

6.10 2.4 M

6.11 39.1 mL

6.12 $H^+(aq)$ and $Cl^-(aq)$ are present.

6.13 HCl

6.14 **(a)** 39.1 g

(b) 79.9 g

(c) 12.2 g

(d) 48.0 g

6.15 **(a)** 39.1 mg

(b) 79.9 mg

(c) 12.2 mg

(d) 48.0 mg

6.16 9 mg

6.17 **(a)** 0.70 Osmol

(b) 0.15 Osmol

6.18 0.88%

Chapter 7

7.1 **(a)** $2\,HNO_3 + Mg(OH)_2 \longrightarrow 2\,H_2O + Mg(NO_3)_2$

(b) $H_2SO_4 + Ba(OH)_2 \longrightarrow 2\,H_2O + BaSO_4$

7.2 $Al(OH)_3 + 3\,HCl \longrightarrow AlCl_3 + 3\,H_2O$

7.3 **(a)** $KHCO_3 + HNO_3 \longrightarrow KNO_3 + CO_2 + H_2O$

(b) $MgCO_3 + H_2SO_4 \longrightarrow MgSO_4 + CO_2 + H_2O$

7.4 $H_2SO_4 + 2\,NH_3 \longrightarrow (NH_4)_2SO_4$

7.5 $CH_3NH_2 + HCl \longrightarrow CH_3NH_3^+ Cl^-$

7.6 **(a)** acidic, $[OH^-] = 3.1 \times 10^{-10}$

(b) basic, $[OH^-] = 3.2 \times 10^{-3}$

7.7 (a) acidic
  (b) basic
  (c) acidic
  (d) acidic;
  Pancreatic juice is least acidic and
  wine is most acidic.

7.8 (a) $[H_3O^+] = 3.2 \times 10^{-7}$ M
  (b) $[H_3O^+] = 1.2 \times 10^{-8}$ M
  (c) $[H_3O^+] = 2.0 \times 10^{-4}$ M
  (d) $[H_3O^+] = 3.2 \times 10^{-4}$ M

7.9 0.73 M

7.10 133 mL

7.11 (a) 0.079 eq
  (b) 0.169 eq
  (c) 0.14 eq

7.12 (a) 0.26 N
  (b) 0.563 N
  (c) 0.47 N

7.13 Chloride ion is too weak a base
  to neutralize any added acid.

Chapter 8

8.1 (a)

$$CH_3-\overset{\displaystyle H}{\underset{\displaystyle OH}{C}}-\overset{\displaystyle O}{C}-OH$$

  Carboxylic acid
  Alcohol

  (b)

  Aromatic ring
  Double bond

8.2 (a)

$$H-\overset{\displaystyle H}{\underset{\displaystyle H}{C}}-\overset{\displaystyle O}{C}-H$$

  (b)

$$H-\overset{\displaystyle H}{\underset{\displaystyle H}{C}}-\overset{\displaystyle H}{\underset{\displaystyle H}{C}}-\overset{\displaystyle O}{C}-O-H$$

8.3

$$H-\overset{\displaystyle H}{\underset{\displaystyle H}{C}}-\overset{\displaystyle H}{\underset{\displaystyle H}{C}}-\overset{\displaystyle H}{\underset{\displaystyle H}{C}}-\overset{\displaystyle H}{\underset{\displaystyle H}{C}}-\overset{\displaystyle H}{\underset{\displaystyle H}{C}}-\overset{\displaystyle H}{\underset{\displaystyle H}{C}}-\overset{\displaystyle H}{\underset{\displaystyle H}{C}}-H$$

8.4 (a) $CH_3CH_2CH_2CH_2CH_3$

(b)

$$CH_3\overset{\displaystyle CH_3}{\underset{\displaystyle |}{CH}}CH_2CH_3$$

(c)

$$CH_3\overset{\displaystyle CH_3}{\underset{\displaystyle CH_3}{C}}CH_3$$

8.5 Structures (a) and (c) are identical
  and are isomers of structure (b).

8.6 $CH_3CH_2CH_2CH_2CH_2CH_3$

$$CH_3CH_2CH_2\overset{\displaystyle CH_3}{\underset{\displaystyle |}{C}}HCH_3$$

$$CH_3CH_2\overset{\displaystyle CH_3}{\underset{\displaystyle |}{C}}HCH_2CH_3$$

$$CH_3CH_2\overset{\displaystyle CH_3}{\underset{\displaystyle CH_3}{C}}CH_3$$

$$CH_3\overset{\displaystyle CH_3}{\underset{\displaystyle |}{C}}H\overset{\displaystyle }{\underset{\displaystyle CH_3}{C}}HCH_3$$

8.8 (a) 2,6-dimethyloctane
  (b) 2,2-diethylheptane

8.9 (a)

$$CH_3CH_2\overset{\displaystyle CH_3}{\underset{\displaystyle |}{C}}HCH_2CH_2CH_3$$

  (b)

$$CH_3CH_2\overset{\displaystyle CH_3}{\underset{\displaystyle CH_3}{C}}H\overset{\displaystyle }{C}HCH_2CH_2CH_3$$

  (c)

$$CH_3\overset{\displaystyle CH_3}{\underset{\displaystyle CH_3}{C}}CH_2\overset{\displaystyle CH_3}{\underset{\displaystyle }{C}}HCH_3$$

8.10 All CH_3's are primary, CH_2's are
  secondary, CH's are tertiary, and
  C's are quaternary.

8.11 (a)

$$CH_3\overset{\displaystyle CH_3}{\underset{\displaystyle |}{C}}HCH_3$$

  Tertiary

  (2-methylpropane)

(b)

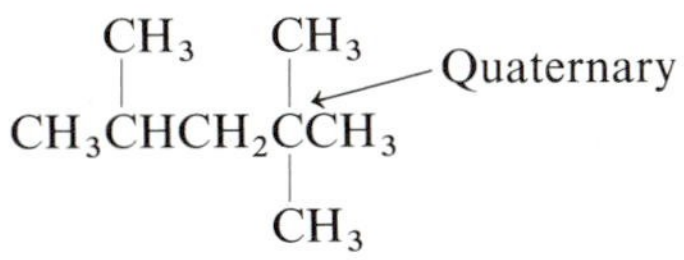

$$CH_3CHCH_2CCH_3$$

CH₃ CH₃ ←— Quaternary

(2,2,4-trimethylpentane)

8.12 (a) 1-ethyl-4-methylcyclohexane
(b) 1-ethyl-3-isopropylcyclopentane

8.13 (a)

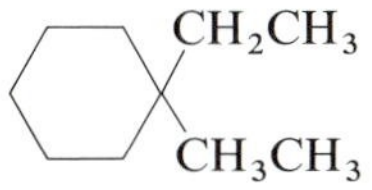

CH₂CH₃
CH₃CH₃

(b) 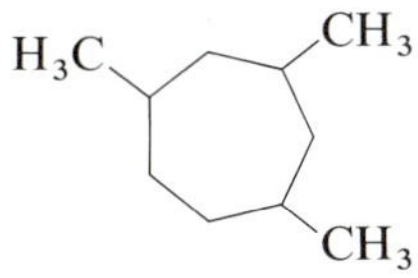 H₃C   CH₃

CH₃

Chapter 9

9.1 (a) 2-methyl-3-heptene

(b) 2-methyl-1,5-hexadiene

9.2 (a)

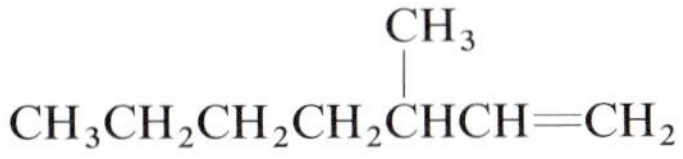

CH₃
CH₃CH₂CH₂CH₂CHCH=CH₂

(b)

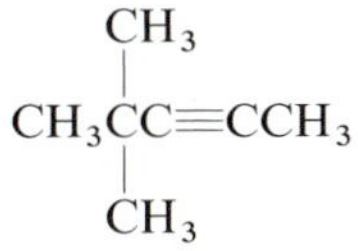

CH₃
CH₃CC≡CCH₃
CH₃

(c)

CH₃
CH₃CH₂CH₂CH=CCH₃

(d)

H₃C   CH₂CH₃
CH₃C—C=CHCH₂CH₃
H₃C

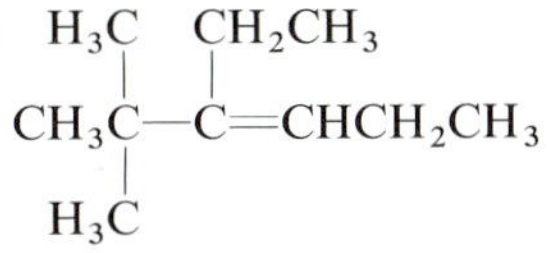

9.3 Compounds (a) and (c) can exist as cis-trans isomers.

9.4

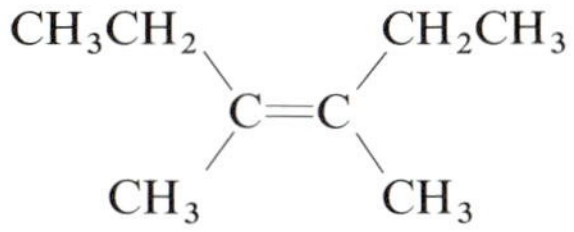

CH₃CH₂   CH₂CH₃
      C=C
CH₃   CH₃

and

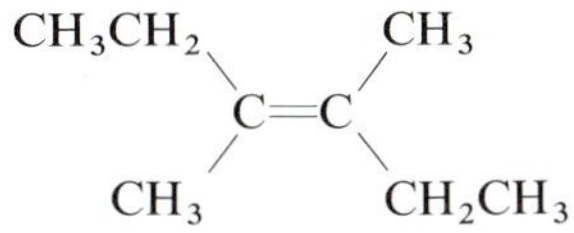

CH₃CH₂   CH₃
      C=C
CH₃   CH₂CH₃

9.5 (a,b,c) CH₃CH₂CH₂CH₃

(d)

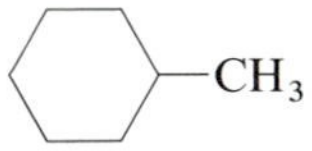

—CH₃

9.6 (a)

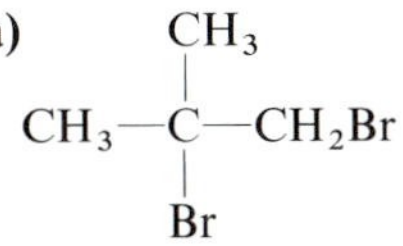

CH₃
CH₃—C—CH₂Br
Br

(b)

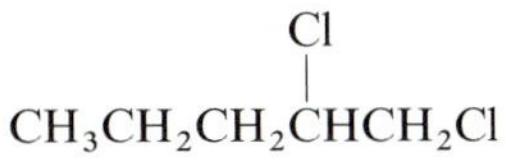

Cl
CH₃CH₂CH₂CHCH₂Cl

9.7 (a)

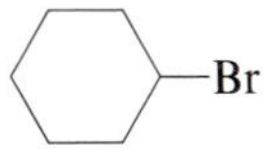

—Br

(b)

Cl
CH₃CH₂CH₂CH₂CHCH₃

(c)

H₃C   I
CH₃CHCHCH₃

(d)

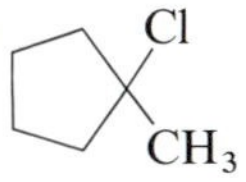

Cl
CH₃

9.8 (a)

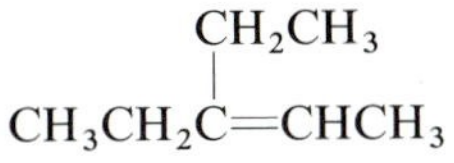

CH₂CH₃
CH₃CH₂C=CHCH₃

(b) (CH₃)₂C=C(CH₃)₂   or

CH₃
(CH₃)₂CHC=CH₂

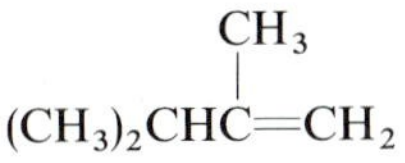

9.9 (a,b)

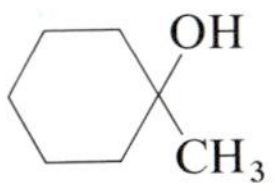

OH
CH₃

9.10

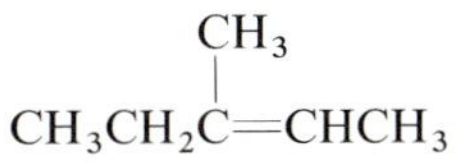

CH₃   or
CH₃CH₂C=CHCH₃

CH₂
CH₃CH₂CCH₂CH₃

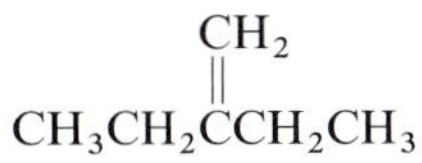

9.11

CH₃
CH₃—C—CH₃
    +

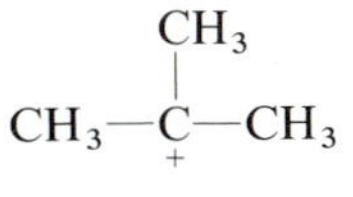

9.12 (a) *ortho*-bromochlorobenzene
(b) butylbenzene
(c) *ortho*-bromomethylbenzene
or *ortho*-bromotoluene

9.13 (a)

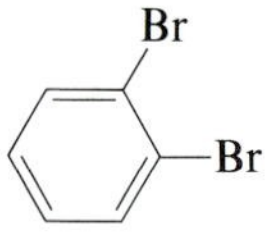

(b) Cl—⟨ ⟩—CH_3

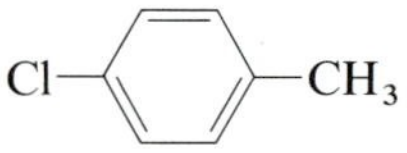

(c) CH_3CH_2

—CH_2CH_3

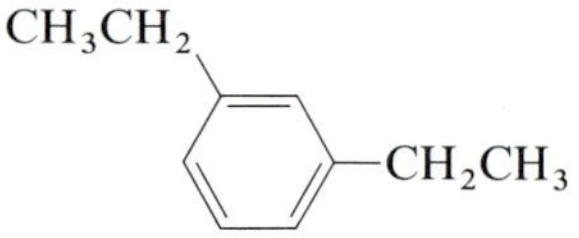

9.14 (a)

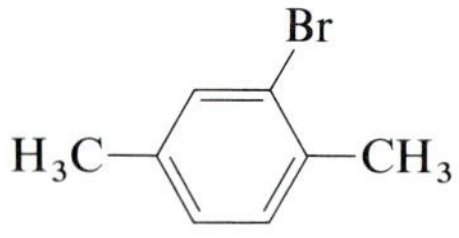

(b)

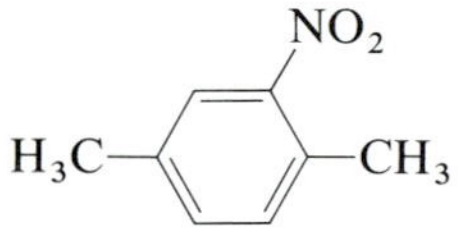

(c)

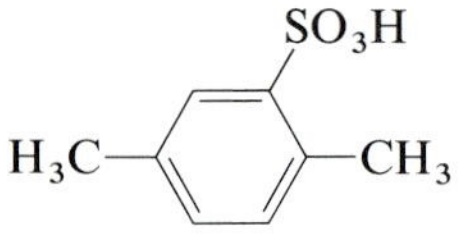

9.15

Br—⟨ ⟩—CH_3

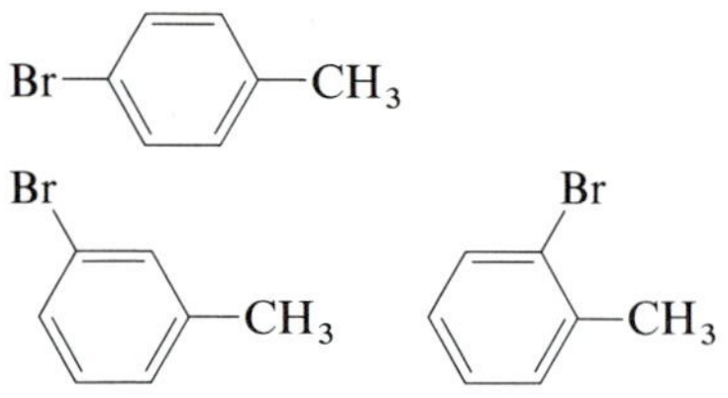

Chapter 10

10.1 **(a)** alcohol
(b) alcohol
(c) phenol
(d) alcohol
(e) ether
(f) ether

10.2 **(a)**

$$CH_3CH_2\overset{\overset{\displaystyle OH}{|}}{\underset{\underset{\displaystyle CH_3}{|}}{C}}CH_2CH_3$$

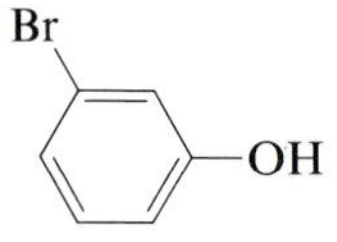

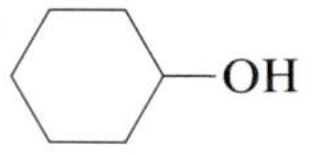

(b)

$$CH_3\overset{\overset{\displaystyle CH_3}{|}}{CH}-O-CH_3$$

(c)

(d)

(e)

$$CH_3CH_2CH_2\overset{\overset{\displaystyle OH}{|}}{CH}CH_2\overset{\overset{\displaystyle CH_3}{|}}{CH}CH_3$$

10.3 **(a)** 3-pentanol
(b) 2-ethyl–1-pentanol
(c) *para*-chlorophenyl methyl ether
(d) *meta*-bromophenol
(e) 4,4-dimethylcyclohexanol

10.4 **(a)** $CH_3CH{=}CH_2$

(b)

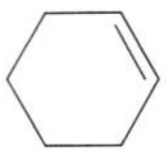

(c) CH_3 and
$$H_2C{=}CHCH_2\overset{\overset{\displaystyle CH_3}{|}}{CH}CH_3$$
$$CH_3CH{=}CHCH\overset{\overset{\displaystyle CH_3}{|}}{}CH_3$$

10.5 **(a)**
$$CH_3\overset{\overset{\displaystyle HO}{|}}{\underset{\underset{\displaystyle CH_3}{|}}{C}}-CH\overset{\overset{\displaystyle CH_3}{}}{}$$

(b) OH and
$$CH_3CH_2\overset{\overset{\displaystyle OH}{|}}{CH}CH_3$$
$$CH_3CH_2CH_2CH_2OH$$

10.6 **(a)** CH_3CH_2COOH

(b)
$$CH_3\overset{\overset{\displaystyle O}{\|}}{C}CH_2CH_2CH_3$$

(c)

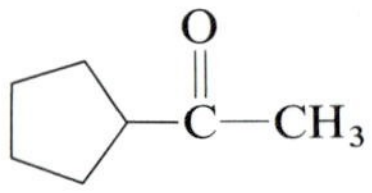

10.7 **(a)**
$$CH_3\overset{\overset{\displaystyle OH}{|}}{CH}CH_3$$

(b)

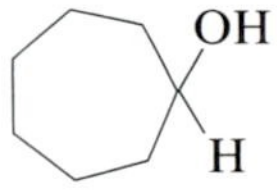

(c) 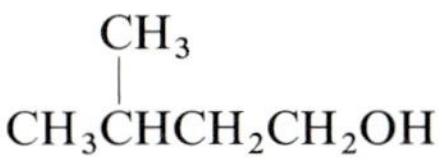   CH₃
        |
    CH₃CHCH₂CH₂OH

10.8 **(a)** CH₃CH₂CH₂S—SCH₂CH₂CH₃

(b) 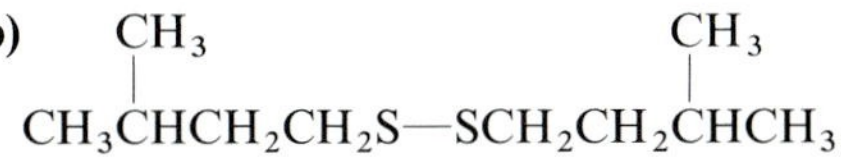   CH₃   CH₃
            | |
    CH₃CHCH₂CH₂S—SCH₂CH₂CHCH₃

10.9 **(a)** primary
(b) secondary
(c) primary
(d) secondary
(e) tertiary
10.10 (a) propylamine
(b) dimethylamine
(c) *N*-ethylaniline

10.11 (a) CH₃CH₂CH₂CH₂NH₂

(b) CH₃CH₂NHCH₃

(c) 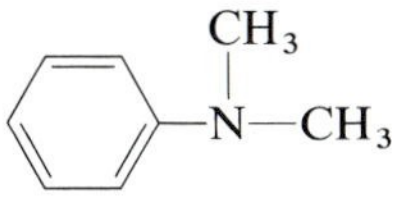   CH₃
                |
         [benzene ring]—N—CH₃

10.12 (a) 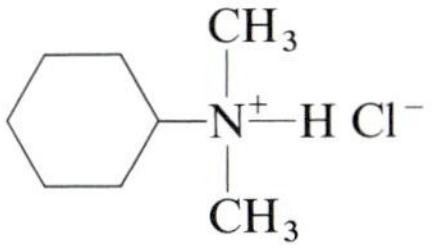   CH₃
                    |
    [cyclohexane ring]—N⁺—H Cl⁻
                    |
                    CH₃

(b) (CH₃CH₂)₂N⁺H₂ Cl⁻

(c) 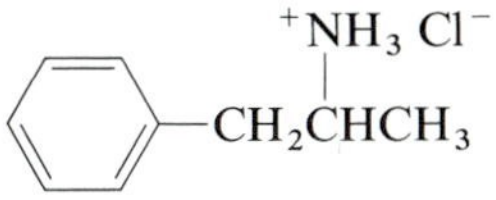   ⁺NH₃ Cl⁻
                    |
    [benzene ring]—CH₂CHCH₃

10.13 (a) (CH₃)₃N

(b) 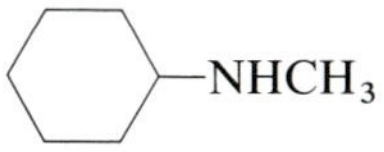 [cyclohexane ring]—NHCH₃

Chapter 11

11.2 **(a)** CH₃CH₂CH₂CH₂CH₂CHO

(b) 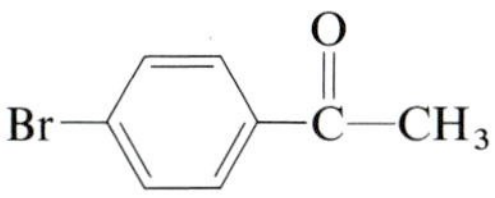   O
                    ‖
    Br—[benzene ring]—C—CH₃

(c)   CH₃   O
            | ‖
    CH₃CH₂CHCH₂CCH₃

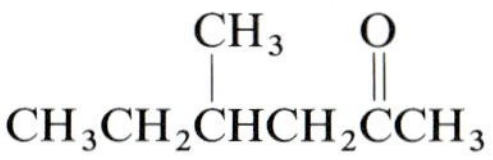

11.3 **(a)** pentanal
(b) 3-pentanone
(c) 4-methylhexanal

11.4 **(a)** 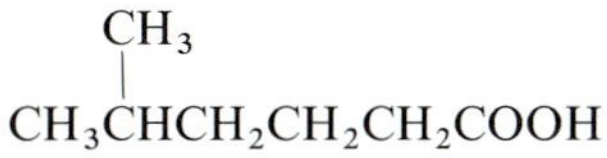   CH₃
                    |
    CH₃CHCH₂CH₂CH₂COOH

(b) 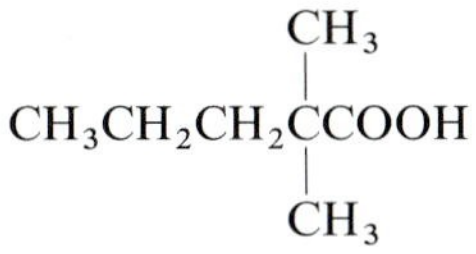   CH₃
                        |
    CH₃CH₂CH₂CCOOH
                        |
                        CH₃

(c) N.R.

11.5 **(a)** 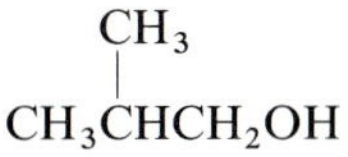   CH₃
                    |
    CH₃CHCH₂OH

(b) 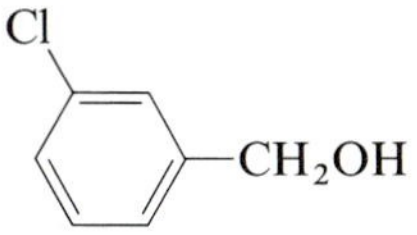 Cl
    [benzene ring]—CH₂OH

(c) 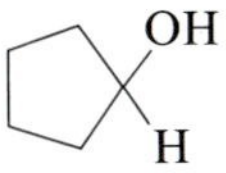 [cyclopentane ring with OH and H]

11.6 **(a)** 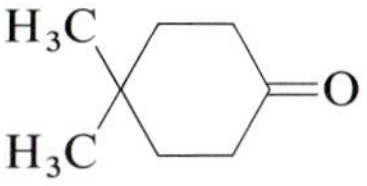 H₃C
                    [cyclohexane ring]=O
        H₃C

(b) 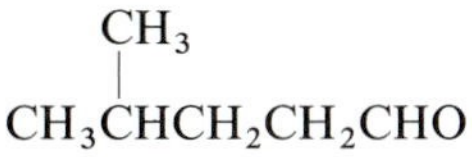   CH₃
            |
    CH₃CHCH₂CH₂CHO

(c) 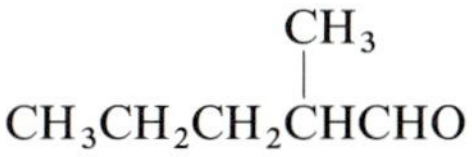   CH₃
                    |
    CH₃CH₂CH₂CHCHO

11.7 **(a)** 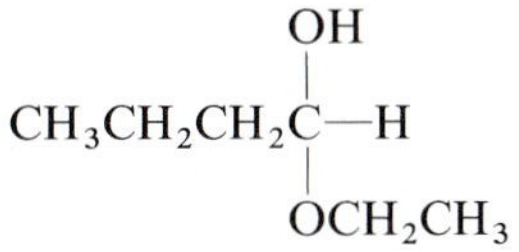   OH
                            |
    CH₃CH₂CH₂C—H
                            |
                        OCH₂CH₃

(b) 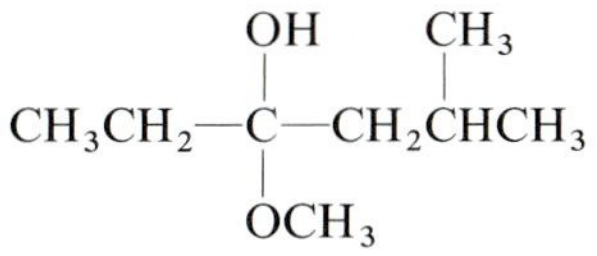   OH   CH₃
                | |
    CH₃CH₂—C—CH₂CHCH₃
                |
                OCH₃

11.8 **(a)** 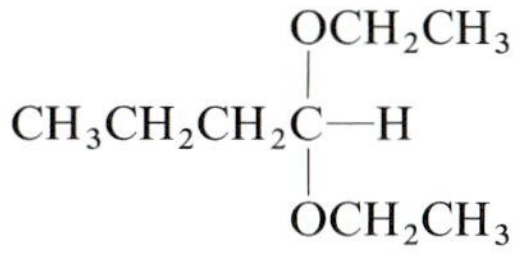   OCH₂CH₃
                            |
    CH₃CH₂CH₂C—H
                            |
                        OCH₂CH₃

(b) 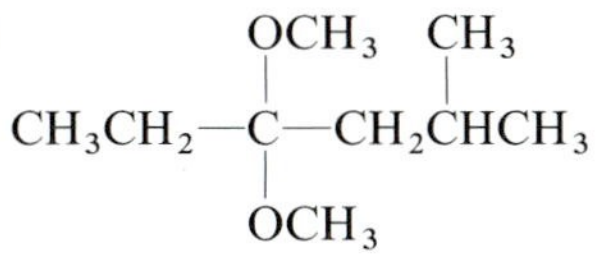   OCH₃   CH₃
                | |
    CH₃CH₂—C—CH₂CHCH₃
                |
                OCH₃

11.9 **(a)** 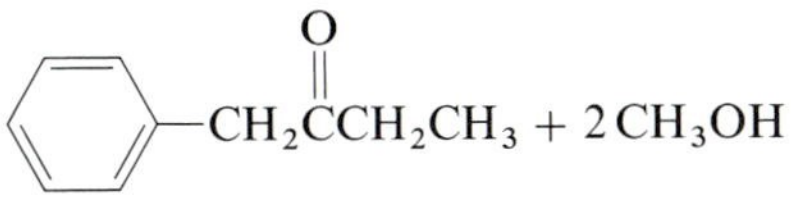   O
                            ‖
    [benzene ring]—CH₂CCH₂CH₃ + 2 CH₃OH

(b)

$$H-\overset{\overset{\displaystyle O}{\|}}{C}-H \ +$$

$$2\,CH_3CH_2CH_2OH$$

11.10 (a)

$$CH_3CH_2\overset{\overset{\displaystyle OH}{|}}{\underset{\underset{\displaystyle CH_3CH_2}{|}}{C}}-\overset{\overset{\displaystyle O}{\|}}{\underset{\underset{\displaystyle CH_3}{|}}{CH}}CH_2CH_3$$

(b)

$$\text{(phenyl)}-CH_2\overset{\overset{\displaystyle OH}{|}}{CH}-\overset{\overset{\displaystyle O}{\|}}{CH}CH\text{(phenyl)}$$

11.11 (a) and (b) cannot undergo aldol reactions.

Chapter 12

12.1 (a) 4-methylpentanoic acid
(b) isopropyl butanoate
(c) *N*-methyl *para*-chlorobenzamide

12.2 (a)

$$CH_3CH_2CH_2\overset{\overset{\displaystyle CH_3}{|}}{CH}CH_2\overset{\overset{\displaystyle O}{\|}}{C}OH$$

(b)

$$CH_3\overset{\overset{\displaystyle CH_3}{|}}{CH}CH_2CH_2\overset{\overset{\displaystyle O}{\|}}{C}NH_2$$

(c)

$$\text{(phenyl)}-\overset{\overset{\displaystyle O}{\|}}{C}OCH_2CH_2CH_3$$

(d)

$$\text{(phenyl with NO}_2)-\overset{\overset{\displaystyle O}{\|}}{C}-OH$$

(e)

$$CH_3CH_2CH_2\overset{\overset{\displaystyle O}{\|}}{C}-NHCH_3$$

(f)

$$CH_3CH_2\overset{\overset{\displaystyle O}{\|}}{C}OCH_2CH_3$$

12.3 (a)

$$CH_3CH_2CH_2\overset{\overset{\displaystyle O}{\|}}{C}-O^-\,K^+ \ + \ H_2O$$

(b)

$$\left(CH_3CH_2CH_2\overset{\overset{\displaystyle CH_3}{|}}{CH}\overset{\overset{\displaystyle O}{\|}}{C}-O^-\right)_2 Ba^{2+} + 2\,H_2O$$

12.4 1.6 g

12.5

$$H-\overset{\overset{\displaystyle O}{\|}}{C}-OCH_2\overset{\overset{\displaystyle CH_3}{|}}{CH}CH_3$$

12.6 (a)

$$\text{(cyclohexyl)}-OH \ +$$

$$HO-\overset{\overset{\displaystyle O}{\|}}{C}CH_2CH_2\overset{\overset{\displaystyle CH_3}{|}}{CH}CH_3$$

(b)

$$CH_3CH_2CH_2CH_2\overset{\overset{\displaystyle O}{\|}}{C}-OH \ +$$

$$HO\overset{\overset{\displaystyle CH_3}{|}}{CH}CH_3$$

12.7 (a)

$$CH_3\overset{\overset{\displaystyle H_3C}{|}}{CH}\overset{\overset{\displaystyle O}{\|}}{C}-OH \ +$$

$$HO\overset{\overset{\displaystyle CH_3}{|}}{CH}CH_3$$

(b)

$$CH_3CH=CH\overset{\overset{\displaystyle O}{\|}}{C}-OH \ +$$

$$HOCH_2CH_3$$

(c)

$$Br-\text{(phenyl)}-\overset{\overset{\displaystyle O}{\|}}{C}-OH \ +$$

$$HOCH_2CH_2CH_3$$

12.8 (a)

$$\text{(cyclopentyl)}-CH_2\overset{\overset{\displaystyle O}{\|}}{C}-\overset{\overset{\displaystyle O}{\|}}{\underset{\underset{\displaystyle \text{(cyclopentyl)}}{|}}{CH}}C-OCH_3$$

(b)

$$CH_3CH_2CH_2\overset{\overset{\displaystyle O}{\|}}{C}-\overset{\overset{\displaystyle O}{\|}}{\underset{\underset{\displaystyle CH_2CH_3}{|}}{CH}}C-OCH_3$$

12.9 Methyl benzoate has no hydrogens on the carbon atom next to the carbonyl group.

12.10 (a)

$$CH_3CHC-NHCH_3$$ (with H_3C and O above)

(b)

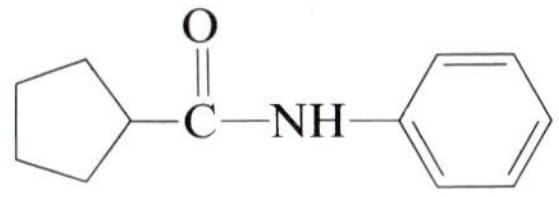

cyclopentyl–C(=O)–NH–phenyl

12.11 (a)

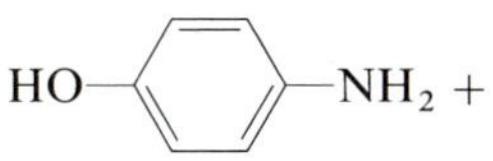

HO–⟨benzene⟩–NH_2 +

HO–CCH_3 (with O above)

12.12 (a)

$$CH_3CH=CHC-OH +$$

$$CH_3NH_2$$

(b)

Cl–⟨benzene⟩–$C-OH$ (with O above) +

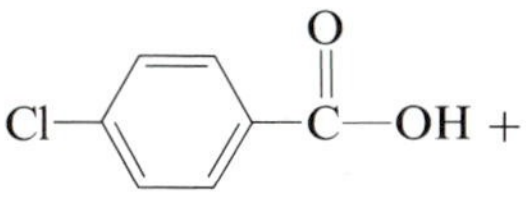

$$(CH_3CH_2)_2NH$$

Chapter 13

13.1 (a) aldopentose
(b) ketotriose
(c) aldotetrose
13.2 (a) and (c) are handed.
13.4 2-Propanol has only three different groups attached to C2, but 2-butanol has four different groups attached to C2.
13.6 (b) and (c) are chiral.

13.7

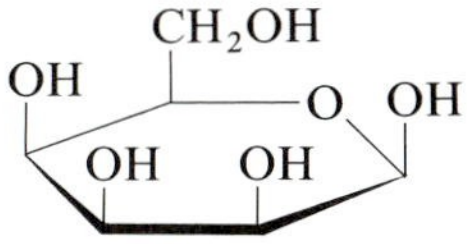

13.8

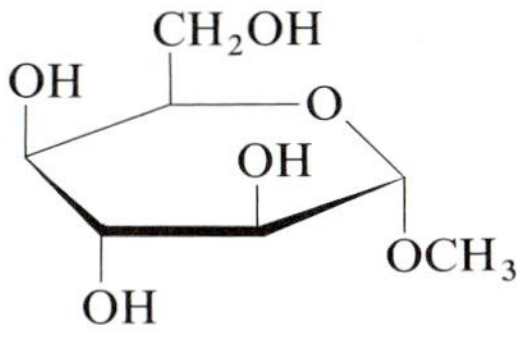

13.9

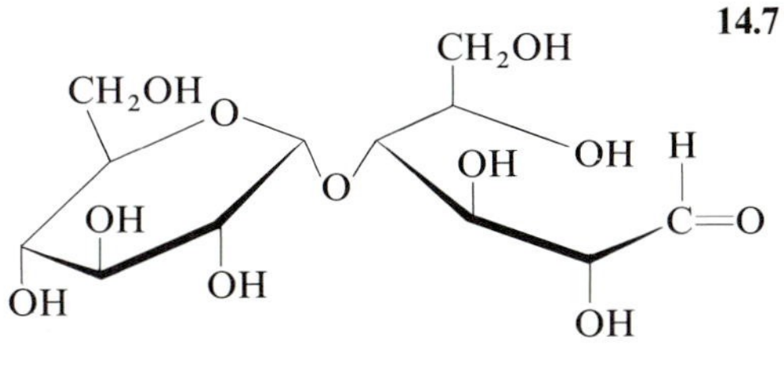

13.10 Cellobiose is a reducing sugar because the right-hand sugar has a hemiacetal linkage.
13.11 Cellobiose yields two molecules of β-D-glucose on hydrolysis.
13.12 Starch is not a reducing sugar.

Chapter 14

14.2

$$CH_3(CH_2)_{18}C-OCH_2(CH_2)_{30}CH_3$$ (with O above)

14.3

$$CH_2O-C(CH_2)_7CH=CH(CH_2)_7CH_3$$
$$CHO-C(CH_2)_7CH=CH(CH_2)_7CH_3$$
$$CH_2O-C(CH_2)_7CH=CH(CH_2)_7CH_3$$

14.4

$$CH_2O-C(CH_2)_7CH=CH(CH_2)_7CH_3$$
$$CHO-C(CH_2)_7CH=CH(CH_2)_7CH_3$$
$$CH_2O-C(CH_2)_7CH=CH(CH_2)_7CH_3$$
$$\xrightarrow[Pd]{3H_2}$$
$$CH_2O-C(CH_2)_{16}CH_3$$
$$CHO-C(CH_2)_{16}CH_3$$
$$CH_2O-C(CH_2)_{16}CH_3$$

14.5

$$\left(CH_3(CH_2)_7CH=CH(CH_2)_7C-O^-\right)_2 Ca^{2+}$$

14.6

$$CH_2-O-C-(CH_2)_{16}CH_3$$
$$CH-O-C-(CH_2)_7CH=CH(CH_2)_7CH_3$$
$$CH_2-O-P-O-CH_2CH_2\overset{+}{N}H_3$$
$$O^-$$

14.7

$$(CH_3)_3\overset{+}{N}CH_2CH_2-O-P-O-CH_2$$
$$O^-$$
$$CH-NH-C(CH_2)_{12}CH_3$$
$$CH-OH$$
$$CH=CH(CH_2)_{12}CH_3$$

14.8

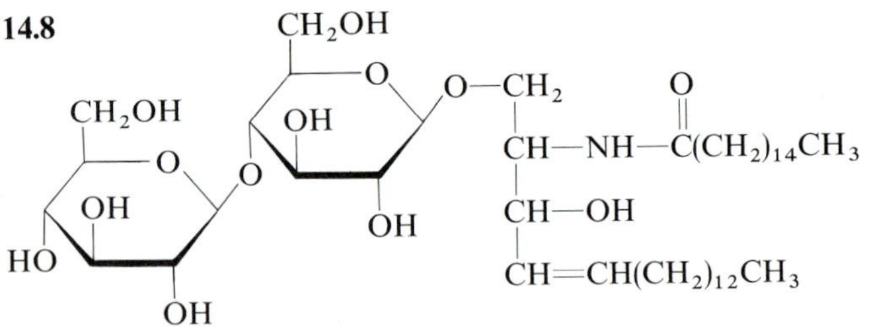

14.9 Estradiol and ethynylestradiol each have an aromatic ring.

Chapter 15

15.1 Phenylalanine, tryptophan, and tyrosine contain aromatic rings; cysteine and methionine contain sulfur; serine and threonine are alcohols; alanine, isoleucine, leucine, and valine have alkyl-group side chains.

15.2

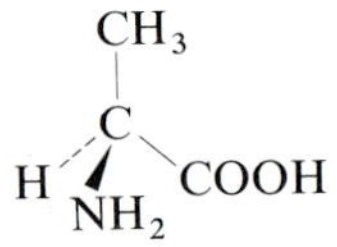

15.3 Alanine is chiral because there are four different groups attached to C2.

15.4 Threonine and isoleucine have two chiral carbon atoms.

15.5 Low pH:

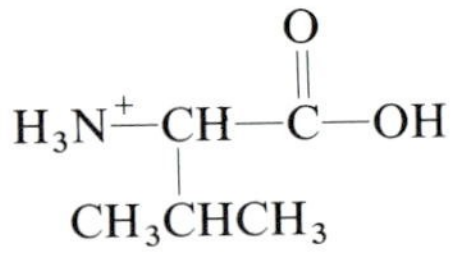

Neutral pH:

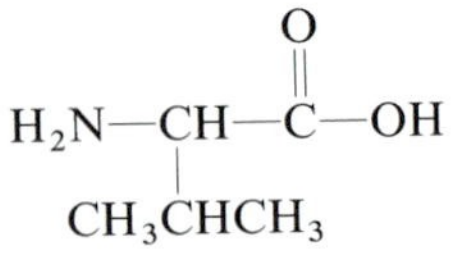

High pH:

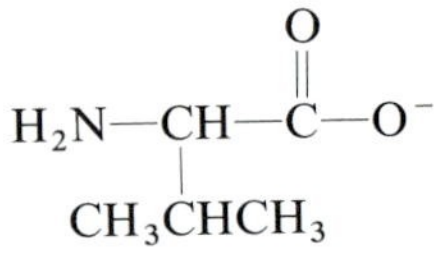

15.6 Val-Cys:

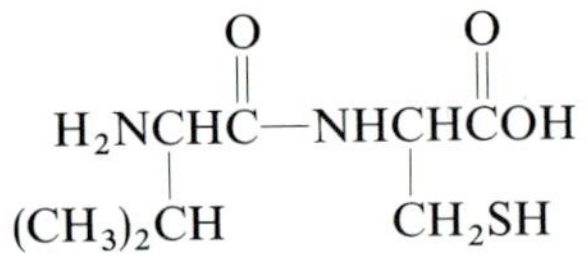

Cys-Val:

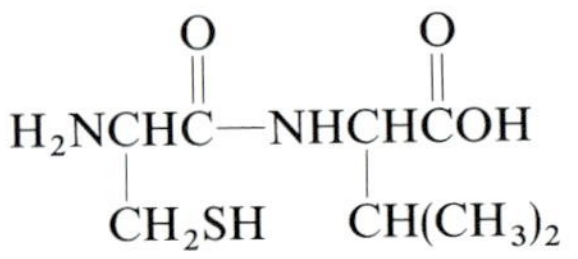

15.7 Val—Tyr—Gly,
Val—Gly—Tyr,
Tyr—Gly—Val,
Tyr—Val—Gly,
Gly—Val—Tyr,
Gly—Tyr—Val

15.8 The eight individual amino acids would be formed.

Chapter 16

16.2 **(a)** hydration of fumaric acid
(b) oxidation of squalene
(c) transfer of a phosphate group to glucose
(d) hydrolysis of cellulose

16.3 lyases

16.4 Vitamin D is a lipid that is structurally similar to steroids.

16.5 Vitamin A is a lipid, and vitamin C has many hydroxyl groups.

Chapter 17

17.1

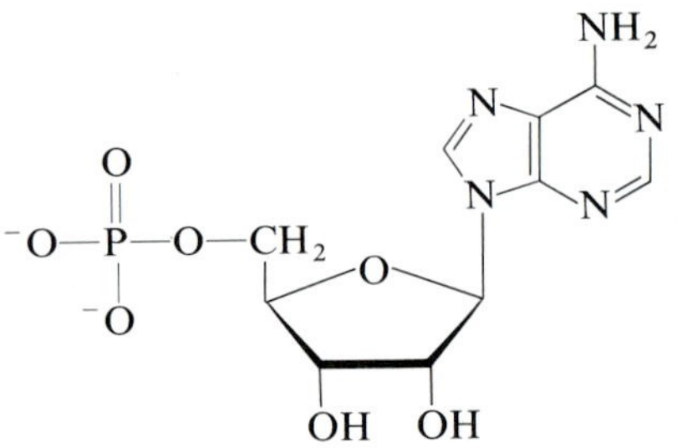

17.2

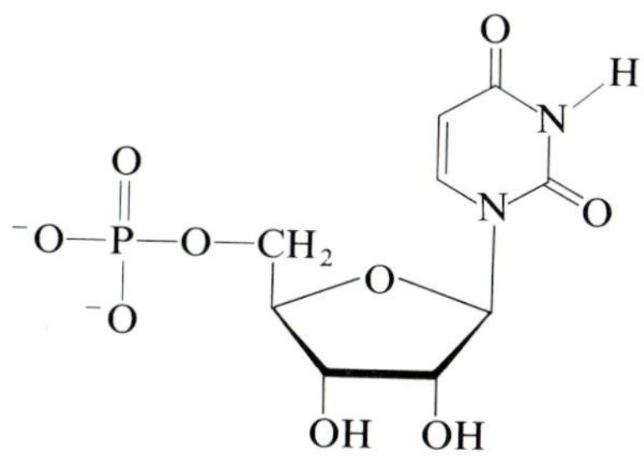

17.3

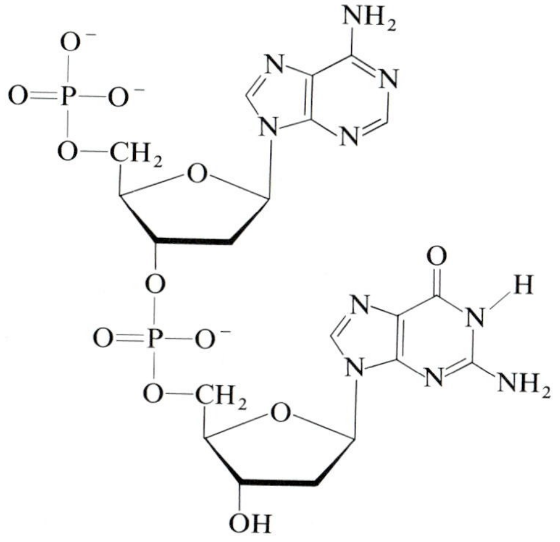

17.4 **(a)** C-G-G-A-T-C-A
(b) T-T-A-C-C-G-A-G-T

17.5 **(a)** C-U-A-A-U-G-G-C-A-U
(b) A-U-A-C-C-G-A-U-C-C-G-U

17.6 **(a)** GCU, GCC, GCA, GCG
(b) UUU, UUC
(c) UUA, UUG, CUU, CUC,
CUA, CUG
(d) GUU, GUC, GUA, GUG
(e) UAU, UAC

17.7 **(a)** Ile **(b)** Ala **(c)** Arg **(d)** Asn

17.8 Leu-Met-Ala-Trp-Pro

17.9 GAA-UAC-CGA-ACC-GGG-AUU

17.10 GAA-TAC-CGA-ACC-GGG-ATT

17.11 X-G-T-A-C-C-G-A-T-T-C-A-A-T-G-C

17.12 A cleavage: ^{32}P-G, ^{32}P-G-A-T, ^{32}P-G-A-T-A-C-G-G
G cleavage: ^{32}P-G-A-T-A-C, ^{32}P-G-A-T-A-C-G, ^{32}P-G-A-T-A-C-G-G-A-T-C
^{32}P-G-A-T-A-C-G-G-A-T-C-G
C cleavage: ^{32}P-G-A-T-A, ^{32}P-G-A-T-A-C-G-G-A-T
T cleavage: ^{32}P-G-A, ^{32}P-G-A-T-A-C-G-G-A

17.14 The informational DNA sequence T-G-A corresponds to the template DNA sequence A-C-T, which corresponds to the mRNA sequence U-G-A, which is a stop codon.

17.15 (a) stop becomes Tyr
(b) Gly becomes Ala

Chapter 18

18.1

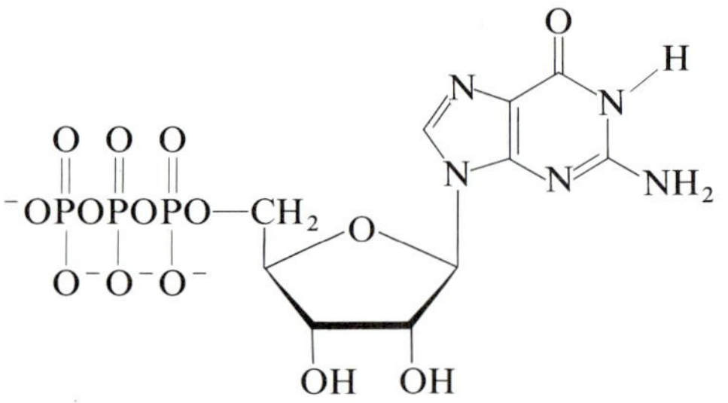

18.2

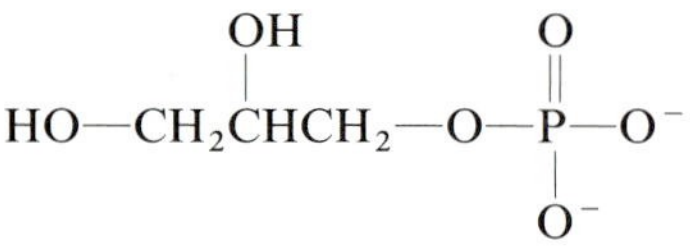

18.3 Activation energy is the difference between the energy level at the top of the curve and that of the starting material. Heat of reaction is the difference between the energy levels of starting material and product.

18.4 Citric acid and isocitric acid are tricarboxylic acids.

18.5

$$CH_3-\overset{O}{\overset{\|}{C}}-CoA + 2O_2 \longrightarrow$$
$$2CO_2 + H-CoA + H_2O$$

18.6 $C_6H_{12}O_6 + 6O_2 \longrightarrow$
$$6CO_2 + 6H_2O$$

18.7 (a) requires NAD^+
(b) requires FAD

18.8

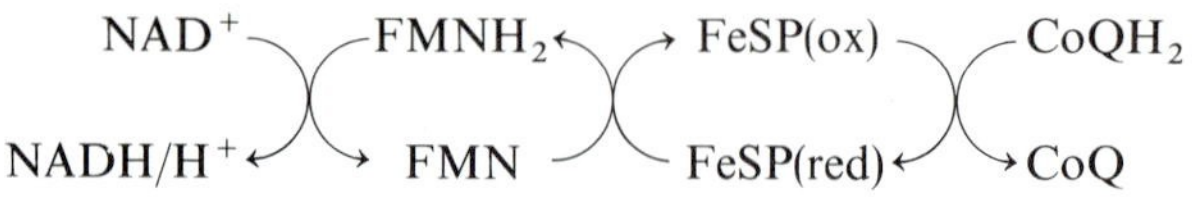

18.9 Each step that converts Fe^{2+} into Fe^{3+} is an oxidation; each step that converts Fe^{3+} into Fe^{2+} is a reduction.

Chapter 19

19.1 glycerin + 3 stearic acid

19.2 Insulin is digested in the stomach to its constituent amino acids.

19.3 $C_6H_{12}O_6 + O_2 \longrightarrow$
$$2C_3H_4O_3 + 2H_2O$$

19.4 3 mol ATP; 2.4 mol ATP

19.5 180 g ATP

19.6 (a) six acetyl CoA from five turns of the spiral
(b) ten acetyl CoA from nine turns of the spiral

19.7 129 ATP

19.8 260 g ATP

19.9

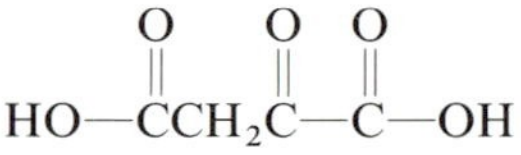

19.10 The oxaloacetic acid produced by oxidative amination of aspartic acid enters the citric-acid cycle.

19.11 seven rounds of the lipigenesis spiral; 7 ATP and 14 NADH

19.12 (a) alanine
(b) aspartic acid

Chapter 20

20.1 $^{218}_{84}Po$

20.2 $^{226}_{88}Ra$

20.3 $^{14}_{6}C \longrightarrow ^{0}_{-1}e + ^{14}_{7}N$

20.4 (a) $^{3}_{1}H \longrightarrow ^{0}_{-1}e + ^{3}_{2}He$

(b) $^{210}_{82}Pb \longrightarrow ^{0}_{-1}e + ^{210}_{83}Bi$

20.5 12%

20.6 4%

20.7 $^{237}_{93}Np$

20.8 $^{241}_{95}Am + ^{4}_{2}He \longrightarrow$
$$2^{1}_{0}n + ^{243}_{97}Bk$$

20.9 $^{40}_{18}Ar + ^{1}_{1}H \longrightarrow ^{1}_{0}n + ^{40}_{19}K$

20.10 $^{235}_{92}U + ^{1}_{0}n \longrightarrow$
$$^{137}_{52}Te + 2^{1}_{0}n + ^{97}_{40}Zr$$

Answers to Selected Review Problems

Chapter 1

1.19 A *physical quantity* is a physical property that can be measured, and consists of a number plus a unit.

1.21 *Mass* measures the amount of matter in an object; *weight* measures the gravitational force that a heavenly body exerts on an object.

1.23 kilogram (kg), cubic meter (m^3), meter (m)

1.25 Temperature in K = temperature in °C + 273.

1.27 *Specific heat* is a physical property since it can be determined without changing the chemical makeup of a substance.

1.29 10^9 pg; 3.5×10^4 pg

1.31 A quart is about 5% smaller than a liter: 1 quart = 0.9464 L

1.33 89 $\dfrac{km}{hr}$

1.35 **(a)** 550 mL = 0.55 L = 0.58 quarts
   (b) 3340 m = 334,000 cm = 3.34 km = 2.08 mi

1.37 10^{-2} L

1.39 102.7°F

1.41 −89.2°C; 184 K

1.43 4,370,000 ft^2

1.45 330 mm

1.47 2×10^{10} cells

1.49 **(a)** 7,926 miles; 7,900 miles; 7,926.38 miles
   (b) 7.926381×10^3 miles

1.51 **(a)** 2.586×10^3
   (b) 4.957500×10^6
   (c) 3.870×10^{-3}

1.53 **(a)** six
   (b) three
   (c) three
   (d) four
   (e) can't tell

1.55 39°C

1.57 0.7856 $\dfrac{g}{mL}$; 0.7856

1.59 554 g; 18.5 oz

1.61 A urinometer will float higher in chloroform than in ethanol.

1.63 0.313 oz (three significant figures)

1.65 4.7×10^4 people/mi^2

1.67 2×10^4 g

1.69 3.9×10^4 g

Chapter 2

2.11 **(a)** nitrogen
   (b) oxygen
   (c) potassium

   (d) chlorine
   (e) calcium
   (f) phosphorus
   (g) magnesium
   (h) manganese

2.13 **(a)** Only the first letter of a chemical symbol is capitalized. Bromine is thus Br.
   (b) The chemical symbol for manganese is Mn; Mg is the symbol for magnesium.
   (c) The symbol for carbon is C; Ca is the symbol for calcium.
   (d) The symbol for potassium is K; po (the "p" should be capitalized: Po) is the symbol for polonium.

2.15 **(a)** Fe
   (b) Cu
   (c) Co
   (d) Mo
   (e) Cr
   (f) F
   (g) S

2.17 proton, 1.007 amu, +1; neutron, 1.009 amu, 0; electron, 5.486 × 10^{-4} amu, −1

2.19 Both isotopes have six protons and six electrons; they differ only in the number of neutrons.

2.21 $^{131}_{53}I$

2.23 first shell: 2 electrons; second

shell: 8 electrons; third shell: 18 electrons

2.25 $P - 1s^2 2s^2 2p^6 3s^2 3p^3$

2.27 Magnesium (atomic number 2) has two electrons in its outer shell.

2.29 Americium (Am) is a metal.

2.31 By the reasoning used in Problem 2.8, you would expect elements in Group 4A to have four electrons in their outer shell.

2.33 Aluminum, which is below boron in the periodic table, is most similar to boron.

2.37 Helium, neon, argon, krypton, xenon, and radon make up the noble-gas family.

2.39 An *element* is a fundamental substance that can't be broken down or chemically changed. A *compound* is composed of two or more elements in definite proportions and can be broken down into its constituent elements. A *mixture* is a blend of substances, in variable proportions, in which the identities of each substance are unchanged.

2.41 Glycine contains ten atoms and is composed of carbon, hydrogen, nitrogen, and oxygen.

2.43 Penicillin V $= C_{16}H_{18}N_2O_5S$

2.47 One atom of carbon weighs more than one atom of hydrogen because the mass of carbon (12 amu) is greater than the mass of hydrogen (1 amu).

2.49 If 10^{23} hydrogen atoms weigh 1 gram, then 10^{23} carbon atoms will weigh 12 grams.

2.51 Strontium occurs directly below calcium in Group 2A and thus has similar chemical behavior.

2.53 The element is a nonmetal, belongs to Group 7A, and has 35 protons. It must therefore be chlorine.

2.55 Zirconium, a metal, has an electron configuration by shell of 2 8 18 10 2.

Chapter 3

3.22 All alkali metals have a single outer-shell s electron, which they can give up to attain a noble-gas configuration. The resulting ion has one less electron than protons, and is thus positively charged.

3.24 Strontium (atomic number 38) has 38 protons and 38 electrons. The strontium ion Sr^{2+} has 38 protons and 36 electrons.

3.25 Calcium $(1s^2 2s^2 2p^6 3s^2 3p^6 4s^2)$ can achieve a noble-gas configuration by losing its two $4s$ electrons.

3.26 Elements in Group 4A are more likely to form covalent compounds than to form ions.

3.29 The reducing agent is Sn, and the oxidizing agent is F_2. Tin loses two electrons, and each fluorine gains one electron.

3.31 (a) $+1$
(b) $+3$
(c) $+2$
(d) $+1$

3.33 The oxidation state of vanadium is $+5$.

3.35 (a) 4
(b) 2
(c) 3
(d) 1
(e) 2

3.37 Tellurium, a Group 6A element, has a covalence of 2 like oxygen, sulfur, and other members of the group.

3.40

3.43 Tetrachloroethylene has a carbon-carbon double bond.

3.45

3.47 A compound with the formula C_2H_8 cannot exist because any

structure drawn would violate the rules of covalence.

3.49

3.51 (a) $NaHCO_3$
(b) KNO_3
(c) $CaCO_3$

3.53 Calcium ion has a charge of $+2$, and phosphate ion has a charge of -3. To form a neutral compound, two phosphate ions (total charge of -6) must bond with three calcium ions (total charge $+6$). Thus, $Ca_3(PO_4)_2$ is the correct formula.

3.57 Either of the two following structures is possible. The first structure is the actual structure for dimethylamine.

Dimethylamine

3.58 Since each oxygen has an oxidation state of -2, four oxygens have a total charge of -8. In order for the permanganate ion to have a charge of -1, the oxidation state of manganese must be $+7$.

3.60 The formula of potassium permanganate is $KMnO_4$.

3.62 (a) $MoCl_5$
(b) $CoBr_3$
(c) Sc_2O_3

3.64 Number of protons: 12 (from carbon) + 16 (from oxygen) + 3 (from hydrogen) = 31
Number of electrons: 8 (inner shell of C and O) + 14 (in bonds) + 10 (in lone pairs) = 32
Acetate ion is negatively charged because it has one more electron than proton.

3.65 Since the charge on Na^+ is $+1$, the charge of hypochlorite ion must be -1. A Lewis structure for the hypochlorite ion is:

$$:\!\ddot{C}\!l\!:\!\ddot{O}\!:^-$$

Chapter 4

4.19 3 Moles of H_2 react with 1 mole of N_2.
9 Moles of H_2 react with 3 moles of N_2.
1 Mole of N_2 reacts with 3 moles of H_2.

4.21 $2\,NaHCO_3 + H_2SO_4 \longrightarrow$
$\quad 2\,CO_2 + Na_2SO_4 + 2\,H_2O$

4.23 $6\,CO_2 + 6\,H_2O \longrightarrow$
$\quad C_6H_{12}O_6 + 6\,O_2$

4.25 A mole of a substance is its formula weight in grams, and has 6.02×10^{23} molecules.

4.27 Diazepam ($C_{16}H_{13}ClN_2O$) has formula weight = 284.5 amu

4.29 **(a)** 0.0641 mol
(b) 0.0595 mol
(c) 0.0418 mol
(d) 0.0143 mol

4.31 151.8 amu; 1.98×10^{-3} mol

4.33 1.8×10^{23} molecules;
6.84×10^{22} molecules

4.35 138 g

4.37 194.0 amu

4.39 2.78×10^{-3} mol

4.41 A molecule has "stability" if it contains a relatively small amount of potential energy and is therefore unreactive.

4.43 The heat of reaction is the exact amount of heat released during a chemical reaction.

4.45 To increase the rate of a chemical reaction, you can (1) increase the reaction temperature, (2) increase the concentration of the reactants, or (3) add a catalyst.

4.47 A smaller E_{act} indicates a lower energy barrier and a faster reaction. Thus, the reaction with $E_{act} = 5$ kcal/mol is faster than the one with $E_{act} = 10$ kcal/mol.

4.51 **(a)** $C_6H_{12}O_6 + 6\,O_2 \longrightarrow$
$\quad 6\,CO_2 + 6\,H_2O) + $ heat
(b) 41 kcal

(c) 180 amu
(d) 738 kcal

4.53 $327 \dfrac{kcal}{mol}$

4.55 56.3 g

4.57 $CH_4 + 2\,Cl_2 \longrightarrow$
$\quad CH_2Cl_2 + 2\,HCl$

4.59 29.9 g; Cl_2 is the limiting reagent.

4.61 4.63 g

4.63 5.6×10^{14} molecules

Chapter 5

5.17 **(a)** 760 mm Hg
(b) 190 mm Hg
(c) 5.7×10^3 mm Hg
(d) 711 mm Hg

5.19 92.6 mm Hg

5.25 22.4 L

5.27 $PV = nRT$

5.29 178 atm

5.31 29.4 atm

5.33 1.0 L of O_2 weighs more than 1.0 L of H_2 because the formula weight of O_2 is greater than the formula weight of H_2, and equal numbers of moles of both are present.

5.35 The CO_2 sample has slightly fewer molecules (3.2×10^{22} molecules CO_2, 3.3×10^{22} molecules N_2). Since the formula weight of CO_2 is much greater than the formula weight of N_2, however, the CO_2 sample weighs more than the N_2 sample.

5.37 63 mL

5.39 41 K

5.41 0.048 mol

5.43 50 kg N_2; 63 Kg air

5.45 Molecules in a liquid are free to move about. Unlike the molecules in a gas, however, the molecules in a liquid are in constant contact with one another.

5.47 Increased pressure raises a liquid's boiling point; decreased pressure lowers a liquid's boiling point.

5.49 When a liquid and a gas are in equilibrium, the number of gas molecules escaping the liquid equal the number of molecules reentering the liquid.

5.51 In a solid, molecules have a fixed geometrical arrangement; in a liquid, molecules are free to move about.

5.53 two moles; 5.0 L

5.55 0.0094 mol

5.57 0.0079 mol; 0.25 g

5.59 At STP, equal volumes of gases have an equal number of moles. However, O_2 has a greater formula weight than H_2, and the vessel containing O_2 will therefore be heavier.

5.61 6×10^{15} L; $10^8 \dfrac{atoms}{L}$

5.63 52 atm

Chapter 6

6.21 The solubility of a gas in a liquid is lower at the higher temperature, and more gas escapes.

6.23 The powdered salt has a much greater surface area than the block of salt and thus comes into greater contact with water and dissolves faster.

6.25 Add water to the mixture. The insoluble aspirin settles out and can be removed by filtration, and the salt can be recovered by evaporation of the water. Alternatively, add chloroform to the mixture and filter off the insoluble salt. The aspirin can then be recovered by evaporating the chloroform.

6.29 400 mL

6.31 Dissolve 2.5 g $B(OH)_3$ in water to a final volume of 500 mL.

6.33 Dissolve 1.5 g $NaCl$ in water to a final volume of 50 mL.

6.35 Dissolve 75 g KBr in water and dilute to 1.0 L.

6.37 **(a)** $6.7\% \ (w/v)$
(b) $4.3\% \ (w/v)$

6.39 230 mL; 1640 mL

6.41 4.5 g

6.43 **(a)** 0.060 mol
(b) 0.38 mol
(c) 1.88 mol

6.45 5.3 mL

6.47 0.38 mL

6.49 If the concentration of Ca^{2+} is 3.0 mEq/L, there are 3.0 mmol of charges due to calcium per liter of blood. Since calcium has a charge of $+2$, there are 1.5 mmol of calcium per liter of blood.

6.51 0.040 Eq

6.52 0.355 g

6.53 The inside of a red blood cell contains dissolved substances and therefore has a higher osmolarity than pure water. Water thus passes through the cell membrane to dilute the inside until pressure builds up and the cell eventually bursts.

6.55 Distilled water separated from a 1.0 M NaCl solution by an osmotic membrane would pass through the membrane to dilute the NaCl until the maximum osmotic pressure of the solution was reached.

6.57 (a) 0.40 M Na_2SO_4
(b) 30% (w/v) NaOH

6.59 Since a 0.40 M NaCl solution has a higher osmolarity than a 0.40 M glucose solution, water would pass through the osmotic membrane from the glucose solution to the NaCl solution until the two were of equal osmolarity.

6.61 9.4 mL

6.63 (a) Dilute 8.3 mL of 12.0 M HCl to a volume of 250 mL.
(b) Dilute 6.7 mL of 12.0 M HCl to a volume of 1.6 L.

6.65 NaCl, 0.147 M; KCl, 0.0040 M; $CaCl_2$, 0.0030 M

6.67 50 mL

6.69 $CuSO_4 \cdot 5H_2O$

Chapter 7

7.15 In water, HBr dissociates almost completely. Water acts as a base to accept the proton, and a solution of H_3O^+ and Br^- results.

7.17 In water, KOH dissociates completely to yield K^+ and OH^- ions.

7.19 Strong acids:
HCl, HBr, HNO_3, H_2SO_4

Weak acids:
CH_3COOH, H_2CO_3

7.21 All aqueous acid solutions contain the hydronium ion, H_3O^+.

7.23 $2HCl(aq) + CaCO_3(s) \longrightarrow$
$H_2O\,(l) + CO_2\,(g) + CaCl_2(aq)$

7.25 Citric acid reacts with sodium bicarbonate to release CO_2 bubbles: $C_6H_5O_7H_3(aq) +$
$3NaHCO_3(aq) \rightarrow C_6H_5O_7Na_3$
$(aq) + 3H_2O\,(l) + 3CO_2(g)$

7.27 When H_3O^+ concentration is expressed as a power of 10, pH is defined as the negative of the exponent of the 10. In mathematical terms, pH is the negative logarithm of H_3O^+ concentration.

7.29 Indicator paper undergoes a series of color changes that correspond to particular pH values. When a drop of test solution is placed on the paper, the color that appears indicates the pH of the solution.

7.31 Water dissociates to give H_3O^+ and OH^- in equal concentrations. Since the concentrations of both acid and base are equal, water is neutral.

7.33 K_w is the product of the concentrations of H_3O^+ and OH^- in any aqueous solution and is numerically equal to 10^{-14}.

7.35 Since the pH of a 0.10 N HCN solution is lower than 7.0, the solution is acidic. If HCN were a strong acid, the pH of the 0.10 N solution would be 1 (Problem 7.34). Thus, HCN is a weak acid.

7.37 (a) $[OH^-] = 10^{-10}$ M
(b) $[OH^-] = 10^{-3}$ M
(c) $[OH^-] = 10^{-14}$ M

7.39 (a) pH = 7.6
(b) pH = 3.3

7.41 (a) 4.5, colorless
(b) 11.2, red
(c) 7.1, pink

7.45 $CH_3COO^-Na^+(aq) + HNO_3(aq)$
$\longrightarrow CH_3COOH(aq)$
$+ NaNO_3(aq).$

7.49 Since one equivalent of any base neutralizes one equivalent of any acid, 0.035 equivalents of NaOH neutralize 0.035 equivalents of H_3PO_4.

7.51 Dilute 2.1 mL of 12.0 M HCl to a final volume of 250 mL.

7.53 $[H_3O^+] = 10^{-12}$ M; $[OH^-] = 10^{-2}$ M

7.55 0.68 g

7.57 (a) 0.50 eq
(b) 0.084 eq
(c) 0.25 eq

7.59 0.34 N

7.61 8.0 mL

7.63 42 mL

7.65 0.35 M

7.69 0.68 N

Chapter 8

8.15 The bonds in organic compounds are covalent bonds.

8.17 Functional groups are groups of atoms that have a characteristic chemical behavior. They are important because most of the chemistry of organic compounds is determined by functional groups. The reactivity of a functional group is similar in all compounds in which it occurs.

8.19 One way is to put a small amount of each sample in a test tube. The two liquids do not mix, and the less dense hexane will lie on top. Alternatively, you might add a small amount of NaCl to each liquid. The salt does not dissolve in hexane.

8.22 (a)

Menthol

(b)

Aspirin

8.23 There are several possible answers to these questions. For example:

(a)

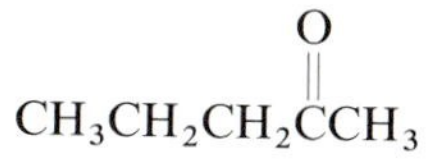

$$CH_3CH_2CH_2\overset{\displaystyle O}{\overset{\|}{C}}CH_3$$

A ketone

(b)

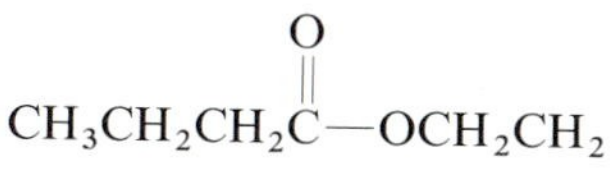

$$CH_3CH_2CH_2\overset{\displaystyle O}{\overset{\|}{C}}-OCH_2CH_2$$

An ester

(c)

$$H_2NCH_2\overset{\displaystyle O}{\overset{\|}{C}}-OH$$

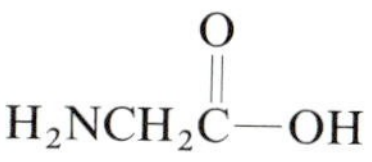

An amine-acid

8.26 In a straight-chain alkane, all carbons are connected in a row and it is possible to draw a path connecting them without retracing or lifting your pen from the paper. A branched-chain alkane has one or more carbon branches, and it is not possible to draw a path connecting all carbons without either retracing your path or lifting your pen from the paper.

8.28 Compounds with the formulas C_5H_{10} and C_4H_{10} are not isomers because they do not have the same molecular formulas.

8.30 A compound can't have a quintary carbon because carbon forms only four bonds, not five.

8.32 $CH_3CH_2CH_2OH$

1-Propanol

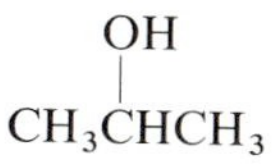

$$\overset{\displaystyle OH}{\underset{}{CH_3CHCH_3}}$$

2-Propanol

$$CH_3CH_2-O-CH_3$$

Ethylmethyl ether

8.34 Since carbon forms only four bonds, the largest number of hydrogens that can be bonded to three carbons is eight: C_3H_8.

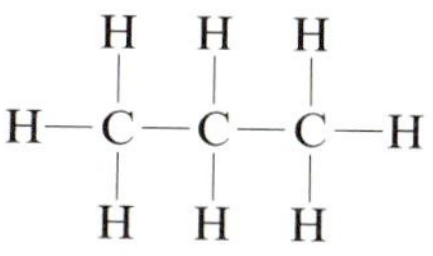

$$\begin{array}{ccc} H & H & H \\ | & | & | \\ H-C-C-C-H \\ | & | & | \\ H & H & H \end{array}$$

8.35 (a) $CH_3CH_2CH_2CH_2OH$

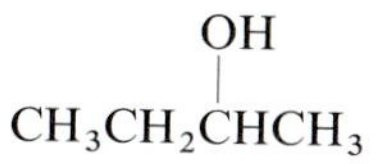

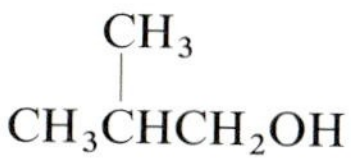

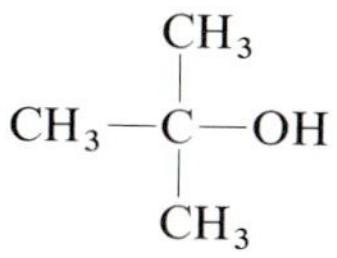

$$\overset{\displaystyle OH}{\underset{}{CH_3CH_2CHCH_3}}$$

$$\overset{}{\underset{\displaystyle CH_3}{CH_3CHCH_2OH}}$$

$$\overset{\displaystyle CH_3}{\underset{\displaystyle CH_3}{CH_3-C-OH}}$$

(c)

$$CH_3CH_2CH_2\overset{\displaystyle O}{\overset{\|}{C}}CH_3$$

$$CH_3CH_2\overset{\displaystyle O}{\overset{\|}{C}}CH_2CH_2$$

$$CH_3\overset{\displaystyle O}{\overset{\|}{\underset{\displaystyle CH_3}{CHCCH_3}}}$$

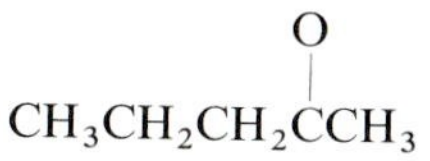

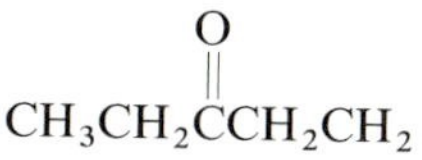

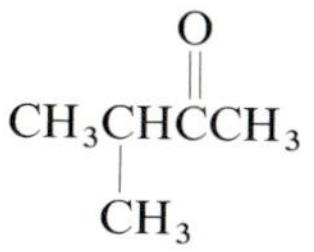

8.36 (a) identical **(d)** isomers
(b) isomers **(e)** isomers
(c) unrelated

8.37 All three structures have a carbon atom with five bonds; not allowed!

8.38 (a) The first two are identical; the third is an isomer.

8.39 (a) 4-Ethyl-3-methyloctane
(b) 5-Isopropyl-3-methyloctane

8.40 (a)

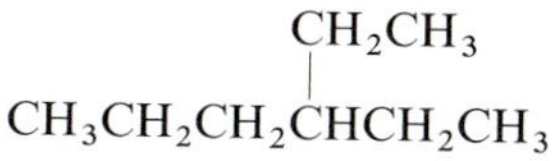

$$\overset{\displaystyle CH_2CH_3}{\underset{}{CH_3CH_2CH_2CHCH_2CH_3}}$$

(b)

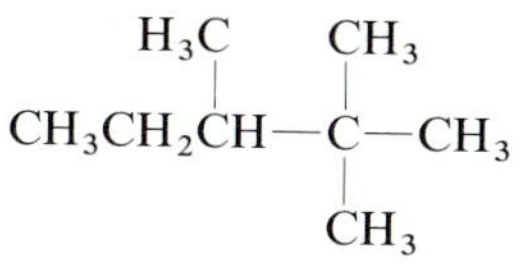

$$\overset{\displaystyle H_3C}{\underset{}{CH_3CH_2CH}}-\overset{\displaystyle CH_3}{\underset{\displaystyle CH_3}{C}}-CH_3$$

8.41

$$\overset{\displaystyle CH_3}{\underset{\displaystyle CH_3}{CH_3-C}}-\overset{\displaystyle CH_3}{\underset{}{CH_2CHCH_3}}$$

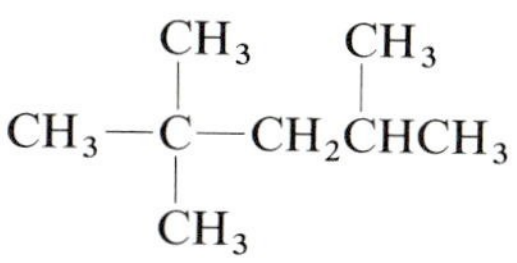

2,2,4-Trimethylpentane

8.43 (a)

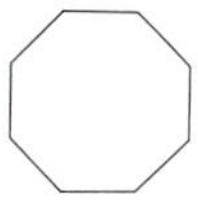

(b) H_3C CH_3

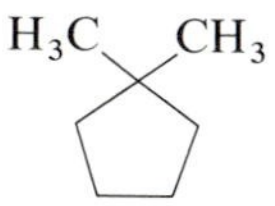

8.44 (a)

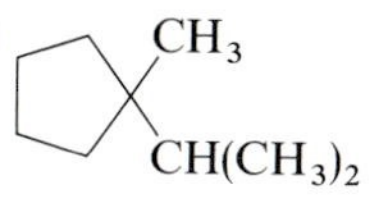

(b)

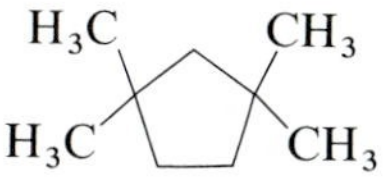

Chapter 9

9.17 The term "aromatic" in chemistry refers to compounds that have a six-membered ring containing three double bonds. The association with aroma is of historical origin but no longer has any meaning.

9.19 alkene: *-ene*; alkyne: *-yne*; aromatic compound: *-benzene*

9.21 (a) 1-Pentene
(b) 5-Methyl-2-hexyne
(c) 2,3-Dimethyl-2-butene

9.23 $CH_3CH_2CH_2C{\equiv}CH$

1-Pentyne

$$CH_3CH_2C{\equiv}CCH_3$$

2-Pentyne

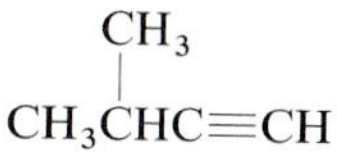

$$\overset{\displaystyle CH_3}{\underset{}{CH_3CHC{\equiv}CH}}$$

3-Methyl-1-butyne

9.27

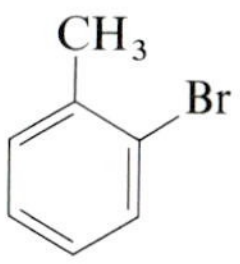

o-Bromotoluene

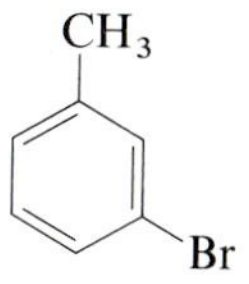

m-Bromotoluene

p-Bromotoluene

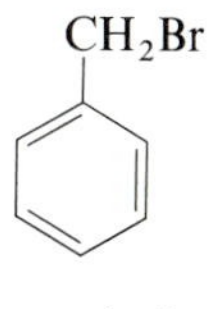

Bromomethylbenzene

9.31 Alkynes don't show cis-trans isomerism because only one group is bonded to each alkyne carbon.

9.34 (a)

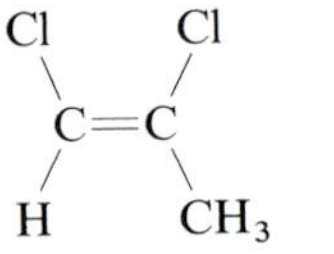

Cl, Cl bonded to C=C, H and CH₃

9.35 (a) These compounds are identical.

9.39 (a)

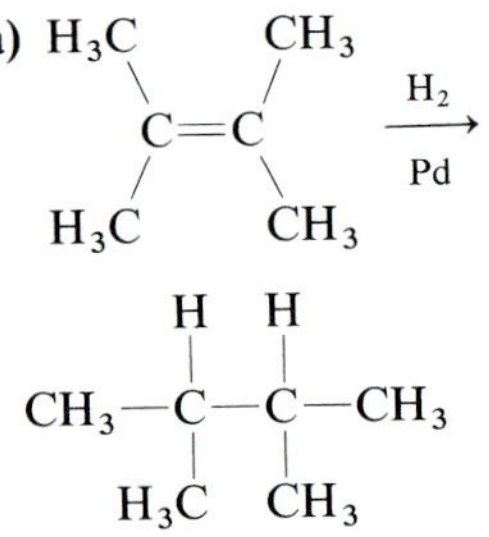

H₃C, CH₃ / C=C / H₃C, CH₃ —H₂, Pd→

CH₃—C(H)(H₃C)—C(H)(CH₃)—CH₃

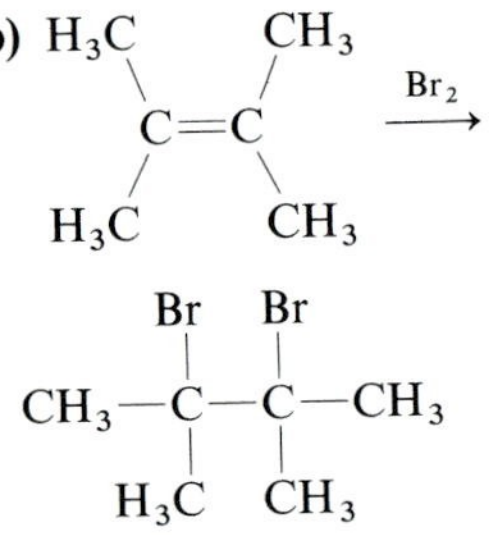

(b)

H₃C, CH₃ / C=C / H₃C, CH₃ —Br₂→

CH₃—C(Br)(H₃C)—C(Br)(CH₃)—CH₃

(c)

H₃C, CH₃ / C=C / H₃C, CH₃ —HBr→

CH₃—C(H)(H₃C)—C(Br)(CH₃)—CH₃

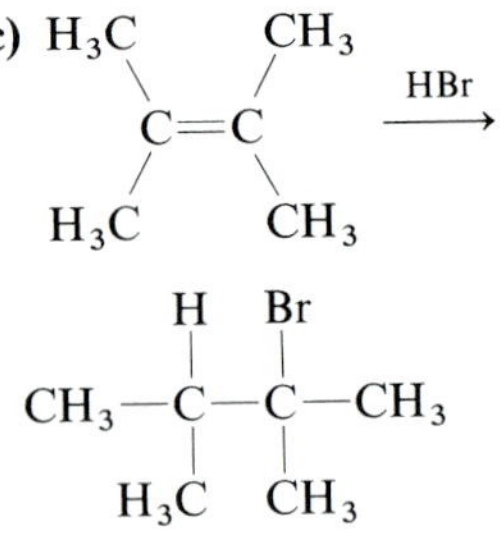

(d)

H₃C, CH₃ / C=C / H₃C, CH₃ —H₂O, H₂SO₄→

CH₃—C(H)(H₃C)—C(OH)(CH₃)—CH₃

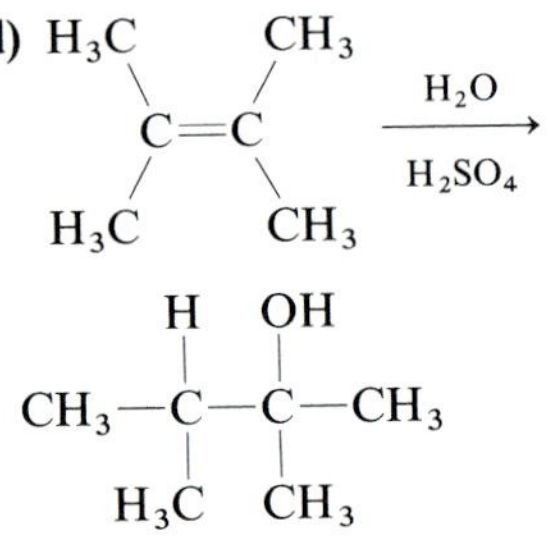

9.43 Under conditions where an alkene would react with all four reagents, benzene reacts only with Br_2 (b) to give

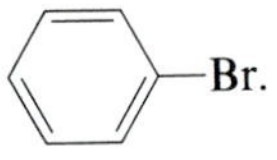

benzene—Br.

9.44 (a)

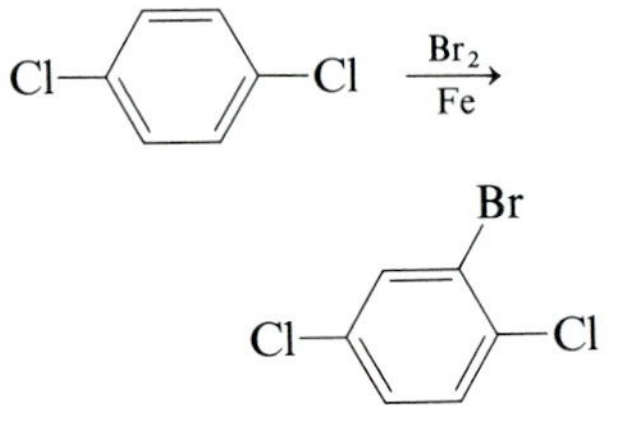

Cl—C₆H₄—Cl —$\frac{Br_2}{Fe}$→ Cl,Cl-substituted benzene with Br

(b)

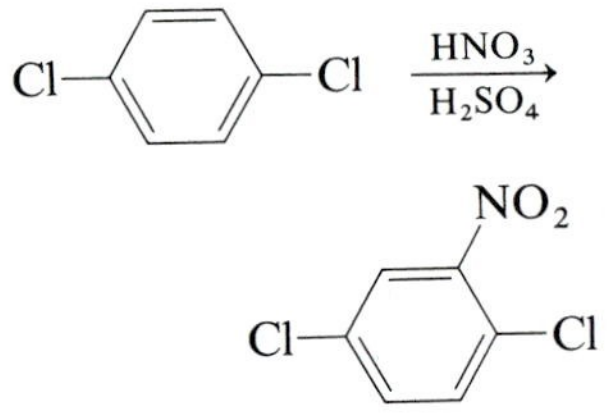

Cl—C₆H₄—Cl —$\frac{HNO_3}{H_2SO_4}$→ Cl,Cl-substituted benzene with NO_2

9.47

O_2N—C₆H₂(—NO_2)(—CH_3)(—NO_2)

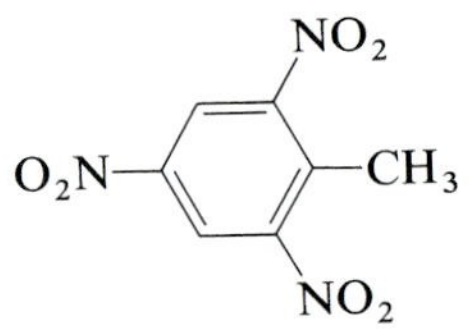

9.49 None of the reagents that add to alkenes react with alkanes like cyclohexane. Thus, you could use one of these reagents, say Br_2, on a sample from each bottle. The cyclohexene reacts with the Br_2, but the cyclohexane does not.

9.51

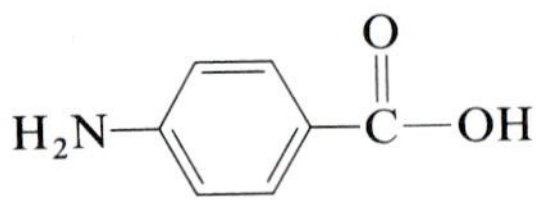

H_2N—C₆H₄—$\overset{O}{\underset{\|}{C}}$—OH

9.53

C₆H₅—CH_2—$CH_2$$\overset{O}{\underset{\|}{C}}$—H

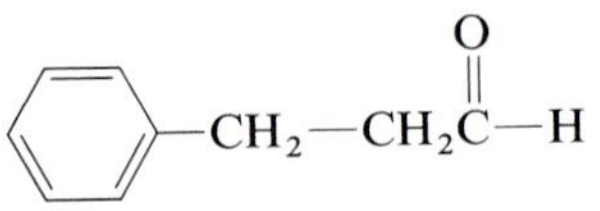

3-Phenylpropanal

Chapter 10

10.16 Alcohols contain —OH groups, which can form hydro-

gen-bonds to each other. Since extra energy (heat) must be supplied to break these hydrogen bonds, alcohols are higher boiling than ethers, which can't hydrogen bond.

10.21 (a) 2-Ethyl-1-pentanol
(b) 3-Methyl-1-butanol
(c) 1,2,4-Butanetriol

10.22 (a)

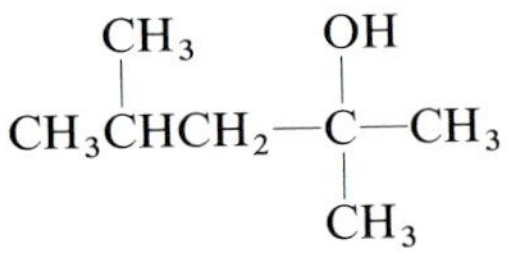

CH_3CHCH_2—C(—CH_3)(—OH)—CH_3 with CH_3 below

(c)

CH_3CH_2—C(—CH_2CH_3)(—CH_2CH_3)—$CH_2CH_2CH_2CH_2OH$

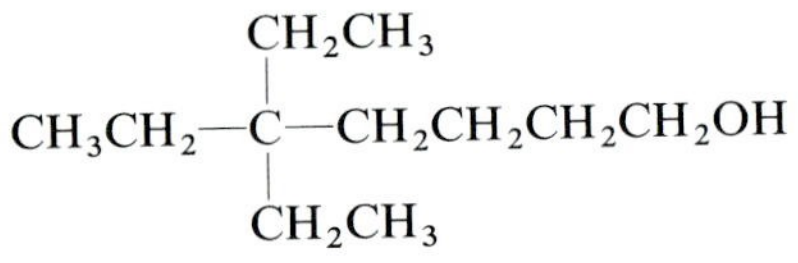

10.24 (b) Isopropyl methyl ether

10.25 (a)

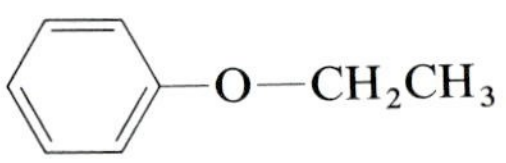

C₆H₅—O—CH_2CH_3

10.29 Tertiary alcohols are not oxidized because they have no —H bonded to the —OH bearing carbon.

10.31 (a) least soluble, forms no hydrogen bonds with water
(b) intermediate, forms some hydrogen bonds with water
(c) most soluble, forms most hydrogen bonds with water

10.33 The simplest way to distinguish the two alcohols is to try to oxidize them. The tertiary alcohol is unreactive toward oxidizing agents, but the secondary alcohol would be converted to a ketone.

10.35

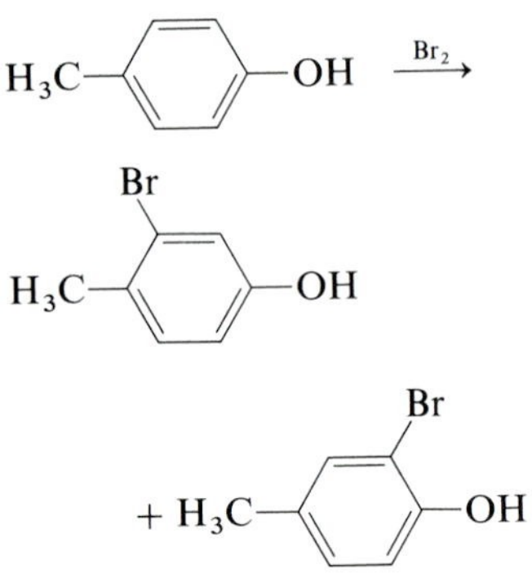

H_3C—C₆H₄—OH —Br_2→

H_3C—C₆H₃(Br)—OH

+ H_3C—C₆H₃(Br)—OH

10.36 (b)

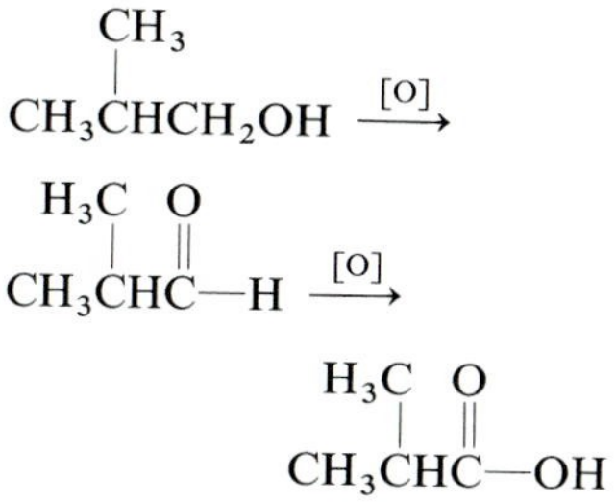

$$CH_3CHCH_2OH \xrightarrow{[O]}$$
(with CH_3 on the CH)

$$CH_3CHC{-}H \xrightarrow{[O]}$$
(with H_3C and O)

$$CH_3CHC{-}OH$$
(with H_3C and O)

10.37 (c)

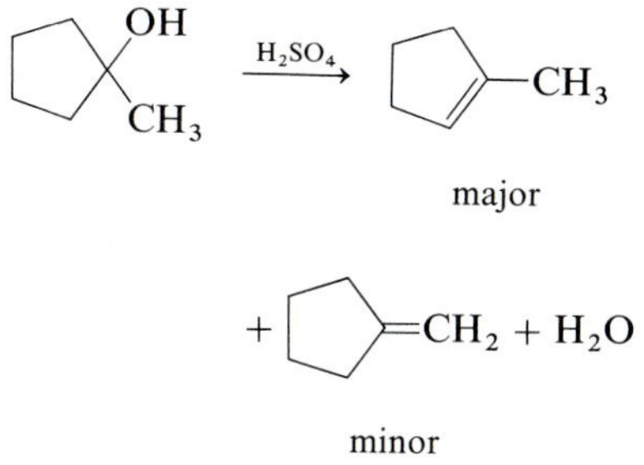

major

$$+ \quad {=}CH_2 + H_2O$$

minor

10.41 (b)

$$CH_3CH_2{-}\overset{H}{\underset{}{N}}{-}CH_2CH_3$$

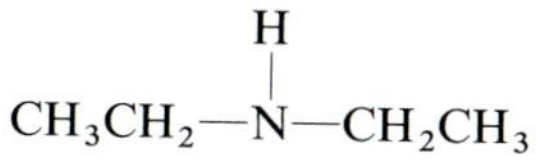

10.42 (b) Isopropylamine
10.45 a tertiary amine
10.48

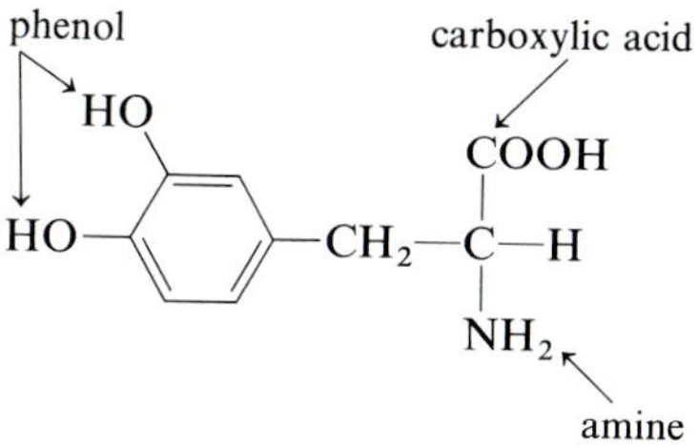

phenol

carboxylic acid

amine

10.51

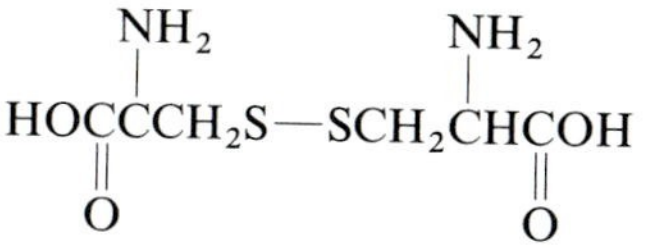

$$HOC\overset{NH_2}{\underset{O}{\overset{|}{C}}}CH_2S{-}SCH_2\overset{NH_2}{\underset{O}{\overset{|}{C}}}HCOH$$

Chapter 11

11.16 There are many possible answers to this question.

(a)

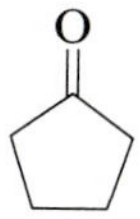

(b)

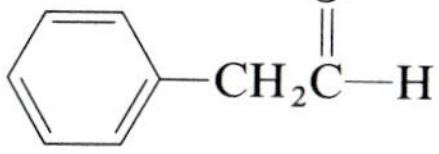

$$-CH_2\overset{O}{\overset{\|}{C}}{-}H$$

11.18 (a) aldehyde
(b) ketone
(c) no carbonyl

11.20 (a)

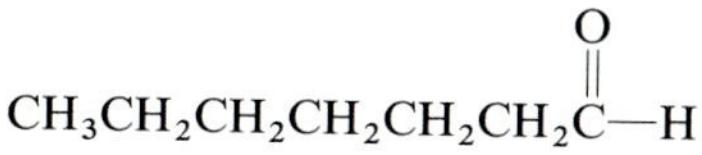

$$CH_3CH_2CH_2CH_2CH_2CH_2\overset{O}{\overset{\|}{C}}{-}H$$

11.21 (a)

$$CH_3CH_2CH_2CH_2\overset{O}{\overset{\|}{C}}CH_2CH_3$$

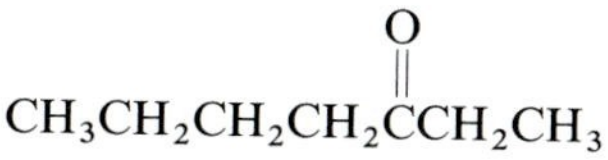

11.22 (a) 2-Methylbutanal
11.23 (a) 2-Butanone
11.27 Tollens' regent oxidizes an aldehyde to a carboxylic acid:

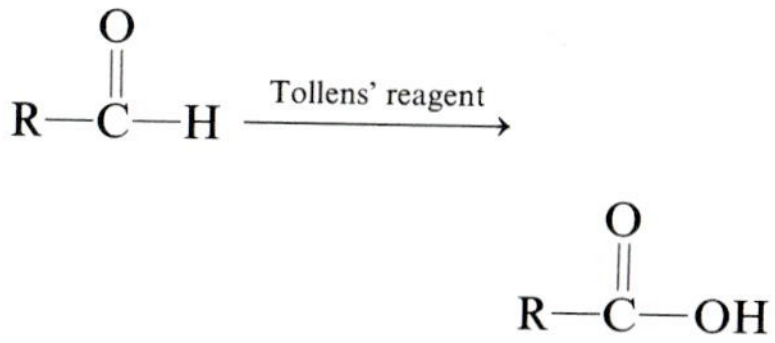

$$R{-}\overset{O}{\overset{\|}{C}}{-}H \xrightarrow{\text{Tollens' reagent}}$$

$$R{-}\overset{O}{\overset{\|}{C}}{-}OH$$

11.28 An aldehyde (RCHO) has only one organic group attached to the carbonyl carbon. When the aldehyde is reduced, it yields a primary alcohol (RCH$_2$OH), which also has only one organic group attached to carbon. A ketone (R$_2$C=O) has two organic groups attached to carbon and therefore yields a secondary alcohol (R$_2$CHOH) when it is reduced.

11.30 (a) no reaction
(b)

$$CH_3CH_2CH_2CH_2CH_2\overset{O}{\overset{\|}{C}}{-}OH + Ag$$

(c)

$$CH_3CH_2CH_2\overset{COOH}{\underset{}{\overset{|}{C}}}HCH_2CH_3 + Ag$$

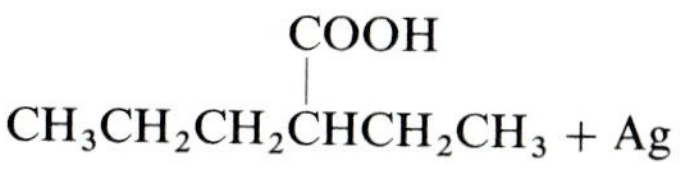

11.31 (b)

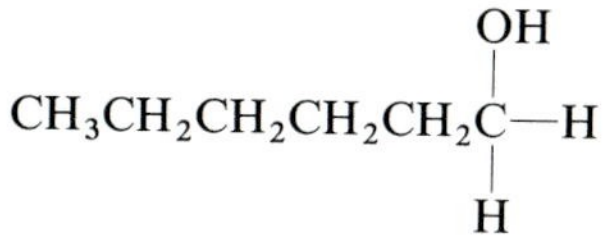

$$CH_3CH_2CH_2CH_2CH_2\overset{OH}{\underset{H}{\overset{|}{C}}}{-}H$$

11.34 (a)

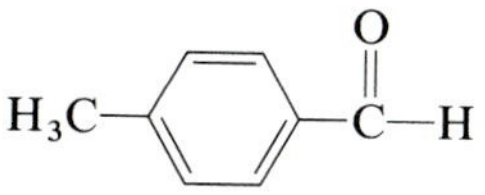

$$H_3C{-} \quad {-}\overset{O}{\overset{\|}{C}}{-}H$$

11.35 (a)

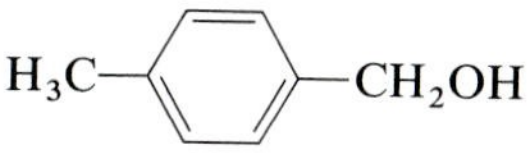

$$H_3C{-} \quad {-}CH_2OH$$

11.36 (a)

$$CH_3CH_2CH_2\overset{CHO}{\underset{}{\overset{|}{C}}}HCH_3$$

11.37 (a)

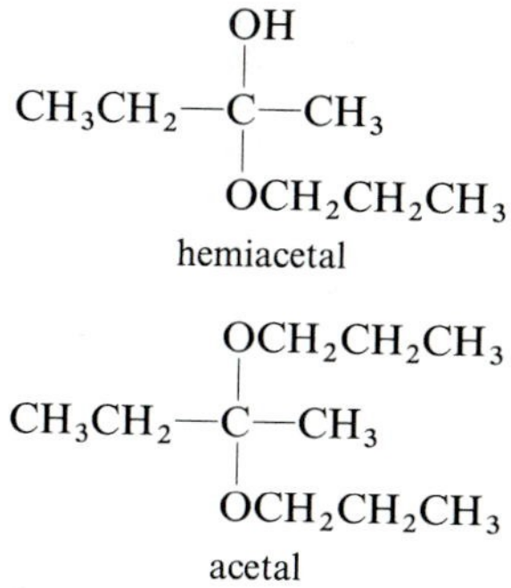

$$CH_3CH_2{-}\overset{OH}{\underset{OCH_2CH_2CH_3}{\overset{|}{C}}}{-}CH_3$$

hemiacetal

$$CH_3CH_2{-}\overset{OCH_2CH_2CH_3}{\underset{OCH_2CH_2CH_3}{\overset{|}{C}}}{-}CH_3$$

acetal

11.38 (a)

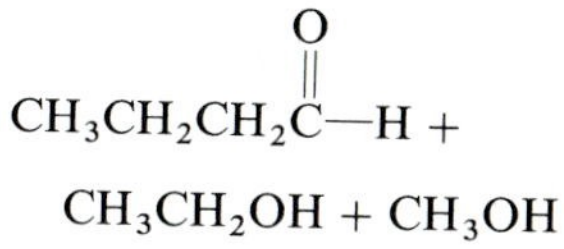

$$CH_3CH_2CH_2\overset{O}{\overset{\|}{C}}{-}H +$$
$$CH_3CH_2OH + CH_3OH$$

11.39

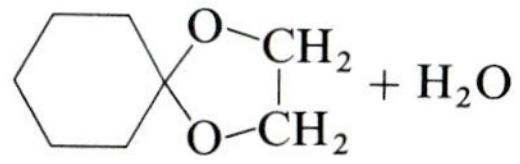

$$+ H_2O$$

11.42

$$HOCH_2\overset{OH}{\underset{}{\overset{|}{C}}}H{-}\overset{OH}{\underset{}{\overset{|}{C}}}H{-}\overset{OH}{\underset{}{\overset{|}{C}}}H{-}\overset{OH}{\underset{}{\overset{|}{C}}}H{-}\overset{O}{\overset{\|}{C}}{-}H$$

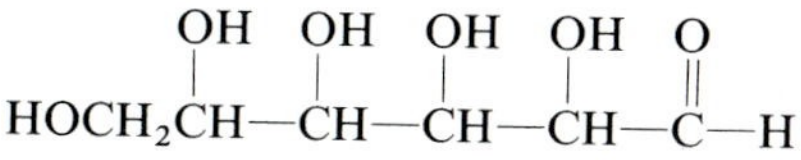

11.47 (a)

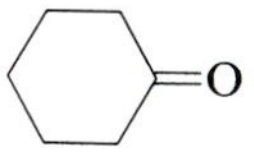

(b)

(c)

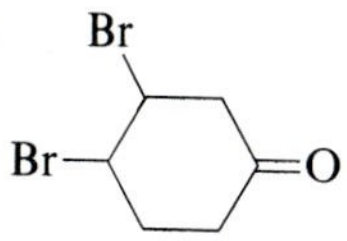

11.48 An aldehyde or ketone must have a hydrogen on the carbon next to the carbonyl group in order to undergo an aldol reaction.

11.49 (a)

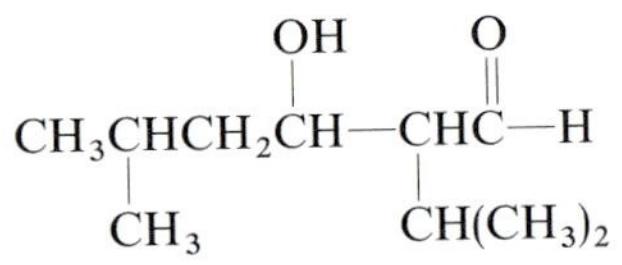

$$CH_3CHCH_2CH-CHC-H$$

with OH and O groups, CH_3 and $CH(CH_3)_2$ substituents

11.50

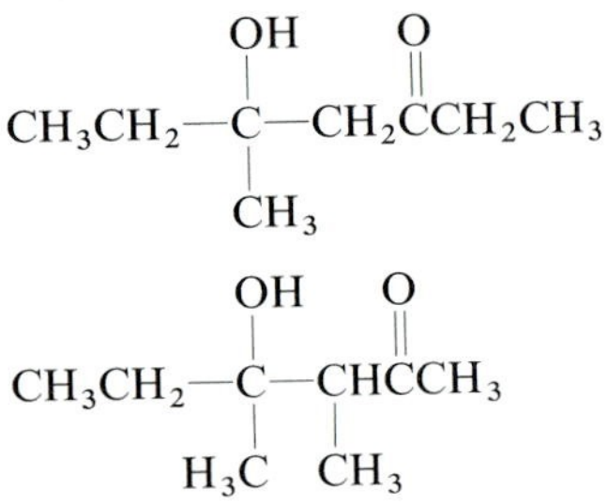

$$CH_3CH_2-\overset{OH}{\underset{CH_3}{C}}-CH_2CCH_2CH_3$$

$$CH_3CH_2-\overset{OH}{\underset{H_3C\ \ CH_3}{C}}-CHCCH_3$$

11.52

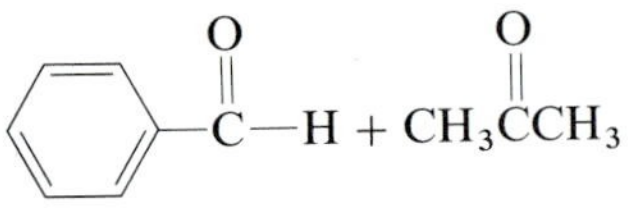

(benzaldehyde) $\overset{O}{C}-H + CH_3\overset{O}{C}CH_3$

Chapter 12

12.15

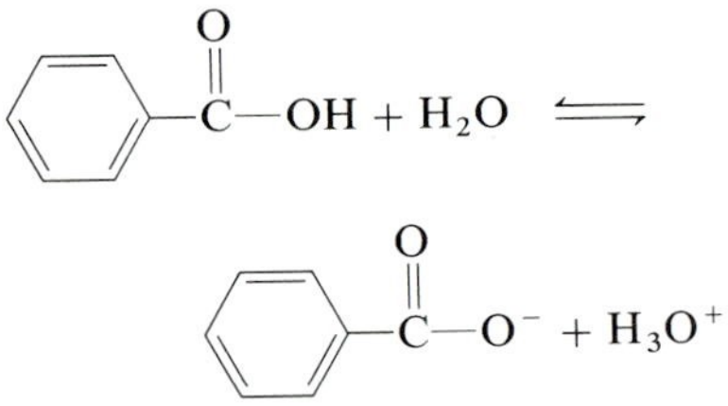

(benzene ring) $\overset{O}{C}-OH + H_2O \rightleftharpoons$

(benzene ring) $\overset{O}{C}-O^- + H_3O^+$

12.17

$$CH_3CH_2CH_2CH_2\overset{O}{C}-OH$$
Pentanoic acid

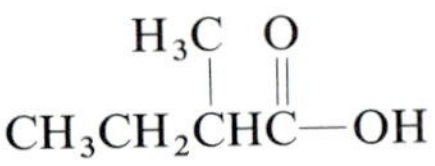

$$CH_3CH_2\overset{H_3C\ \ O}{\underset{}{CHC}}-OH$$
2-Methylbutanoic acid

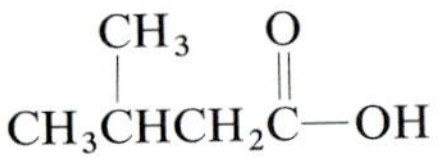

$$CH_3\overset{CH_3}{\underset{}{CHCH_2}}\overset{O}{C}-OH$$
3-Methylbutanoic acid

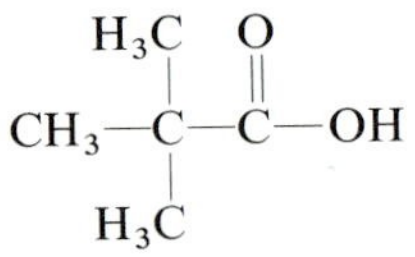

$$CH_3-\overset{H_3C\ \ O}{\underset{H_3C}{C}}-\overset{O}{C}-OH$$
2,3-Dimethylpropanoic acid

12.19 (a) Hexanoic acid **(c)** 2-Ethylbutanoic acid
(e) 4-Bromo-2-methylbutanoic acid

12.21 (a)

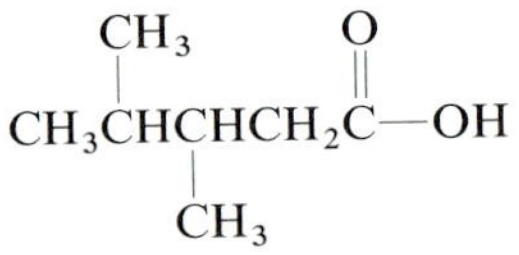

$$CH_3CHCHCH_2\overset{O}{C}-OH$$
with CH_3 and CH_3 substituents

(c) CH_3CH_2 (benzene ring) $\overset{O}{C}-OH$

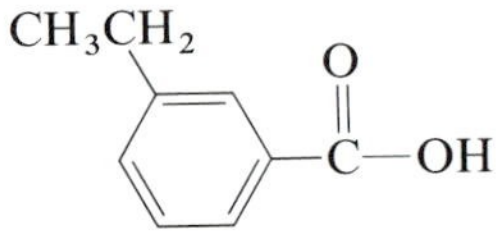

12.23

$$HO-\overset{O}{C}-CHCH_2\overset{O}{C}-OH$$
with OH substituent

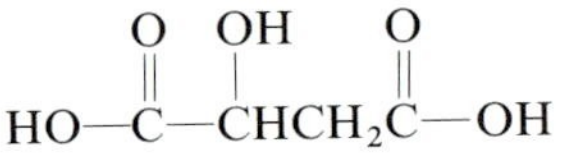

12.25 (a) 9.37 g   **(b)** 28.0 g
12.27 (b) Methyl 4-methylpentanoate   **(d)** Ethyl benzoate

12.28–12.29 (a)

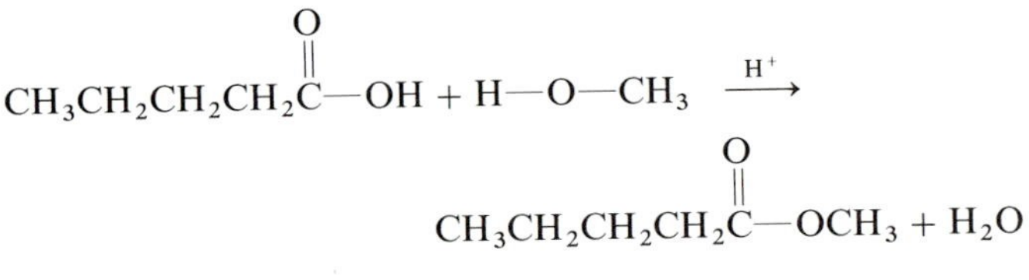

$$CH_3CH_2CH_2CH_2\overset{O}{C}-OH + H-O-CH_3 \xrightarrow{H^+}$$

$$CH_3CH_2CH_2CH_2\overset{O}{C}-OCH_3 + H_2O$$
Methylpentanoate

(c)

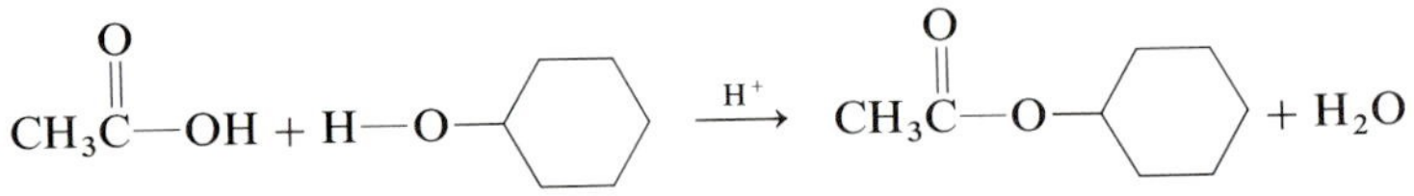

$$CH_3\overset{O}{C}-OH + H-O-\text{(cyclohexyl)} \xrightarrow{H^+} CH_3\overset{O}{C}-O-\text{(cyclohexyl)} + H_2O$$
Cyclohexyl acetate

12.30–12.31 (a)

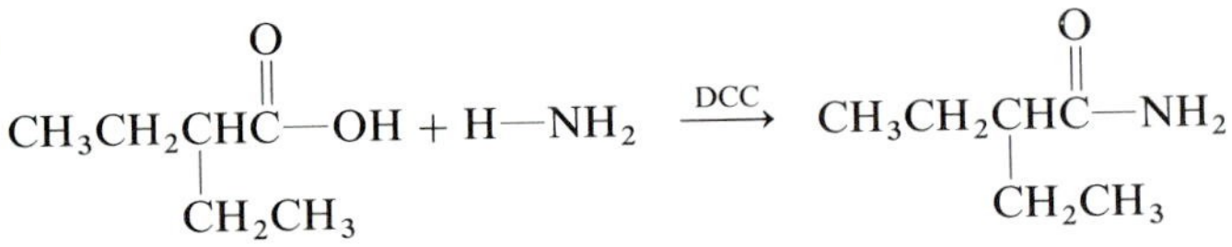

$$CH_3CH_2\overset{O}{\underset{CH_2CH_3}{CHC}}-OH + H-NH_2 \xrightarrow{DCC} CH_3CH_2\overset{O}{\underset{CH_2CH_3}{CHC}}-NH_2$$
2-Ethylbutanamide

12.32–12.33 (a)

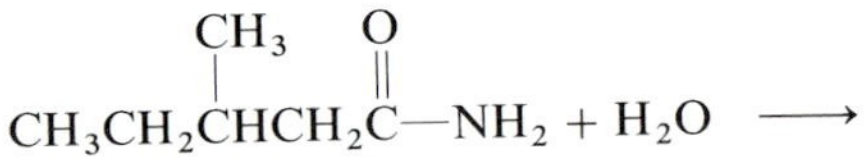

$$CH_3CH_2\overset{CH_3}{\underset{}{CHCH_2}}\overset{O}{C}-NH_2 + H_2O \longrightarrow$$
3-Methylpentanamide

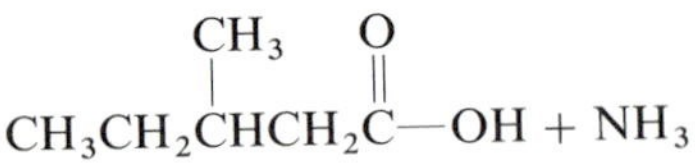

$$CH_3CH_2\overset{CH_3}{\underset{}{CHCH_2}}\overset{O}{C}-OH + NH_3$$

(c)

(benzene ring) $\overset{O}{C}-\overset{CH_3}{\underset{}{NCH_2CH_3}} + H_2O \longrightarrow$
N-Ethyl-N-methylbenzamide

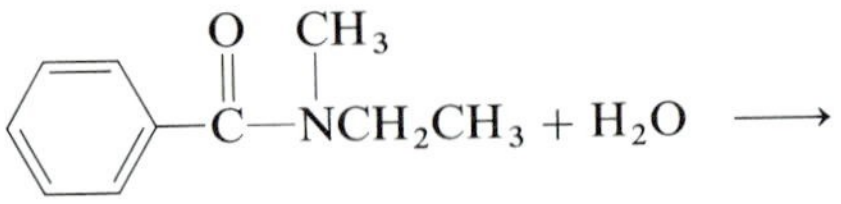

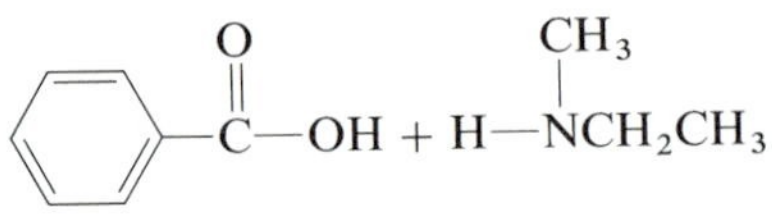

(benzene ring) $\overset{O}{C}-OH + H-\overset{CH_3}{\underset{}{NCH_2CH_3}}$

12.35

Methylanthranilate

12.37

4-Hydroxybutanoic acid

12.44

$$HO-CH_2-\overset{\displaystyle O}{\overset{\|}{C}}-CH_2-OH$$

$$+ \; H-O-\overset{\displaystyle O}{\underset{\displaystyle O^-}{\overset{\|}{P}}}-O^-$$

12.46 Esters (a) and (b) can't undergo Claisen condensation reactions because they have no hydrogen atoms bonded to the carbon next to the ester group.

12.47 (a)

$$\underset{\overset{\displaystyle |}{CH_3}}{CH_3CHCH_2}\overset{\displaystyle O}{\overset{\|}{C}}-\underset{\overset{\displaystyle |}{CH(CH_3)_2}}{CHC}\overset{\displaystyle O}{\overset{\|}{}}-OCH_3$$

$$+ \; CH_3OH$$

(b)

$$CH_3\overset{\displaystyle O}{\overset{\|}{C}}-CH_2\overset{\displaystyle O}{\overset{\|}{C}}-\underset{\overset{\displaystyle |}{CH_3}}{OCHCH_3}$$

$$+ \; CH_3\underset{\overset{\displaystyle |}{CH_3}}{CHOH}$$

Chapter 13

13.17 (a) an aldotetrose
(b) a ketopentose

13.19

13.21 glucose
13.23 the α form
13.27-13.28

13.31 Chiral: **(a)** shoe,
(c) light bulb
Achiral: **(b)** bed,
(d) flower pot
13.34 2-Bromo-2-chloropropane has a symmetry plane and is therefore achiral, but 2-bromo-2-chlorobutane does not have a symmetry plane and is chiral.
13.37 A hemiacetal has a carbon atom bonded to one —OH group and one —OR group. An acetal has a carbon bonded to two —OR groups.

13.41

13.46 Lactose and maltose have hemiacetal linkages that give positive Tollens' test results. Sucrose, however, has no hemiacetal group and is thus unreactive toward Tollens' reagent.

13.47–13.48

β-D-Glucose

β-D-Glucose

Gentiobiose is a reducing sugar because it has a hemiacetal linkage on the right-hand sugar unit. The left-hand sugar has an acetal linkage.

Chapter 14

14.16 A fat or oil is a triester of glycerol with three fatty acids—a triacylglycerol.
14.18 The fatty acids in an animal fat are mostly saturated; those in a vegetable oil are mostly unsaturated.
14.19

$$CH_3(CH_2)_{14}\overset{\displaystyle O}{\overset{\|}{C}}-OCH_2(CH_2)_{14}CH_3$$

14.21 Simple lipids:
(a) prostaglandin E$_1$,
(c) progesterone
Complex lipids:
(b) a lecithin,
(d) a sphingomyelin,
(e) a cerebroside,
(f) glyceryl trioleate
14.26 A vegetable oil can be converted into a solid cooking fat by hydrogenating its double bond(s).
14.28 0.13 g
14.29 (a)

$$\underset{\overset{\displaystyle |}{Br}}{CH_3(CH_2)_7CH}-\underset{\overset{\displaystyle |}{Br}}{CH(CH_2)_7}\overset{\displaystyle O}{\overset{\|}{C}}OH$$

(b)

$$CH_3(CH_2)_7CH-CH(CH_2)_7COH$$
with H and H below the two CH carbons, and O double-bonded above the C.

(c)

$$CH_3(CH_2)_7CH=CH(CH_2)_7COCH_3$$
with O double-bonded above the C.

14.33 Phosphoglycerides are more soluble in water than triglycerides because they have an ionic phosphate group that is solvated by water.

14.35 In addition to phospholipids, glycolipids, cholesterol, and proteins are also present in cell membranes.

14.43 11.5 g

14.44 Cholesterol regulates membrane fluidity.

14.47 Thromboxane A_2 is closely related to the prostaglandins.

Chapter 15

15.10 When referring to an amino acid, the prefix "α" means that the amino group is bonded to the carbon next to the —COOH carbon. In other words, the —NH_2 and —COOH groups are bonded to the same carbon.

15.14 (a)

$$H_2N-CH-C-OH$$
with CH below, bonded to H_3C and CH_3; O double-bonded above the C.

Valine

or

$$H_2N-CH-C-OH$$
with CH_2 below, then CH bonded to H_3C and CH_3; O double-bonded above the C.

Isoleucine

(b)

$$H_2N-CH-C-OH$$
with H—C—OH and CH_3 below; O double-bonded above the C.

Threonine

(c)

$$H_2N-CH-C-OH$$
with CH_2SH below; O double-bonded above the C.

Cysteine

(d)

$$H_2N-CH-C-OH$$
with CH_2 below, bonded to a benzene ring with OH.

Tyrosine

15.16 L-alanine and D-alanine are mirror images.

15.18 (a) basic
(b) neutral
(c) acidic
(d) neutral

15.19 (a) pH = 3 (its isoelectric point)
(b) pH = 13

15.26 In the presence of a mild oxidizing agent, two cysteine thiol groups can form a disulfide bridge linking them together.

15.28 Fibrous proteins consist of polypeptide chains arranged side by side. Globular proteins are coiled into nearly spherical shapes.

15.31 hydrogen bonding

15.32 The disulfide bridges that cysteine forms help to stabilize a protein's tertiary structure.

15.34 Met—Ile—Lys
Ile—Met—Lys
Lys—Met—Ile
Met—Lys—Ile
Ile—Lys—Met
Lys—Ile—Met

15.37 If a diabetic took insulin orally, digestive enzymes would hydrolyze it, and the individual amino acids would be absorbed as food.

15.40 (a)

$$H_2N-CH_2-C-OCH_3$$
with O double-bonded above the C.

an ester

(b)

$$Cl^-\ H_3\overset{+}{N}-CH_2-C-OH$$
with O double-bonded above the C.

an ammonium salt

15.44 A peptide rich in Asp and Lys residues in more soluble in water than a peptide rich in Val and Ala residues. The side chains of Asp and Lys are polar and are better solvated by water than the nonpolar, hydrophobic side chains of Val and Ala.

15.46

$$H_2N-CH-C-OH$$
with CH_2 then COOH below; O double-bonded above the C.

Aspartic acid

$$+\ H_2N-CH-C-OH\ +\ CH_3OH$$
with CH_2 below bonded to a benzene ring; O double-bonded above the C.

Phenylalanine

Chapter 16

16.7 An enzyme is a large protein molecule that catalyzes biochemical reactions.

16.9 Catalysts speed up reactions by lowering the energy barrier to product formation.

16.13 A coenzyme is a type of cofactor. A cofactor can be either an inorganic ion or a small organic molecule called a coenzyme.

16.14 An enzyme is a large three-dimensional molecule with a crevice into which its substrate can fit. Enzymes are specific in their action because only one or a few molecules have the appropriate shape to fit into the crevice.

16.16 (a) A *protease* catalyzes this reaction, which is the hydrolysis of an amide.

(b) Either a *dehydrase* or a *decarboxylase* catalyzes this reaction.

(c) A *dehydrogenase* catalyzes the introduction of a double bond into a molecule.

16.19 Dimethylurea inhibits urease by competitive inhibition since its structure is similar to that of urea.

16.24 Since the apoenzyme is the protein part of a holoenzyme, it is more likely to be denatured than the coenzyme.

16.26 (a) slow down the reaction

(b) probably slow down the reaction

(c) stop or slow down the reaction

(d) probably speed up reaction

16.31 Irreversible inhibition is the most difficult to treat medically because of the strength of the covalent bond between the inhibitor and the enzyme.

16.32 Papain is effective as a meat tenderizer because it hydrolyzes peptide bonds and partially digests the proteins in the meat.

16.35 A hormone transmits a chemical message from a gland to a target tissue. A neurotransmitter carries an impulse between neighboring nerve cells.

16.36 A vitamin acts as a cofactor for an enzyme, enabling an enzyme to catalyze biochemical reactions.

16.37 11 L

16.39 vitamins A, D, C, and K

16.41 A synapse is the gap between two nerve cells (neurons). Neurotransmitters released by one neuron cross the synapse to receptors on a second neuron and transmit the nerve impulse.

16.44 The body's endocrine system manufactures and secretes hormones, which regulate biochemical activities.

16.45 the hypothalamus

16.49 The same type of lock-and-key specificity exists for both an enzyme/substrate complex and a hormone/receptor complex.

16.51 Epinephrine raises the blood sugar level by increasing the rate of glycogen breakdown in the liver.

16.52 Vitamins are required in the diet; hormones and enzymes are synthesized in the body.

Chapter 17

17.19 "Deoxy" means that the oxygen atom attached to carbon 2 of ribose is missing and that there is an —H group on carbon 2 rather than an —OH group.

17.23 Messenger RNA (mRNA) carries the genetic message from DNA to ribosomes. Ribosomal RNA (rRNA) bonds to protein to constitute the physical makeup of ribosomes. Transfer RNA (tRNA) transports specific amino acids to the ribosomes, where they are incorporated into proteins.

17.26 A gene carries the DNA code needed to synthesize a specific protein.

17.28 hydrogen bonds between base pairs

17.31 The 5′ end of a nucleic acid has a free phosphoric acid group, and the 3′ end has a free —OH group.

17.35 The two DNA strands are complementary. The template strand is used for transcription of mRNA and is complementary to mRNA. Thus, mRNA is a copy of the information strand (with U replacing T).

17.37 An anticodon is a sequence of three nucleotides on a tRNA molecule that is complementary to the sequence of a codon.

17.39 Exons and introns are sections of DNA. Exons carry the genetic message, whereas introns do not code for any part of the protein being synthesized. Introns constitute about 90% of DNA and exons 10%.

17.41 32% T, 18% C

17.43 (a) Thr

(b) Gly

(c) Leu

17.45 A—T—G—G—C—T

17.47 Tyr—Arg

17.49 A mutation in RNA affects only one molecule of RNA; other intact molecules of RNA can still carry out protein synthesis. A mutation in DNA is much more drastic, however, because there is only one molecule of DNA per gene, and any error will be copied into all subsequent DNA molecules during replication.

17.50 Thr replaces Ile

17.51 A mutation must occur in a germ cell (sperm or egg) in order for it to be passed down to future generations.

17.52 A restriction endonuclease is an enzyme that cleaves DNA at a specific base sequence. Restriction endonucleases produce fragments of DNA that are of manageable size for sequencing.

17.54 If codons were made up of two, rather than three nucleotides, only $4^2 = 16$ nucleotide combinations would be possible. Since there are 20 amino acids, it would not be possible to code for all of them with only two nucleotides.

17.57 UAU–GGU–GGU–UUU– AUG–UAA (one of several)

Chapter 18

18.12 Organelles are subcellular structures that perform specialized tasks within the cell.

are the inner folds of a mitochondrion. On the cristae are protuberances where the energy production of the cell takes place.

18.16 Metabolism is the total of all reactions that take place in cells. Digestion is a part of metabolism in which food is broken down into small organic molecules.

18.18 acetyl-CoA

18.19 Digestion, citric-acid cycle, respiratory chain, oxidative phosphorylation

18.21 ATP has a triphosphate group bonded to C5 of ribose, and ADP has a diphosphate group in that position.

18.25 The energy of the reaction products must be lower than that of the reactants.

18.26 ATP is a high-energy molecule because energy is released when ATP reacts with most other molecules to transfer a phosphate group.

18.30 in mitochondria

18.32 Oxaloacetic acid,

$$\overset{O}{\underset{\parallel}{HOOCCH_2C}}COOH$$

18.33 (a) steps 3, 4, 6, and 8
(b) steps 3 and 4
(c) step 7

18.38 NAD^+ and FMN

18.39 water, $6 H^+$ ions, and energy

18.41 iron

Chapter 19

19.13 Digestion occurs in the mouth, stomach, and small intestine; it involves the enzyme-catalyzed hydrolysis of food components into small molecules.

19.15 aerobic conditions: acetyl CoA; anaerobic conditions: lactic acid; fermentation: ethanol.

19.17 pyruvic acid,

$$\overset{O}{\underset{\parallel}{CH_3C}}COOH$$

19.18 in the cytoplasm of muscle, fat, and liver cells

19.21 pyruvic acid and oxaloacetic acid

19.24 (a) steps 1, 3, and 6
(b) steps 2, 5, and 8
(c) step 6

19.26 507 g

19.31 in mitochondria

19.33 In the fatty-acid spiral, reaction occurs on a continually shortened fatty acid until the entire fatty acid is consumed. In a cycle, the product of the final step is a reactant in the first step.

19.35 in the cytoplasm of liver cells

19.37 17 mol

19.40 5

19.44

$$CH_3\overset{O}{\underset{\parallel}{C}}-CH_2-\overset{O}{\underset{\parallel}{C}}-CoA \xrightarrow[\substack{AcetylCoA \\ transferase}]{CoA}$$

$$2 CH_3\overset{O}{\underset{\parallel}{C}}-CoA$$

19.45 (a) 61
(b) 112

19.46 caprylic acid, 3; myristic acid, 6

19.49 1400 g

19.51 Amino acids in the amino-acid pool can be used to form tissue proteins or nucleic acids, or they can be catabolized for energy.

19.53 (a)

$$\underset{\text{(benzene ring)}}{}-CH_2-\overset{O}{\underset{\parallel}{C}}-\overset{O}{\underset{\parallel}{C}}-OH$$

19.54 α-Ketoglutaric acid

Chapter 20

20.12 transmutation

20.14 ^4_2He

20.17 Gamma radiation has the highest penetrating power, alpha radiation the lowest.

20.19 A high-energy reactive ion is produced.

20.22 A neutron in the nucleus decomposes to a proton and an electron, which is emitted as a beta particle.

20.23 An alpha particle is a helium nucleus, ^4_2He; a helium atom is a helium nucleus plus two electrons.

20.25 25%

20.27 a Geiger counter and a film badge

20.29 Rems indicate the amount of tissue damage from any type of radiation, and allow comparisons between different types of radiation to be made.

20.31 (1) c;
(2) b;
(3) d;
(4) a

20.33 alpha particle: atomic number decreases by two, mass number deceases by four;
beta particle: atomic number increases by one, mass number is unchanged;
gamma ray: both atomic number and mass number are unchanged.

20.36 A transuranium element is an element with an atomic number greater than 92; it is produced by bombardment of a lighter element with high-energy particles.

20.39 If the mass of $^{235}_{92}U$ is too small, the product neutrons will escape before encountering another nucleus, and the reaction will stop.

20.41 $^{75}_{34}Se \rightarrow {}^{0}_{-1}e + {}^{75}_{35}Br$

20.43 21/3 years

20.44 $^{35}_{17}Cl$

20.45 approx 1 ng; approx 5×10^{-3} ng

20.46 (a) $^{140}_{55}Cs$
(b) $^{246}_{96}Cm$

20.47 (a) $^{186}_{76}Os$

20.49 $^{198}_{80}Hg + {}^{1}_{0}n \rightarrow$
$^{198}_{79}Au + {}^{1}_{1}H$ (a proton)

Photo Credits
and Acknowledgments

opening photo Image generated on the Evans & ...and PS 390 computer graphics system using Tripos® ...BYL computational chemistry software. **Fig. 16.2(b)** Courtesy of Dr. Christine Humblet, Parke-Davis. **Fig. 16.11** Dr. E. R. Degginger. **Fig. 16.13** Manfred Kage/Peter Arnold, Inc. **Interlude** Dr. Edward J. Bottone, Dept. of Microbiology, Mount Sinai Hospital, New York.

Chapter 17 opening photo CDC, Science Source/Photo Researchers. **Fig. 17.1** Biophoto Associates, Science Source/Photo Researchers. **Fig. 17.4** Richard Feldman/National Institutes of Health. **Fig. 17.8** O. L. Miller, Jr., and B. R. Beatty, *J. Cell Physiol.*, Vol. 74 (1969). **Fig. 17.10(c)** Image generated on the Evans & Sutherland PS 390 computer graphics system using Tripos® SYBYL computational chemistry software. **Fig. 17.13** O. L. Miller, Jr., Barbara A. Hamkalo, and C. A. Thomas, Jr., *Science*, 169: 392–395, 1970. ©1970 by the American Association for the Advancement of Science. **Fig. 17.15** Bill Longcore/Photo Researchers.

Chapter 18 opening photo Lennart Nilsson from *The Body Victorious* ©Delacorte Press. **Fig. 18.8(b)** Image generated on the Evans & Sutherland PS 390 computer graphics system using Tripos® SYBYL computational chemistry software.

Chapter 19 opening photo Food for Thought. **Fig. 19.3** Jack Fields/Photo Researchers.

Chapter 20 opening photo Courtesy of the International Atomic Energy Agency. **Application 1** Courtesy of Marshall/University of California, Irvine. **Application 2** CNRI, Science Photo Library/Photo Researchers. **Interlude** Tom McHugh/Photo Researchers.

Index

Note: A page number set in boldface type indicates that the entry is defined on that page.

inorganic ion or a small organic molecule called a coenzyme.

16.14 An enzyme is a large three-dimensional molecule with a crevice into which its substrate can fit. Enzymes are specific in their action because only one or a few molecules have the appropriate shape to fit into the crevice.

16.16 **(a)** A *protease* catalyzes this reaction, which is the hydrolysis of an amide.

(b) Either a *dehydrase* or a *decarboxylase* catalyzes this reaction.

(c) A *dehydrogenase* catalyzes the introduction of a double bond into a molecule.

16.19 Dimethylurea inhibits urease by competitive inhibition since its structure is similar to that of urea.

16.24 Since the apoenzyme is the protein part of a holoenzyme, it is more likely to be denatured than the coenzyme.

16.26 **(a)** slow down the reaction

(b) probably slow down the reaction

(c) stop or slow down the reaction

(d) probably speed up reaction

16.31 Irreversible inhibition is the most difficult to treat medically because of the strength of the covalent bond between the inhibitor and the enzyme.

16.32 Papain is effective as a meat tenderizer because it hydrolyzes peptide bonds and partially digests the proteins in the meat.

16.35 A hormone transmits a chemical message from a gland to a target tissue. A neurotransmitter carries an impulse between neighboring nerve cells.

16.36 A vitamin acts as a cofactor for an enzyme, enabling an enzyme to catalyze biochemical reactions.

16.37 11 L

16.39 vitamins A, D, C, and K

16.41 A synapse is the gap between two nerve cells (neurons). Neurotransmitters released by one

neuron cross the synapse to receptors on a second neuron and transmit the nerve impulse.

16.44 The body's endocrine system manufactures and secretes hormones, which regulate biochemical activities.

16.45 the hypothalamus

16.49 The same type of lock-and-key specificity exists for both an enzyme/substrate complex and a hormone/receptor complex.

16.51 Epinephrine raises the blood sugar level by increasing the rate of glycogen breakdown in the liver.

16.52 Vitamins are required in the diet; hormones and enzymes are synthesized in the body.

Chapter 17

17.19 "Deoxy" means that the oxygen atom attached to carbon 2 of ribose is missing and that there is an —H group on carbon 2 rather than an —OH group.

17.23 Messenger RNA (mRNA) carries the genetic message from DNA to ribosomes.
Ribosomal RNA (rRNA) bonds to protein to constitute the physical makeup of ribosomes.
Transfer RNA (tRNA) transports specific amino acids to the ribosomes, where they are incorporated into proteins.

17.26 A gene carries the DNA code needed to synthesize a specific protein.

17.28 hydrogen bonds between base pairs

17.31 The 5′ end of a nucleic acid has a free phosphoric acid group, and the 3′ end has a free —OH group.

17.35 The two DNA strands are complementary. The template strand is used for transcription of mRNA and is complementary to mRNA. Thus, mRNA is a copy of the information strand (with U replacing T).

17.37 An anticodon is a sequence of three nucleotides on a tRNA molecule that is complementary to the sequence of a codon.

17.39 Exons and introns are sections of DNA. Exons carry the genetic message, whereas introns do not code for any part of the protein being synthesized. Introns constitute about 90% of DNA and exons 10%.

17.41 32% T, 18% C

17.43 **(a)** Thr

(b) Gly

(c) Leu

17.45 A—T—G—G—C—T

17.47 Tyr—Arg

17.49 A mutation in RNA affects only one molecule of RNA; other intact molecules of RNA can still carry out protein synthesis. A mutation in DNA is much more drastic, however, because there is only one molecule of DNA per gene, and any error will be copied into all subsequent DNA molecules during replication.

17.50 Thr replaces Ile

17.51 A mutation must occur in a germ cell (sperm or egg) in order for it to be passed down to future generations.

17.52 A restriction endonuclease is an enzyme that cleaves DNA at a specific base sequence. Restriction endonucleases produce fragments of DNA that are of manageable size for sequencing.

17.54 If codons were made up of two, rather than three nucleotides, only $4^2 = 16$ nucleotide combinations would be possible. Since there are 20 amino acids, it would not be possible to code for all of them with only two nucleotides.

17.57 UAU–GGU–GGU–UUU–AUG–UAA (one of several)

Chapter 18

18.12 Organelles are subcellular structures that perform specialized tasks within the cell.

18.15 *Cristae* are the inner folds of a mitochondrion. On the cristae are protuberances where the energy production of the cell takes place.

18.16 Metabolism is the total of all reactions that take place in cells. Digestion is a part of metabolism in which food is broken down into small organic molecules.

18.18 acetyl-CoA

18.19 Digestion, citric-acid cycle, respiratory chain, oxidative phosphorylation

18.21 ATP has a triphosphate group bonded to C5 of ribose, and ADP has a diphosphate group in that position.

18.25 The energy of the reaction products must be lower than that of the reactants.

18.26 ATP is a high-energy molecule because energy is released when ATP reacts with most other molecules to transfer a phosphate group.

18.30 in mitochondria

18.32 Oxaloacetic acid,

$$
\underset{\text{HOOCCH}_2\text{CCOOH}}{\overset{\overset{\text{O}}{\|}}{}}
$$

18.33 (a) steps 3, 4, 6, and 8
(b) steps 3 and 4
(c) step 7

18.38 NAD$^+$ and FMN

18.39 water, 6 H$^+$ ions, and energy

18.41 iron

Chapter 19

19.13 Digestion occurs in the mouth, stomach, and small intestine; it involves the enzyme-catalyzed hydrolysis of food components into small molecules.

19.15 aerobic conditions: acetyl CoA; anaerobic conditions: lactic acid; fermentation: ethanol.

19.17 pyruvic acid,

$$
\underset{\text{CH}_3\text{CCOOH}}{\overset{\overset{\text{O}}{\|}}{}}
$$

19.18 in the cytoplasm of muscle, fat, and liver cells

19.21 pyruvic acid and oxaloacetic acid

19.24 (a) steps 1, 3, and 6
(b) steps 2, 5, and 8
(c) step 6

19.26 507 g

19.31 in mitochondria

19.33 In the fatty-acid spiral, reaction occurs on a continually shortened fatty acid until the entire fatty acid is consumed. In a cycle, the product of the final step is a reactant in the first step.

19.35 in the cytoplasm of liver cells

19.37 17 mol

19.40 5

19.44

$$
\underset{\text{CH}_3\text{C}-\text{CH}_2-\text{C}-\text{CoA}}{\overset{\overset{\text{O}}{\|}\qquad\qquad\overset{\text{O}}{\|}}{}} \xrightarrow[\text{transferase}]{\overset{\text{CoA}}{\text{AcetylCoA}}}
$$

$$
2\,\underset{\text{CH}_3\text{C}-\text{CoA}}{\overset{\overset{\text{O}}{\|}}{}}
$$

19.45 (a) 61
(b) 112

19.46 caprylic acid, 3; myristic acid, 6

19.49 1400 g

19.51 Amino acids in the amino-acid pool can be used to form tissue proteins or nucleic acids, or they can be catabolized for energy.

19.53 (a)

$$
\text{C}_6\text{H}_5-\text{CH}_2-\underset{}{\overset{\overset{\text{O}}{\|}}{\text{C}}}-\underset{}{\overset{\overset{\text{O}}{\|}}{\text{C}}}-\text{OH}
$$

19.54 α-Ketoglutaric acid

Chapter 20

20.12 transmutation

20.14 ^4_2He

20.17 Gamma radiation has the highest penetrating power, alpha radiation the lowest.

20.19 A high-energy reactive ion is produced.

20.22 A neutron in the nucleus decomposes to a proton and an electron, which is emitted as a beta particle.

20.23 An alpha particle is a helium nucleus, ^4_2He; a helium atom is a helium nucleus plus two electrons.

20.25 25%

20.27 a Geiger counter and a film badge

20.29 Rems indicate the amount of tissue damage from any type of radiation, and allow comparisons between different types of radiation to be made.

20.31 (1) c;
(2) b;
(3) d;
(4) a

20.33 alpha particle: atomic number decreases by two, mass number deceases by four;
beta particle: atomic number increases by one, mass number is unchanged;
gamma ray: both atomic number and mass number are unchanged.

20.36 A transuranium element is an element with an atomic number greater than 92; it is produced by bombardment of a lighter element with high-energy particles.

20.39 If the mass of $^{235}_{92}\text{U}$ is too small, the product neutrons will escape before encountering another nucleus, and the reaction will stop.

20.41 $^{75}_{34}\text{Se} \rightarrow {}^{0}_{-1}\text{e} + {}^{75}_{35}\text{Br}$

20.43 21/3 years

20.44 $^{35}_{17}\text{Cl}$

20.45 approx 1 ng; approx 5×10^{-3} ng

20.46 (a) $^{140}_{55}\text{Cs}$
(b) $^{246}_{96}\text{Cm}$

20.47 (a) $^{186}_{76}\text{Os}$

20.49 $^{198}_{80}\text{Hg} + {}^{1}_{0}\text{n} \rightarrow$
$^{198}_{79}\text{Au} + {}^{1}_{1}\text{H}$ (a proton)

Photo Credits and Acknowledgments

Chapter 16 opening photo Image generated on the Evans & Sutherland PS 390 computer graphics system using Tripos® SYBYL computational chemistry software. **Fig. 16.2(b)** Courtesy of Dr. Christine Humblet, Parke-Davis. **Fig. 16.11** Dr. E. R. Degginger. **Fig. 16.13** Manfred Kage/Peter Arnold, Inc. **Interlude** Dr. Edward J. Bottone, Dept. of Microbiology, Mount Sinai Hospital, New York.

Chapter 17 opening photo CDC, Science Source/Photo Researchers. **Fig. 17.1** Biophoto Associates, Science Source/ Photo Researchers. **Fig. 17.4** Richard Feldman/National Institutes of Health. **Fig. 17.8** O. L. Miller, Jr., and B. R. Beatty, *J. Cell Physiol.*, Vol. 74 (1969). **Fig. 17.10(c)** Image generated on the Evans & Sutherland PS 390 computer graphics system using Tripos® SYBYL computational chemistry software. **Fig. 17.13** O. L. Miller, Jr., Barbara A. Hamkalo, and C. A. Thomas, Jr., *Science*, 169: 392–395, 1970. ©1970 by the American Association for the Advancement of Science. **Fig. 17.15** Bill Longcore/Photo Researchers.

Chapter 18 opening photo Lennart Nilsson from *The Body Victorious* ©Delacorte Press. **Fig. 18.8(b)** Image generated on the Evans & Sutherland PS 390 computer graphics system using Tripos® SYBYL computational chemistry software.

Chapter 19 opening photo Food for Thought. **Fig. 19.3** Jack Fields/Photo Researchers.

Chapter 20 opening photo Courtesy of the International Atomic Energy Agency. **Application 1** Courtesy of Marshall/ University of California, Irvine. **Application 2** CNRI, Science Photo Library/Photo Researchers. **Interlude** Tom McHugh/ Photo Researchers.

Index

Note: A page number set in boldface type indicates that the entry is defined on that page.

Some Important Families of Organic Molecules

Family Name	Functional Group Structure[a]	Simple Example	Name Ending
Alkane	(contains only C—H and C—C single bonds)	CH_3CH_3 ethane	-ane
Alkene	$\diagdown C = C \diagup$	$H_2C = CH_2$ ethylene	-ene
Alkyne	$—C \equiv C—$	$H—C \equiv C—H$ acetylene (ethyne)	-yne
Arene	(benzene ring structure)	(benzene ring) benzene	none
Alcohol	$—\overset{\vert}{\underset{\vert}{C}}—O—H$	$CH_3—OH$ methyl alcohol (methanol)	-ol
Ether	$—\overset{\vert}{\underset{\vert}{C}}—O—\overset{\vert}{\underset{\vert}{C}}—$	$CH_3—O—CH_3$ dimethyl ether	none
Amine	$—N—H, —\overset{\vert}{N}—H, —\overset{\vert}{\underset{\vert}{N}}—$ with H	$CH_3—NH_2$ methylamine	-amine
Aldehyde	$—\overset{O}{\overset{\Vert}{C}}—H$	$CH_3—\overset{O}{\overset{\Vert}{C}}—H$ acetaldehyde (ethanal)	-al
Ketone	$C—\overset{O}{\overset{\Vert}{C}}—C$	$CH_3—\overset{O}{\overset{\Vert}{C}}—CH_3$ acetone	-one
Carboxylic acid	$—\overset{O}{\overset{\Vert}{C}}—OH$	$CH_3—\overset{O}{\overset{\Vert}{C}}—OH$ acetic acid	-ic acid
Ester	$—\overset{O}{\overset{\Vert}{C}}—O—$	$CH_3—\overset{O}{\overset{\Vert}{C}}—O—CH_3$ methyl acetate	-ate
Amide	$—\overset{O}{\overset{\Vert}{C}}—NH_2, —\overset{O}{\overset{\Vert}{C}}—\overset{\vert}{N}—H, —\overset{O}{\overset{\Vert}{C}}—\overset{\vert}{\underset{\vert}{N}}—$	$CH_3—\overset{O}{\overset{\Vert}{C}}—NH_2$ acetamide	-amide

[a] The bonds whose connections aren't specified are assumed to be attached to carbon or hydrogen atoms in the rest of the molecule.